Published for
OXFORD INTERNATIONAL
AQA EXAMINATIONS

International A Level
BIOLOGY

AS and
A LEVEL

Fran Fuller
Glenn Toole
Susan Toole

OXFORD
UNIVERSITY PRESS

OXFORD
UNIVERSITY PRESS

Great Clarendon Street, Oxford, OX2 6DP, United Kingdom

Oxford University Press is a department of the University of Oxford. It furthers the University's objective of excellence in research, scholarship, and education by publishing worldwide. Oxford is a registered trade mark of Oxford University Press in the UK and in certain other countries

First published in 2016

British Library Cataloguing in Publication Data
Data available

978-0-19-837601-9

10 9 8 7

Paper used in the production of this book is a natural, recyclable product made from wood grown in sustainable forests.
The manufacturing process conforms to the environmental regulations of the country of origin.

Printed in United Kingdom

Acknowledgements

The publishers would like to thank the following for permissions to use their photographs:

Cover: PrimePhoto/Shutterstock

p3: Alamy/Martin Shields; **p6:** Science Photo Library/Andrew Syred; **p18:** Science Photo Library/Chris Priest; **p20:** Science Photo Library/Mauro Fermariello; **p21:** Science Photo Library; **p31:** Science Photo Library/Dr Gopal Murti; **p36:** Science Photo Library/J.C. Revy; **p39:** Science Photo Library/ George D. Lepp; **p52l:** Science Photo Library/Prof. P. Motta/Dept. of Anatomy/ Univeristy "La Sapienza", Rome; **p52r:** Science Photo Library/J.C. Revy; **p65:** Science Photo Library/Microfi eld Scientifi c Ltd; **p66:** Sidney Moulds/Science Photo Library; **p69t:** Science Photo Library/CNRI; **p69b:** Science Photo Library/ Prof. Motta, Correr and Nottola/Univeristy "La Sapienza", Rome; **p73:** Science Photo Library/Alfred Pasieka; **p75:** Science Photo Library/Frances Leroy, Biocosmos; **p79t:** Alamy/FLPA; **p79b:** Glenn and Susan Toole; **p80t:** Alamy/ Ron Steiner; **p80b:** Alamy/Natural Visions; **p83t:** Science Photo Library/ George Ranalli; **p83m:** Science Photo Library/Tony Wood; **p83b:** Science Photo Library/Bob Gibbons; **p84:** Science Photo Library/Jeanne White; **p85:** Angela Mariana Zamalloa Vargas/EyeEm; **p112:** Laguna Design/Getty Images; **p120t:** Science Photo Library/Claude Nuridsany & Marie Perennou; **p120m:** Oxford Scientific/Bontica; **p120b:** Glenn and Susan; **p122:** Alamy/Mark J. Barrett; **p126:** PhyloWin; **p130:** Glenn and Susan Toole; **p133l:** Science Photo Library/Barbara Strnadova; **p133r:** Science Photo Library/BSIP, Fife; **p134b:** Science Photo Library/Simon Fraser; **p136:** Alamy/Art Kowalsky; **p137:** Glenn and Susan Toole; **p138:** Alamy/Worldwide Picture Library; **p139l:** Photodisc 6; **p139r:** Science Photo Library/Leslie J Borg; **p142l:** Oxford Scientific/PHOTOTAKE Inc; **p142r:** Science Photo Library/NIBSC; **p143:** Science Photo Library/Eye of Science; **p147t:** Alamy/PHOTOTAKE Inc; **p147b:** Alamy/Medical-on-line; **p151:** Photodisc 18; **p158:** Glenn and Susan Toole; **p159l:** Alamy/Ron Steiner; **p159r:** Alamy/Natural Visions; **p166:** Science Photo Library/Dr Gopal Murti; **p167:** Shutterstock; **p168:** Science Photo Library/Custom Medical Stock Photo; **p170:** Science Photo Library/GustoImages; **p176:** Shutterstock; **p192:** Alamy/ imagebroker; **p195t:** Science Photo Library/CNRI; **p195m:** Science Photo Library/Steve Gschmeissner; **p195b:** Science Photo Library/ Tony Wood; **p202:** Science Photo Library/Prof. Motta/G. Macchiarelli/ Univeristy "La Sapienza", Rome; **p205t:** Universal Images Group/ Getty Images; **p205b:** Peter Dazeley/Getty Images; **p206:** Science Photo Library/GJLP; **p207:** Science Photo Library/Prof. P.M. Motta, G. Macchiarelli, S.A Nottola; **p213:** Oxford Scientific/PHOTOTAKE Inc; **p216:** Oxford Scientific/Carolina Bio. Supply C; **p218:** Alamy/ Celeste Daniels; **p228:** Nigel Cattlin/Visuals Unlimited, Inc/Getty Images; **p229t:** Thomas Splettstoesser, Visuals Unlimited /Science Photo Library; **p229m:** Centre For Bioimaging, Rothamsted Research/Science Photo Library; **p229b:** Nigel Cattlin/Science Photo Library; **p230r:** Geoff Kidd/Science Photo Library; **p230l:** Nigel Cattlin/Science Photo Library; **p236:** Science Photo Library/Prof. G Gimenez-Martin; **p237:** Steve Gschmeissner/Science Photo Library; **p238:** Centre For Infections/Public Health England/Science Photo Library; **p239:** Oxford Scientific/PHOTOTAKE Inc; **p242:** Alamy/ Bruce Coleman Inc; **p244:** Shutterstock; **p248:** iStockphoto; **p249:** Glenn and Susan Toole; **p251:** Shutterstock; **p252:** Alamy/Chris Johnson; **p254:** iStockphoto; **p255t:** Alamy/Juniors Bildharchiv; **p255b:** Alamy/Chris Gomersall; **p256 - p259r:** iStockphoto; **p263:** Photolibrary; **p266:** Glenn and Susan Toole; **p267:** Science Photo Library/Simon Fraser; **p268:** Glenn and Susan Toole; **p269:** Science Photo Library/Simon Fraser; **p270:** Science Photo Library/ Bob Gibbons; **p275:** Science Photo Library/Dr Jeremy Burgess; **p276:** Oxford Scientific/Herve Conge Phototake Inc; **p278:** Science Photo Library/Dr Kenneth R. Miller; **p287:** Science Photo Library/ Martyn Chillmaid; **p293:** Science Photo Library/Eye of Science; **p294:** Shutterstock; **p300:** Science Photo Library/ISM; **p309:** Science Photo Library/Suzanne L.& Joseph T. Collins; **p312:** Oxford Scientific/Raymond Blythe; **p316:** Oxford Scientific/Phototake Inc; **p317t:** Alamy/Acro Images GmbH; **p317b:** Alamy/AGStockUSA, Inc; **p319:** Alamy/22DigiTal; **p321t:** Nigel Cattlin/Science Photo Library; **p321b:** Alamy/Paul Glendell; **p323:** Science Photo Library/Martin Dohrn; **p325:** Science Photo Library/Scott Sinklier/AGStockUSA; **p329:** Alamy/guatebrian; **p330:** Glenn and Susan Toole; **p332 & p333:** Alamy/David Noton Photography; **p336:** Science Photo Library/Hugh Spencer; **p337:** Alamy/Chris Gomersall; **p338 - p340t:** Glenn and Susan Toole; **p341:** Alamy/geogphotos; **p346:** Science Photo Library/Alfred Pasieka; **p350:** Science Photo Library/Biophoto Associates; **p353:** iStockphoto; **p359:** Dr. John Brackenbury/Science Photo Library; **p364:** Oxford Scientific/David Fox; **p368:** Alamy/ David Hosking; **p375:** Science Photo Library; **p377:** Science Photo Library/Anatomical Travelogue; **p379:** Science Photo Library/ Omikron; **p382:** Science Photo Library/Jean Clauderevy, ISM; **p384:** Science Photo Library/Steve Gschmeissner; **p389:** Science Photo Library/CRNI; **p392:** Science Photo Library/Steve Gschmeissner; **p400:** Science Photo Library/Steve Percival; **p406 & p407:** Science Photo Library; **p412t:** iStockphoto; **p412b:** Leonello Calvetti/Getty Images; **p421t:** bfrontlineservices.com.au; **p424:** Gilles Mermet/ Science Photo Library; **p428 - p430:** Shutterstock; **p435:** Science Photo Library/Astrid & Hanns-Frieder Michler; **p437:** Science Photo Library/J.C. Revy; **p440:** Science Photo Library/Medi-Mation; **p452:** Science Photo Library/Rosenfeld Images Ltd; **p454:** Science Photo Library/Dr Yorgos Nikas; **p458:** Science Photo Library/Dr Gopal Murti; **p461:** Alamy/Victorio Castellani; **p467:** Alamy/F1online digitale Bildagentur GmbH.

Contents

Answers to the Practice Questions are available at www.oxfordsecondary.com/oxfordaqaexams-alevel-biology

Learning objectives

- → At the beginning of each topic, there is a list of learning objectives.
- → These are matched to the specification and allow you to monitor your progress.
- → A specification reference is also included.
  Specification reference: 3.1.1

This book contains many different features. Each feature is designed to support and develop the skills you will need for your exams, as well as foster and stimulate your interest in biology.

Terms that you will need to be able to define and understand are shown in **bold type** within the text.

Where terms are not explained within the same topic, they are highlighted in **bold orange text**. You can look these words up in the glossary.

Application features

These features contain important and interesting applications of biology in order to emphasise how scientists and engineers have used their scientific knowledge and understanding to develop new applications and technologies. There are also application features to develop your maths skills, and to develop your practical skills. In addition, there is a feature on each of the ten Required Practicals clearly sign-posted within each relevant chapter.

Synoptic link

These highlight how the sections relate to each other. Linking different areas of biology together becomes increasingly important, as many exam questions (particularly at A Level) will require you to bring together your knowledge from different areas.

Extension features

These features contain material that is beyond the specification, which is designed to stretch you and provide you with a broader knowledge and understanding and lead the way into the types of thinking and areas you might study in further education. As such, neither the detail nor the depth of questioning will be required for the exams. But this book is about more than getting through the exams.

1 Extension and application features have questions that link the material with concepts that are covered in the specification. There are also extension features to develop your maths skills, and to develop your practical skills.

Study tip

Study tips contain prompts to help you with your revision.

Hint

Hint features give other information or ways of thinking about a concept to support your understanding.

Summary questions

1 These are short questions that test your understanding of the topic and allow you to apply the knowledge and skills you have acquired. The questions are ramped in order of difficulty.

Mathematical Skills in A level Biology

Biology students are often less comfortable with the application of mathematics compared with students such as physicists, for whom complex maths is a more obvious everyday tool. Nevertheless, it is important to realise that biology does require competent maths skills in many areas. It is important to practise these skills so you are familiar with them as part of your routine study of the subject.

Confidence with mental arithmetic is very helpful, but among the most important skills is that of taking care and checking calculations. We may not be required to understand the detailed theory of the maths we use, but we do need to be able to apply the skills accurately, whether simply calculating percentages or means, or substituting numbers into complex-looking algebraic equations, such as in statistical tests.

This chapter is designed to help with some of the regularly encountered mathematical problems in biology.

Working with the correct units

In biology it is very important to be secure in the use of correct units. These must always be written clearly in calculations.

Maths link
MS 0.1

Base units

The units we use are from the Système Internationale – the SI units. In biology we most commonly use the SI base units:

- metre (m) for length, height, distance
- kilogram (kg) for mass
- second (s) for time
- mole (mol) for the amount of a substance.

You should develop good habits right from the start, being careful to use the correct abbreviation for each unit used. For example, seconds should be abbreviated to s, not 'sec' or 'S'.

Derived units

Biologists also use SI derived units, such as:

- square metres (m^2) for area
- cubic metre (m^3) for volume
- cubic centimetre (cm^3), also written as millilitre (ml), for volume
- degree Celsius (°C) for temperature
- mole per litre (mol/L, mol dm^{-3}) is usually used for concentration of a substance in solutions (although the official SI derived unit is moles per cubic metre)
- joule (J) for energy
- pascal (Pa) for pressure
- volt (V) for electrical potential.

265

Mathematical section to support and develop your mathematical skills required for your course. Remember, at least 10% of your exam will involve mathematical skills.

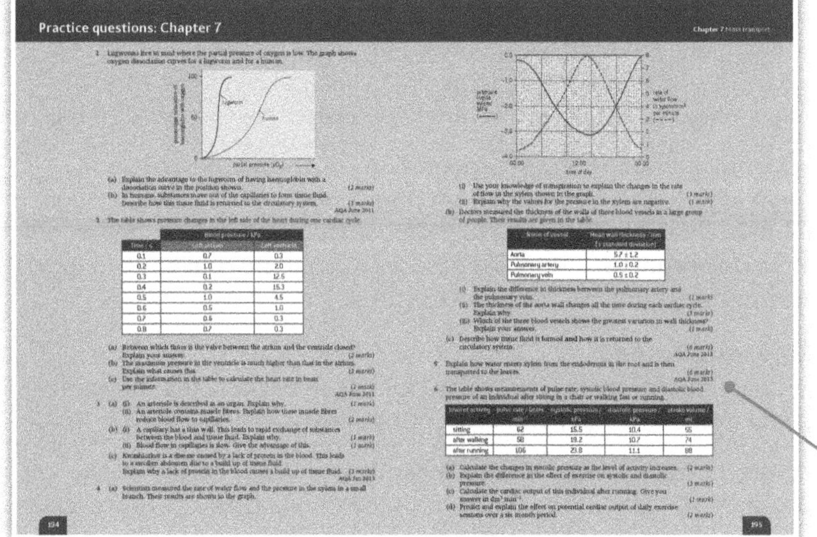
Practice questions: Chapter 7

Practice questions at the end of each chapter including questions that cover practical and maths skills.

1 Biological molecules
1.1 Carbohydrates

Learning objectives:

- → Describe how large molecules like carbohydrates are constructed.
- → Describe the structure of a monosaccharide.
- → Describe how to carry out the Benedict's test for reducing sugars.

Specification reference: 3.1.1.1 and 3.1.1.2

As the word suggests, carbohydrates are carbon molecules (carbo) combined with water (hydrate). Some carbohydrate molecules are small whilst others are large.

Life based on carbon

Carbon atoms have an unusual feature. They very readily form bonds with other carbon atoms. This allows a sequence of carbon atoms of various lengths to be built up. These form a backbone along which other atoms can be attached. This permits an immense number of different types and sizes of molecule, all based on carbon. The variety of life that exists on Earth is a consequence of living organisms being based on the versatile carbon atom. As a result, carbon-containing molecules are known as organic molecules. In living organisms, there are relatively few other atoms that attach to carbon. Life is therefore based on a small number of chemical elements.

Hint

In biology certain prefixes are commonly used to indicate numbers. There are two systems, one based on Latin and the other on Greek. The Greek terms that are used when referring to chemicals are:

- mono – one
- di – two
- tri – three
- tetra – four
- penta – five
- hexa – six
- poly – many

The making of large molecules

Many organic molecules, including carbohydrates, are made up of a chain of individual molecules. Each of the individual molecules that make up these chains is given the general name **monomer**. The carbon atoms of these monomers join to form longer chains. These longer chains of repeating monomer units are called **polymers**. How this happens is explained in Topic 1.2. Biological molecules such as carbohydrates and proteins are often polymers. These polymers are based on a surprisingly small number of chemical elements. Most are made up of just four elements: carbon, hydrogen, oxygen, and nitrogen.

In carbohydrates, the basic monomer unit is a sugar, otherwise known as a saccharide. A single monomer is therefore called a **monosaccharide**. A pair of monosaccharides can be combined to form a **disaccharide**. Monosaccharides can also be combined in much larger numbers to form **polysaccharides**.

Monosaccharides

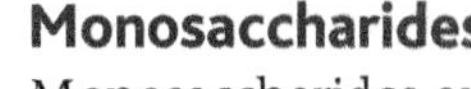

Monosaccharides are sweet-tasting, soluble substances that have the general formula $(CH_2O)_n$, where n can be any number from 3 to 7.

Perhaps the best-known monosaccharide is glucose. This molecule is a hexose (6-carbon) sugar and has the formula $C_6H_{12}O_6$. However, the atoms of carbon, hydrogen, and oxygen can be arranged in many different ways, forming **isomers**. Although the molecular structure is often shown as a straight chain for convenience, the atoms actually form a ring, as in Figure 1, which can take a number of forms.

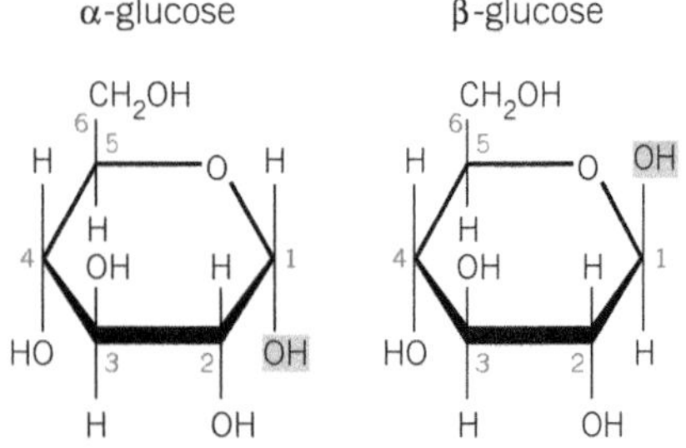

▲ **Figure 1** *Molecular arrangements of α- and β-glucose (five carbon atoms at the intersection of the lines have been omitted for simplicity. Each line represents a covalent bond).*

Test for reducing sugars

All monosaccharides and some disaccharides (e.g., maltose) are reducing sugars. Reduction is a chemical reaction involving the gain of electrons. A reducing sugar is therefore a sugar that can donate electrons to (or reduce) another chemical, in this case Benedict's reagent. The test for a reducing sugar is therefore known as the Benedict's test. Benedict's reagent is an alkaline solution of copper(II) sulfate. When a reducing sugar is heated with Benedict's reagent it forms an insoluble orange-brown to red precipitate of copper(I) oxide. The test is carried out as follows:

- Add 2 cm^3 of the food sample to be tested to a test tube. If the sample is not already in liquid form, first grind it up in water.
- Add an equal volume of Benedict's reagent.
- Heat the mixture in a gently boiling water bath for 5 minutes.

1 Food sample dissolved in water

2 Equal volume of Benedict's reagent added

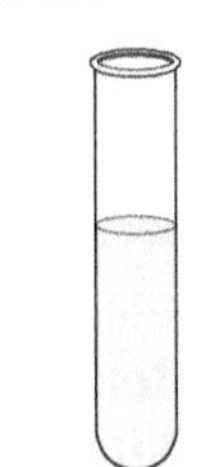

3 Heated in water bath – if reducing sugar present, solution turns orange–brown

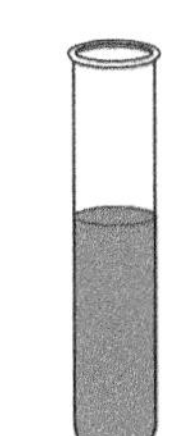

▲ **Figure 2** *The Benedict's test*

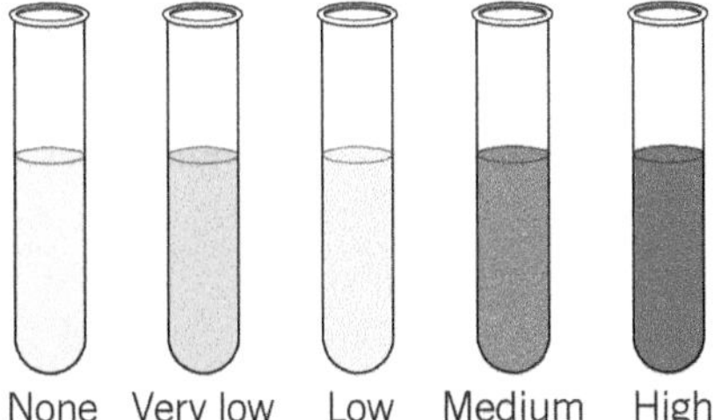

▲ **Figure 3** *Results of Benedict's test according to the concentration of reducing sugar present*

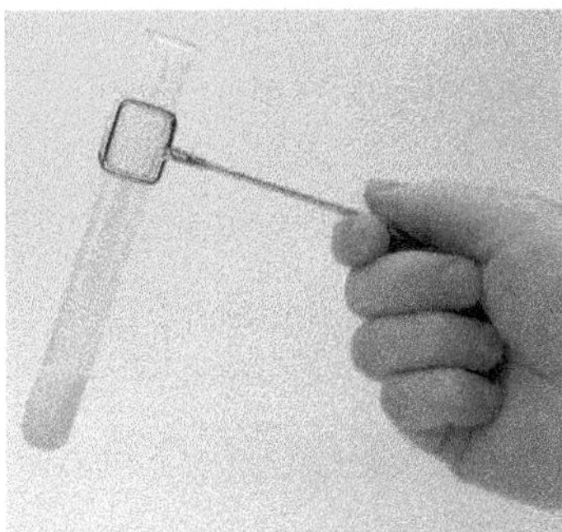

▲ **Figure 4** *If a reducing sugar is present, an orange-brown colour is formed*

Semi-quantitative nature of the Benedict's test

Table 1 shows the relationship between the concentration of reducing sugar and the colour of the solution and precipitate formed during the Benedict's test. The differences in colour mean that the Benedict's test is semi-quantitative, that is it can be used to estimate the approximate amount of reducing sugar in a sample.

The Benedict's test was carried out on five food samples. The results are shown in Table 2.

1. Place the letters in sequence of the increasing amount of reducing sugar in each sample.
2. Suggest a way, other than comparing colour changes, in which different concentrations of reducing sugar could be estimated.
3. Explain why it is not possible to distinguish between very concentrated samples, even though their concentrations are different.

Study tip

The Benedict's test may be a practical exercise but make sure that you have a detailed knowledge of the procedure.

Summary questions

1. Large molecules often contain carbon. Why is this?
2. What is the general name for a molecule that is made up of many similar repeating units?
3. Why does Benedict's reagent turn red when heated with a reducing sugar?

▼ **Table 1** *The Benedict's test*

Concentration of reducing sugar	Colour of solution and precipitate
None	Blue
Very low	Green
Low	Yellow
Medium	Brown
High	Red

▼ **Table 2**

Sample	Colour of solution
A	Yellowish brown
B	Green
C	Red
D	Dark brown
E	Yellowish green

1.2 Carbohydrates – disaccharides and polysaccharides

Learning objectives:

- → Describe how monosaccharides are linked together to form disaccharides.
- → Describe how α-glucose molecules are linked to form starch.
- → Describe the test for non-reducing sugars.
- → Describe the test for starch.

Specification reference: 3.1.1.2

You have seen that, in carbohydrates, the monomer unit is called a monosaccharide. Pairs of monosaccharides can be combined to form a **disaccharide**. Monosaccharides can also be combined in much larger numbers to form **polysaccharides**.

Disaccharides

When combined in pairs, monosaccharides form a disaccharide. For example:

- Glucose linked to glucose forms maltose.
- Glucose linked to fructose forms sucrose.
- Glucose linked to galactose forms lactose.

When the monosaccharides join, a molecule of water is removed and the reaction is therefore called a **condensation reaction**. The bond that is formed is called a **glycosidic bond**.

When water is added to a disaccharide under suitable conditions, it breaks the glycosidic bond releasing the constituent monosaccharides. This is called **hydrolysis** (addition of water that causes breakdown).

Figure 1a illustrates the formation of a glycosidic bond by the removal of water (condensation reaction). Figure 1b shows the breaking of the glycosidic bond by the addition of water (hydrolysis reaction).

a *Formation of glycosidic bond by removal of water (condensation reaction)*

b *Breaking of glycosidic bond by addition of water (hydrolysis reaction)*

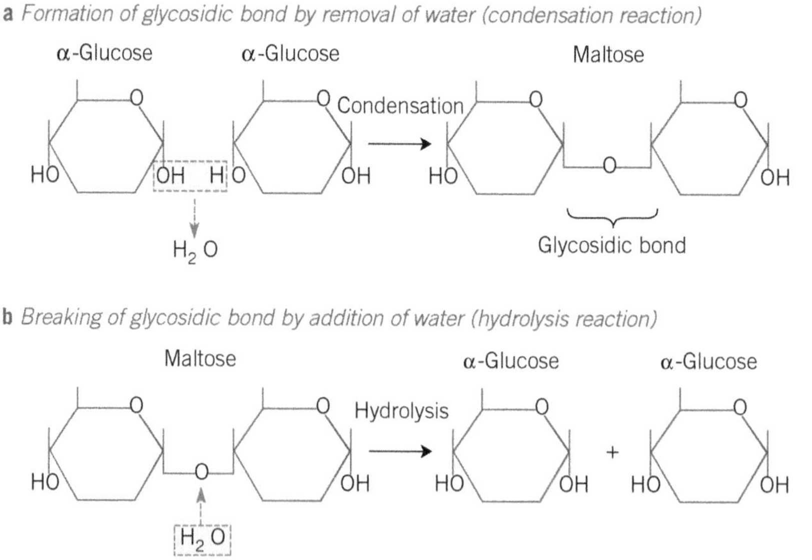

▲ **Figure 1** *Formation and breaking of a glycosidic bond by condensation and hydrolysis*

Study tip

Be clear about the difference between the terms condensation and hydrolysis. Both involve the use of water in reactions. However, condensation is the *giving out* of water in reactions whilst hydrolysis is the *taking in* of water to split molecules in reactions.

Hint

To help you remember that condensation is *giving out* water, think of condensation on your bedroom window on cold mornings. This is water that you have *given out* in your breath.

Hint

Polysaccharides illustrate an important principle – that a few basic monomer units can be combined in a number of different ways to give a large range of different biological molecules.

Test for non-reducing sugars

Some disaccharides (e.g., maltose) are reducing sugars. To detect these you would use the **Benedict's test**, as described in Topic 1.1. Other disaccharides, such as sucrose, are known as non-reducing sugars because they do not change the colour of Benedict's reagent when they are heated with it. In order to detect a non-reducing sugar it must first be broken down into its monosaccharide components by hydrolysis. The process is carried out as follows:

- If the sample is not already in liquid form, you must first grind it up in water.
- Add 2 cm³ of the food sample being tested to 2 cm³ of Benedict's reagent in a test tube.
- Place the test tube in a gently boiling water bath for 5 minutes. If the Benedict's reagent does not change colour (the solution remains blue), then a reducing sugar is *not* present.
- Add another 2 cm³ of the food sample to 2 cm³ of dilute hydrochloric acid in a test tube and place the test tube in a gently boiling water bath for 5 minutes. The dilute hydrochloric acid will hydrolyse any disaccharide present into its constituent monosaccharides.
- Slowly add some sodium hydrogencarbonate solution to the test tube in order to neutralise the hydrochloric acid. (Benedict's reagent will not work in acidic conditions.) Test with pH paper to check that the solution is alkaline.
- Retest the resulting solution by heating it with 2 cm³ of Benedict's reagent in a gently boiling water bath for 5 minutes.
- If a non-reducing sugar was present in the original sample, the Benedict's reagent will now turn orange-brown. This is due to the reducing sugars that were produced from the hydrolysis of the non-reducing sugar.

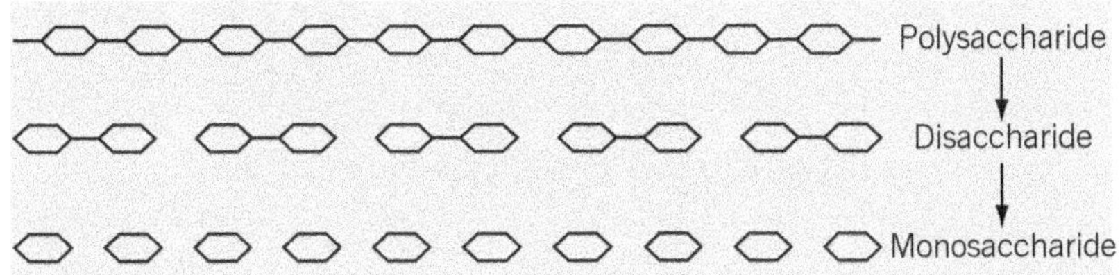

▲ **Figure 2** *The breakdown of a polysaccharide into disaccharides and monosaccharides*

Polysaccharides

Polysaccharides are polymers, formed by combining together many monosaccharide molecules. The monosaccharides are joined by glycosidic bonds that were formed by **condensation reactions**. As polysaccharides are very large molecules, they are usually insoluble. This feature makes them suitable for storage. When they are hydrolysed, polysaccharides break down into disaccharides or monosaccharides (Figure 2). Some polysaccharides, such as cellulose (see Topic 1.3), are not used for storage but give structural support to plant cells.

Starch is a polysaccharide that is found in many parts of plants in the form of small granules or grains, for example, starch grains in chloroplasts. It is formed by the linking of between 200 and 100 000 α-glucose molecules by glycosidic bonds in a series of condensation reactions. More details of starch and its functions are given in Topic 1.3.

Test for starch

Starch is easily detected by its ability to change the colour of the iodine in potassium iodide solution from yellow to blue-black (Figure 3). The test is carried out at room temperature. The method is as follows:

- Place 2 cm³ of the sample being tested into a test tube (or add two drops of the sample into a depression on a spotting tile).
- Add two drops of iodine solution and shake or stir.
- The presence of starch is indicated by a blue-black coloration.

▲ **Figure 3** *Test for starch*

Summary questions

1. Which one, or more, monomer units make up each of the following carbohydrates?
   a. lactose
   b. sucrose
   c. starch
2. Glucose ($C_6H_{12}O_6$) combines with fructose ($C_6H_{12}O_6$) to form the disaccharide sucrose. From your knowledge of how disaccharides are formed, work out the formula of sucrose.
3. To hydrolyse a disaccharide it can be boiled with hydrochloric acid, but if hydrolysis is carried out by an enzyme a much lower temperature (40 °C) is used. Why is this?

1.3 Starch and cellulose

Learning objectives:

- → Describe how α-glucose monomers are arranged to form the polymers of starch and glycogen.
- → Describe how β-glucose monomers are arranged to form the polymer cellulose.
- → Describe how the molecular structures of starch, glycogen, and cellulose relate to their functions.

Specification reference: 3.1.1.2

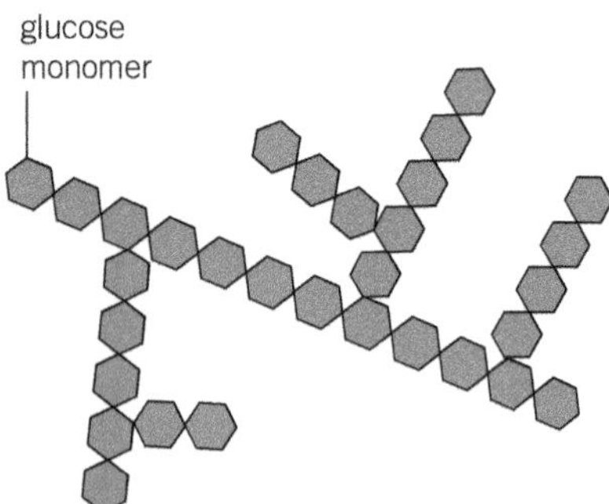

▲ **Figure 2** *Branched polysaccharides have both 1,4 and 1,6 glycosidic bonds*

▲ **Figure 3** *False colour scanning electron micrograph (SEM) of starch grains (blue) in the cells of a potato. Starch is a compact storage material.*

In organisms, a wide range of different molecules with very different properties can be made from a limited range of smaller molecules. What makes the larger molecules different is the various ways in which the smaller molecules are combined to form them. Let us look at some of these larger molecules by considering three important polysaccharides.

Starch

Starch is a polysaccharide that is found in many parts of a plant in the form of small grains. Especially large amounts occur in seeds and storage organs, such as potato tubers. It forms an important component of food and is the major energy source in most diets. Starch is made up of two polysaccharides, amylose and amylopectin. Amylose is made up of chains of α-glucose monosaccharides linked by glycosidic bonds that are formed by **condensation** reactions. The unbranched chain is wound into a tight coil that makes the molecule very compact and so an ideal storage molecule. The structure of an amylose molecule is shown in Figure 1. Amylopectin is a branched polysaccharide formed from α-glucose molecules linked by both 1,4 and 1,6 glycosidic bonds.

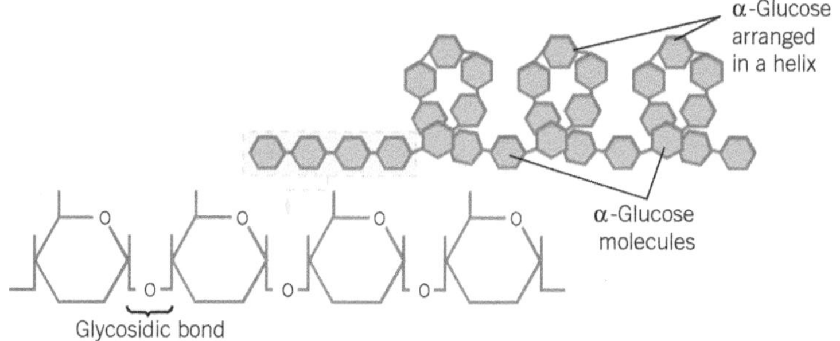

▲ **Figure 1** *Structure of an amylose molecule*

The main role of starch is energy storage, something it is especially suited for because:

- it is insoluble and therefore does not tend to draw water into the cells by **osmosis**
- being insoluble, it does not easily diffuse out of cells
- it is compact, so a lot of it can be stored in a small space
- when hydrolysed it forms α-glucose, which is both easily transported and readily used in respiration.

Starch is never found in animal cells. Instead a similar polysaccharide, called glycogen, serves the same role.

Amylopectin and glycogen

Glycogen is very similar in structure to amylopectin in starch but has shorter chains and is more highly branched. It is sometimes called 'animal starch' because it is the major carbohydrate storage product of animals. In animals it is stored as small granules mainly in the muscles and the liver. Its structure suits it for storage for the same reasons as those given for starch. However, because it is made up of smaller chains, it is even more readily hydrolysed to α-glucose. Glycogen is found in animal cells but never in plant cells. The roles of the hormones insulin and glucagon will be covered in Chapter 27.

Cellulose

Cellulose differs from starch and glycogen in one major respect: it is made of monomers of β-glucose rather than α-glucose. This seemingly small variation produces fundamental differences in the structure and function of this polysaccharide. The main reason for this is that, in the β-glucose units, the positions of the —H group and the —OH group on a single carbon atom are reversed. In β-glucose the —OH group is above, rather than below, the ring. This means that to form glycosidic links, each β-glucose molecule must be rotated by 180° compared to its neighbour. The result is that the —CH_2OH group on each β-glucose molecule alternates between being above and below the chain (Figure 4).

Rather than forming a coiled chain like starch, cellulose has straight, unbranched chains. In cell walls these chains run parallel to one another, allowing **hydrogen bonds** to form cross-linkages between adjacent chains. Whilst each individual hydrogen bond adds very little to the strength of the molecule, the sheer overall number of them makes a considerable contribution to strengthening the cellulose cell wall. This makes cellulose a valuable structural material in plants. The arrangement of β-glucose chains in a cellulose molecule is shown in Figure 4.

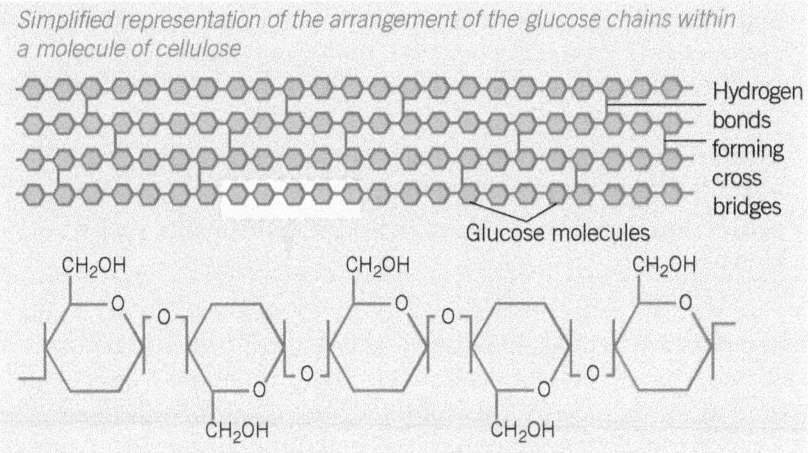

Simplified representation of the arrangement of the glucose chains within a molecule of cellulose

Hydrogen bonds forming cross bridges

Glucose molecules

CH_2OH CH_2OH CH_2OH CH_2OH CH_2OH

◀ **Figure 4** *Structure of a cellulose molecule*

The cellulose chain, unlike that of starch, has adjacent glucose molecules rotated by 180°. This allows hydrogen bonds to be formed between the hydroxyl (—OH) groups on adjacent parallel chains, which help to give cellulose its structural stability.

The cellulose molecules are grouped together to form microfibrils which, in turn, are arranged in parallel groups called fibres (Figure 5).

Cellulose is a major component of plant cell walls and provides rigidity to the plant cell. The cellulose cell wall also prevents the cell from bursting as water enters it by osmosis. It does this by exerting an inward pressure that stops any further influx of water. As a result, living plant cells are turgid and push against one another, making herbaceous parts of the plant semi-rigid. This is especially important in maintaining stems and leaves in a turgid state so that they can provide the maximum surface area for photosynthesis.

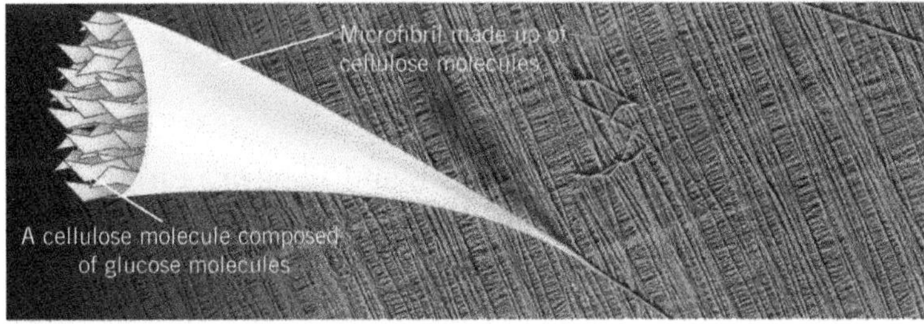

▲ **Figure 5** *Structure of a cellulose microfibril*

Summary questions

From the following list of carbohydrates choose one or more that most closely fits each of the statements below. Each carbohydrate may be used once, more than once, or not at all.

α-glucose β-glucose starch glycogen cellulose

1. Stains deep blue with iodine solution.
2. Is known as 'animal starch'.
3. Found in plants.
4. Are polysaccharides.
5. Monosaccharide found in starch.
6. Has a structural function.
7. Can be hydrolysed.

1.4 Lipids

Learning objectives:

- → Describe how triglycerides are formed.
- → Describe how fatty acids can vary.
- → Describe the structure of a phospholipid.
- → Describe how you can identify the presence of a lipid.

Specification reference: 3.1.1.3

Lipids are a varied group of substances that share the following characteristics:

- They contain carbon, hydrogen, and oxygen.
- The proportion of oxygen to carbon and hydrogen is smaller than in carbohydrates.
- They are insoluble in water.
- They are soluble in organic solvents such as alcohols and acetone.

The three main groups of lipids are **triglycerides** (fats and oils), phospholipids, and waxes.

Roles of lipids

The main role of lipids is in **plasma membranes**. Phospholipids contribute to the flexibility of membranes and the transfer of lipid-soluble substances across them. Other roles of lipids include:

- **an energy source**. When oxidised, lipids provide more than twice the energy as the same mass of carbohydrate.
- **waterproofing**. Lipids are insoluble in water and are therefore useful as a waterproofing material. Both plants and insects have waxy cuticles that conserve water, whilst mammals produce an oily secretion from the sebaceous glands in the skin.
- **insulation**. Fats are slow conductors of heat and when stored beneath the body surface help to retain body heat.
- **protection**. Fat is often stored around delicate organs, such as the kidney.

Fats are solid at room temperature (10–20 °C), whereas oils are liquid. Triglycerides are so called because they have three (tri) fatty acids combined with glycerol (glyceride). Each fatty acid forms a bond with glycerol in a **condensation** reaction (Figure 1). **Hydrolysis** of a triglyceride therefore produces glycerol and three fatty acids.

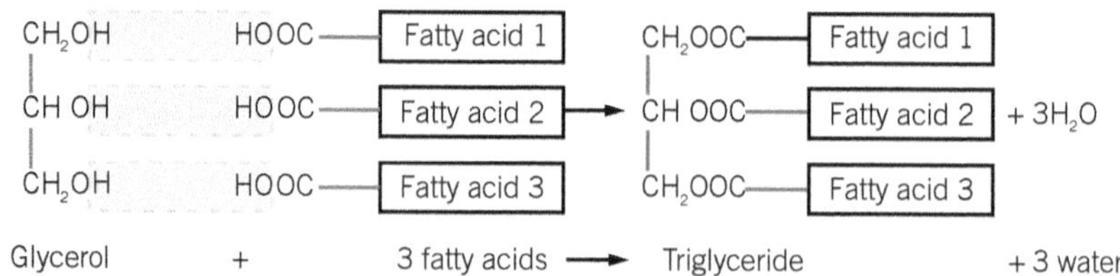

▲ **Figure 1** *Formation of a triglyceride*

The three fatty acids may all be the same, thereby forming a simple triglyceride, or they may be different, in which case a mixed triglyceride is produced. In either case it is a condensation reaction.

As the glycerol molecule in all triglycerides is the same, the differences in the properties of different fats and oils come from variations in the fatty acids. There are over 70 fatty acids and all have a carboxyl (—COOH) group with a hydrocarbon chain attached and so can be represented by the formula RCOOH, where the hydrocarbon chain

Hint

Fats are generally made of saturated fatty acids, whereas oils are made of unsaturated ones.

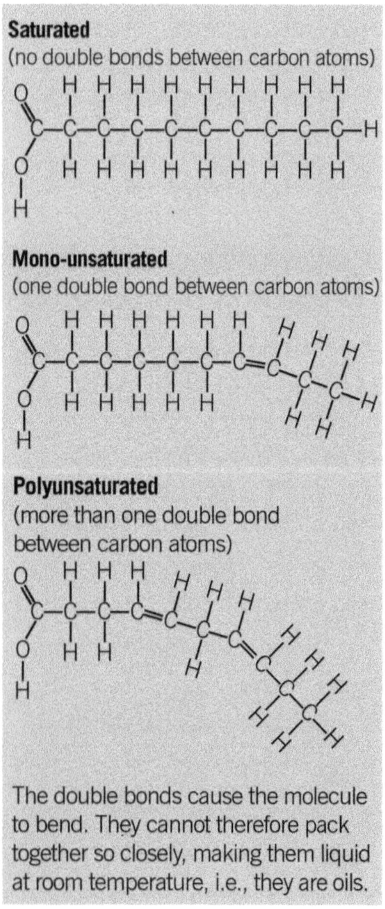

Saturated
(no double bonds between carbon atoms)

Mono-unsaturated
(one double bond between carbon atoms)

Polyunsaturated
(more than one double bond between carbon atoms)

The double bonds cause the molecule to bend. They cannot therefore pack together so closely, making them liquid at room temperature, i.e., they are oils.

▲ **Figure 2** *Saturated and unsaturated fatty acids*

R can be saturated or unsaturated. If this chain has no carbon–carbon double bonds, the fatty acid is then described as **saturated**, because all the carbon atoms are linked to the maximum possible number of hydrogen atoms, that is, they are saturated with hydrogen atoms. If there is a single double bond, it is **mono-unsaturated**; if more than one double bond is present, it is **polyunsaturated**. These differences are illustrated in Figure 2.

Phospholipids

Phospholipids are similar to lipids except that one of the fatty acid molecules is replaced by a phosphate molecule (Figure 3). Whereas fatty acid molecules repel water (are hydrophobic), phosphate groups attract water (are hydrophilic). A phospholipid is therefore made up of two parts:

- **a hydrophilic 'head'**, which interacts with water (is attracted to it) but not with fat
- **a hydrophobic 'tail'**, which orients itself away from water but mixes readily with fat.

This means that when these phospholipid molecules are placed in water they position themselves so that the hydrophilic heads are as close to the water as possible and the hydrophobic tails are as far away from the water as possible (Figure 4).

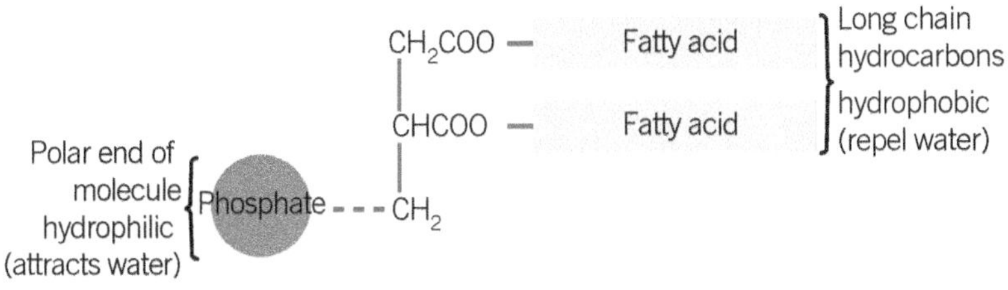

▲ **Figure 3** *Structure of a phospholipid*

Test for lipids

The test for lipids is known as the emulsion test and is carried out as follows:

1. Take a completely dry and grease-free test tube.
2. To 2 cm^3 of the sample being tested, add 5 cm^3 of ethanol.
3. Shake the tube thoroughly to dissolve any lipid in the sample.
4. Add 5 cm^3 of water and shake gently.
5. A cloudy, white solution indicates the presence of a lipid.
6. As a control, repeat the procedures using water instead of the sample; the final solution should remain clear.

The cloudiness is due to any lipid in the sample being finely dispersed in the water to form an emulsion. Light passing through this emulsion is refracted as it passes from oil droplets to water droplets, making it appear cloudy.

Study tip

Do not use terms like water-loving and water-hating. Use the correct scientific terms hydrophilic and hydrophobic.

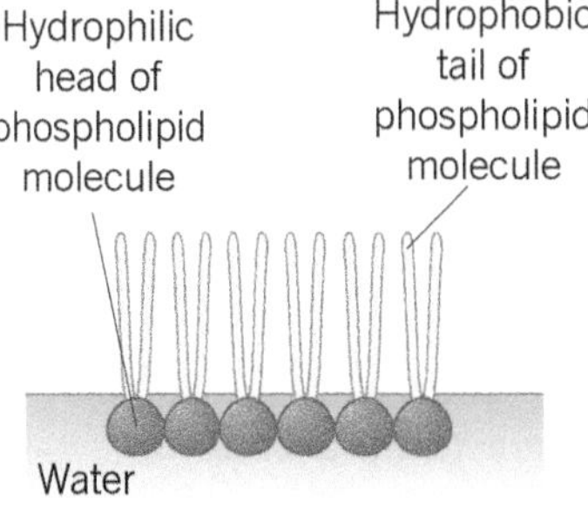

a *Monolayer*

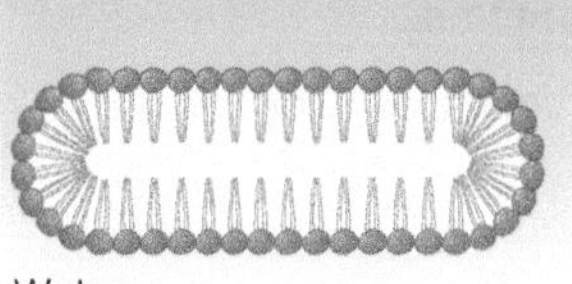

b *Bilayered sheet*

▲ **Figure 4** *Two arrangements of phospholipid molecules in water*

Summary questions

1. In the following passage give the most suitable word for each of the letters **a** to **e**.

   Fats and oils make up a group of lipids called **a** which, when hydrolysed, form **b** and fatty acids. A fatty acid with more than one carbon–carbon double bond is called **c**. In a phospholipid, the number of fatty acids is **d**; these are described as **e** because they repel water.

2. State **two** differences between a triglyceride molecule and a phospholipid molecule.
3. Organisms that move frequently use lipids rather than carbohydrates as an energy store. Why might this be an advantage?

1.5 Proteins

Learning objectives:

- → Describe how amino acids are linked to form polypeptides – the primary structure of proteins.
- → Describe how polypeptides are arranged to form the secondary structure and then the tertiary structure of a protein.
- → Describe how the quaternary structure of a protein is formed.
- → Describe how proteins can be identified.

Specification reference: 3.1.1.4

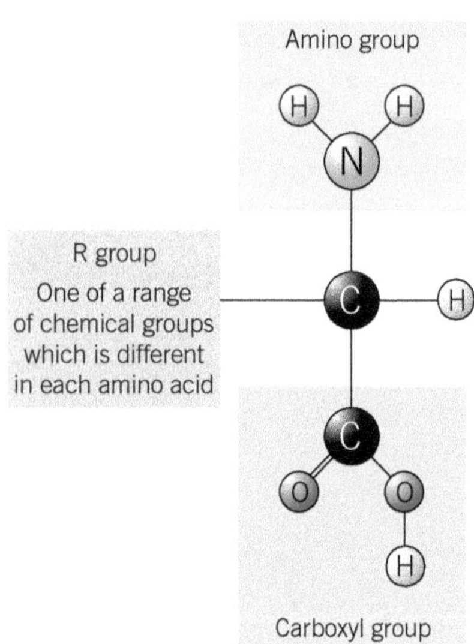

▲ **Figure 1** *The general structure of an amino acid*

> **Study tip**
>
> Distinguish between **condensation reactions** (molecules combine producing water) and **hydrolysis reactions** (molecules are split up by taking in water).

The word 'protein' is a Greek word meaning 'of first importance'. Proteins are very large molecules. The types of carbohydrates and lipids in all organisms are relatively few and they are very similar. However each organism has numerous proteins that differ from species to species. The shape of any one type of protein molecule differs from that of all other types of proteins. One group of proteins, enzymes, is involved in almost every living process. There is a vast range of different enzymes that between them perform a very diverse number of functions.

Structure of an amino acid

Amino acids are the basic **monomer** units that combine to make up a **polymer** called a polypeptide. Only around 100 amino acids have been identified, of which 20 occur naturally in proteins.

Every amino acid has a central carbon atom to which are attached four different chemical groups:

- amino group (—NH_2) – a basic group from which the amino part of the name amino acid is derived
- carboxyl group (—COOH) – an acidic group which gives the amino acid the acid part of its name
- hydrogen atom (—H)
- R group – a variety of different chemical groups. Each amino acid has a different R group.

The general structure of an amino acid is shown in Figure 1.

The formation of a peptide bond

In the same way that monosaccharide monomers combine to form disaccharides (see Topic 1.2), so amino acid monomers can combine to form a dipeptide. The process is essentially the same – namely, the removal of a water molecule in a **condensation** reaction. The water is made by combining an —OH from the carboxyl group of one amino acid with an —H from the amino group of another amino acid. The two amino acids then become linked by a new **peptide bond** between the carbon atom of one amino acid and the nitrogen atom of the other. The formation of a peptide bond is illustrated in Figure 2. In the same way as a glycosidic bond of a disaccharide can be broken by the addition of water (**hydrolysis**), so the peptide bond of a dipeptide can also be broken by hydrolysis to give its two constituent amino acids.

The primary structure of proteins – polypeptides

Through a series of condensation reactions, many amino acid monomers can be joined together in a process called **polymerisation**. The resulting chain of many hundreds of amino acids is called a **polypeptide**. The sequence of amino acids in a polypeptide chain forms the primary structure of any protein. As polypeptides have many (usually hundreds) of the 20 naturally occurring amino acids joined together in a particular sequence, it follows that there is an almost limitless number of possible combinations, and therefore types, of primary protein structure.

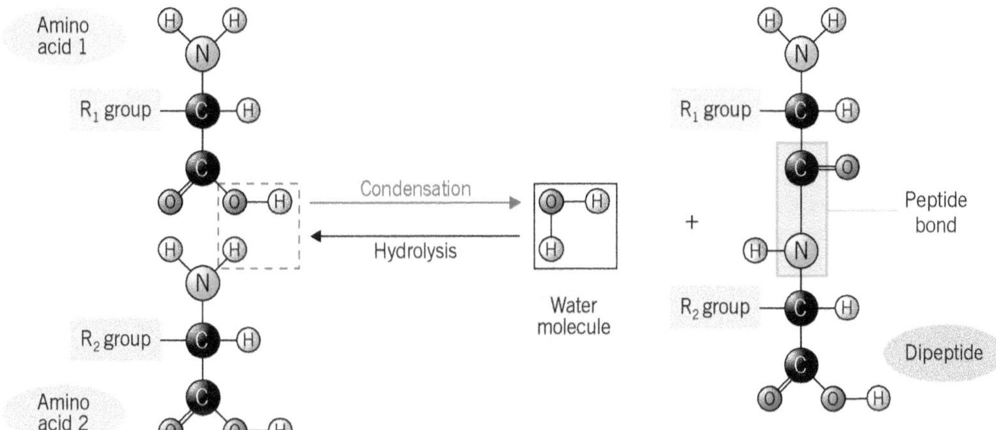

▲ **Figure 2** *The formation of a peptide bond*

It is the primary structure of a protein that determines its ultimate shape and hence its function. A change in just a single amino acid in this primary sequence can lead to a change in the shape of the protein and may stop it carrying out its function. In other words, a protein's shape is very specific to its function. Change its shape and it will function less well, if at all.

Some proteins consist of a single polypeptide. More commonly, however, a protein is made up of a number of polypeptide chains.

The secondary structure of proteins

The linked amino acids that make up a polypeptide possess both —NH and —C=O groups on either side of every peptide bond. The hydrogen of the —NH group has an overall positive charge whilst the O of the —C=O group has an overall negative charge. These two groups therefore readily form weak bonds, called **hydrogen bonds**. This causes the long polypeptide chain to be twisted into a 3-D shape, such as the coil known as an α-helix. Figure 3 illustrates the structure of an α-helix.

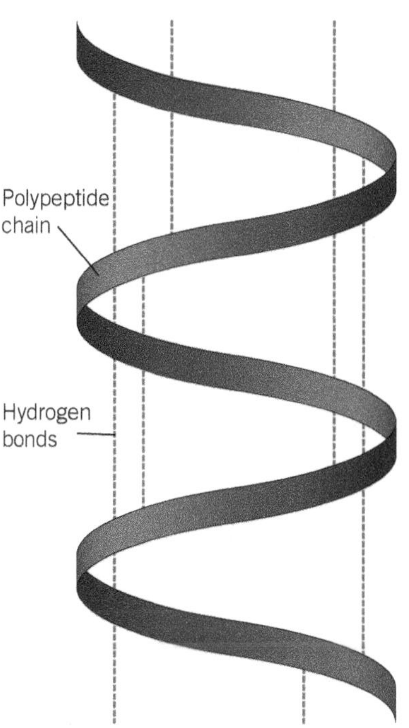

▲ **Figure 3** *Structure of the α-helix*

Tertiary structure of proteins

The α-helices of the secondary structure can be twisted and folded even more to give the polypeptide chain a complex, and often unique, 3-D structure (Figure 4). This is known as the tertiary structure. This structure is maintained by a number of different bonds, including:

- **disulfide bonds** – which are fairly strong and therefore not easily broken down.
- **ionic bonds** – which are formed between any carboxyl and amino groups that are not involved in forming peptide bonds. They are weaker than disulfide bonds and are easily broken by changes in pH.
- **hydrogen bonds** – which are numerous but easily broken.

It is the 3-D shape of a protein that is important when it comes to how it functions. It makes each protein distinctive and allows it to recognise, and be recognised by, other molecules. It can then interact with them in a very specific way.

Hint

Remember that, although the 3-D structure is important to how a protein functions, it is the sequence of amino acids (primary structure) that determines the 3-D shape in the first place.

a The primary structure of a protein is the sequence of amino acids in the polypeptide. This sequence determines its properties and shape. Following the elucidation of the amino acid sequence of the hormone insulin by Frederick Sanger in 1954, the primary structure of many other proteins is now known.

b The secondary structure is the shape that the polypeptide chain forms as a result of hydrogen bonding. This is most often a spiral known as the α-helix, although other configurations occur.

c The tertiary structure is due to the bending and twisting of the polypeptide helix into a compact structure. All three types of bond – disulfide, ionic, and hydrogen – contribute to the maintenance of the tertiary structure.

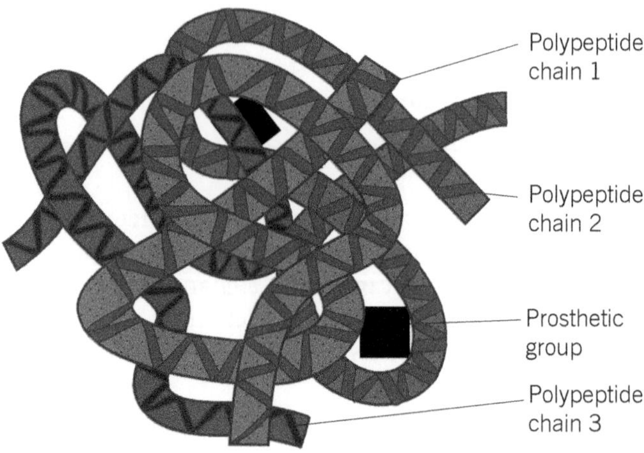

d The quaternary structure arises from the combination of a number of different polypeptide chains and associated non-protein (prosthetic) groups into a large, complex protein molecule.

▲ **Figure 4** *Structure of proteins*

Quaternary structure of proteins

Large proteins often form complex molecules containing a number of individual polypeptide chains that are linked in various ways. There may also be non-protein (prosthetic) groups associated with the molecules (Figure 4d), such as the iron-containing haem group in haemoglobin.The structure and function of haemoglobin are considered in Topic 5.7.

Test for proteins

The most reliable protein test is the biuret test, which detects peptide bonds. It is performed as follows:

- Place a sample of the solution to be tested in a test tube and add an equal volume of sodium hydroxide solution at room temperature.
- Add a few drops of very dilute (0.05%) copper(II) sulfate solution and mix gently.
- A purple coloration indicates the presence of peptide bonds and hence a protein. If no protein is present, the solution remains blue.

Study tip

You can simply refer to adding **biuret** reagent to test for protein. A purple colour shows protein is present; a blue colour indicates that protein is absent.

Hint

Think of the polypeptide chain as a piece of string. In a fibrous protein many pieces of the string are twisted together into a rope, whilst in a globular protein the pieces of string, usually fewer, are rolled into a ball.

Summary questions

1 What type of bond links amino acids together?

2 What type of reaction is involved in linking amino acids together?

3 What **four** different components make up an amino acid?

Protein shape and function

Proteins perform many different roles in living organisms. Their roles depend on their molecular shape, which can be of two basic types.

- Fibrous proteins, such as collagen, have structural functions.
- Globular proteins, such as enzymes and haemoglobin, carry out metabolic functions. You will learn more about enzymes in Topic 3.1 and about haemoglobin in Topic 5.7.

It is the very different structure and shape of each of these types of protein that enables them to carry out their functions.

Fibrous proteins

Fibrous proteins form long chains that run parallel to one another. These chains are linked by cross-bridges and so form very stable molecules. One example is **collagen**. Its molecular structure is as follows:

- The primary structure is an unbranched polypeptide chain.
- In the secondary structure the polypeptide chain is very tightly wound.
- Lots of the amino acid glycine helps close packing.
- In the tertiary structure the chain is twisted into a second helix.
- Its quaternary structure is made up of three such polypeptide chains wound together in the same way that individual fibres are wound together in a rope.

Collagen is found in tendons. Tendons join muscles to bones. When a muscle contracts the bone is pulled in the direction of the contraction.

1 Explain why the quaternary structure of collagen makes it a suitable molecule for a tendon.

2 Suggest how the cross-linkages between the amino acids of polypeptide chains increase the strength and stability of a collagen fibre.

3 Explain why the arrangement of collagen molecules is necessary for the efficient functioning of a tendon.

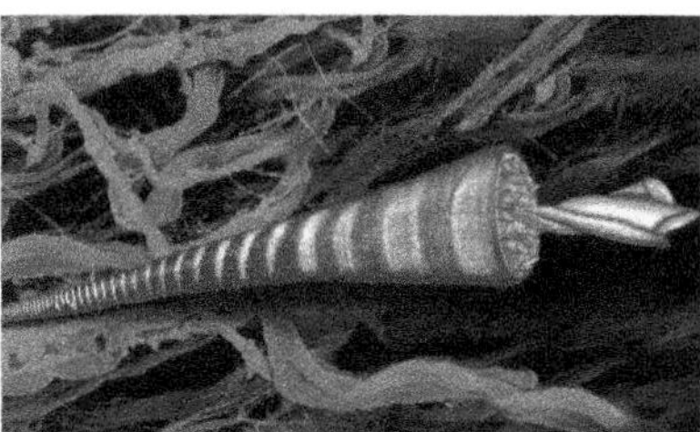

Figure 5 *Fine structure of the fibrous protein collagen*

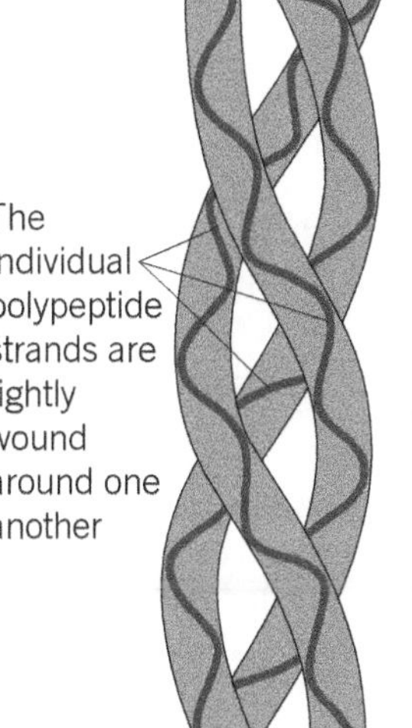

Figure 6 *Structure of fibrous proteins*

Practice questions: Chapter 1

1 **(a)** The table shows some substances found in cells. Complete the table to show the properties of these substances. Put a tick in the box if the statement is correct.

	Substance			
Statement	Starch	Glycogen	Deoxyribose	DNA helicase
Substance contains only the elements carbon, hydrogen, and oxygen				
Substance is made from amino acid monomers				
Substance is found in both animal cells and plant cells				

(4 marks)

(b) The diagram shows two molecules of β-glucose.

On the diagram, draw a box around the atoms that are removed when the two β-glucose molecules are joined by condensation. *(2 marks)*

(c) (i) Hydrogen bonds are important in cellulose molecules. Explain why. *(2 marks)*

(ii) A starch molecule has a spiral shape. Explain why this shape is important to its function in cells. *(1 mark)*

AQA Jan 2011

2 **(a)** Omega-3 fatty acids are unsaturated. What is an *unsaturated* fatty acid? *(2 marks)*

(b) Scientists investigated the relationship between the amount of omega-3 fatty acids eaten per day and the risk of coronary heart disease. The graph shows their results. Do the data show that eating omega-3 fatty acids prevents coronary heart disease? Explain your answer. *(3 marks)*

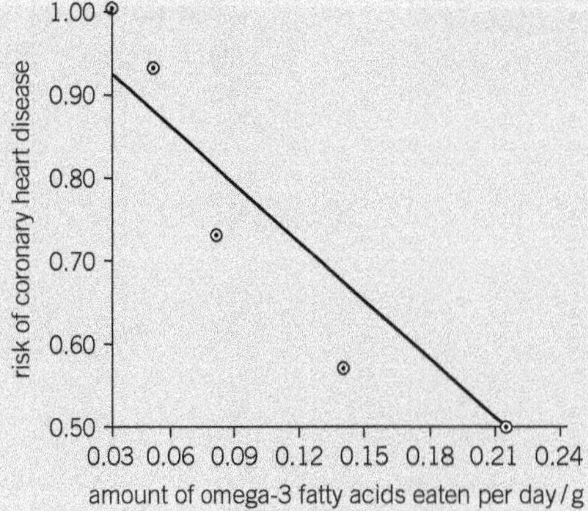

(c) Olestra is an artificial lipid. It is made by attaching fatty acids, by condensation, to a sucrose molecule. The diagram below shows the structure of olestra. The letter R shows where a fatty acid molecule has attached.

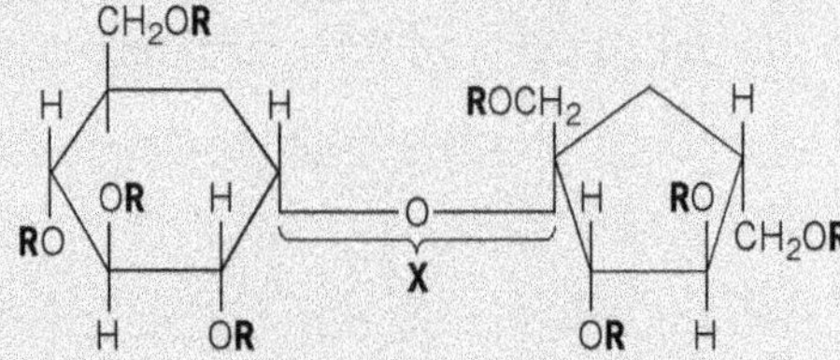

(i) Name bond **X**. *(1 mark)*

(ii) A triglyceride does *not* contain sucrose or bond **X**. Give **one** other way in which the structure of a triglyceride is different to olestra. *(1 mark)*

(iii) Starting with separate molecules of glucose, fructose, and fatty acids, how many molecules of water would be produced when one molecule of olestra is formed? *(1 mark)*

AQA Jan 2011

3 **(a)** (i) The equation shows the reaction catalysed by the enzyme lactase. Complete this equation.

Lactose + ... ⟶ Glucose + ... *(2 marks)*

(b) (ii) Name the type of chemical reaction shown in this equation. *(1 mark)*

(c) Lactase is an enzyme. Lactose is a reducing sugar.

(i) Describe how you could use the biuret test to distinguish a solution of the enzyme lactase from a solution of lactose. *(1 mark)*

(ii) Explain the result you would expect with the enzyme. *(1 mark)*

AQA Jan 2010

4 Some seeds contain lipids. Describe how you could use the emulsion test to show that a seed contains lipids. *(3 marks)*

AQA Jan 2012

Answers to the Practice Questions are available at
www.oxfordsecondary.com/oxfordaqaexams-alevel-biology

2 Cells and cell structure

2.1 Investigating the structure of cells

Learning objectives:

- → Define magnification and resolution.
- → Define cell fractionation.
- → Describe how ultracentrifugation works.

Specification reference: 3.1.2.1

▼ **Table 1** *Units of length*

Unit	Symbol	Equivalent in metres
Kilometre	km	10^3
Metre	m	1
Millimetre	mm	10^{-3}
Micrometre	µm	10^{-6}
Nanometre	nm	10^{-9}

Study tip

Careless language costs marks. Candidates sometimes state that optical microscopes have a longer wavelength than electron microscopes. What they mean to say is that *light* has a longer wavelength than a beam of *electrons*.

The cell is the basic unit of life. However, with a few exceptions, cells are not visible to the naked eye and their structure is only apparent when seen under a microscope.

Microscopy

Microscopes are instruments that magnify the image of an object. A simple convex glass lens can act as a magnifying glass but such lenses work more effectively if they are in a compound light microscope. The relatively long wavelength of light rays means that a light microscope can only distinguish between two objects if they are 0.2 µm, or further, apart. This limitation can be overcome by using beams of **electrons** rather than beams of light. With their shorter wavelengths, the beam of electrons in the electron microscope can distinguish two objects as close together as 0.1 nm.

Magnification

The material that is put under a microscope is referred to as the **object**. The appearance of this material when viewed under the microscope is referred to as the **image**.

The magnification of an object is how many times bigger the image is when compared to the object.

$$\textbf{magnification} = \frac{\textbf{size of image}}{\textbf{size of object}}$$

In practice, it is more likely that you will be asked to calculate the size of an object when you know the size of the image and the magnification. In this case:

$$\textbf{size of object} = \frac{\textbf{size of image}}{\textbf{magnification}}$$

The important thing to remember when calculating the magnification is to ensure that the units of length (Table 1) are the same for both the object and the image.

Imagine, for example, that you know an object is actually 100 nm in length and you are asked how much it is magnified in a photograph. You should first measure the object in the photograph. Suppose it is 10 mm long. The magnification is:

$$\frac{\textbf{size of image}}{\textbf{size of object}} = \frac{\textbf{10 mm}}{\textbf{100 nm}}$$

Now convert the measurements to the same units – normally the smallest – which in this case is nanometres. There are 10 000 000 nanometres in 10 millimetres and therefore the magnification is:

$$\frac{\textbf{size of image}}{\textbf{size of object}} = \frac{\textbf{10 000 000 nm}}{\textbf{100 nm}} = \frac{\textbf{100 000}}{\textbf{1}} = \textbf{×100 000 times}$$

Resolution

The resolution, or resolving power, of a microscope is the minimum distance apart that two objects can be in order for them to appear as separate items. Whatever the type of microscope, the resolving power depends on the wavelength or form of radiation used. In a light microscope it is about 0.2 µm. This means that any two objects that are 0.2 µm or more apart will be seen separately, but any objects closer together than 0.2 µm will appear as a single item. In other words, greater resolution means greater clarity – that is, the image produced is clearer and more precise.

Increasing the magnification will increase the size of an image, but does not always increase the resolution. Every microscope has a limit of resolution. Up to this point increasing the magnification will reveal more detail but beyond this point increasing the magnification will not do this. The object, whilst appearing larger, will just be more blurred.

Hint

Practise working out actual sizes from diagrams and photographs with a given scale. Practice makes it easy.

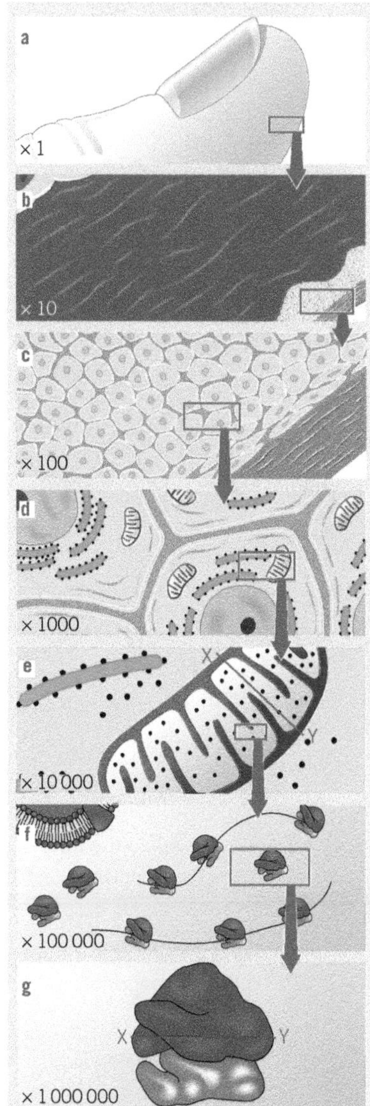

▲ **Figure 1** *The effect of progressive magnification of a portion of human skin*

Cell fractionation

In order to study the structure and function of the various organelles that make up cells, it is necessary to obtain large numbers of isolated organelles.

Cell fractionation is the process by which cells are broken up and the different components they contain, including organelles, are separated out.

Before cell fractionation can begin, the tissue is placed in a cold, **isotonic** buffered solution. The solution is:

- cold – to reduce enzyme activity that might break down the organelles
- isotonic – to prevent organelles bursting or shrinking as a result of osmotic gain or loss of water. An isotonic solution is one that has the same water potential as the original tissue.
- buffered – to maintain a constant pH.

There are two stages to cell fractionation:

Homogenation

Cells are broken up by a homogeniser (blender). This releases the organelles from the cells. The resultant fluid, known as homogenate, is then filtered to remove any complete cells and large pieces of debris.

Ultracentrifugation

Ultracentrifugation is the process by which the fragments in the filtered homogenate are separated in a machine called an ultracentrifuge. This spins tubes of homogenate at very high speed in order to create a centrifugal force. For animal cells, the process is as follows:

- The tube of filtrate is placed in the ultracentrifuge and spun at a slow speed.
- The heaviest organelles, the nuclei, are forced to the bottom of the tube, where they form a thin sediment or pellet.
- The fluid at the top of the tube (supernatant) is removed, leaving just the sediment of nuclei.
- The supernatant is transferred to another tube and spun in the ultracentrifuge at a faster speed than before.
- The next heaviest organelles, the mitochondria, are forced to the bottom of the tube.
- The process is continued in this way so that, at each increase in speed, the next heaviest organelle is sedimented and separated out (see Table 2 on the next page).

Study tip

Remember that the isotonic solution used during cell fractionation prevents *organelles* bursting or shrinking as a result of osmotic gain or loss of water. Don't refer to *cells* bursting or shrinking.

Summary questions

1 Distinguish between magnification and resolution.

2 An organelle that is 5 µm in diameter appears under a microscope to have a diameter of 1 mm. How many times has the organelle been magnified?

3 A cell organelle called a ribosome is typically 25 nm in diameter. If viewed under an electron microscope that magnifies it ×400 000, what would the diameter of the ribosome appear to be in millimetres?

4 At a magnification of ×12 000 a structure appears to be 6 mm long. What is its actual length?

5 Chloroplasts have a greater mass than mitochondria but a smaller mass than nuclei. Starting with a sample of plant cells, describe briefly how you would obtain a sample rich in chloroplasts. Use Table 2 to help you.

6 Using the magnifications given in Figure 1, calculate the actual size of the following organelles as measured along the line labelled X–Y. In your answer, use the most appropriate units from Table 1.

- a The organelle in box E.
- b The organelle in box G.

7 Using Table 2 and Figure 2, suggest which organelle or organelles (there may be more than one) would most likely be found in each of the following:

- a sediment 1
- b sediment 3
- c supernatant 1
- d supernatant 3.

▼ **Table 2** *Separation of organelles by ultracentrifugation*

Organelles to be separated out	Speed of centrifugation / gravitational force	Duration of centrifugation / min
Nuclei	1000	10
Mitochondria	3500	10
Lysosomes	16 500	20
Ribosomes	100 000	60

A summary of cell fractionation is given in Figure 2.

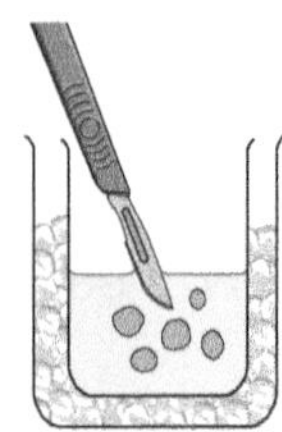
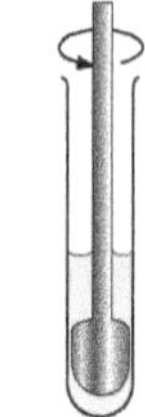
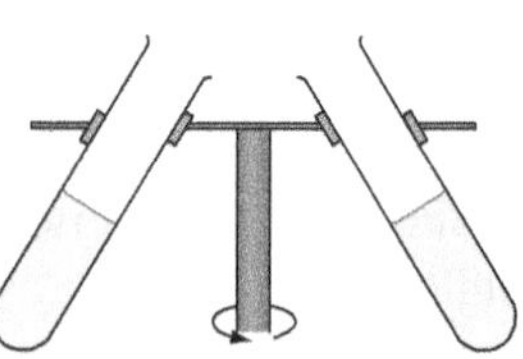

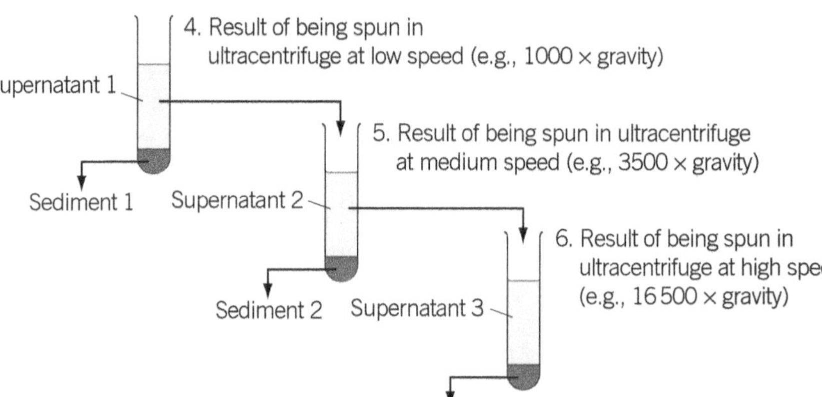

▲ **Figure 2** *Summary of cell fractionation*

The techniques of cell fractionation and ultracentrifugation have enabled considerable advances in biological knowledge. They have allowed a detailed study of the structure and function of organelles.

▲ **Figure 3** *An ultracentrifuge used to separate the various components of cell homogenate*

2.2 The electron microscope

Light microscopes have poor resolution as a result of the relatively long wavelength of light. In the 1930s, however, a microscope was developed that used a beam of electrons instead of light. This is called an electron microscope and it has two main advantages:

- The electron beam has a very short wavelength and the microscope can therefore resolve objects well – it has a high resolving power.
- As electrons are negatively charged the beam can be focused using electromagnets (Figure 1).

The best modern electron microscopes can resolve objects that are just 0.1 nm apart – 2000 times better than a light microscope. Because electrons are absorbed by the molecules in air, a near-vacuum has to be created within the chamber of an electron microscope in order for it to work effectively.

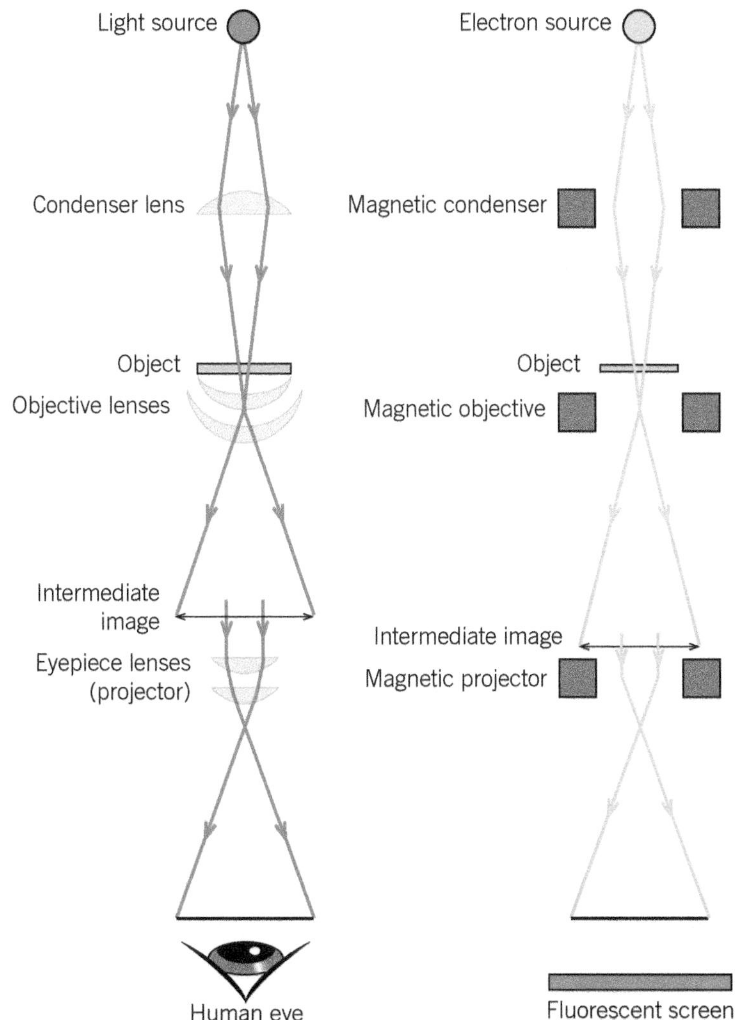

▲ **Figure 1** *Comparison of radiation pathways in light and electron microscopes*

Learning objectives:

- → Describe how electron microscopes work.
- → Describe the differences between a transmission electron microscope and a scanning electron microscope.
- → Describe the limitations of the transmission and the scanning electron microscopes.

Specification reference: 3.1.2.1

Study tip

Remember that the greater resolving power of an electron microscope compared to a light microscope is due to the electron beam having a shorter wavelength than light.

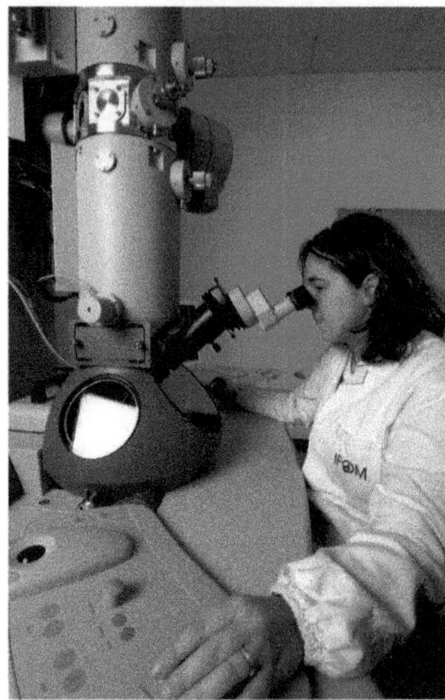

▲ **Figure 2** *Scientist looking at a sample using a TEM*

There are two types of electron microscope:

- the transmission electron microscope (TEM)
- the scanning electron microscope (SEM).

The transmission electron microscope

The transmission electron microscope (TEM) consists of an electron gun that produces a beam of electrons that is focused onto the specimen by a condenser electromagnet. In a TEM, the beam passes through a thin section of the specimen. Parts of this specimen absorb electrons and therefore appear dark. Other parts of the specimen allow the electrons to pass through and so appear bright. An image is produced on a screen and this can be photographed to give a **photomicrograph**. The resolving power of a TEM is 0.1 nm, although problems with specimen preparation mean that this cannot always be achieved. The main limitations of the TEM are:

- The whole system must be in a vacuum and therefore living specimens cannot be observed.
- A complex 'staining' process is required and even then the image is only in black and white.
- The specimen must be extremely thin.
- The image may contain artefacts. Artefacts are things that result from the way the specimen is prepared. Artefacts may appear on the finished photomicrograph but are not part of the natural specimen. It is therefore not always easy to be sure that what is seen on a photomicrograph really exists in that form.

Study tip

Look at photographs taken with an SEM and a TEM and make sure you can identify cell organelles. Don't just rely on diagrams.

In the TEM the specimens must be extremely thin to allow electrons to penetrate. The result is therefore a flat, 2-D image. We can partly get over this by taking a series of sections through a specimen. We can then build up a 3-D image of the specimen by looking at the series of photomicrographs produced. However, this is a slow and complicated process. One way in which this problem has been overcome is the development of the scanning electron microscope (SEM).

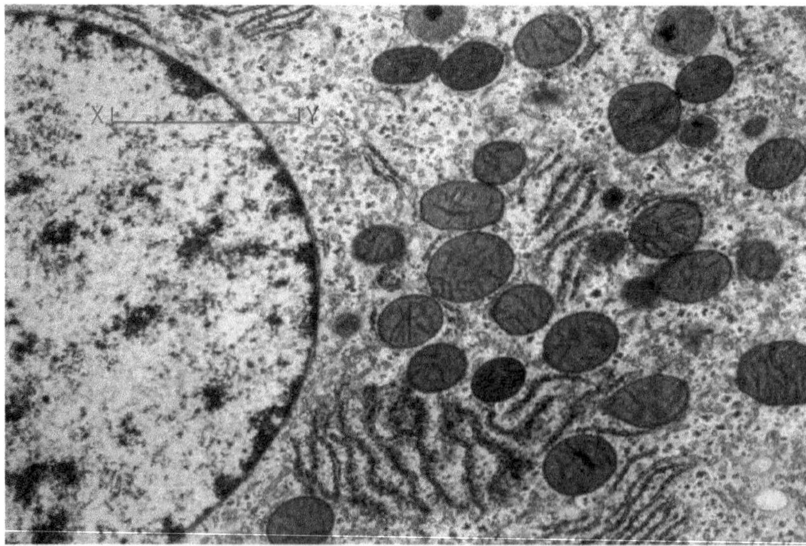

▲ **Figure 3** *Part of an animal cell seen under a TEM*

The scanning electron microscope

All the limitations of the TEM also apply to the scanning electron microscope (SEM), except that specimens need not be extremely thin as electrons do not need to penetrate the specimen. Basically similar to a TEM, the SEM directs a beam of electrons on to the surface of the specimen from above, rather than penetrating it from below. The beam is then passed back and forth across a portion of the specimen in a regular pattern. The electrons are scattered by the specimen and the pattern of this scattering depends on the contours of the specimen surface. We can build up a 3-D image by computer analysis of the pattern of scattered electrons and secondary electrons produced. The basic SEM has a lower resolving power than a TEM, around 20 nm, but is still 10 times better than a light microscope.

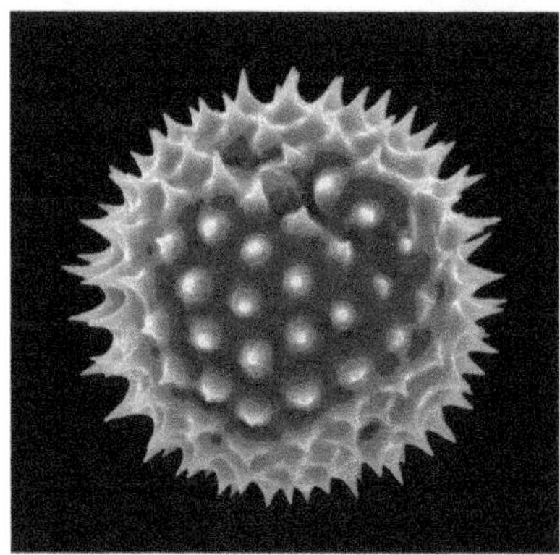

▲ **Figure 4** *False-colour SEM of a pollen grain from a marigold plant*

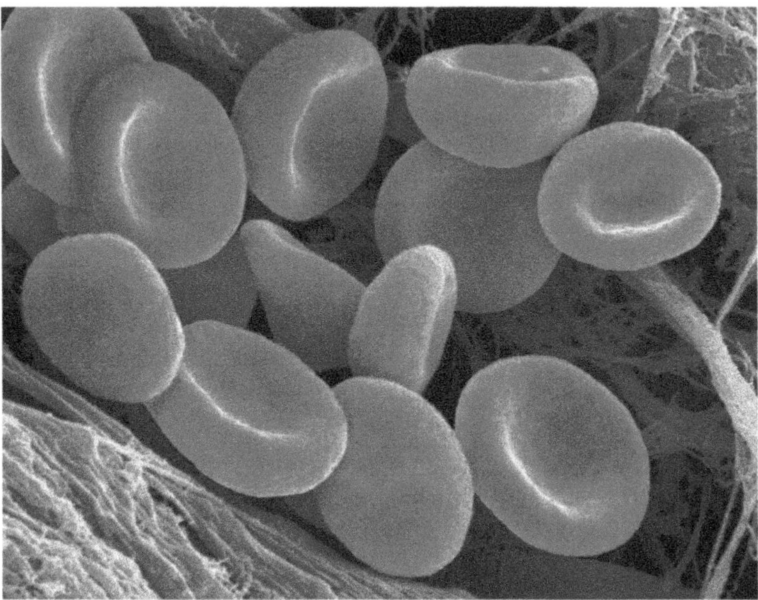

▲ **Figure 5** *False-colour SEM of human red blood cells*

Summary questions

1. Why is the electron microscope able to resolve objects better than the light microscope?
2. Why do specimens have to be kept in a near-vacuum in order to be viewed effectively using an electron microscope?
3. Which of the biological structures listed here:
   plant cell (100 µm) DNA molecule (2 nm) virus (100 nm)
   actin molecule (3.5 nm) a bacterium (1 µm)
   can, in theory, be resolved by the following?
   a a light microscope
   b a transmission electron microscope
   c a scanning electron microscope.
4. In practice, the theoretical resolving power of an electron microscope cannot always be achieved. Why not?
5. In the photomicrograph of the animal cell in Figure 3, the line X–Y represents a length of 5 µm. What is the magnification of this photomicrograph?

2.3 Eukaryotic cell structure

Learning objectives:

→ Describe the structure and functions of the nucleus, mitochondria, chloroplasts, rough and smooth endoplasmic reticulum, Golgi apparatus, Golgi vesicles, and lysosomes.

→ Describe the structure and function of the cell wall in plants, algae, and fungi.

→ Describe the structure and function of the cell vacuole in plants.

Specification reference: 3.1.2.1

Hint

When you look at a group of animal cells, such as epithelial cells, under a light microscope, you cannot see the cell-surface membrane because it is too thin to be observed. What you actually see is the boundary between cells.

Synoptic link

The cell-surface membrane is covered in Topic 4.1, and DNA is covered in Chapter 7.

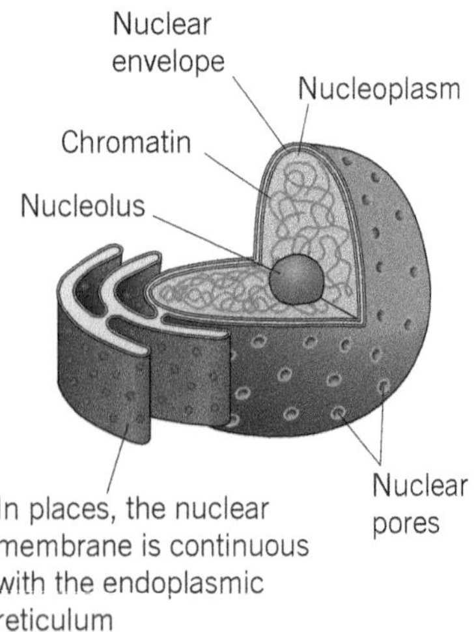

▲ **Figure 1** *The nucleus*

Each cell can be regarded as a metabolic compartment, a separate place where the chemical processes of that cell occur. Cells in multicellular organisms are often adapted to perform a particular function. Depending on that function, each cell type has an internal structure that suits it for its job. This is known as the **ultrastructure** of the cell. **Eukaryotic** cells have a distinct nucleus and possess membrane-bound organelles. They differ from prokaryotic cells, such as bacteria. More details of these differences are given in Topic 2.5. Using an electron microscope, you can see the structure of organelles within cells. The most important of these organelles are described below, with the exception of the cell-surface membrane.

The nucleus

The nucleus (Figure 1) is the most prominent feature of a eukaryotic cell, such as an epithelial cell. The nucleus contains the organism's hereditary material and controls the cell's activities. Usually spherical and between 10 μm and 20 μm in diameter, the nucleus has a number of parts.

- The **nuclear envelope** is a double membrane that surrounds the nucleus. Its outer membrane is continuous with the endoplasmic reticulum of the cell and often has ribosomes on its surface. It controls the entry and exit of materials in and out of the nucleus and contains the reactions taking place within it.
- **Nuclear pores** allow the passage of large molecules, such as messenger RNA, out of the nucleus. There are typically around 3000 pores in each nucleus, each 40–100 nm in diameter.
- **Nucleoplasm** is the granular, jelly-like material that makes up the bulk of the nucleus.
- **Chromosomes** consist of protein-bound, linear DNA.
- The **nucleolus** is a small spherical region within the nucleoplasm. It manufactures ribosomal RNA and assembles the ribosomes. There may be more than one nucleolus in a nucleus.

The functions of the nucleus are to:

- act as the control centre of the cell through the production of mRNA and tRNA and hence protein synthesis (see Chapter 8)
- retain the genetic material of the cell in the form of DNA and chromosomes
- manufacture ribosomal RNA and ribosomes.

The mitochondrion

Mitochondria (Figures 2 and 3) are usually rod-shaped and 1–10 μm in length. They are made up of the following structures:

- Around the organelle is a **double membrane** that controls the entry and exit of material. The inner of the two membranes is folded to form extensions known as cristae.

- **Cristae** are extensions of the inner membrane, which in some species extend across the whole width of the mitochondrion. These provide a large surface area for the attachment of enzymes and other proteins involved in respiration.
- The **matrix** makes up the remainder of the mitochondrion. It contains proteins, lipids, ribosomes, and DNA, which allow the mitochondria to control the production of some of their own proteins. Many enzymes involved in respiration are found in the matrix.

Mitochondria are the sites of the aerobic stages of respiration (the Krebs cycle and the oxidative phosphorylation pathway, which are covered in Topics 20.3 and 20.4). They are therefore responsible for the production of the energy-carrier molecule, ATP, using substances derived from respiratory substrates, such as glucose. Because of this, the number and size of the mitochondria, and the number of their cristae, are high in cells that have a high level of metabolic activity and therefore require a plentiful supply of ATP. Examples of metabolically active cells include muscle and epithelial cells. Epithelial cells in the intestines require a lot of ATP in the process of absorbing substances from the intestines by active transport.

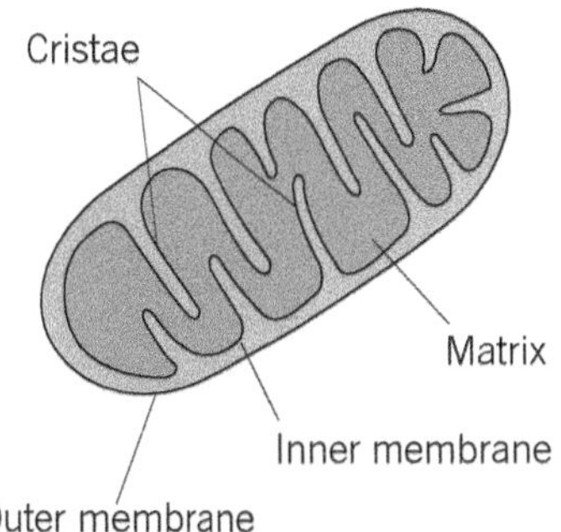

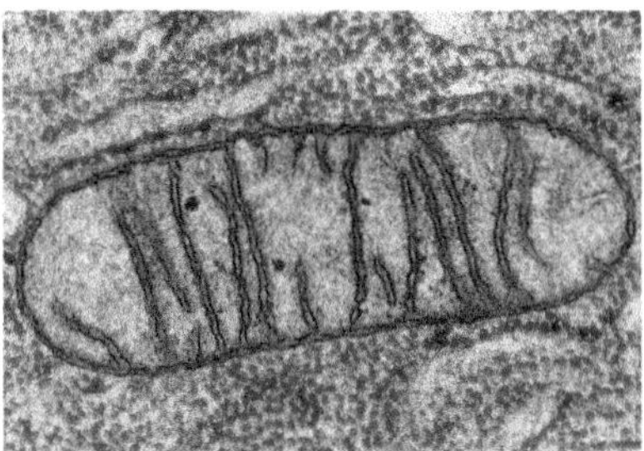

▲ **Figure 2** *The basic structure of a mitochondrion (top); false-colour TEM of a mitochondrion (bottom)*

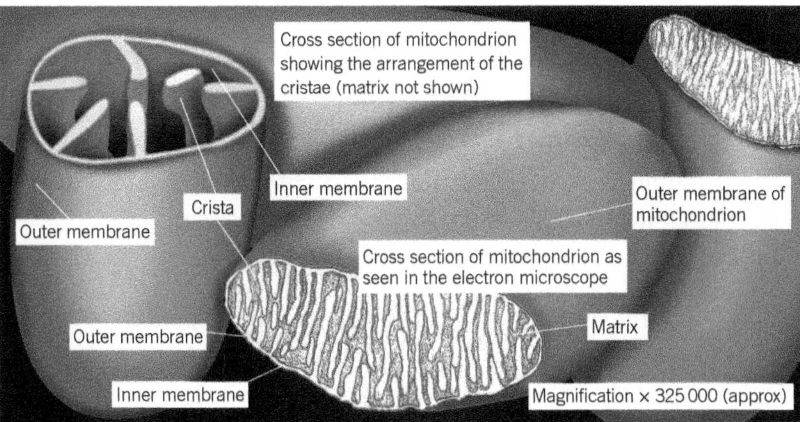

▲ **Figure 3** *Mitochondria*

Chloroplasts

Chloroplasts (Figure 4) are the organelles that carry out photosynthesis (see Topic 19.1). They vary in shape and size but are typically disc-shaped, 2–10 μm long, and 1 μm in diameter. Their main features are as follows:

- **The chloroplast envelope** is a double plasma membrane that surrounds the organelle. It is highly selective in what it allows to enter and leave the chloroplast.
- **The grana** are stacks of up to 100 disc-like structures called **thylakoids**. Within the thylakoids is the photosynthetic pigment called **chlorophyll**. Some thylakoids have tubular extensions that join up with thylakoids in adjacent grana. The grana are where the first stage of photosynthesis (light absorption) takes place.
- **The stroma** is a fluid-filled matrix where the second stage of photosynthesis (synthesis of sugars) takes place. Within the stroma are a number of other structures, such as starch grains.

Hint

Chloroplasts have DNA and may have evolved from free-living prokaryotic cells, but they are organelles, not cells.

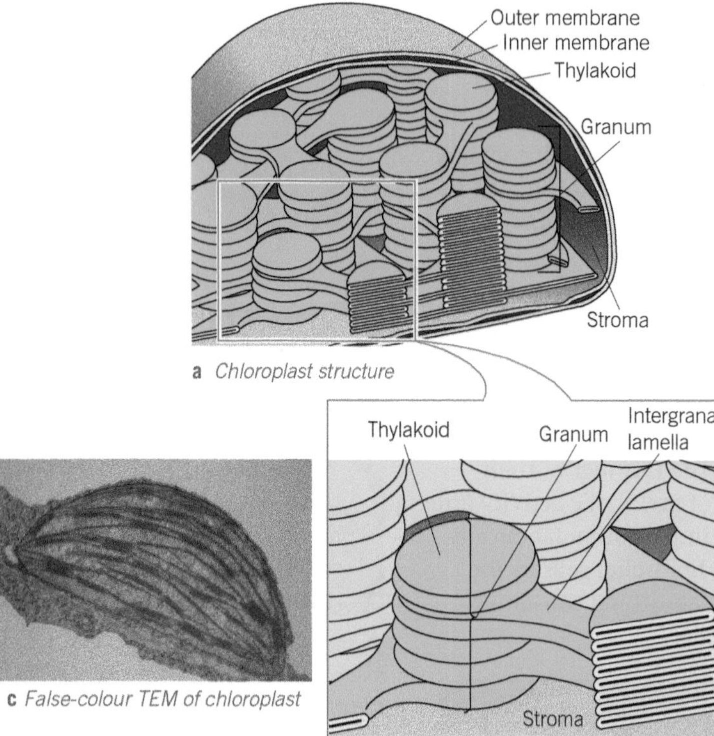

a *Chloroplast structure*

b *Grana and thylakoids*

c *False-colour TEM of chloroplast*

Figure 4 *Chloroplast structure*

> **Study tip**
>
> Not all plant cells have chloroplasts. Think about root cells. These are below the soil surface where light rarely penetrates and so no photosynthesis is possible.

Chloroplasts are adapted in the following ways to their function of harvesting sunlight and carrying out photosynthesis:

- The granal membranes provide a large surface area for the attachment of chlorophyll, electron carriers, and enzymes that carry out the first stage of photosynthesis. These chemicals are attached to the membrane in a highly ordered fashion.
- The fluid of the stroma possesses all the enzymes needed to make sugars in the second stage of photosynthesis.
- Chloroplasts contain both DNA and ribosomes so that they can quickly and easily manufacture some of the proteins needed for photosynthesis.

Endoplasmic reticulum

The endoplasmic reticulum (ER) is an elaborate 3-D system of sheet-like membranes spreading through the cytoplasm of the cells. It is continuous with the outer nuclear membrane. The membranes enclose a network of tubules and flattened sacs called cisternae (see Figure 5). There are two types of ER:

- **Rough endoplasmic reticulum (RER)** has ribosomes present on the outer surfaces of the membranes. Its functions are to:
  - provide a large surface area for the synthesis of proteins and glycoproteins
  - provide a pathway for the transport of materials, especially proteins, throughout the cell.
- **Smooth endoplasmic reticulum (SER)** lacks ribosomes on its surface and is often more tubular in appearance. Its functions are to:
  - synthesise, store, and transport lipids
  - synthesise, store, and transport carbohydrates.

It follows that cells that manufacture and store large quantities of carbohydrates, proteins, and lipids have a very extensive ER. Such cells include liver and secretory cells, for example the epithelial cells that line the intestines.

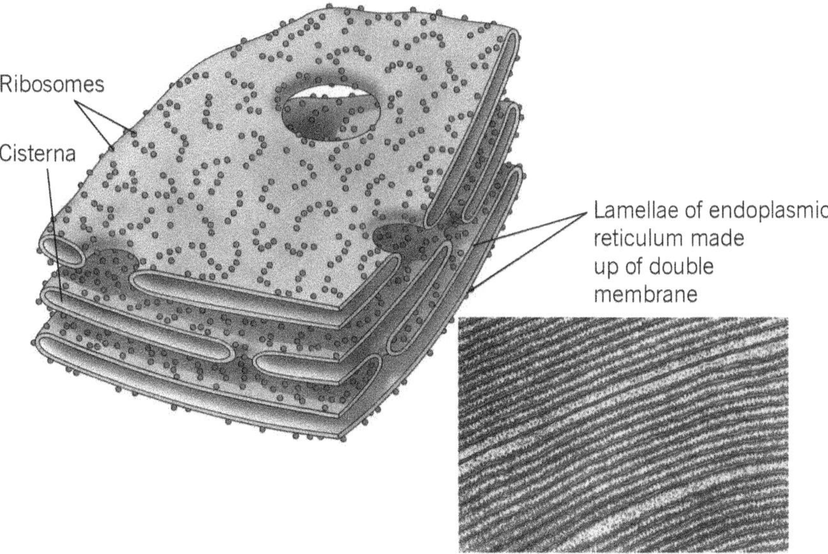

▲ **Figure 5** *Structure of RER (above); false-colour TEM of a section through RER (red) (right)*

Golgi apparatus

The Golgi apparatus occurs in almost all eukaryotic cells and is similar to the SER in structure except that it is more compact. It consists of a stack of membranes that make up flattened sacs, or **cisternae**, with small, rounded, hollow structures called vesicles. The proteins and lipids produced by the ER are passed through the Golgi apparatus in strict sequence. The Golgi modifies these proteins, often adding non-protein components, such as carbohydrate, to them. It also 'labels' them, allowing them to be accurately sorted and sent to their correct destinations. Once sorted, the modified proteins and lipids are transported in Golgi vesicles, which are regularly pinched off from the ends of the Golgi cisternae (Figure 6). These vesicles may move to the cell surface, where they fuse with the membrane and release their contents to the outside.

The functions of the Golgi apparatus are to:

- add carbohydrate to proteins to form glycoproteins
- produce secretory enzymes, such as those secreted by the pancreas
- synthesise and transport carbohydrates such as pectin which are used in making cell walls in plants
- transport, modify, and store lipids
- form lysosomes.

The Golgi apparatus is especially well developed in secretory cells, such as the epithelial cells that line the intestines.

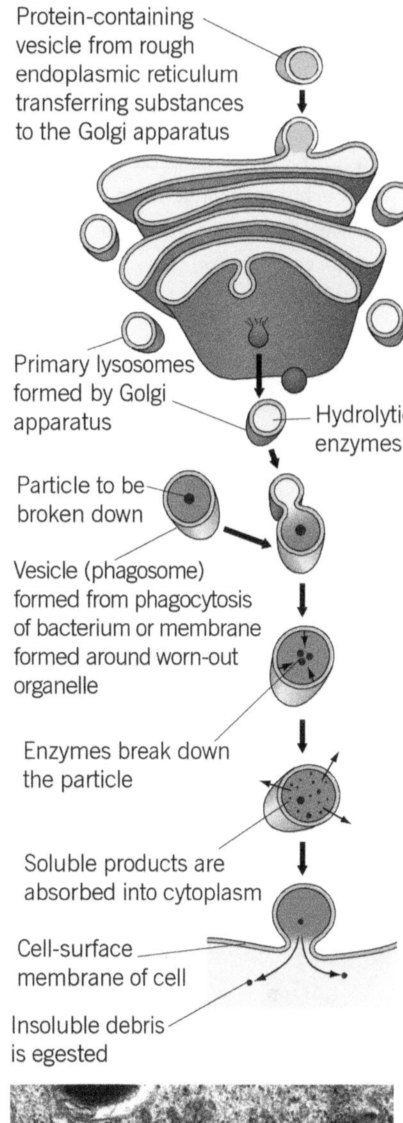

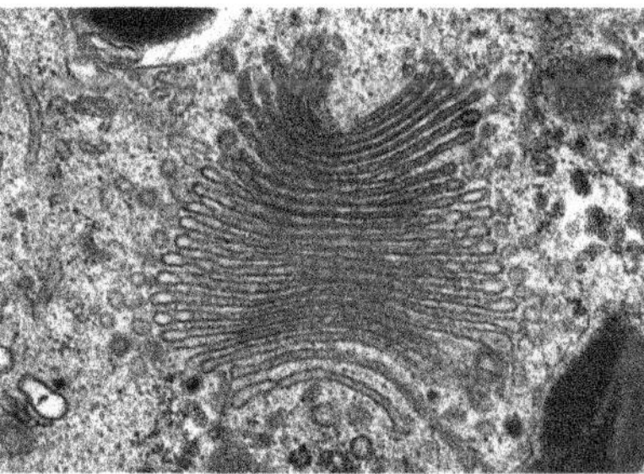

▲ **Figure 6** *The Golgi apparatus and the formation and functioning of a lysosome (top); false-colour TEM of a Golgi apparatus (orange) (bottom)*

Hint

To help you understand the functions of the Golgi apparatus, think of it as the cell's post office, but receiving, sorting, and delivering proteins and lipids, rather than letters.

Hint

Lysosomes can be thought of as refuse disposal operatives. They remove useless and potentially dangerous material (e.g., bacteria) and reuse the useful parts, disposing of only that which cannot be recycled.

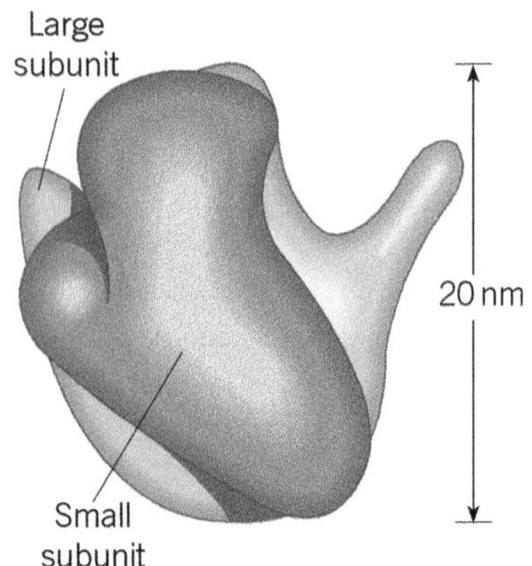

▲ **Figure 7** *Structure of a ribosome*

Synoptic link

Look back to Topic 1.3, to refresh your knowledge of cellulose. Osmosis will be covered in Topic 4.3.

Study tip

Plant cells have a cell-surface membrane *and* a cell wall, not just a cell wall.

Lysosomes

Lysosomes are formed when the vesicles produced by the Golgi apparatus contain enzymes such as proteases and lipases. They also contain lysozymes, enzymes that hydrolyse the cell walls of certain bacteria. As many as 50 different types of such enzymes may be contained in a single lysosome. Up to 1.0 μm in diameter, lysosomes isolate these enzymes from the rest of the cell before releasing them, either to the outside or into a **phagocytic** vesicle within the cell (Figure 6).

The functions of lysosomes are to:

- hydrolyse material ingested by phagocytic cells, such as white blood cells and bacteria
- release enzymes to the outside of the cell (exocytosis) in order to destroy material around the cell
- digest worn out organelles so that the useful chemicals they are made of can be reused
- completely break down cells after they have died (autolysis).

Given the roles that lysosomes perform, it is not surprising that they are especially abundant in secretory cells, such as epithelial cells, and in phagocytic cells.

Ribosomes

Ribosomes are small cytoplasmic granules found in all cells. They may occur in the cytoplasm or be associated with the RER. There are two types, depending on the cells in which they are found:

- **80S** – found in eukaryotic cells, are around 25 nm in diameter.
- **70S** – found in prokaryotic cells, mitochondria, and chloroplasts, are slightly smaller.

Ribosomes have two subunits – one large and one small (Figure 7) – each of which contains ribosomal RNA and protein. Despite their small size, they occur in such vast numbers that they can account for up to 25 per cent of the dry mass of a cell. Ribosomes are the site of protein synthesis.

Cell wall

Characteristic of all plant cells, the cell wall consists of microfibrils of the polysaccharide cellulose, embedded in a matrix. Cellulose microfibrils have considerable strength and so contribute to the overall strength of the cell wall. Cell walls have the following features:

- They consist of a number of polysaccharides, such as cellulose.
- There is a thin layer, called the **middle lamella**, which marks the boundary between adjacent cell walls and cements adjacent cells together.

The functions of the cellulose cell wall are:

- to provide mechanical strength in order to prevent the cell bursting under the pressure created by the osmotic entry of water

- to give mechanical strength to the plant as a whole
- to allow water to pass along it and so contribute to the movement of water through the plant.

The cell walls of algae are made up of either cellulose or glycoproteins, or a mixture of both.

The cell walls of fungi and bacteria (see Topic 2.5) do not contain cellulose. In eukaryotic fungi the cell wall comprises of a mixture of a nitrogen-containing polysaccharide called **chitin**, a polysaccharide called glycan, and glycoproteins.

Vacuoles

A fluid-filled sac bounded by a single membrane may be termed a vacuole. Within mature plant cells there is usually one large central vacuole. The single membrane around it is called the **tonoplast**. A plant vacuole contains a solution of mineral salts, sugars, amino acids, wastes, and sometimes pigments such as anthocyanins.

Plant vacuoles serve a variety of functions:

- They support herbaceous plants, and herbaceous parts of woody plants, by making cells turgid.
- The sugars and amino acids may act as a temporary food store.
- The pigments may colour petals to attract pollinating insects.

Relating cell ultrastructure to function

As each organelle has its own function, it is possible to deduce, with reasonable accuracy, the role of a cell by looking at the number and size of the organelles it contains. For example, as mitochondria produce ATP that is used as a temporary energy store, it follows that cells with many mitochondria are likely to require a lot of ATP and therefore have a high rate of metabolism. Even within each mitochondrion, the more dense and numerous the cristae, the greater the metabolic rate of the cell possessing these mitochondria.

Summary questions

1 State in which process ribosomes are important.

2 List **three** carbohydrates that are absorbed by an epithelial cell of the small intestine.

3 State the organelle that is being referred to in each of the following descriptions:

- a It possesses structures called cristae.
- b It contains chromatin.
- c It synthesises glycoproteins.
- d It digests worn out organelles.

4 The following sentences give the name of a cell type and a brief description of its role. Suggest **two** organelles that might be numerous and/or well developed in each of the cells.

- a A sperm cell swims a considerable distance carrying the male chromosomes.
- b One type of white blood cell engulfs and digests foreign material.
- c Liver cells manufacture proteins and lipids at a rapid rate.

2.4 Cell differentiation and organisation

Learning objectives:

- Describe the advantages of cellular differentiation.
- Describe how cells are arranged into tissues.
- Describe how cells are arranged into organs.
- Describe how organs are arranged into organ systems.

Specification reference: 3.1.2.1

In multicellular organisms, cells are specialised to perform specific functions. Similar cells are then grouped together into tissues, tissues into organs, and organs into organ systems for increased efficiency.

Cell differentiation

Single-celled organisms perform all essential life functions inside the boundaries of a single cell. Although they perform all functions adequately, they cannot be totally efficient at all of them, because each function requires a different type of cellular structure. One activity may be best carried out by a long, thin cell, whilst another might suit a spherically shaped cell. No one cell can provide the best conditions for all functions. The cells of multicellular organisms are each adapted in different ways to perform a particular role. All cells in an organism are initially identical. As it matures, each cell takes on its own individual characteristics that suit it to the function that it will perform when it is mature. In other words, each cell becomes specialised in structure to suit the role that it will carry out. This is known as cell differentiation.

All the cells in an organism, such as a human, are derived by mitotic divisions of the fertilised egg. It follows that they all contain exactly the same genes. How then does the cell differentiate? Every cell contains the genes needed for it to develop into any one of the many different cells in an organism. But not all these genes are switched on (expressed) in any one cell. Different genes are switched on in each type of differentiated cell. Other genes are either switched off or expressed to a lesser extent.

It is not just the shape of different cells that varies, but also the numbers of each of their organelles. For example, a muscle or sperm cell will have many mitochondria, whilst a bone cell has very few. White blood cells have many lysosomes whereas a muscle cell has very few.

The cells of a multicellular organism have therefore evolved to become more and more suited to one specialised function. In doing so, they have lost the ability to carry out other functions. They are dependent on other cells to perform these activities for them. These other cells are specially adapted to their own particular function and perform it more effectively. As a result, the whole organism functions efficiently.

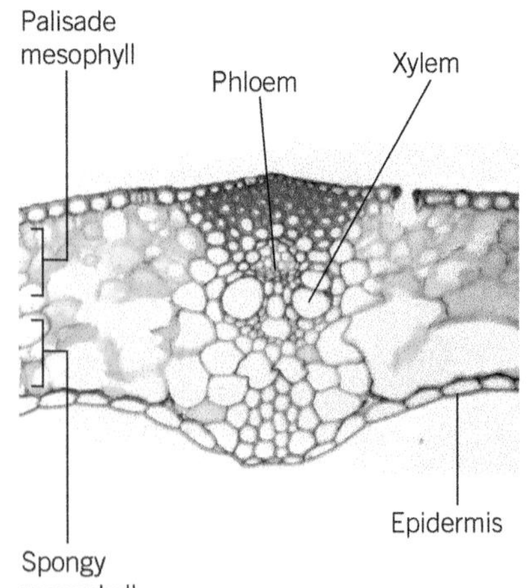

▲ **Figure 1** *Some of the various tissues that make up the organ called a leaf*

Tissues

For working efficiency, cells are normally aggregated together. Such a collection of similar cells that perform a specific function is known as a **tissue**. Examples of tissues include:

- **epithelial tissues**, which are found in animals and consist of sheets of cells. They line the surfaces of organs and often have a protective or secretory function. There are many types, including those made up of thin, flat cells that line organs where diffusion takes place, for example, the alveoli of the lungs (see Topics 5.4 and 5.6), and the ciliated epithelium that lines a duct such as the trachea (see Topic 5.4). The cilia are used to move mucus over the epithelial surface.

- **xylem** (see Topic 16.1), which occurs in plants and is made up of a number of cell types. It is used to transport water and mineral ions throughout the plant and also gives mechanical support.

Organs

Just as cells are aggregated into tissues, so tissues are aggregated into **organs**. An organ is a combination of tissues that are coordinated to perform a variety of functions, although they often have one predominant major function. In animals, for example, the stomach is an organ that carries out the digestion of certain types of food (see Topic 12.1). It is made up of tissues such as:

- muscle to churn and mix the stomach contents
- epithelium to protect the stomach wall and produce secretions
- connective tissue to hold together the other tissues.

In plants, a leaf (Figure 1) is an organ made up of the following tissues:

- palisade mesophyll made up of leaf palisade cells (see Topic 19.1) that carry out photosynthesis
- spongy mesophyll adapted for gaseous diffusion
- epidermis to protect the leaf and allow gaseous diffusion
- phloem to transport organic materials away from the leaf
- xylem to transport water and ions into the leaf.

It is not always easy to determine which structures are organs. Blood capillaries, for example, are not organs, whereas arteries and veins are both organs. All three structures have the same major function, namely the transport of blood. However, capillaries are made up of just one tissue – epithelium – whereas arteries and veins are made up of many tissues, including epithelial, muscle, and connective tissues.

Organ systems

Organs work together as a single unit known as an **organ system**. These systems may be grouped together to perform particular functions more efficiently. There are a number of organ systems in humans.

- The **digestive system** digests and processes food. It is made up of organs that include the salivary glands, oesophagus, stomach, duodenum, ileum, pancreas, and liver.
- The **respiratory system** is used for breathing and gas exchange. It is made up of organs that include the trachea, bronchi, and lungs.
- The **circulatory system** (Figure 2) pumps and circulates blood. It is made up of organs that include the heart, arteries, and veins.

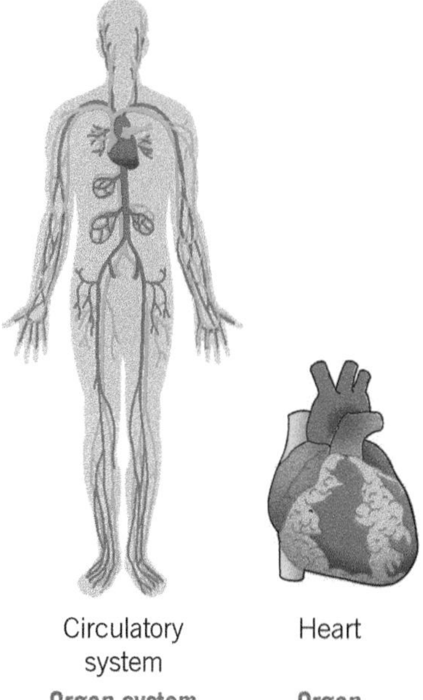

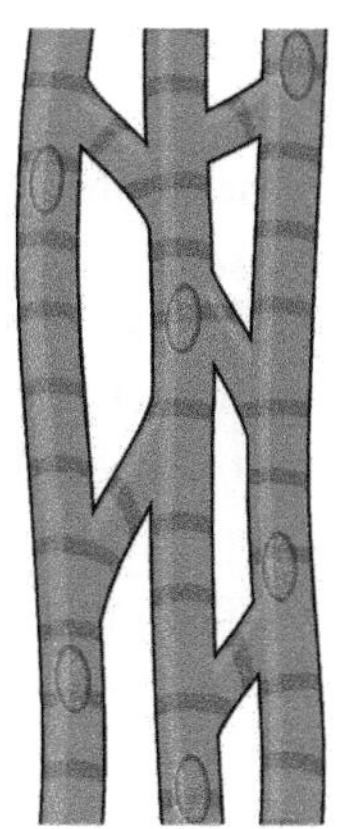

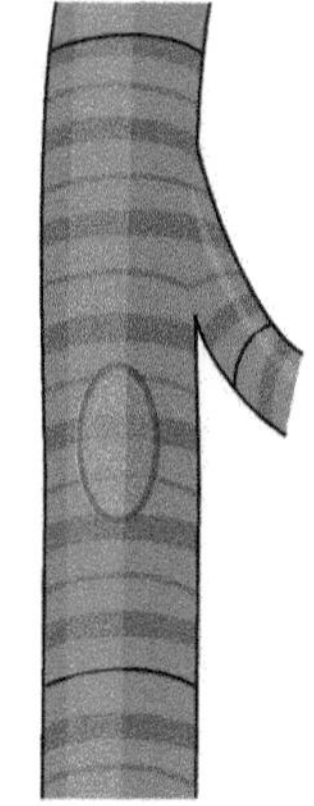

▲ **Figure 2** *The circulatory system as an example of an organ system*

Summary questions

1. What is a tissue?
2. Why is an artery described as an organ whereas a blood capillary is not?
3. State whether each of the following is a tissue or an organ:
   a heart **b** xylem **c** lungs **d** epithelium.

2.5 The structure of prokaryotic cells

Learning objectives:

- → Describe prokaryotic cells.
- → Explain how prokaryotes differ from eukaryotes.

Specification reference: 3.1.2.2

Globally, cholera is of great significance, killing an estimated 120 000 people each year. The agent that causes the disease is a curved, rod-shaped bacterium called *Vibrio cholerae*. Bacteria are examples of prokaryotic cells. The structure of a generalised prokaryotic cell is shown in Figure 1. The differences between prokaryotic and eukaryotic cells are listed in Table 1.

▼ **Table 1** *Comparison of prokaryotic and eukaryotic cells*

Prokaryotic cells	Eukaryotic cells
No true nucleus, only a diffuse area of nuclear material with no nuclear envelope	Distinct nucleus, with a nuclear envelope
No nucleolus	Nucleolus is present
Circular molecules of DNA but no chromosomes	Chromosomes present, in which DNA is located
No membrane-bound organelles	Membrane-bound organelles, such as mitochondria, are present
No chloroplasts, only photosynthetic regions in some bacteria	Chloroplasts present in plants and algae
Ribosomes are smaller (70S type)	Ribosomes are larger (80S type)
No endoplasmic reticulum or associated Golgi apparatus and lysosomes	Endoplasmic reticulum present along with Golgi apparatus and lysosomes
Cell wall made of peptidoglycan	Where present, cell wall is made mostly of cellulose (or chitin in fungi)

▼ **Table 2** *Roles of structures found in a bacterial cell*

Cell structure	Role
Cell wall	Physical barrier that protects against mechanical damage and excludes certain substances
Capsule	Protects bacterium from other cells and helps groups of bacteria to stick together for further protection
Cell-surface membrane	Acts as a differentially permeable layer that controls the entry and exit of chemicals
Flagellum (not in all bacterial cells)	Aids movement of bacterium because its rigid corkscrew shape and rotating base help the cell spin through fluids
Circular DNA	Possesses the genetic information for the replication of bacterial cells
Plasmid	Possesses genes that aid the survival of bacteria in adverse conditions, e.g., produces enzymes that break down antibiotics

Structure of a bacterial cell

Bacteria occur in every habitat in the world – they are versatile, adaptable, and very successful. Much of their success is a result of their small size, normally ranging from 0.1 µm to 10 µm in length. Their cellular structure is relatively simple (Figure 1). All bacteria possess a **cell wall**, which is made up of the glycoprotein murein. This is a polymer of polysaccharides and peptides. Many bacteria further protect themselves by secreting a **capsule** of mucilaginous slime around this wall. **Flagella** occur in certain types of bacteria. Their rigid corkscrew shape and rotating base enable bacteria to spin through fluids.

Inside the cell wall is the **cell-surface membrane**, within which is the cytoplasm that contains ribosomes of the 70S type. These ribosomes are smaller than those of eukaryotic cells (80S type), but nevertheless still synthesise proteins. Bacteria store food reserves as glycogen granules and oil droplets. The genetic material in bacteria is in the form of a **circular molecule of DNA**. Separate from this are smaller circular molecules of DNA, called **plasmids**. These can reproduce themselves independently and can carry genes that give the bacterium resistance to harmful chemicals, such as antibiotics. Plasmids are used extensively as vectors (carriers of genetic information) in genetic engineering. The roles of the main structures in a bacterial cell are summarised in Table 2.

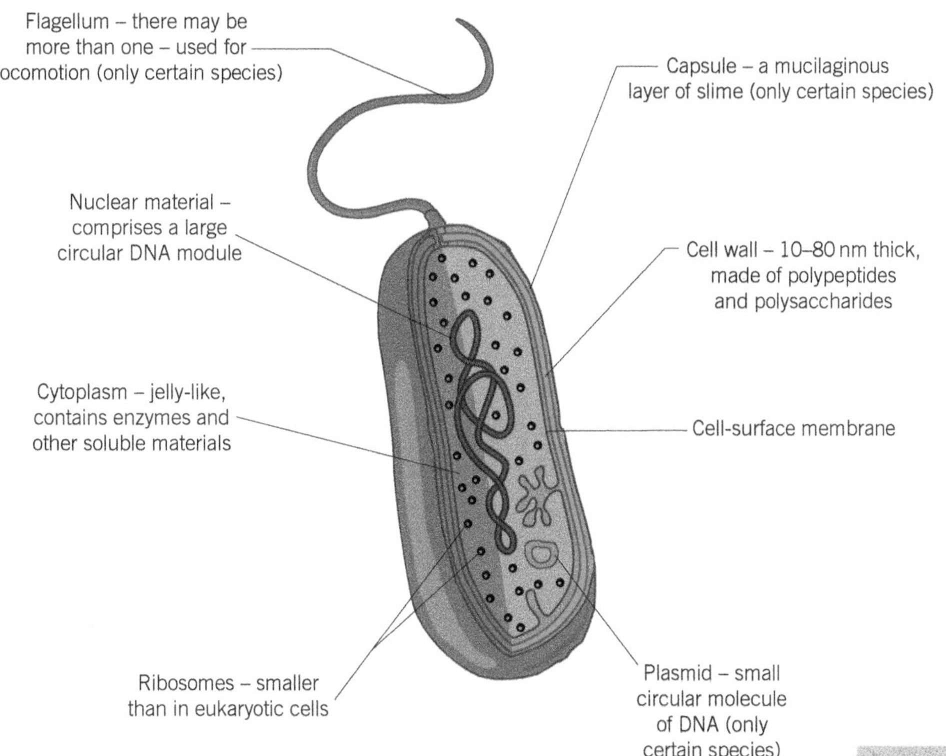

▲ **Figure 1** *Structure of a generalised bacterial cell*

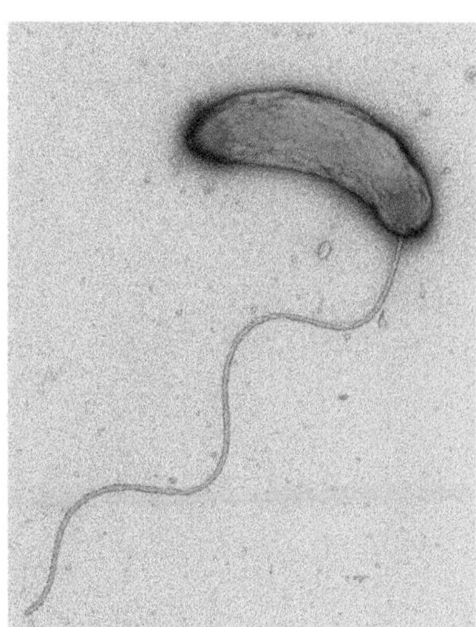

▲ **Figure 2** *False-colour transmission electron micrograph (TEM) of the cholera bacterium,* Vibrio cholerae

Summary question

1 Table 3 lists some of the features of cells. For the letter in each box, write down **one** of the following:
'present' if the feature always occurs
'absent' if it never occurs
'sometimes' if it occurs in some cells but not others.

▼ **Table 3** *Features of prokaryotic and eukaryotic cells*

Feature	Prokaryotic cell	Eukaryotic cell
Nuclear envelope	A	B
Cell wall	C	D
Flagellum	E	F
Ribosomes	G	H
Plasmid	I	J
Cell-surface membrane	K	L
Mitochondria	M	N

Study tip

Prokaryotic cells have no membrane-bound organelles. You should be able to relate this to their small size and your knowledge of the sizes of the membrane-bound organelles in Topic 2.3.

Practice questions: Chapter 2

1 An amoeba is a single-celled eukaryotic organism. Scientists used a transmission electron microscope to study an amoeba. The diagram shows its structure.

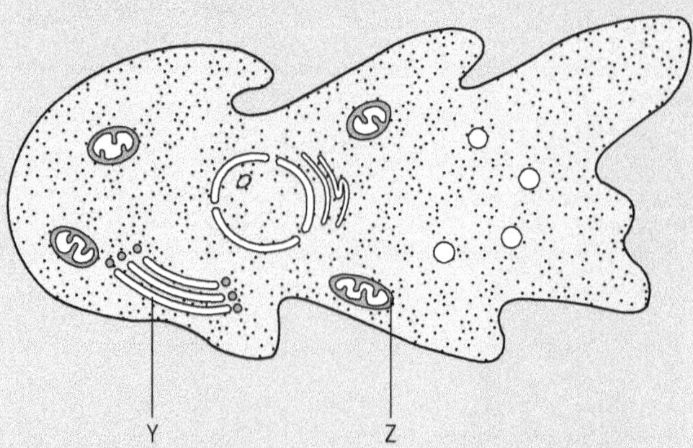

(a) (i) Name organelle **Y**. (*1 mark*)
(ii) Name **two** other structures in the diagram which show that the amoeba is a eukaryotic cell. (*2 marks*)
(b) What is the function of organelle **Z**? (*1 mark*)
(c) The scientists used a transmission electron microscope to study the structure of the amoeba. Explain why. (*2 marks*)

AQA June 2012

2 The photograph shows part of the cytoplasm of a cell.

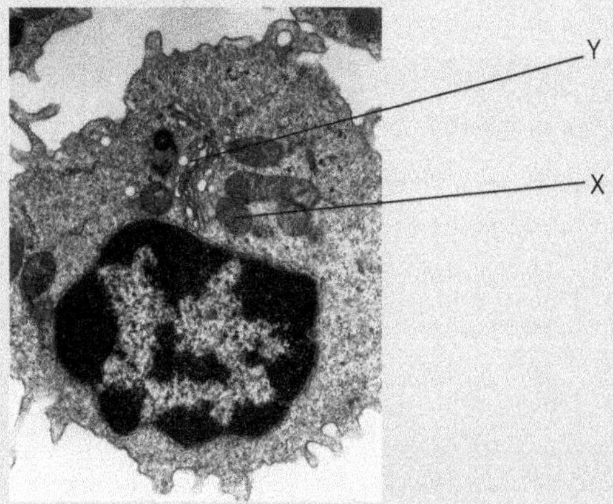

(a) (i) Organelle X is a mitochondrion. What is the function of this organelle? (*1 mark*)
(ii) Name organelle Y. (*1 mark*)
(b) This photograph was taken using a transmission electron microscope. The structure of the organelles visible in the photograph could not have been seen using an optical (light) microscope. Explain why. (*2 marks*)

AQA Jan 2013

3 The diagram shows a chloroplast as seen with an electron microscope.

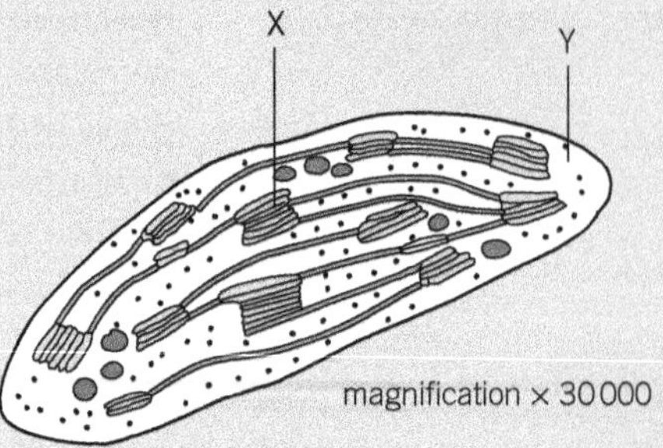

(a) Name **X** and **Y**. *(2 marks)*
(b) Describe the function of a chloroplast. *(2 marks)*
(c) Calculate the maximum length of this chloroplast in micrometres (μm). Show your working. *(2 marks)*

AQA Jan 2012

4 **(a)** The table shows some features of cells. Complete the table by putting a tick in the box if the feature is present in the cell.

Feature	Cell		
	Cholera bacterium	Epithelial cell from intestine	Epithelial cell from alveolus of lung
Cell-surface membrane			
Flagellum			
Nucleus			

(3 marks)

(b) The diagram shows part of an epithelial cell from an insect's gut.

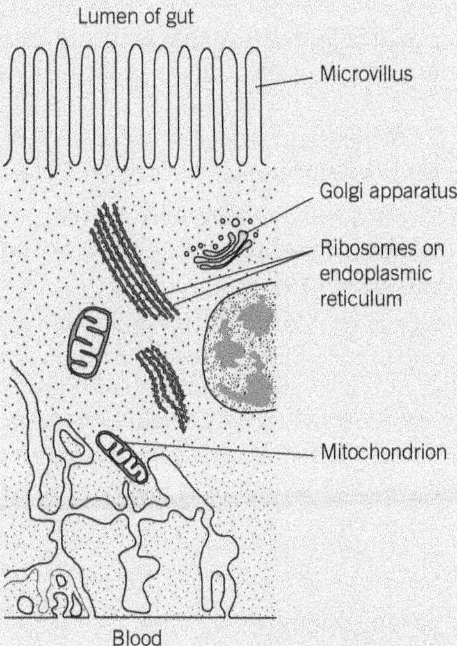

This cell is adapted for the three functions listed below. Using a **different** feature from the diagram for each answer, explain how this cell is adapted for each of these functions:

(i) the active transport of substances from the cell into the blood *(2 marks)*
(ii) the synthesis of enzymes *(2 marks)*
(iii) rapid diffusion of substances from the lumen of the gut into the cytoplasm, *(1 mark)*

AQA Jan 2012

Answers to the Practice Questions are available at
www.oxfordsecondary.com/oxfordaqaexams-alevel-biology

3 Biochemical reactions in cells

3.1 Enzymes and enzyme action

Learning objectives:

- → Describe how enzymes speed up chemical reactions.
- → Describe how the structures of enzyme molecules relate to their function.
- → Describe the lock and key model of enzyme action.
- → Describe the induced fit model of enzyme action.

Specification reference: 3.1.3.1 and 3.1.3.2

Enzymes are globular proteins that act as catalysts. Catalysts alter the rate of a chemical reaction without undergoing permanent changes themselves. They can be reused repeatedly and are therefore effective in small amounts. Enzymes do not make reactions happen, they simply alter the speed of reactions that already occur, sometimes by a factor of many millions.

Enzymes as catalysts lowering activation energy

Let us consider a typical chemical reaction:

sucrose + water ⟶ glucose + fructose
(substrates) (products)

For reactions like this to take place naturally a number of conditions must be satisfied:

- The sucrose and water molecules must collide with sufficient energy to alter the arrangement of their atoms to form glucose and fructose.
- The energy of the products (glucose and fructose) must be less than that of the substrates (sucrose and water).
- An initial boost of energy is needed to kick start the reaction. The minimum amount of energy needed to activate the reaction in this way is called the **activation energy**.

There is an activation energy level, like an energy hill or barrier, which must initially be overcome before the reaction can proceed. Enzymes work by lowering this activation energy level (Figure 1). In this way, enzymes allow reactions to take place at a lower temperature than normal. This enables some metabolic processes to occur rapidly at the human body temperature of 37 °C, which is relatively low in terms of chemical reactions. Without enzymes these reactions would proceed too slowly to sustain life as we know it.

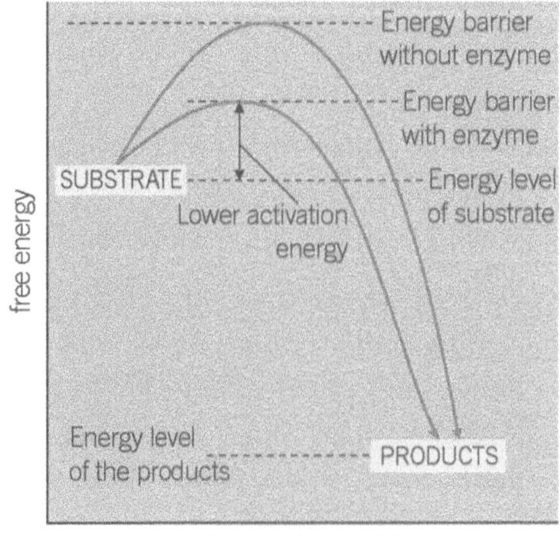

▲ **Figure 1** *How enzymes lower activation energy*

Enzyme structure

From Topic 1.5 you will be aware that enzymes, being globular proteins, have a specific 3-D shape that is the result of their sequence of amino acids (primary stucture). Although an enzyme molecule is large overall, only a small region of it is functional. This is known as the **active site** and is made up of a relatively small number of amino acids. The active site forms a small, hollow depression within the much larger enzyme molecule.

The molecule on which the enzyme acts is called the **substrate**. This fits neatly into this depression to form an **enzyme–substrate complex** (Figure 2). The substrate molecule is held within the active site by bonds that temporarily form between certain amino acids of the active site and groups on the substrate molecule.

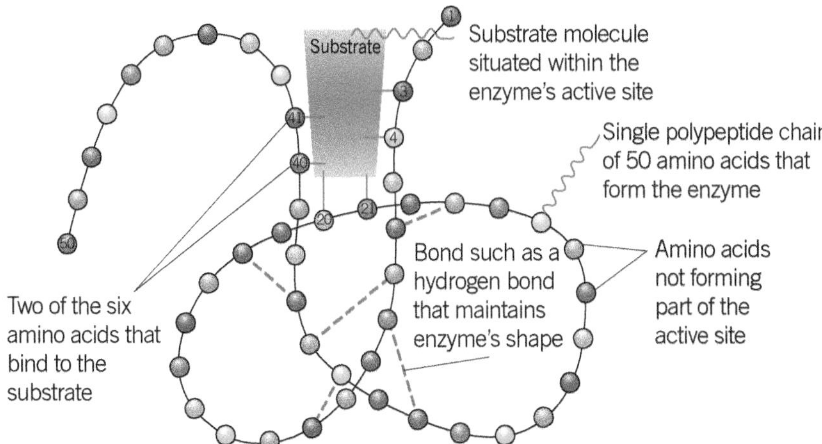

▲ **Figure 2** *Enzyme–substrate complex showing the six out of 50 amino acids that form the active site*

Hint

If a stone is lying behind a mound, we need to expend energy to move it down a hillside, either by pushing the stone over the mound or reducing the height of the mound. Once it starts to move, the stone gathers momentum and rolls to the bottom. Hence an initial input of energy (**activation energy**) starts a reaction that then continues of its own accord. Enzymes achieve the equivalent of lowering the mound of earth.

Lock and key model of enzyme action

Scientists often try to explain their observations by producing a representation of how something works. This is known as a scientific model. Examples include the physical models used to explain enzyme action. One model proposes that enzymes work in the same way as a key operates a lock – each key has a specific shape that fits and operates only a single lock. In a similar way, a substrate will only fit the active site of one particular enzyme. This model is supported by the observation that enzymes are specific in the reactions that they catalyse. The shape of the substrate (key) exactly fits the active site of the enzyme (lock). This is known as the **lock and key model** (Figure 3).

One limitation of this model is that the enzyme, like a lock, is considered to be a rigid structure. However, scientists observed that other molecules could bind to enzymes at sites other than the active site. In doing so, they altered the activity of the enzyme. This suggested that the enzyme's shape was being altered by the binding molecule. In other words, its structure was not rigid but flexible. In true scientific fashion this led to an alternative model being proposed, one that better fitted the observations. This was the called the induced fit model.

Study tip

The substrate does *not* have the 'same shape' as the active site. The substrate has a **complementary shape** to the active site.

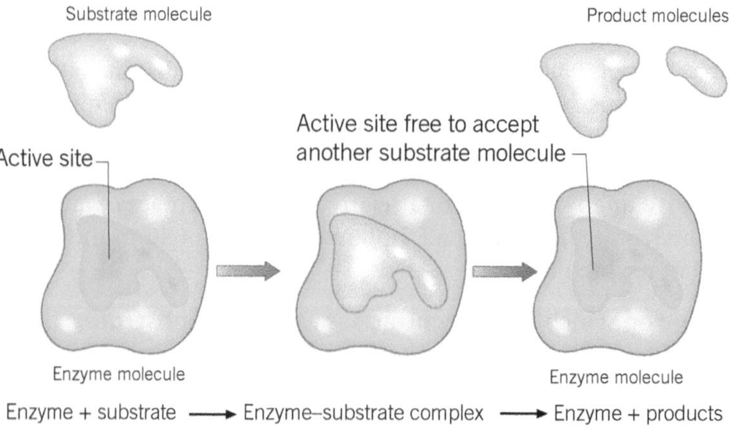

▲ **Figure 3** *Mechanism of enzyme action*

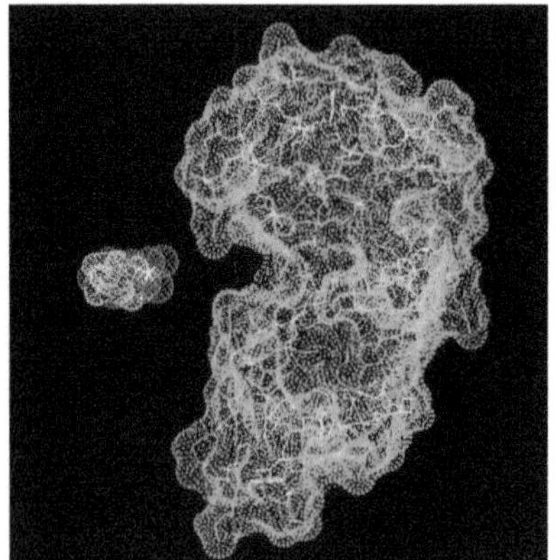

▲ **Figure 4** *Molecular computer graphics image of the enzyme ribonuclease A (right) and its substrate (left) approaching the enzyme's active site*

Induced fit model of enzyme action

Although the lock and key model describes the basic method of enzyme action, the induced fit model is more refined. It proposes that, rather than being a rigid lock, the enzyme actually changes its shape slightly to fit the profile of the substrate. In other words, the enzyme is flexible and can mould itself around the substrate in the way that a glove moulds itself to the shape of the hand. The enzyme has a certain general shape, just as a glove has, but this alters in the presence of the substrate. As it changes its shape, the enzyme puts a strain on the substrate molecule. This strain distorts a particular bond and consequently lowers the activation energy needed to break the bond.

The induced fit model is therefore a modified version of the lock and key model. It is a better explanation of the scientific observations because it explains:

- how other molecules can affect enzyme activity
- how the activation energy is lowered.

Any change in an enzyme's environment is likely to change its shape. The very act of colliding with its substrate is a change in its environment and so its shape changes – induced fit.

Summary questions

1 What is a catalyst?

2 Why are enzymes effective in tiny quantities?

3 Explain why changing one of the amino acids that make up the active site could prevent the enzyme from functioning.

4 Why might changing certain amino acids that are *not* part of the active site also prevent the enzyme from functioning?

3.2 Factors affecting enzyme action

Before considering how pH and temperature affect enzymes, it is worth bearing in mind that, for an enzyme to work, it must:

- come into physical contact with its **substrate**
- have an **active site** which fits the substrate.

Almost all factors that influence the rate at which an enzyme works do so by affecting one or both of the above. In order to investigate how enzymes are affected by various factors we need to be able to measure the rate of the reactions they catalyse.

Measuring enzyme-catalysed reactions

To measure the progress of an enzyme-catalysed reaction we usually measure its time course, that is, how long it takes for a particular event to run its course. The two 'events' most frequently measured are:

- the formation of the products of the reaction, for example, the volume of oxygen produced when catalase acts on hydrogen peroxide (Figure 1)
- the disappearance of the substrate, for example, the reduction in concentration of starch when it is acted upon by amylase (Figure 2).

Although the graphs in Figures 1 and 2 differ, the explanation for their shapes is the same:

- At first there is a lot of substrate (hydrogen peroxide or starch) but no product (water and oxygen, or maltose).
- It is therefore very easy for substrate molecules to come into contact with the empty active sites on the enzyme molecules.
- All enzyme active sites are filled and the substrate is rapidly broken down into its products.
- The amount of substrate decreases as it is broken down, resulting in an increase in the amount of product.
- As the reaction proceeds, there is less and less substrate and more and more product.
- It becomes more difficult for the substrate molecules to come into contact with the enzyme molecules because there are fewer substrate molecules and also the product molecules may 'get in the way' of substrate molecules and prevent them reaching an active site.
- It therefore takes longer for the substrate molecules to be broken down by the enzyme and so its rate of disappearance slows, and consequently the rate of formation of product also slows. Both graphs 'tail off'.
- The rate of reaction continues to slow until there is so little substrate that any further decrease in its concentration cannot be measured.
- The graphs flatten out because all the substrate has been used up and so no new product can be produced.

Effect of temperature on enzyme action

A rise in temperature increases the **kinetic energy** of molecules. As a result, the molecules move around more rapidly and collide with each

Learning objectives:

- → Describe how the rate of an enzyme-controlled reaction is measured.
- → Describe how temperature affects the rate of an enzyme-controlled reaction.
- → Describe how pH affects the rate of an enzyme-controlled reaction.
- → Describe how substrate concentration affects the rate of reaction.

Specification reference: 3.1.3.2

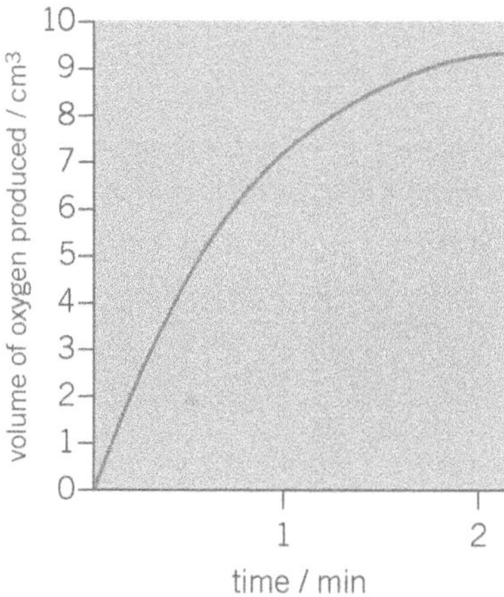

▲ **Figure 1** *Measurement of the formation of oxygen due to the action of catalase on hydrogen peroxide*

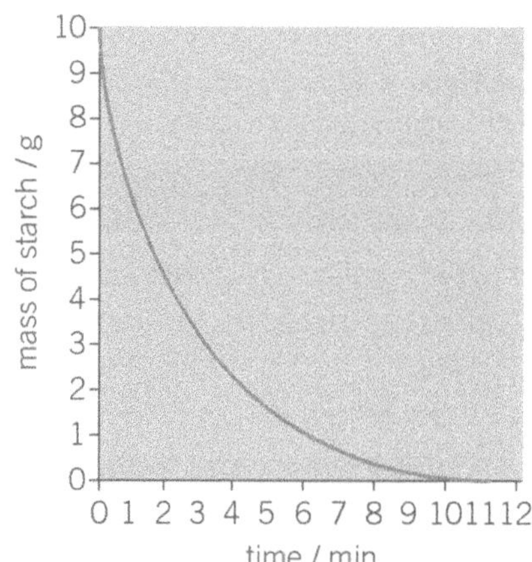

▲ **Figure 2** *Measurement of the disappearance of starch due to the action of amylase*

Study tip

Rate is always expressed per unit time.

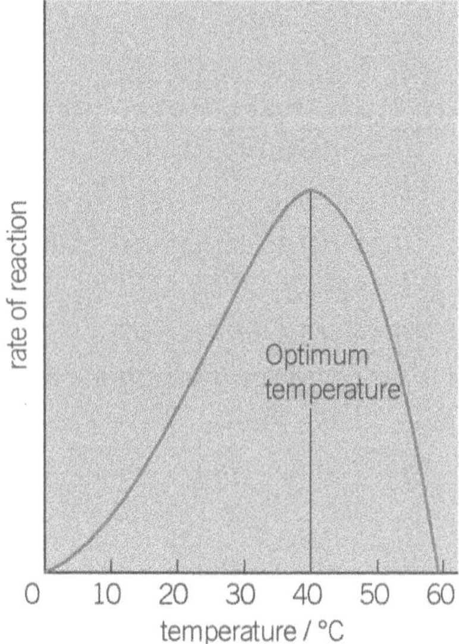

▲ **Figure 3** *Effect of temperature on the rate of an enzyme-controlled reaction*

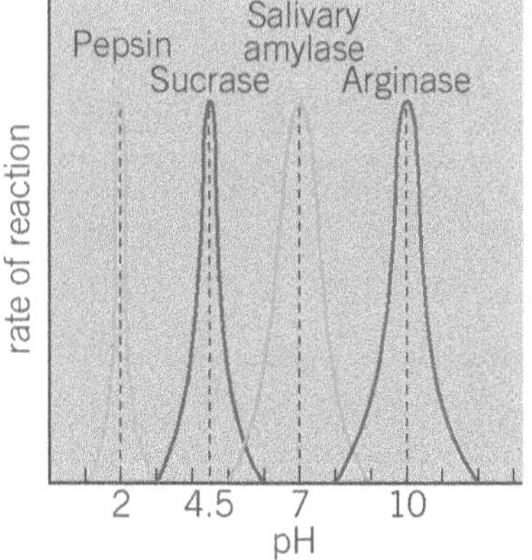

▲ **Figure 4** *Effect of pH on the rate of an enzyme-controlled reaction*

Hint

When considering how factors affect enzyme action, think shape change.

Study tip

Enzymes are not alive and so cannot be killed. Use the correct term, **denatured**.

other more often. In an enzyme-catalysed reaction, this means that the enzyme and substrate molecules come together more often in a given time, so that the rate of reaction increases.

Shown on a graph, this gives a rising curve. However, once the temperature has risen past a certain point, any continued rise will begin to cause the hydrogen and other bonds in the enzyme molecule to break. This results in the enzyme, including its active site, changing shape. At first, the substrate fits less easily into this changed active site, slowing the rate of reaction. For many human enzymes this may begin at temperatures of around 45 °C.

At some point, usually around 60 °C, the enzyme is so disrupted that it stops working altogether. It is said to be denatured. **Denaturation** is a permanent change and, once it has occurred, the enzyme does not function again. Shown on a graph, the rate of the reaction follows a falling curve. The actual effect of temperature on the rate of an enzyme reaction is a combination of these two factors (Figure 3). The optimum working temperature differs from enzyme to enzyme. Some work fastest at around 10 °C, whereas others continue to work rapidly at 80 °C. Many enzymes in the human body have an optimum temperature of about 40 °C. Our body temperatures have, however, evolved to be 37 °C. This is advantageous for several reasons:

- Although higher body temperatures would increase the metabolic rate slightly, the advantages are offset by the additional energy (food) that would be needed to maintain the higher temperature.
- Other proteins, apart from enzymes, may be denatured at higher temperatures.
- At higher body temperatures, any further rise in temperature, for example, during illness, might denature the enzymes.

Effect of pH on enzyme action

The pH of a solution is a measure of its hydrogen ion concentration. Each enzyme has an optimum pH, that is, a pH at which it works fastest (Figure 4). In a similar way to a rise in temperature, a change in pH reduces the effectiveness of an enzyme and may eventually cause it to become denatured and to stop working altogether.

The pH affects how an enzyme works in the following ways:

- A change in pH alters the charges on the amino acids that make up the active site of the enzyme. As a result, the substrate can no longer become attached to the active site and so the enzyme–substrate complex cannot be formed.
- A change in pH can cause the bonds that maintain the enzyme's tertiary structure to break. The enzyme therefore changes shape. These changes can alter the shape of the active site and the substrate may therefore no longer fit it. The enzyme has been denatured.

Even small changes in pH can change the arrangement of the active site of an enzyme. The arrangement of the active site is partly determined by the hydrogen and ionic bonds between $—NH_2$ and —COOH groups of the polypeptides that make up the enzyme. The change in H^+ ions affects this bonding, causing the active site to change shape.

As pH fluctuations inside organisms are usually small, they are far more likely to reduce an enzyme's activity than to denature it.

Effects of substrate concentration on the rate of enzyme action

If the amount of enzyme is fixed at a constant level and substrate is slowly added, the rate of reaction increases in proportion to the amount of substrate that is added. This is because, at low substrate concentrations, the enzyme molecules have only a limited number of substrate molecules to collide with, and therefore the active sites of the enzymes are not working to full capacity. As more substrate is added, the active sites gradually become filled, until the point where all of them are working as fast as they can. The rate of reaction is at its maximum (V_{max}). After that, the addition of more substrate will have no effect on the rate of reaction. In other words, when there is an excess of substrate, the rate of reaction levels off. A summary of the effect of substrate concentration on the rate of enzyme action is given in Figure 6.

Hint

The active site and the substrate are not 'the same', any more than a key and a lock are the same – in some senses they are more like opposites. The correct term is **complementary**.

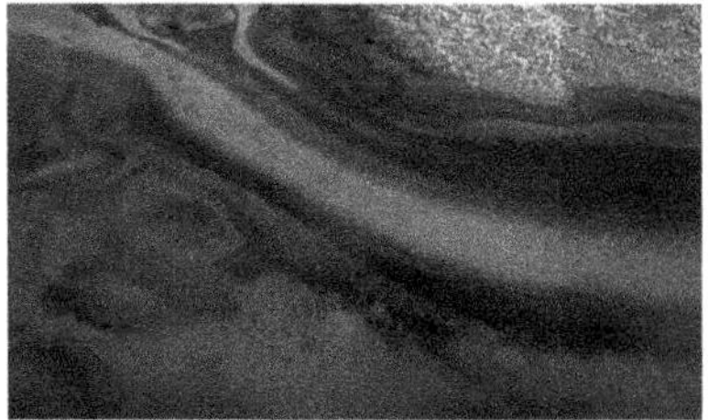

▲ **Figure 5** *Enzymes in the algae in this hot spring remain functional at temperatures of 80 °C whereas in most organisms they would be denatured at this temperature*

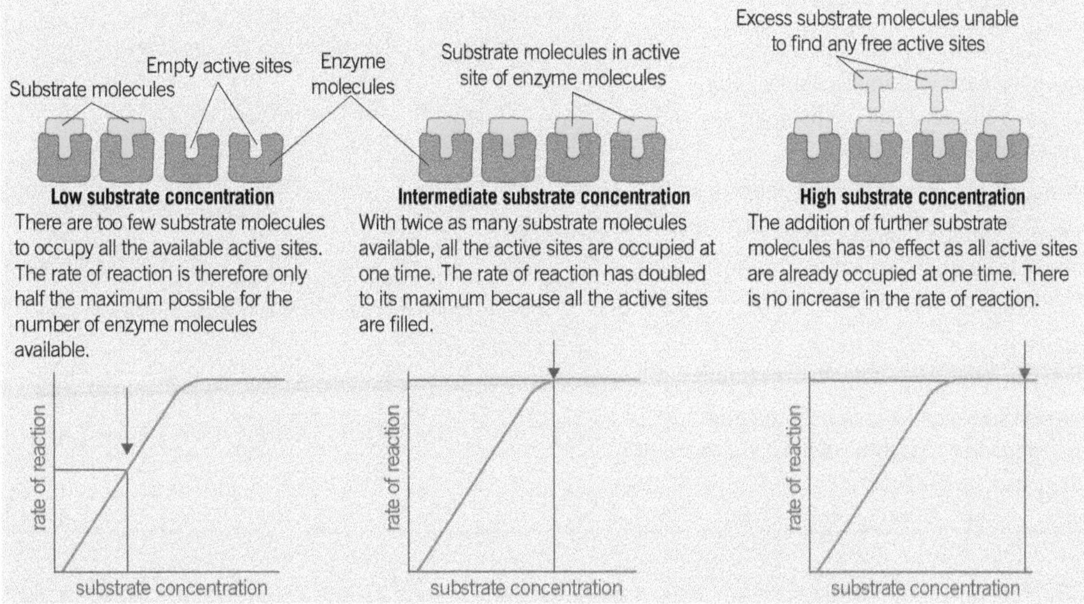

▲ **Figure 6** *Effect of substrate concentration on the rate of an enzyme-controlled reaction*

Summary questions

1. Explain why enzymes function less well at lower temperatures.
2. Explain how high temperatures may completely prevent enzymes from functioning.
3. Enzymes produced by microorganisms are responsible for spoiling food. Using this fact and your knowledge of enzymes, can you suggest a reason why the following procedures are carried out?
   a. Food is heated to a high temperature before being canned.
   b. Some foods, such as onions, are preserved in vinegar.

Required practical 1

Investigating the effect of substrate concentration on the activity of an enzyme.

Catalase is an intracellular enzyme which breaks down hydrogen peroxide. Hydrogen peroxide is produced as a by product of **aerobic** respiration and its presence in cells can be damaging if it is not removed. The equation for the breakdown of hydrogen peroxide is as follows:

$$2H_2O_2 \xrightarrow{\text{Catalase}} 2H_2O + O_2$$

The rate of this reaction could be determined by collecting the volume of oxygen released following the breakdown by catalase of its substrate hydrogen peroxide.

An alternative approach uses the fact that discs of paper soaked in a catalase solution give off bubbles of oxygen when they are dropped into hydrogen peroxide. These bubbles cause the discs to rise to the surface. The time it takes for the discs to rise correlates to the rate of production of oxygen. In this experiment, homogenised potato tubers are used as a source of catalase.

Method

- Peel a potato and chop it into small chunks before homogenising in a blender. Filter the homogenate and store the filtrate on ice until required.
- Using a hole punch, cut several discs of filter paper.
- Prepare a range of different concentrations of hydrogen peroxide by a simple dilution of a stock solution of hydrogen peroxide.
- Place a solution of the lowest hydrogen peroxide concentration in a test tube to a depth of exactly 6 cm.
- Remove a small sample of the potato filtrate and place this in a petri dish. Add three paper discs to the filtrate so they soak up the liquid.
- Taking one disc, transfer it into the hydrogen peroxide solution. Push the disc to the bottom of the solution using a glass rod. Remove the rod and immediately start the timer.
- Record the time taken for the disc to rise to the surface.
- Repeat the experiment three times for each hydrogen peroxide concentration using a freshly soaked disc each time.

1. Why are you advised to store the potato extract on ice until it is required?
2. What safety precautions should you carry out when using substances such as hydrogen peroxide in the laboratory?
3. The uncertainty in time for the filter paper to reach the surface is assumed to be 0.21 seconds. Suggest how this figure has been derived.
4. Explain how you would calculate the absolute uncertainty for the rate of reaction in a low concentration of hydrogen peroxide where the time taken to for the disc to reach the surface was 21 seconds.
5. What happens to the absolute uncertainty as the concentration of hydrogen peroxide is increased?
6. Sketch a graph to show the predicted outcome of the effect of increasing the substrate concentration on the rate of this enzyme-controlled reaction.
7. Name two variables which have not been controlled in this experiment.

3.3 Enzyme inhibition

Enzyme inhibitors are substances that directly or indirectly interfere with the functioning of the **active site** of an enzyme and so reduce its activity. Sometimes the inhibitor binds itself so strongly to the active site that it cannot be removed and so permanently prevents the enzyme functioning. Most inhibitors only make temporary attachments to the enzyme molecule. These are called reversible inhibitors and are of two types:

- competitive inhibitors – which bind to the active site of the enzyme
- non-competitive inhibitors – which bind to the enzyme at a position other than the active site.

Learning objectives:

→ Describe how competitive inhibitors and non-competitive inhibitors affect the active site.

→ Define enzyme inhibition.

Specification reference: 3.1.3.2

Competitive inhibitors

Competitive inhibitors have a molecular shape similar to that of the substrate. This allows them to occupy the active site of an enzyme. They therefore compete with the substrate for the available active sites (Figure 1). It is the difference between the concentration of the inhibitor and the concentration of the substrate that determines the effect that this has on enzyme activity. If the substrate concentration is increased, the effect of the inhibitor is reduced. The inhibitor is not permanently bound to the active site and so, when it leaves, another molecule can take its place. This could be a substrate or inhibitor molecule, depending on how much of each type is present. Sooner or later, all the substrate molecules will occupy an active site, but the greater the concentration of inhibitor, the longer this will take. An example of competitive inhibition occurs with an important respiratory enzyme that acts on succinic acid. Another compound, called malonic acid, can inhibit the enzyme because it has a very similar molecular shape to succinic acid. It therefore easily combines with the enzyme and blocks succinic acid from combining with the enzyme's active site.

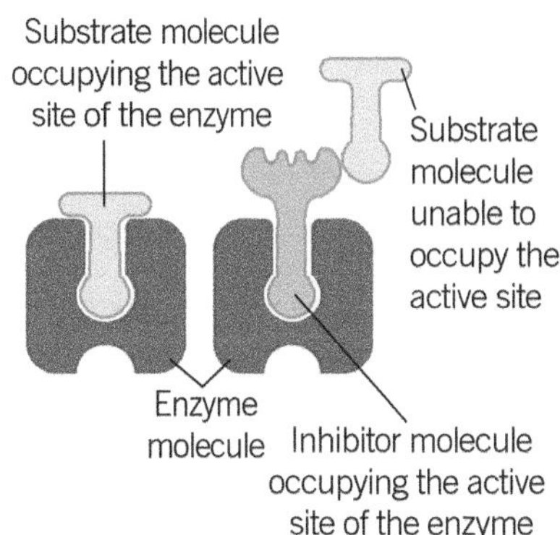

▲ **Figure 1** *Competitive inhibition*

Non-competitive inhibitors

Non-competitive inhibitors attach themselves to the enzyme at a site which is not the active site. Upon attaching to the enzyme, the

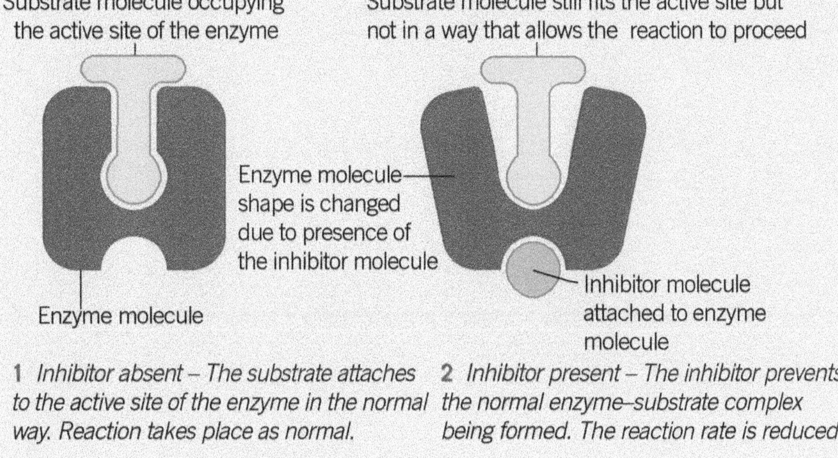

▲ **Figure 2** *Non-competitive inhibition*

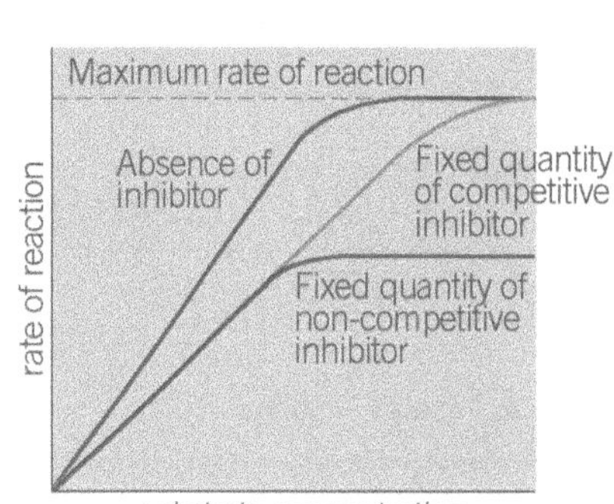

▲ **Figure 3** *Comparison of competitive and non-competitive inhibition on the rate of an enzyme-controlled reaction at different substrate concentrations*

inhibitor alters the shape of the enzyme's active site in such a way that substrate molecules can no longer occupy it, and so the enzyme cannot function (Figure 2). As the substrate and the inhibitor are not competing for the same site, an increase in substrate concentration does not decrease the effect of the inhibitor (Figure 3).

Summary questions

1 Distinguish between a competitive and a non-competitive inhibitor.

2 An enzyme-controlled reaction is inhibited by substance X. Suggest a simple way in which you could tell whether substance X is acting as a competitive or a non-competitive inhibitor.

Control of metabolic pathways

A metabolic pathway is a series of reactions in which each step is catalysed by an enzyme. In the tiny space inside a single cell, there are many hundreds of different metabolic pathways. The pathways are not at all haphazard, but highly structured. The enzymes that control a pathway are often attached to the inner membrane of a cell organelle in a very precise sequence. Inside each organelle optimum conditions for the functioning of particular enzymes may be provided. To keep a steady level of a particular chemical in a cell, the same chemical often acts as an inhibitor of an enzyme at the start of a reaction.

Let us look at the example illustrated in Figure 4. The end product inhibits enzyme A. If for some reason the quantity of end product increases above normal, then there will be greater inhibition of enzyme A. As a result, less end product will be produced and its concentration will return to normal. If the quantity of the end product falls below normal there will be less of it to inhibit enzyme A. Consequently, more end product will be produced and, again, its concentration will return to normal. In this way, the concentration of any chemical can be maintained at a relatively constant level.This is known as **end-product inhibition**. This type of inhibition is usually non-competitive.

1 Different conditions affect how enzymes work. Name one that might vary between one organelle and another.

2 Suggest why enzymes are attached to the inner membrane of an organelle 'in a very precise sequence'.

3 If an end product inhibits enzyme B rather than enzyme A, what would be:

   a the initial effect on the quantity of intermediate 1?

   b the overall longer-term effect on the level of the end product?

4 What is the advantage of end-product inhibition being non-competitive rather than competitive? Explain your answer in terms of how the two types of inhibition take place.

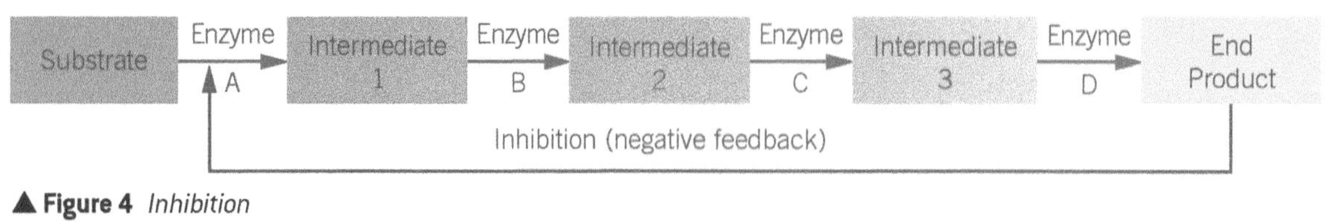

▲ **Figure 4** *Inhibition*

Practice questions: Chapter 3

1 Figure 1 represents an enzyme molecule and three other molecules that could combine with it.

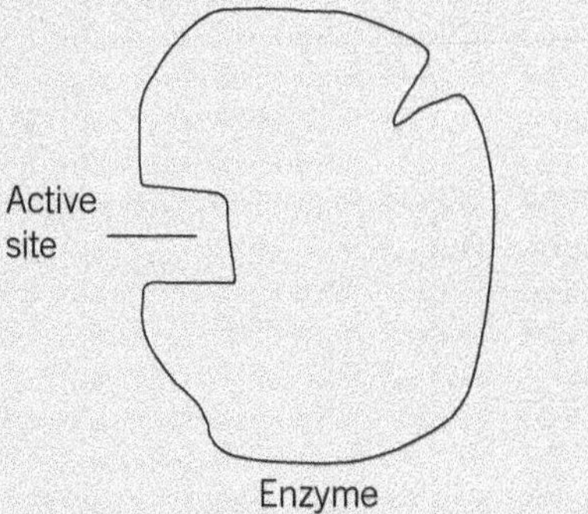

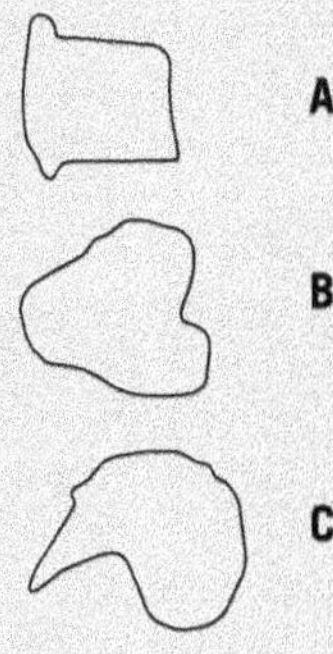

▲ **Figure 1**

(a) Which molecule is the substrate for the enzyme? Give a reason for your answer. *(1 mark)*

(b) Use the diagram to explain how a **non-competitive** inhibitor would decrease the rate of the reaction catalysed by this enzyme. *(3 marks)*

(c) Lysozyme is an enzyme. A molecule of lysozyme is made up of 129 amino acid molecules joined together. In the formation of its active site, the two amino acids that are at positions 35 and 52 in the amino acid sequence need to be close together.

(i) Name the bonds that join amino acids in the primary structure.

(ii) Suggest how the amino acids at positions 35 and 52 are held close together to form the active site. *(3 marks)*

AQA 2006

2 A student carried out an investigation into the mass of product formed in an enzyme-controlled reaction at three different temperatures. Only the temperature was different for each experiment. The results are shown in Figure 2.

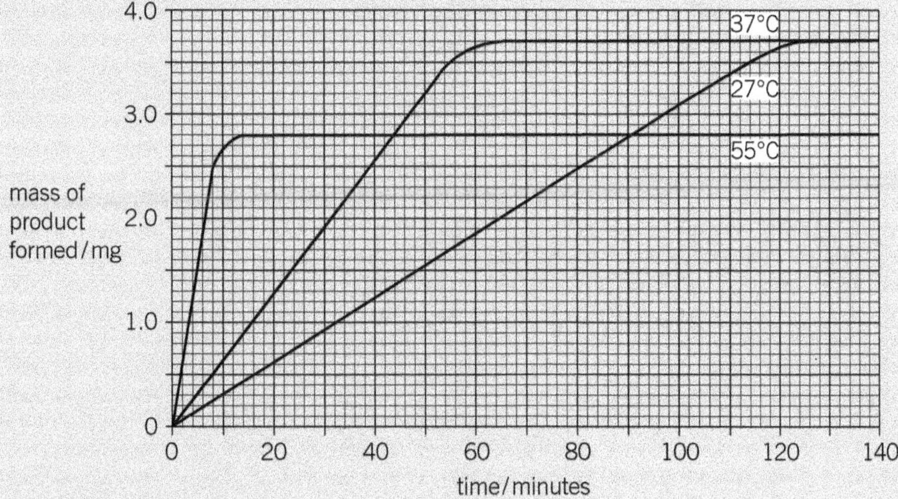

▲ **Figure 2**

(a) Use your knowledge of enzymes to explain:

(i) why the initial rate of reaction was highest at 55 °C

(ii) the shape of the curve for 55 °C after 20 minutes. *(5 marks)*

(b) Explain why the curves for 27 °C and 37 °C level out at the same value. *(2 marks)*

AQA 2006

3 **(a)** Many reactions take place in living cells at temperatures far lower than those required for the same reactions in a laboratory.
Explain how enzymes enable this to happen. *(3 marks)*

(b) Figure 3 shows the results of tests to determine the optimum temperature for the activity of this amylase.

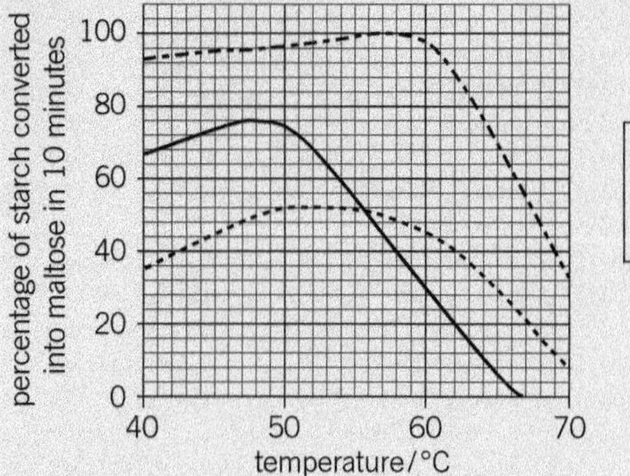

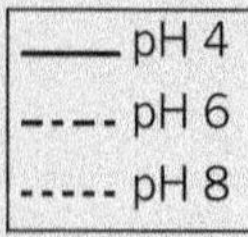

▲ **Figure 3**

(i) Copy and complete the table with the optimum temperature for the activity of amylase at each pH value.

(ii) Describe and explain the effect of temperature on the rate of reaction of this enzyme at pH 4.

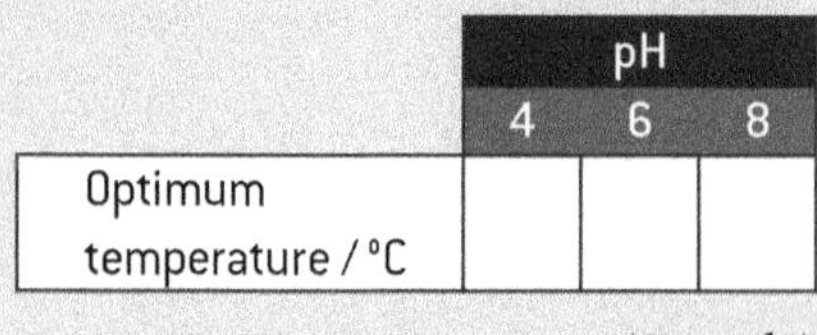

	pH		
	4	6	8
Optimum temperature / °C			

(*7 marks*)
AQA 2004

4 In an investigation, the rate at which phenol was broken down by the enzyme phenol oxidase was measured in solutions with different concentrations of phenol. The experiment was then repeated with a non-competitive inhibitor added to the phenol solutions. Figure 4 shows the results.

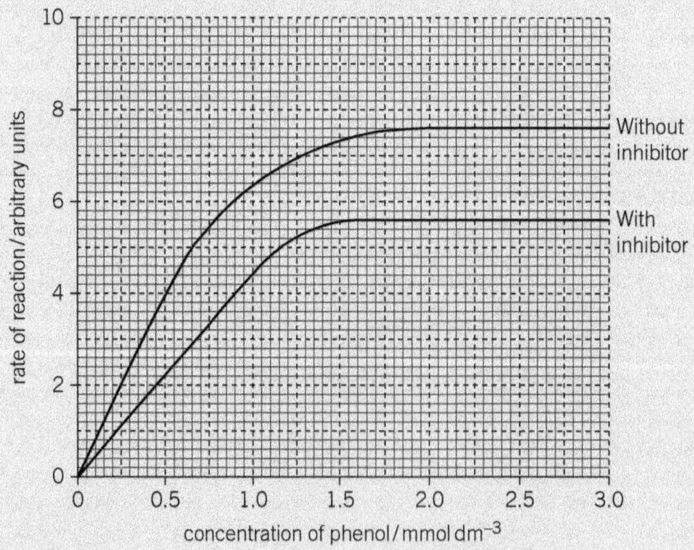

▲ **Figure 4**

(a) Explain why an increase in concentration of phenol solution from 2.0 to 2.5 mmol dm^{-3} has no effect on the rate of the reaction without the inhibitor. (*2 marks*)

(b) Explain the effect of the non-competitive inhibitor. (*2 marks*)

(c) Calculate the percentage decrease in the maximum rate of the reaction when the inhibitor was added. Show your working. (*2 marks*)

(d) Make a copy of the graph and draw a curve on it to show the results expected if a competitive inhibitor instead of a non-competitive inhibitor had been used. (*1 mark*)

AQA 2005

Answers to the Practice Questions are available at
www.oxfordsecondary.com/oxfordaqaexams-alevel-biology

4 Transport in and out of cells

4.1 Plasma membranes

All membranes around and within cells (including those around and within cell organelles) have the same basic structure and are known as **plasma membranes**.

The cell-surface membrane is the plasma membrane that surrounds cells and forms the boundary between the cell cytoplasm and the environment. It allows different conditions to be established inside and outside a cell. It controls the movement of substances in and out of the cell. Before you learn about how the cell-surface membrane achieves this, you need first to look in more detail at the molecules that form its structure.

Learning objectives:

→ Describe the structure of the cell-surface membrane.

→ Describe the functions of the various components of the cell-surface membrane.

→ Describe the fluid-mosaic model.

Specification reference: 3.1.4.1

Phospholipids

You looked at the molecular structure of a phospholipid in Topic 1.4. Phospholipids form a bilayer sheet. They are important components of cell-surface membranes for the following reasons:

- One layer of phospholipids has its hydrophilic heads pointing inwards (interacting with the water in the cell cytoplasm).
- The other layer of phospholipids has its hydrophilic heads pointing outwards (interacting with the water that surrounds all cells).
- The hydrophobic tails of both phospholipid layers point into the centre of the membrane – protected, as it were, from the water on both sides.

Lipid-soluble material moves through the membrane via the phospholipid portion. The functions of phospholipids in the membrane are to:

- allow lipid-soluble substances to enter and leave the cell
- prevent water-soluble substances entering and leaving the cell
- make the membrane flexible.

Hint

Organelles such as mitochondria and chloroplasts are surrounded by two plasma membranes. The term cell-surface membrane is reserved only for the plasma membrane around the cell.

Study tip

When representing a phospholipid it is important to be accurate. It has a *single* phosphate head and *two* fatty acid tails. Don't show too many heads and/or too many tails.

Proteins

The proteins of the cell-surface membrane are arranged less symmetrically than the regular pattern of phospholipids. They are embedded in the phospholipid bilayer in two main ways:

- **Extrinsic proteins** occur either on the surface of the bilayer or only partly embedded in it, but they never extend completely across it. They act either to give mechanical support to the membrane or, in conjunction with glycolipids, as cell receptors for molecules such as hormones.
- **Intrinsic proteins** completely span the phospholipid bilayer from one side to the other. Some act as carriers to transport water-soluble material across the membrane whereas others are enzymes.

The functions of the proteins in the membrane are to:

- provide structural support
- allow the transport of water-soluble substances by forming channels across the membrane

Hint

All plasma membranes found around and inside cells have the same phospholipid bilayer structure. What gives plasma membranes their different properties are the different substances that they contain – especially proteins.

- allow **active transport** across the membrane by acting as ion carriers for sodium, potassium, etc.
- form recognition sites by identifying cells
- help cells adhere together
- act as receptors, for example, for hormones.

Fluid-mosaic model of the cell-surface membrane

The way in which all the various molecules are combined into the structure of the cell-surface membrane is shown in Figure 1. This arrangement is known as the **fluid-mosaic model** for the following reasons:

- fluid because the individual phospholipid molecules can move relative to one another. This gives the membrane a flexible structure that is constantly changing in shape.
- mosaic because the proteins that are embedded in the phospholipid bilayer vary in shape, size, and pattern in the same way as the stones or tiles of a mosaic.

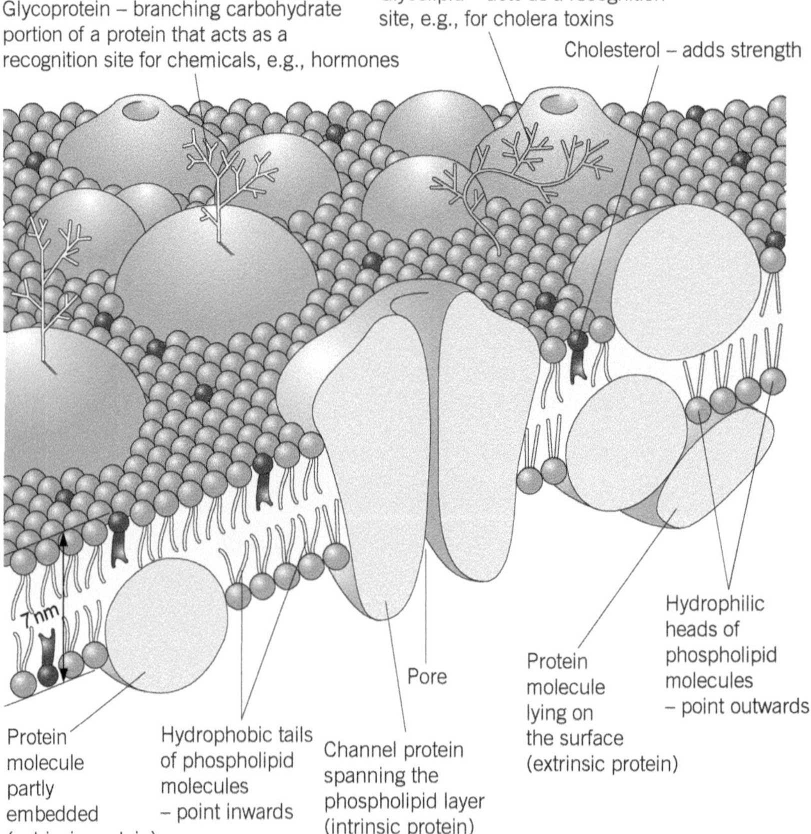

▲ **Figure 1** *The fluid-mosaic model of the cell-surface membrane*

Summary questions

1 What is the overall function of the cell-surface membrane?

2 Which end of the phospholipid molecule lies towards the inside of the cell-surface membrane?

3 Through which molecule in the cell-surface membrane are each of the following likely to pass in order to get in or out of a cell?

  a a molecule that is soluble in lipids

  b a mineral ion.

4 From your knowledge of the cell-surface membrane, suggest **two** properties that a drug should possess if it is to enter a cell rapidly.

4.2 Diffusion

The exchange of substances between cells and the environment occurs in ways that require metabolic energy (active transport) and in ways that do not (passive transport). Diffusion is an example of passive transport.

Explanation of diffusion

As all movement requires energy, it is possibly confusing to describe diffusion as passive transport. In this sense, 'passive' means that the energy comes from the natural, inbuilt motion of particles, rather than from ATP. To help understand diffusion and other passive forms of transport it is necessary to understand that:

- all particles in liquids and gases are constantly in motion due to the kinetic energy that they possess
- this motion is random, with no set pattern to the way the particles move around
- particles are constantly bouncing off one another as well as off other objects, for example, the sides of the vessel in which they are contained.

Given these facts, Figure 1 shows how particles that are concentrated together in part of a closed vessel will, of their own accord, distribute themselves evenly throughout the vessel as a result of diffusion.

Diffusion is therefore defined as:

the net movement of molecules or ions from a region where they are more highly concentrated to one where their concentration is lower.

1 If 10 particles occupying the left-hand side of a closed vessel are in random motion, they will collide with each other and the sides of the vessel. Some particles from the left-hand side move to the right, but initially there are no available particles to move in the opposite direction, so the movement is in one direction only. There is a large concentration gradient and diffusion is rapid.

2 After a short time the particles (still in random motion) have spread themselves more evenly. Particles can now move from right to left as well as from left to right. However, with a higher concentration of particles (seven) on the left than on the right (three), there is a greater probability of a particle moving to the right than in the reverse direction. There is a smaller concentration gradient and diffusion is slower.

3 Some time later, the particles will be evenly distributed throughout the vessel and the concentrations will be equal on each side. The system is in equilibrium. However, the particles are not static but remain in random motion. With equal concentrations on each side, the probability of a particle moving from left to right is equal to the probability of one moving in the opposite direction. There is no concentration gradient and no net movement.

4 Now the particles remain evenly distributed and will continue to be so. Although the number of particles on each side remains the same, individual particles are continuously changing position. This situation is called **dynamic equilibrium**.

Learning objectives:

- → Define diffusion and how it occurs.
- → Describe what affects the rate of diffusion.
- → Describe how facilitated diffusion differs from diffusion.

Specification reference: 3.1.4.2

Hint

Remember that diffusion is the *net* movement of particles. All particles move at random in diffusion, it is just that more move in one direction than in the other. This is due to concentration differences.

1
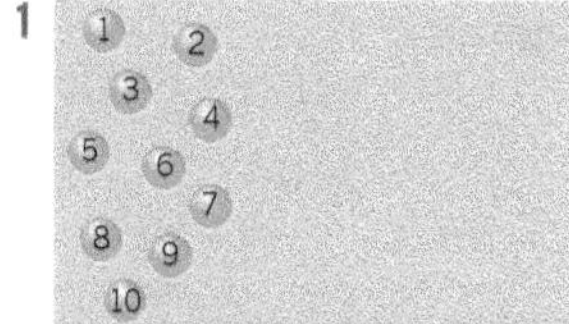

2

3
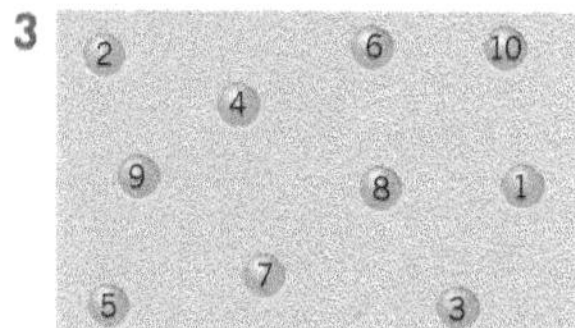

4

▲ **Figure 1** *Diffusion*

Study tip

Diffusion is proportional to the *difference* in concentration between two regions (the concentration gradient).

Hint

Diffusion only occurs between different concentrations of the *same* substance. For example, it may occur between different concentrations of oxygen or between different concentrations of carbon dioxide. It *never* occurs between different concentrations of oxygen and carbon dioxide.

Rate of diffusion

There are a number of factors that affect the rate at which molecules or ions diffuse across an exchange surface:

- **concentration gradient**. The greater the difference in the concentration of molecules or ions on either side of an exchange surface, the faster the rate of diffusion.
- **area over which diffusion takes place**. The larger the area of an exchange surface, the faster the rate of diffusion. See more about microvilli on page 154.
- **thickness of exchange surface**. The thinner an exchange surface, the faster the rate of diffusion.

The relationship between these three factors can be expressed as follows:

$$\textbf{diffusion} \propto \frac{\textbf{surface area} \times \textbf{difference in concentration}}{\textbf{length of diffusion path}}$$

Although this expression gives a good guide to the rate of diffusion, it is not wholly applicable to cells because diffusion is also affected by:

- the nature of the plasma membrane – its composition
- the size and nature of the diffusing molecule, for example, small molecules diffuse faster than large ones, and fat-soluble molecules diffuse faster than water-soluble ones.

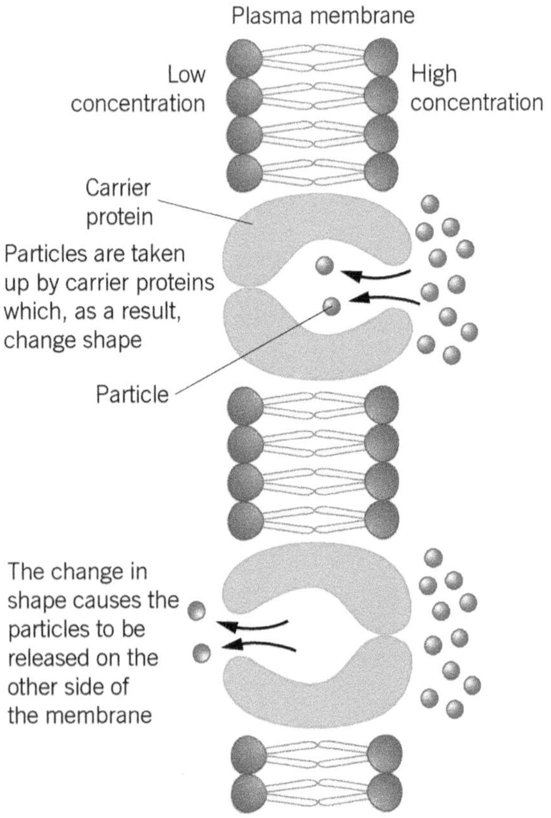

▲ **Figure 2** *Facilitated diffusion involving carrier proteins*

Facilitated diffusion

Facilitated diffusion is a passive process. It relies only on the inbuilt motion (kinetic energy) of the diffusing molecules. There is no external input of energy. Like diffusion, it occurs down a concentration gradient, but it differs in that it occurs at specific points on the plasma membrane where there are special protein molecules. These proteins form water-filled channels (protein channels) across the membrane. These allow water-soluble ions to pass through. Such ions would usually diffuse only very slowly through the phospholipid bilayer of the plasma membrane. The channels are selective, each opening only in the presence of a specific ion. If the particular molecule is not present, the channel remains closed. In this way, there is some control over the entry and exit of substances.

An alternative form of facilitated diffusion involves carrier proteins that span the plasma membrane. When a molecule that is specific to the protein is present outside the cell, it binds with the protein. This causes the carrier protein to change shape in such a way that the molecule is released to the inside of the membrane (Figure 2). No external energy is needed for this. The molecules move from a region where they are highly concentrated to one of lower concentration, using only the kinetic energy of the molecules themselves.

Diffusion in action

Starch in the diet is digested by the enzymes amylase and then maltase to form glucose. Glucose must be absorbed into the body so that it can be used by cells as a substrate for respiration. The glucose is absorbed from the exchange surface of the small intestine into the epithelial cells that line it. This absorption occurs partly by diffusion.

1 Glucose molecules mostly diffuse into cells through the pores in the proteins that span the phospholipid bilayer. Why do they not pass easily through the phospholipid bilayer?

2 State **two** changes to the structure of cell-surface membranes that would increase the rate at which glucose diffuses into a cell.

3 The other molecule required by cells for respiration is oxygen. This diffuses into the blood through the epithelial layers of the alveoli and blood capillaries. By how much would each of the following changes increase or decrease the rate of diffusion of oxygen?
   a The surface area of the alveoli is doubled.
   b The surface area of the alveoli is halved and the oxygen concentration gradient is doubled.
   c The oxygen concentration gradient is halved and the total thickness of the epithelial layers is doubled.
   d The oxygen concentration of the blood is halved and the carbon dioxide concentration of the alveoli is doubled.

Summary questions

1 State **three** factors that affect the rate of diffusion.

2 How does facilitated diffusion differ from diffusion?

3 Explain why facilitated diffusion is a passive process.

4.3 Osmosis

Learning objectives:

- → Define osmosis.
- → Define the water potential of pure water.
- → Describe the effect of solutes on water potential.
- → Describe how water potential affects water movement.
- → Describe the result of placing animal cells and plant cells into pure water.

Specification reference: 3.1.4.2

In the last topic you learnt about diffusion. You will now learn about a special case of diffusion, known as osmosis. Osmosis only involves the movement of water molecules.

What is osmosis?

Osmosis is defined as:

the movement of water from a region where it has a higher water potential to a region where it has a lower water potential through a partially permeable membrane.

Cell-surface membranes and other plasma membranes such as those around organelles are partially permeable, that is, they are permeable to water molecules and a few other small molecules, but not to larger molecules.

Solutions and water potential

A solute is any substance that is dissolved in a solvent, for example, water. The solute and the solvent together form a solution.

Water potential is represented by the Greek letter psi (Ψ), and is measured in units of pressure, usually kilopascals (kPa). Water potential is the pressure created by water molecules. Under standard conditions of temperature and pressure (25 °C and 100 kPa), pure water is said to have a water potential of zero.

It follows that:

- the addition of a solute to pure water will lower its water potential
- the water potential of a solution (water + solute) must always be less than zero, that is, a negative value
- the more solute that is added (i.e., the more concentrated a solution), the lower (more negative) its water potential
- water will move by osmosis from a region of higher (less negative) water potential (e.g., −20 kPa) to one of lower (more negative) water potential (e.g., −30 kPa).

One way of finding the water potential of cells or tissues is to place them in a series of solutions of different water potentials. Where there is no net gain or loss of water from the cells or tissues, the water potential inside the cells or tissues must be the same as that of the external solution.

Explanation of osmosis

Consider the hypothetical situation in Figure 1 overleaf, in which a partially permeable plasma membrane separates two solutions.

- The solution on the left has a low concentration of solute molecules whilst the solution on the right has a high concentration of solute molecules.
- Both the solute and water molecules are in random motion due to their **kinetic energy**.
- The partially permeable plasma membrane, however, only allows water molecules across it and not solute molecules.

Hint

Remember that, whilst diffusion can be the movement of *any* molecule, osmosis is the movement of water molecules *only*.

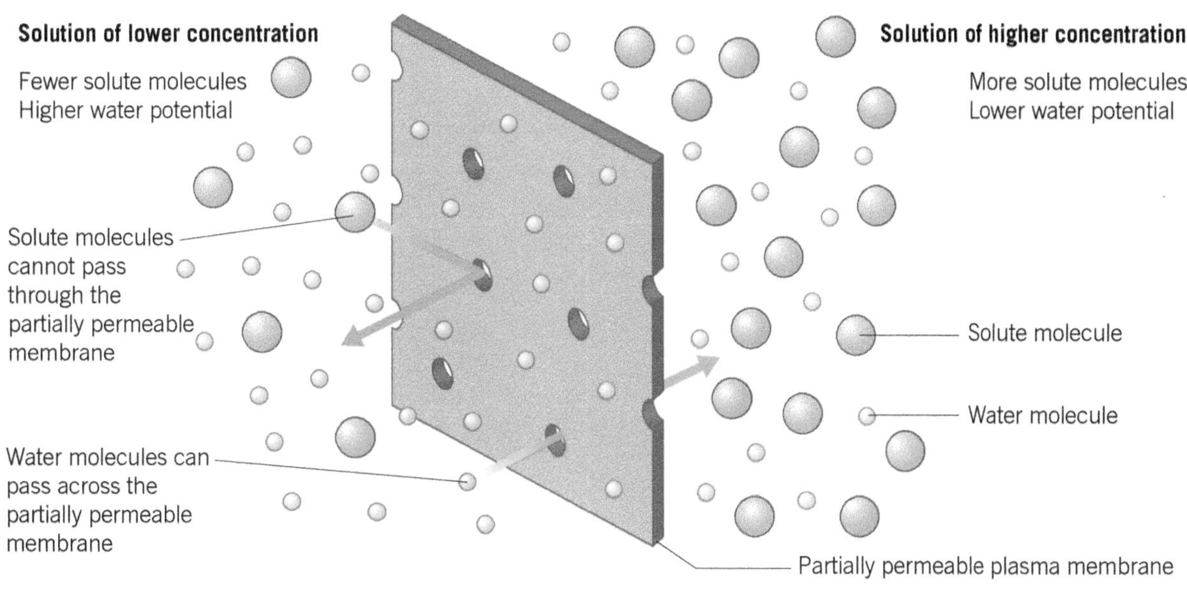

▲ **Figure 1** *Osmosis*

- The water molecules diffuse from the left-hand side, which has the higher water potential, to the right-hand side, which has the lower water potential, that is, along a water potential gradient (Figure 2).
- At the point where the water potentials on either side of the plasma membrane are equal, a dynamic equilibrium is established and there is no net movement of water.

Understanding water potential

The highest value of water potential, that of pure water, is zero, and so all other values are negative. The more negative the value, the lower the water potential. Think of water potential as an overdraft at a bank. The bigger the overdraft, the more negative is the amount of money you have. The smaller the overdraft, the less negative is the amount of money you have.

Osmosis and animal cells

Animal cells, such as red blood cells, contain a variety of solutes dissolved in their watery cytoplasm. If a red blood cell is placed in pure water it will absorb water by osmosis because it has a lower water potential. Cell-surface membranes are very thin (7 nm) and, although they are flexible, they cannot stretch to any great extent. The cell-surface membrane will therefore break, bursting the cell and releasing its contents (in red blood cells this is called haemolysis). This does not happen because animal cells are normally bathed in a liquid that has the same water potential as the cells. In our example, the liquid is the blood plasma. The plasma and red blood cells have the same water potential. If a red blood cell is placed in a solution with a water potential lower than its own, water leaves by osmosis and the cell shrinks and becomes shrivelled (see Table 1 on the next page).

Study tip

Remember that all water potential values are negative. The highest water potential is zero. Therefore the lower the water potential, the more negative it becomes.

Key

x kPa Water potential of cell

Direction of water movement

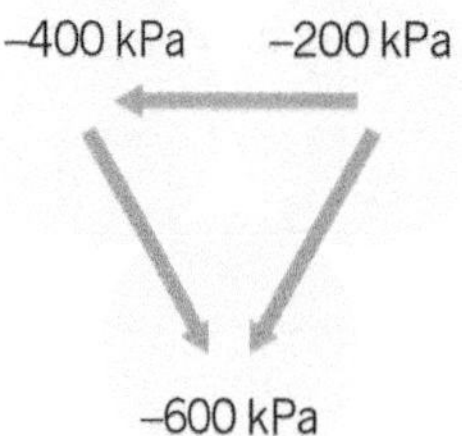

Water moves from higher water potential to lower water potential. The highest water potential is zero.

▲ **Figure 2** *Movement of water between cells along a water potential gradient*

▼ **Table 1** *Summary of osmosis in an animal cell, for example, a red blood cell*

Water potential (ψ) of external solution compared to cell solution	higher (less negative)	equal	lower (more negative)
Net movement of water	into cell	no overall movement	out of cell
State of cell	swells and bursts	no change	shrinks
	Contents, including haemoglobin, are released Remains of cell surface membrane	Normal red blood cell	Haemoglobin is more concentrated, giving cell a darker appearance Cell shrunken and shrivelled

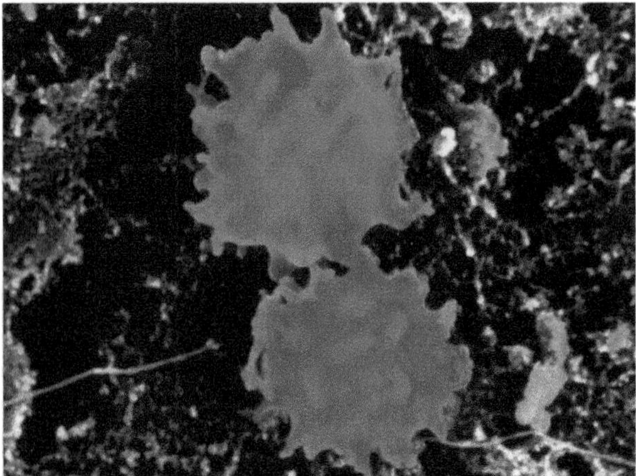

▲ **Figure 3** *Scanning electron micrograph (SEM) of red blood cells that have been placed in a solution of lower water potential. Water has left by osmosis and the cells have become shrunken and shrivelled.*

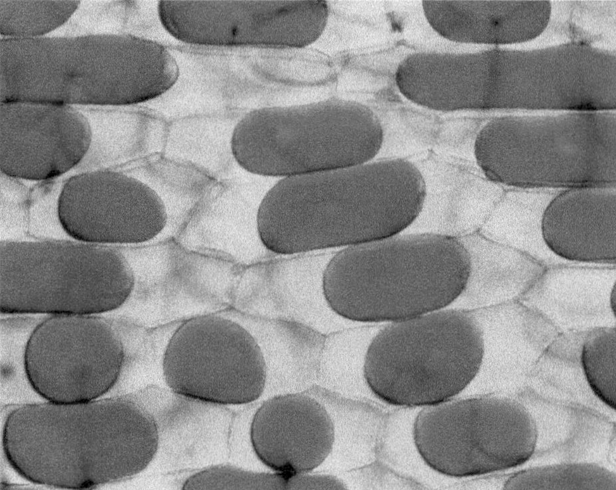

▲ **Figure 4** *Onion epidermal cells showing plasmolysis. The protoplasts, with their vacuoles containing red liquid, have shrunk and pulled away from the cell walls.*

Summary questions

1. What is meant by a partially permeable membrane?
2. Under standard conditions of pressure and temperature, what is the water potential of pure water?
3. Four cells have the following water potentials:
   Cell A = −200 kPa
   Cell B = −250 kPa
   Cell C = −100 kPa
   Cell D = −150 kPa.
   In what order would the cells have to be placed for water to pass from one cell to the next if they are arranged in a line?

Osmosis and plant cells

▼ **Table 2** *Summary of osmosis in a plant cell*

Water potential (ψ) of external solution compared to cell solution	higher (less negative)	equal	lower (more negative)
Net movement of water	into cell	no overall movement	out of cell
Protoplast	swells	no change	shrinks
Condition of cell	turgid	incipient plasmolysis	plasmolysed
	Protoplast pushed against cell wall; Nucleus; Cellulose cell wall; Protoplast	Protoplast beginning to pull away from the cell wall	Protoplast completely pulled away from the cell wall

For the purposes of the following explanations, the plant cell can be divided into three parts:

- the **central vacuole**, which contains a solution of salts, sugars, and organic acids in water
- the **protoplast**, consisting of the outer cell-surface membrane, nucleus, cytoplasm, and the inner vacuole membrane
- the **cellulose cell wall**, a tough, inelastic covering that is permeable to even large molecules.

Like animal cells, plant cells also contain a variety of solutes, mainly dissolved in the water of the large cell vacuole that each possesses. When placed in pure water they also absorb water by osmosis because of their lower (more negative) water potential. Unlike animal cells, however, they are unable to control the composition of the fluid around their cells. Indeed, plant cells are normally permanently bathed in almost pure water, which is constantly absorbed from the plant's roots. Water entering a plant cell by osmosis causes the protoplast to swell and press on the cell wall. Because the cell wall is capable of only very limited expansion, a pressure builds up on it that resists the entry of further water. In this situation, the protoplast of the cell is kept pushed against the cell wall and the cell is said to be **turgid**.

If the same plant cell is placed in a solution with a lower water potential than its own, water leaves by osmosis. The volume of the cell decreases. A stage is reached where the protoplast no longer presses on the cellulose cell wall. At this point the cell is said to be at **incipient plasmolysis**. Further loss of water will cause the cell contents to shrink further and the protoplast to pull away from the cell wall. In this condition the cell is said to be **plasmolysed**. These events are summarised in Table 2.

1 Explain why an animal cell placed in pure water bursts whereas a plant cell placed in pure water does not.

2 Plant cells that have a water potential of −600 kPa are placed in solutions of different water potentials. State in each of the following cases whether, after 10 minutes, the cells would be turgid, plasmolysed or at incipient plasmolysis.
   a Solution A = −400 kPa
   b Solution B = −600 kPa
   c Solution C = −900 kPa
   d Solution D = pure water

3 If an animal cell with a water potential of −700 kPa was placed in each of the solutions, in which solutions is it likely to burst?

Required practical 2

Investigating the effect of solute concentration on water uptake

Potato tubers are a useful source of plant tissues for investigating the effect of solute concentration on the uptake or loss of water.

- A series of five different concentrations of sodium chloride was made up using simple dilution of a 1 M stock solution as shown in Table 1.

▼ **Table 1**

1 M Sodium chloride solution (cm^3)	Water (cm^3)	Final sodium chloride concentration (M)
100		1.00
75		0.75
50		0.50
25		0.25
0		0.00

1 Complete Table 1 to show the volume of water used to obtain the final sodium chloride concentration.

- A number of cylinders of potato tissue were cut using a cork borer. The cylinders were each cut to a length of 40 mm. The cylinders were rinsed briefly to remove any surface starch, dried, and then weighed.
- Three cylinders were then placed in each of the solutions shown in Table 1. The mass of each cylinder was recorded before being submerged in the solution. Cylinders were each labelled using a piece of different coloured thread. The cylinders were left in solution for 24 hours and then reweighed and the lengths remeasured.

2 Why was it important to label each cylinder?

3 Suggest two precautions that should be carried out in order to obtain accurate results from weighing the cylinders.

- The change in mass and length for each cylinder was recorded and a mean percentage change was calculated for each concentration of sodium chloride. The results are shown in Table 2.

▼ Table 2

Sodium chloride concentration (M)	Mean percentage change in mass	Mean percentage change in length
1.00	−26.4	−10.8
0.75	−30.3	−8.3
0.50	−28.6	−6.7
0.25	−3.00	0.0
0.00	+18.7	10.8

4 Plot a graph of the results in Table 2.

5 Using your understanding of osmosis, explain the general pattern shown by the results for the 0.00 and 0.75 M solutions.

6 Suggest an explanation for differences in the pattern of results for mass and length seen between the 0.75 and the 1.00 M solutions.

4.4 Active transport

Learning objectives:

- → Define active transport.
- → Describe what is needed for active transport to take place.

Specification reference: 3.1.4.3

Study tip

Carrier proteins have a specific tertiary structure and will only transport particular substances across a membrane. They are not enzymes and do not have an active site.

Active transport allows cells to exchange molecules and / or ions against a concentration gradient. Metabolic energy is required for the process. Once inside the cell the molecules are prevented from leaking back by the barrier of the cell-surface membrane's bilayer. In this way, a different environment is maintained on either side of the membrane.

What is active transport?

Active transport is defined as:

the movement of molecules or ions into or out of a cell from a region of lower concentration to a region of higher concentration using energy from ATP and carrier molecules.

It differs from passive forms of transport in the following ways:

- Metabolic energy in the form of ATP is needed.
- Materials are moved against a concentration gradient, that is, from a lower to a higher concentration.
- Carrier protein molecules which act as 'pumps' are involved.
- The process is very selective, with specific substances being transported.

Active transport uses ATP in one of two ways:

- used directly by carrier proteins to move molecules
- by using a concentration gradient that has already been set up by direct active transport. This is also known as **co-transport** and is explained in Topic 12.3.

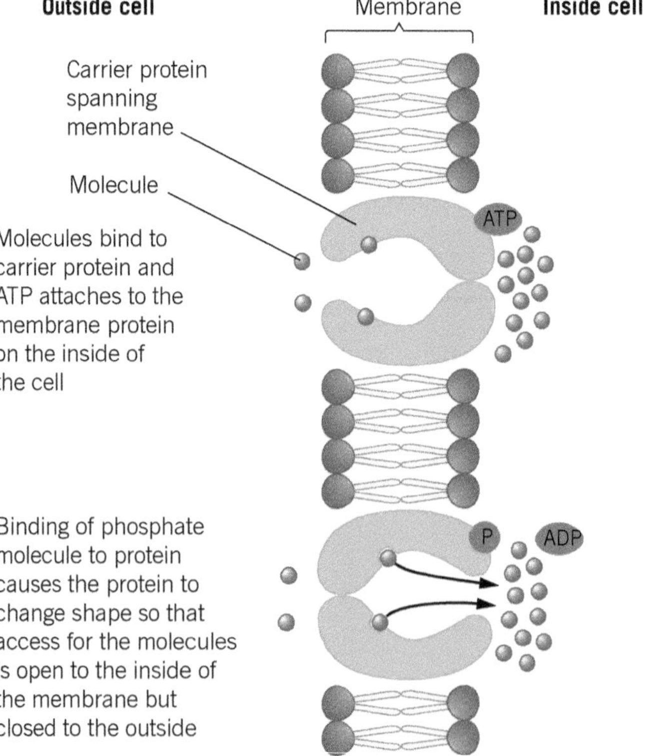

▲ **Figure 1** *Active transport*

Direct active transport of a single molecule or ion is described below.

- The carrier proteins span the cell-surface membrane and on one side accept the molecules or ions to be transported.
- The molecules or ions bind to receptors on the channels of the carrier protein.
- On the inside of the cell, ATP binds to the protein, causing it to split into ADP and a phosphate molecule. As a result, the protein molecule changes shape and opens to the opposite side of the membrane.
- The molecules or ions are then released to the other side of the membrane.
- The phosphate molecule is released from the protein and recombines with the ADP to form ATP during respiration.
- This causes the protein to revert to its original shape, ready for the process to be repeated.

These events are illustrated in Figure 1. It is important to distinguish between active transport and facilitated diffusion. Both use carrier proteins but facilitated diffusion occurs *down* a concentration gradient, whereas active transport occurs *against* a concentration gradient. This means that facilitated diffusion does not require metabolic energy, whereas active transport does. The metabolic energy is provided in the form of ATP.

Sometimes more than one molecule or ion may be moved in the same direction at the same time by active transport. Occasionally, the molecule or ion is moved into a cell at the same time a different one is being removed from it. One example of this is the **sodium–potassium pump**.

In the sodium–potassium pump, sodium ions are actively removed from the cell whilst potassium ions are actively taken in from the surroundings. This process is essential to a number of important processes in the organism, including the creation of a nerve impulse.

How ATP acts as a source of energy

Adenosine triphosphate (ATP) is a nucleotide and as the name suggests, has three phosphate groups (see Figure 2). These are the key to how ATP acts as a source of energy. The bonds between these phosphate groups are unstable and so have a low **activation energy**, which means that they are easily broken. When they do break they release a considerable amount of energy. Usually in living cells it is only the terminal phosphate that is removed, according to the equation:

ATP	+	**(H_2O)**	→	**ADP**	+	**P_I**	+	**E**
adenosine triphosphate		**water**		**adenosine diphosphate**		**inorganic phosphate**		**energy**

How ATP is synthesised

The conversion of ATP to ADP is a reversible reaction and therefore energy can be used to add an inorganic phosphate to ADP to re-form ATP according to the reverse of the equation above (see Figure 3). This reaction is catalysed by the enzyme **ATP synthase**.

Hint

Carrier proteins are involved in both facilitated diffusion and active transport *but* any given protein carrier is very specific about what it carries and by which method.

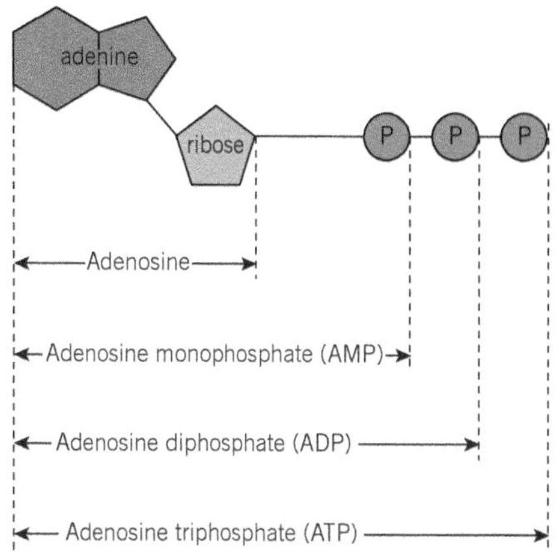

▲ **Figure 2** *Structure of ATP*

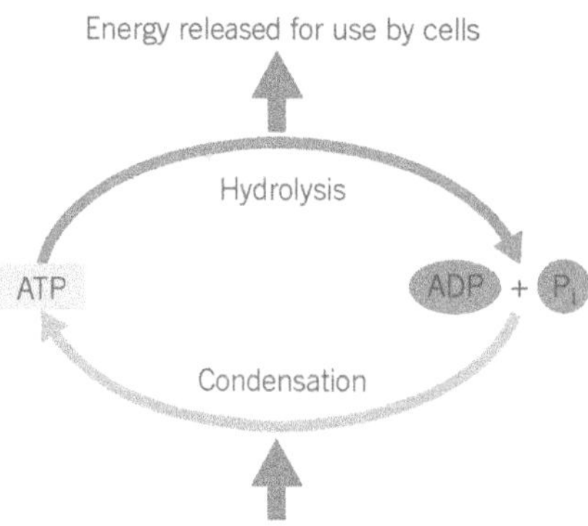

▲ **Figure 3** *Interconversion of ATP and ADP*

ATP and biological processes

ATP cannot be stored and so has to be continuously made within the mitochondria of cells that need it. Cells such as muscle fibres and the epithelium of the small intestine, which require energy for movement and active transport respectively, possess many large mitochondria.

ATP is used in energy-requiring processes in cells including:

- **metabolic processes**. ATP provides the energy needed to build up macromolecules from their basic units. For example, making starch from glucose or polypeptides from amino acids.
- **movement**. ATP provides the energy for muscle contraction. In muscle contraction, ATP provides the energy for the filaments of muscle to slide past one another and therefore shorten the overall length of a muscle fibre.
- **active transport**. ATP provides the energy to change the shape of carrier proteins in plasma membranes. This allows molecules or ions to be moved against a concentration gradient.
- **secretion**. ATP is needed to form the lysosomes necessary for the secretion of cell products.
- **activation of molecules**. The inorganic phosphate released during the hydrolysis of ATP can be used to phosphorylate other compounds in order to make them more reactive, thus lowering the activation energy in enzyme-catalysed reactions. For example, the addition of phosphate to glucose molecules at the start of glycolysis.

Hint

ATP is synthesised during reactions that *release* energy and it is hydrolysed to provide energy for reactions that *require* it.

Study tip

Don't think about ATP as a high-energy substance. ATP is an intermediate energy substance that is used to transfer energy.

Summary questions

1 ATP is sometimes referred to as an immediate energy source. Explain why.

2 Explain how ATP can make an enzyme-catalysed reaction take place more readily.

3 State **three** roles of ATP in plant cells.

4 State **one** similarity and **one** difference between active transport and facilitated diffusion.

5 The presence of many mitochondria is typical of cells that carry out active transport. Explain why this is so.

6 In the production of urine, glucose is initially lost from the blood but is then reabsorbed into the blood by cells in the kidneys. Explain why it is important that this reabsorption occurs by active transport rather than by diffusion.

Practice questions: Chapter 4

1 A student investigated the effect of putting cylinders cut from a potato into sodium chloride solutions of different concentrations. He cut cylinders from a potato and weighed each cylinder. He then placed each cylinder in a test tube. Each test tube contained a different concentration of sodium chloride solution. The tubes were left overnight. He then removed the cylinders from the solutions and reweighed them.

(a) Before reweighing, the student blotted dry the outside of each cylinder. Explain why. *(2 marks)*

The student repeated the experiment several times at each concentration of sodium chloride solution. His results are shown in the graph.

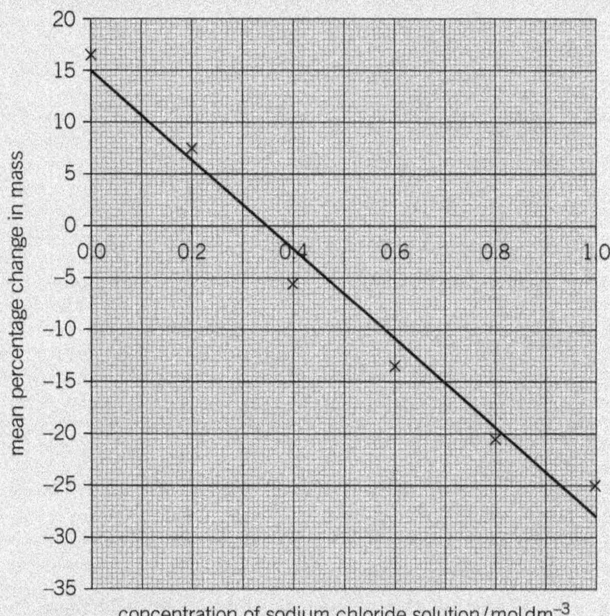

(b) The student made up all the sodium chloride solutions using a 1.0 / mol dm^{-3} sodium chloride solution and distilled water.

Complete Table 1 to show how he made 20 cm^3 of a 0.2 mol dm^{-3} sodium chloride solution.

▼ **Table 1**

Volume of 1.0 mol dm^{-3} sodium chloride solution	Volume of distilled water

(1 mark)

(c) The student calculated the *percentage* change in mass rather than the change in mass. Explain the advantage of this. *(2 marks)*

(d) The student carried out several repeats at each concentration of sodium chloride solution. Explain why the repeats were important. *(2 marks)*

(e) Use the graph to find the concentration of sodium chloride solution that has the same water potential as the potato cylinders. *(1 mark)*

AQA Jan 2011

2 Some substances can cross the cell-surface membrane of a cell by simple diffusion through the phospholipid bilayer. Describe other ways by which substances cross this membrane. *(5 marks)*

AQA Jan 2013

3 **(a)** Give **two** ways in which active transport is different from facilitated diffusion. *(2 marks)*

Scientists investigated the effect of a drug called a proton pump inhibitor. The drug is given as a tablet to people who produce too much acid in their stomach. It binds to a carrier protein in the surface membrane of cells lining the stomach. This carrier protein usually moves hydrogen ions into the stomach by active transport.

The scientists used two groups of people in their investigation. All the people produced too much acid in their stomach. People in group **P** were given the drug. Group **Q** was the control group.

The graph shows the results.

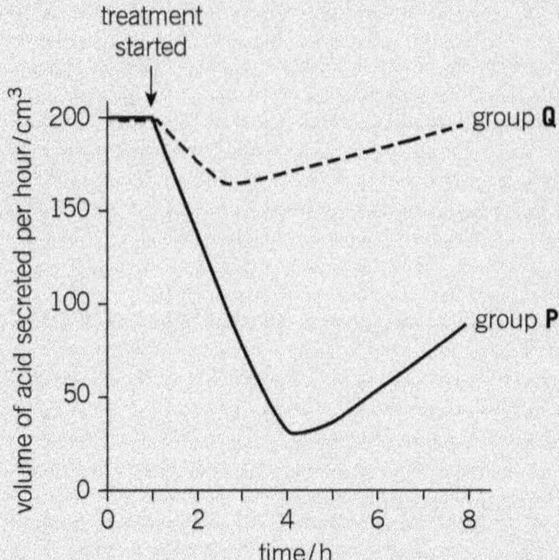

(b) (i) The scientists used a control group in this trial. Explain why. *(1 mark)*

(ii) Suggest how the control group would have been treated. *(2 marks)*

(c) Describe the effect that taking the drug had on acid secretion. *(1 mark)*

(d) Calculate the percentage decrease in acid secretion of group **P** compared to group **Q** after 8 hours. *(2 marks)*

AQA June 2011

4 Scientists investigated the percentages of different types of lipid in plasma membranes from different types of cell. **Table 2** shows some of their results.

▼ **Table 2**

Type of lipid	Percentage of lipid in plasma membrane by mass		
	Cell lining ileum of mammal	Red blood cell of mammal	The bacterium *Escherichia coli*
Cholesterol	17	23	0
Glycolipid	7	3	0
Phospholipid	54	60	70
Others	22	14	30

(a) The scientists expressed their results as *percentage of lipid in plasma membrane by mass*. Explain how they would find these values. *(2 marks)*

(b) Cholesterol increases the stability of plasma membranes. Cholesterol does this by making membranes less flexible.
Suggest **one** advantage of the different percentage of cholesterol in red blood cells compared with cells lining the ileum. *(1 mark)*

(c) *E. coli* has no cholesterol in its cell-surface membrane. Despite this, the cell maintains a constant shape. Explain why. *(2 marks)*

AQA SAMS AS PAPER 1

Answers to the Practice Questions are available at
www.oxfordsecondary.com/oxfordaqaexams-alevel-biology

5 Gas exchange and the transport of oxygen

5.1 Exchange between organisms and their environment

Learning objectives:

- → Explain how the size of an organism and its structure relate to its surface area to volume ratio.
- → Describe how larger organisms increase their surface area to volume ratio.

Specification reference: 3.1.5.1

The external environment is different from the internal environment found within an organism and within its cells. To survive, organisms transfer materials between the two environments. This transfer takes place at exchange surfaces and always involves crossing cell plasma membranes. The environment around the cells of multicellular organisms is called **tissue fluid**. The majority of cells are too far from exchange surfaces for diffusion alone to supply or remove the various materials that tissue fluid needs to keep its composition relatively constant. Therefore, once absorbed, materials are rapidly distributed to the tissue fluid and the waste products are returned to the exchange surface for removal. This involves a mass transport system. It is this mass transport system that maintains the diffusion gradients that bring materials to and from the cell-surface membranes.

The size and metabolic rate of an organism will affect the amount of each material that is exchanged. For example, organisms with a high metabolic rate exchange more materials and so require a larger surface area to volume ratio. In turn this is reflected in the type of exchange surface and transport system that has evolved to meet the requirements of each organism. In this chapter you will investigate the adaptations of exchange surfaces and transport systems in a variety of organisms.

Examples of things that need to be interchanged between an organism and its environment include respiratory gases (oxygen and carbon dioxide), nutrients (glucose, fatty acids, amino acids, vitamins, minerals), excretory products (urea and carbon dioxide), and heat.

Except for heat, these exchanges can take place in two ways:

- passively (no metabolic energy is required), by diffusion and osmosis
- actively (metabolic energy is required), by active transport.

Surface area to volume ratio

Exchange takes place at the surface of an organism, but the materials absorbed are used by the cells that mostly make up its volume. For exchange to be effective, the exchange surface(s) of the organism must be large compared with its volume.

Small organisms have a surface area that is large enough, compared with their volume, to allow efficient exchange across their body surface. However, as organisms become larger, their volume increases at a faster rate than their surface area (Table 1). Because of this, simple diffusion of substances across the outer surface can only meet the needs of relatively inactive organisms. Even if the outer surface could supply enough of a substance, it would still take too long for it to reach the middle of the organism if diffusion alone was the method of transport. Organisms have therefore evolved one or more of the following features:

Study tip

In a cell the lowest oxygen concentration is inside the mitochondria, where oxygen is used up in respiration. Mitochondria also contain the highest concentration of carbon dioxide. This maintains the diffusion gradient for these gases to diffuse in and out of the cell.

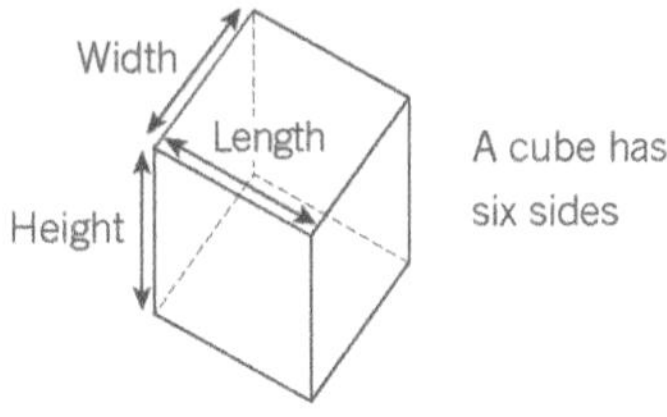

▲ **Figure 1** *Calculating volume*

- a flattened shape, so that no cell is ever far from the surface (e.g., a flatworm or a leaf)
- specialised exchange surfaces with large areas to increase the surface area to volume ratio (e.g., lungs in mammals, gills in fish).

You may be asked to calculate the surface area to volume ratio of cells with different shapes. To make these calculations reasonably straightforward, cells or organisms may have to be assumed to have a uniform shape, although in practice they almost never do.

▼ **Table 1** *How the surface area to volume ratio gets smaller as an object becomes larger*

Length of edge of a cube / cm	Surface area of whole cube (area of one side × 6 sides) / cm^2	Volume of cube (length × width × height) / cm^3	Ratio of surface area to volume (surface area ÷ volume)
1	1 × 6 = **6**	1 × 1 × 1 = **1**	$\frac{6}{1}$ = **6.0 : 1**
2	4 × 6 = **24**	2 × 2 × 2 = **8**	$\frac{24}{8}$ = **3.0 : 1**
3	9 × 6 = **54**	3 × 3 × 3 = **27**	$\frac{54}{27}$ = **2.0 : 1**
4	16 × 6 = **96**	4 × 4 × 4 = **64**	$\frac{96}{64}$ = **1.5 : 1**
5	25 × 6 = **150**	5 × 5 × 5 = **125**	$\frac{150}{125}$ = **1.2 : 1**
6	36 × 6 = **216**	6 × 6 × 6 = **216**	$\frac{216}{216}$ = **1.0 : 1**

Worked example: Calculating the surface area to volume ratio of cells with different shapes

For example, let us assume that a cell has the shape of a sphere that is 10 μm in diameter. The surface area of a sphere is calculated using the formula: $4\pi r^2$

In this example: r = 5 μm (radius = half the diameter) and we will use the value of π as 3.14.

Therefore the surface area of the cell = 4 × 3.14 × (5 × 5) = 314 μm^2

The volume of a sphere is calculated using the formula: $\frac{4}{3}\pi r^3$

Therefore the volume of the cell = $\frac{4}{3}$ × 3.14 × (5 × 5 × 5) = 523.33 μm^3

The surface area to volume ratio is therefore 314 ÷ 523.33 = 0.6 : 1

Hint

Remember that substances not only have to move into cells through the cell-surface membrane but also into organelles like mitochondria through the plasma membrane that surrounds them. All plasma membranes are therefore thin.

Features of specialised exchange surfaces

To allow effective transfer of materials across specialised exchange surfaces by diffusion or active transport, exchange surfaces show the following characteristics:

- a large surface area relative to the volume of the organism, which increases the rate of exchange
- very thin so that the diffusion distance is short and therefore materials cross the exchange surface rapidly
- selectively permeable to allow selected materials to cross

- movement of the environmental medium, for example, air, to maintain a diffusion gradient
- a transport system to ensure the movement of the internal medium, for example, blood, in order to maintain a diffusion gradient.

You saw in Topic 4.2 that the relationship between certain of these factors can be expressed as:

$$\textbf{diffusion} \propto \frac{\textbf{surface area} \times \textbf{difference in concentration}}{\textbf{length of diffusion path}}$$

Being thin, specialised exchange surfaces are easily damaged and dehydrated. They are therefore often located inside an organism. Where an exchange surface is located inside the body, the organism needs to have a means of moving the external medium over the surface, for example, a means of ventilating the lungs in a mammal.

Summary questions

1 Name **four** general things that need to be exchanged between organisms and their environment.

2 Calculate the surface area to volume ratio of a cube that has sides 10 mm long.

3 Name **three** factors that affect the rate of diffusion of substances into cells.

Significance of the surface area to volume ratio in organisms

The graph in Figure 2 shows the surface area to volume ratios of different-sized cubes. The ratios are actually 1:1, 2:1,3:1 etc. but are shown as single numbers for ease of plotting.

1 Microscopic organisms obtain their oxygen by diffusion across their body surface. Using the graph, explain how they are able to obtain sufficient oxygen for their needs.

2 The blue whale is the largest animal on the planet. It spends much of its life in cold waters with temperatures between 0 °C and 6 °C. Use the graph to explain one way in which large size is an advantage to blue whales.

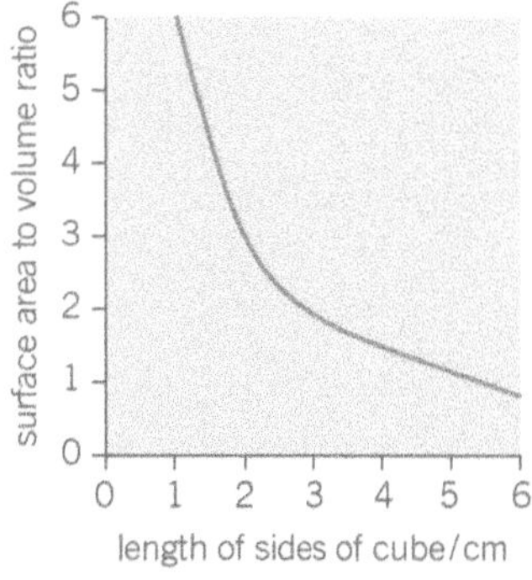

▲ **Figure 2** *Surface area to volume ratios*

Calculating a surface area to volume ratio

Consider the shape shown in Figure 3, which has dimensions marked on it. Use the information below to calculate the ratio of surface area to volume of this shape (to two decimal places).

The area of a disc (like those at the ends of an enclosed cylinder) is calculated using the formula πr^2.

The external surface area of an enclosed cylinder is calculated using the formula $2\pi rh + 2\pi r^2$.

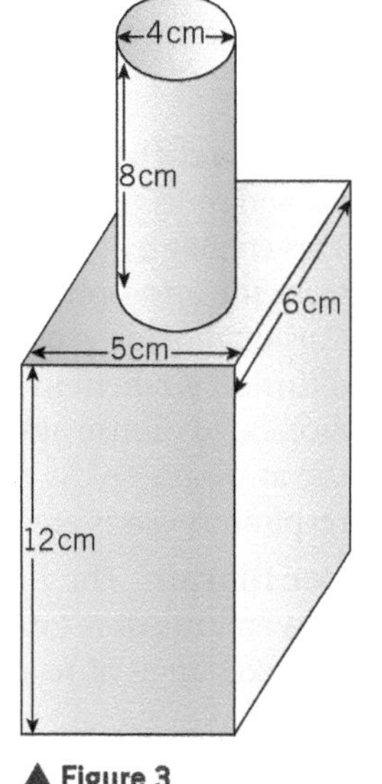

▲ **Figure 3**

5.2 Gas exchange in single-celled organisms and insects

Learning objectives:

- → Describe how single-celled organisms exchange gases.
- → Describe how terrestrial insects balance the need to exchange gases with the need to conserve water.
- → Describe how insects exchange gases.

Specification reference: 3.1.5.2

Gas exchange in single-celled organisms

Single-celled organisms are small and therefore have a large surface area to volume ratio. Oxygen is absorbed by diffusion across their body surface, which is covered only by a cell-surface membrane. In the same way, carbon dioxide from respiration diffuses out across their body surface. When a living cell is surrounded by a cell wall, this is completely permeable and so there is no barrier to the diffusion of gases.

Gas exchange in insects

Most insects are terrestrial (live on land). The problem for all terrestrial organisms is that water easily evaporates from the surface of their bodies and they can become dehydrated. They have evolved adaptations to conserve water. However, efficient gas exchange requires a thin, permeable surface with a large area. These features conflict with the need to conserve water. Overall, as a terrestrial organism, the insect has to balance the opposing needs of exchanging respiratory gases with reducing water loss.

To reduce water loss, terrrestrial organisms usually exhibit two features:

- **Waterproof coverings** over their body surfaces. In the case of insects this covering is a rigid outer skeleton that is covered with a waterproof cuticle.
- **Small surface area to volume ratio** to minimise the area over which water is lost.

These features mean that insects cannot use their body surface to diffuse respiratory gases in the way that a single-celled organism does. Instead they have evolved an internal network of tubes called **tracheae**. The tracheae are supported by strengthened rings to prevent them from collapsing. The tracheae divide into smaller tubes called **tracheoles**. The tracheoles extend throughout all the body tissues of the insect. Atmospheric air, with the oxygen it contains, is therefore brought directly to the respiring tissues.

Respiratory gases move in and out of the tracheal system in two ways:

- **Down a diffusion gradient**. When cells are respiring, oxygen is used up and so its concentration towards the ends of the tracheoles falls. This creates a diffusion gradient that causes gaseous oxygen to diffuse from the atmosphere along the tracheae and tracheoles to the cells. Carbon dioxide is produced by cells during respiration. This creates a diffusion gradient in the opposite direction. This causes gaseous carbon dioxide to diffuse along the tracheoles and tracheae from the cells to the atmosphere. As diffusion in air is much more rapid than in water, respiratory gases are exchanged quickly by this method.
- **Ventilation**. The movement of muscles in insects can create mass movements of air in and out of the tracheae. This further speeds up the exchange of respiratory gases.

Hint

Every cell of an insect is only a very short distance from one of the tracheae or tracheoles and so the diffusion pathway is always short.

Gases enter and leave tracheae through tiny pores, called **spiracles**, on the body surface. The spiracles may be opened and closed by a valve. When the spiracles are open, water can evaporate from the insect. For much of the time insects keep their spiracles closed to prevent this water loss. Periodically they open the spiracles to allow gas exchange. Part of an insect tracheal system is illustrated in Figure 1.

The tracheal system is an efficient method of gas exchange. It does, however, have some limitations. It relies mostly on diffusion to exchange gases between the environment and the cells. For diffusion to be effective, the diffusion pathway needs to be short. As a result this limits the size that insects can attain. Not that being small has hindered insects. They are one of the most successful groups of organisms on Earth.

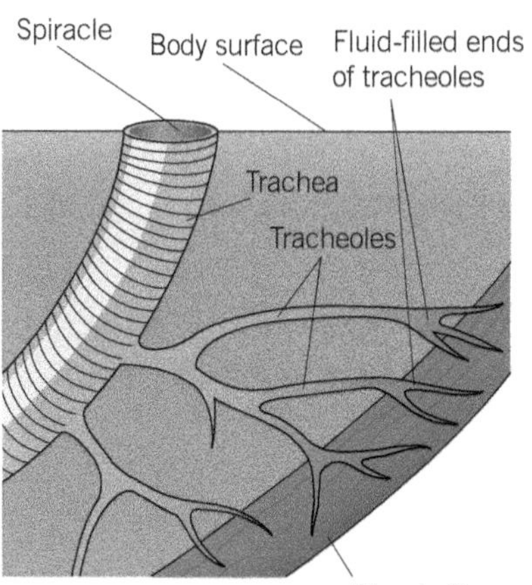

▲ **Figure 1** *Part of an insect tracheal system*

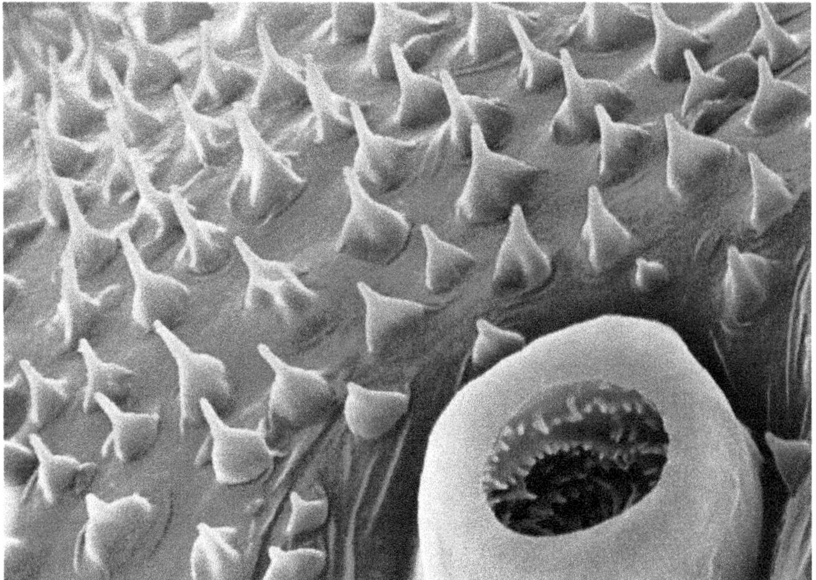

▲ **Figure 2** *Scanning electron micrograph (SEM) of a spiracle (or air pore, bottom right) of an insect*

Summary questions

1. By what process is carbon dioxide removed from a single-celled organism?
2. How do insects prevent excessive water loss from their tracheal system?
3. Explain why there is a conflict in terrestrial insects between the need for gas exchange and the need to conserve water.
4. Why does the tracheal system limit the size of insects?

Spiracle movements

An experiment was carried out to measure the concentrations of oxygen and carbon dioxide in the tracheal system of an insect over a period of time. During the experiment the opening and closing of the insect's spiracles was observed and recorded. The results are shown in Figure 3.

1. Describe what happens to the concentration of oxygen in the tracheae when the spiracles are closed.
2. Suggest an explanation for this change in the concentration of oxygen when the spiracles are closed.
3. From the information provided by the graph, suggest what causes the spiracles to open.
4. What is the advantage of these spiracle movements to a terrestrial insect?
5. Fossil insects have been discovered that are larger than insects that occur on Earth today. What does this suggest about the composition of the atmosphere at the time when these fossil insects lived?

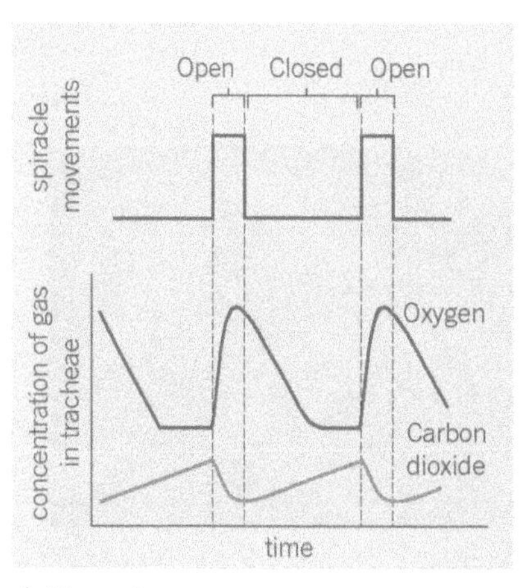

▲ **Figure 3**

5.3 Gas exchange in the leaf of a plant

Learning objectives:

- → Describe how plants exchange gases.
- → Describe the structure of a dicotyledonous plant leaf.
- → Describe how the leaf is adapted for efficient gas exchange.

Specification reference: 3.1.5.2

Like animal cells, all plant cells take in oxygen and produce carbon dioxide during respiration. When it comes to gas exchange, however, plants show one important difference from animals. Some plant cells carry out photosynthesis. During photosynthesis, plant cells take in carbon dioxide and produce oxygen. At times the gases produced in one process can be used for the other. This reduces the need for gas exchange with the external air. Overall, this means that the volumes and types of gases that are being exchanged by a plant leaf change. This depends on the balance between the rates of photosynthesis and respiration.

- When photosynthesis is taking place, although some carbon dioxide comes from respiration of cells, most of it has to be obtained from the external air. In the same way, some oxygen from photosynthesis is used in respiration but most of it **diffuses** out of the plant.
- When photosynthesis is not occurring, for example, in the dark, oxygen diffuses into the leaf because it is constantly being used by cells during respiration. In the same way, carbon dioxide produced during respiration diffuses out.

Study tip

The diffusion gradients in and out of the leaf are maintained by mitochondria carrying out respiration and chloroplasts carrying out photosynthesis.

Structure of a plant leaf and gas exchange

In some ways, gas exchange in plants is not unlike that of insects (see Topic 5.2).

- No living cell is far from the external air, and therefore from a source of oxygen and carbon dioxide.
- Diffusion takes place in the gas phase (air), which makes it more rapid than if it were in water.

Hint

Remember that plant cells respire all the time, but only plant cells with chloroplasts photosynthesise – and then only when the conditions are right.

Overall, therefore, there is a short, fast diffusion pathway. In addition, a plant leaf has a very large surface area compared with the volume of living tissue. For these reasons, no specialised transport system is needed for gases, which simply move in and through the plant by diffusion. Most gaseous exchange occurs in the leaves, which show the following adaptations for rapid diffusion:

- a thin, flat shape that provides a large surface area and short diffusion pathways
- many small pores, called **stomata**, mostly in the lower epidermis (Figure 1)
- numerous interconnecting air spaces that occur throughout the mesophyll.

The structure of a leaf is shown in Figure 2.

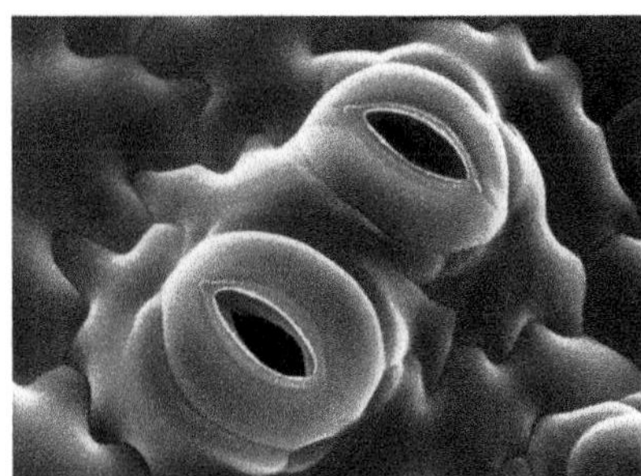

▲ **Figure 1** *Coloured scanning electron micrograph (SEM) showing two open stomata each surrounded by guard cells.*

Stomata

Stomata are minute pores that occur mainly, but not exclusively, on the leaves, especially the undersides. Each stoma (singular) is surrounded by a pair of special cells (guard cells). These cells can open and close the stomatal pore (Figure 3). In this way, they can control the rate of gaseous exchange. This is important because terrestrial organisms lose water by evaporation. Plants have to balance the conflicting needs of gas exchange and control of water loss. They do this by completely or partly closing stomata at times when water loss would be excessive. Plants

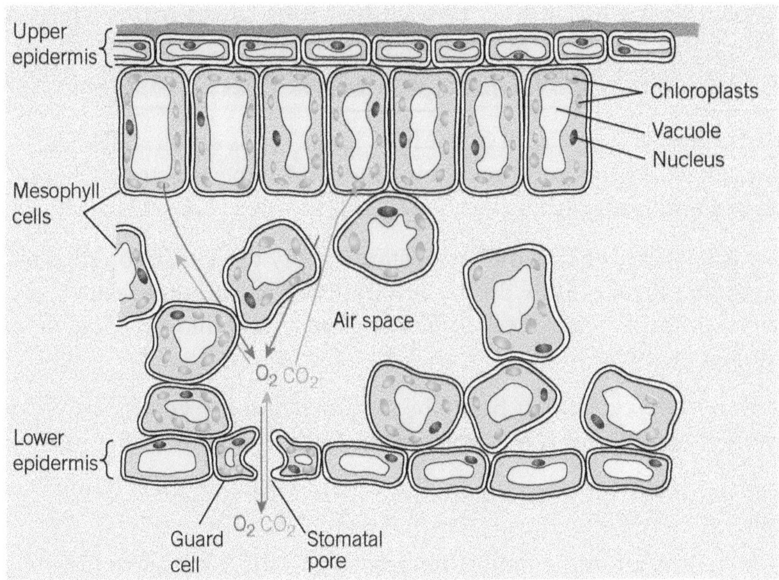

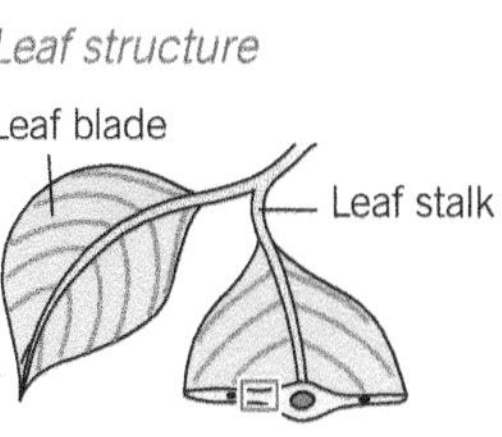

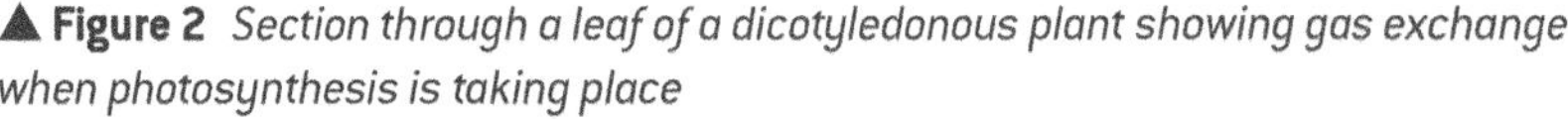
▲ **Figure 2** *Section through a leaf of a dicotyledonous plant showing gas exchange when photosynthesis is taking place*

that are adapted to dry conditions (xerophytic plants) have additional features, such as rolled leaves or sunken stomata, which reduce water loss when stomata are open.

Summary questions

1. State **two** similarities between gas exchange in a plant leaf and gas exchange in a terrestrial insect.
2. State **two** differences between gas exchange in a plant leaf and gas exchange in a terrestrial insect.
3. What is the advantage to a plant of being able to control the opening and closing of stomata?

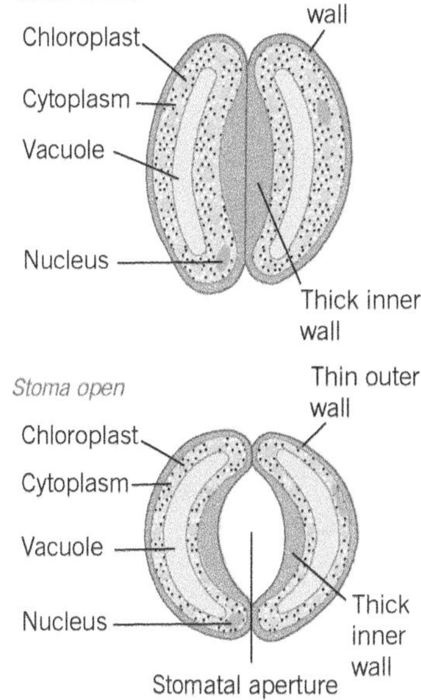

▲ **Figure 3** *Surface view of a stoma closed and open*

Exchange of carbon dioxide

The graph in Figure 4 shows the volume of carbon dioxide produced by a sample of tomato plants at different light intensities.

1. Which process produces carbon dioxide in the tomato plants?
2. Which process uses up carbon dioxide in the tomato plants?
3. Explain why, at point X, carbon dioxide is neither taken up nor given out by the tomato plants.
4. Some herbicides cause the stomata of plants to close. Suggest how these herbicides might lead to the death of a plant.

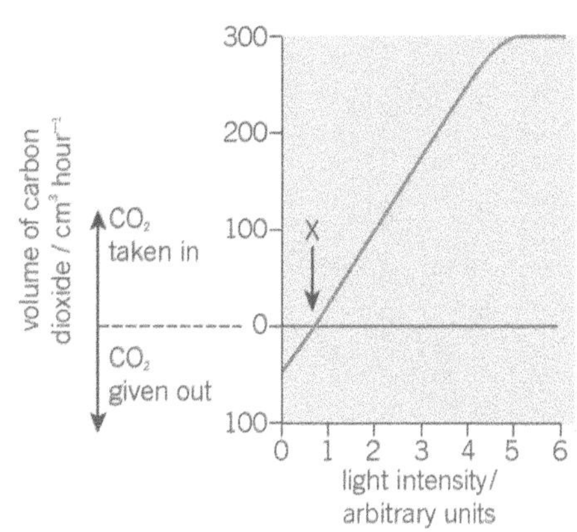

▲ **Figure 4**

5.4 Structure of the human gas-exchange system

Learning objectives:

- → Describe how the human gas-exchange system is arranged.
- → Describe the functions of its main parts.

Specification reference: 3.1.5.2

In this chapter you will read about how lungs act as an interface for the exchange of gases and how their function can be affected by both pathogens and lifestyle.

All aerobic organisms require a constant supply of oxygen to release energy in the form of ATP during respiration. The carbon dioxide produced in the process needs to be removed as its build-up could be harmful to the body.

The volume of oxygen that has to be absorbed and the volume of carbon dioxide that must be removed are large in mammals because:

- they are relatively large organisms with a large volume of living cells
- they maintain a high body temperature and therefore have high metabolic and respiratory rates.

As a result mammals have evolved specialised surfaces, called **lungs**, to ensure efficient gas exchange between the air and their blood.

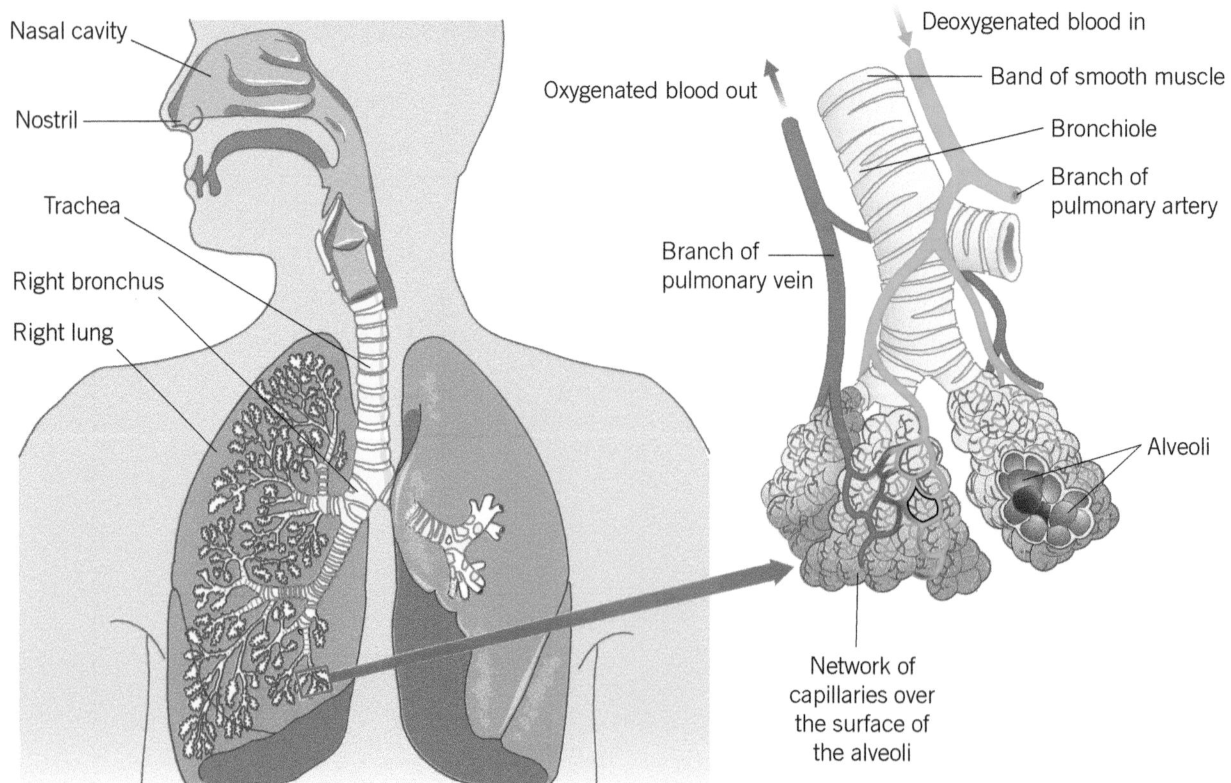

▲ **Figure 1** *The gross structure of the human gas-exchange system*

Mammalian lungs

The lungs are the site of gas exchange in mammals. They are located inside the body because:

- air is not dense enough to support and protect these delicate structures
- they would otherwise lose a great deal of water and dry out.

The lungs are supported and protected by a bony box called the **rib cage**. The ribs can be moved by the muscles between them. This enables the lungs to be ventilated by a tidal stream of air, thereby ensuring that the air within them is constantly replenished. The main parts of the human gas-exchange system and their structures and functions are described below.

- The **lungs** are a pair of lobed structures made up of a series of highly branched tubules, called bronchioles, which end in tiny air sacs called alveoli.
- The **trachea** is a flexible airway that is supported by rings of cartilage. The cartilage prevents the trachea collapsing as the air pressure inside falls when breathing in. The tracheal walls are made up of muscle, lined with ciliated epithelium and goblet cells. The goblet cells produce mucus that traps dirt particles and bacteria from the air breathed in. The cilia move the mucus, laden with dirt and microorganisms, up to the throat, from where it passes down the oesophagus into the stomach.
- The **bronchi** are two divisions of the trachea, each leading to one lung. They are similar in structure to the trachea and, like the trachea, they also produce mucus to trap dirt particles and have cilia that move the dirt-laden mucus towards the throat. The larger bronchi are supported by cartilage, although the amount of cartilage is reduced as the bronchi get smaller.
- The **bronchioles** are a series of branching subdivisions of the bronchi. Their walls are made of muscle lined with epithelial cells. This muscle allows them to constrict so that they can control the flow of air in and out of the alveoli.
- The **alveoli** are minute air sacs, with a diameter of between 100 μm and 300 μm, at the end of the bronchioles. They contain some collagen and elastic fibres, and they are lined with epithelium. The elastic fibres allow the alveoli to stretch as they fill with air when breathing in. They then spring back during breathing out in order to expel the carbon dioxide-rich air. The alveolar membrane is the gas-exchange surface.

Summary questions

1 State **two** reasons why humans need to absorb large volumes of oxygen from the lungs.

2 List in the correct sequence all the structures that air passes through on its journey from the gas-exchange surface of the lungs to the nose.

3 Explain how the cells lining the trachea and bronchus protect the alveoli from damage.

Hint

The ending '-ioles' is commonly used in biology to denote a smaller version of a structure. Hence 'bronchioles' are small bronchi, and 'arterioles' are small arteries.

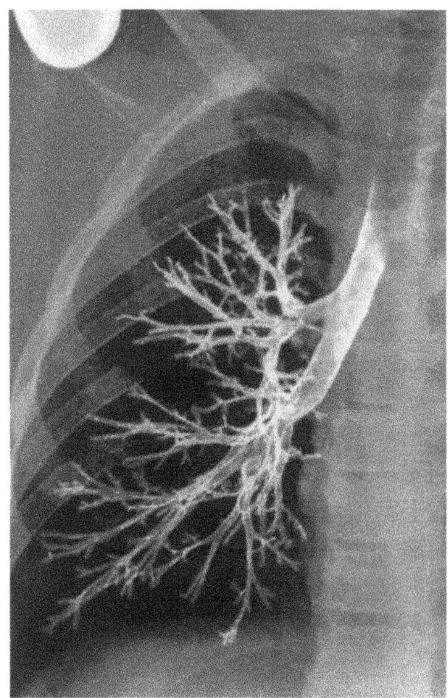

▲ **Figure 2** *False-colour X-ray of the bronchus and bronchioles of a healthy human lung*

▲ **Figure 3** *False-colour scanning electron micrograph (SEM) of a section of the epithelium of the trachea showing ciliated cells (green)*

5.5 The mechanism of breathing

Learning objectives:

→ Describe how air is moved into the lungs when breathing in.

→ Describe how air is moved out of the lungs when breathing out.

→ Define pulmonary ventilation and how to calculate it.

Specification reference: 3.1.5.2

To maintain diffusion of gases across the alveolar epithelium, air must be constantly moved in and out of the lungs. We call this process breathing, or **ventilation**. When the air pressure of the atmosphere is greater than the air pressure inside the lungs, air is forced into the alveoli. This is called **inspiration** (inhalation). When the air pressure in the lungs is greater than that of the atmosphere, air is forced out of the lungs. This is called **expiration** (exhalation). The pressure changes within the lungs are brought about by the movement of two sets of muscles:

- the diaphragm, which is a sheet of muscle that separates the thorax from the abdomen
- the intercostal muscles, which lie between the ribs. There are two sets of intercostal muscles:
  - the **internal intercostal muscles**, whose contraction leads to expiration
  - the **external intercostal muscles**, whose contraction leads to inspiration.

Hint

There are two basic physical laws that will help you to understand the movement of air during breathing:

- Within a closed container, if the volume of the container increases, the pressure decreases. If the volume of the container decreases, the pressure increases.
- Gases move from a region where their pressure is higher to a region where their pressure is lower.

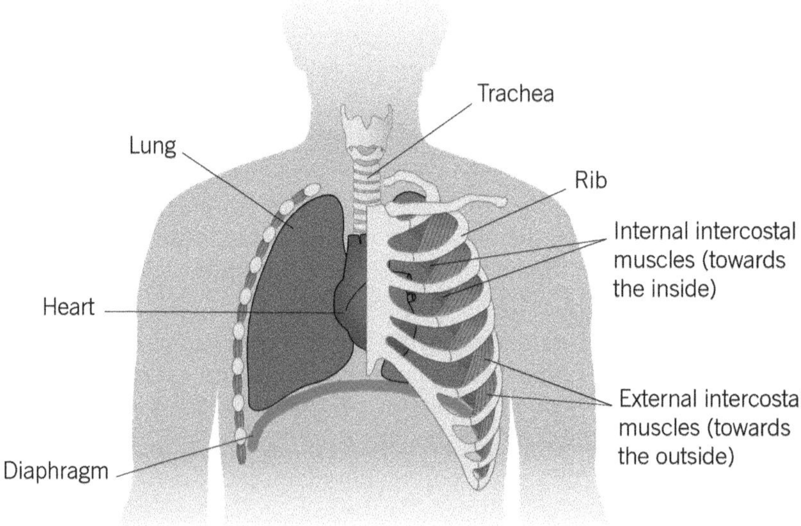

▲ **Figure 1** *The arrangement of the diaphragm and intercostal muscles*

Inspiration

Breathing in is an active process (it uses energy) and occurs as follows:

- The external intercostal muscles contract, whilst the internal intercostal muscles relax.
- The ribs are pulled upwards and outwards, increasing the volume of the thorax.
- The diaphragm muscles contract, causing it to flatten, which also increases the volume of the thorax.
- The increased volume of the thorax results in reduction of pressure in the lungs.
- Atmospheric pressure is now greater than pulmonary pressure, and so air is forced into the lungs.

Study tip

The volume of blood pumped by the heart in one minute is called the 'cardiac output'. Do not confuse this with 'pulmonary output'. Remember cardiac = heart and pulmonary = lungs.

Expiration

Breathing out is a largely passive process (it does not require much energy) and occurs as follows:

- The internal intercostal muscles contract, whilst the external intercostal muscles relax.
- The ribs move downwards and inwards, decreasing the volume of the thorax.
- The diaphragm muscles relax, making it return to its upwardly domed position, again decreasing the volume of the thorax.
- The decreased volume of the thorax increases the pressure in the lungs.
- The pulmonary pressure is now greater than that of the atmosphere, and so air is forced out of the lungs.

During normal quiet breathing, the recoil of the elastic lungs is the main cause of air being forced out (like air being expelled from a partly inflated balloon). Only under more strenuous conditions, such as exercise, do the various muscles play a part.

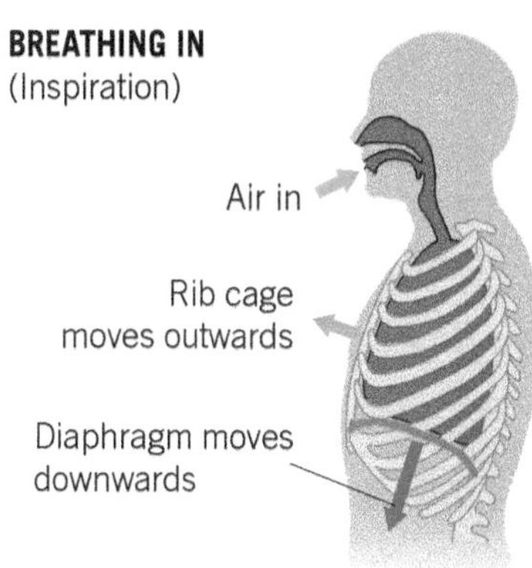

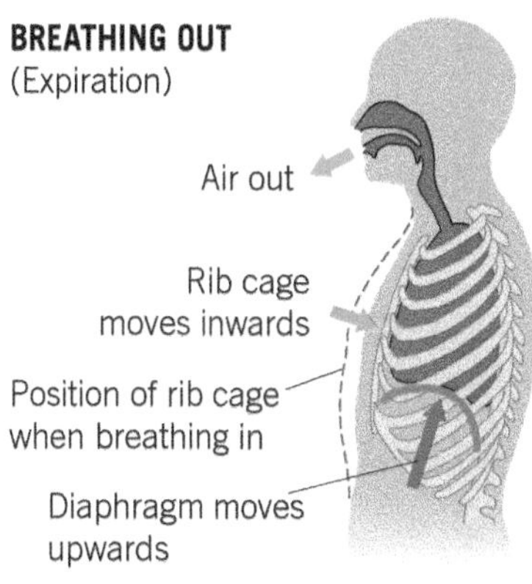

▲ **Figure 2** *Position of ribs and diaphragm during inspiration and expiration*

Pulmonary ventilation

It is sometimes useful to know how much air is taken in and out of the lungs in a given time. To do this you can use a measure called pulmonary ventilation. Pulmonary ventilation is the total volume of air that is moved into the lungs during 1 minute. To calculate it you multiply together two factors:

- tidal volume, which is the volume of air normally taken in at each breath when the body is at rest. This is usually around 0.5 dm^3.
- ventilation (breathing) rate, that is, the number of breaths taken in 1 minute. This is normally 12–20 breaths in a healthy adult.

Pulmonary ventilation is expressed as $dm^3 min^{-1}$.

To summarise:

pulmonary ventilation rate = tidal volume × breathing rate
($dm^3 min^{-1}$) (dm^3) (min^{-1})

Summary questions

1. From the graphs in Figure 3, determine the tidal volume of this person.
2. From the graphs in Figure 3, calculate the breathing rate of this person. Give your answer in breaths per minute. Show how you arrived at your answer.
3. If the volume of air in the lungs when the person inhaled was 3000 cm^3, what would the volume of air in the lungs be after the person had exhaled? Show your working.
4. Explain how muscles create the change of pressure in the alveoli over the period 0 s to 0.5 s.

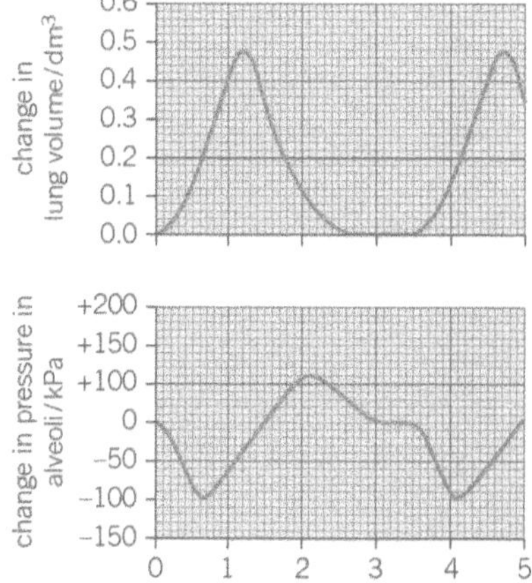

▲ **Figure 3** *The volume and pressure changes that occurred in the lungs of a person when breathing whilst at rest*

5.6 Exchange of gases in the lungs

Learning objectives:

- → Define the essential features of exchange surfaces.
- → Describe how gases are exchanged in the alveoli of humans.

Specification reference: 3.1.5.2

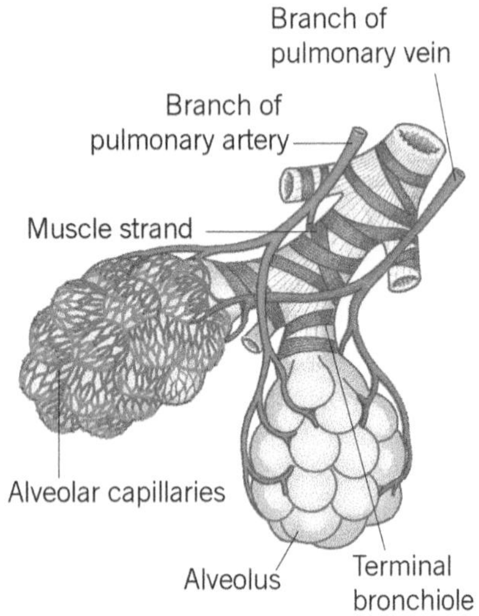

▲ **Figure 1** *Alveoli*

The site of gas exchange in mammals is the epithelium of the alveoli. These alveoli are minute air sacs about 100–300 μm in diameter and situated in the lungs. To ensure a constant supply of oxygen to the body, a diffusion gradient must be maintained at the alveolar surface.

Essential features of exchange surfaces

To enable efficient transfer of materials across them by diffusion or active transport, exchange surfaces have the following characteristics:

- They have a **large surface area to volume ratio** – to speed up the rate of exchange.
- They are **very thin** – to keep the diffusion pathway short and so allow materials to cross rapidly.
- They are **partially or selectively permeable** – to allow selected materials to diffuse easily.
- There is **movement of the environmental medium**, for example, air – to maintain a diffusion gradient.
- There is **movement of the internal medium**, for example, blood – to maintain a diffusion gradient.

We saw in Topic 4.2 that the relationship between some of these factors is described in the following expression, known as Fick's Law:

$$\textbf{diffusion} \propto \frac{\textbf{surface area} \times \textbf{difference in concentration}}{\textbf{length of diffusion path}}$$

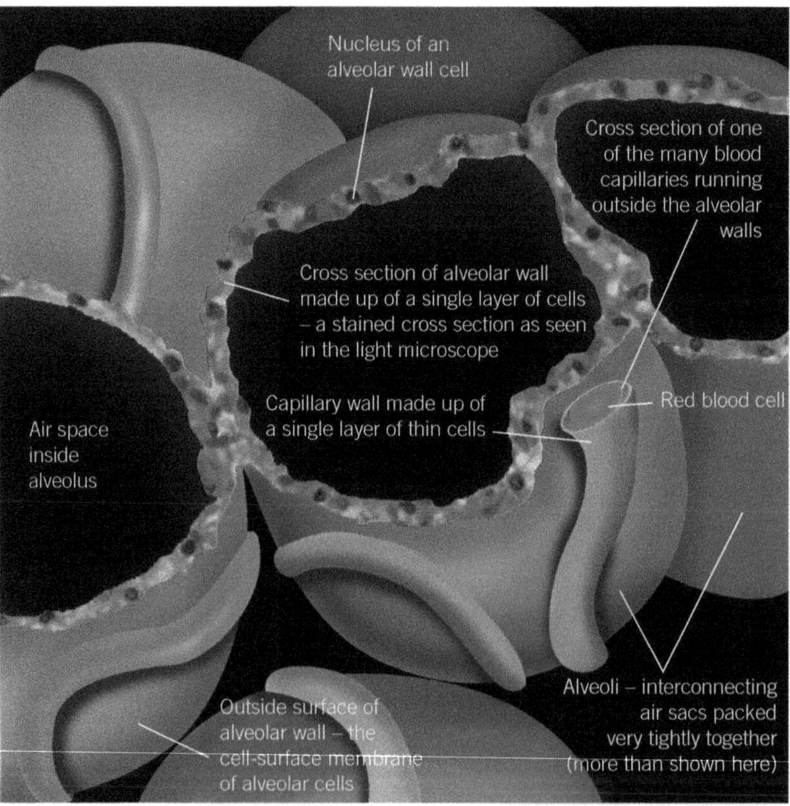

▶ **Figure 2** *External appearance and exchange surface features of a group of alveoli*

Being thin, these specialised exchange surfaces are easily damaged and therefore are often located inside an organism for protection. Where an exchange surface, such as the lungs, is located inside the body, the organism needs to have some means of moving the external medium over the surface, for example, a means of ventilating the lungs in a mammal.

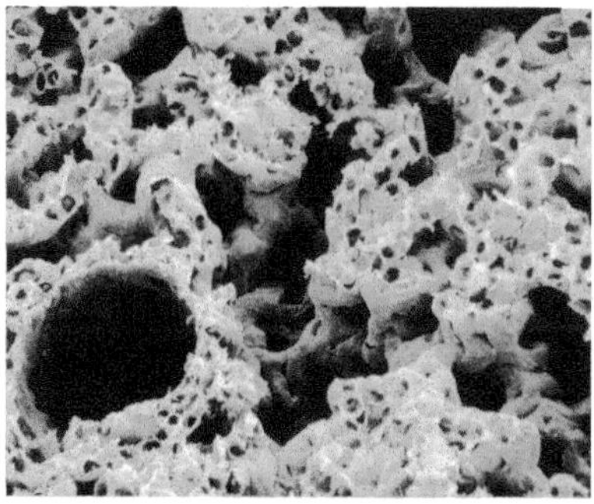

▲ **Figure 3** *False-colour scanning electron micrograph (SEM) of a section of human lung tissue showing alveoli surrounded by blood capillaries*

Role of the alveoli in gas exchange

There are about 300 million alveoli in each human lung. Their total surface area is around 70 m² – about half the area of a tennis court. Their structure is shown in Figures 1 and 2. Each alveolus is lined mostly with epithelial cells only 0.05–0.3 μm thick. Around each alveolus is a network of pulmonary capillaries that are so narrow (7–10 μm) that red blood cells are flattened against the thin capillary walls in order to squeeze through. These capillaries have walls that are only a single layer of cells thick (0.04–0.2 μm). Diffusion of gases between the alveoli and the blood will be very rapid because:

- red blood cells are slowed as they pass through pulmonary capillaries, allowing more time for diffusion
- the distance between the alveolar air and red blood cells is reduced as the red blood cells are flattened against the capillary walls
- the walls of both alveoli and capillaries are very thin and therefore the distance over which diffusion takes place is very short
- alveoli and pulmonary capillaries have a very large total surface area
- breathing movements constantly ventilate the lungs, and the action of the heart constantly circulates blood around the alveoli. Together, these actions ensure that a steep concentration gradient of the gases to be exchanged is maintained.
- blood flow through the pulmonary capillaries maintains a concentration gradient.

The diffusion of gases in an alveolus is illustrated in Figure 4.

Hint

The diffusion pathway is short because the alveoli have only a single layer of epithelial cells and the blood capillaries have only a single layer of endothelial cells.

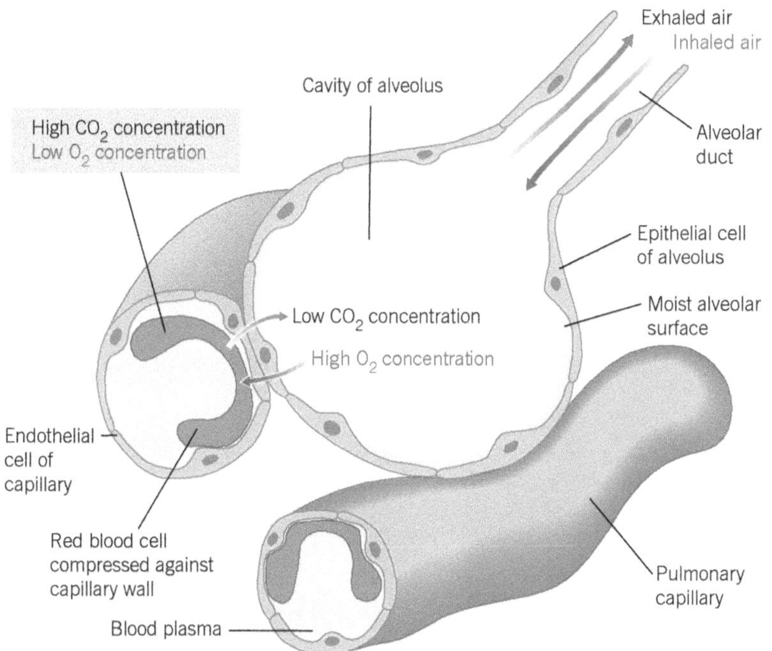

▲ **Figure 4** *Diffusion of gases in an alveolus*

Summary questions

1 How does each of the following features contribute to the efficiency of gas exchange in alveoli?

   a The wall of each alveolus is not more than 0.3 µm thick.

   b There are 300 million alveoli in each lung.

   c Each alveolus is covered by a dense network of pulmonary blood capillaries.

   d Each pulmonary capillary is very narrow.

2 If the number of alveoli in each lung was increased to 600 million and the pulmonary ventilation was doubled, how many times greater would the rate of diffusion be?

Interpreting lung function measurements

Various tests of lung function are used to monitor the changes that occur in the lungs due to different diseases.

1 Pulmonary fibrosis is a lung disease that causes the epithelium of the lungs to become irreversibly thickened. It also leads to reduced elasticity of the lungs. One symptom of the disease is shortness of breath, especially when exercising. Suggest why this symptom arises.

2 One measure of lung function is forced expiratory volume (FEV). This is the volume of air that can forcibly be blown out in 1 second, after full inspiration. Suggest how pulmonary fibrosis might affect FEV and explain why.

5.7 Haemoglobin

Once oxygen has diffused into the bloodstream, it is the role of the circulatory system to deliver it to respiring cells. You will learn more about the circulatory system in Chapter 15. The transport of oxygen is the role of the protein **haemoglobin**. In Topic 1.5 you looked at the structure of proteins and how the shape of a protein is important to how it functions. You have seen that the primary structure of a protein is the sequence of amino acids, determined by DNA, that makes up a polypeptide chain. It is this sequence that determines how the polypeptide is shaped into its tertiary structure. Linking together a number of polypeptides, sometimes along with non-protein groups, gives rise to a protein's quaternary structure. Let us now look at the haemoglobins – a group of protein molecules that have a quaternary structure.

Learning objectives:

- → Define haemoglobins and their role.
- → Describe how haemoglobins from different organisms differ and why.
- → Define oxygen loading and unloading.

Specification reference: 3.1.5.3

Haemoglobin molecules

The haemoglobins are a group of chemically similar molecules found in a wide variety of organisms. The structure of a haemoglobin molecule is shown in Figure 1. It is made up as follows:

- **primary structure**, consisting of four polypeptide chains
- **secondary structure**, in which each of these polypeptide chains is coiled into a helix
- **tertiary structure**, in which each polypeptide chain is folded further into a precise shape – an important factor in its ability to carry oxygen
- **quaternary structure**, in which all four polypeptides are linked together to form an almost spherical molecule. Each polypeptide is associated with a haem group – which contains a ferrous (Fe^{2+}) ion. Each Fe^{2+} ion can combine with a single oxygen molecule (O_2), making a total of four O_2 molecules that can be carried by a single haemoglobin molecule in humans.

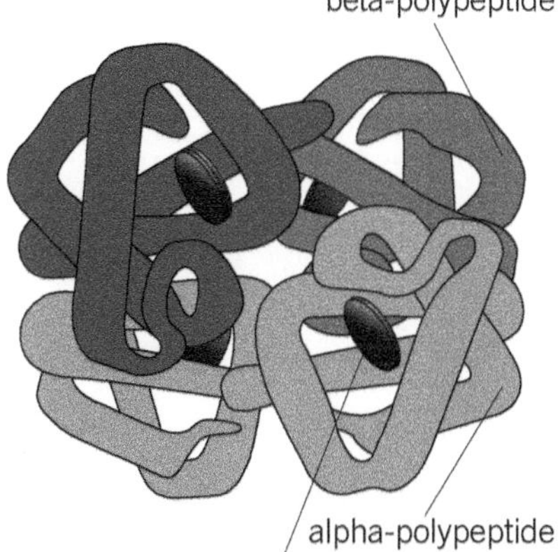

▲ **Figure 1** *Quaternary structure of a haemoglobin molecule*

The role of haemoglobin

The role of haemoglobin is to transport oxygen. To be efficient at transporting oxygen, haemoglobin must:

- readily associate with oxygen at the surface where gas exchange takes place
- readily dissociate from oxygen at those tissues requiring it.

These two requirements may appear to contradict each other, but they are achieved by a remarkable property of haemoglobin. It changes its affinity for oxygen under different conditions (Table 1). It achieves this because its shape changes in the presence of certain substances, such as carbon dioxide. In the presence of carbon dioxide, the new shape of the haemoglobin molecule binds more loosely to oxygen. As a result, haemoglobin releases its oxygen.

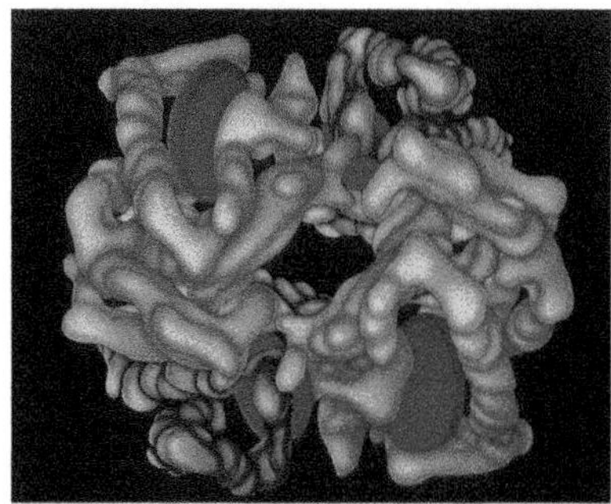

▲ **Figure 2** *Computer graphic representation of a haemoglobin molecule showing two pairs of polypeptide chains (orange and blue) and each chain associated with a haem group (red)*

▼ **Table 1** *Affinity of haemoglobin for oxygen under different conditions*

Region of body	Oxygen concentration	Carbon dioxide concentration	Affinity of haemoglobin for oxygen	Result
Gas exchange surface	High	Low	High	Oxygen is attached
Respiring tissues	Low	High	Low	Oxygen is released

Why have different haemoglobins?

Scientists long ago observed that many organisms possessed haemoglobin. They proposed that it carried oxygen from the gas-exchange surface to the tissues that required it for respiration. If so, this meant that it must readily combine with oxygen. Consequently they investigated the ability of haemoglobin from different organisms to combine with oxygen. Results showed that there were different types of haemoglobins. These exhibited different properties relating to the way they took up and released oxygen. At two ends of the range are:

- **haemoglobins with a high affinity for oxygen**. These take up oxygen more easily and release it less readily.
- **haemoglobins with a low affinity for oxygen**. These take up oxygen less easily and release it more readily.

Scientists questioned why this should happen. Further observation and experimentation showed a correlation between the type of haemoglobin in an organism and factors such as the environment in which it lived or its metabolic rate. Explanations for some of these correlations are as follows:

- An organism living in an environment with little oxygen requires a haemoglobin that readily combines with oxygen if it is to absorb enough of it. Provided that the organism's metabolic rate is not very high, the fact that this form of haemoglobin does not release its oxygen as readily into the tissues will not be a problem.
- An organism with a high metabolic rate needs to release oxygen readily into its tissues. Provided that there is plenty of oxygen in the organism's environment, it is more important to have a haemoglobin that releases its oxygen easily than one that takes it up easily.

Why do different haemoglobins have different affinities for oxygen?

The answer, scientists discovered, lies in the shape of the molecule. Different haemoglobin molecules have slightly different sequences of amino acids and therefore slightly different shapes. Depending on the shape, haemoglobin molecules range from those that have a high affinity for oxygen to those that have a low affinity for oxygen.

Loading and unloading oxygen

The process by which haemoglobin combines with oxygen is called **loading**, or **associating**. In humans this takes place in the lungs.

The process in which haemoglobin releases its oxygen is called **unloading**, or **dissociating**. In humans this takes place in the tissues.

Summary questions

1. Describe the quaternary structure of haemoglobin.
2. Explain how DNA leads to different haemoglobin molecules having different affinities for oxygen.
3. When the body is at rest, only one of the four oxygen molecules carried by haemoglobin is normally released into the tissues. Suggest why this could be an advantage when the organism becomes more active.
4. Carbon monoxide binds permanently to haemoglobin in preference to oxygen. Suggest a reason why a person breathing in carbon monoxide might lose consciousness.

5.8 Oxygen-haemoglobin dissociation curves

Having looked at haemoglobin in Topic 5.7, let us now consider its properties. How does it load and unload oxygen and what effect does carbon dioxide have on this process?

Learning objectives:

- → Define an oxygen dissociation curve.
- → Describe the effect of carbon dioxide concentration on the curve.
- → Describe how the properties of the haemoglobins in different organisms relate to the organism's environment and way of life.

Specification reference: 3.1.5.3

Oxygen-haemoglobin dissociation curves

When haemoglobin is exposed to different partial pressures of oxygen (see Hint), it does not absorb the oxygen evenly. At very low concentrations of oxygen, the four polypeptides of the haemoglobin molecule are closely united, and so it is difficult to absorb the first oxygen molecule. However, binding changes the shape of haemoglobin, making binding of the next oxygen molecule easier/possible. Once loaded, this oxygen molecule causes the polypeptides to load the remaining three oxygen molecules very easily. The graph of this relationship is known as the **oxygen dissociation curve** (see Figure 1). You will notice from this graph that a very small decrease in the partial pressure of oxygen leads to a lot of oxygen becoming dissociated from haemoglobin. The graph tails off at very high oxygen concentrations simply because the haemoglobin is almost saturated with oxygen.

You saw in Topic 5.7 that there are a number of different types of haemoglobin molecules, each with a different shape and hence a different affinity for oxygen. In addition, the shape of any one type of haemoglobin molecule can change under different conditions. These facts both mean that there are a large number of different oxygen-haemoglobin dissociation curves. They all have a roughly similar shape but differ in their position on the axes.

The many different oxygen dissociation curves are better understood if two facts are always kept in mind:

- the further to the left the curve, the greater is the affinity of haemoglobin for oxygen (so it takes up oxygen readily and releases it less easily)
- the further to the right the curve, the lower is the affinity of haemoglobin for oxygen (so it takes up oxygen less readily and releases it more easily).

Hint

Measuring oxygen concentration
The amount of a gas that is present in a mixture of gases is measured by the pressure it contributes to the total pressure of the gas mixture. This is known as the **partial pressure** of the gas and, in the case of oxygen, is written as pO_2. It is measured in kilopascals (kPa). Normal atmospheric pressure is 100 kPa. As oxygen makes up 21 per cent of the atmosphere, its partial pressure is normally 21 kPa.

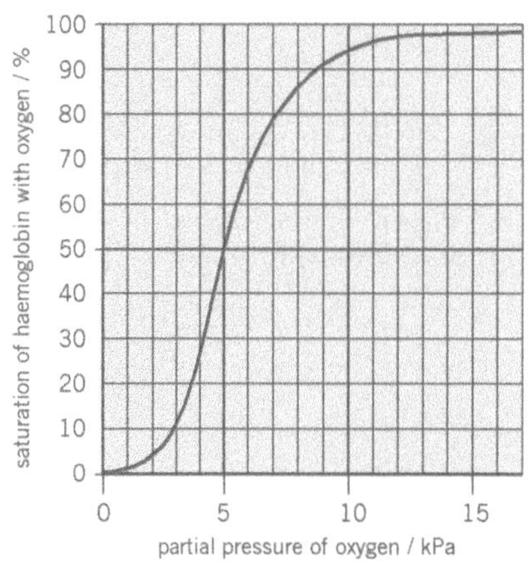

▲ **Figure 1** *Oxygen-haemoglobin dissociation curve for adult human haemoglobin*

Effects of carbon dioxide concentration

Haemoglobin has a reduced affinity for oxygen in the presence of carbon dioxide. The greater the concentration of carbon dioxide, the more readily the haemoglobin releases its oxygen (the Bohr effect). This explains the changes in the affinity of haemoglobin for oxygen in different regions of the body.

- At the gas-exchange surface (e.g., lungs), the concentration of carbon dioxide is low because it diffuses across the exchange surface and is expelled from the organism. The affinity of haemoglobin for oxygen is increased, which, coupled with the high concentration of oxygen in the lungs, means that oxygen is readily loaded by haemoglobin. The reduced carbon dioxide concentration has shifted the oxygen dissociation curve to the left (Figure 2).
- In rapidly respiring tissues (e.g., muscles), the concentration of carbon dioxide is high. The affinity of haemoglobin for oxygen is therefore

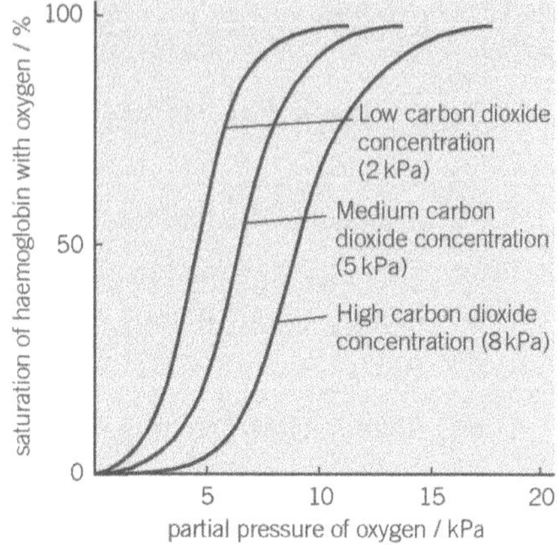

▲ **Figure 2** *The effect of carbon dioxide concentration on the oxygen dissociation curve*

reduced, which, coupled with the low concentration of oxygen in the muscles, means that oxygen is readily unloaded from the haemoglobin into the muscle cells. The increased carbon dioxide level has shifted the oxygen dissociation curve to the right (Figure 2).

You have just seen that the greater the concentration of carbon dioxide, the more readily haemoglobin releases its oxygen. This is because dissolved carbon dioxide is acidic and the low pH causes haemoglobin to change shape. Let us see how this works in the transport of oxygen by haemoglobin.

Loading, transport, and unloading of oxygen

- At the gas-exchange surface carbon dioxide is constantly being removed.
- The pH is raised due to the low concentration of carbon dioxide.
- The higher pH changes the shape of haemoglobin into one that enables it to load oxygen readily.
- This shape also increases the affinity of haemoglobin for oxygen, so it is not released whilst being transported in the blood to the tissues.
- In the tissues, carbon dioxide is produced by respiring cells.
- Carbon dioxide is acidic in solution, so the pH of the blood within the tissues is lowered.
- The lower pH changes the shape of haemoglobin into one with a lower affinity for oxygen.
- Haemoglobin releases its oxygen into the respiring tissues.

The above process is a flexible way of ensuring that there is always sufficient oxygen for respiring tissues. The more active a tissue, the more oxygen is unloaded. This works as follows:

The higher the rate of respiration → the more carbon dioxide the tissues produce → the lower the pH → the greater the haemoglobin shape change → the more readily oxygen is unloaded → the more oxygen is available for respiration.

In humans, haemoglobin normally becomes saturated with oxygen as it passes through the lungs. In other words, most of the haemoglobin molecules are loaded with their maximum four oxygen molecules. When this haemoglobin reaches a tissue with a low respiratory rate, only one of these molecules will normally be released. The blood returning to the lungs will therefore contain haemoglobin that is still 75 per cent saturated with oxygen. If a tissue is very active, for example, an exercising muscle, then three oxygen molecules will usually be unloaded from each haemoglobin molecule. These events are shown in Figure 3.

Key
- Haemoglobin molecule is loaded with oxygen in the lungs
- Haemoglobin molecule in a resting tissue unloads only 25% of its oxygen
- Haemoglobin molecule in an active tissue unloads 75% of its oxygen

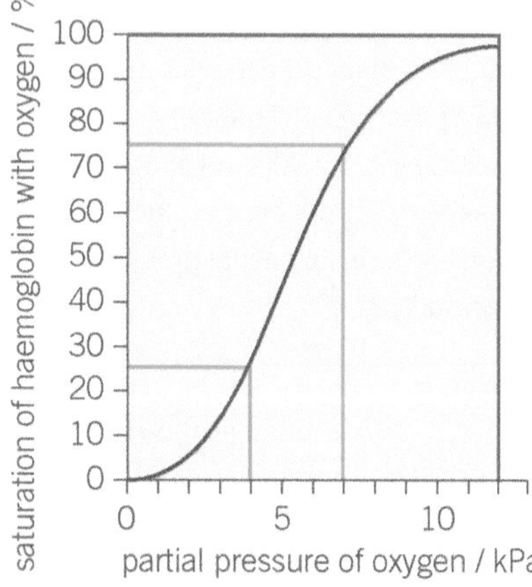

▲ **Figure 3** *The loading and unloading of haemoglobin with oxygen*

Different lives – different haemoglobins

Where you live is important

If you walk along a sandy seashore when the tide is out, you may come across worm casts. Beneath these probably lies an organism called a lugworm. The lugworm is not very active, spending almost all its life in a U-shaped burrow. Most of the time the lugworm is covered by sea water, which it circulates through its burrow. Oxygen diffuses into the lugworm's blood from the water and it uses haemoglobin to transport oxygen to its tissues. When the tide goes out, the lugworm can no longer circulate a fresh supply of oxygenated water through its burrow. As a result, the water in the burrow contains progressively less oxygen as the lugworm uses it up. The lugworm needs to extract as much oxygen as possible from the water in the burrow if it is to survive until the tide covers it again. Figure 4 shows the oxygen dissociation curve of lugworm haemoglobin compared to that of adult human haemoglobin.

1 In Figure 4, line A is drawn at a partial pressure of oxygen of 2 kPa. This is the partial pressure of oxygen found in lugworm burrows after the sea no longer covers them. Using figures from the graph, explain why a lugworm can survive at these concentrations of oxygen whereas a human could not.
2 Using the graphs in Figure 4, explain how the lugworm is able to obtain sufficient oxygen from an environment that contains so little.
3 Suggest one feature of a lugworm's way of life described in the passage that helps it to survive in an environment that has very little oxygen. Explain how this feature aids survival.
4 Haemoglobin usually loads oxygen less readily when the concentration of carbon dioxide is high (the Bohr effect). The haemoglobin of lugworms does not exhibit this effect. Explain why to do so could be harmful.
5 Suggest a reason why lugworms are not found higher up the seashore.
6 Llamas are animals that live at high altitudes. At these altitudes the atmospheric pressure is lower and so the partial pressure of oxygen is also lower. It is therefore difficult to load haemoglobin with oxygen. Suggest where the oxygen dissociation curve of llama haemoglobin is shifted to, relative to human haemoglobin. A sketch graph might help to clarify your answer.

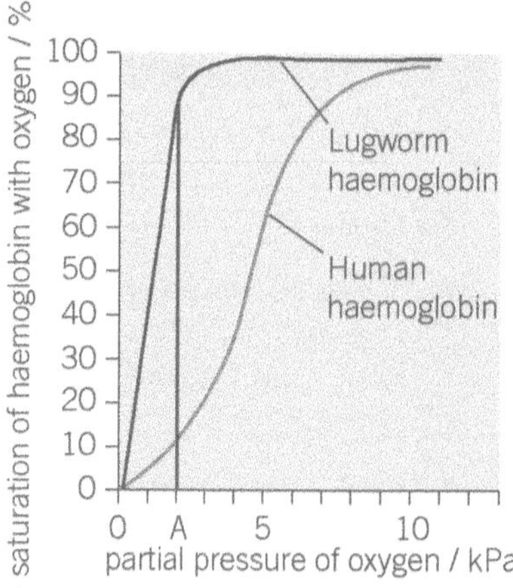

▲ **Figure 4** *Comparison of the oxygen dissociation curves of lugworm and human haemoglobin*

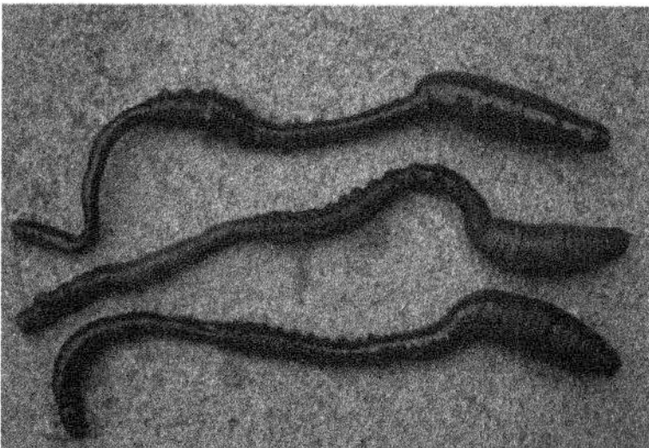

▲ **Figure 5** *Three lugworms lying on sand (top); lugworm casts at the entrances to their burrows (bottom)*

Size matters

Mice are small mammals and therefore have a large surface area to volume ratio. As a result they tend to lose heat rapidly when the environmental temperature is lower than their body temperature. To compensate for this they have a high metabolic rate that generates heat and helps them to maintain their normal body temperature. Figure 6 shows the oxygen dissociation curve for the haemoglobin of a mouse compared to that of adult human haemoglobin.

7 The partial pressure of oxygen at which haemoglobin is 50 per cent saturated is known as the unloading pressure. Calculate the difference between the unloading pressure of oxygen for human haemoglobin and that for mouse haemoglobin.
8 The oxygen dissociation curve of the mouse is shifted to the right of that for a human.
   a What difference does this make to the way oxygen is unloaded from mouse haemoglobin compared with human haemoglobin?
   b What advantage does this have for the maintenance of body temperature in mice?
   c The position of the oxygen dissociation curve for mouse haemoglobin means that its haemoglobin loads oxygen less readily than human haemoglobin. Use the graph to explain why this is of no disadvantage to the mouse, given that the partial pressure of oxygen in air is normally 21 kPa.

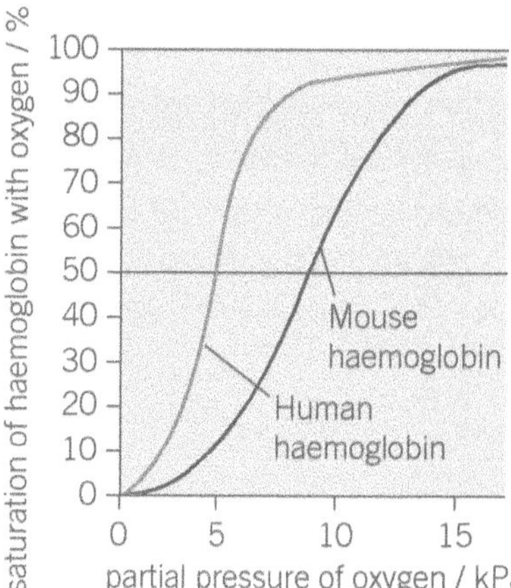

▲ **Figure 6** *Oxygen dissociation curves of mouse and human haemoglobin*

Summary question

1 Study Figure 1 on page 77 and answer the following questions:

   a At what partial pressure of oxygen is the haemoglobin 50 per cent saturated with oxygen?

   b What is the percentage saturation of haemoglobin with oxygen when the partial pressure of oxygen is 9 kPa?

   c In an exercising muscle the partial pressure of oxygen is 4 kPa whilst in the lungs it is 12 kPa. What percentage of the oxyhaemoglobin from the lungs will have released its oxygen to an exercising muscle?

2 a What is the effect of increased carbon dioxide concentration on oxygen dissociation?

   b How does this change the saturation of haemoglobin with oxygen?

3 A rise in temperature shifts the oxygen dissociation curve to the right. Suggest how this enables an exercising muscle to work more efficiently.

Hint

Remember that oxygen in blood is transported in combination with haemoglobin as oxyhaemoglobin.

9 Using a sketch graph suggest the shapes and relative positions of the oxygen-haemoglobin dissociation curves in the following mammals:

   a a human

   b an elephant

   c a shrew.

Activity counts

Flight in birds and swimming in fish are both energy-demanding processes. The muscles that move a bird's wings are powerful and require a lot of oxygen to enable them to respire at a sufficient rate to keep the body airborne. Flight muscles have a very high metabolic rate and, during flight, much of the blood pumped by the heart goes to these muscles. Whereas birds use a great deal of energy opposing gravity in a medium that gives little support, fish have a different problem. They expend considerable energy swimming through a medium that is very dense and therefore difficult to move through.

10 Suggest whether the oxygen dissociation curve of pigeon haemoglobin is shifted to the right or left of the curve for human haemoglobin. Explain your answer.

11 The mackerel is a type of fish that swims freely in the surface waters of the sea. These fish rely on their ability to swim very fast in order to escape from predators. The plaice is a marine fish that uses a different strategy. These fish spend much of their lives stationary or moving very slowly on the sea bed, where they are camouflaged by their skin colour. The two fish are of relatively similar mass. Sketch a graph to show the relative positions of the oxygen-haemoglobin dissociation curves in these two fish.

12 Ice fish live in the Antarctic and are the only vertebrates to completely lack haemoglobin. Suggest a reason why they can survive without haemoglobin.

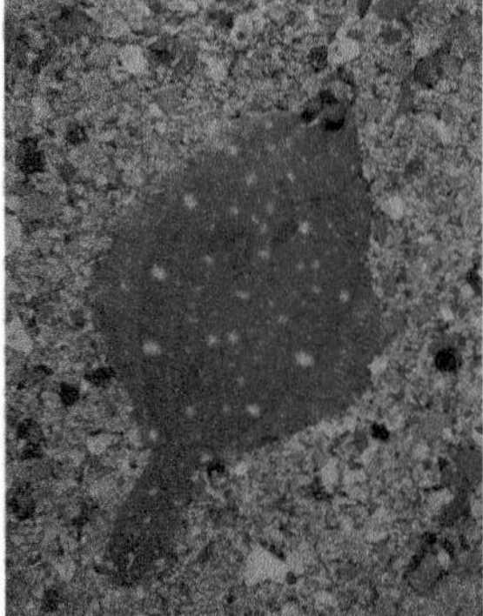

▲ **Figure 7** *Mackerel (left) live in surface waters and swim rapidly. Plaice (right) live on the sea bed and move very slowly.*

Practice questions: Chapter 5

1 The diagram shows the position of the diaphragm at times **P** and **Q**.

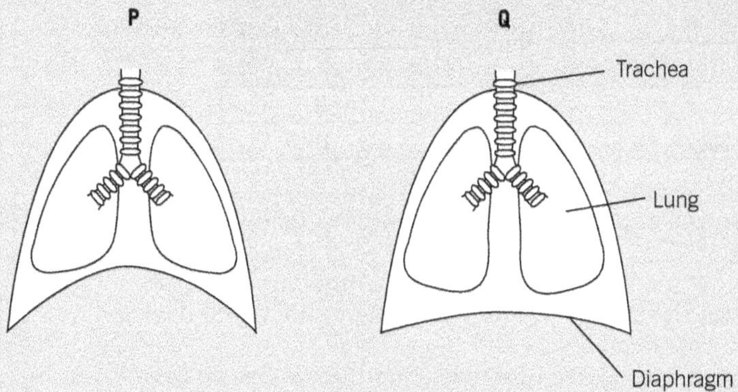

(a) Describe what happens to the diaphragm between times **P** and **Q** to bring about the change in its shape. *(2 marks)*

(b) Air moves into the lungs between times **P** and **Q**. Explain how the diaphragm causes this. *(3 marks)*

(c) Describe how oxygen in air in the alveoli enters the blood in capillaries. *(2 marks)*

AQA June 2012

2 Scientists investigated the effect of tuberculosis (TB) on breathing. They obtained data from African miners aged 20 to 65 years.

They divided the miners into groups based on how many times they had had TB.

- Group **P**, never had TB
- Group **Q**, had TB once
- Group **R**, had TB twice

The data were for forced expiratory volume (FEV). FEV is the maximum volume a person can breathe out in 1 second.

Their results are shown in this graph:

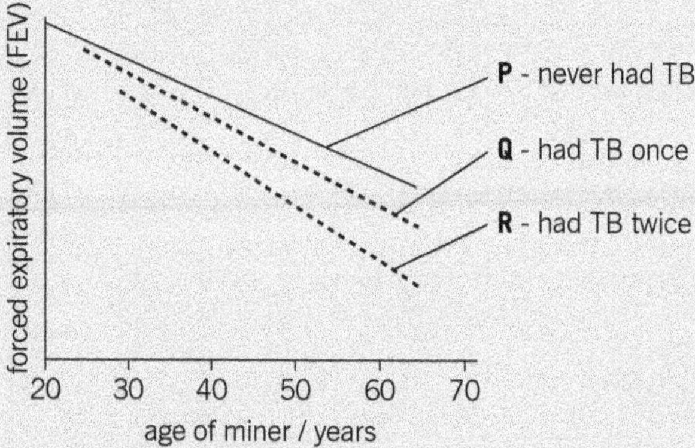

(a) Describe the results. *(3 marks)*

(b) TB leads to permanent changes in the gas-exchange system.
These changes include fibrosis.
Explain how fibrosis caused by TB could have produced the changes in FEV of the miners. *(3 marks)*

3 Insects such as beetles obtain oxygen by drawing air into their tracheae through spiracles. Diving beetles live in ponds. They carry a bubble of air under their wing cases when they swim underwater. The bubble supplies air to the spiracles. When the bubble has been used up, the beetle comes to the surface to collect a new bubble.

An investigation was carried out into the effect of temperature on diving beetles. Three beetles, **A**, **B**, and **C**, of the same species were observed in thermostatically controlled water baths. The number of times each beetle surfaced to renew its air bubble was counted at three different temperatures.

The results are shown in the table below.

Temperature / °C	Number of times air bubble was renewed per hour		
	Beetle A	Beetle B	Beetle C
10	10	12	8
20	18	22	18
30	44	48	38

(a) Calculate the mean number of times the air bubble was renewed per hour at each temperature. *(1 mark)*

(b) Sketch a graph to show the relationship between temperature and the mean number of times the air bubble was renewed per hour, and name the shape of the line obtained. *(2 marks)*

(c) The number of times the air bubble is renewed per hour is related to a beetle's need for oxygen to carry out aerobic respiration, which is catalysed by enzymes. Explain what the data reveal about the size of the effect of each 10 °C rise in temperature on the rate of respiration. *(2 marks)*

4 Lugworms live in mud where the partial pressure of oxygen is low. The graph shows oxygen-haemoglobin dissociation curves for a lugworm and for a human.

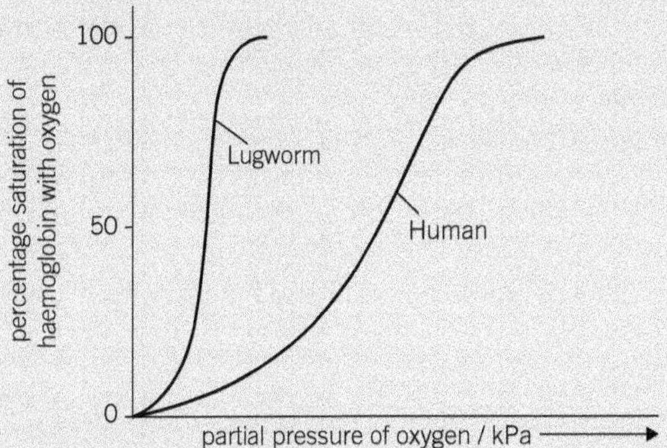

(a) Explain the advantage to the lugworm of having haemoglobin with a dissociation curve in the position shown. *(2 marks)*

(b) In humans, substances move out of the capillaries to form tissue fluid. Describe how this tissue fluid is returned to the circulatory system. *(3 marks)*

AQA June 2011

5 **(a)** Flatworms are small animals that live in water. They have no specialised gas-exchange or circulatory systems. The drawing shows one type of flatworm.

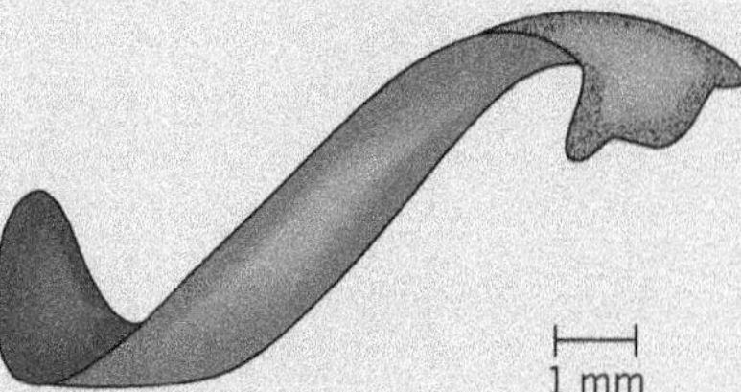

(i) Name the process by which oxygen reaches the cells inside the body of this flatworm. *(1 mark)*

(ii) The body of a flatworm is adapted for efficient gas exchange between the water and the cells inside the body. Using the diagram, explain how **two** features of the flatworm's body allow efficient gas exchange. *(2 marks)*

(b) (i) A leaf is an organ. What is an organ? *(1 mark)*

(ii) Describe how carbon dioxide in the air outside a leaf reaches mesophyll cells inside the leaf. *(3 marks)*

AQA June 2012

Answers to the Practice Questions are available at
www.oxfordsecondary.com/oxfordaqaexams-alevel-biology

6 Variation

6.1 Investigating variation

One look around you and it is clear that living things differ. If one species differs from another this is called **interspecific variation** (Figure 1). But members of the same species also differ from each other. This is called **intraspecific variation** (Figure 2). Every one of the billions of organisms on planet Earth is unique. Even identical twins, who are born with the same DNA, vary as a result of their different experiences. How then do we measure the differences between these characteristics?

Learning objectives:

- → Describe how variation is measured.
- → Describe sampling and why it is used.
- → Describe the causes of variation.

Specification reference: 3.1.6

Making measurements

All scientists measure things, but this is a particular problem for biologists. This is because they are usually measuring some aspect of living organisms and all living organisms are different. For this reason, biologists have to take many measurements of the same thing. They cannot reliably determine the height of buttercups or the number of red cells in 1 mm^3 of human blood by taking a single measurement. Equally, they cannot measure every buttercup or human being in existence. What they do is take samples.

Sampling

Sampling involves taking measurements of individuals, selected from the population of organisms which is being investigated. In theory, if these individuals are representative of the population as a whole, then the measurements can be relied upon. But are the measurements representative? There are several reasons why they might not be, including:

- **sampling bias**. The selection process may be biased. The investigators may be making unrepresentative choices, either deliberately or unwittingly. Are they as likely to take samples of buttercups from a muddy area as a dry one? Will they avoid areas covered in cow dung or rich in nettles?
- **chance**. Even if sampling bias is avoided, the individuals chosen may, by pure chance, not be representative. The 50 buttercup plants selected might just happen to be the 50 tallest in the population.

The best way to prevent sampling bias is to eliminate, as far as possible, any human involvement in choosing the samples. This can be achieved by carrying out **sampling at random**. One method is to:

1. Divide the study area into a grid of numbered lines, for example, by stretching two long tape measures at right angles to each other.
2. Using random numbers, from a table or generated by a computer, obtain a series of coordinates.
3. Take samples at the intersection of each pair of coordinates.

We cannot completely remove chance from the sampling process but we can minimise its effect by:

- **using a large sample size**. The more individuals that are selected, the smaller is the probability that chance will influence the result. If we sample only five buttercups there is a high probability that they

▲ **Figure 1** *Variation between species (interspecific variation)*

Study tip

Remember that intraspecific variation is variation within a species.

▲ **Figure 2** *Variation within a species (intraspecific variation): despite the immense variety that they show, all dogs, including this Pug and Great Dane, belong to the same species – Canis familiaris*

Synoptic link

Details of the process of meiosis are covered in Topic 9.1.

Summary questions

1. State **three** ways in which genetic variation can be increased in sexually reproducing organisms.
2. How is genetic variation increased in asexually reproducing organisms?
3. Give **two** reasons why a sample may not be representative of the population as a whole.
4. How may sampling bias be prevented?

may all be taller than average. If we sample 500 there is a much lower probabilty that they will all be taller than average. The greater the sample size the more reliable the data will be.

- **analysis of the data collected**. Accepting that chance will play a part, the data collected can be analysed using statistical tests to determine the extent to which chance may have influenced the data. These tests allow us to decide whether any variation observed is the result of chance or is more likely to have some other cause.

Causes of variation

Variation is the result of two main factors: genetic differences and environmental influences. In most cases it is a combination of both.

Genetic differences

Genetic differences are due to the different alleles of the genes that each individual organism possesses. These differences not only arise in living individuals but also change from generation to generation. Genetic variation arises as a result of:

- **mutations**. The production of new forms, alleles, or genes and chromosomes may or may not be passed on to the next generation.
- **meiosis**. This special form of nuclear division forms the gametes. This mixes up the genetic material before it is passed into the gametes, all of which are therefore different (see Topic 9.1).
- **fusion of gametes**. In sexual reproduction the offspring inherit some characteristics from each parent and are therefore different from both of them. Which gamete fuses with which at fertilisation is a random process, further adding to the variety of offspring that two parents can produce.

Variation in asexually reproducing organisms can only be caused by mutation. In contrast, variation in sexually reproducing organisms can be caused by all three methods. It follows that populations of sexually reproducing organisms are more varied than asexually reproducing organisms.

Environmental influences

The environment exerts an influence on all organisms. These influences affect the way the organism's genes are expressed. The genes set limits, but it is largely the environment that determines where, within those limits, an organism lies. In buttercups, for example, the genes of one plant may determine that it will grow much taller than other plants. If, however, the seed germinated in an environment of poor light or low soil nitrogen, the plant may not grow properly and it will be short. Environmental influences include climatic conditions (e.g., temperature, rainfall, and sunlight), soil conditions, pH, and food availability.

In most cases variation is due to the combined effects of genetic differences and environmental influences. It is very hard to distinguish between the effects of the many genetic and environmental influences that combine to produce differences between individuals. As a result, it is very difficult to draw conclusions about the causes of variation in any particular case. Any conclusions that are drawn are usually tentative and should be treated with caution.

Required practical 3

Investigating variation in plant pigments

Plant species such as *Coleus* can show variation in the pigmentation of their leaves. One way to investigate the pigments present would be to use thin-layer chromatography (TLC). TLC plates can be obtained from various scientific suppliers.

▲ **Figure 3** *Variation in the pigmentation of* Coleus *leaves*

- Cut strips from a TLC plate such that they are the right size to fit the chromatography chamber.
- Set up the chromatography chamber (e.g., a beaker sealed with foil) and place the running solvent (5 parts cyclohexane, 3 parts propanone, and 2 parts petroleum ether) in the bottom of the chamber to a depth of approximately 1 cm. Cover with the foil and leave in a fume cupboard to equilibrate.
- Using a hole punch, remove three or four leaf discs from the ***Coleus*** leaf being investigated and place these in a mortar with some silver sand.
- Break up the tissue with a pestle and then add a small volume of the solvent propanone (0.5–1.0 cm^3), mix, and allow the debris to settle.
- Carefully draw off the supernatant and store in a stoppered container.
- Taking care not to touch the surface of the TLC strip, make two marks at the edges of the strip approximately 1.5 cm from the bottom. The supernatant will be loaded on a line between these two points
- Using a fine Pasteur pipette, add small drops of the supernatant in a line across the plate. Allow these to dry and then repeat until there is a concentrated line across the plate.
- Place the TLC strip into the chromatography chamber and replace the cover.
- Remove from the chamber once the running solvent is close to the top of the strip and immediately mark the position of the solvent – the 'solvent front'.

R_f values for each of the pigments separated can be calculated by dividing the distance moved by the pigment by the distance of the solvent front from the loading line.

1. Why is it important to allow the chromatography chamber to equilibrate before placing the TLC strip in place?
2. The yellow pigment carotene has an R_f value of 0.91 whereas another yellow pigment xanthophyll has an R_f value of 0.47. Which of the two pigments would be closest to the solvent front?
3. How could you identify the pigments you have separated from the ***Coleus*** leaves?

6.2 Types of variation

Learning objectives:

- → Describe the types of variation.
- → Define the mean of a normal distribution.
- → Describe standard deviation and how is it calculated.

Specification reference: 3.1.6

Variation is the result of either genetic differences or the influence of the environment. In many cases it is a combination of both genetic and environmental factors.

Variation due to genetic factors

Where variation is the result of genetic factors organisms fit into a few distinct forms and there are no intermediate types. This is called **discontinuous variation**. In the ABO blood grouping system, for example, there are four distinct groups A, B, AB, and O (Figure 1). A character displaying this type of variation is usually controlled by a single gene. This variation can be represented on a bar chart or pie graph. Environmental factors have little influence on this type of variation.

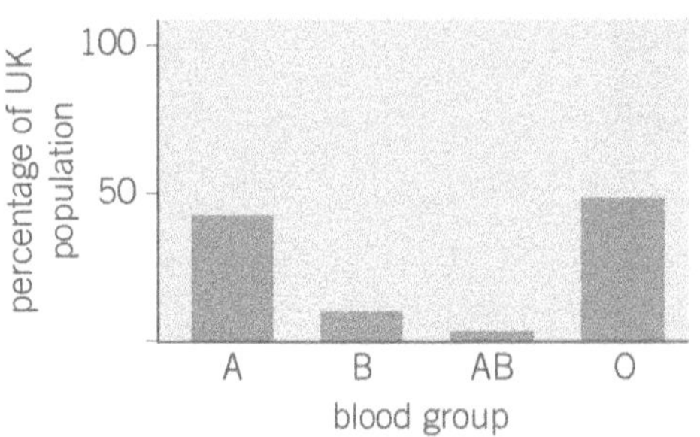

◀ **Figure 1** *Variation due to genetic factors, illustrated by the percentage of the UK population that has blood group A, B, AB, or O*

▼ **Table 1** *Frequency of heights (measured to the nearest 2 cm)*

Height / cm	Frequency
140	0
144	1
148	23
152	90
156	261
160	393
164	458
168	413
172	177
176	63
180	17
184	4
188	1
190	0
192	0

Variation due to environmental influences

Some characteristics of organisms grade into one another, forming a continuum. This is called **continuous variation**. In humans, two examples are height and mass. Characteristics that display this type of variation are not controlled by a single gene, but by many genes (polygenes). Environmental factors also play a major role in determining where on the continuum an organism actually lies. For example, individuals who are genetically predetermined to be the same height actually grow to different heights due to variations in environmental factors, such as diet. This type of variation is the product of polygenes and the environment. Table 1 shows the number of people (frequency) in a particular sample with various heights. If we take these data and plot them on a graph we obtain a bell-shaped curve known as a **normal distribution curve** (Figure 2).

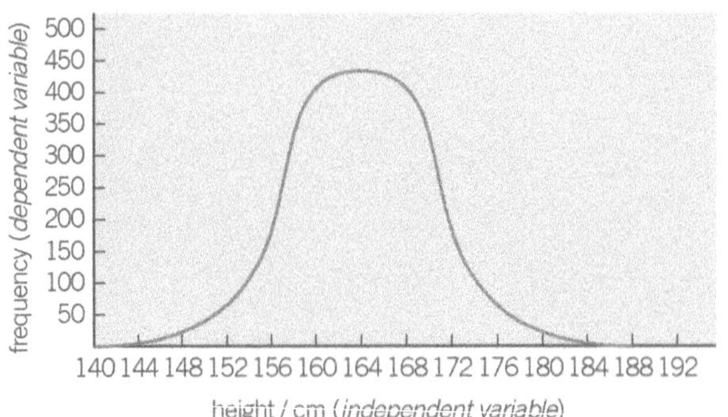

▲ **Figure 2** *Graph of frequency against height for a sample of humans*

Mean and standard deviation

A normal distribution curve always has the same basic shape (Figure 2). It differs in two measurements: its maximum height and its width.

- The **mean** is the measurement at the maximum height of the curve. The mean of a sample of data provides an average value and is useful information when comparing one sample with another. It does not, however, provide any information about the range of values within the sample. For example, the mean number of children in a sample of eight families may be two. However, this could be made up of eight families each with two children or six families with no children and two families with eight children each.
- The **standard deviation** is a measure of the width of the curve. It gives an indication of the range of values either side of the mean. A standard deviation is the distance from the mean to the point where the curve changes from being convex to concave (the point of inflection). In a normal distribution, **68 per cent** of all the measurements lie within ± 1.0 standard deviation of the mean. Increasing this width to almost ± 2.0 (actually ± 1.96) standard deviations takes in **95** of all measurements. These measurements are illustrated in Figure 3.

Study tip

A large standard deviation means a lot of variety, whereas a small standard deviation means little variety.

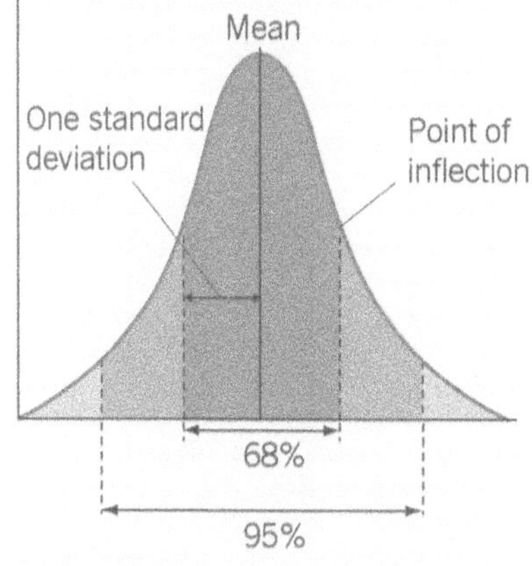

▲ **Figure 3** *The normal distribution curve showing values for standard deviation*

Calculating standard deviation

At first sight, the formula for standard deviation can look complex:

$$\textbf{standard deviation} = \sqrt{\frac{\Sigma(x - \bar{x})^2}{n - 1}}$$

where:

Σ = **the sum of**

x = **measured value (from the sample)**

$\bar{x}$ = **mean value**

n = **total number of values in the sample.**

Hint

You don't need to learn the standard deviation formula by heart as it is provided in your exams.

Hint

Remember to square **all** the numbers, not just the negative ones.

However, it is straightforward to calculate and less frightening if you take it step by step. The following very simple example, using the six measured values (x) 4, 1, 2, 3, 5, and 0, illustrates each step in the process.

- Calculate the mean value ($\bar{x}$):

$$\mathbf{4 + 1 + 2 + 3 + 5 + 0 = 15}$$

$$\mathbf{15 \div 6 = 2.5}$$

- Subtract the mean value from each of the measured values ($x - \bar{x}$). This gives:

+1.5, −1.5, −0.5, +0.5, +2.5, −2.5

- As some of these numbers are negative, we need to make them positive. To do this, square **all** the numbers $(x - \bar{x})^2$. Remember to square all the numbers and not just the negative ones. This gives:

2.25, 2.25, 0.25, 0.25, 6.25, 6.25

- Add all these squared numbers together:

$$\Sigma(x - \bar{x})^2 = \mathbf{17.5}$$

- Divide this number by the original number of measurements less one, that is, 5:

$$\frac{\Sigma(x - \bar{x})^2}{n - 1} = \frac{17.5}{5} = 3.5$$

- As all the numbers have been squared, the final step is to take the square root in order to get back to the same units as the mean:

$$\sqrt{\frac{\Sigma(x - \bar{x})^2}{n - 1}} = \sqrt{3.5} = 1.87$$

Summary questions

1 In the following list of statements, decide whether each refers to variation due to genetic or environmental factors.

  a An example is the ABO blood grouping system in humans.

  b It can be represented by a line graph.

  c It is usually controlled by a single gene.

  d It can be represented as a bar graph.

  e A mean can be calculated.

  f An example is the length of the body in rats.

2 In a population of men the systolic blood pressure shows a normal distribution. The mean of the population is 125 (measured in mm Hg) and the standard deviation is 10. If the population was 1000, how many of them have a blood pressure between 115 and 135 mm Hg?

Practice questions: Chapter 6

1 Twin studies have been used to determine the relative effects of genetic and environmental factors on the development of a type of diabetes. The table shows the concordance (where both twins have the condition) in genetically identical and genetically non-identical twins.

Concordance in genetically identical twins / %	Concordance in genetically non-identical twins / %
85	35

(a) What do the data show about the relative effects of environmental and genetic factors on the development of diabetes? *(1 mark)*

(b) Suggest **two** factors which should be taken into account when collecting the data in order to draw valid conclusions. *(2 marks)*

AQA 2005

2 Maize seeds were an important food crop for the people who lived in Peru. The seeds could be kept for long periods. Each year, some were sown to grow the next crop. Archaeologists have found well-preserved stores. The graph shows the lengths of seeds collected from three stores of different ages.

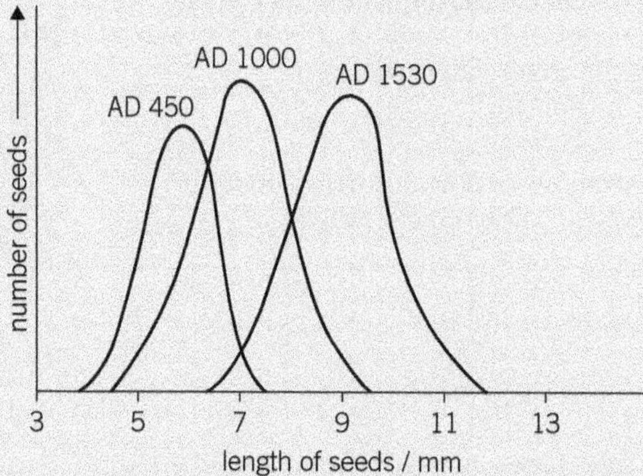

Within each store the maize seeds showed a range of different lengths. Give **two** causes of this variation and an explanation for each. *(4 marks)*

AQA 2004

3 The graph shows the variation in length of 86 Atlantic salmon.

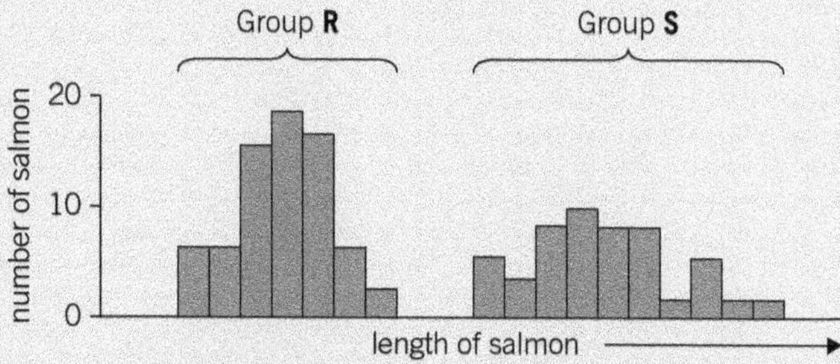

(a) Give **two** possible causes of variation. *(2 marks)*

(b) When comparing variation in size between two groups of organisms, it is often considered more useful to compare standard deviations rather than ranges. Explain why. *(2 marks)*

AQA 2005

4 The graph shows variation in the number of spots on the wing cases of a species of ladybird.

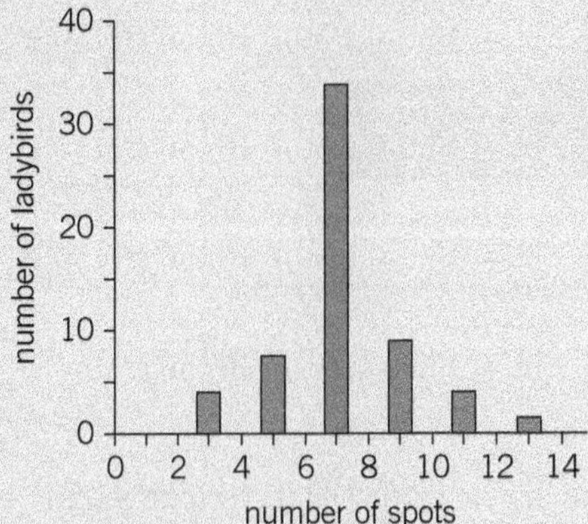

(a) The number of spots on the wing cases of this species of ladybird is determined by genes. What does the graph suggest about the genetic control of spot number in this species? *(1 mark)*

(b) Give **one** piece of evidence from the graph that variation in the number of spots is normally distributed. *(1 mark)*

AQA 2003

5 ABO blood groups in humans are an example of discontinuous variation, whereas height in humans is an example of continuous variation.

(a) Describe how discontinuous variation differs from continuous variation in terms of:

(i) genetic control

(ii) the effect of the environment

(iii) the range of phenotypes. *(3 marks)*

(b) Genetically identical twins often show slight differences in their appearance at birth.
Suggest **one** way in which these differences may have been caused. *(1 mark)*

AQA 2006

Answers to the Practice Questions are available at
www.oxfordsecondary.com/oxfordaqaexams-alevel-biology

7 DNA, genes, and chromosomes

7.1 Structure of RNA and DNA

Nucleic acids are a group of important molecules, of which the best known are **ribonucleic acid** (**RNA**) and **deoxyribonucleic acid** (**DNA**). The double helix structure of DNA makes it immediately recognisable. DNA carries genetic information. The identification of this extraordinary molecule as the material that passes on the features of organisms from one generation to the next is one of the most remarkable feats of experimental biology. The discovery of the precise molecular arrangement of DNA was no less remarkable. Despite its complex structure, DNA is made up of nucleotides that have just three basic components.

Learning objectives:

→ Describe the structure of a nucleotide.

→ Describe the structure of RNA.

→ Describe the structure of DNA.

Specification reference: 3.1.7.1

Nucleotide structure

Individual nucleotides are made up of three components:

- a pentose sugar (so called because it has five carbon atoms)
- a phosphate group
- a nitrogen-containing organic base. These are: cytosine **C**, thymine **T**, uracil **U**, adenine **A**, and guanine **G**.

Study tip

Spelling can make a difference. 'Thymine' is a base in DNA but 'thiamine' is vitamin B_1.

The pentose sugar, phosphate group, and organic base are joined, as a result of condensation reactions, to form a single nucleotide (**mononucleotide**) as shown in Figure 1. Two mononucleotides may, in turn, be joined as a result of a condensation reaction between the deoxyribose sugar of one mononucleotide and the phosphate group of another. The bond formed between them is called a **phosphodiester bond** (Figure 2). The new structure is called a **dinucleotide**. The continued linking of mononucleotides in this way forms a long chain known as a **polynucleotide**. In addition to DNA and RNA, some other biologically important molecules contain nucleotides. For simplicity the various components of nucleotides are represented by symbols, as shown in Table 1.

▼ **Table 1** *Components of nucleotides*

Name of molecule	Symbol
Phosphate	(circle)
Pentose sugar	(pentagon)
Adenine	adenine
Guanine	guanine
Cytosine	cytosine
Thymine	thymine
Uracil	uracil

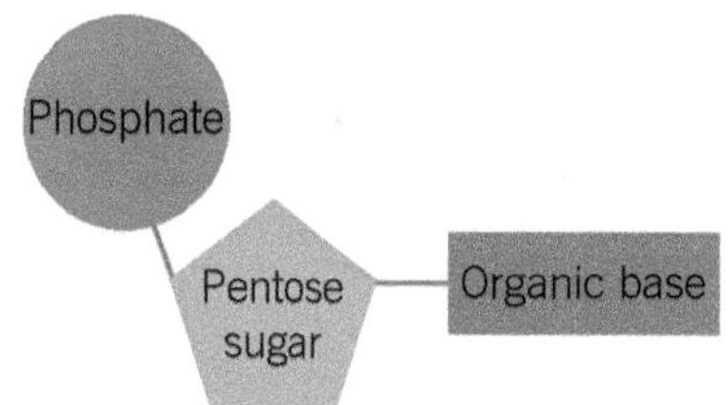

▲ **Figure 1** *Simplified structure of a nucleotide*

Study tip

Do not get confused between DNA and proteins. DNA is a sequence of nucleotides and proteins are sequences of *amino acids*. Nucleotides join to form a *polynucleotide*, amino acids join to form a *polypeptide*.

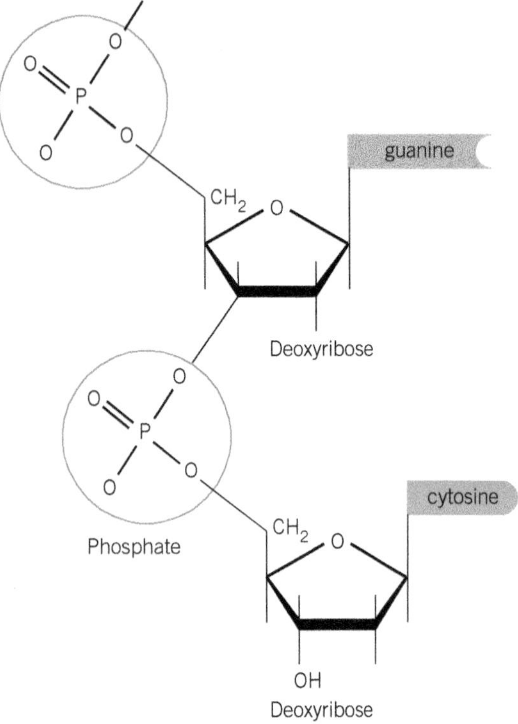

▲ **Figure 2** *The structure of a phosphodiester bond between a guanine nucleotide and a cytosine nucleotide*

Ribonucleic acid (RNA) structure

Ribonucleic acid is a polymer made up of nucleotides. It is a single, relatively short polynucleotide chain in which the pentose sugar is always **ribose** and the organic bases are adenine, guanine, cytosine, and uracil (Figure 3). Three types of RNA are important in every cell. Transfer RNA (tRNA) transfers genetic information from DNA to the ribosomes. The ribosomes themselves are made up of proteins and another type of RNA (rRNA). A third type of RNA (mRNA) is involved in protein synthesis. You will learn more about the structure and functions of mRNA and tRNA in the section on protein synthesis (Topics 8.2 and 8.3).

DNA structure

In 1953, James Watson and Francis Crick worked out the structure of DNA, following pioneering work by Rosalind Franklin on the X-ray diffraction patterns of DNA. This opened the door for many of the major developments in biology over the next half-century.

In DNA the pentose sugar is deoxyribose and the organic bases are adenine, thymine, guanine and cytosine. DNA is made up of two strands of nucleotides (polynucleotides). Each of the two strands is extremely long, and they are joined together by hydrogen bonds formed between certain bases. In its simplified form, DNA can be thought of as a ladder in which the phosphate and deoxyribose molecules alternate to form the uprights and the organic bases pair together to form the rungs (Figure 4).

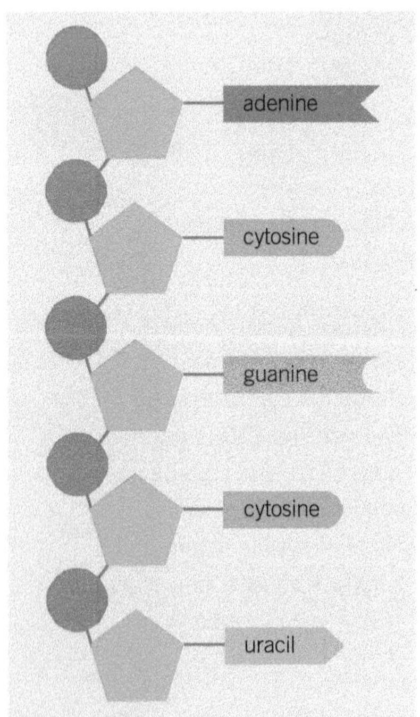

▲ **Figure 3** *Section of an RNA molecule*

Base pairing

The bases on the two strands of DNA attach to each other by hydrogen bonds. It is these hydrogen bonds that hold the two strands together. The base pairing is specific:

- Adenine always pairs with thymine.
- Guanine always pairs with cytosine.

As a result of these pairings, adenine is said to be **complementary** to thymine, and guanine is said to be complementary to cytosine.

It follows that the quantities of adenine and thymine in DNA are always the same, and so are the quantities of guanine and cytosine. However, the ratio of adenine and thymine to guanine and cytosine varies from species to species.

The double helix

In order to appreciate the structure of DNA, you need to imagine the ladder-like arrangement of the two polynucleotide chains being twisted. In this way, the uprights of phosphate and deoxyribose wind around one another to form a double helix. They form the structural backbone of the DNA molecule.

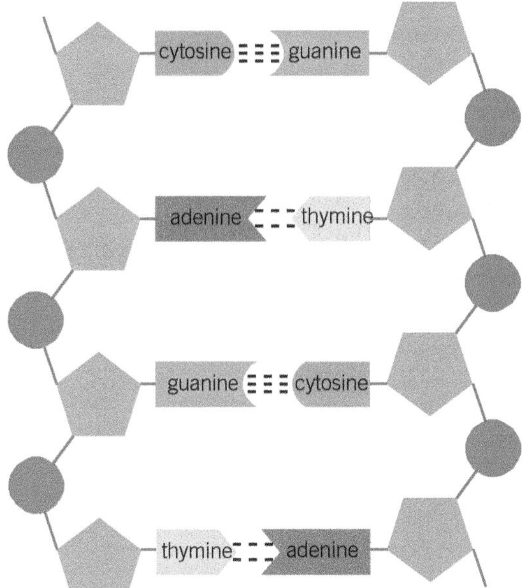

▲ **Figure 4** *Basic structure of DNA. DNA structure may be likened to a ladder in which alternating phosphate and deoxyribose molecules make up the 'uprights' and pairs of organic bases comprise the 'rungs'. Note the base pairings are always cytosine–guanine and adenine–thymine. This ensures a standard 'rung' length. Note also that the 'uprights' run in the opposite direction to each other (i.e., are antiparallel).*

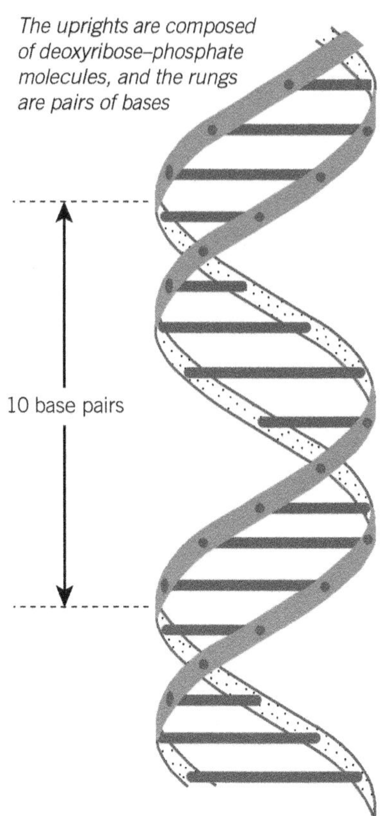

▲ **Figure 5** *The double helix structure of DNA*

The stability of DNA

DNA is a stable molecule because:

- The phosphodiester backbone protects the more chemically reactive organic bases inside the double helix.
- Hydrogen bonds link the organic base pairs forming bridges (rungs) between the phosphodiester uprights. As there are three hydrogen bonds between cytosine and guanine, compared with two hydrogen bonds between adenine and thymine, the higher the proportion of C–G pairings, the more stable the DNA molecule.

There are other interactive forces between the base pairs that hold the molecule together, such as base stacking.

Function of DNA

DNA is the hereditary material responsible for passing genetic information from cell to cell and from generation to generation. In total, there are around 3.2 billion base pairs in the DNA of a typical mammalian cell. This vast number means that there is an almost infinite variety of sequences of bases along the length of a DNA molecule. It is this variety that provides the genetic diversity within living organisms.

The DNA molecule is adapted to carry out its functions in a number of ways:

- It is a very stable structure that normally passes from generation to generation without change. Only rarely does it mutate (but see Topic 9.1).

Hint

In every molecule of DNA, the phosphate group, the deoxyribose, and the four bases are always the same. What differs between one DNA molecule and another are the proportions and, more importantly, the sequence of each of the four bases.

Synoptic link

More detail on mutations is given in Topic 17.3.

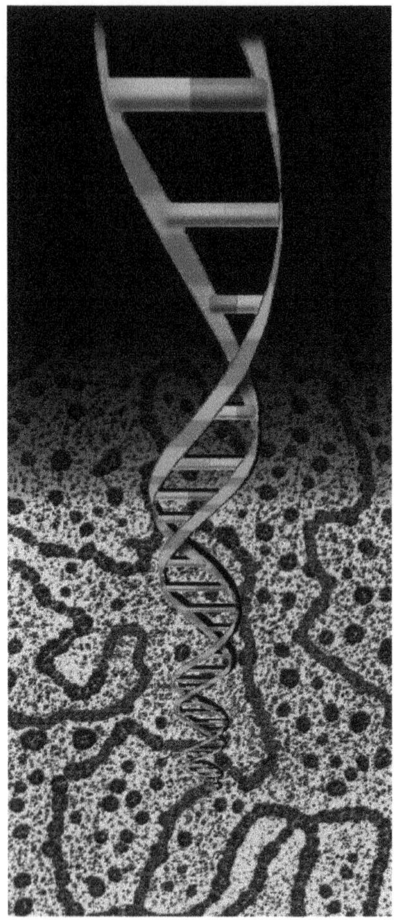

▲ **Figure 6** *Computer generated image of DNA molecules*

- Its two separate strands are joined only with hydrogen bonds, which allow them to separate during DNA replication (Topic 7.3) and protein synthesis.
- It is an extremely large molecule and therefore carries an immense amount of genetic information.
- By having the base pairs within the helical cylinder of the deoxyribose–phosphate backbone, the genetic information is to some extent protected from being corrupted by outside chemical and physical forces.
- Complementary base pairing leads to DNA being able to replicate and to transfer information as mRNA (see Topic 8.2).

The function of the remarkable molecule that is DNA depends on the sequence of base pairs that it possesses. This sequence is important to everything it does and, indeed, to life itself.

Summary questions

1 List the three basic components of a nucleotide.

2 Suggest why the base pairings of adenine with cytosine and guanine with thymine do not occur.

3 If the bases on one strand of DNA are TGGAGACT, determine the base sequence on the other strand.

4 If 19.9 per cent of the base pairs in human DNA are guanine, calculate what percentage of human DNA is thymine. Show your reasoning.

Unravelling the role of DNA

We now take for granted that DNA is the hereditary material that passes genetic information from cell to cell and from generation to generation. However, we haven't always known that this was the case because there were other contenders for this role, in particular proteins.

With the knowledge available at the time, scientists thought that proteins were the more likely candidate because of their considerable chemical diversity. DNA was considered to have too few components and to be chemically too simple to fulfil the role. However, not all scientists were convinced and so they set about finding experimental evidence to determine the true nature of hereditary material.

1 Assess the advantages of scientists questioning the validity of a current theory rather than automatically accepting it.

Scientists work by using **observations** and current knowledge to form a **hypothesis**. From this, they make **predictions** about the outcome of a particular **investigation**. By carrying out this investigation a number of times, they collect the experimental evidence that allows them to accept or reject their hypothesis.

2 Explain what is meant by the term hypothesis in the scientific sense.

Investigations were needed to test the hypothesis that DNA was the hereditary material.

One such investigation involved experiments using mice and a bacterium that can cause pneumonia. The bacterium exists in two forms:

- a safe form that does not cause pneumonia, known as the R-strain

- a harmful form that causes pneumonia, known as the S-strain.

Mice were separately injected with living bacteria of the safe form and dead bacteria from the harmful form.

The group of mice injected with the living safe form of bacteria remained healthy, as did the group injected with the dead harmful form of bacteria. So, when mice were injected with both types together, it would not have been surprising to get a similar result. These mice, however, developed pneumonia. The experiment and the results are summarised in Figure 7.

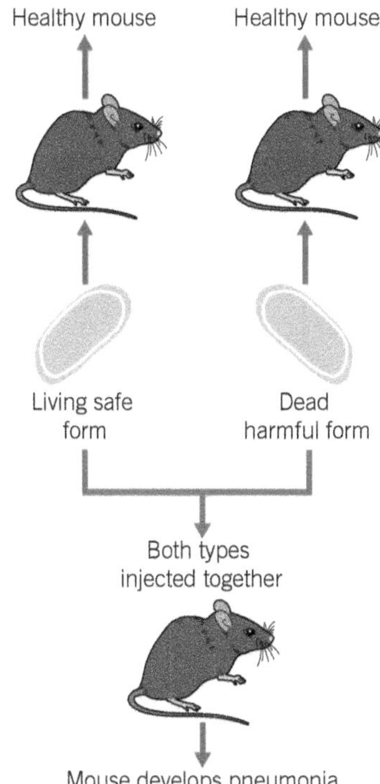

▲ **Figure 7** *Summary of an experiment to determine the nature of hereditary material in an organism*

Living bacteria of the harmful form were isolated from the mice with pneumonia. There are three possible explanations for this:

- Experimental error, for example, the harmful forms in the mixture were not all killed.
- The living safe form had mutated into the harmful form. This is possible but extremely unlikely, especially given that the experiment was repeated many times with the same result.
- Pneumonia is caused by a toxin. The harmful form of the bacterium has the information on how to make the toxin but, being dead, cannot do so. The safe form has the means of making the toxin but lacks the information on how to do so. The information on how to make the toxin may have been transferred from the harmful form to the safe form, which then produced it.

3 State what simple procedure could be carried out to discount the first explanation.

4 Mutations happen very rarely. Explain why this helps to discount the second explanation.

The third explanation was considered worthy of further investigation and so a series of experiments was designed and carried out as follows:

- The living harmful bacteria that were found in the mice with pneumonia were collected.
- Various substances were isolated from these bacteria and purified.
- Each substance was added to suspensions of living safe bacteria to see whether it would transform them into the harmful form.
- The only substance that produced this transformation was purified DNA.
- When an enzyme that breaks down DNA was added, the ability to carry out the transformation ceased.

Other experiments provided further proof that DNA was the hereditary material and also suggested a mechanism by which it could be transferred from one bacterial cell to another.

- It had been observed that viruses infect bacteria, causing the bacteria to make more viruses.
- As the virus is made up of just protein and DNA, one or the other must possess the instructions that the bacteria use to make new viruses.
- The protein and DNA in the viruses were each labelled with a different radioactive element.
- One sample of bacteria was infected by viruses with radioactive protein whilst another sample was infected by viruses with radioactive DNA.
- After giving the bacteria time to replicate the viruses and bacteria in both samples were separated from one another.
- Only the sample with bacteria that had been infected by viruses labelled with radioactive DNA produced radioactive viruses.

This was evidence that DNA was the material that had provided the bacteria with the genetic information needed to make the viruses.

5 A new scientific discovery often presents moral, economic, and ethical issues. Justify why it is necessary for society to analyse the risks and benefits of these discoveries before they are developed.

A prime location

In order to understand how nucleotides are arranged in nucleic acids, it is necessary to know how the carbon atoms in the pentose molecule are numbered. Of particular importance is the numbering of the 3-prime (3′) and 5-prime (5′) carbon atoms. The 5′ carbon has an attached phosphate group, whereas the 3′ has a hydroxyl group.

Figure 8 shows a nucleotide with the 3′ and 5′ carbon atoms marked on its pentose sugar.

When nucleotides are organised into the double strands of a DNA molecule, one strand runs in the 5′ to 3′ direction whilst the other runs the opposite way – in the 3′ to 5′ direction. The two strands are therefore said to be antiparallel.

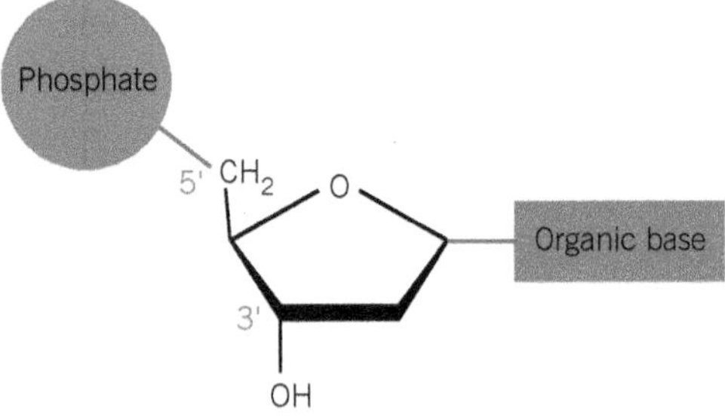

▲ **Figure 8** *Nucleotide showing positions of the 3-prime (3′) and 5-prime (5′) carbon atoms on the pentose sugar*

Nucleic acids can only be synthesised *in vivo* in the 5′ to 3′ direction. This is because the enzyme DNA polymerase that assembles nucleotides into a DNA molecule can only attach nucleotides to the hydroxyl (OH) group on the 3′ carbon molecule.

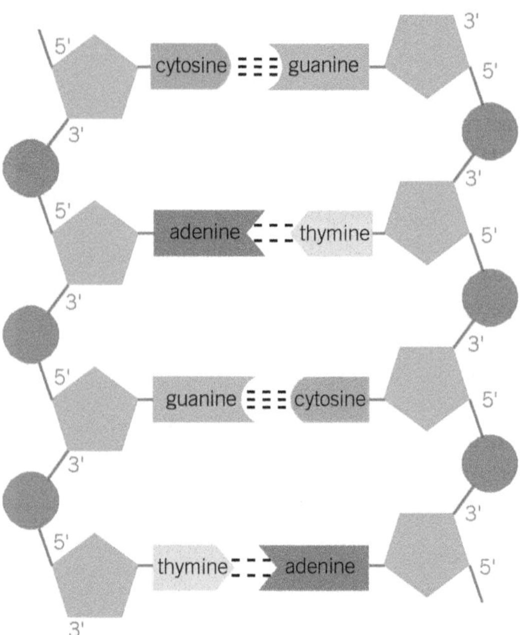

▲ **Figure 9** *DNA molecule showing the 3-prime and 5-prime carbon atoms labelled. Notice that one strand runs 5′ to 3′ whilst the other runs 3′ to 5′. They are* ***antiparallel.***

1. Suggest what the term *in vivo* means in the context of synthesising DNA.
2. From your knowledge of the way enzymes work, explain why DNA polymerase can only attach nucleotides to the hydroxyl (OH) group on the 3′ carbon molecule.

7.2 DNA, genes, and chromosomes

In Topics 2.3 and 2.5 you saw that, according to their organisation, there are two types of cells: **eukaryotic cells** and **prokaryotic cells**. You looked at some of the differences between the two. These differences extend to their DNA:

- In prokaryotic cells, such as bacteria, the DNA molecules are shorter, form a circle, and are not associated with protein molecules. Prokaryotic cells therefore do not have chromosomes.
- In eukaryotic cells, the DNA molecules are longer, form a line (are linear) rather than a circle, and occur in association with proteins called **histones** to form structures called **chromosomes**. The mitochondria and chloroplasts of eukaryotic cells also contain DNA that, like the DNA of prokaryotic cells, is short, circular, and not associated with proteins.

Learning objectives:

- → Distinguish between the DNA in prokaryotic cells and the DNA in eukaryotic organisms.
- → Describe the structure of a chromosome.
- → Explain how genes are arranged on a DNA molecule.
- → Describe the nature of homologous chromosomes.
- → Explain what is meant by an allele.

Specification reference: 3.1.7.2

Chromosome structure

Chromosomes are only visible as distinct structures when a cell is dividing. For the rest of the time they are widely dispersed throughout the nucleus. When they first become visible at the start of cell division, chromosomes appear as two threads, joined at a single point (Figure 1). Each thread is called a **chromatid** because DNA has already replicated to give two identical DNA molecules (see Topic 7.3). The DNA in chromosomes is held by histones. The considerable length of DNA found in each cell (around 2 m in every human cell) is highly coiled and folded, as illustrated in Figure 2.

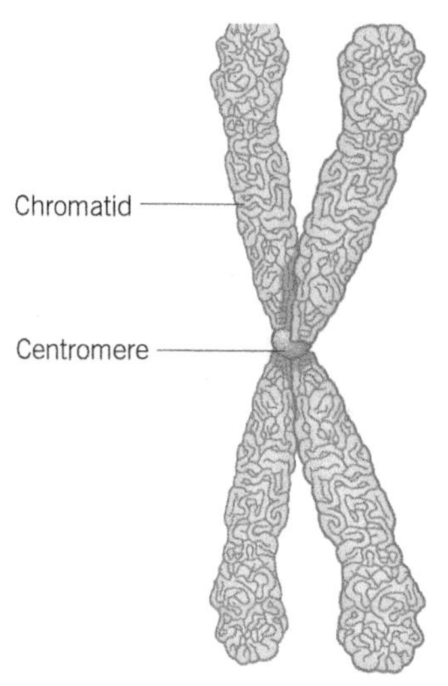

▲ **Figure 1** *Structure of a chromosome*

Study tip

Do not confuse the two threads (chromatids) of a chromosome with the two strands of the DNA double helix.

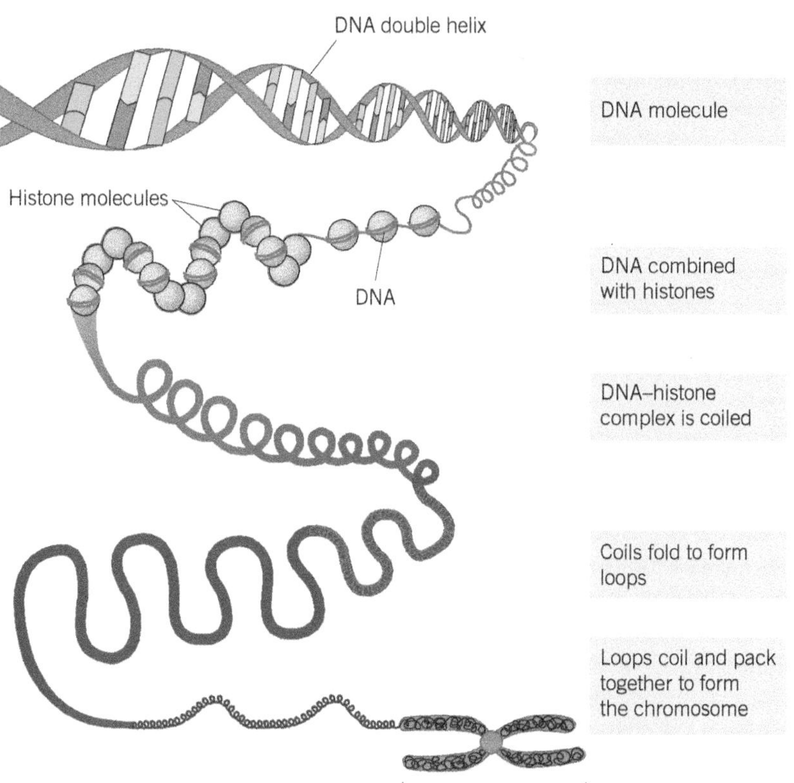

▲ **Figure 2** *How DNA is packed into a chromosome*

Hint

It is often mistakenly thought that humans have just 46 chromosomes throughout the body, rather than 46 in every single cell (except sperm and eggs).

You already know that DNA is a double helix. From Figure 2 you can see that this helix is wound around histones to fix it in position. This DNA–histone complex is then coiled. The coil, in turn, is looped and further coiled before being packed into the chromosome. In this way, a lot of DNA is condensed into a single chromosome. If you follow the diagram carefully you will see that a chromosome contains just a single molecule of DNA, although this is very long. This single DNA molecule has many genes along its length (see Topic 8.1).

Although the number of chromosomes is always the same for normal individuals of the same species, it varies from one species to another. For example, whilst humans have 46 chromosomes, potato plants have 48 and dogs have 78. In most species, there is an even number of chromosomes in the cells of adults.

Homologous chromosomes

Sexually produced organisms, such as humans, are the result of the fusion of a sperm and an egg, each of which contributes one complete set of chromosomes to the offspring. Therefore, each cell will contain two copies of each chromosome. One of each pair is derived from the chromosomes provided by the mother in the egg (maternal chromosomes) and the other is derived from the chromosomes provided by the father in the sperm (paternal chromosomes). These are known as **homologous pairs** and the total number is referred to as the diploid number. In humans this is 46.

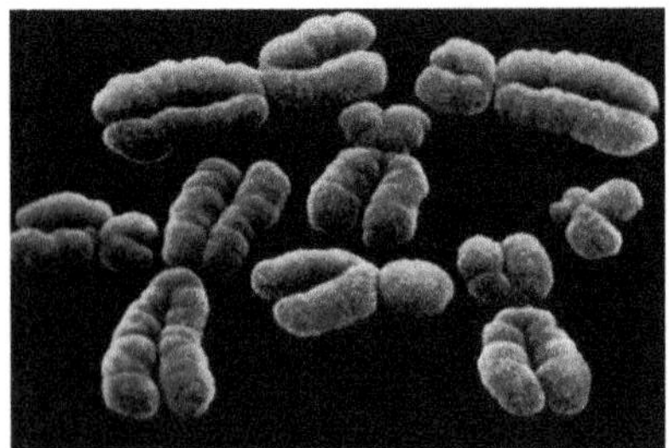

▲ **Figure 3** *False-colour scanning electron micrograph (SEM) of a group of human chromosomes*

A homologous pair is always two chromosomes that carry the same genes but not necessarily the same alleles of the genes.

For instance, a homologous pair of chromosomes may each possess genes for tongue rolling and blood group, but one chromosome may carry the allele for non-roller and blood group A, whilst the other carries the allele for roller and blood group B. During meiosis, the halving of the number of chromosomes is done in a manner which ensures that each daughter cell receives only one chromosome from each homologous pair. In this way, each cell receives just one gene for each characteristic of the organism. Such cells are referred to as haploid (see Topic 9.1). When these haploid cells combine, the diploid state, with paired homologous chromosomes, is restored.

What is a gene?

A gene is a section of DNA that contains the coded information for making polypeptides and functional RNA. The coded information is in the form of a specific sequence of bases along the DNA molecule.

Hint

Do not confuse genes and alleles. A *gene* refers to a sequence of DNA that controls a particular characteristic, such as blood groups. Genes can exist in two or more different forms called *alleles*.

Polypeptides make up proteins and so genes determine the proteins of an organism. Enzymes are proteins. As enzymes control chemical reactions they are responsible for an organism's development and activities. In other words, genes, along with environmental factors, determine the nature and development of all organisms. Each gene occupies a specific position (**locus**) along the DNA molecule. A genome is the complete set of genes present in a cell.

The gene is a base sequence of DNA that codes for:

- the amino acid sequence of a polypeptide
- or a functional RNA, including rRNA and tRNAs (Topic 8.3).

In eukaryotes, much of the nuclear DNA does not code for polypeptides and there are non-coding regions called introns both between genes and within genes (Topic 8.1).

What is an allele?

An **allele** is one of a number of alternative forms of a gene. You have seen that genes are sections of DNA that contain coded information in the form of specific sequences of bases. A gene may exist in two or more different forms. Each of these forms is called an allele. Each individual inherits one allele from each of its parents. These two alleles may be the same or they may be different. When they are different, each allele has a different base sequence, therefore a different amino acid sequence, so produces a different polypeptide.

Any changes in the base sequence of a gene produces a new allele of that gene (mutation) and results in a different sequence of amino acids being coded for.

Summary questions

1 Contrast the DNA of a prokaryotic cell with that of a eukaryotic cell.

2 State the function of the protein found in chromosomes.

3 Explain how the considerable length of a DNA molecule is compacted into a chromosome.

7.3 DNA replication

Learning objectives:

- → Describe what happens during DNA replication.
- → Describe how a new polynucleotide strand is formed.
- → Explain why the process of DNA replication is called semi-conservative.

Specification reference: 3.1.7.3

The cells that make up organisms are always derived from existing cells by the process of division. Cell division occurs in two main stages:

- **Nuclear division** is the process by which the nucleus divides. There are two types of nuclear division, mitosis and meiosis.
- **Cell division** follows nuclear division and is the process by which the whole cell divides.

Before a nucleus divides its DNA must be replicated (copied). This is to ensure that all the daughter cells have the genetic information to produce the enzymes and other proteins that they need.

The process of DNA replication is clearly very precise because, with the exception of gametes, all the new cells are more or less identical to the original one. This is a remarkable achievement when one considers the complexity of the DNA molecule. How then does DNA replication take place? Of the possible methods, it is the semi-conservative model that is universally accepted.

Hint

It is a basic scientific principle that when smaller molecules are built up into larger ones, energy is required. When larger molecules are broken down into smaller ones, energy is released.

Semi-conservative replication

There are four requirements for semi-conservative replication to take place:

- The four types of nucleotide, each with their bases of adenine, guanine, cytosine, and thymine, must be present.
- Both strands of the DNA molecule must act as a template for the attachment of these nucleotides.
- The enzyme DNA polymerase is needed to catalyse the reaction.
- A source of chemical energy is required to drive the process.

The process of semi-conservative replication is illustrated in Figure 1. It takes place as follows:

- The enzyme **DNA helicase** breaks the hydrogen bonds linking the base pairs of DNA.
- As a result the double helix separates into its two strands and unwinds.
- Each exposed polynucleotide strand then acts as a template to which complementary nucleotides are attracted.
- Energy is used to activate these nucleotides.
- The activated nucleotides are joined together by the enzyme **DNA polymerase** to form the 'missing' polynucleotide strand on each of the two original polynucleotide strands of DNA.

Each of the new DNA molecules contains one of the original DNA strands, that is, half the original DNA has been saved and built into each of the new DNA molecules (Figure 2). The process is therefore termed semi-conservative replication.

Synoptic link

The structure of DNA is covered in Topic 7.1.

Study tip

Remember that DNA replication uses complementary base pairings to produce two identical copies.

a A representative portion of DNA, which is about to undergo replication.

b An enzyme, DNA helicase, causes the two strands of the DNA to separate.

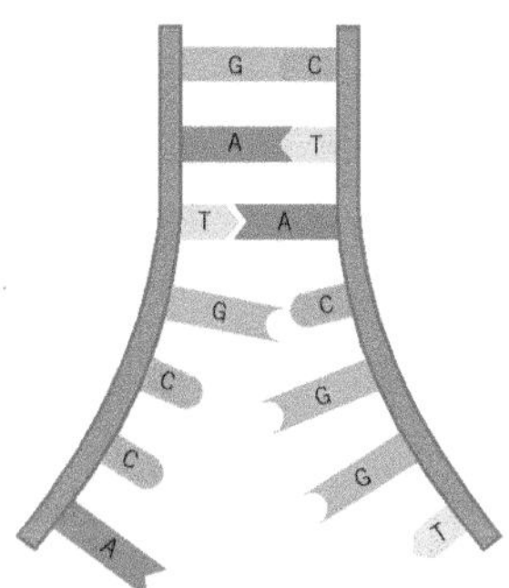

c DNA helicase completes the splitting of the two strands. Meanwhile, free nucleotides that have been activated are attracted to their complementary bases.

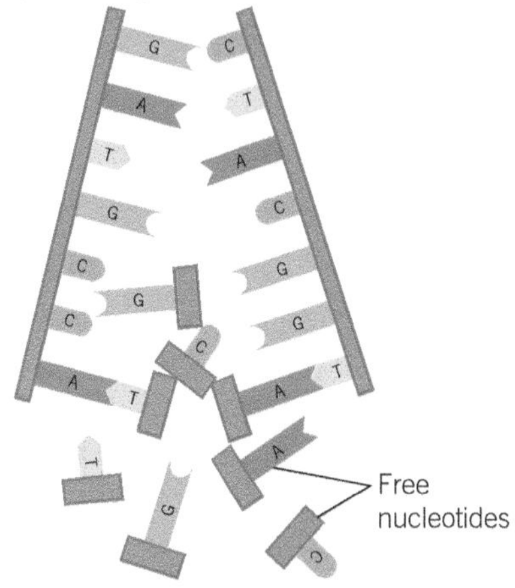

d Once the activated nucleotides are lined up, they are joined together by DNA polymerase (bottom three nucleotides). The remaining unpaired bases continue to attract their complementary nucleotides.

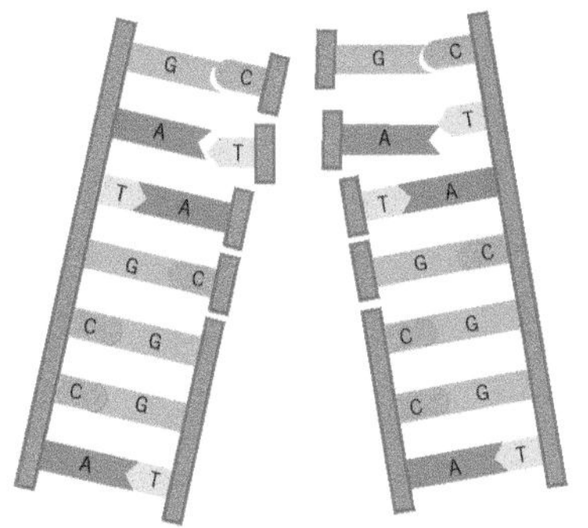

e Finally, all the nucleotides are joined using DNA polymerase to form a complete polynucleotide chain. In this way, two identical double strands of DNA are formed. As each double strand retains half of the original DNA material, this method of replication is called the semi-conservative method.

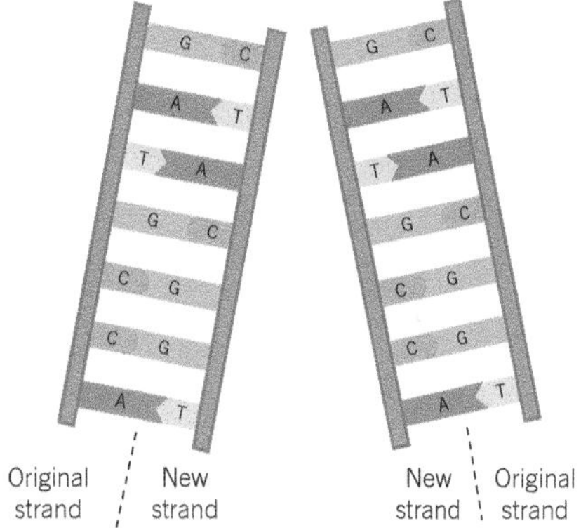

▲ **Figure 1** *The semi-conservative replication of DNA*

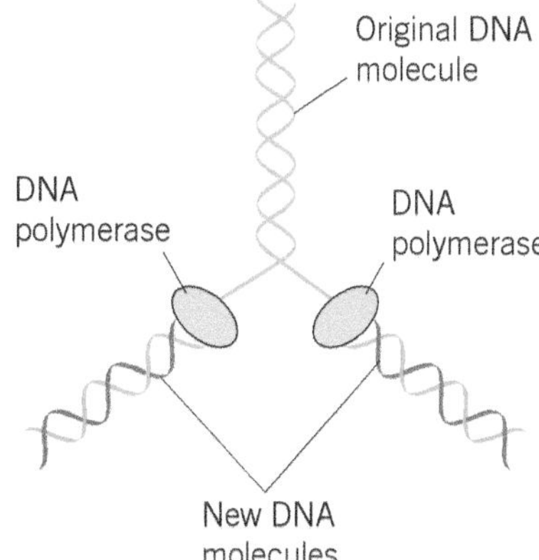

▲ **Figure 2** *Role of DNA polymerase in the semi-conservative replication of DNA*

Summary questions

1. If the bases on a portion of the original strand of DNA are ATGCTACG, what would the equivalent sequence of bases be on the newly formed strand?
2. Why is the process of DNA replication described as semi-conservative?
3. If an inhibitor of DNA polymerase were introduced into a cell, explain what the effect would be on DNA replication.

Practice questions: Chapter 7

1 Below is a short sequence of DNA bases:

TTTGTATACTAGTCTACTTCGTTAATA

(a) (i) What is the maximum number of amino acids for which this sequence of DNA bases could code? *(1 mark)*

(ii) The number of amino acids coded for could be fewer than your answer to part **(a)(i)**.
Give **one** reason why. *(1 mark)*

(b) Explain how a change in the DNA base sequence for a protein may result in a change in the structure of the protein. *(3 marks)*

(c) A piece of DNA consisted of 74 base pairs. The two strands of the DNA, strands **A** and **B**, were analysed to find the **number** of bases of each type that were present. Some of the results are shown in Table 1.

▼ **Table 1**

	Number of bases			
	C	G	A	T
Strand **A**	26			
Strand **B**	19		9	

Complete the table by writing in the missing values. *(2 marks)*

AQA June 2011

2 **Figure 1** shows a short section of a DNA molecule.

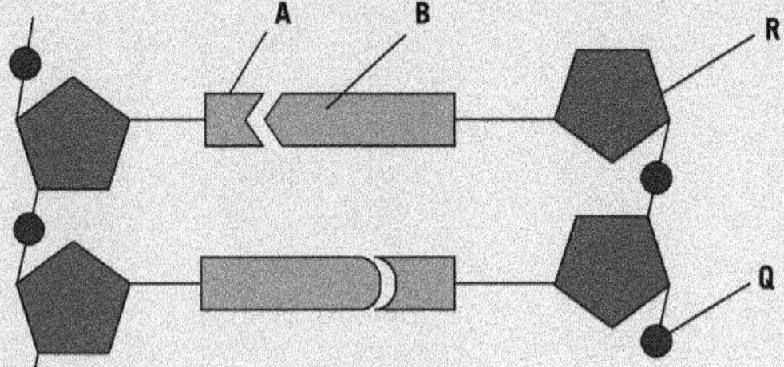

▲ **Figure 1**

(a) Name parts **R** and **Q**. *(2 marks)*

(b) Name the bonds that join **A** and **B**. *(1 mark)*

(c) Ribonuclease is an enzyme. It is 127 amino acids long.
What is the minimum number of DNA bases needed to code for ribonuclease? *(1 mark)*

(d) Shown below is the sequence of DNA bases coding for seven amino acids in the enzyme ribonuclease.

GTTTACTACTCTTCTTCTTTA

The number of each type of amino acid coded for by this sequence of DNA bases is shown in Table 2.

▼ **Table 2**

Amino acid	Number present
Arg	3
Met	2
Gln	1
Asn	1

Use Table 2 and sequence of bases to work out the sequence of amino acids in this part of the enzyme. Write your answer in the boxes below.

Gln						

(1 mark)

(e) Explain how a change in a sequence of DNA bases could result in a non-functional enzyme. *(3 marks)*

3 Figure 1 shows part of a DNA molecule.

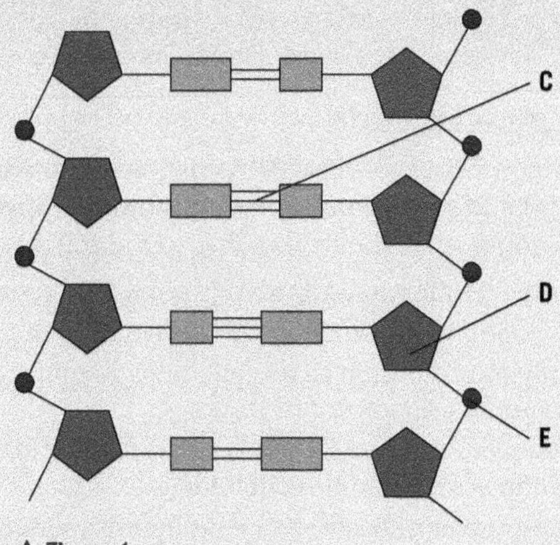

▲ Figure 1

(a) (i) DNA is a polymer. What is the evidence from the diagram that DNA is a polymer? *(1 mark)*

(ii) Name the parts of the diagram labelled **C**, **D**, and **E**. *(3 marks)*

(iii) In a piece of DNA, 34% of the bases were thymine. Complete Table 3 to show the names and percentages of the other bases. *(2 marks)*

▼ Table 3

Name of base	Percentage
Thymine	34
....	34
....	....
....	....

AQA June 2012

4 How does DNA replicate? *(6 marks)*

AQA Jan 2013

5 (a) DNA helicase is important in DNA replication. Explain why. *(2 marks)*

(b) Table 4 shows the types of DNA molecule that could be present in samples **1** to **3**. Use your knowledge of semi-conservative replication to complete the table with a tick if the DNA molecule is present in the sample. *(3 marks)*

▼ Table 4

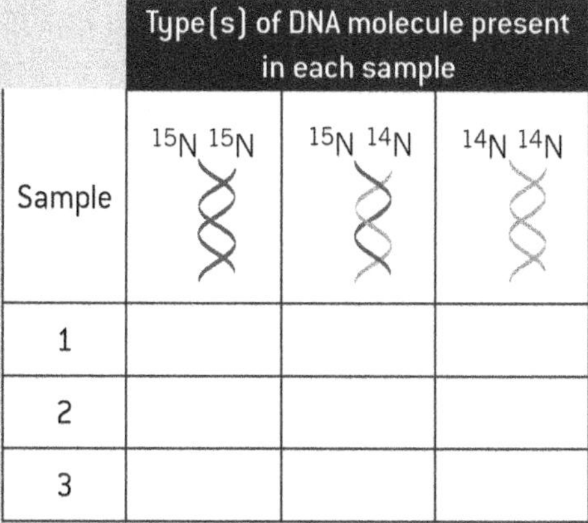

Sample	Type(s) of DNA molecule present in each sample		
	^{15}N ^{15}N	^{15}N ^{14}N	^{14}N ^{14}N
1			
2			
3			

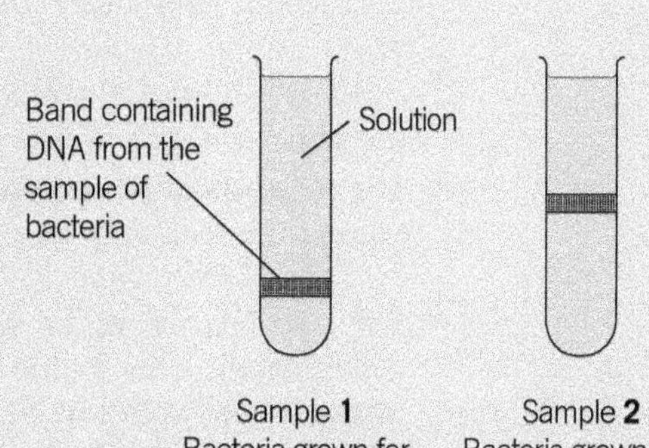

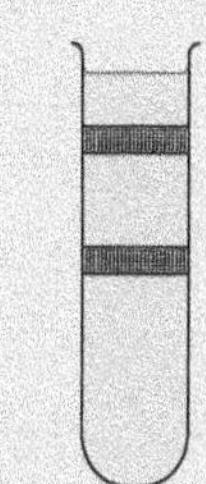

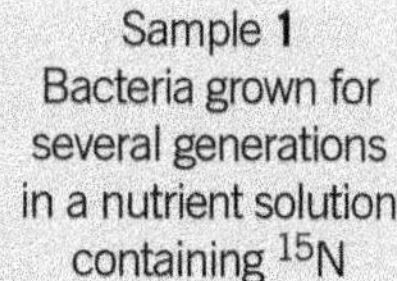

Sample **1**
Bacteria grown for several generations in a nutrient solution containing ^{15}N

Sample **2**
Bacteria grown in a nutrient solution containing ^{14}N for one cell division

Sample **3**
Bacteria grown in a nutrient solution containing ^{14}N for two cell divisions

▲ Figure 4

Answers to the Practice Questions are available at
www.oxfordsecondary.com/oxfordaqaexams-alevel-biology

8 Protein synthesis

8.1 The genetic code

Learning objectives:

→ Explain that the genetic code consists of three bases coding for one amino acid.

→ Explain why the triplet code is universal, non-overlapping, and degenerate.

Specification reference: 3.1.7.2 and 3.1.8.1

The genetic code

In trying to discover how DNA bases coded for amino acids, scientists suggested that there must be a minimum of three bases that coded for each amino acid. Their reasoning was as follows:

- Only 20 different amino acids regularly occur in proteins.
- Each amino acid must have its own code of bases on the DNA.
- Only four different bases (adenine, guanine, cytosine, and thymine) are present in DNA.
- If each base coded for a different amino acid, only four different amino acids could be coded for.
- Using a pair of bases, 16 (4^2) different codes are possible, which is still inadequate.
- Three bases produce 64 (4^3) different codes, more than enough to satisfy the requirements of 20 amino acids.

As the code has three bases for each amino acid, each sequence of three bases is called a triplet. As there are 64 possible triplets and only 20 amino acids, it follows that some amino acids are coded for by more than one triplet.

Features of the genetic code

Further experiments have revealed the following features of the genetic code:

- A few amino acids are each coded for by only a single triplet.
- The remaining amino acids are coded for by between two and six triplets each.
- The code is known as a '**degenerate code**' because most amino acids are coded for by more than one triplet.
- A triplet is always read in one particular direction along the DNA strand.
- The start of a DNA sequence that codes for a polypeptide is always the same triplet. This codes for the amino acid methionine. If this first methionine molecule does not form part of the final polypeptide, it is later removed.
- Three triplets do not code for any amino acid. These are called 'stop codons' and mark the end of a polypeptide chain. They act in much the same way as a full stop at the end of a sentence.
- The code is **non-overlapping**, in other words each base in the sequence is read only once. Thus six bases numbered 123456 are read as triplets 123 and 456, rather than as triplets 123, 234, 345, 456.
- The code is **universal** – with a few minor exceptions each triplet codes for the same amino acid in all organisms. This is indirect evidence for evolution.

Much of the DNA in eukaryotes does not code for polypeptides. For example, between genes there are non-coding sequences made up of multiple repeats of base sequences. Even within genes, only certain sequences code for amino acids. These coding sequences are called **exons**. Within the gene these exons are separated by further non-coding sequences called **introns**. Some genes code for ribosomal RNA (rRNA) and transfer RNAs (tRNAs).

Summary questions

1 Describe what a gene is.
2 Calculate the minimum number of bases required to code for a chain of six consecutive amino acids.
3 Explain how a change in one base along a DNA molecule may result in an enzyme becoming non-functional.
4 A section of DNA has the following sequence of bases along it:
TAC GCT CCG CTG TAC. All of the bases are part of the code for amino acids. The first base in the sequence is the start of the code.
  a Calculate the number of amino acids that the section of DNA codes for.
  b Determine which two sequences code for the same amino acid.
  c It is possible that this sequence codes for many different amino acids or many copies of the same amino acid. From your knowledge of the genetic code explain how this can happen.

Interpreting the genetic code

Table 1 is a genetic code table showing the amino acids that each codon (set of three nucleotides in mRNA) is translated into during protein synthesis. An amino acid is indicated by three letters of its name, for example Arg = **arg**inine, Ile = **isole**ucine. To find the code for any amino acid you find the relevant three letters (usually the first three) of its name in Table 1 and then read:

▼ **Table 1** *The genetic code. The base sequences shown are those on mRNA*

First position	Second position				Third position
	U	C	A	G	
U	Phe	Ser	Tyr	Cys	U
	Phe	Ser	Tyr	Cys	C
	Leu	Ser	Stop	Stop	A
	Leu	Ser	Stop	Trp	G
C	Leu	Pro	His	Arg	U
	Leu	Pro	His	Arg	C
	Leu	Pro	Gln	Arg	A
	Leu	Pro	Gln	Arg	G
A	Ile	Thr	Asn	Ser	U
	Ile	Thr	Asn	Ser	C
	Ile	Thr	Lys	Arg	A
	Met	Thr	Lys	Arg	G
G	Val	Ala	Asp	Gly	U
	Val	Ala	Asp	Gly	C
	Val	Ala	Glu	Gly	A
	Val	Ala	Glu	Gly	G

- the first base in the sequence from the column on the left
- the second base in the sequence from the row at the top
- the third base in the sequence from the column to the right.

You can also use the table to find an amino acid that is coded for by a particular codon. For example, UGC codes for the amino acid Cys (cysteine):

- the first letter (U) is in the column on the left
- the second letter (G) is in the row at the top
- the third letter (C) is in the column to the right.

You will notice that most amino acids have more than one codon, for example, alanine (Ala) has four codons, GCU, GCC, GCA, and GCG.

Using Table 1, answer the following questions. In each case identify amino acids by their three-letter codon(s).

1 List the two amino acids that have only one codon and state what it is in each case.
2 Name the amino acids that have each of the following codons:
  a CUC
  b AAA
  c GAU
3 For each of the following base sequences on a DNA molecule, deduce the sequence of amino acids in the order in which they would occur in the resultant polypeptide.
  a ATGCGTTAAGGCAGT
  b GCTAAGTTTCCAGAT

8.2 Polypeptide synthesis – transcription and splicing

Learning objectives:

- → Explain how pre-messenger RNA is produced from DNA in the process called transcription.
- → Describe how pre-messenger RNA is modified to form messenger RNA.

Specification reference: 3.1.8.2

Hint

The bakery analogy is to help understanding but this should not be used when you are writing a scientific explanation.

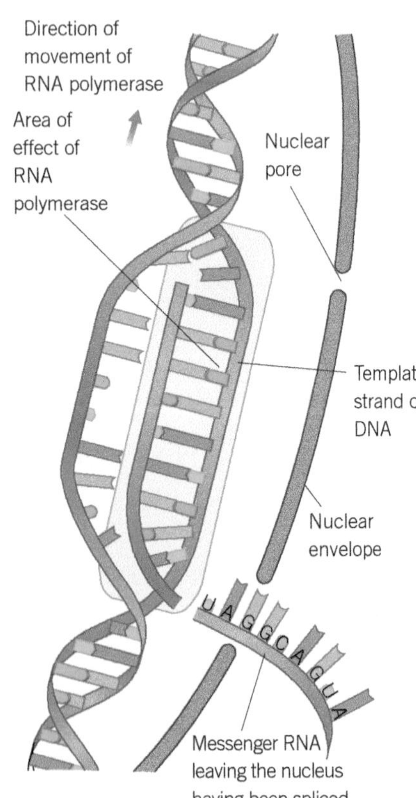

▲ **Figure 1** *Summary of transcription*

We saw in Topic 1.5 that proteins are made up of one or more polypeptides. Proteins, especially enzymes, are essential to all aspects of life. Every organism needs to make its own unique proteins. The biochemical machinery in the cytoplasm of each cell has the capacity to make every protein from just 20 amino acids. Exactly which proteins it manufactures depends upon the instructions that are provided, at any given time, by the DNA in the cell's nucleus. The basic process is as follows:

- DNA provides the instructions in the form of a long sequence of bases.
- A complementary section of part of this sequence is made in the form of a molecule called pre-mRNA – a process called **transcription**.
- The pre-mRNA is spliced to form mRNA.
- The mRNA is used as a template to which complementary tRNA molecules attach and the amino acids they carry are linked to form a polypeptide – a process called **translation**.

The process can be likened to a bakery, where the basic equipment and ovens (cell organelles) can manufacture any variety of cake (protein) from relatively few basic ingredients (amino acids). Which particular variety of cake is made depends on the recipe (genetic code) that the baker uses on any particular day. By choosing different recipes at different times, rather than making everything all the time, the baker can meet seasonal demands, adapt to changing customer needs, and avoid waste.

DNA replication can be likened to the publication of many copies of a recipe book (genome) – making a photocopy of a recipe to use in the bakery is therefore transcription. Making the cakes, using the photocopied recipe, is translation. If the book is not removed from the library, many copies of the recipe can be made, and the same cakes can be produced in many places at the same time or over many years.

Transcription in eukaryotic cells

Transcription is the process of making pre-mRNA using part of the DNA as a template. The process, which is illustrated in Figure 1, is as follows:

- An enzyme acts on a specific region of the DNA, causing the two strands to separate and expose the nucleotide bases in that region.
- The nucleotide bases on one of the two DNA strands, known as the **template strand**, pair with their complementary RNA nucleotides from the pool that is present in the nucleus. The enzyme **RNA polymerase** then moves along the strand and joins the nucleotides together to form a pre-mRNA molecule.

- In this way, an exposed guanine base on the DNA binds to the cytosine base of a free nucleotide. Similarly, cytosine links to guanine, and thymine joins to adenine. The exception is adenine on DNA, which links to uracil rather than thymine.
- As the RNA polymerase joins the nucleotides one at a time to build a strand of pre-mRNA, the DNA strands rejoin behind it. As a result, only about 12 base pairs on the DNA are exposed at any one time.
- When the RNA polymerase reaches a particular sequence of bases on the DNA that it recognises as a stop codon, it detaches, and the production of pre-mRNA is then complete.

Splicing of pre-mRNA

In prokaryotic cells, transcription results directly in the production of mRNA from DNA. In eukaryotic cells transcription results in the production of pre-mRNA, which is then spliced to form mRNA. The DNA of a gene sequence in eukaryotic cells is made up of sections called exons that code for proteins, and sections called introns that do not. These intervening introns would prevent the synthesis of a polypeptide. In the pre-mRNA of eukaryotic cells, the base sequences corresponding to the introns are removed and the functional exons are joined together during a process called **splicing**. As most prokaryotic cells do not have introns, splicing of their DNA is unnecessary. The process of splicing is shown in Figure 2.

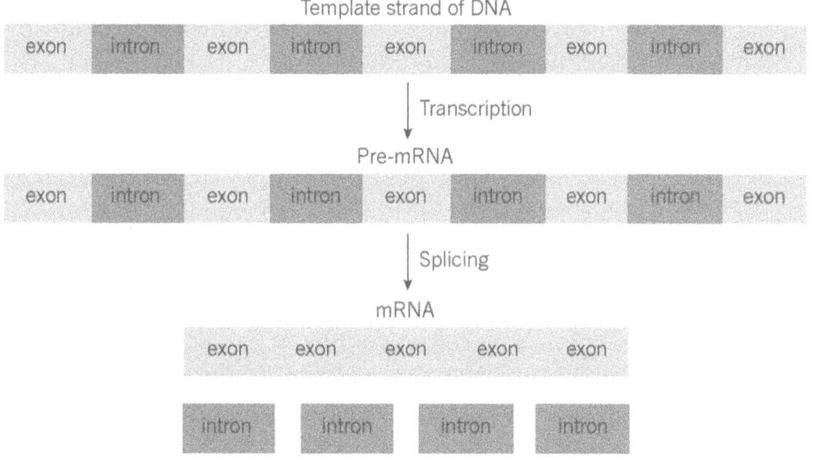

▲ **Figure 2** *Splicing of pre-mRNA to form mRNA*

The mRNA molecules are too large to diffuse out of the nucleus and so, once they have been spliced, they leave via a nuclear pore. Outside the nucleus, the mRNA is attracted to the ribosomes to which it becomes attached, ready for the next stage of the process – translation.

Summary questions

1. Describe the role of RNA polymerase in transcription.
2. State which other enzyme is involved in transcription and describe its role.
3. Explain why splicing of pre-mRNA is necessary.
4. A sequence of bases along the template strand of DNA is ATGCAAGTCCAG.
   a. Deduce the sequence of bases on a pre-messenger RNA molecule that has been transcribed from this part of the DNA molecule.
   b. Calculate how many amino acids the sequence potentially codes for.
5. A gene is made up of 756 base pairs. The mRNA that is transcribed from this gene is only 524 nucleotides long. Explain why there is this difference.

8.3 Polypeptide synthesis – translation

Learning objectives:

- → Explain how a polypeptide is synthesised during the process of translation.
- → Describe the roles of messenger RNA and transfer RNA in translation.

Specification reference: 3.1.7.1 and 3.1.8.2

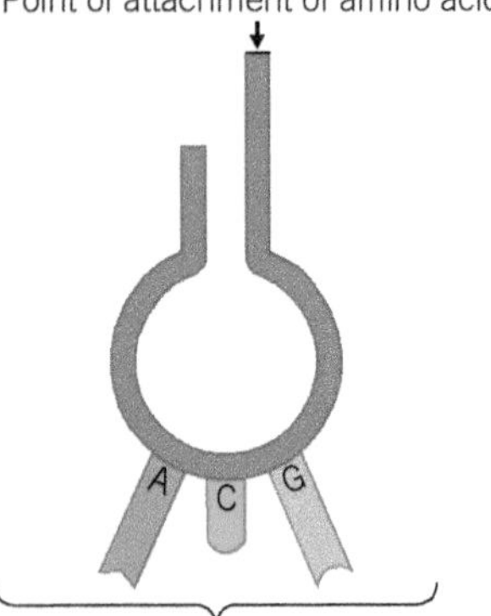

Anticodon – this sequence of ACG means that the amino acid cysteine will attach to the other end of this tRNA molecule. This anticodon will combine with the codon UGC on an mRNA molecule during the formation of a polypeptide. The mRNA codon UGC therefore translates into the amino acid cysteine.

▲ **Figure 1** *Simplified structure of one type of tRNA*

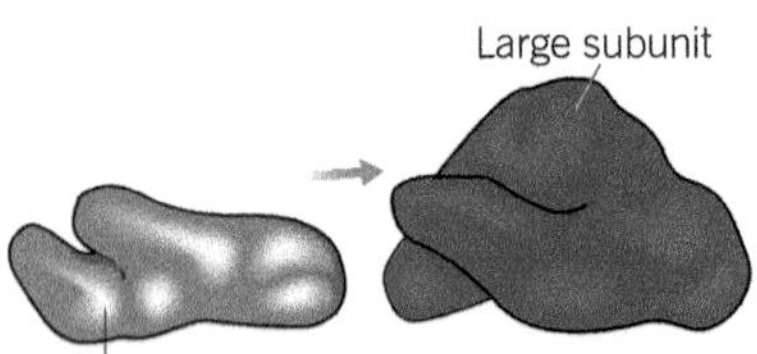

▲ **Figure 2** *Structure of a ribosome. The smaller subunit fits into a depression on the surface of the larger one.*

Hint

Remember that there is no thymine in any RNA molecule. It is uracil in RNA that pairs with adenine.

In Topic 8.2 you looked at how the triplet code of DNA is transcribed into a sequence of **codons** (genetic code) on messenger RNA (mRNA). The next stage is to translate the codons on the mRNA into a sequence of amino acids that make up a polypeptide.

There are about 60 different transfer RNAs (tRNAs). A particular tRNA has a specific **anticodon** and attaches to a specific amino acid (see Figure 1). Each amino acid therefore has one or more tRNA molecule, each with its own anticodon of bases.

Synthesising a polypeptide

Once mRNA has passed out of the nuclear pore it determines the synthesis of a polypeptide. The following explanation of how a polypeptide is made is illustrated in Figures 3 and 4. (You don't need to learn the codons or amino acids given in brackets below, this information is only to help you follow the process.)

- A ribosome (Figure 4, part 1) becomes attached to the starting codon (AUG) at one end of the mRNA molecule.
- The tRNA molecule with the complementary anticodon sequence (UAC) moves to the ribosome and pairs up with the codon on the mRNA. This tRNA carries a specific amino acid (methionine).
- A tRNA molecule with a complementary anticodon (UGC) pairs with the next codon on the mRNA (ACG). This tRNA molecule carries another amino acid (threonine).
- The two amino acids (methonine and threonine) on the tRNA are joined by a **peptide bond** using an enzyme and **ATP**, which is hydrolysed to provide the required energy.
- The ribosome moves on to the third codon (GAU) in the sequence on the mRNA, thereby linking the amino acids (threonine and aspartic acid) on the second and third tRNA molecules (Figure 4, part 2).
- As this happens, the first tRNA is released from its amino acid (methionine) and is free to collect another of its specific amino acids (methionine) from the amino acid pool in the cell.
- The process continues in this way, with up to 15 amino acids being added each second, until a polypeptide chain is built up (Figure 4, part 3).
- Up to 50 ribosomes can pass immediately behind the first, so that many identical polypeptides can be assembled simultaneously (Figure 3).
- The synthesis of a polypeptide continues until a ribosome reaches a stop codon. At this point, the ribosome, mRNA, and the last tRNA molecule all separate and the polypeptide chain is complete.

In summary, the DNA sequence of triplets that make up a gene determines the sequence of codons on mRNA. The sequence of codons on mRNA determine the order in which the tRNA molecules line up.

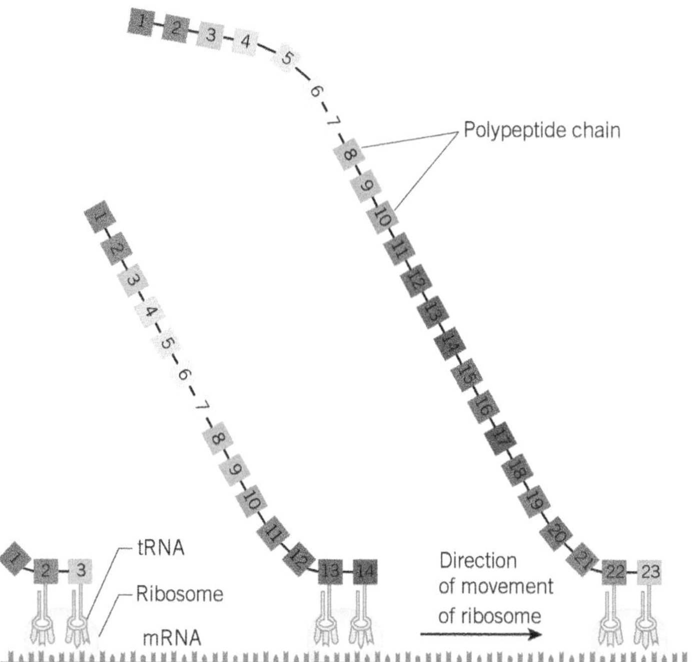

▲ **Figure 3** *Polypeptide formation*

They, in turn, determine the sequence of amino acids in the polypeptide. In this way, genes precisely determine which proteins a cell manufactures. As many of these proteins are enzymes, genes effectively control the activities of cells.

Summary questions

1. Name the cell organelle that is involved in translation.
2. A codon found on a section of mRNA has the sequence of bases AUC. List the sequence of bases found on:
   a. the tRNA anticodon that attaches to this codon
   b. the template strand of DNA that formed the mRNA codon.
3. Describe the role of tRNA in the process of translation.
4. A strand of mRNA has 64 codons but the protein produced from it has only 63 amino acids. Suggest a reason for this difference.

Study tip

ATP has two roles in translation. It is required to provide energy to attach amino acids to tRNA and also to attach amino acids together.

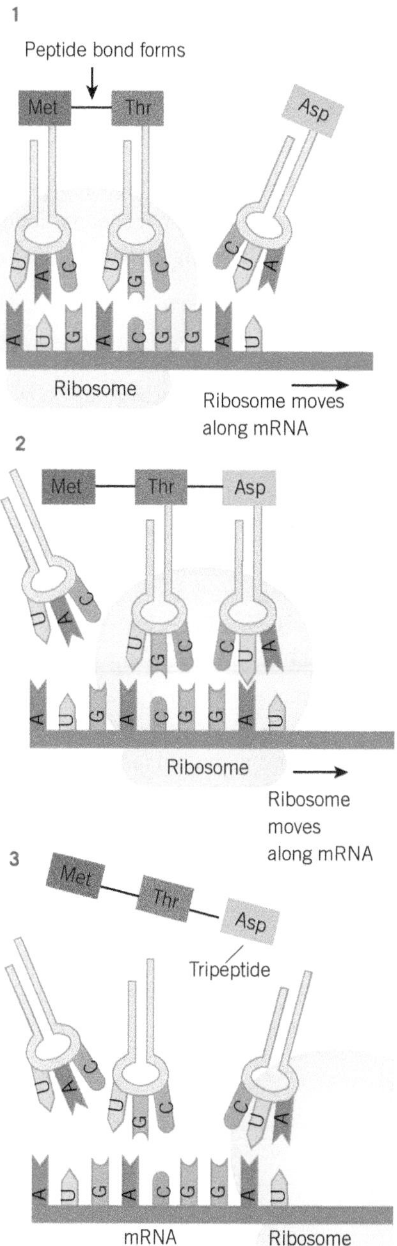

▲ **Figure 4** *Translation*

Protein synthesis

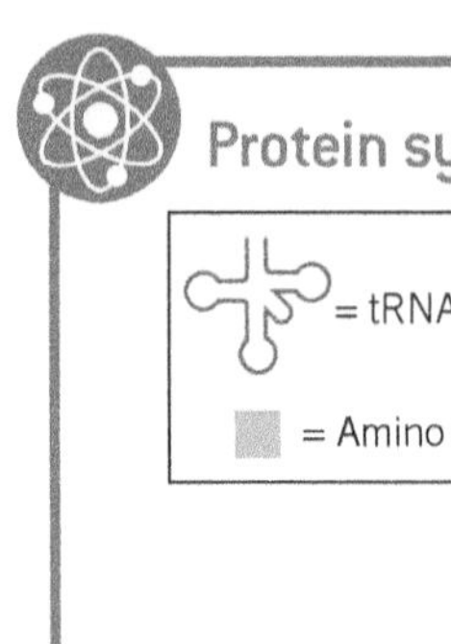

▲ **Figure 5**

Figure 5 shows the formation of part of a polypeptide along a section of eight codons. Codons 4 and 5 have been left blank. Using Figure 5 and Table 1, answer the following questions.

1. Name the structures **X** and **Y**.
2. State the chemical group shown on the end of the polypeptide chain.
3. Determine the anticodon sequence on tRNA molecule 4.
4. Deduce the sequence of the first five amino acids in the polypeptide.
5. Determine the sequence of bases on the portion of DNA from which codons 1 – 3 are transcribed.
6. A DNA mutation results in the base cytosine being replaced by uracil in codon 8. Explain the significance of this change.
7. Another mutant form of a gene causes the inversion (reversal) of the code for the amino acid glutamine (Glu).
   a. Consider all possible outcomes from this change and explain the effect on the polypeptide in each case.
   b. If the polypeptide formed from this mutant gene forms part of an enzyme, suggest **two** reasons why it might fail to function. Explain your answer.

▼ **Table 1** *The base sequences shown are those on mRNA*

First position	Second position				Third position
	U	C	A	G	
U	Phe	Ser	Tyr	Cys	U
	Phe	Ser	Tyr	Cys	C
	Leu	Ser	Stop	Stop	A
	Leu	Ser	Stop	Trp	G
C	Leu	Pro	His	Arg	U
	Leu	Pro	His	Arg	C
	Leu	Pro	Gln	Arg	A
	Leu	Pro	Gln	Arg	G
A	Ile	Thr	Asn	Ser	U
	Ile	Thr	Asn	Ser	C
	Ile	Thr	Lys	Arg	A
	Met	Thr	Lys	Arg	G
G	Val	Ala	Asp	Gly	U
	Val	Ala	Asp	Gly	C
	Val	Ala	Glu	Gly	A
	Val	Ala	Glu	Gly	G

8.4 Protein folding

Sometimes a single polypeptide chain is a functional protein. Often, a number of polypeptides are linked together to give a functional protein (see Topics 1.5 and 5.7). What happens to the newly synthesised polypeptide next depends on the protein being made.

Learning objectives:

- → Explain how polypeptide chains are folded into a tertiary structure.
- → Explain the role of chaperone proteins.

Specification reference: 3.1.8.3

Assembling a protein

The primary structure of a polypeptide is coiled or folded to produce a secondary structure. Then, interactions between R-groups further coil and fold the polypeptide to produce the final tertiary structure that gives the protein its characteristic 3-D shape. Different polypeptide chains, along with non-protein groups, can be linked to form functional proteins, such as haemoglobin, that have a quaternary structure.

The position of the R groups in the polypeptide determines which folded protein structure is possible. For example, the amino acid cysteine has an R group consisting of a sulfhydryl (—SH) group. This can form a disulfide bond with another cysteine molecule at a different position in the polypeptide chain. The formation of a disulfide bond would fold the chain into a particular shape. Similarly, some R groups are acidic whereas others are basic, allowing ionic bonds to form between these groups, which again will coil or fold the chain. However, in any given polypeptide chain, there are still many thousands of possible folding formations. Furthermore, as the polypeptide chain is formed on the ribosome, those regions where the amino acids have **hydrophobic** R groups will tend to aggregate together or join with other hydrophobic molecules spontaneously. This could lead to a 'mis-folded' protein (Figure 1), which could mean that, even if other bonds such as disulfide or ionic bonds do form, the final shape will not be correct. Not only will this mis-folded protein not function, but the build-up of mis-folded protein aggregates in cells could cause disease, as occurs with Alzheimer's disease.

Study tip

Remember there are 20 different amino acids found in proteins, which differ only in their R groups. Some amino acids have neutral, non-polar R groups that would be hydrophobic. Like the hydrophobic regions of the phospholipids you studied in Topics 1.4 and 4.1, hydrophobic parts of a molecule will interact with each other.

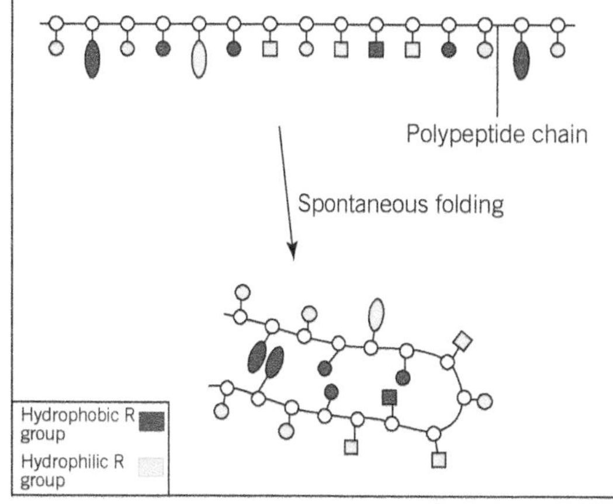

▲ **Figure 1** *The interaction of the hydrophobic R groups in the growing polypeptide chain forms a fold that could prevent the protein achieving its correct final 3-D shape*

Chaperone proteins

Cells produce chaperone proteins which ensure that the polypeptide chains synthesised on ribosomes fold correctly. One example is a group of proteins called HSP 70. These bind to the hydrophobic regions of polypeptides as they are being formed and prevent incorrect hydrophobic interactions occurring before the polypeptide chain can be completed.

Other polypeptides need a different type of protein to ensure that they fold correctly. HSP 60 proteins are known as **chaperonins**. They are large, cylindrical proteins with a central compartment. The polypeptide chain fits into this compartment, where it is isolated from other protein molecules and so is prevented from interacting with them. Interactions between the chaperonin molecule and the polypeptide then enable the protein to fold correctly.

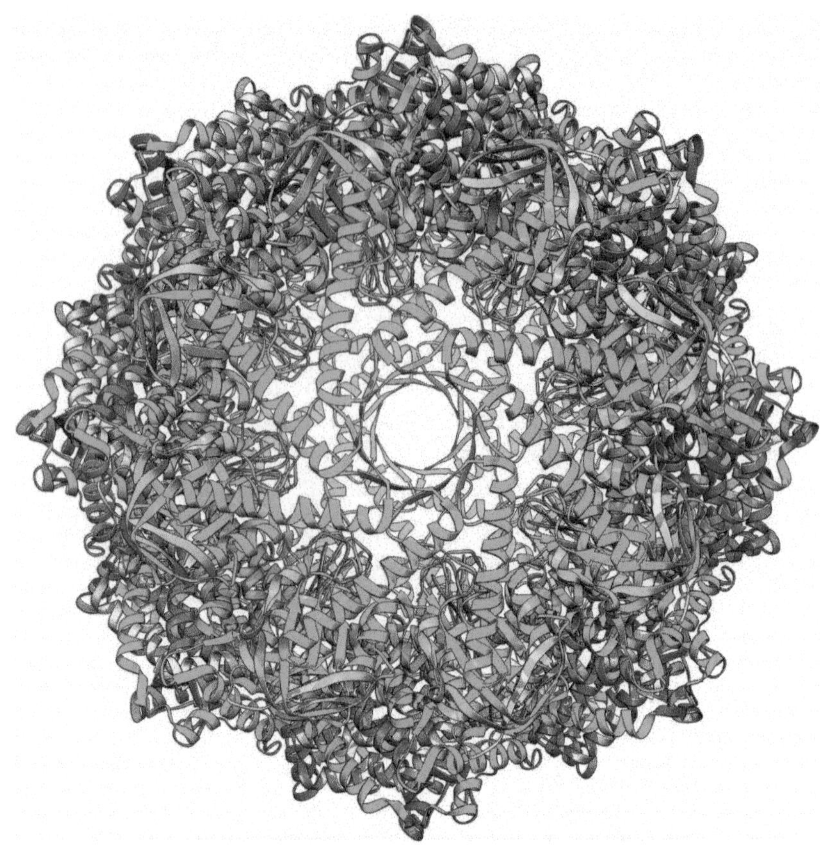

▲ **Figure 2** *Molecular model of a chaperonin protein. The clear area in the centre of the molecule is referred to as an Anfinsen cage and is named after Christian Anfinsen, who was awarded the Nobel Prize for Chemistry in 1972 for this discovery*

Some chaperone proteins are found in the endoplasmic reticulum, where they facilitate the folding and assembly of membrane and secretory proteins.

Summary questions

1. Why is it essential that proteins such as enzymes are folded correctly?
2. Suggest why incorrectly folded proteins might build up in cells?
3. The letters HSP stand for 'Heat Shock Protein'. Suggest why proteins such as HSP 60 and HSP 70 might also be important in cells that have been subjected to a rise in temperature.

Practice questions: Chapter 8

1 Figure 1 shows part of a pre-mRNA molecule.

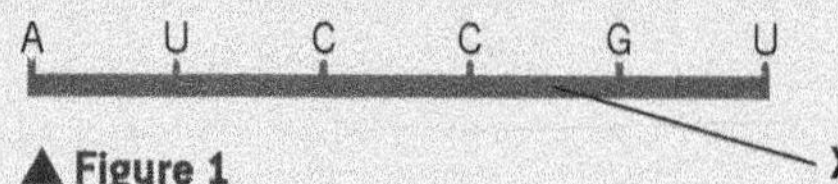

▲ **Figure 1**

(a) (i) Name the two substances that make up part **X**. *(1 mark)*

(ii) Give the sequence of bases on the DNA strand from which this pre-mRNA has been transcribed. *(1 mark)*

(b) (i) Give **one** way in which the structure of an mRNA molecule is different from the structure of a tRNA molecule. *(1 mark)*

(ii) Explain the difference between pre-mRNA and mRNA. *(1 mark)*

(c) Table 1 shows the percentage of different bases in two pre-mRNA molecules.

The molecules were transcribed from the DNA in different parts of a chromosome.

▼ **Table 1**

Part of chromosome	Percentage of base			
	A	G	C	U
Middle	38	20	24	
End	31	22	26	

(i) Complete the table by writing the percentage of uracil (U) in the appropriate boxes. *(1 mark)*

(ii) Explain why the percentages of bases from the middle part of the chromosome and the end part are different. *(2 marks)*

AQA June 2011

2 (a) What name is used for the non-coding sections of a gene? *(1 mark)*

Figure 2 shows a DNA base sequence. It also shows the effect of two mutations on this base sequence. Table 2 shows DNA triplets that code for different amino acids.

Original DNA base sequence	A	T	T	G	G	C	G	T	G	T	C	T
Amino acid sequence												
Mutation **1** DNA base sequence	A	T	T	G	G	A	G	T	G	T	C	T
Mutation **2** DNA base sequence	A	T	T	G	G	C	C	T	G	T	C	T

▲ **Figure 2**

▼ **Table 2**

DNA triplets	Amino acid
GGT, GGC, GGA, GGG	Gly
GTT, GTA, GTG, GTC	Val
ATC, ATT, ATA	Ile
TCC, TCT, TCA, TCG	Ser
CTC, CTT, CTA, CTG	Leu

(b) Complete Figure 1 to show the sequence of amino acids coded for by the original DNA base sequence. *(1 mark)*

(c) Some gene mutations affect the amino acid sequence. Some mutations do not. Use the information from Figure 1 and Table 2 to explain:

(i) whether mutation 1 affects the amino acid sequence *(2 marks)*

(ii) how mutation 2 could lead to the formation of a non-functional enzyme. *(3 marks)*

(d) Gene mutations occur spontaneously.

(i) During which part of the cell cycle are gene mutations most likely to occur? *(1 mark)*

(ii) Suggest an explanation for your answer. *(1 mark)*

AQA June 2010

3 Figure 3 represents one process that occurs during protein synthesis.

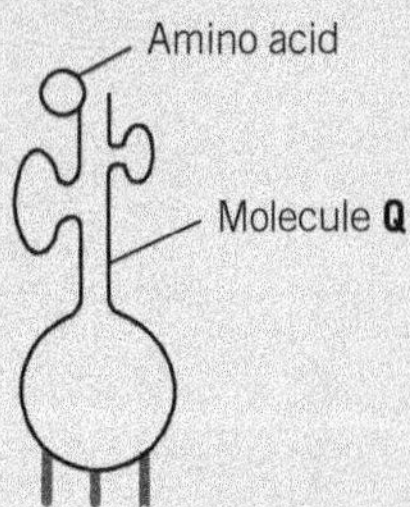

A U G C C G U A C C G A C U

(a) Name the process shown. *(1 mark)*

(b) Identify the molecule labelled **Q**. *(1 mark)*

(c) In Figure 3, the first codon is AUG. Give the base sequence of the complementary DNA base sequence and the missing anticodon. *(2 marks)*

Table 3 shows the base triplets that code for two amino acids.

▼ Table 3

Amino acid	Encoding base triplet
Aspartic acid	GAC, GAU
Proline	CCA, CCG, CCC, CCU

(d) Describe how two amino acids differ from one another. You may use a diagram to help your description. *(1 mark)*

(e) Deletion of the sixth base (G) in the sequence shown in Figure 3 would change the nature of the protein produced but substitution of the same base would not. Use the information in Table 3 and your own knowledge to explain why. *(3 marks)*

AQA SAMS PAPER 2

4 Read the following passage:

The sequence of bases in a molecule of DNA codes for proteins. Different sequences of bases code for different proteins. The genetic code, however, is degenerate. Although the base sequence for AGT codes for serine, other sequences may also code for this same amino acid. There are four base sequences which code for the amino acid glycine. These are CCA, CCC, CCG, and CCT. There are also four base sequences coding for the amino acid proline. These are GGA, GGC, GGG, and GGT.

Pieces of DNA which have a sequence where the same base is repeated many times are called slippery. When slippery DNA is copied during replication, errors may occur in copying.

Individual bases may be copied more than once. This may give rise to differences in the protein which is produced by the piece of DNA containing the errors.

Use information in the passage and your own knowledge to answer the following questions.

(a) Different sequences of bases code for different proteins (lines 1–2). Explain how. *(2 marks)*

(b) The base sequence AGT codes for serine (lines 2–3). Give the mRNA codon transcribed from this base sequence. *(2 marks)*

(c) Glycine-proline-proline is a series of amino acids found in a particular protein. Give the sequence of DNA bases for these three amino acids which contains the longest slippery sequence. *(2 marks)*

(d) Explain how copying bases more than once may give rise to differences in the protein (lines 9–10). *(2 marks)*

(e) Starting with mRNA in the nucleus of a cell, describe how a molecule of protein is synthesised. *(6 marks)*

AQA 2005

Answers to the Practice Questions are available at
www.oxfordsecondary.com/oxfordaqaexams-alevel-biology

9 Genetic diversity, species, and taxonomy

9.1 Meiosis and genetic variation

Learning objectives:

- → Explain the importance of meiosis.
- → Explain how meiosis leads to genetic variation.

Specification reference: 3.1.9.1

The division of cells involves, firstly, the division of the nucleus and, secondly, the division of the cell as a whole. The division of the nucleus of cells occurs in one of two ways:

- **Mitosis** produces two daughter nuclei with the same number of chromosomes as the parent cell and as each other. We shall learn more about mitosis in Topic 17.1.
- Meiosis produces four daughter nuclei, each with half the number of chromosomes as the parent cell.

Why has meiosis evolved?

In sexual reproduction, two gametes fuse to give rise to new offspring. If each gamete has a full set of chromosomes (diploid number), then the fused cell that they produce has double this number. In humans, the diploid number of chromosomes is 46, which means that this cell would have 92 chromosomes. This doubling of the number of chromosomes would continue at each generation. It follows that, in order to maintain a constant number of chromosomes in the adults of a species, the number of chromosomes must be halved at some stage in the life cycle. This halving occurs as a result of meiosis.

Every diploid cell of an organism has two sets of chromosomes (see Topic 7.2), with one set provided by each parent. During meiosis, the chromosome pairs separate, so that only one chromosome from each pair enters each cell. This is known as the **haploid** number of chromosomes, which, in humans, is 23. When two haploid gametes fuse at fertilisation, the diploid number of chromosomes is restored.

The process of meiosis

Meiosis involves two nuclear divisions that normally occur one after the other:

1. In the **first meiotic division (meiosis 1)** the homologous chromosomes pair up and their chromatids wrap around each other. Equivalent portions of these chromatids may be exchanged in a process called **crossing over**. The significance of this is discussed later. By the end of this stage the homologous pairs have separated, with one chromosome from each pair going into one of the two daughter cells.
2. In the **second meiotic division (meiosis 2)** the chromatids move apart. At the end of meiosis 2, four cells have been formed. In humans, each of these cells contains 23 chromatids.

This is summarised in Figure 1.

In addition to halving the number of chromosomes, meiosis also produces genetic variation among the offspring, allowing species to adapt and survive in a changing world. Meiosis brings about this genetic variation in the following two ways:

- independent segregation of homologous chromosomes
- recombination of homologous chromosomes by crossing over.

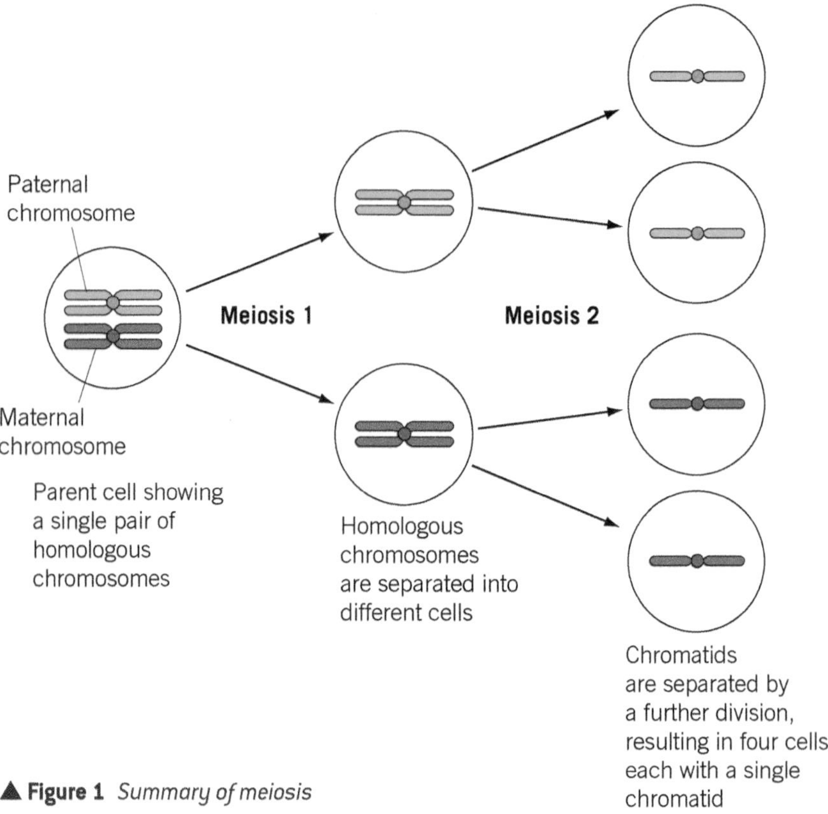

▲ **Figure 1** *Summary of meiosis*

Before we look at these two processes in more detail, let us remind ourselves of the meaning of three important terms:

- **gene** – a section of DNA that codes for a polypeptide
- **locus** – the position of a gene on a chromosome or DNA molecule
- **allele** – one of the different forms of a particular gene.

Independent segregation of homologous chromosomes

During meiosis 1, each chromosome lines up alongside its homologous partner (see Figure 2). In humans, for example, this means that there will be 23 pairs of homologous chromosomes lying side by side. When these homologous pairs arrange themselves in this line they do so randomly. One of each pair will pass to each daughter cell. Which one of the pair goes into the daughter cell, and with which one of any of the other pairs, depends on how the pairs are lined up in the parent cell. Since the pairs are lined up at random, the combination of chromosomes that goes into the daughter cell at meiosis 1 is also random. This is called **independent segregation**.

Variety from new genetic combinations

Each member of a homologous pair of chromosomes has exactly the same genes and therefore determines the same characteristics (e.g., eye colour and blood group). However, the alleles of these genes may differ (e.g., they may code for brown or blue eyes, or blood group A or B). The random distribution, and consequent independent assortment, of these chromosomes therefore produces new or different combinations of alleles. An example is shown in Figure 2. This diagram focuses on just two homologous pairs. The stages shown on the figure are:

Hint

Imagine your chromosomes as two packs of 23 cards, red and blue, in which the cards are labelled from A to W. You were given the red pack by your mother and the blue pack by your father. Independent segregation is like dealing a card, of each letter in turn, at random from either of these two packs. Your final hand of 23 cards could contain any proportion of red and blue cards. In fact there are 2^{23} (over 8 million) different possible combinations.

- **Stage 1**. One pair of homologous chromosomes carries the gene for eye colour – with one of the pair carrying the allele for brown eyes and the other carrying the allele for blue eyes. The other pair of chromosomes carries the gene for blood group. One of these homologous chromosomes carries the allele for blood group A, and the other carries the allele for blood group B. There are two possible arrangements, P and Q, of the two chromosomes at the start of meiosis. Both are equally probable, but each produces a different outcome in terms of the characteristics that may be passed on via the gametes.
- **Stage 2**. At the end of meiosis 1, the homologous chromosomes have segregated into two separate cells.
- **Stage 3**. At the end of meiosis 2, the chromosomes have segregated into chromatids, producing four gametes for each arrangement. The actual gametes are different, depending on the original arrangement (P or Q) of the chromosomes at stage 1.

Hint

Imagine your packs of blue and red cards again. Recombination is like taking a red card and a blue card, each with the same letter, and tearing an identical portion from each card and attaching it to the other card, so that you have new cards that are part red and part blue. You can do this in an almost infinite number of ways and all before you start to deal them as before!

Arrangement P produces the following types of gamete:

- Brown eyes and blood group B.
- Blue eyes and blood group A.

Arrangement Q produces the following types of gamete:

- Blue eyes and blood group B.
- Brown eyes and blood group A.

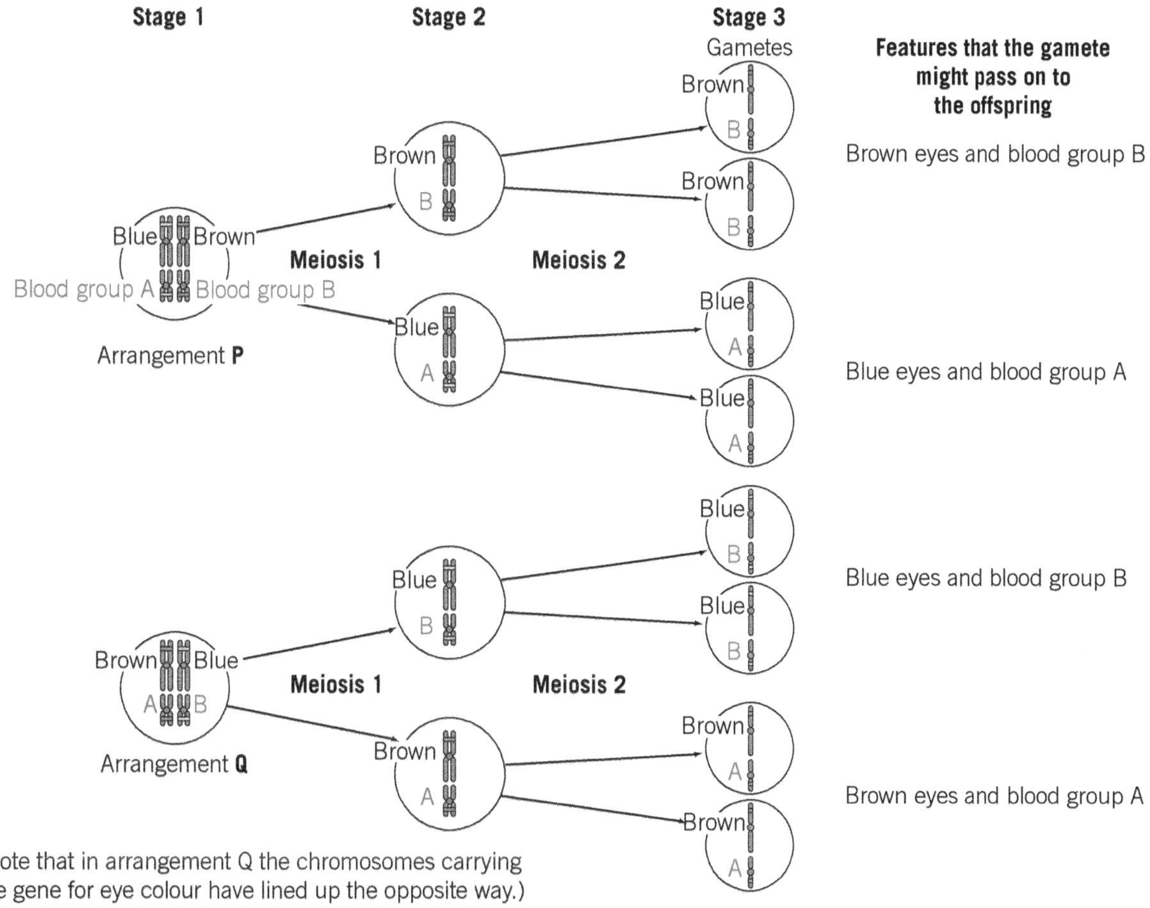

(Note that in arrangement Q the chromosomes carrying the gene for eye colour have lined up the opposite way.)

▲ **Figure 2** *Genetic variation produced as a result of independent segregation of chromosomes during meiosis. This diagram illustrates the independent segregation of alleles of genes responsible for eye colour and blood group, which are carried on separate chromosomes.*

Where the cells produced in meiosis are gametes these will be genetically different as a result of the different combinations of the maternal and paternal chromosomes that they contain. These haploid gametes fuse randomly at fertilisation. The haploid gametes produced by meiosis must fuse to restore the diploid state. Each gamete has a different genetic make-up and their random fusion therefore produces variety in the offspring. Where the gametes come from different parents (as is usually the case), two different genetic make-ups are combined and even more variety results. Sexual reproduction produces a **lot** of genetic variation – because two individuals contribute to the offspring. The evolution of meiosis allowed for successful sexual reproduction in eukaryotes. Meiosis as a process also produces some genetic variation.

Genetic recombination by crossing over

You saw above that, during meiosis 1, each chromosome lines up alongside its homologous partner. The following events then take place:

- The chromatids of each pair become twisted around one another.
- During this twisting process tensions are created and portions of the chromatids break off.
- These broken portions may then rejoin with the chromatids of their homologous partner.
- Usually it is the equivalent portions of homologous chromosomes that are exchanged.
- In this way, new genetic combinations are produced (Figure 3).

The chromatids cross over one another many times and so the process is known as **crossing over**. The broken-off portions of chromatid recombine with another chromatid, so this process is called **recombination**.

The effect of this recombination by crossing over on the cells produced at the end of meiosis is illustrated in Figure 4. Compare the four cells that result with those shown in Figure 1. If there is no recombination by crossing over only two different types of cell are produced. However, if recombination does occur, four different cell types are produced, each (probably) with different allele combinations. Crossing over therefore increases genetic variety even further.

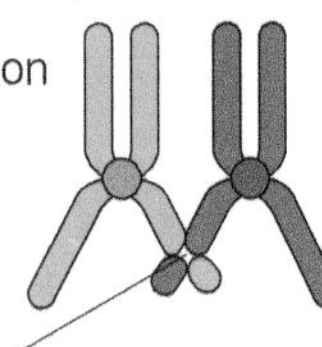

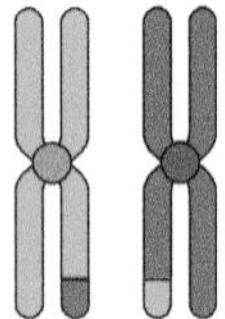

▲ **Figure 3** *Crossing over*

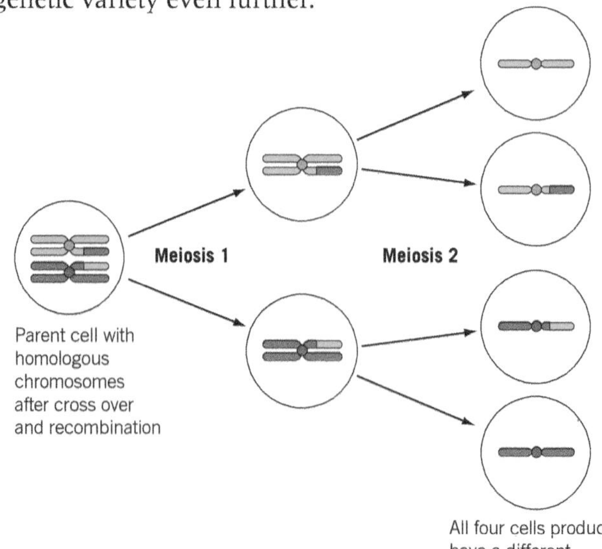

▲ **Figure 4** *Genetic variation as a result of recombination by crossing over*

Summary questions

1. A cell is examined and found to have 27 chromosomes. Is it likely to be haploid or diploid? Explain your answer.
2. In which **two** ways does meiosis lead to an increase in genetic variety?
3. Study Figure 2. Imagine that both alleles of the gene on the smaller pair of homologous chromosomes are for blood group A (rather than blood groups A and B). List all the different combinations of alleles in the gametes.
4. A mule is a cross between a horse (64 chromosomes) and a donkey (62 chromosomes). Mules therefore have 63 chromosomes. From your knowledge of meiosis, suggest why mules cannot produce gametes and are therefore sterile.

9.2 Classification

Scientists have identified and named around 1.8 million different living organisms. No one knows how many types remain to be identified. Estimates for the total number of species on Earth vary from 10 million to 100 million. The figure is likely to be around 14 million. These represent only the species that exist today. Some scientists have estimated that 99 per cent of the species that have existed on Earth are now extinct, and almost all of them have left no fossil record. With such a vast number of organisms it is clearly important for scientists to name them and sort them into groups.

Classification is the organisation of living organisms into groups. This process is not random but is based on a number of accepted principles. Before we examine how organisms are grouped according to these principles, let us look at how scientists distinguish one type of organism from another.

Learning objectives:

- → Explain what is meant by a species and how they are named.
- → Explain how organisms are classified.
- → Understand how classification relates to evolution.

Specification reference: 3.1.10.1 and 3.1.10.2

The concept of a species

A species is the basic unit of classification. A definition of a species is not easy, but members of a single species have certain things in common:

- **They are similar to one another but different from members of other species**. They have very similar genes and therefore closely resemble one another physically and biochemically. They have similar patterns of development and similar immunological features and they occupy the same **ecological niche**.
- **They are capable of breeding to produce living, fertile offspring**. They are therefore able to successfully produce more offspring. This means that, when a species reproduces sexually, any of the genes of its individuals can, in theory, be combined with any other, that is, they belong to the same **gene pool**.

Naming species – the binomial system

Historically, scientists often gave new organisms a name that described their features, for example, blackbird, rainbow trout. This practice resulted in the same names being used in different parts of the world for very different species. Therefore, it was difficult for scientists to be sure they were referring to the same organism. Over 200 years ago the Swedish botanist Linnaeus overcame this problem by devising a common system of naming organisms. This system is still in use today.

Organisms are identified by two names and hence the system is called the **binomial system**. Its features are as follows:

- It is a universal system based upon Latin or Greek names.
- The first name, called the **generic name**, denotes the genus to which the organism belongs. This is equivalent to the surname that is used to identify people and is shared by their close relatives.

Study tip

Remember that members of the same species are capable of breeding to produce *fertile* offspring, not *viable* offspring.

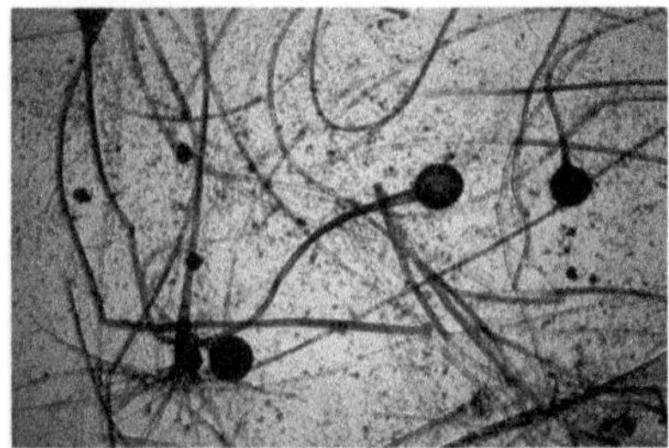

▲ **Figure 1** *(From top to bottom) the fungus* Mucor mucedo *(bread mould); the plant* Lathyrus odoratus *(sweet pea); the animal* Panthera tigris *(tiger). The classification of these organisms is shown in Table 1 on the next page.*

- The second name, called the **specific name**, denotes the species to which the organism belongs. This is equivalent to the first (or given) name used to identify people. However, unlike in humans, it is never shared by other species within the genus.

There are a number of rules that are applied to the use of the binomial system in scientific writing:

- The names are printed in italics or, if handwritten, they are underlined to indicate that they are scientific names.
- The first letter of the generic name is in upper case (capitals), but the specific name is in lower case (small letters).
- If the specific name is not known, it can be written as 'sp.', for example, *Panthera* sp.

The naming of organisms is in a constant state of change. Current names reflect the present state of scientific knowledge and understanding. In the same way, the classification of species is regularly changing as our knowledge of their evolution, physical features, biochemistry, and behaviour increases.

Grouping species together – the principles of classification

With so many species, past and present, it makes sense to organise them into manageable groups. This allows better communication between scientists and avoids confusion. The grouping of organisms is known as **classification**, whilst the theory and practice of biological classification is called **taxonomy**.

There are two main forms of biological classification, each used for a different purpose.

- **Artificial classification** divides organisms according to differences that are useful at the time. Such features may include colour, size, number of legs, leaf shape, etc. These are described as analogous characteristics, where they have the same function but do not have the same evolutionary origins. For example, the wings of butterflies and birds are both used for flight but they originated in different ways.
- **Natural classification**:
  - is based upon the evolutionary relationships between organisms and their ancestors
  - classifies species into groups using shared features derived from their ancestors
  - arranges the groups into a hierarchy, in which the groups are contained within larger composite groups with no overlap.

Relationships in a natural classification are based upon homologous characteristics. Homologous characteristics have similar evolutionary origins regardless of their functions in the adult of a species. For example, the wing of a bird, the arm of a human, and the front leg of a horse all have the same basic structure and all evolved from a common ancestor and are therefore homologous.

It must be remembered that all systems of classification are human inventions. They are developed for our convenience. The natural world does not follow any system of classification, nor is it is bound by our ideas.

Organising the groups of species – taxonomy

Each group within a natural biological classification is called a taxon (plural taxa). Taxonomy is the study of these groups and their positions in a hierarchical order, where they are known as taxonomic ranks. These are based upon the evolutionary line of descent of the group members. The largest group is the **domain**. All living organisms are grouped into one of three **domains**: Archaea (Archaebacteria), Bacteria (Eubacteria), and Eukarya. Within each domain are a number of **kingdoms** and each organism is placed into one of these. Within each kingdom the largest groups are known as **phyla**. Organisms in each phylum (singular) have a body plan that is radically different from organisms in any other phylum. Diversity within each phylum allows it to be divided into **classes**. Each class is divided into **orders** of organisms that have additional features in common. Each order is divided into **families** and at this level the differences are less obvious. Each family is divided into **genera** and each genus (singular) into **species**. As examples of how the system works (rather than names to be learnt), the classification of three organisms is given in Table 1.

▼ **Table 1** *Classification of three organisms from different kingdoms*

Rank	Pin mould	Sweet pea	Tiger
Domain	Eukarya	Eukarya	Eukarya
Kingdom	Fungi	Plantae	Animalia
Phylum	Zygomycota	Magnoliophyta	Chordata
Class	Zygomycetes	Magnoliopsida	Mammalia
Order	Mucorales	Fabales	Carnivora
Family	Mucoraceae	Fabaceae	Felidae
Genus	*Mucor*	*Lathyrus*	*Panthera*
Species	*mucedo*	*odoratus*	*tigris*

Phylogeny

We saw in the last section that the hierarchical order of taxonomic ranks is based upon the evolutionary line of descent of the group members. This evolutionary relationship between organisms is known as **phylogeny**. The term is derived from the word phylum, which, in classification, is a group of related or similar organisms. The phylogeny of an organism reflects the evolutionary branch that led up to it. The phylogenetic relationships of different species are usually represented by a tree-like diagram called a phylogenetic tree. In these diagrams, the oldest species is at the base of the tree whilst the most recent ones are represented by the ends of the branches. An example is shown in Figure 2.

Hint

A useful mnemonic for remembering the order of the taxonomic ranks is '**D**ecide **K**ing **P**rawn **C**urry **O**r **F**at **G**reasy **S**ausages'.

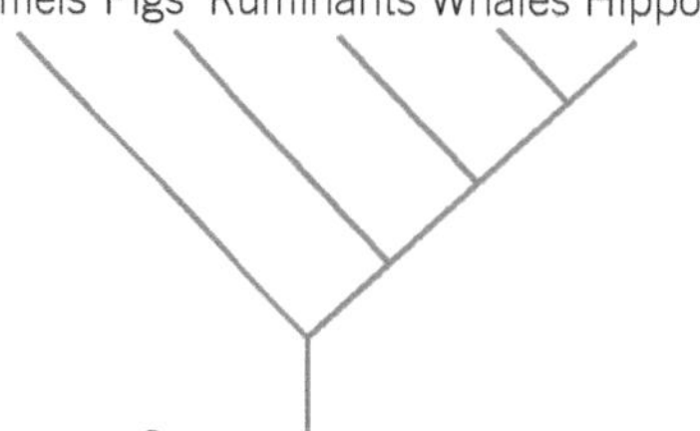

▲ **Figure 2** *A phylogenetic tree showing the evolutionary relationship between certain mammals*

Summary questions

1. What are the **two** main things that all members of a species share?
2. What are the **three** features of a natural system of classification?
3. *Rana temporaria* is the frog commonly found in Britain. Table 2, which is incomplete, shows part of its classification. Give the most appropriate name for each of the blanks represented by the numbers 1–7.

▼ **Table 2**

Kingdom	Animalia
1	Chordata
2	Amphibia
3	Anura
4	Ranidae
Genus	5
6	7

The difficulties of defining species

A species may be defined in terms of observable similarities and the ability to produce fertile offspring. There are, however, certain difficulties with this definition. These include:

- Species are not fixed forever, but change and evolve over time. In time, some individuals may develop into a new species.
- Within a species there can be considerable variation among individuals. All dogs, for example, belong to the same species, but **artificial selection** has led to a variety of different breeds.
- Many species are extinct and most of these have left no fossil record.
- Some species rarely, if ever, reproduce sexually.
- Members of different populations of the same species may be isolated, for example, by oceans, and so never meet and therefore never interbreed.
- Populations of organisms that are isolated from one another may be classified as different species. These populations may turn out to be of the same species when their ability to interbreed is tested.
- Some species are sterile (see below).

1. Even where groups of extinct organisms have left fossil records, it is very difficult to distinguish different species. Suggest two reasons why.
2. Suggest reasons why it is often difficult to classify organisms as distinct species.

▲ **Figure 3** *A horse (right) and a donkey (left), although different species, are capable of mating and producing offspring called mules*

A horse and a donkey (Figure 3) are capable of mating and producing offspring, which are known as mules. A horse and a donkey are, however, different species and the resulting mules are infertile, that is, they almost never produce offspring when mated with each other. Why are mules infertile? It is all down to the number of **chromosomes** and the first stage of **meiosis**. A horse has 64 chromosomes (32 pairs) and a donkey has 62 chromosomes (31 pairs). The **gametes** of a horse and a donkey therefore have 32 and 31 chromosomes respectively. When the gametes of a horse and a donkey fuse, the offspring (the mule) has 63 chromosomes. These gametes are formed by meiosis, which cannot take place if there is an odd number of chromosomes. So a mule cannot produce gametes and is therefore normally infertile. However, mitosis can take place and therefore a mule grows and develops normally.

There have been occasional cases of a fertile female mule. This event is very rare, so much so that the Romans had a saying that meant 'when a mule foals', which was the equivalent of our modern 'once in a blue moon'.

3. From your knowledge of the events during meiosis 1 (see Topic 9.1), suggest a reason why the cells of a mule with their 63 chromosomes are unable to undergo meiosis and so cannot produce gametes.
4. Does the fact that fertile mules occasionally occur make a mule a distinct species? Give reasons for your answer.

Phylogenetic relationships

Figure 4 shows a phylogenetic tree for birds and certain reptiles.

1. Which group is the closest relative of the snakes?
2. Are dinosaurs more closely related to crocodiles or birds?
3. Suggest a reason why dinosaurs are not shown along the time line like all the other groups.

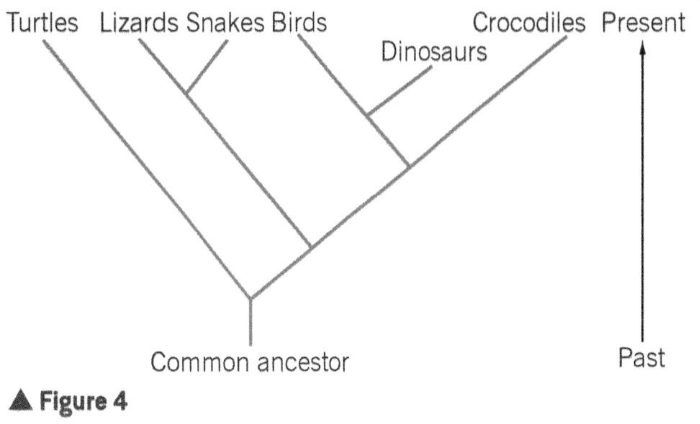

▲ **Figure 4**

9.3 Genetic comparisons using DNA and proteins

Learning objectives:

- ➔ Explain how comparisons of base sequences in DNA can be used to investigate how closely related organisms are.
- ➔ Explain what DNA hybridisation is and how is it used to determine relationships between organisms.
- ➔ Explain how comparisons of amino acid sequences in proteins can be used to investigate the relationships between organisms.
- ➔ Explain how immunological comparisons are used to investigate variations in proteins.

Specification reference: 3.1.10.2 and 3.1.11.1

In Topic 9.2, you saw that classification systems were originally based on features that could easily be observed. As science has developed it has become possible to use a wider range of evidence to determine the evolutionary relationships between organisms.

Evolution affects not only the visible internal and external features that adapt and change, but also the molecules of which they are made. DNA determines the proteins of an organism, including **enzymes** (see Chapter 8), and proteins determine the features of an organism. It follows that changes in the features of an organism are due to changes in its DNA. Comparing the DNA and proteins of different **species** helps scientists to determine the evolutionary relationships between them.

Comparison of DNA base sequences

When one species gives rise to another species during evolution, the DNA of the new species will initially be very similar to that of the species that gave rise to it. Due to **mutations**, the sequences of nucleotide bases in the DNA of both will change. Consequently, over time, the new species will accumulate more and more differences in its DNA. As a result, we would expect species that are more closely related to show more similarity in their DNA base sequences than species that are more distantly related. As there are millions of base sequences in every organism, DNA contains a vast amount of information about the evolutionary history of all organisms.

One way to determine similarities between the DNA of different organisms is to use a technique called DNA hybridisation.

DNA hybridisation

DNA hybridisation depends upon a particular property of the DNA double helix. When DNA is heated, its double strand separates into its two complementary single strands. When cooled, the complementary bases on each strand recombine with each other to reform the original double strand. Given sufficient time, all strands in a mixture of DNA will pair up with their partners.

Using this property, DNA hybridisation can be used to compare the DNA of two species in the following manner:

- DNA from two species is extracted, purified, and cut into short pieces.
- The DNA from one of the species is labelled by attaching a radioactive or fluorescent marker to it. It is then mixed with unlabelled DNA from the other species.
- The mixture of both sets of DNA is heated to separate their strands.
- The mixture is cooled to allow the strands to combine with other strands that have a complementary sequence of bases.

Hint

As DNA determines the features of an organism, using the similarities in DNA as evidence for a close evolutionary relationship between species provides a direct record. Using similarities in the features themselves provides an indirect record. However, some DNA is non-functional and does not code for proteins. Analysis of this DNA can provide new evidence of relationships between organisms.

Synoptic link

An understanding of DNA hybridisation requires information found in Topic 7.1.

- Some of the double strands that reform will be made up of one strand from each species. This is called **hybridisation** and the new strands are called hybrid strands. These can be identified because they are 50 per cent labelled.
- These hybrid strands are separated out and the temperature is increased in stages.
- At each temperature stage the degree to which the two strands are still linked together is measured.
- If the two species are closely related they will share many complementary nucleotide bases.
- There will therefore be more hydrogen bonds linking them together in the hybrid strand.
- The greater the number of hydrogen bonds, the stronger the hybrid strand will be.
- The stronger the hybrid strand, the higher the temperature needed to separate it into its two single strands.
- The higher the temperature at which the hybrid strand splits, the more closely the two species are related.
- The lower the temperature at which it splits, the more distantly the species are related.

The process of DNA hydridisation is summarised in Figure 1.

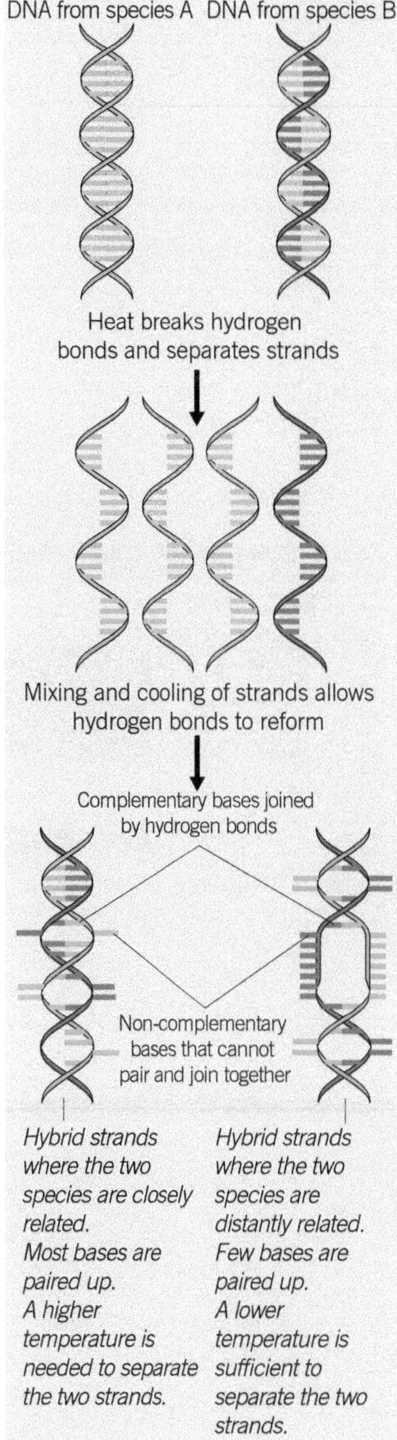

▲ **Figure 1** *DNA hybridisation*

Use of DNA base sequencing in classifying plants

Until recently the classification of flowering plants had been based on the appearance of a plant's physical features. This led to flowering plants being placed in one of two groups: the monocotyledons, which have a single seed leaf (and generally have thin, narrow leaves), and the dicotyledons, which have two seed leaves (and generally have broad leaves).

A team of scientists at The Royal Botanical Gardens, Kew recently devised a new classification of the families of flowering plants. This was based on the DNA sequences of three genes found in all plants. Their work was carried out as follows:

- They used 565 species that between them represented all the known families of flowering plants in the world.
- For each plant, the DNA sequences of all three genes were determined.
- The sequences for each species were compared using computer analysis.
- A phylogenetic tree for the families of flowering plants was devised based upon the DNA sequences of the species used.

The phylogenetic tree that the scientists produced showed how species have evolved into natural groups. These groupings represent evolutionary relationships better than any previous form of classification has ever done.

Genome sequencing

Advances in the techniques for sequencing DNA now mean that whole genomes can be analysed relatively quickly and cheaply using what is termed next-generation sequencing. The data generated (nucleotide sequences) is stored as bioinformatics databases. One example of such a database is hosted by the European Molecular Biology Laboratory (EMBL). Genomes can be compared using a BLAST search. This stands for Basic Local Alignment Search Tool. This is essentially a computer programme which identifies regions of similarity between different DNA sequences.

Genome sequences of known organisms are stored on the database. Comparing how similar the DNA of an unidentified organism is to that of the known organisms lets us classify it, because organisms which are closely related in evolutionary terms have a greater degree of similarity between their DNA sequences.

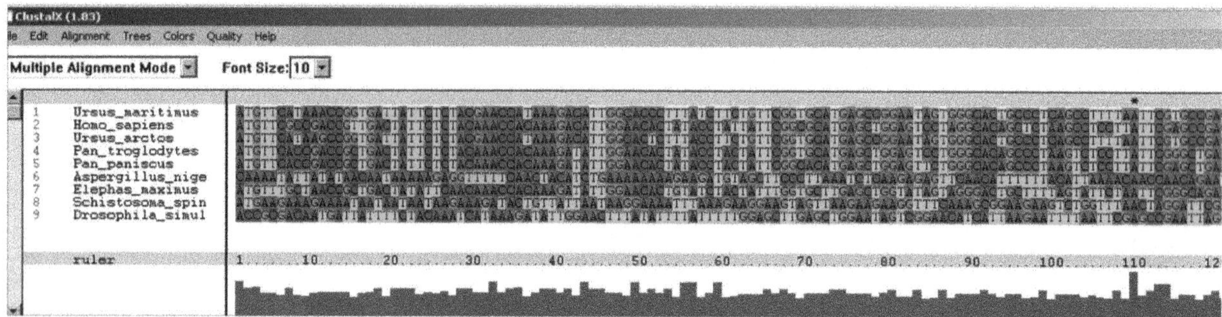

▲ **Figure 2** *Results of a DNA sequence analysis of nine different organisms*

One application of this technique has been in the analysis of microbes present in environmental samples. The species, genera, and even phyla of the organisms in the sample may be unknown at the time of collecting. The goal of the sequencing is to determine precisely the species present without the need to isolate and culture each individual species.

Comparison of amino acid sequences in proteins

The sequence of amino acids in proteins is determined by DNA. The degree of similarity in the amino acid sequence of the same protein in two species will therefore reflect how closely related the two species are.

Once the amino acid sequence for a chosen protein has been determined for two species, the two sequences are compared. This can be done by counting either the number of similarities or the number of differences in each sequence. An example is shown in Figure 3. Here there is a short sequence of seven amino acids of the same protein in six different species. The table on the right of the figure shows both the number of differences and the number of similarities.

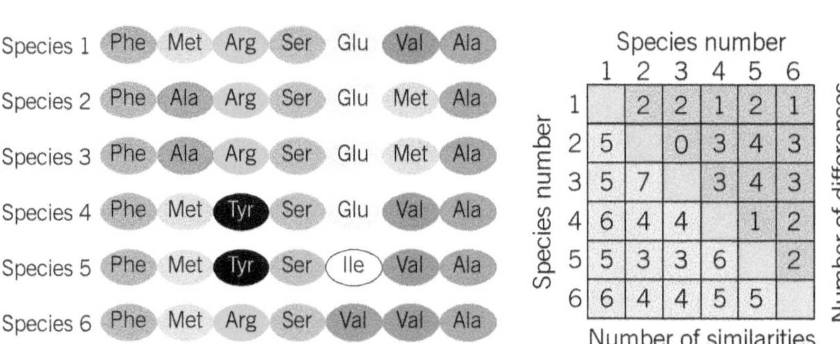

Species number	1	2	3	4	5	6
1		2	2	1	2	1
2	5		0	3	4	3
3	5	7		3	4	3
4	6	4	4		1	2
5	5	3	3	6		2
6	6	4	4	5	5	

▲ **Figure 3** *Comparison of amino acid sequence in part of the same protein in six species*

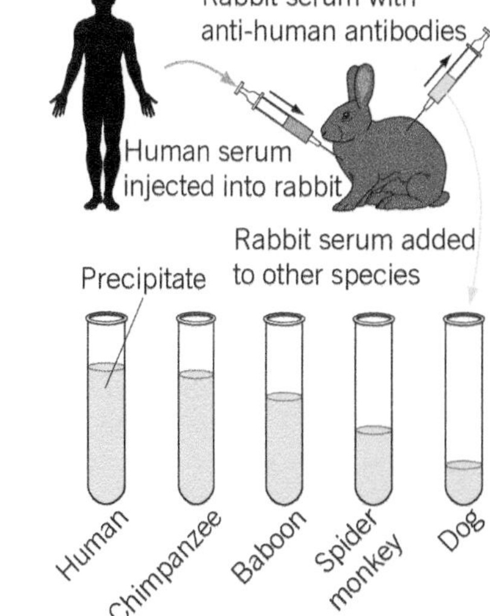

▲ **Figure 4** *Immunological comparison of human serum with the sera of other species*

Immunological comparisons of proteins

The proteins of different species can also be compared using immunological techniques. The principle behind this method is the fact that **antibodies** of one species will respond to specific **antigens** on proteins, such as albumin, in the blood **serum** of another. The process is carried out as follows:

- Serum albumin from species A is injected into species B.
- Species B produces antibodies specific to all the antigen sites on the albumin from species A.
- Serum is extracted from species B – this serum contains antibodies specific to the antigens on the albumin from species A.
- Serum from species B is mixed with serum from the blood of a third species C.
- The antibodies respond to their corresponding antigens on the albumin in the serum of species C.
- The response is the formation of a precipitate.
- The greater the number of similar antigens, the more precipitate is formed and the more closely species A and C are related.
- The fewer the number of similar antigens, the less precipitate is formed and the more distantly the species are related.

An example of this technique is illustrated in Figure 4. In this case, species A is a human, species B is a rabbit, and species C is represented by a variety of other mammals.

Synoptic link

An understanding of the technique described in Figure 4 requires information found in Topic 14.4.

Summary questions

1 During the process of DNA hybridisation, explain the following:
   a why DNA is heated
   b why some hybrid strands require a higher temperature to separate the two strands than others
   c the significance of the difference, described in **b**, in determining the relationships between species.
2 Using the information in Figure 4 state, with reasons, which two species are most closely related.

Practice questions: Chapter 9

1 Figure 1 shows a pair of chromosomes at the start of meiosis. The letters represent alleles.

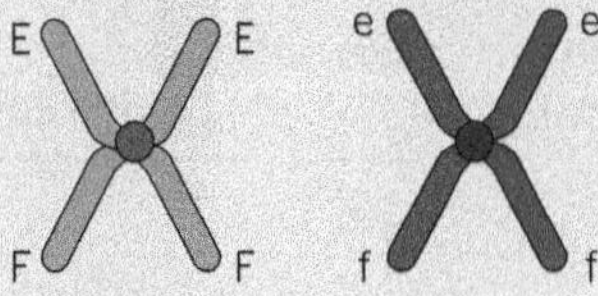

▲ **Figure 1**

(a) What is an allele? *(1 mark)*

(b) Explain the appearance of one of the chromosomes in Figure 1. *(2 marks)*

(c) The cell containing this pair of chromosomes divided by meiosis. Figure 2 shows the distribution of chromosomes from this pair in four of the gametes produced.

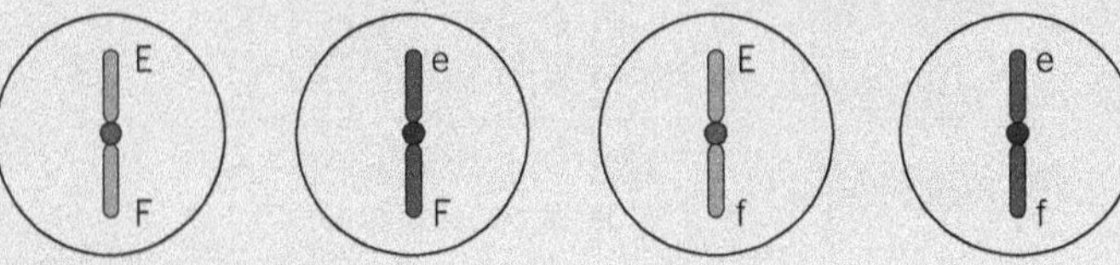

▲ **Figure 2**

(i) Some of the gametes formed during meiosis have new combinations of alleles. Explain how the gametes with the combinations of alleles Ef and eF have been produced. *(2 marks)*

(ii) Only a few gametes have the new combination of alleles Ef and eF. Most gametes have the combination of alleles EF and ef. Suggest why only a few gametes have the new combination of alleles, Ef and eF. *(1 mark)*

(d) Figure 3 shows a cell with six chromosomes.

▲ **Figure 3**

(i) This cell produces gametes by meiosis. Draw a diagram to show the chromosomes in one of the gametes. *(2 marks)*

(ii) How many different types of gametes could be produced from this cell as a result of different combinations of maternal and paternal chromosomes? *(1 mark)*

(e) (i) Calculate the number of different types of gametes that can be produced in a species with a diploid number of 24. *(1 mark)*

(ii) Assuming random fertilisation, calculate the number of different combinations of maternal and paternal chromosomes in the zygotes of this species. *(1 mark)*

AQA June 2010

2 The diagram shows part of a pre-mRNA molecule.

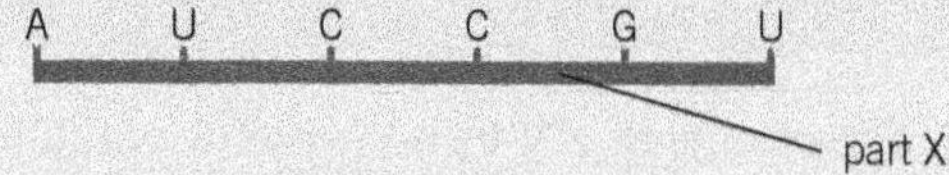

(a) (i) Name the two substances that make up part **X**. *(1 mark)*
(ii) Give the sequence of bases on the DNA strand from which this pre-mRNA has been transcribed. *(1 mark)*
(b) (i) Give **one** way in which the structure of an mRNA molecule is different from the structure of a tRNA molecule. *(1 mark)*
(ii) Explain the difference between pre-mRNA and mRNA. *(1 mark)*
(c) The table shows the percentage of different bases in two pre-mRNA molecules.

The molecules were transcribed from the DNA in different parts of a chromosome.

Part of chromosome	Percentage of base			
	A	G	C	U
Middle	38	20	24	
End	31	22	26	

(i) Complete the table by writing the percentage of uracil (U) in the appropriate boxes. *(1 mark)*
(ii) Explain why the percentages of bases from the middle part of the chromosome and the end part are different. *(2 marks)*

AQA June 2011

3 Organisms can be classified using a hierarchy of phylogenetic groups.
(a) Explain what is meant by:
(i) a hierarchy *(2 marks)*
(ii) a phylogenetic group. *(1 mark)*
(b) Cytochrome c is a protein involved in respiration. Scientists determined the amino acid sequence of human cytochrome c. They then:

- determined the amino acid sequences in cytochrome c from five other animals
- compared these amino acid sequences with that of human cytochrome c
- recorded the number of differences in the amino acid sequence compared with human cytochrome c.

The table shows their results.

Animal	Number of differences in the amino acid sequence compared with human cytochrome c
A	1
B	12
C	12
D	15
E	21

(i) Explain how these results suggest that animal **A** is the most closely related to humans. *(2 marks)*
(ii) A student who looked at these results concluded that animals **B** and **C** are more closely related to each other than to any of the other animals. Suggest **one** reason why this might **not** be a valid conclusion. *(1 mark)*
(iii) Cytochrome c is more useful than haemoglobin for studying how closely related different organisms are. Suggest **one** reason why. *(1 mark)*

AQA June 2013

Answers to the Practice Questions are available at
www.oxfordsecondary.com/oxfordaqaexams-alevel-biology

10 Biodiversity within a community

10.1 Genetic diversity

Learning objectives:

- → Explain why organisms are different from one another.
- → Describe what factors influence genetic diversity.

Specification reference: 3.1.11.1

In Topic 9.3, you saw how comparisons of DNA and proteins could be used to measure genetic diversity. In this chapter you will learn how similarities and differences in DNA result in genetic diversity. Genetic diversity is just one aspect of **biodiversity**.

Organisms are varied. Around 1.8 million **species** of organisms on Earth have been identified and named. Many more are unnamed or undiscovered. Estimates of the total number of species on this planet range from 5 million to 100 million. All of these species are different.

Even between members of the same species there are a multitude of differences. Almost every one of the 6.5 billion people alive in 2008 were similar enough to be recognised as humans and yet different enough to be distinguished from one another. What makes us and other species similar and yet different?

▲ **Figure 1** *Examples of genetic diversity (from top to bottom): anemone; lichens; mountain goat; fritillary butterfly*

Genetic diversity

You learnt in Topic 9.3 that it is proteins which make organisms different and, in Topic 8.1, that it is DNA which determines the considerable variety of proteins that make up each organism. Therefore, similarities and differences between organisms may be defined in terms of variation in DNA. Hence it is differences in DNA that lead to the vast genetic diversity we find on Earth.

You also saw in Topic 8.1 that a section of DNA that codes for one or more polypeptides is called a gene. All members of the same species have the same genes. For example, all humans have a gene for blood group, just as all snapdragons *(Antirrhinum majus)* have a gene for petal colour. Which blood group humans have depends on which two **alleles** of the gene they possess. Likewise, the colour of a snapdragon's petals depends on which two alleles for petal colour it possesses. Organisms therefore differ in their alleles, not their genes. It is the combination of alleles they possess that makes species (and individuals within that species) different from one another.

The greater the number of different alleles that all members of a species possess, the greater the genetic diversity of that species. The greater the genetic diversity, the more likely that a species will be able to adapt to some environmental change. This is because it will have a wider range of alleles and therefore a wider range of characteristics. There is therefore a greater probability that some individual will possess a characteristic that suits it to the new environmental conditions. Genetic diversity is reduced when a species has fewer different alleles. Let us now look at some factors that influence genetic diversity.

Hint

Remember that an allele is one alternative form of a gene and, as such, is a length of DNA on one chromosome of a homologous pair.

Selective breeding

Selective breeding is also known as **artificial selection**. It involves identifying individuals with the desired characteristics and using them to parent the next generation. Offspring that do not exhibit the desired characters are killed, or at least prevented from breeding. In this way, alleles for unwanted characteristics are bred out of the population. The variety of alleles in the population is deliberately restricted to a small number of desired alleles. Over many generations, this leads to a population in which all individuals possess the desired qualities, but which has reduced genetic diversity.

Selective breeding is commonly carried out in order to produce high-yielding breeds of domesticated animals and strains of plants. For example, in plants such as wheat, the features selected for include large grains with a high gluten content, short stems, and resistance to disease.

The founder effect

The founder effect occurs when just a few individuals from a population colonise a new region. These few individuals will carry with them only a small fraction of the alleles of the population as a whole. These alleles may not be representative of the larger population. The new population that develops from the few colonisers will therefore show less genetic diversity than the population from which they came. The founder effect is seen when new volcanic islands rise out of the sea. The few individuals that colonise these barren islands give rise to populations that are genetically distinct from the populations they left behind. The new populations may, in time, develop into a separate species. As these species have fewer alleles, they are less able to adapt to changing conditions.

Genetic bottlenecks

Populations of a species may from time to time suffer a dramatic drop in numbers. Sometimes the reason for this drop is a chance event, such as a volcanic eruption or interference by man. The few survivors will possess a much smaller variety of alleles than the original population. In other words, their genetic diversity will be less. As these few individuals breed and become re-established, the genetic diversity of the new population will remain restricted. This effect can be illustrated by elephant seals. The population of northern elephant seals was hunted by humans until, by about 1900, just 20 remained in a colony on the coast of Mexico. Their population has since increased but, when compared with the population of southern elephant seals, it shows considerably less genetic diversity. Less diversity means fewer alleles, making it less likely that the population can adapt to any change in its environment.

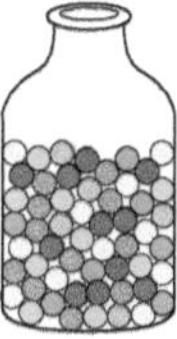

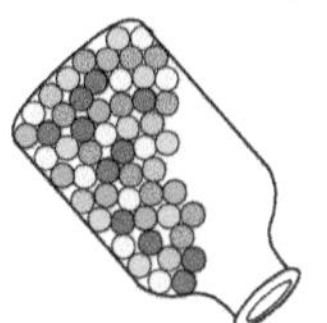

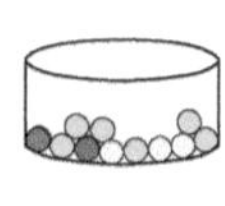

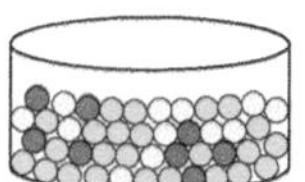

▲ **Figure 2** *The bottleneck effect*

Synoptic link

While you should be aware of the factors which can affect genetic diversity a detailed knowledge of this factors is not required at AS level. This material forms part of Topic 3.3.7 and you will meet it again in Chapter 23.

Ethics of selective breeding in domesticated animals

Having outlined the background to selective breeding and the techniques involved, let us now consider some of the ethical issues surrounding it.

Almost all scientific progress has both benefits and risks. Scientists carry out research into various techniques of selective breeding, from which they can report on the benefits and risks of each. Interested parties, such as government agencies, farmers, food producers, supermarkets, and shopkeepers, and organisations such as animal welfare groups can then debate the issues surrounding these reports. As part of this debate it is always worth bearing in mind that:

- Scientific research requires funding. Who funds the research may have an influence on its outcome. Consider how the type of research undertaken might differ according to whether a farming body, a food producer, a supermarket, or an animal welfare group was funding the work.
- Scientists are largely self-regulating. They help to make decisions on what experiments are carried out and how. Different scientists may have different codes of ethics depending on their personal, moral, and religious beliefs. This may affect the experiments they perform.

Ethical implications

There is no doubt that selective breeding has produced livestock that gives greater yields. In turn, this gives us a reliable source of cheap food throughout the year. It has raised our standard of living. At the same time, selective breeding of domesticated animals raises some ethical issues.

- Is selective breeding interfering with nature? Do other species have 'rights'? Should the development of new varieties be allowed to take place naturally, as it has successfully done for millions of years? Do domesticated animals have value in themselves or only in terms of their usefulness to humans? Should we accept a lower standard of living in return for natural rather than artificial breeding of animals?
- What features should be selected for and who decides? Is selective breeding of pets equally as acceptable as selecting high-yielding food animals? Is it acceptable to select for features in cats, dogs, and other pets simply because the features are fashionable or desirable to their owners?
- How do we balance increased yield with animal welfare? At what point does a further increase in udder size become disabling for the cow rather than just inconvenient? Is it acceptable to inconvenience a cow anyway? Is it reasonable to selectively breed animals that are better suited to living in environmentally regulated sheds for long periods rather than in open fields?
- Producing genetically uniform livestock by selective breeding leads to varieties of animals with a narrower range of alleles and hence less genetic diversity. Could we be losing forever some alleles that might be of benefit to the animals and mankind in the future?
- In an effort to control global warming, should we select animals that produce less methane (a major greenhouse gas) and/or grow well at lower environmental temperatures and so do not need overwintering in heated sheds in colder climates?
- Are we driven too much by consumerism? Should we stop trying to select increasingly higher-yielding varieties and simply be prepared to pay more for our food? Is this fair on the poorest people in our society?
- Should we stop the traditional methods of selective breeding as these are still artificial? Is selective breeding an acceptable alternative to producing required characteristics by genetic engineering?

Scientists will continue to explore, and try to explain, aspects of the world about them, including research into selective breeding. It is up to society at large to decide how to use the information they provide. Whilst scientists will inform us about what *can* be done, it is answering ethical questions like those above that will help us to decide what *ought* to be done.

1. Those who make decisions about whether a new scientific or technological discovery should be developed have a very difficult task. Suggest some of the factors that they need to take into account before making a decision.
2. Apart from ethical principles, suggest what other guidance any decision must comply with.

Selective breeding in cattle

The breeding of farm animals, such as cattle, has never been a random affair. For many thousands of years, farmers have chosen the most suitable domesticated animals and plants to parent the next generation. Selective breeding is therefore not new, but traditional forms of selective breeding are slow and imprecise. What has changed is the pace and extent of selective breeding that has occurred over the past 50 years. What has prompted this change?

As consumers we want a reliable supply of a wide range of foods at minimum cost. We therefore create a highly competitive market that puts farmers under pressure to cut costs and supply cheap food in order to stay in business. As a result, food production has become ever more intensive. To meet our demand for cheap, plentiful food, scientists and farmers have worked together to breed higher-yielding varieties of domesticated plants and animals.

One method of increasing the pace of change by selective breeding is to use **artificial insemination (AI)**. AI is the collection of semen and its introduction into the vagina by artificial means. The semen of a single bull can be used to inseminate hundreds of cows. In the UK 80 of insemination in cattle is by artificial means.

Cattle have been selectively bred for two main purposes:

- for meat – beef breeds (e.g., Herefords and Aberdeen Angus). Desirable characteristics include a high muscle to bone ratio, and rapid growth and weight gain.
- for milk – dairy breeds (e.g., Friesians and Holstein). Desirable characteristics include high production of milk with a high fat and protein content, an udder that suits a milking machine, and rapid delivery of milk.

The rapid change in cattle characteristics as a result of selection has raised a number of issues. For example, in dairy cattle it has had the following results:

- The genetic diversity of cattle has been reduced. Specialist breeds of cow, suited to local conditions, have disappeared, to be replaced by the highly bred Holstein–Friesian (the typical black-and-white cow) that now makes up 90 per cent of European dairy herds.
- The doubling of milk yield per cow over the past 50 years has put a strain on the animals' welfare. Mastitis (inflammation of the udder), lameness, and infertility are all more common now.
- The natural lifespan of a cow is up to 25 years but most cows now go for slaughter after 5 years.
- Calves would normally suckle for 6–12 months but, so that the milk can be used by humans, they are removed from their mothers within 1–2 days.
- From the age of 2 years, dairy cows produce calves, and hence milk, continuously throughout their lives.

3 Traditional forms of selective breeding involve farmers choosing the most suitable animals to breed the next generation. Suggest reasons why this process is described as 'slow and imprecise'.
4 Suggest one possible danger of allowing the semen of a single bull to inseminate hundreds of cows.
5 Suggest two advantages of selective breeding in dairy cattle.
6 The Vegan Society opposes many aspects of the modern dairy industry. Suggest two ethical objections that groups like the Vegan Society might have to the selective breeding of dairy cattle.

◀ **Figure 3** *Cattle, such as the Highland variety (left) and Friesian variety (right), have been bred to produce different characteristics*

Hint

Ethics are a set of standards that are followed by a particular group of individuals and are designed to regulate their behaviour. They determine what is acceptable and legitimate in pursuing the aims of the group.

Summary questions

1 State whether each of the following is likely to **increase** or **decrease** genetic diversity:
  a increasing the variety of alleles within a population
  b breeding together closely related cats to develop varieties with longer fur
  c a few seeds from a plant community on Iceland reaching the new volcanic island of Surtsey and establishing a new population
  d mutation (permanent change to the DNA) of an allele.
2 Explain how a difference in an organism's DNA might lead to it having a different appearance and hence the species showing greater genetic diversity.
3 Explain how genetic bottlenecks reduce genetic diversity.

10.2 Species diversity

Learning objectives:

→ Explain what is meant by species diversity.

→ Explain how an index of diversity can be used as a measure of species diversity.

Specification reference: 3.1.11.2

Throughout this unit we have considered the variety of living organisms. **Biodiversity** is the general term used to describe variety in the living world. It refers to the number and variety of living organisms in a particular area and has three components:

- **Genetic diversity** refers to the variety of genes possessed by the individuals that make up any one species.
- **Species diversity** refers to the number of different species and the number of individuals of each species within any one community.
- **Ecosystem diversity** refers to the range of different habitats within a particular area.

One measure of biodiversity is species diversity. It has two components:

- the number of different species in a given area. This is referred to as the **species richness**.
- the proportion of the community that is made up of an individual species.

▲ **Figure 1** *In a tropical rainforest there is high species diversity*

Two communities may have the same number of species but the proportions of the community made up of each species may differ markedly. For example, a natural meadow and a field of wheat may both have 25 species. However, in the meadow, all 25 species might be equally abundant, whereas, in the wheat field, over 95 per cent of the plants may be a single species of wheat.

Measuring species diversity

Consider the data shown in Table 1 about two different habitats. It does not tell us much about the differences between the two habitats because, in both cases, the total number of species and the total number of individuals are identical. However, if we measure the species diversity, we get a different picture.

▲ **Figure 2** *In the sub-arctic tundra there is low species diversity*

▼ **Table 1** *Number and types of species found in two different habitats within the same ecosystem*

Species found	Numbers found in habitat X	Numbers found in habitat Y
A	10	3
B	10	5
C	10	2
D	10	36
E	10	4
No. of species	5	5
No. of individuals	50	50

One way of measuring species diversity is to use an index that is calculated as follows:

$$d = \frac{N(N-1)}{\Sigma n(n-1)}$$

Where:

d = species diversity index

N = total number of organisms of all species

n = total number of organisms of each species

Σ = the sum of

To use the index to calculate the species diversity of the two habitats, you must first calculate $n(n-1)$ for each species in each habitat. You can then calculate the sum of $n(n-1)$ for each species. These calculations are shown in Table 2.

▼ Table 2 *Calculation of $n(n-1)$ and $\Sigma n(n-1)$ for habitats X and Y*

Species	Numbers (n) found in habitat X	$n(n-1)$	Numbers (n) found in habitat Y	$n(n-1)$
A	10	10(9) = **90**	3	3(2) = **6**
B	10	10(9) = **90**	5	5(4) = **20**
C	10	10(9) = **90**	2	2(1) = **2**
D	10	10(9) = **90**	36	36(35) = **1260**
E	10	10(9) = **90**	4	4(3) = **12**
	$\Sigma n(n-1)$	**450**	$\Sigma n(n-1)$	**1300**

You can now calculate the species diversity index for each habitat.

Habitat X: $$d = \frac{50(49)}{450} = \frac{2450}{450} = 5.44$$

Habitat Y: $$d = \frac{50(49)}{1300} = \frac{2450}{1300} = 1.88$$

The higher the value d, the greater is the species diversity. So, in this case, although the total number of species and the total number of individuals are the same in both habitats, the species diversity of habitat X is much greater.

Species diversity and ecosystems

Biodiversity reflects how well an **ecosystem** functions. The higher the species diversity index, the more stable an ecosystem usually is and the less it is affected by climate change. For example, if there is a drought, a community with a high species diversity index is much more likely to have at least one species that is able to tolerate drought than a community with a low species diversity index. At least some members are therefore likely to survive the drought and maintain the community.

In extreme environments, such as hot deserts, only a few species have the necessary adaptations to survive the harsh conditions. The species diversity

Hint

Calculating a species diversity index provides a number that makes it easier to compare the variety in different habitats. It would be so much harder, and less precise, if we had to rely on descriptions of different habitats to make these comparisons.

Summary questions

1 What is meant by species diversity?

2 Table 3 shows the numbers of each of six species of plant found in a salt-marsh community. Calculate the species diversity index for this salt-marsh community using the formula shown earlier. Show your working.

▼ Table 3

Species	Numbers in salt marsh
Salicornia maritima	24
Halimione portulacoides	20
Festuca rubra	7
Aster tripolium	3
Limonium humile	3
Suaeda maritima	1

3 Explain why it is more useful to calculate a species diversity index than just to record the number of species present.

index is therefore normally low. This usually results in an unstable ecosystem in which the make-up of the community is dominated by climatic factors rather than by the organisms within the community. In less hostile environments, the species diversity index is normally high. This usually results in a stable ecosystem in which the make-up of the community is dominated by living organisms rather than by climate.

1 Scientists believe that the production of greenhouse gases by human activities is contributing to climate change. Explain why an increase in greenhouse gases is more likely to result in damage to communities with a low species diversity index than to communities with a high index.

2 The graph in Figure 3 shows the effect of environmental change on the stability and the functioning of ecosystems.
   a Describe the relationship between environmental change and the community with a low species diversity index.
   b Explain the different responses to environmental change in communities with a low and a high species diversity.

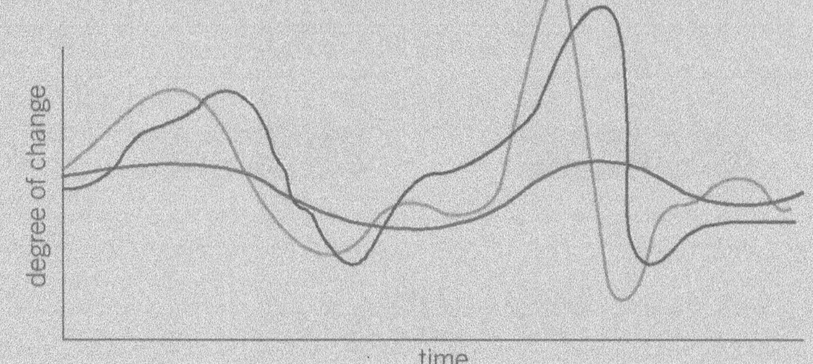

▲ **Figure 3**

▶ **Figure 4** *In harsh environments, like this hot desert, only a few species are adapted to survive the extreme conditions and therefore species diversity is low*

10.3 Species diversity and human activities

In our efforts to provide enough food for the human population at a low cost, mankind has had a considerable impact on the natural world. This impact has led to a reduction in biodiversity. In this topic you will look at how two human activities, agriculture and deforestation, have reduced species diversity.

Learning objectives:

- → Describe the influence of deforestation and the impact of agriculture on species diversity.

Specification reference: 3.1.11.2

Impact of agriculture

As natural ecosystems develop over time, they may become complex communities with many individuals of a large number of different species. In other words, these communities have a high species diversity index. Agricultural ecosystems are controlled by humans and are different. You saw in Topic 10.1 that farmers select species for particular qualities that make them more productive. As a result, the number of species, and the genetic variety of alleles they possess, is reduced to the few that exhibit the desired features. To be economic, the numbers of these desirable species needs to be large. Any particular area can only support a certain amount of biomass. If most of the area is taken up by the one species that the farmer considers desirable, it follows that there is a smaller area available for all the other species. These many other species have to compete for what little space and resources are available. Many will not survive this competition. In addition, pesticides are used to exclude these species because they compete for the light, mineral ions, water, and food required by the farmed species. The overall effect is a reduction in species diversity. The species diversity index is therefore low in agricultural ecosystems.

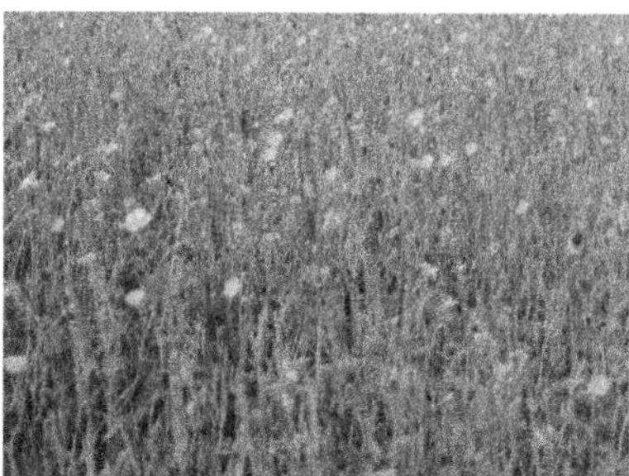

▲ **Figure 1** *High species diversity in a hay meadow*

▲ **Figure 2** *Low species diversity in a field grown for silage*

Balancing food production with conservation

With the human population set to increase from around 7 billion in 2014 to 11 billion by the end of this century, clearly food production has to keep pace. So how can this be done in a way which conserves species diversity? There are several examples of incentives or restrictions which may be imposed on farmers to encourage sustainable agriculture:

- financial incentives from governments to introduce features such as buffer strips (areas around crops or water courses which are left uncultivated and where native species can survive)
- financial incentives such as tax relief for restoring and managing areas of woodland in regions previously deforested to make way for crops
- restricting the expansion of the area of agricultural land in favour of improving productivity in current areas
- removing subsidies for production in favour of financial incentives such as those above.

Education has a large part to play. Each year enormous areas of land are lost from food production due to soil erosion by rain and wind. This is usually a result of poor farming practices. Changing these practices and rescuing degraded farm land reduces the need to cultivate more land and allows conservation of wild areas with high levels of biodiversity. Some of these practices, such as the sowing of a mixture of plant species for grazing (a herbal ley) rather than grass land consisting of just one or two species, increase the quality of the soil whilst immediately increasing the biodiversity. This is a good example of sustainable practice in farming.

Impact of deforestation

Forests are the natural vegetation over much of the Earth. Without human intervention, they would, and once did, cover much of the planet. As forests form many layers between the ground and the tops of the trees, there are numerous **habitats** available. Many different species are adapted to living in these different habitats and species diversity is therefore high. Indeed, the tropical rainforests have the highest species diversity of any ecosystem. Whilst some deforestation is the result of accidental fires, the vast majority is due to deliberate human actions. Deforestation is the permanent clearing of forests and the conversion of the land to other uses, such as agriculture, grazing, housing, and reservoirs. In addition, some forests have been destroyed as a result of man-made pollutants producing acid rain.

The most serious consequence of deforestation is the loss of biodiversity. Some estimates suggest that up to 50 000 species are being lost each year due to deforestation. It is in the tropical rainforests that the loss is greatest. Despite covering only 7 per cent of the Earth's surface, tropical rainforests account for half of all its species. The replacement of these and other forests by agriculture, housing, or reservoirs has considerably reduced species diversity. Even where areas are reforested there is still an overall loss of species diversity as the new forests grown for commercial purposes have just a few predominant tree types.

Summary questions

1 Explain how agriculture has reduced species diversity.

2 Why is there a reduction in species diversity when a forest is replaced by grassland for grazing sheep or cattle?

3 Why does the loss of tropical rainforest have a greater effect on global biodiversity than the loss of any other ecosystem?

Human activity and loss of species in the UK

▲ **Figure 3** *Deforestation*

The present rate of species extinction is thought to be between 100 and 1000 times greater than at any other time in evolutionary history. The main cause of species loss is the clearance of land in order to grow crops and meet the demand for food from an ever-increasing human population. An area of rainforest roughly the size of the UK is cleared every year. Throughout the world, habitats are being lost. Most of this habitat loss has entailed the replacement of natural communities of high species diversity with agricultural ones of low species diversity. The conservation agencies in the UK have made estimates of the percentage of various habitats that have been lost in the UK since 1900. These estimates are shown in Table 1.

▲ **Figure 4** *Heathland (left) and mixed woodland (right)*

▼ **Table 1**

Habitat	Habitat loss since 1900 / %	Main reason for habitat loss
Hay meadow	95	Conversion to highly productive grass and silage
Chalk grassland	80	Conversion to highly productive grass and silage
Lowland fens and wetlands	50	Drainage and reclamation of land for agriculture
Limestone pavements in England	45	Removal for sale as rockery stone
Lowland heaths on acid soils	40	Conversion to grasslands and commercial forests
Lowland mixed woodland	40	Conversion to commercial conifer plantations and farmland
Hedgerows	30	To make larger fields to accommodate farm machinery

1 There are currently approximately 350 000 km of hedgerow in the UK. How many kilometres were there in 1900?
2 Some lowland mixed woodlands have been replaced by other woodland. Explain how this change might still result in a lower species diversity.
3 Suggest **one** benefit and **one** risk associated with the conversion of hay meadows and chalk grasslands to highly productive grass and silage.
4 In what ways might the information in the table be used to inform decision-making on preserving habitats and biodiversity?
5 The European Union gives grants to farmers to replant hedges. Explain how replanting hedges might affect the species diversity found on farms.

Practice questions: Chapter 10

1 **(a)** What is a *species*? *(2 marks)*

(b) Scientists investigated the diversity of plants in a small area within a forest. The table shows their results.

Plant species	Number of individuals
Himalayan raspberry	20
Heartwing sorrel	15
Shala tree	9
Tussock grass	10
Red cedar	4
Asan tree	6
Spanish needle	8
Feverfew	8

The index of diversity can be calculated by the formula:

$$d = \frac{N(N-1)}{\sum n(n-1)}$$

where
d = index of diversity
N = total number of organisms of all species
n = total number of organisms of each species

(i) Use the formula to calculate the index of diversity of plants in the forest. Show your working. *(2 marks)*

(ii) The forest was cleared to make more land available for agriculture. After the forest was cleared, the species diversity of insects in the area decreased. Explain why. *(3 marks)*

AQA June 2013

2 **(a)** What information is required to calculate an index of diversity for a particular community? *(1 mark)*

(b) Farmers clear tropical forest and grow crops instead. Explain how this causes the diversity of insects in the area to decrease. *(3 marks)*

(c) Farmers manage the ditches that drain water from their fields. If they do not, the ditches will become blocked by plants. Biologists investigated the effects of two different ways of managing ditches on farmland birds:

- ditch **A** was cleared of plants on both banks
- ditch **B** was cleared of plants on one bank.

The graph shows the number of breeding birds of all species along the two ditches, before and after management.

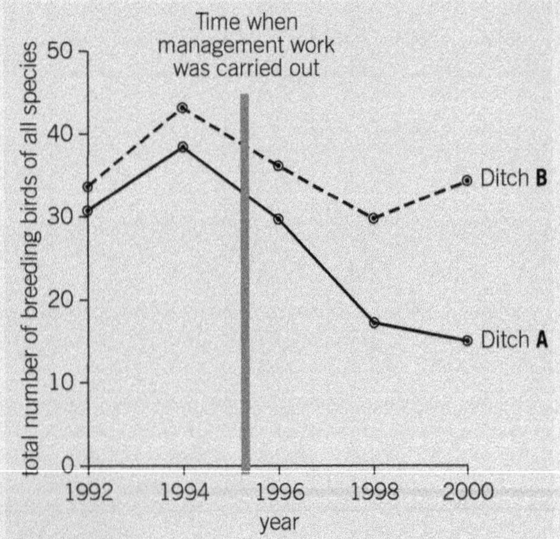

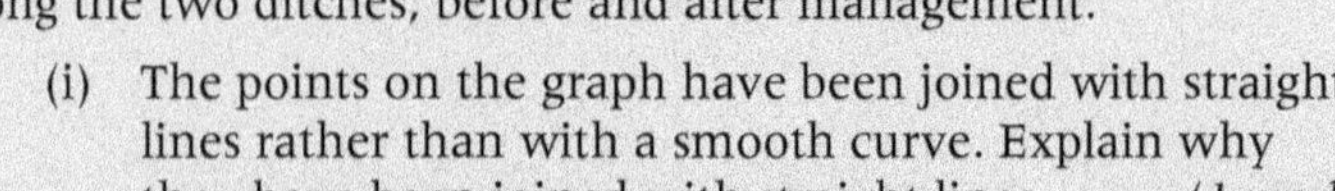

(i) The points on the graph have been joined with straight lines rather than with a smooth curve. Explain why they have been joined with straight lines. *(1 mark)*

(ii) It would have been useful to have had a control ditch in this investigation. Explain why. *(1 mark)*

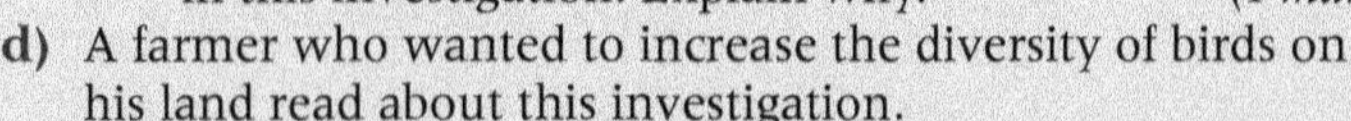

(d) A farmer who wanted to increase the diversity of birds on his land read about this investigation.

He concluded that clearing the plants from one bank would not decrease diversity as much as clearing the plants from both banks. Evaluate this conclusion. *(3 marks)*

AQA Jan 2011

3 Costa Rica is a Central American country. It has a high level of species diversity.

(a) There are over 12 000 species of plants in Costa Rica. Explain how this has resulted in a high species diversity of animals. *(2 marks)*

(b) The number of species present is one way to measure biodiversity. Explain why an index of diversity may be a more useful measure of biodiversity. *(2 marks)*

(c) Crops grown in Costa Rica are sprayed with pesticides. Pesticides are substances that kill pests. Scientists think that pollution of water by pesticides has reduced the number of species of frog.

(i) Frogs lay their eggs in pools of water. These eggs are small. Use this information to explain why frogs' eggs are very likely to be affected by pesticides in the water. *(2 marks)*

(ii) An increase in temperature leads to evaporation of water. Suggest how evaporation may increase the effect of pesticides on frogs' eggs. *(1 mark)*

AQA June 2011

4 To reduce the damage caused by insect pests, some farmers spray their fields of crop plants with pesticide. Many of these pesticides have been shown to cause environmental damage.

Bt plants have been genetically modified to produce a toxin that kills insect pests. The use of Bt crop plants has led to a reduction in the use of pesticides.

Scientists have found that some species of insect pest have become resistant to the toxin produced by the Bt crop plants.

The figure shows information about the use of Bt crops and the number of species of insect pest resistant to the Bt toxin in one country.

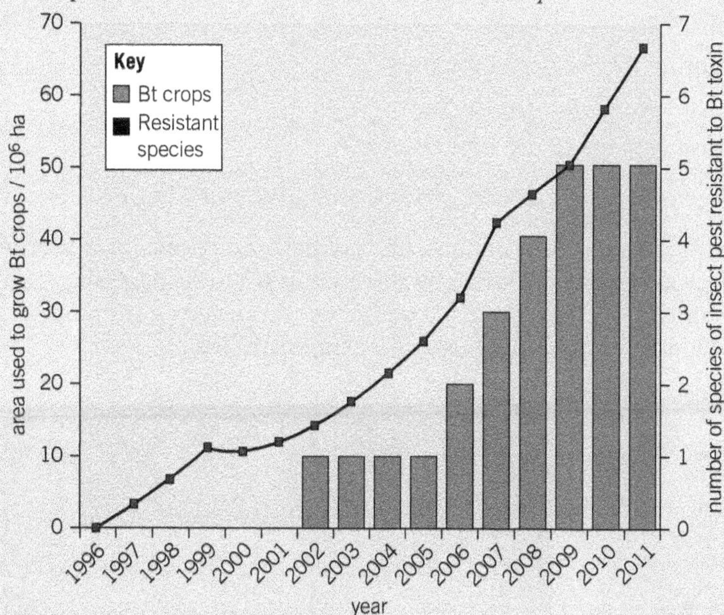

(a) Can you conclude that the insect pest resistant to Bt toxin found in the years 2002 to 2005 was the same insect species? Explain your answer. *(1 mark)*

(b) One farmer stated that the increase in the use of Bt crop plants had caused a mutation in one of the insect species and that this mutation had spread to other species of insect. Was he correct? Explain your answer. *(4 marks)*

(c) There was a time lag between the introduction of Bt crops and the appearance of the first insect species that was resistant to the Bt toxin. Explain why there was a time lag. *(3 marks)*

AQA SAMS AS PAPER 2

(d) Calculate the actual increase and the percentage increase in the area used to grow Bt crops between 2000 and 2010. *(2 marks)*

Answers to the Practice Questions are available at

www.oxfordsecondary.com/oxfordaqaexams-alevel-biology

11 Causes of disease

11.1 Pathogens

Learning objectives:

- → Explain what is meant by a pathogen.
- → Describe how pathogens enter the body.
- → Describe different ways in which pathogens cause disease.

Specification reference: 3.2.1.1

Microorganism is a general term for an organism that is too small to be seen without a microscope. Microorganisms include bacteria and viruses. Many microorganisms live more or less permanently in our bodies, benefiting from doing but causing us no harm. Some of these microorganisms are beneficial to us. Other microorganisms, however, cause **disease** – these are called **pathogens**. Before you look at these it is worth considering what is meant by disease.

What is disease?

It is difficult to say what is meant by 'disease'. Disease is not a single thing, but rather a description of certain symptoms, either physical or mental, or both. Disease suggests a malfunction of body or mind that has an adverse effect on good health. It has mental, physical, and social aspects.

Study tip

Be clear whether you are referring to a **bacterium** or a **virus**. Do not interchange the two terms.

Microorganisms as pathogens

For a microorganism to be considered a pathogen, it must:

- gain entry to the host
- colonise the tissues of the host
- resist the defences of the host
- cause damage to the host tissues.

Pathogens include bacteria, viruses, and fungi (see Figures 1–3).

If a pathogen gets into the host and colonises its tissue, an **infection** results. Disease occurs when an infection leads to recognisable symptoms in the host. When a pathogen is transferred from one individual to another, it is known as **transmission.**

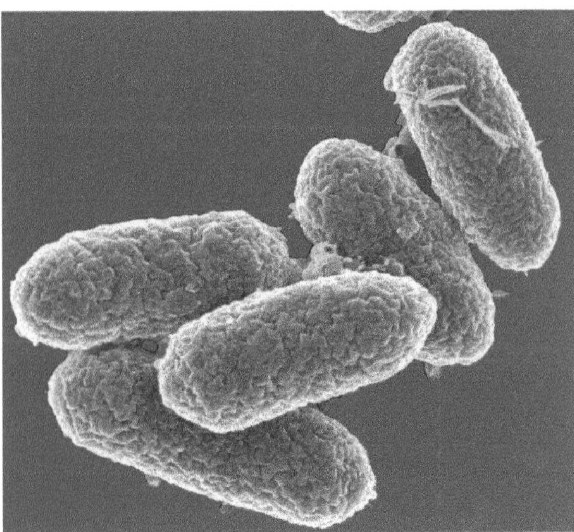

▲ **Figure 1** *Bacterial pathogen that causes salmonella food poisoning*

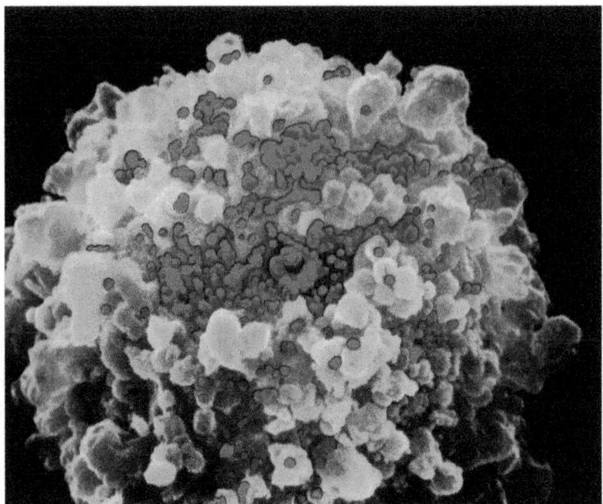

▲ **Figure 2** *False-colour scanning electron micrograph (SEM) of a T lymphocyte blood cell (green) infected with human immunodeficiency virus (HIV; red), the agent that causes acquired immune deficiency syndrome (AIDS)*

How do microorganisms get into the body?

Pathogens normally get into the body by penetrating one of the organism's interfaces with the environment. An interface is a surface or boundary linking two systems, in this case linking the external environment with the internal environment of the body. One of these interfaces is the skin. However, as the skin forms a thick, continuous layer, it is an effective barrier to infection. Invasion therefore normally occurs only when the skin is broken. This may happen as a result of cuts and abrasions or through the bites of insects and other animals. Some interfaces of the body have evolved to allow exchange of material between the internal and external environments. As a result the body linings at these points are thin, moist (and therefore sticky), have a large surface area, and are well supplied with blood vessels. Just as these features make for easy entry of molecules, so they also make for easy entry of pathogenic microorganisms. Interfaces of the body are hence common points of entry and include:

- the **gas-exchange system**. Many pathogens enter the body through the gas-exchange surfaces. Pathogens that cause influenza, tuberculosis, and bronchitis infect in this way.
- the **digestive system**. Food and water may carry pathogens into the stomach and intestines via the mouth. Cholera, typhoid, and dysentery pathogens enter the body by this route.
- the **reproductive system**. Bacteria such as *Treponema pallidum* (which causes syphilis) and viruses such as the herpes virus can be transmitted through intimate sexual contact (see Topic 13.3).

To help prevent the entry of pathogens, the body has a number of natural defences. These include:

- a mucus layer that covers exchange surfaces and forms a thick, sticky barrier that is difficult to penetrate
- the production of enzymes that break down the pathogens
- the production of stomach acid, which kills microorganisms.

How do pathogens cause disease?

Pathogens affect the body in two main ways:

- **by damaging host tissues**. Sometimes the sheer number of pathogens causes damage by, for example, preventing tissues functioning properly. Viruses inhibit the synthesis of DNA, RNA, and proteins by the host cells and divert cell resources towards making new virus particles. Many pathogens break down the membranes of the host cells.
- **by producing toxins**. Most bacterial pathogens produce toxins. The cholera bacterium produces a toxin that leads to excessive water loss from the lining of the intestines.

Some diseases, like malaria, have a single cause, but others, like heart disease, have a number of causes. Pathogens, lifestyle, and genetic factors can all cause disease.

How quickly a pathogen causes damage, and hence the onset of symptoms, is related to how rapidly the pathogen divides. Pathogens like those causing gastroenteritis divide about every 30 minutes and so symptoms of diarrhoea and vomiting become apparent within 24 hours of infection. The gastroenteritis pathogen also causes damage only when present in very large numbers. Other pathogens, such as the typhoid bacterium, cause harm when their numbers are relatively small.

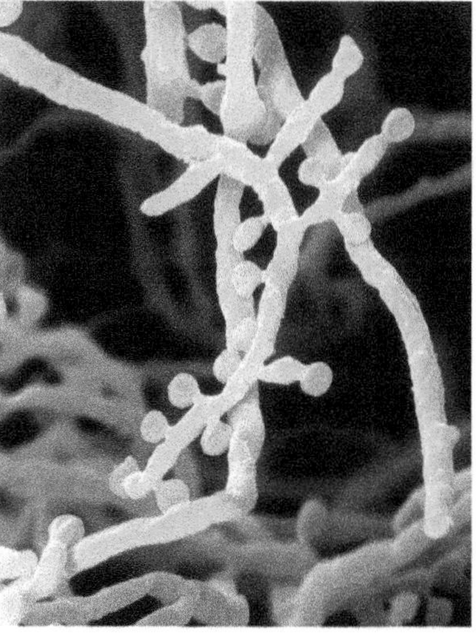

▲ **Figure 3** *False-colour scanning electron micrograph (SEM) of the fungus that causes athlete's foot*

Study tip

Remember that not all microorganisms cause disease. Therefore, when referring to a disease-causing organism, use the term pathogen, not microorganism.

Hint

It is the pathogen that enters the body and causes the symptoms, etc., *not* the disease. Do not write 'Cholera enters the body and this leads to diarrhoea', but rather write '*The bacterium that causes cholera* enters the body and this leads to diarrhoea'.

Summary questions

1. What is a pathogen?
2. Why are the digestive and respiratory systems often the sites of entry for pathogens?
3. In which **two** ways do pathogens cause disease?
4. Suggest **one** reason why oral antibiotics are not normally used to treat gastroenteritis and other diarrhoeal diseases.

11.2 Data and disease

Learning objectives:

→ Analyse and interpret data on disease.

→ Explain what is meant by a correlation.

→ Explain how a causal link is established.

Specification reference: 3.2.1.2

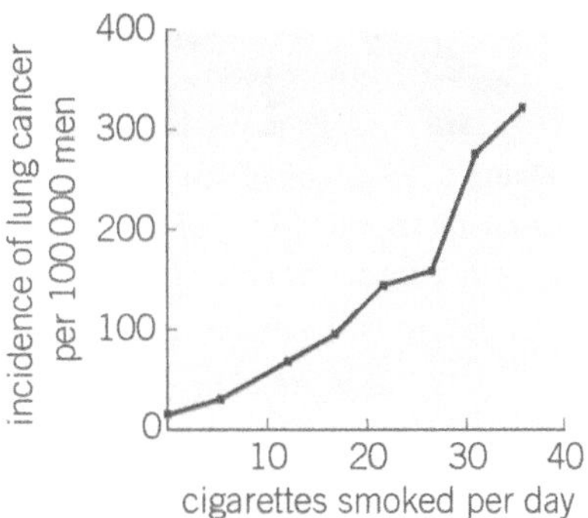

▲ **Figure 1** *Annual incidence of lung cancer per 100 000 men in the USA correlated to the daily consumption of cigarettes*

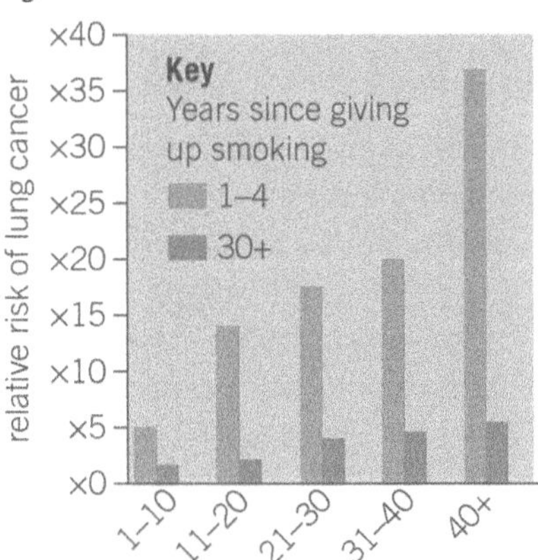

▲ **Figure 2** *The bar chart shows the risk of developing cancer in relation to the number of cigarettes smoked per day before stopping and the number of years since giving up smoking.*
Source: AQA, 2003

Study tip

It is important to be clear that a correlation does *not* mean that there is a causal link.

There is a considerable amount of information concerning disease and its possible causes. Newspapers, magazines, radio, TV, and the Internet bombard us with the latest statistics and research concerning the connection between various factors and the incidence of disease. Some of the information appears contradictory. So what can we believe? How can we tell if something is good or bad for us?

Analysing and interpreting data on disease

Epidemiology is the study of the incidence (number of cases) and pattern of a disease with a view to finding the means of preventing and controlling it. To do this, epidemiologists collect data on diseases and then look for a pattern or a relationship between these diseases and various factors in the lives of people who have them.

Correlations and causal relationships

A **correlation** occurs when a change in one of two variables is reflected by a change in the other variable.

The interpretation of the data in Figure 1 shows that there is a correlation between the number of cigarettes smoked a day and the incidence of lung cancer. What we *cannot* do, however, is to conclude that smoking is the **cause** of lung cancer. The data immediately seem to suggest this is the case but there is no actual evidence here to prove it. There needs to be a clear causal connection between smoking cigarettes and lung cancer before we can say that the case is proven. These data alone show only a correlation and not a cause. It could be that people who smoke exercise less and that it is the lack of exercise rather than the actual smoking which causes lung cancer. To prove that smoking cigarettes is the cause of lung cancer we would need experimental evidence to show that some component of the tobacco smoke led directly to people getting lung cancer. Recognising the distinction between a correlation and a causal relationship is a necessary and important skill.

Figure 1 shows how the incidence of lung cancer changes with the number of cigarettes smoked a day. What can we conclude? Well, nothing really. We can see that the more cigarettes that are smoked, the greater are the number of deaths from lung cancer. In other words, there is a positive correlation between the two factors. However, we cannot conclude that it is the cigarette smoke that causes lung cancer. Even though this graph does not itself establish a link, scientists have produced compelling experimental evidence to show that smoking tobacco definitely can cause lung cancer (see Topic 11.3).

Looking critically at data

It is easy to accept data and other scientific information at face value, but it should be looked at critically. To do this, consider the following questions when deciding how reliable the data are.

- Has the right factor been measured and have the correct questions been asked?
- How were the data gathered, were the methods reliable, and was the right apparatus used?

- Do those collecting the data have a vested interest in the outcome of the research?
- Has the study been repeated, with the same results and conclusions, by other people?
- Are there still unanswered questions?

Summary questions

Study Figure 2 and answer the following questions.

1 State **two** correlations shown by the information in this bar chart.

2 Explain why the information provided does not show a causal relationship between the correlations you have identified.

3 The *y*-axis of the bar chart is labelled 'relative risk of lung cancer'. Explain what this means. (It may help to refer to Topic 11.3.)

Hill's Criteria of Causation

Sir Austin Bradford Hill, a medical statistician, set out nine criteria which have to be met in order to establish a causal relationship between a specific factor and a disease. Hill stated:

'...None of these nine viewpoints can bring indisputable evidence for or against a cause and effect hypothesis... What they can do, with greater or lesser strength, is to help answer the fundamental question - is there any other way of explaining the set of facts before us? Is there any other answer equally or more likely than cause and effect?'

The nine criteria are as follows:

1 **Temporality**. This means that exposure to the factor must occur before the disease develops. This is the only essential criteria and it means, for example, that in order to say that exposure to UV radiation can cause skin cancer, those people who develop the disease must be shown to have been exposed before the disease developed.

2 **Strength of Association.** This can be established by statistical tests such as correlation coefficients. The greater the correlation between, for example, a diet high in salt and high blood pressure, the more likely it is the a high salt diet is causing high blood pressure.

3 **Dose Response Relationship.** This means that the more people are exposed to the factor, the more chance they have of developing the disease. Figure 1 shows that this is clearly the case with cigarette smoking and lung cancer.

4 **Consistency.** Similar correlations can be seen in different groups of people under different circumstances. Correlations between smoking and lung cancer have been demonstrated in different populations in different countries using different investigative techniques.

5 **Theoretical Plausibility.** There is an underlying biological reason to explain the link between the factor and the disease. For example, high salt levels in blood plasma lower the water potential leading to more water moving into the plasma by osmosis. This increases blood volume which in turn will increase blood pressure.

6 **Coherence.** When many different types of experimental data provide evidence supporting a causal relationship and there is no plausible conflicting theory. The conclusion that smoking causes lung cancer is supported by population studies (epidemiological studies), experimental studies on animals and cells and other biological data. This is discussed further in Topic 11.3.

7 **Specificity.** Ideally there is only one cause. This is frequently not the case, For example, high blood pressure can also be due to other environmental factors such as increased levels of stress.

8 **Experimental Evidence**. Experimental evidence that supports findings from epidemiological studies increases the validity of conclusions about causal relationships. For example, tar is a component of tobacco smoke and this was shown to cause cancers when applied to the skin of laboratory animals under controlled conditions.(Topic 11.3).

9 **Analogy.** When something is suspected of causing a disease, then other factors which are similar to this supposed cause should be considered. For example, if tar is cigarettes is a proposed cause of lung cancer, does exposure to tar in other forms produce cancers? In fact it has been shown to be the case that, for example, car mechanics exposed to the tars in used engine oil are at increased risk of developing skin cancer.

1 Use the Hill Criteria to discuss the difficulties in establishing a cause and effect relationship between cigarette smoking and coronary heart disease (see Topics 11.3 and 15.5).

11.3 Lifestyle and health

Learning objectives:

- → Explain what is meant by risk and how it is measured.
- → Assess the factors that affect the risk of contracting cancer and coronary heart disease.

Specification reference: 3.2.1.2

There are a number of disorders that result from an individual's lifestyle and the decisions they make. In some cases the harmful consequences of their behaviour are known at the outset. Most people who begin smoking are aware of the increased risk of lung **cancer** and **emphysema**. In other cases, the damage may only become apparent later. Exercise, normally beneficial, can lead to **osteoarthritis** if it is excessive or inappropriate.

What is risk?

Before we consider specific risk factors associated with cancer and coronary heart disease (CHD), let us consider what is meant by risk. Risk has many definitions depending on the context in which it is used. In respect of health, perhaps the simplest definition is:

a measure of the probability that damage to health will occur as a result of a given hazard

The concept of risk has two elements:

- the probability that a hazardous event will occur
- the consequences of that hazardous event.

This affects how we view risk. As an example, we may have a high probability of catching a cold, but, as the consequences are minor, we do not worry too much. The consequences of being struck by lightning are very severe, but, as the probability of this occurring is very low, again it does not worry us much. It is when the probability is high and the consequences severe that we become concerned.

Measurement of risk

Risk can be measured as a value that ranges from 0 per cent (no harm will occur) to 100 per cent (harm will certainly occur).

Health risks need a timescale

To tell someone that their risk of dying is 100 per cent is meaningless because every one of us will die sometime. To state that their risk of dying *in the next month* is 100 per cent has an altogether different meaning.

Risk is often relative

Risk is measured by comparing the likelihood of harm occurring in those exposed to a hazard with those who are not exposed to it, for example, smokers may be 15 times more likely to develop lung cancer than non-smokers.

Even when a risk is quantified, there are so many factors to consider that it is difficult to understand the risk. For example, take the figure above: smokers are 15 times more likely to develop lung cancer. To be able to understand the risk we need to know many other things:

- Over what time period does this occur?
- How does the number of cigarettes smoked a day affect the figure?
- Do stress levels, alcohol intake, occupation, gender, pollution, or other factors have an influence?
- Does it change according to where the smokers live, for example, in different countries, or in the city or countryside?

Hint

Remember that risk is about probabilities, *not* certainties. Some people may lead a lifestyle that would seem to put them at considerable risk and yet live well into old age. Conversely, others with few risk factors may become ill.

Misleading statistics

There is often so much more to a statistic, but reports in the media may be very misleading because they focus on a single figure. The impression given is that this figure applies to everyone, when often this is far from the case. In 2007 there were headlines such as 'HRT alert after more than 1000 women die' in national newspapers. This was certainly a disturbing statistic for the million or more women on hormone replacement therapy. A look behind the statistic showed that these extra deaths were over a 14-year period. The number of extra deaths each year was therefore 72. This is still a cause for concern but not nearly as alarming as the headlines would have us believe.

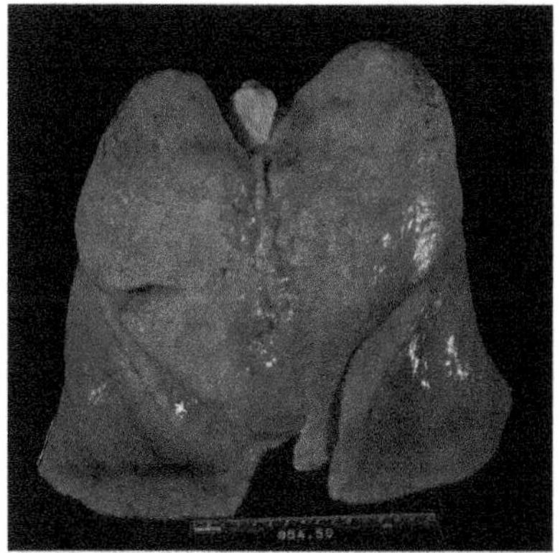

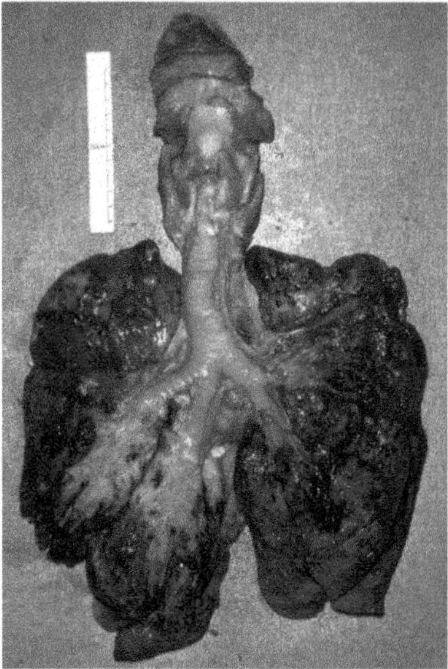

▲ **Figure 1** *Normal healthy lungs (top); smoker's lungs affected by cancer (bottom)*

Risk factors and cancer

Cancer is not a single disease and, likewise, does not have a single cause. Some causal factors are beyond our individual control, for example, age and genetic factors. Others are lifestyle factors and therefore within our power to change.

Lifestyle choices and cancer

We can do nothing about our genes or our age but our lifestyle can expose us to environmental and **carcinogenic** factors that put us at risk of contracting cancer. It is thought that about half the people who are diagnosed with cancer in the UK could have avoided getting the disease if they had changed their lifestyle. The specific lifestyle factors that contribute to cancer include:

- **smoking**. Not only smokers are in danger – those who passively breathe tobacco smoke also have an increased risk of getting cancer.
- **diet**. What we eat and drink affects our risk of contracting cancer. There is strong evidence that a low-fat, high-fibre diet, rich in fruit and vegetables, reduces the risk.
- **obesity**. Being overweight increases the risk of cancer.
- **physical activity**. People who take regular exercise are at lower risk from some cancers than those who take little or no exercise.
- **sunlight**. The more that someone is exposed to sunlight or light from sunbeds, the greater is their risk of skin cancer.

Smoking and lung cancer

Life insurance companies have calculated that, on average, smoking a single cigarette lowers an individual's life expectancy by 10.7 minutes – longer than it takes to smoke the cigarette! Whilst this is a statistical deduction rather than a scientific one, there is now clear scientific evidence to support the view that smoking cigarettes damages your health and reduces life expectancy. One type of evidence comes from correlations between cigarette smoking and certain diseases.

Figure 2 on the next page shows deaths from lung cancer in the UK correlated to the number of cigarettes smoked per year during a period in the last century. Study it carefully and then answer the questions overleaf.

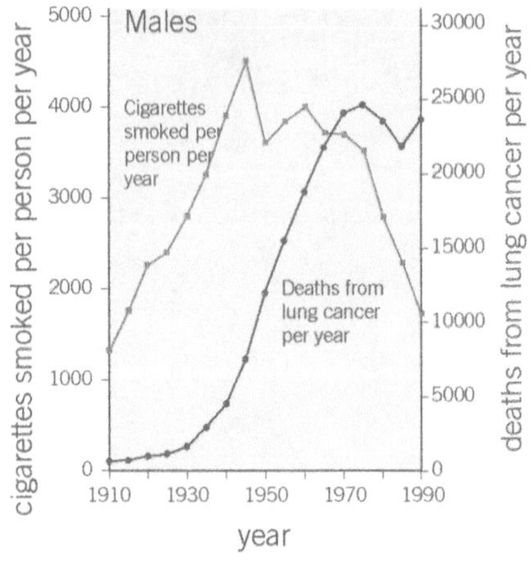

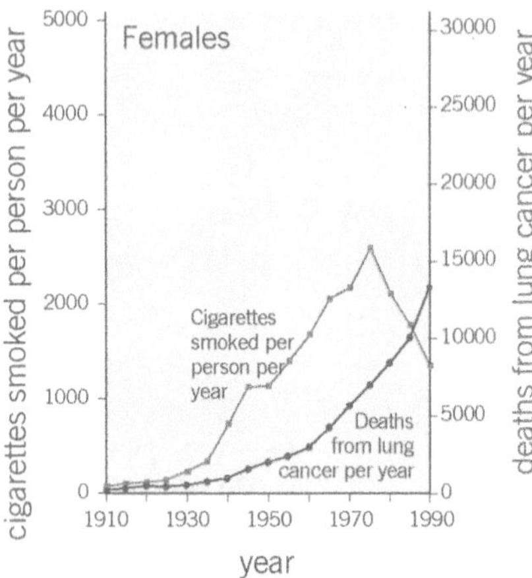

▲ **Figure 2** *Incidence of deaths from lung cancer in the UK correlated to cigarettes smoked per year (1910–90)*

1 In which decade did smoking reach its peak for the following?
   a males
   b females.
2 Explain how the graphs show that there is a correlation between the number of cigarettes smoked and deaths from lung cancer in both sexes.
3 In both sexes, the number of deaths per year from lung cancer increased over the period 1910 to 1970. Suggest **three** possible reasons for this.
4 Suggest a reason why there is a time lag between the number of cigarettes smoked and a corresponding change in the number of deaths from lung cancer.

Risk factors and coronary heart disease

Coronary heart disease (CHD) is the largest cause of death globally. There are a number of factors that increase the risk of an individual developing CHD. When combined together, four or five of these factors produce a disproportionately greater risk. Some factors, such as our genes, age, and sex, are beyond our control, but there are others that we can do something about.

Factors we can control (lifestyle factors)

There are certain factors in our lives that we can control, such as:

- **smoking.** Smokers are between two and six times more likely to develop CHD than non-smokers. Giving up smoking is the single most effective way of increasing life expectancy.
- **high blood pressure**. Excessive prolonged stress, certain diets, and lack of exercise all increase blood pressure and hence the risk of CHD.
- **blood cholesterol levels**. These can be kept lower by including fewer saturated fatty acids in the diet.
- **obesity**. A body mass index of over 25 is associated with an increased risk of CHD.
- **diet**. A high concentration of salt in the diet raises blood pressure, whereas high levels of saturated fatty acids increase blood cholesterol concentration. Both therefore increase the risk of CHD. By contrast, foods such as dietary fibre reduce the risk of CHD by lowering blood cholesterol levels.
- **physical activity**. Aerobic exercise can lower blood pressure and blood cholesterol as well as help avoid obesity – all of which reduce the risk of CHD.

Study tip

Graphical data may be presented using a dual scale as shown in Figure 2. Look carefully at the labelling and units of each axis when analysing this type of data.

Reducing the risk of cancer and CHD

There are measures that we can all take to reduce our chances of getting both cancer and CHD. These include:

- giving up or not taking up smoking
- avoiding becoming overweight
- reducing salt intake in the diet

- reducing intake of cholesterol and **saturated fats** in the diet
- taking regular aerobic exercise
- keeping alcohol consumption within safe limits
- increasing the intake of dietary fibre and **antioxidants** in the diet.

Smoking and disease

Sixty years ago smoking was a highly popular pastime. Today, both individuals and governments worldwide are making strenuous efforts to eliminate it. How such a change was brought about illustrates the role that scientific knowledge can play in changing personal perceptions and public policy.

History of smoking

When tobacco was first introduced to Britain in the 16th century it was usually smoked in pipes. At the end of the 19th century the invention of the cigarette-making machine made tobacco readily available to all. Initially, only men smoked, but women took up smoking in the 1920s. By 1945, the equivalent of 12 cigarettes a day for every British male were being smoked. At the time, the public regarded smoking as a harmless pleasure. Doctors, however, were alarmed by a phenomenal increase in deaths from lung cancer. At a 1947 conference, a number of scientists suggested tobacco smoke as a possible cause of the increase. Their problem was how to convince the public, governments, and the rest of the scientific world.

1 Scientists need to look at all possible explanations for the correlations that they have recognised. Suggest another possible cause of lung cancer, other than smoking, that they could have investigated.

Epidemiological evidence linking smoking to disease

Epidemiologists collect data on diseases and then look for correlations between these diseases and various factors in the lives of those who have them. The world's longest-running survey of smoking began in the UK in 1951. This survey, and others elsewhere in the world, has revealed a number of statistical facts about smokers.

- A regular smoker is three times more likely to die prematurely than a non-smoker.
- The more cigarettes smoked per day, the earlier, on average, a smoker dies.
- Smokers who give up the habit improve their life expectancy compared to those who continue to smoke.
- One in two long-term smokers will die early as a result of smoking.
- The incidence of pulmonary disease increases with the number of cigarettes smoked.
- Smokers make up 98 per cent of emphysema sufferers.

Data like those in Figure 3 were used to help establish a link between disease and smoking.

2 What correlation is shown by the data in Figure 3?

Summary questions

1 What single lifestyle change within the population of the UK would bring about the greatest reduction in cancer rates?

2 In what **three** ways would 30 minutes of brisk exercise each day reduce your chances of developing coronary heart disease?

3 A friend asks how she can change her diet in order to live longer. What advice do you give her?

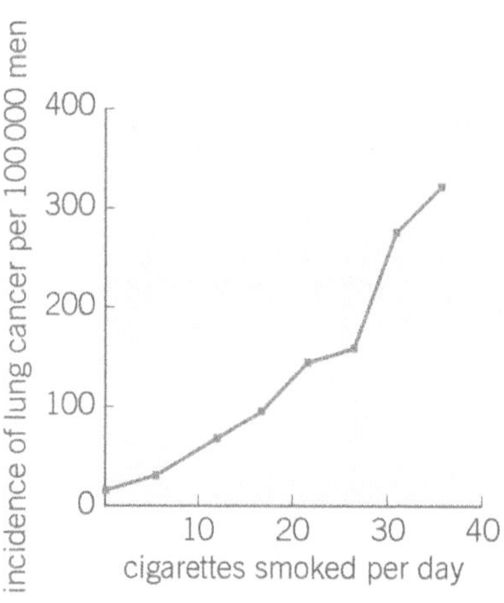

▲ **Figure 3** *Annual incidence of lung cancer per 100 000 men in the USA correlated to daily consumption of cigarettes*

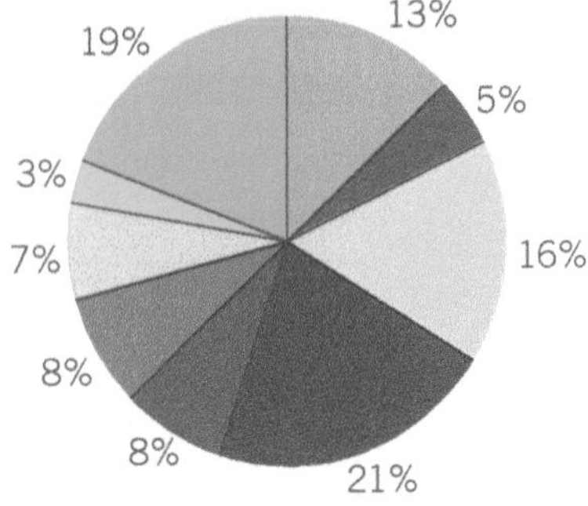

Key

- Respiratory disease
- Injuries and poisoning
- All other causes
- Coronary heart disease
- Stroke
- Other cardiovascular diseases
- Lung cancer
- Colo-rectal cancer
- Other cancer

▲ **Figure 4** *Causes of death in men in the UK in 2004. Source: National Statistics website: www.statistics.gov.uk. Crown copyright material is reproduced with the permission of the Controller, Office of Public Sector Information (OPSI).*

Epidemiological statistics show correlations between lung cancer and smoking. These include:

- A man smoking 25 cigarettes a day is 25 times more likely to die of lung cancer than a non-smoker.
- The longer a person smokes, the greater the risk of developing lung cancer. Smoking 20 cigarettes a day for 40 years increases the risk of lung cancer eight times more than smoking 40 cigarettes a day for 20 years.
- When a person stops smoking, their risk of developing lung cancer decreases and approaches that of a non-smoker after around 10–15 years (depending on age and amount of tobacco consumed).
- The death rate from lung cancer is 18 times greater in a smoker than in a non-smoker.

Cigarette manufacturers and some smokers argued that these epidemiological correlations were coincidental.

3 Much of the data linking smoking to lung cancer were collected from very large samples of the population. Suggest why this weakens the argument that the link is coincidental.

4 Do the data provide evidence of a causal link between lung cancer and smoking? Explain your answer.

Experimental evidence linking smoking to disease

Scientists carried out experiments in the 1960s in which dogs were made to inhale cigarette smoke. The smoke was either inhaled directly or first passed through a filter tip. Those dogs that inhaled the filtered smoke remained generally healthy. Those inhaling unfiltered smoke developed pulmonary disease and early signs of lung cancer. Scientists then carried out a further series of experiments that allowed them to formulate a new hypothesis from each result, which they could then test experimentally.

- Machines were used to simulate the action of smoking and to collect the harmful constituents that accumulated in the filters.
- These were then analysed chemically and each constituent was tested in the laboratory for its ability to damage epithelial cells and mutate the genes they contain. This was done by adding tar to the skin of mice or to cells that had been grown in culture.
- As a result of such tests it was shown that the tar found in cigarette smoke contained **carcinogens**.
- The constituent chemicals of the tar were each tested and one, benzopyrene (BP), was shown to mutate DNA.
- The scientists still had to demonstrate precisely how it caused cancer. They carried out experiments that showed that BP is absorbed by epithelial cells and converted to a derivative. This then binds with a **gene** and mutates it.
- Another experiment showed that this **mutation** led to uncontrolled cell division of epithelial cells and hence the growth of a **tumour**.
- Even this was not proof. In further experiments, scientists showed that the mutations of the gene in a cancer cell occurred at three specific

points on the DNA. When the derivative of BP from tobacco smoke was used to mutate the gene, it caused changes to the DNA at precisely the same points.

5 What is the **key** evidence that smoking is a cause of lung cancer?

The evidence was now conclusive. Smoking tobacco could cause lung cancer. This is not to say that it always does, but simply that there is an increased risk – it is about probabilities not certainties.

This case study illustrates how scientists can suspect a correlation, collect the epidemiological evidence to demonstrate that the correlation exists, and then design and carry out a series of experiments to establish a causal link. We can never say absolutely that something is proven, only that there is proof within the bounds of our current scientific knowledge. That knowledge, and the theories based on it, are constantly being adapted in the light of new scientific evidence and discoveries.

These experiments convinced the public of the health risks of smoking and led to reduced use of tobacco in the UK (from 82 per cent of the male population in 1948 to 30 per cent in 2002). This changed view in turn persuaded the government to take measures that were designed to reduce smoking. These included:

- progressively raising taxes on tobacco
- banning tobacco advertising
- placing health warnings on tobacco products
- banning smoking in work and public places, including bars, pubs, and clubs.

6 'My father smoked 30 cigarettes a day and lived to be 95.' This type of argument is sometimes used to suggest that smoking is not harmful. Explain why scientists do not accept this reasoning.

The ethics of animal experimentation

The experiments described here will have indirectly prevented millions of premature deaths. However, the involvement of beagle dogs and mice in these experiments provoked a public outcry. This led to the 1986 Animal Act, which established a three-tier licensing system to limit and control animal experiments. The Act laid down a set of ethical standards to be followed, and restricted the use of animals to cases where there is no realistic alternative.

▲ **Figure 5** *Smoking these 20 cigarettes would, on average, reduce your life expectancy by 3½ hours*

Practice questions: Chapter 11

1 The graph in Figure 1 gives information about the effects of cigarette smoking, plasma cholesterol concentrations, and high blood pressure on the incidence of heart disease in American men.

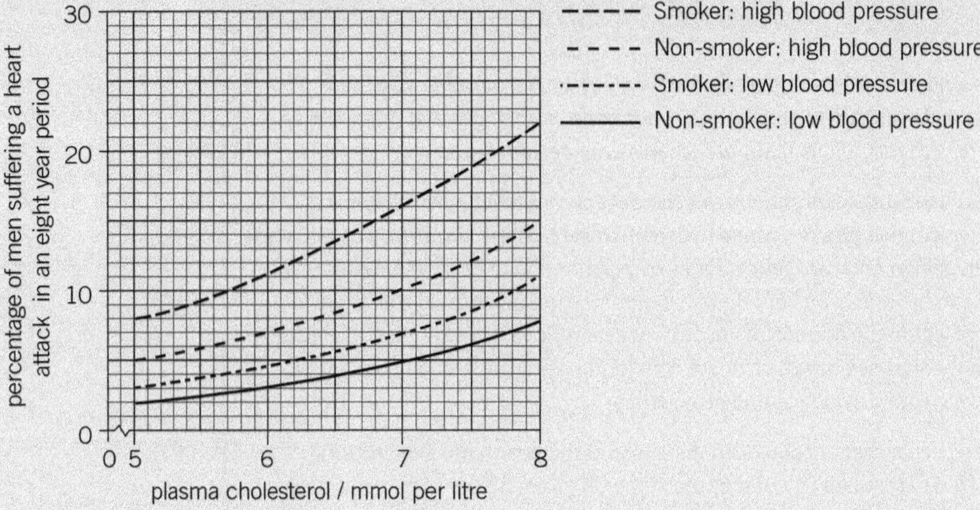

▲ **Figure 1**

(a) A non-smoker with low blood pressure has a plasma cholesterol concentration of 5 mmol per litre. Over a period of time this concentration increases to 8 mmol per litre. By how many times has his risk of heart disease increased? Show your working. *(2 marks)*

(b) Two non-smoking men with low blood pressure both have plasma cholesterol concentrations of 5 mmol per litre. One of them starts to smoke and the plasma cholesterol concentration of the other increases to 7 mmol per litre. Which man is now at the greater risk of heart disease? Explain your answer. *(3 marks)*

AQA 2001

2 The table shows the number of deaths from various causes in a group of individuals of the same age. Individuals were identified as smokers or non-smokers.

Cause of death	Number of deaths among smokers	Number of deaths among non-smokers
Total deaths (all causes)	7316	4651
Coronary artery disease	3361	1973
Strokes	556	428
Aneurysm	86	29
Lung cancer	397	37
Other causes	2916	2184

(a) Why was it necessary for the smokers and the non-smokers to be the same age? *(2 marks)*

(b) Do the figures in the table show that smokers were more likely to have died from a stroke than non-smokers? Use suitable calculations to support your answer. *(3 marks)*

(c) Figures 2 and 3 show information from one study of lung cancer and lung diseases in adults of all ages in the UK.

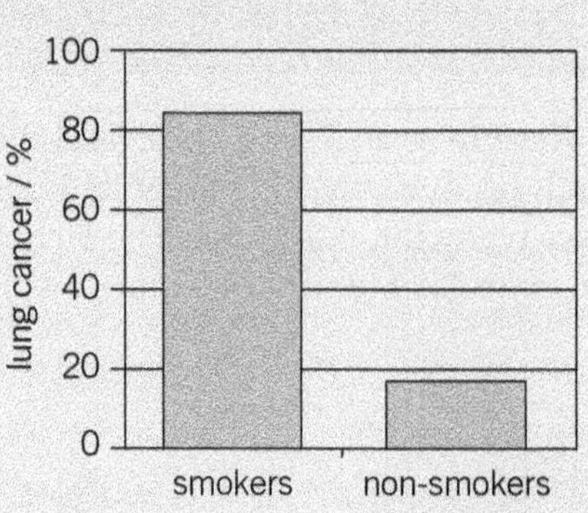

▲ **Figure 2** *Proportion of people with lung cancer who are smokers or non-smokers*

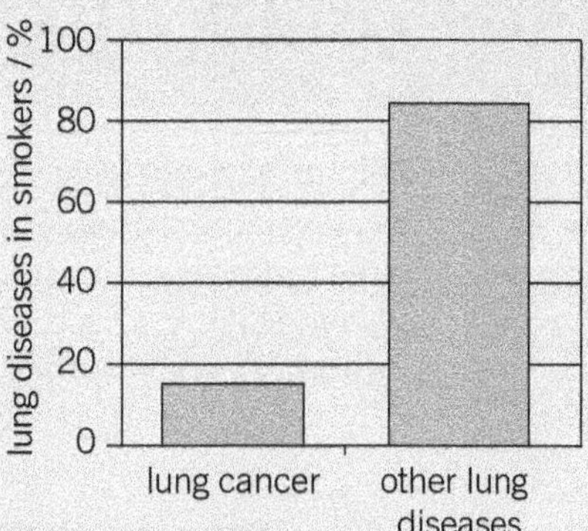

▲ **Figure 3** *Proportion of types of lung disease in smokers with lung disease*

(i) Give **three** conclusions that can be drawn from the results of this study. *(3 marks)*

(ii) Suggest **two** reasons why conclusions made only on the basis of these data may not be reliable. *(2 marks)*

AQA 2003; AQA 2002

3 Figure 4 shows the influence of different risk factors on the incidence of coronary heart disease in women (7.5 mm Hg is equal to 1 kilopascal).

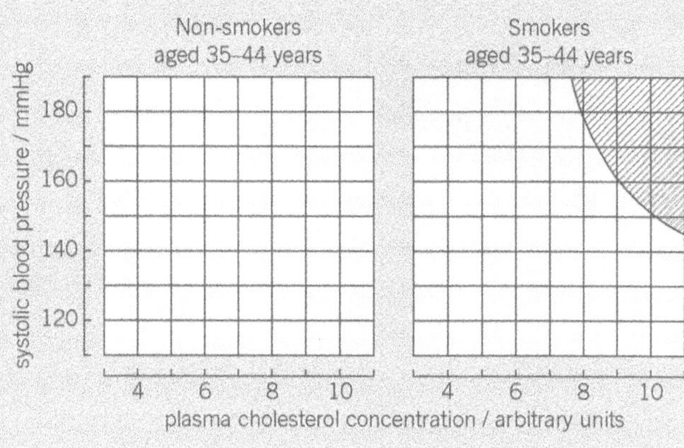

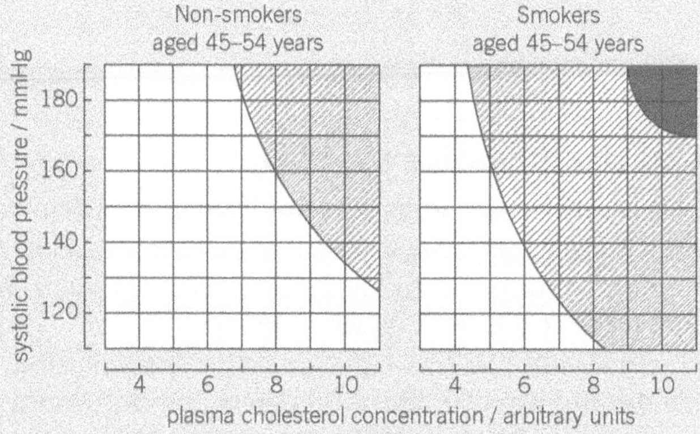

Key

Risk of developing coronary heart disease during next 10 years

less than 15% 15–30% greater than 30%

▲ **Figure 4**

(i) Use Figure 4 to give the characteristics of women with the highest risk of developing coronary heart disease.

(ii) Figure 4 only has limited value in predicting whether a particular woman might develop coronary heart disease. Explain why. *(5 marks)*

Answers to the Practice Questions are available at
www.oxfordsecondary.com/oxfordaqaexams-alevel-biology

12 Digestion and absorption

12.1 Enzymes and digestion

Learning objectives:

- → Describe the structure and function of the major parts of the digestive system.
- → Explain how the digestive system breaks down food both physically and chemically.

Specification reference: 3.2.1.1

Study tip

Digestion is the process in which *large* molecules are hydrolysed by enzymes into *small* molecules, which can be absorbed and assimilated.

The human digestive system is made up of a long muscular tube and its associated glands. The glands produce **enzymes** that hydrolyse large molecules into small ones ready for absorption. The digestive system (Figure 1) is therefore an exchange surface through which food substances are absorbed.

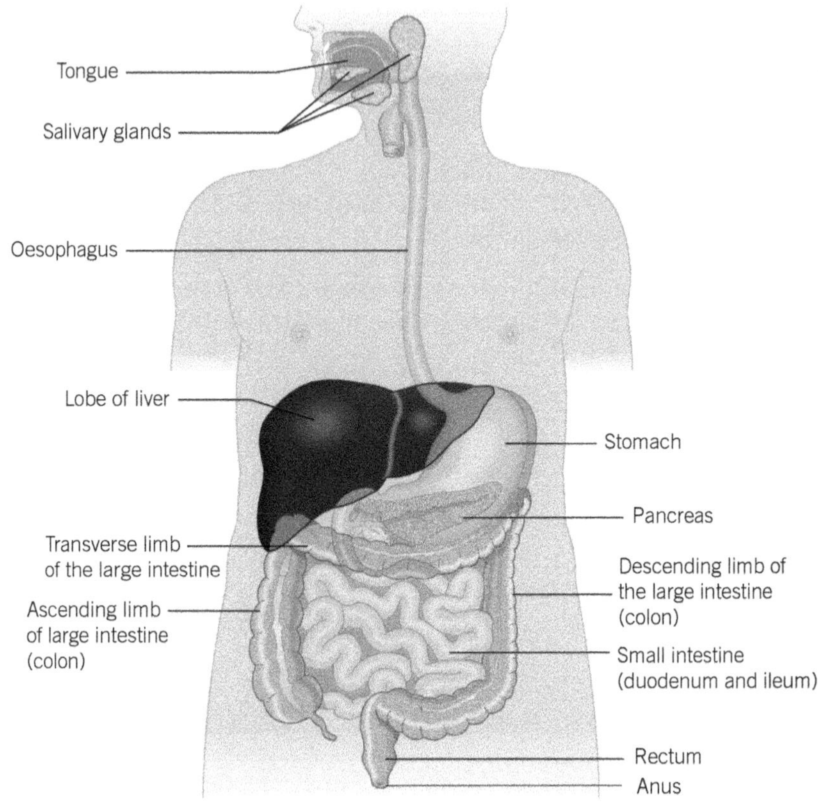

▲ **Figure 1** *Human digestive system*

Major parts of the digestive system

- The **oesophagus** carries food from the mouth to the stomach.
- The **stomach** is a muscular sac with an inner layer that produces enzymes. Its role is to store and digest food, especially proteins. It has glands that produce enzymes which digest protein.
- The stomach empties into the small intestine. The small intestine consists of the **duodenum** and ileum. Digestion continues in the duodenum using the enzymes from the pancreas which enter via the pancreatic duct. Secretions from the liver via the gall bladder assist the activity of these enzymes.
- The **ileum** is a long muscular tube. Food is further digested in the ileum by enzymes that are produced by its walls and by glands that pour their secretions into it. The inner walls of the ileum are folded into villi, which gives them a large surface area. The surface area of these villi is further increased by millions of tiny projections, called microvilli, on the epithelial cells of each villus. This adapts the ileum for its purpose of absorbing the products of digestion into the bloodstream.

- The **large intestine** absorbs water. Most of the water that is absorbed is water from the secretions of the many digestive glands.
- The **rectum** is the final section of the intestines. The faeces are stored here before periodically being removed via the anus in a process called **egestion**.
- The **salivary glands** are situated near the mouth. They pass their secretions via a duct into the mouth. These secretions contain the enzyme amylase, which **hydrolyses** starch into maltose.
- The **pancreas** is a large gland situated below the stomach. It produces a secretion called pancreatic juice. This secretion contains proteases to hydrolyse proteins, lipase to hydrolyse lipids, and amylase to hydrolyse starch.

What is digestion?

In humans, as with many organisms, digestion takes place in two stages:

1 physical breakdown
2 chemical digestion.

Physical breakdown

If the food is large, it is broken down into smaller pieces by means of structures such as the teeth. This not only makes it possible to ingest the food but also provides a large surface area for chemical digestion. Food is churned by the muscles in the stomach wall and this also physically breaks it up.

Chemical digestion

Chemical digestion hydrolyses large, insoluble molecules into smaller, soluble ones. It is carried out by enzymes. All digestive enzymes function by **hydrolysis**. Hydrolysis is the splitting up of molecules by adding water to the chemical bonds that hold them together. Enzymes are specific and so it follows that more than one enzyme is needed to hydrolyse a large molecule. Usually one enzyme hydrolyses a large molecule into sections and these sections are then hydrolysed into smaller molecules by one or more additional enzymes. There are different types of digestive enzymes, three of which are particularly important:

- **Carbohydrases** hydrolyse carbohydrates, ultimately to monosaccharides.
- **Lipases** hydrolyse lipids (fats and oils) into glycerol and fatty acids.
- **Proteases** hydrolyse proteins, ultimately to amino acids.

You will look at these three groups of digestive enzymes in more detail in the next topic.

Summary questions

1 List the parts of the digestive system that are responsible for the physical breakdown of food.

2 The pancreas can be described as an exocrine and an endocrine gland – explain why.

3 What is meant by egestion?

Hint

The contents of the intestines are *not* inside the body. Molecules and ions only truly enter the body when they cross the cells and cell-surface membranes of the epithelial lining of the intestines.

Hint

All organisms are made up of the same biological molecules and therefore your food consists almost entirely of other organisms, or parts of them. You must first hydrolyse them into molecules that are small enough to pass across cell-surface membranes.

Synoptic link

It will help you understand this topic if you revisit Chapter 1.

12.2 Digestion

Learning objectives:

- → Explain the role of enzymes in the breakdown of carbohydrates, lipids, and proteins.
- → Explain that enzymes are specific to their substrates, and that different macromolecules require different enzymes for their breakdown.

Specification reference: 3.2.1.2

Carbohydrate digestion

It usually takes more than one enzyme to completely hydrolyse a large molecule. Typically, one enzyme hydrolyses the molecule into smaller sections and then other enzymes further hydrolyse these sections into their monomers. These enzymes are usually produced in different parts of the digestive system. It is obviously important that enzymes are added to the food in the correct sequence. This is true of starch digestion, for example.

First, the enzyme amylase is produced in the mouth and by the pancreas. Amylase hydrolyses the alternate glycosidic bonds of the starch molecule to produce the disaccharide maltose. Maltose is in turn hydrolysed into the monosaccharide α-glucose by a second enzyme, a disaccharidase called maltase. Maltase is produced by the lining of the ileum.

In humans the process takes place as follows:

- Saliva enters the mouth from the salivary glands and is thoroughly mixed with the food during chewing.
- Saliva contains **salivary amylase**. This starts hydrolysing any starch in the food to maltose. It also contains mineral salts that help to maintain the pH at around neutral. This is the optimum pH for salivary amylase to work.
- The food is swallowed and enters the stomach, where the conditions are acidic. This acid **denatures** the amylase and prevents further hydrolysis of the starch.
- After a time the food is passed into the small intestine, where it mixes with the secretion from the pancreas called pancreatic juice.
- The pancreatic juice contains **pancreatic amylase**. This continues the hydrolysis of any remaining starch to maltose. Alkaline salts are produced by both the pancreas and the intestinal wall to maintain the pH at around neutral so that the amylase can function.
- Muscles in the intestine wall push the food along the ileum. Its epithelial lining produces the enzyme **maltase**. Maltase is not released into the lumen of the ileum but is attached to the cell-surface membranes of the epithelial cells that line the ileum. It is therefore referred to as a **membrane-bound disaccharidase**. The maltase hydrolyses the maltose from starch breakdown into α-glucose.

In addition to maltose, there are two other common disaccharides in the diet that are hydrolysed – sucrose and lactose.

Hint

Remind yourself of the test for starch. The disaccharide maltose is a reducing sugar – how would you test for this?

Study tip

In the test for a non-reducing sugar such as sucrose, the sugar is first hydrolysed by boiling it in dilute hydrochloric acid. The presence of the enzyme sucrase lowers the activation energy needed for hydrolysis to occur, so hydrolysis occurs rapidly at body temperature.

Sucrose is found in many natural foods, especially fruits. Lactose is found in milk, and hence in milk products, such as yoghurt and cheese. Each disaccharide is hydrolysed by a membrane-bound disaccharidase as follows:

- **Sucrase** hydrolyses the single glycosidic bond in the sucrose molecule. This hydrolysis produces the two monosaccharides glucose and fructose.
- **Lactase** hydrolyses the single glycosidic bond in the lactose molecule. This hydrolysis produces the two monosaccharides glucose and galactose.

Hint

Enzyme names usually end in '-ase' and start with the the first part of the name of their substrate (the substance on which they act). Hence maltase hydrolyses maltose, and sucrase hydrolyses sucrose.

Lipid digestion

Lipids are hydrolysed by enzymes called **lipases**. Lipases are enzymes produced in the pancreas that hydrolyse the ester bond in triglycerides to form fatty acids and monoglycerides. A monoglyceride is a glycerol molecule with a single fatty acid molecule attached. Lipids (fats and oils) are first split up into tiny droplets called **micelles** (Topic 12.3) by **bile salts**, which are produced by the liver. This process is called **emulsification** and increases the surface area of the lipids so that the action of lipases is speeded up.

Protein digestion

Proteins are large, complex molecules that are hydrolysed by a group of enzymes called **peptidases** (proteases). There are a number of different peptidases:

- **Endopeptidases** hydrolyse the peptide bonds between amino acids in the central region of a protein molecule, forming a series of peptide molecules.
- **Exopeptidases** hydrolyse the peptide bonds on the terminal amino acids of the peptide molecules formed by endopeptidases. In this way they progressively release dipeptides and single amino acids.
- **Dipeptidases** hydrolyse the bond between the two amino acids of a dipeptide. Dipeptidases are membrane bound, being part of the cell-surface membrane of the epithelial cells lining the ileum.

Summary questions

1. Define hydrolysis.
2. List **two** structures that produce amylase.
3. Suggest why the stomach does not have villi or microvilli.
4. Name the final product of starch digestion in the gut.
5. List **three** enzymes produced by the epithelium of the ileum.

Lactose intolerance

Milk is the only food of human babies and so they produce a relatively large amount of lactase, the enzyme that hydrolyses lactose, the sugar in milk. As milk forms a less significant part of the diet in adults, the production of lactase diminishes as children get older. This reduction can be so great in some adults that they produce little, or no, lactase at all.

This was not a problem to our ancestors but can be to humans of today. Humans that produce no lactase cannot hydrolyse the lactose they consume. When the undigested lactose reaches the large intestines, microorganisms hydrolyse it. This gives rise to small soluble molecules and a large volume of gas. This can result in diarrhoea because the soluble molecules lower the water potential of the material in the colon. The condition is known as lactose intolerance. Some people with the condition cannot consume milk or milk products at all whereas others can consume them only in small amounts.

▲ **Figure 1** *Milk and milk products*

1. a. Suggest the process by which micoorganisms produce 'a large volume of gas' in lactose-intolerant individuals.
   b. Suggest a reason why this gas is unlikely to be carbon dioxide.
2. Suggest an explanation for why lactose intolerance is a problem for modern-day humans but wasn't for our ancestors.
3. Explain how the lowering of water potential in the colon can cause diarrhoea.

12.3 Absorption of the products of digestion

You saw in Topic 12.2 how enzymes hydrolyse carbohydrates, fats, and proteins. The products of this hydrolysis are monosaccharides, amino acids, monoglycerides, and fatty acids. You will now see how these products are absorbed by the ileum.

Learning objectives:

- → Describe the structure of the ileum.
- → Explain how the ileum is adapted for the function of absorption.
- → Explain how monosaccharides and amino acids are absorbed.
- → Explain how triglycerides are absorbed.

Specification reference: 3.2.1.3

Structure of the ileum

The ileum is adapted to the function of absorbing the products of digestion. The wall of the ileum is folded and possesses finger-like projections, about 1 mm long, called **villi** (Figure 1). They have thin walls lined with epithelial cells, on the other side of which is a rich network of blood capillaries. The villi considerably increase the surface area of the ileum and therefore accelerate the rate of absorption.

Villi are situated at the interface between the lumen (cavity) of the intestines (in effect outside the body) and the blood and other tissues inside the body. They are part of a specialised exchange surface adapted for the absorption of the products of digestion. Their properties increase the efficiency of absorption in the following ways:

- They increase the surface area for diffusion.
- They are very thin walled, thus reducing the distance over which diffusion takes place.
- They contain muscle and so are able to move. This helps to maintain diffusion gradients because their movement mixes the contents of the ileum. This ensures that as the products of digestion are absorbed from the food adjacent to the villi, new material rich in the products of digestion replaces it.
- They are well supplied with blood vessels so that blood can carry away absorbed molecules and hence maintain a diffusion gradient.
- The epithelial cells lining the villi possess **microvilli** (Figure 2). These are finger-like projections of the cell-surface membrane that further increase the surface area for absorption.

Synoptic link

You will better understand the contents of this topic if you first read through Topics 4.2, 4.4, and 5.1.

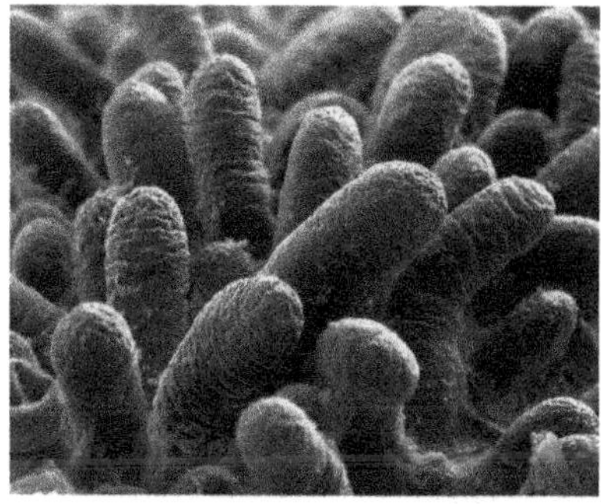

▲ **Figure 1** *False-colour scanning electron micrograph (SEM) of villi in the lining of the ileum*

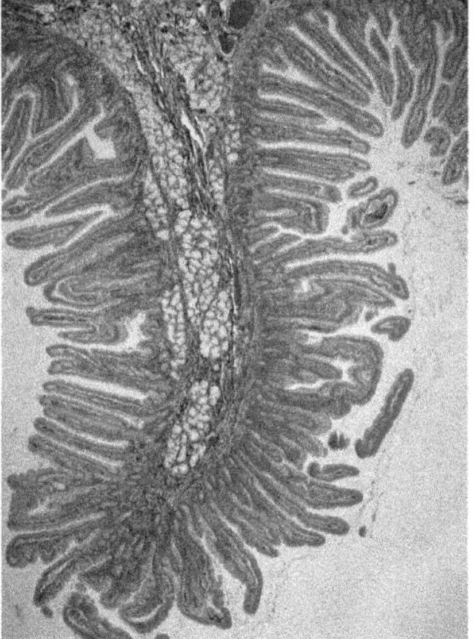

◀ **Figure 2** *Light micrograph of a section through a villus in the small intestine. Villi are projections that increase the surface area for the absorption of food. They are covered in microvilli (smaller, finger-like projections) that further increase this surface area.*

Absorption of amino acids and monosaccharides

The digestion of proteins produces amino acids, whereas that of carbohydrates produces monosaccharides such as glucose, fructose, and galactose. The methods of absorbing these products are the same, namely facilitated diffusion and co-transport.

The role of diffusion in absorption

Diffusion (Topic 4.2) is the net movement of molecules or ions from a region where they are highly concentrated to a region where their concentration is lower.

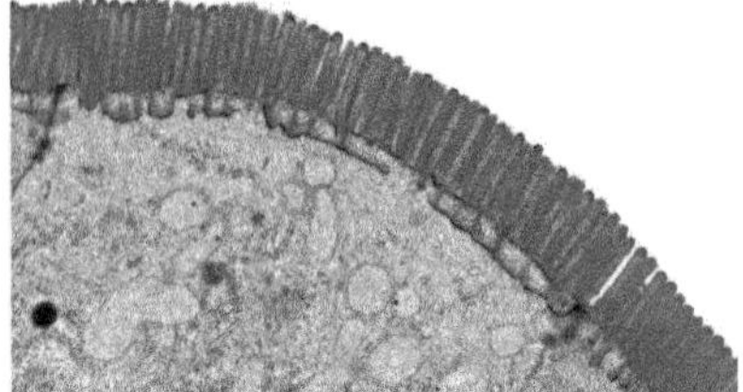

▲ **Figure 3** *Microvilli on an epithelial cell from the small intestine*

As carbohydrates and proteins are being digested continuously, there is normally a greater concentration of glucose and amino acids within the ileum than in the blood. There is therefore a concentration gradient down which glucose moves by facilitated diffusion from inside the ileum into the blood. Given that the blood is constantly being circulated by the heart, the glucose absorbed into it is continuously being removed by the cells as they use it up during respiration. This helps to maintain the concentration gradient between the inside of the ileum and the blood. This means the rate of movement by facilitated diffusion across epithelial cell-surface membranes is increased.

Role of active transport in absorption

At best, diffusion only results in the concentrations either side of the intestinal epithelium becoming equal. This means that not all the available glucose and amino acids can be absorbed in this way and some may pass out of the body. The reason why this does not happen is because glucose and amino acids are also being absorbed by **active transport** (see Topic 4.4). This means that all the glucose and amino acids should be absorbed into the blood.

Study tip

Do not confuse villi and microvilli. Villi (Topic 6.10) are 1 mm projections of the wall of the ileum whereas microvilli are 0.6 µm projections of the cell-surface membrane of the epithelial cells that line this wall. Microvilli are therefore more than one thousand times smaller than villi.

The actual mechanism by which they are absorbed from the small intestine is an example of **co-transport**. This term is used because either glucose or amino acids are drawn into the cells along with sodium ions that have been actively transported out by the sodium–potassium pump (see Topic 4.4). It takes place in the following manner (see Figure 4):

1. Sodium ions are actively transported out of epithelial cells by the sodium–potassium pump into the blood. This takes place through one type of **protein molecule** found in the cell-surface membrane of the epithelial cells.
2. This maintains a much higher concentration of sodium ions in the lumen of the intestine than inside the epithelial cells.
3. Sodium ions diffuse into the epithelial cells down this concentration gradient through a different type of protein carrier

(co-transport protein) in the cell-surface membrane. As the sodium ions diffuse in through this second carrier protein, they carry either amino acid molecules or glucose molecules into the cell with them.

4 The glucose/amino acids pass into the blood plasma by facilitated diffusion using another type of carrier.

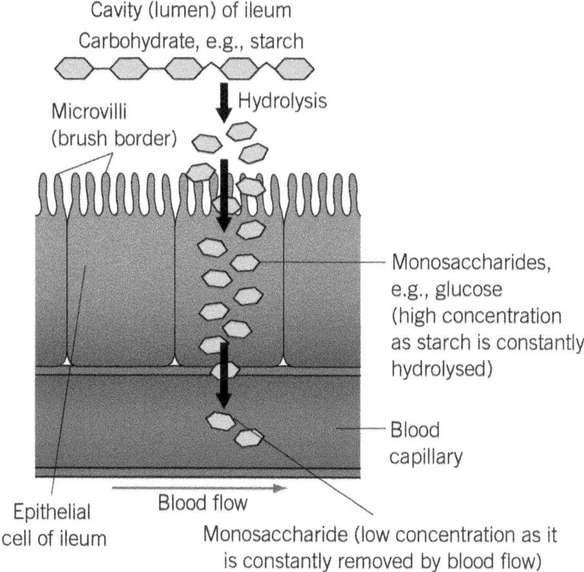

▲ **Figure 4** *Absorption of monosaccharides (e.g., glucose) by diffusion in the ileum*

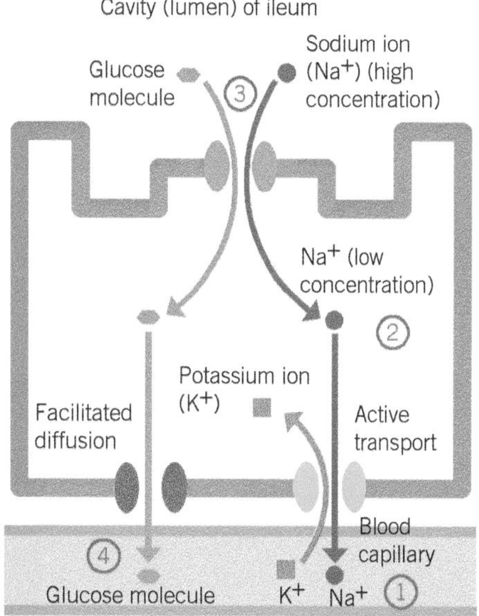

▲ **Figure 5** *Co-transport of a glucose molecule*

Both sodium ions and glucose/amino acid molecules move into the cell, but whilst the sodium ions move *down* their concentration gradient, the glucose molecules move *against* their concentration gradient. It is the sodium ion concentration gradient, rather than ATP directly, that powers the movement of glucose and amino acids into the cells. This makes it an indirect rather than a direct form of active transport.

Absorption of triglycerides

Once formed during digestion, monoglycerides and fatty acids remain in association with the bile salts that initially emulsified the lipid droplets (see Topic 12.2). The structures formed are called **micelles**. They are tiny, being around 4–7 nm in diameter. Through the movement of material within the lumen of the ileum, the micelles come into contact with the epithelial cells lining the villi of the ileum. Here the micelles break down, releasing the monoglycerides and fatty acids. As these are non-polar molecules, they easily diffuse across the cell-surface membrane into the epithelial cells.

Once inside the epithelial cells, monoglycerides and fatty acids are transported to the endoplasmic reticulum where they are recombined to form triglycerides. Starting in the endoplasmic reticulum and continuing in the Golgi apparatus, the triglycerides associate with cholesterol and lipoproteins to form structures called **chylomicrons**. Chylomicrons are special particles adapted for the transport of lipids.

Chylomicrons move out of the epithelial cells by **exocytosis**. They enter lymphatic capillaries called **lacteals** that are found at the centre of each villus. The process is illustrated in Figure 6.

From here, the chylomicrons pass, via lymphatic vessels, into the blood system. The triglycerides in the chylomicrons are hydrolysed by an enzyme in the endothelial cells of blood capillaries, from where they diffuse into cells.

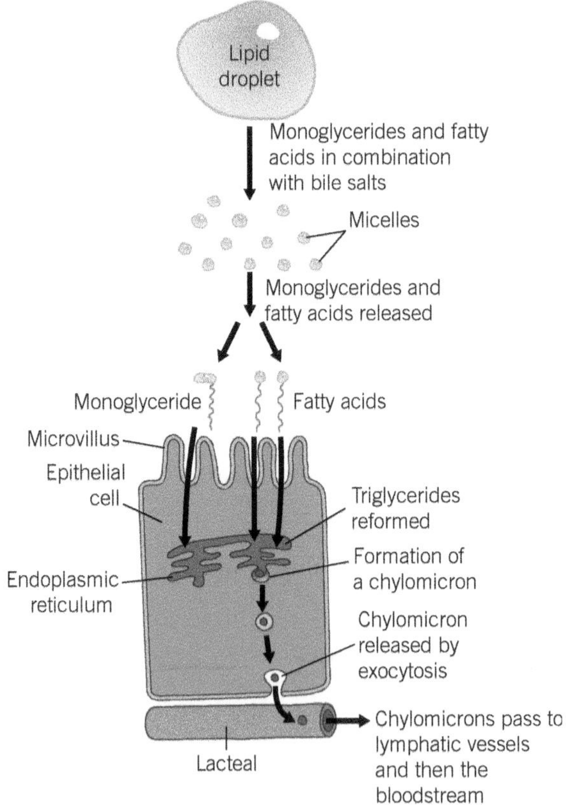

▲ **Figure 6** *The absorption of triglycerides*

Absorption of fatty acids

Bile salts play a role in the digestion and absorption of fatty acids. One end of the bile salt molecule is soluble in fat (lipophilic) but not in water (hydrophobic). The other end is soluble in water (hydrophilic) but not in fat (lipophobic). Bile salt molecules therefore arrange themselves with their lipophilic ends in fat droplets, leaving their lipophobic ends sticking out. In this way they prevent fat droplets from sticking to each other to form large droplets, leaving only tiny ones (micelles). It is in this form that fatty acids reach the epithelial cells of the ileum where they break down, releasing the fatty acids for absorption.

An experiment was carried out to investigate the absorption of fatty acids. Six sections of intestine were filled with a fatty acid called oleic acid. To each section were added different mixtures of other contents, as shown in Table 1.

Iodoacetate inhibits an enzyme involved in glycolysis – a stage of the respiratory process in cells that involves phosphorylation.

▼ **Table 1**

Contents of section of intestine					Relative amount of oleic acid absorbed in 10 hours
Bile salts	Glycerol	Phosphate	Glycerol phosphate	Iodoacetate	
✓	✗	✗	✗	✗	2.9
✓	✗	✓	✗	✗	1.1
✓	✓	✗	✗	✗	2.6
✓	✓	✓	✗	✗	5.8
✓	✗	✗	✓	✗	8.5
✓	✗	✗	✓	✓	0.0

✓ = substance present ✗ = substance absent

From the information in Table 1:

1 List **three** pieces of evidence that support the idea that the absorption of fatty acids in the intestine is increased if they are combined with a compound of glycerol and phosphate.

2 What is the evidence supporting the view that the absorption of fatty acids involves phosphorylation.

Summary questions

1 List **three** organelles that you would expect to be numerous and/or well developed in an epithelial cell of the ileum, giving a reason for your choice in each case.

2 Name the other chemical that moves across epithelial cells with glucose molecules during co-transport.

3 In addition to having microvilli, state **one** other feature of the epithelial cells of the ileum that would increase the rate of absorption of amino acids.

4 In each of the following events in the glucose co-transport system, state whether the movements are active or passive:

  a sodium ions move out of the epithelial cell

  b sodium ions move into the epithelial cell

  c glucose molecules move into the epithelial cell.

Practice questions: Chapter 12

1 Figure 1 shows an epithelial cell from the small intestine.

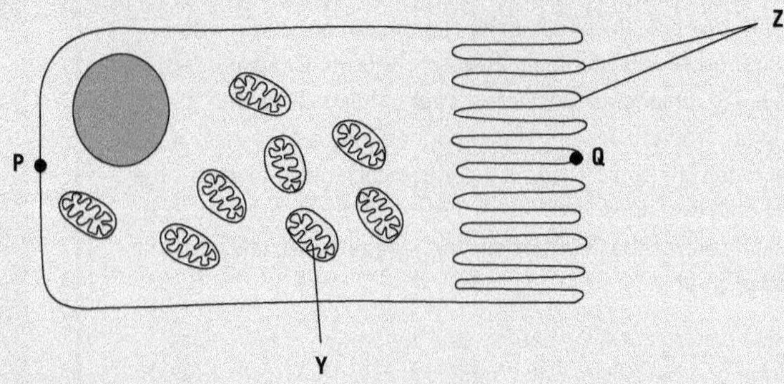

▲ Figure 1

(a) (i) Name organelle **Y**. *(1 mark)*

(ii) There are large numbers of organelle Y in this cell. Explain how these organelles help the cell to absorb the products of digestion. *(2 marks)*

(b) This diagram shows the cell magnified 1000 times. Calculate the actual length of the cell between points **P** and **Q**. Give your answer in µm. Show your working. *(2 marks)*

(c) Coeliac disease is a disease of the human digestive system. In coeliac disease, the structures labelled **Z** are damaged.

Although people with coeliac disease can digest proteins, they have low concentrations of amino acids in their blood.

Explain why they have low concentrations of amino acids in their blood. *(2 marks)*

AQA Jan 2010

2 Gluten is a protein found in wheat. When gluten is digested in the small intestine, the products include peptides. Peptides are short chains of amino acids. These peptides cannot be absorbed by facilitated diffusion and leave the gut in faeces. Some people have coeliac disease. The epithelial cells of people with coeliac disease do not absorb the products of digestion very well. In these people, some of the peptides from gluten can pass between the epithelial cells lining the small intestine and enter the intestine wall. Here, the peptides cause an immune response that leads to the destruction of microvilli on the epithelial cells.

Use the information in the passage and your own knowledge to answer the following questions.

(a) Name the type of chemical reaction which produces amino acids from proteins. *(1 mark)*

(b) The peptides released when gluten is digested cannot be absorbed by facilitated diffusion (lines 2–3). Suggest why. *(3 marks)*

(c) The epithelial cells of people with coeliac disease do not absorb the products of digestion very well (lines 4–5). Explain why *(3 marks)*

(d) Explain why the peptides cause an immune response *(1 mark)*

(e) Scientists have identified a drug which might help people with coeliac disease. It reduces the movement of peptides between epithelial cells. They have carried out trials of the drug with patients with coeliac disease.
Suggest two factors that should be considered before the drug can be used on patients with the disease. *(2 marks)*

AQA June 2012

3 **(a)** Name the monosaccharides of which the following disaccharides are composed:
  (i) sucrose *(1 mark)*
  (ii) lactose. *(1 mark)*

(b) Amylase and maltase are involved in the digestion of starch in the small intestine. Complete the table by identifying where these enzymes are produced and the product of the reaction they catalyse. *(2 marks)*

AQA Jan 2013

Name of enzyme	Where the enzyme is produced	Product of the reaction catalysed by the enzyme
Amylase		
Maltase		

4 Figure 2 shows a cell from the kidney. This cell rapidly absorbs glucose.

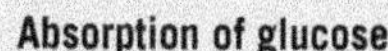

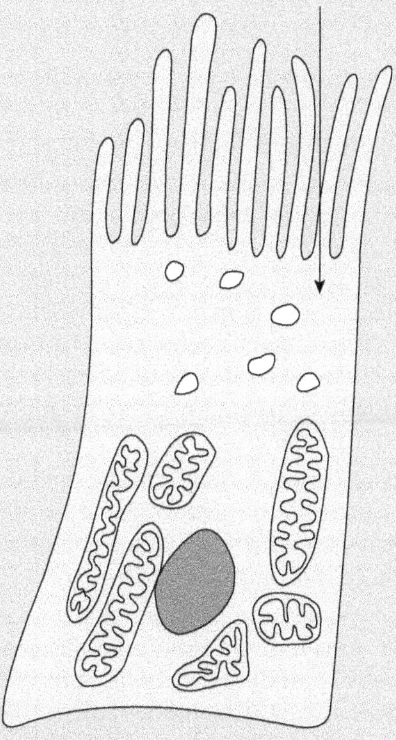

▲ **Figure 2**

(a) (i) Give two structures shown in Figure 2 that are adaptations of this cell which make the rapid absorption of glucose possible. *(1 mark)*
  (ii) For each structure you identified in part 5(a)(i), explain how it makes the rapid absorption of glucose possible. *(2 marks)*

(b) The absorption of glucose into the cell leads to the movement of water into the cell. Explain how. *(2 marks)*

AQA June 2014

Answers to the Practice Questions are available at
www.oxfordsecondary.com/oxfordaqaexams-alevel-biology

13 Human diseases

13.1 Cholera

Learning objectives:

→ Explain how *Vibrio cholerae* causes the symptoms of cholera.

Specification reference: 3.2.3.1

Globally, cholera is of great significance, killing an estimated 120 000 people each year. The agent that causes the disease is a curved, rod-shaped bacterium called *Vibrio cholerae*. It is characterised by the presence of a flagellum at one end. Bacteria are examples of prokaryotic cells (Topic 2.5).

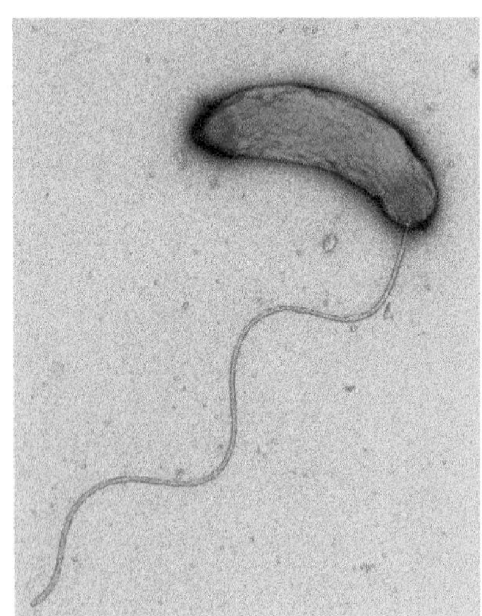

▲ **Figure 1** *False-colour transmission electron micrograph (TEM) of the cholera bacterium,* Vibrio cholerae

How the cholera bacterium causes disease

The main symptoms of cholera are diarrhoea and, consequently, dehydration. Some people infected with the cholera bacterium show few if any symptoms. Some act as carriers, unwittingly spreading the disease.

Vibrio cholerae is transmitted by the ingestion of water or, more rarely, food that has been contaminated with faecal material containing this pathogen.

Almost all the *Vibrio cholerae* bacteria ingested by humans are killed by the acidic conditions in the stomach. However a few may survive, especially if the pH is above 4.5. They then cause disease as follows:

- When the surviving bacteria reach the small intestine they use their flagella to propel themselves, in a corkscrew-like fashion, through the mucus lining of the intestinal wall.
- They then start to produce a toxic protein. This protein has two parts. One part binds to specific carbohydrate receptors on the cell-surface membrane. Only the epithelial cells of the small intestine have these specific receptors, which explains why the cholera toxin only affects this region of the body. The other, toxic part of the protein enters the epithelial cells. This causes the **ion channels** of the cell-surface membrane to open, so that the chloride **ions** that are normally contained within the epithelial cells flood into the lumen of the intestine.
- The loss of chloride ions from the epithelial cells raises their **water potential**, whilst the increase of chloride ions in the lumen of the intestine lowers its water potential. Water therefore flows from the cells into the lumen.
- The loss of ions from the epithelial cells establishes a concentration gradient. Ions therefore move by diffusion into the epithelial cells from the surrounding tissues, including the blood. This, in turn, establishes a water potential gradient that causes water to move by **osmosis** from the blood and other tissues into the intestine.
- It is this loss of water from the blood and other tissues into the intestine that causes the symptoms of cholera, namely severe diarrhoea and dehydration.

Cholera is treated by restoring the water and ions that have been lost using oral rehydration therapy (see Topic 13.2).

Study tip

Always write in *precise* scientific terms rather than in vague, generalised terms. For example, state that cholera may be spread 'by faecal contamination of the hands' rather than 'by not making sure your hands are clean after using the toilet'.

Transmission of cholera

Cholera is transmitted by the ingestion of water or, more rarely, food that has been contaminated with faecal material containing the pathogen. Such contamination can arise because:

- drinking water is not properly purified
- untreated sewage leaks into water courses
- food is contaminated by people who prepare and serve it
- organisms, especially shellfish, have fed on untreated sewage released into rivers or the sea.

1 Given that cholera is transmitted by food and water that is contaminated with faecal matter, suggest three measures that may be used to limit the spread of the disease.
2 Suggest a reason why, in countries where cholera is common, babies who are breast-fed are affected by cholera far less often than babies who are bottle-fed.
3 Suggest how inhibiting the development of a flagellum in the bacterium that causes cholera may prevent the disease.
4 Suggest a reason why injecting antibiotics into the blood can be effective in killing the cholera bacterium whereas the same antibiotics taken orally (by mouth) are not effective.

▲ **Figure 2** *Cholera is most easily transmitted where there is a lack of clean water, sanitation is poor, and houses lack basic facilities*

Summary questions

1 Which domain and genus would you classify the bacterium which causes cholera in?
2 How does the cholera toxin cause diarrhoea?

13.2 Oral rehydration therapy

Learning objectives:

- → Describe the action of oral rehydration therapy in the treatment of cholera.
- → Discuss the role of science in developing improved oral rehydration therapies.
- → Discuss the ethical procedures that govern drug trials.

Specification reference: 3.2.3.2

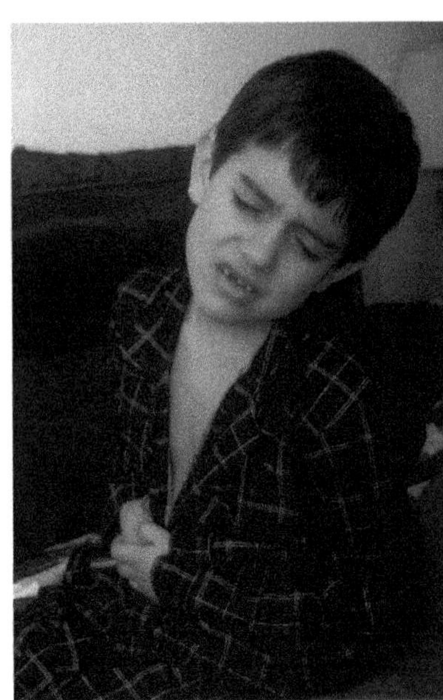

▲ **Figure 1** *Abdominal pain is a symptom of diarrhoea*

Cholera is just one of a number of diarrhoeal diseases that infect the intestines. Treatments for these diseases take a number of forms, of which oral rehydration therapy is one of the most important.

What causes diarrhoea?

Diarrhoea is an intestinal disorder in which watery faeces are produced frequently. The causes include:

- damage to the epithelial cells lining the intestine
- loss of microvilli due to toxins
- excessive secretion of water due to toxins, for example, cholera toxin.

As a result of diarrhoea, excessive fluid is lost from the body and/or insufficient fluid is taken in to make up for this loss. Either way, dehydration results and this may be fatal.

What is oral rehydration therapy?

To treat diarrhoeal diseases it is vital to rehydrate the patient. Just drinking water is ineffective for two reasons:

- Water is not being absorbed from the intestine. Indeed, as in the case of cholera, water is actually being lost from cells.
- The drinking of water does not replace the electrolytes (ions) that are being lost from the epithelial cells of the intestine.

It is possible to replace the water and electrolytes intravenously by a drip, but this requires trained personnel. What is needed is a suitable mixture of substances that can safely be taken by mouth and which will be absorbed by the intestine. But how can the patient be rehydrated if the intestine is not absorbing water? As it happens, there is more than one type of carrier protein in the cell-surface membranes of the epithelial cells that absorb sodium ions. The trick is to develop a rehydration solution that uses these alternative pathways. As sodium ions are absorbed, so the water potential of the cells falls and water enters the cells by osmosis. Therefore, a rehydration solution needs to contain:

- water – to rehydrate the tissues
- sodium – to replace the sodium ions lost from the epithelium of the intestine and to make optimum use of the alternative sodium-glucose carrier proteins
- glucose – to stimulate the uptake of sodium ions from the intestine and to provide energy
- potassium – to replace lost potassium ions and to stimulate appetite
- other **electrolytes** – such as chloride and citrate ions, to help prevent electrolyte imbalance.

These ingredients can be mixed and packaged as a powder, which can be made up into a solution with boiled water as needed. This can then be administered by people with minimal training. The solution must be given regularly, and in large amounts, throughout the illness.

Developing and testing improved oral rehydration solutions

The development of oral rehydration solutions resulted from a long process of scientific experimentation.

- Early rehydration solutions led to side effects, especially in children. These were caused by excess sodium.
- Mixtures with a lower sodium content but more glucose were tested. Unfortunately, the additional glucose lowered the water potential in the lumen of the small intestine so much that it started to draw even more water from the epithelial cells. This made the dehydration even worse.
- Lowering the glucose content reduced this effect but, as glucose also acted as a respiratory substrate, it reduced the amount of energy being supplied to the patient. The problem then was how to supply the glucose without it having an osmotic effect.
- One answer was to use starch in place of some of the glucose. Starch is a large, insoluble molecule that consequently has no osmotic effect. It is, however, broken down steadily by amylase and maltase in the small intestine into its glucose **monomers** (see Topic 12.2). By experimenting with different concentrations of starch, a rehydration solution was developed that released glucose at the optimum rate, so it was taken up as it was produced, without adversely affecting the water potential. Further scientific research is being carried out to find the best source of starch.

Rice starch is a popular choice for two main reasons:

- It is readily available in many parts of the world, especially where diarrhoeal diseases are common.
- It provides other nutrients, such as amino acids. Not only are these nutrients nutritionally valuable, but they also help the uptake of sodium ions from the small intestine.

As rice flour produces a very viscous solution, it is hard to swallow. One answer to this problem is to partly digest the starch with amylase. The smaller, and hence more soluble, starch components produce a less viscous drink.

Testing new drugs, including oral rehydration solutions

The development of any medicine takes place in stages, each of which must be tested for its safety. Whilst initial testing can be done on tissue cultures and animals, to be sure of a drug's effectiveness and safety, it must eventually be tested on humans. This is normally carried out in four phases.

1. A small number (20–80) of usually healthy people are given a tiny amount of the drug to test for side effects rather than to see if the drug is effective. The dose may be increased gradually in a series of such trials. This stage takes around 6 months.
2. The drug is then given to a slightly larger number of people (100–300) who have the condition that the drug is designed to treat. This is to check that the drug works and to look at any safety issues. This stage takes up to 2 years.

Summary questions

1. The following inexpensive home-made rehydration solution is recommended when commercial products are not available:
   8 level teaspoons of sugar +
   1 level teaspoon of table salt dissolved in 1 litre of boiled water.
   - **a** Give **two** reasons why sugar (glucose) is included in the mixture.
   - **b** Table salt is sodium chloride. Give **two** reasons why it is included in the mixture.
   - **c** Why is it essential that the water is boiled?
2. Bananas are rich in potassium. It is sometimes recommended that mashed banana is added to the mixture. Give **two** reasons why this might help the patient recover.
3. Suggest another advantage of adding mashed banana to the mixture before drinking, especially in the case of children.
4. Sports drinks contain a high proportion of glucose to help replace the glucose used during strenuous exercise. Explain in terms of water potential why these drinks are therefore not suitable to rehydrate people with diarrhoea.

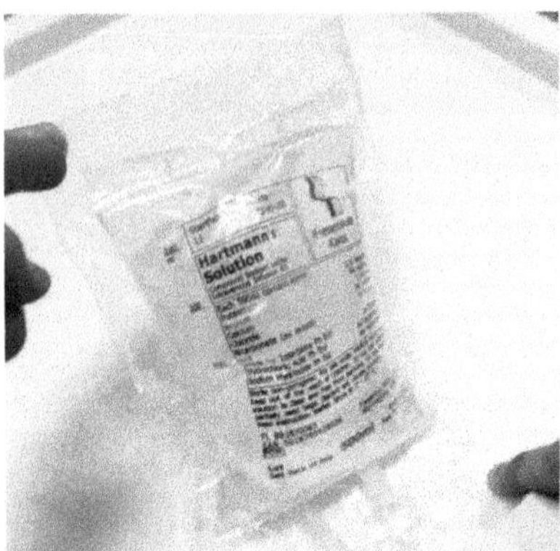

▲ **Figure 2** *Bag containing rehydration solution*

Hint

It is important to remember that oral rehydration solutions do not prevent or cure diarrhoea. They simply rehydrate and nourish the patient until the diarrhoea is cured by some other means.

3 A large-scale trial of many thousands of patients then takes place. Many are given a dummy drug, called a **placebo**. Often neither the scientists nor the patients know who has taken the real drug and who has taken the placebo until after the trial. This type of trial is known as a double-blind trial. These trials take many years.

4 If the drug passes all these stages it may be granted a licence, but its use and effects are still monitored over many years to check for any long-term effects.

1 Why must drugs ultimately be tested on humans?
2 Suggest a reason why a placebo is necessary to ensure that the results of a drug trial are reliable.
3 Suggest why the results of a double-blind trial may be more reliable than those of a trial in which the patients know whether they are taking the real drug or a placebo.

Before a licence is granted, the results of the trials will be published in a scientific journal, such as *The Lancet*. This helps to ensure the validity of the results in three ways:

- To be considered worthy of publication by the editors of such journals, any research must conform to accepted scientific standards.
- Other scientists, especially those in the same field of research, are able to critically review the findings and challenge them if they feel that they are inaccurate or misleading. This is known as peer review.
- The experiments/trials can be replicated by others to see if the same results can be obtained.

These trials raise a number of ethical issues. Who should take part in such trials and how should they be recompensed? How can the participants be made aware of the dangers, especially when the trials themselves are designed to expose these dangers? What happens when things go wrong (as happened in March 2006, during a drug trial in London, when six young men became seriously ill)?

13.3 The human immunodeficiency virus (HIV)

Learning objectives:

- → Describe the structure of HIV.
- → Explain how HIV replicates.
- → Explain how HIV causes AIDS.
- → Describe the treatment and control of AIDS.
- → Explain why antibiotics are ineffective against viruses.

Specification reference: 3.2.4.1 and 3.2.4.2

The human immunodeficiency virus (HIV) causes the disease **acquired immune deficiency syndrome (AIDS)**. Among contagious diseases it is a relative newcomer, having been first diagnosed in 1981. In this topic you will look at the structure of HIV and how it leads to the symptoms of AIDS.

Structure of HIV

The structure of HIV is shown in Figure 1. On the outside is a **lipid envelope**, embedded in which are peg-like **attachment proteins**. Inside the envelope is a protein layer called the **capsid** that encloses two single strands of **RNA** and some enzymes. One of these enzymes is **reverse transcriptase**, so called because it catalyses the production of DNA from RNA – the reverse reaction to that carried out by transcriptase. The presence of reverse transcriptase, and consequent ability to make DNA from RNA, means that HIV belongs to a group of viruses called **retroviruses**.

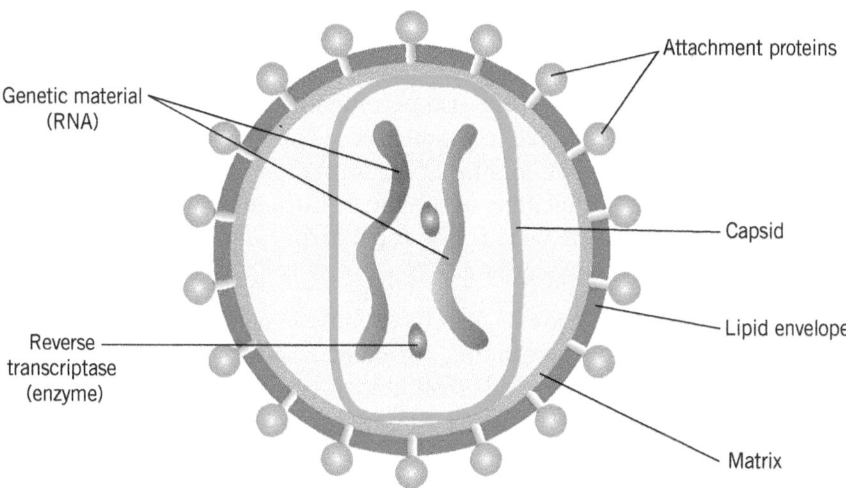

▲ **Figure 1** *Structure of HIV*

Replication of HIV

Being a virus, HIV cannot replicate itself. Instead it uses its genetic material to instruct the host cell's biochemical mechanisms to produce the components required to make new HIV. It does so as follows:

- Following infection HIV enters the bloodstream and circulates around the body.
- A protein on the HIV readily binds to a protein called CD4. Although this protein occurs on a number of different human cells, HIV most frequently attaches to helper T cells (see Topic 14.3).
- The protein capsid fuses with the cell-surface membrane and releases its contents into the helper T cell, including the HIV RNA and enzymes.
- The HIV reverse transcriptase converts the virus's RNA into double-stranded DNA.

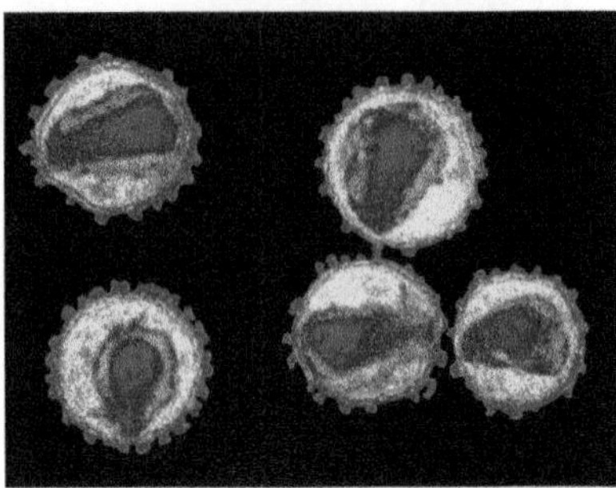

▲ **Figure 2** *False-colour transmission electron micrograph (TEM) of HIV*

- The newly made DNA is moved into the helper T cell's nucleus where it is inserted into the cell's own DNA.
- The HIV DNA in the nucleus creates **messenger RNA** (mRNA), using the cell's enzymes. This mRNA contains the instructions for making new viral proteins and the RNA to go into the new HIV.
- The mRNA passes out of the nucleus through a nuclear pore and uses the cell's protein synthesis mechanisms to make HIV particles.
- The HIV particles break away from the helper T cell using a piece of its cell-surface membrane to forms their lipid envelope.

Once infected with HIV a person is said to be **HIV positive**. However, the replication of HIV often goes into dormancy and only recommences, leading to AIDS, many years later.

How HIV causes the symptoms of AIDS

HIV specifically attacks helper T cells. HIV causes AIDS by killing or interfering with the normal functioning of helper T cells. An uninfected person normally has between 800 and 1200 helper T cells in each cubic millimetre of blood. In a person with AIDS this number can be as low as 200 mm^{-3}. You will see (Topic 14.3) that helper T cells are important in cell-mediated immunity. Without a sufficient number of helper T cells, the immune system cannot stimulate **B cells** to produce antibodies or the cytotoxic T cells that kill cells infected by pathogens. Memory cells may also become infected and destroyed. As a result, the body is unable to produce an adequate immune response and becomes susceptible to other infections and cancers. Many people with AIDS develop infections of the lungs, intestines, brain, and eyes, as well as experiencing weight loss and diarrhoea. It is these secondary diseases that ultimately cause death.

HIV does not kill individuals directly. By infecting the immune system, HIV prevents it from functioning normally. As a result those infected by HIV are unable to respond effectively to other pathogens. It is these infections, rather than HIV, that ultimately cause ill health and eventual death.

Treating HIV

Although there is no cure for HIV, infections can be treated using a range of different drugs. The treatment is known as anti-retroviral therapy (ART). Each different type of drug targets a different stage in the replication and spread of HIV within the body. Table 1 lists the different types of drugs and the stage in viral replication that each drug targets.

▼ **Table 1**

Type of drug	Site of action
Attachment and entry inhibitors	Block the attachment points for the HIV viral protein onto the helper T cell. Some bind to proteins on the virus whereas others bind to receptors on the cell.
Reverse transcriptase inhibitors	There are two types. The non-nucleoside inhibitors act as non-competitive inhibitors, binding to reverse transcriptase at a region away from the active site. Nucleoside inhibitors act as alternative nucleotides in the synthesis of the viral DNA copy. When they are incorporated into the viral DNA copy, the polynucleotide chain is terminated.
Integrase inhibitors	Once reverse transcriptase has synthesised a DNA copy of the viral RNA, this copy must become part of the genomic DNA. This is called integration. Integrase inhibitors act on the enzyme responsible for this process.
Protease inhibitors	Inhibit the enzymes responsible for completing the modification of the proteins that are incorporated into new virus particles.

Various combinations of the above drugs are used in ART and the World Health Organisation (WHO) now recommends that ART is started as soon as a person tests positive for HIV, even if the helper T cell count is still high. WHO is also now recommending the use of some reverse transcriptase inhibitors by people who are at high risk of HIV infection, as part of a combined programme of prevention.

Why antibiotics are ineffective against viral diseases like AIDS

Antibiotics work in a number of different ways. One way is by preventing bacteria from making normal cell walls.

In bacterial cells, as in plant cells, water constantly enters by osmosis. This entry of water would normally cause the cell to burst. That it doesn't burst is due to the wall that surrounds all bacterial cells. This wall is made of **murein** (peptidoglycan), a tough material that is not easily stretched. As water enters the cell by osmosis, the cell expands and pushes against the cell wall. Being relatively inelastic, the cell wall resists expansion and so halts further entry of water. Antibiotics like penicillin inhibit certain enzymes required for the synthesis and assembly of the peptide cross-linkages in bacterial cell walls. This weakens the walls, making them unable to withstand pressure. As water enters naturally by osmosis, the cell bursts and the bacterium dies.

Viruses rely on the host cells to carry out their metabolic activities and therefore lack their own metabolic pathways and cell structures. As a result antibiotics are ineffective because there are no metabolic mechanisms or cell structures for them to disrupt. Viruses also have a protein coat rather than a murein cell wall and so do not have sites where antibiotics can work. In any case, when viruses are within an organism's own cells, antibiotics cannot reach them.

Hint

The use of drugs to prevent disease is called prophylaxis.

Synoptic link

Osmosis was covered in Topic 4.3 and the structure of RNA in Topic 7.1.

Summary questions

1. Explain why HIV is called a retrovirus.
2. Distinguish between HIV and AIDS.
3. Tuberculosis (TB) is a lung disease spread through the air. Suggest a possible reason why the widespread use of condoms might help reduce the incidence of TB in a population.
4. Outline the role of enzyme inhibitors in the control of HIV.

Practice questions: Chapter 13

1 The diagram shows a human immunodeficiency virus (HIV).

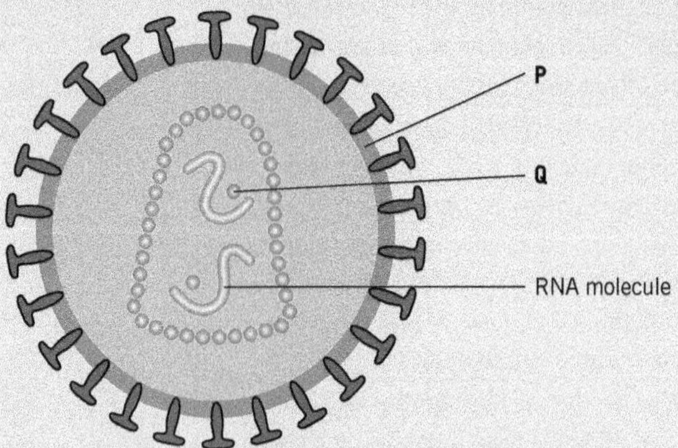

(a) (i) Name structure P and enzyme Q. *(2 marks)*
(ii) What is the function of the RNA molecules in this virus? *(1 mark)*
(b) Describe how new viruses are produced after HIV has infected a T cell. *(3 marks)*

AQA June 2010

2 If people become infected with *Salmonella* bacteria, this can cause food poisoning. People taking the drug morphine are more likely to develop *Salmonella* food poisoning. Scientists investigated the effect of morphine on the development of *Salmonella* food poisoning in mice. They fed a large number of mice with food infected with *Salmonella* bacteria. Half of the mice were then given morphine and the others were given a placebo which did not contain morphine. All of the mice given morphine died of *Salmonella* food poisoning within 5 days. Only half of the mice given the placebo died of *Salmonella* food poisoning. *Salmonella* food poisoning is 20 times more frequent in people with AIDS than in healthy people. Doctors used to assume that this was only due to the immune system not functioning effectively. The scientists working with mice think that morphine might be partly responsible for the high rate of *Salmonella* food poisoning in people with AIDS. Some people with AIDS take morphine, either for pain or because they are addicted to the drug. HIV is often transmitted between addicts who inject drugs such as morphine.
Explain how people become infected with *Salmonella* bacteria and how this causes food poisoning. *(6 marks)*

AQA Jan 2012

3 (a) Infection by the cholera bacterium can cause acute diarrhoea. Explain how. *(2 marks)*
(b) The bacteria that cause cholera can be found in seawater. Outbreaks of cholera often begin in populations living near the coast. Scientists in Bangladesh investigated the relationship between outbreaks of cholera and sea temperature. They used the number of people admitted to hospital with cholera as a measure of the number of cases of the disease. The graph shows their results.

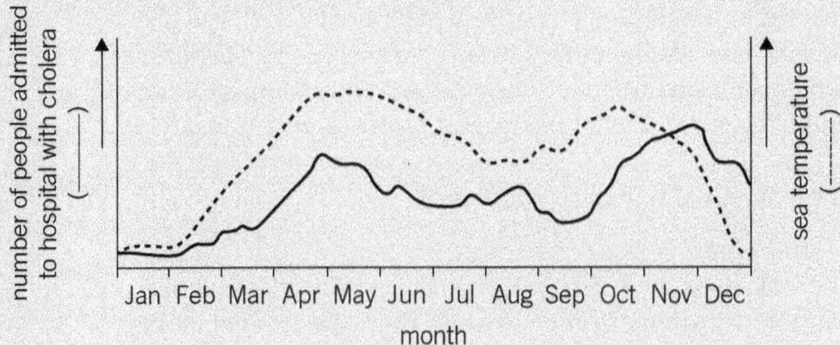

Describe the relationship between sea temperature and the number of people admitted to hospital with cholera between January and June. *(2 marks)*

(c) Some scientists have suggested that a rise in sea temperatures could lead to an increase in outbreaks of cholera. Do these data support this suggestion? Give reasons for your answer. *(2 marks)*

(d) In areas where there are repeated outbreaks of cholera, most people who become infected by cholera bacteria do not become ill. Suggest and explain one reason why *(2 marks)*

AQA June 2013

4 (a) Cholera bacteria produce toxins which increase secretion of chloride ions into the lumen of the intestine. Explain why this results in severe diarrhoea (watery faeces). *(3 marks)*

(b) Scientists investigated how effective two oral rehydration solutions, A and B, were in treating patients with diarrhoea caused by cholera.

Solution A contained glucose.

Solution B was identical to A, except that glucose was replaced by starch.

The graph shows their results.

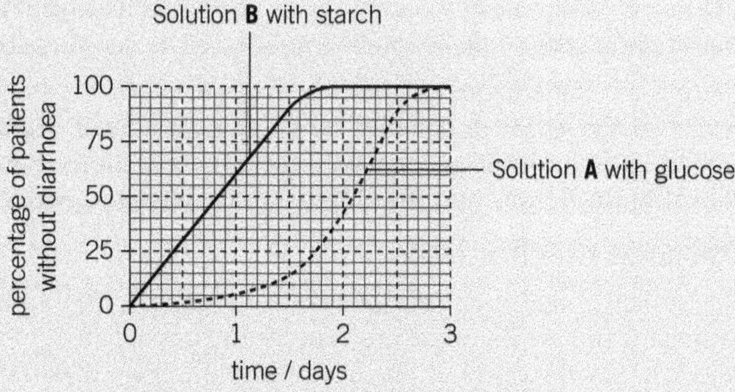

(i) The water potential of solution B was higher (less negative) than the water potential of solution A. Explain why *(1 mark)*

(ii) In this study, 100 patients were treated with solution A and 100 patients were treated with solution B.

Calculate the difference in the number of patients without diarrhoea after 1 day's treatment with solution A and those without diarrhoea after 1 day's treatment with solution B.

Show your working. *(2 marks)*

AQA Jan 2013

14 Mammalian blood – defensive mechanisms

14.1 Cell recognition and the cells of the immune system

Learning objectives:

- → Describe the cells of the immune system.
- → Explain how the body distinguishes between its own cells and foreign cells.

Specification reference: 3.2.5.1

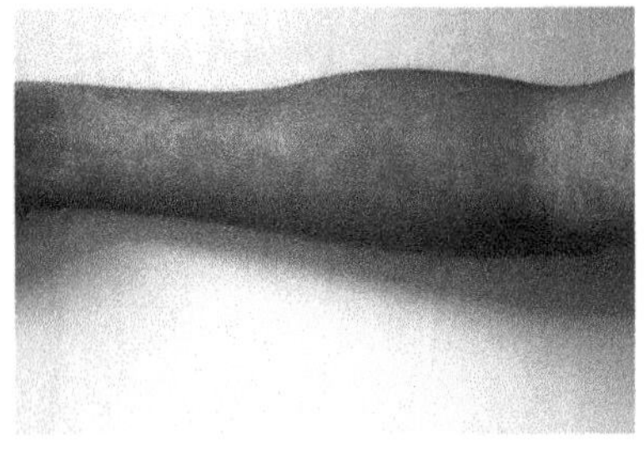

▲ **Figure 1** *Measles is a highly infectious viral disease that mainly affects young children before they have acquired immunity to it*

Tens of millions of humans die each year from infectious diseases. Many more survive and others appear never to be affected in the first place. Why are there these differences?

Any infection is, in effect, an interaction between the **pathogen** and the body's various defence mechanisms. Sometimes the pathogen overwhelms the defences and the individual dies. Sometimes the body's defence mechanisms overwhelm the pathogen and the individual recovers from the disease. Having overwhelmed the pathogen once, however, the body's defences then seem to be better prepared for a second infection by the same pathogen and can destroy it before it can cause any harm. This is known as **immunity** and is the main reason why some people are unaffected by certain pathogens.

There is a complete range of intermediates between the stages described above. Much depends on the overall state of health of an individual. A fit, healthy adult will rarely die of an infection. Those in ill health, the young, and the elderly are usually more vulnerable.

Defence mechanisms

The human body has a range of defences to protect itself from pathogens (Figure 2). Some are general and immediate non-specific defences, like the skin forming a barrier to the entry of pathogens, and **phagocytosis** (see Topic 14.2). Others are more specific, less rapid, but longer lasting. These responses involve a type of white blood cell called a lymphocyte and take two forms:

- cell-mediated responses involving T lymphocytes
- humoral responses involving B lymphocytes.

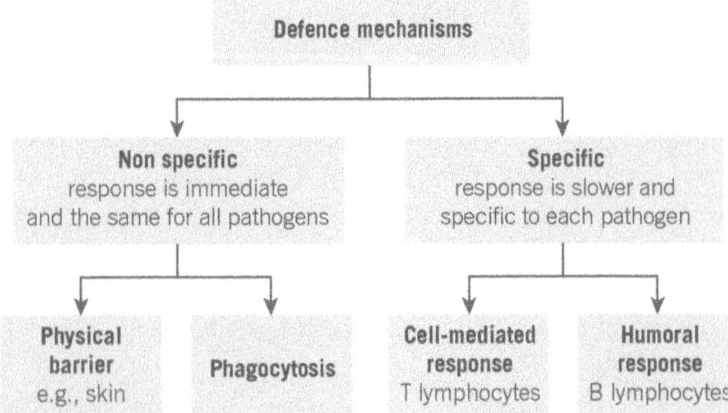

▲ **Figure 2** *Summary of defence mechanisms*

Before we look in detail at these defence mechanisms, let us first consider how the body distinguishes its own cells from foreign material.

Recognising your own cells

To defend the body from invasion by foreign material, lymphocytes must be able to distinguish the body's own cells and molecules (**self**) from those that are foreign (**non-self**). If they could not do this, the lymphocytes would destroy the organism's own tissues.

Each type of cell, self or non-self, has specific molecules on its surface that identify it. These molecules are known as **antigens** and, whilst they can be of various types, it is the proteins that are the most important (see Topic 14.3). This is because proteins have enormous variety and a highly specific tertiary structure. It is this variety of specific 3-D structure of proteins that distinguishes one cell from another. It is these protein molecules that usually allow the immune system to identify:

- pathogens, for example, the human immunodeficiency virus (see Topic 13.3)
- non-self material, such as cells from other organisms of the same species
- toxins, including those produced by certain pathogens, such as the bacterium that causes cholera
- abnormal body cells, such as cancer cells.

All of the above are potentially harmful and their identification is the first stage in removing the threat they pose. Although this response is clearly advantageous to the organism, it has implications for humans who have had tissue or organ transplants. The immune system recognises these as non-self even though they have come from individuals of the same species. It therefore attempts to destroy the transplant. To minimise the effect of this tissue rejection, donor tissues for transplant are normally matched as closely as possible to those of the recipient. The best matches often come from relatives who are genetically close. In addition, immunosuppressant drugs are often administered to reduce the level of the immune response that still occurs.

How lymphocytes recognise cells belonging to the body

- There are probably around 10 million different lymphocytes present at any time, each capable of recognising a different chemical shape.
- In the fetus, these lymphocytes are constantly colliding with other cells.
- Infection in the fetus is rare because it is protected from the outside world by the mother and, in particular, the placenta.
- Lymphocytes will therefore collide almost exclusively with the body's own material (self).
- Some of the lymphocytes will have receptors that exactly fit those of the body's own cells.
- These lymphocytes either die or are suppressed.
- The only remaining lymphocytes are those that might fit foreign material (non-self), and therefore only respond to foreign material.
- In adults, lymphocytes produced in the bone marrow initially only encounter self-antigens.
- Any lymphocytes that show an immune response to these self-antigens undergo programmed cell death (apoptosis) before they can differentiate into mature lymphocytes.
- No clones of these anti-self lymphocytes will appear in the blood, leaving only those that might respond to non-self antigens.

It is important to remember that specific lymphocytes are not produced in response to an infection, but that they already exist – all 10 million different types. Given that there are so many different types of lymphocytes, there is a high probability that, when a pathogen gets into the body, one of these lymphocytes will have a protein on its surface that is complementary to one of the proteins of the pathogen. In other words, the lymphocyte will 'recognise' the pathogen. Not surprisingly with so many different lymphocytes, there are very few of each type. When an infection occurs, the one type already present that has the complementary proteins to those of the pathogen is stimulated to divide to build up its numbers to a level where it can be effective in destroying it. This is called clonal selection and you will learn more about it in Topic 14.4. This explains why there is a time lag between exposure to the pathogen and the body's defences bringing it under control.

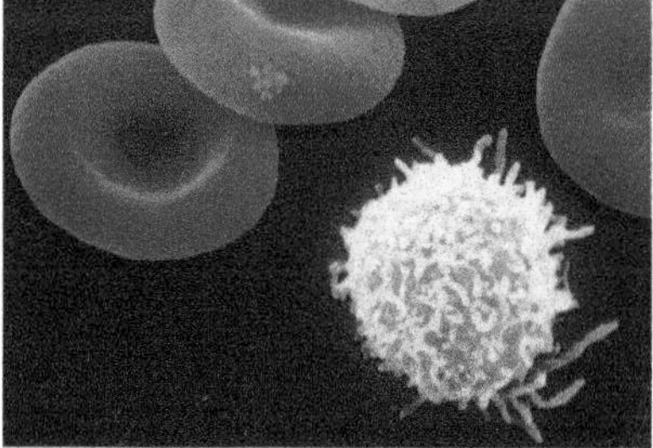

▲ **Figure 3** *False-colour scanning electron micrograph (SEM) of a single human lymphocyte (blue) and red blood cells (red)*

Summary questions

1. State **two** differences between a specific and a non-specific defence mechanism.
2. After a pathogen gains entry to the body it is often a number of days before the body's immune system begins to control it. Suggest a possible reason why this is so.
3. In the above case, suggest why it would be inaccurate to say that the body takes days to 'respond' to the pathogen.

14.2 Phagocytosis

Learning objectives:

- → Describe the first line of defence against disease.
- → Explain the process of phagocytosis.
- → Describe the role of lysosomes in phagocytosis.

Specification reference: 3.2.5.1

If a pathogen is to infect the body it must first gain entry. Clearly, then, the body's first line of defence is to form a physical or chemical barrier to entry. Should this fail, the next line of defence is the white blood cells. There are two types of white blood cell: phagocytes and lymphocytes. Phagocytes ingest and destroy the pathogen by a process called phagocytosis before it can cause harm. Lymphocytes are involved in immune responses (Topics 14.3 and 14.4).

Despite various barriers, such as the skin or sticky mucus in the respiratory tract, pathogens still frequently gain entry and the next line of defence is then phagocytosis.

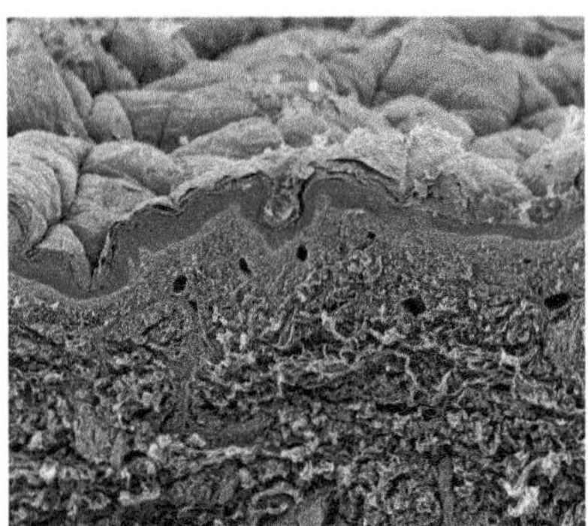

▲ **Figure 1** *Human skin forms a tough outer layer that acts as a barrier to the entry of pathogens*

Phagocytosis

Large particles, such as some types of bacteria, can be engulfed by cells in vesicles formed from the cell-surface membrane. This process is called phagocytosis and is a non-specific defence mechanism. In the blood, the types of white blood cells that carry out phagocytosis are known as **phagocytes**. They provide an important defence against the pathogens that manage to enter the body. Some phagocytes travel in the blood but can move out of blood vessels into other tissues. Phagocytosis is illustrated in Figure 2 and is summarised below and in Figure 3.

- Chemical products of pathogens or dead, damaged, and abnormal cells act as attractants, causing phagocytes to move towards the pathogen (e.g., a bacterium).
- Phagocytes have several receptors on their cell-surface membrane that recognise, and attach to, chemicals or **antigens** on the surface of the pathogen.
- They engulf the pathogen to form a vesicle, known as a **phagosome**.
- Lysosomes move towards the vesicle and fuse with it.
- Enzymes called **lysozymes** are present within the lysosome. These lysozymes destroy ingested bacteria by hydrolysis of their cell walls. The process is the same as that for the digestion of food in the intestines, namely the hydrolysis of larger, insoluble molecules into smaller, soluble ones.
- The soluble products from the breakdown of the pathogen are absorbed into the cytoplasm of the phagocyte.

Synoptic link

Look back at Topic 2.3 to remind yourself of the role of lysosomes in the cell.

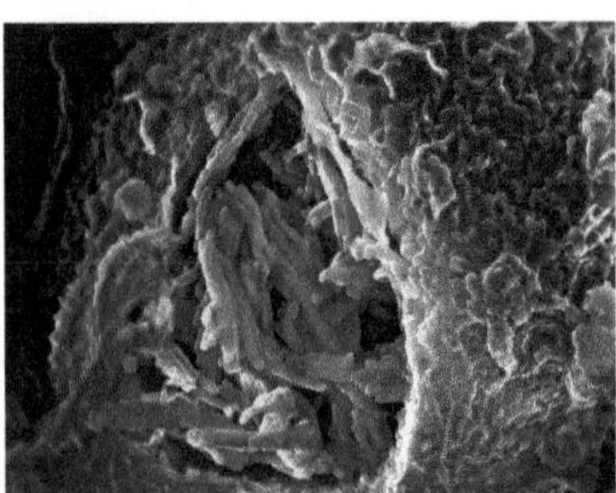

▲ **Figure 2** *False-colour scanning electron micrograph (SEM) of a phagocyte (red) engulfing tuberculosis bacteria (yellow), a process known as phagocytosis*

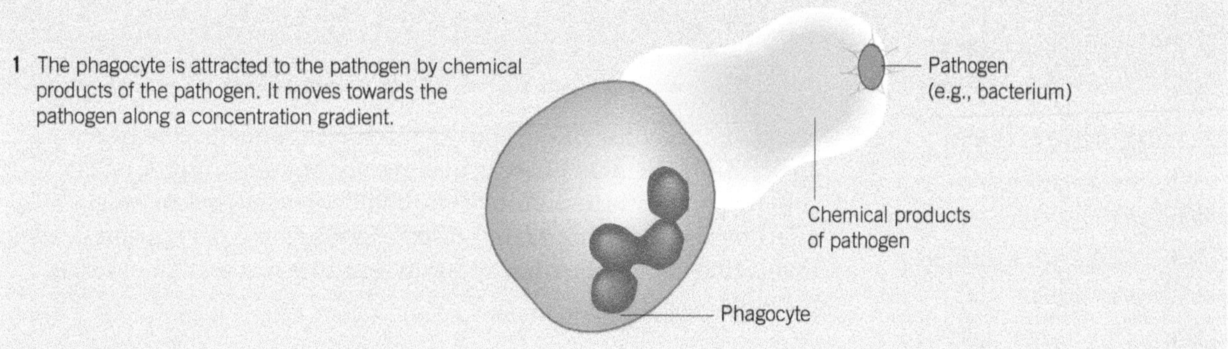

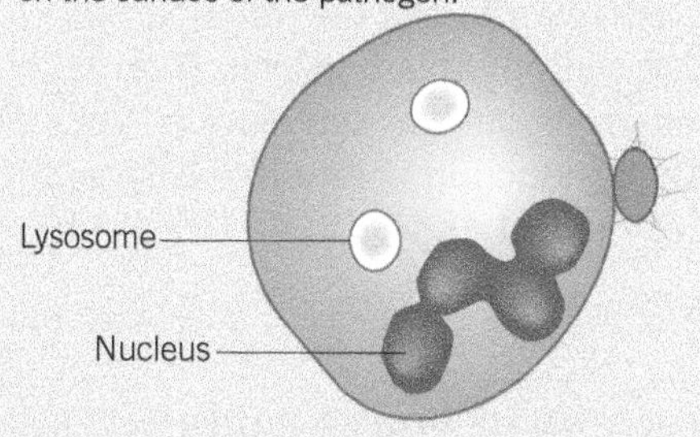

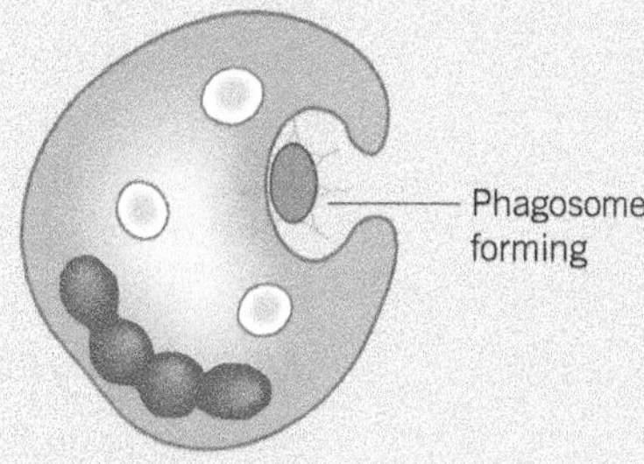

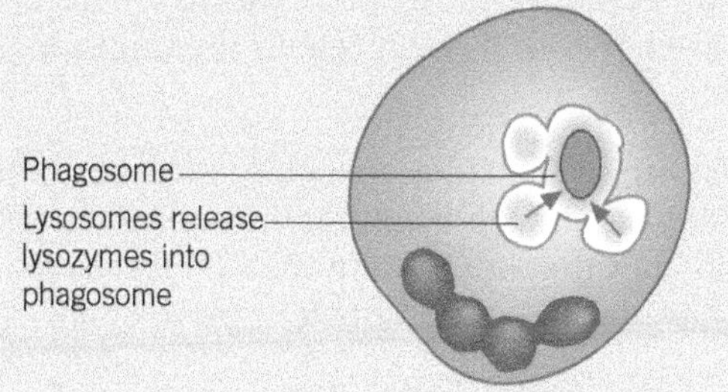

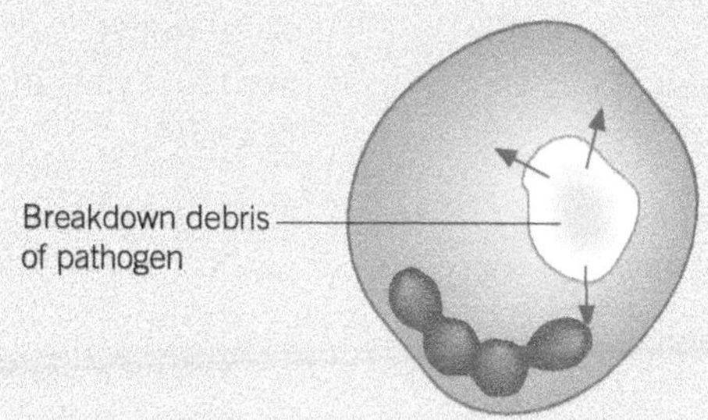

▲ **Figure 3** *Summary of phagocytosis*

Summary questions

1 In the following passage, state the missing word indicated by each letter **a–d**.

Pathogens that invade the body may be engulfed by cells that carry out **a**. The engulfed pathogen forms a vesicle known as a **b**. Once engulfed the pathogen is broken down by enzymes called **c** released from organelles called **d**.

2 Among other places, the enzymes from **1c** are found in tears. Suggest a reason why this is so.

14.3 T lymphocytes and cell-mediated immunity

Learning objectives:

- → State the definition of an antigen.
- → Describe the two main types of lymphocyte.
- → Explain the role of T cells (T lymphocytes) in cell-mediated immunity.

Specification reference: 3.2.5.1 and 3.2.5.2

The initial response of the body to infection is non-specific (see Topic 14.1). The next phase is the primary immune response that confers immunity. Immunity is the ability of organisms to resist infection by protecting against disease-causing microorganisms or their toxins that invade their bodies. It involves the recognition of foreign material (antigens).

Antigens

An **antigen** is any part of an organism or substance that is recognised as non-self (foreign) by the immune system and stimulates an immune response. Antigens are usually proteins that are part of the cell-surface membranes or cell walls of invading cells, such as microorganisms, or abnormal body cells, such as cancer cells. The presence of an antigen triggers the production of an **antibody** as part of the body's defence system (see Topic 14.4).

Lymphocytes

Immune responses such as phagocytosis are **non-specific** (see Topic 14.2) and occur whatever the infection. The body also has **specific** responses that react to specific antigens. These are slower in action at first, but they can provide long-term immunity. This specific immune response depends on a type of white blood cell called a **lymphocyte**. Lymphocytes are produced by stem cells in the bone marrow. There are two types of lymphocyte, each with its own role in the immune response:

- **B lymphocytes (B cells)** are so called because they mature in the bone marrow. They are associated with humoral immunity, that is, immunity involving antibodies that are present in body fluids, or 'humour' such as blood plasma. This is described in more detail in Topic 14.4.
- **T lymphocytes (T cells)** are so called because they mature in the thymus gland. They are associated with cell-mediated immunity, that is, immunity involving body cells.

Cell-mediated immunity

Lymphocytes respond to an organism's own cells that have been infected by non-self material from a different species, for example, a virus. They also respond to cells from other individuals of the same species because these are genetically different. These cells therefore have antigens on their cell-surface membrane that are different from the antigens on the organism's own cells. T lymphocytes can distinguish these invader cells from normal cells because:

- phagocytes that have engulfed and hydrolysed a pathogen present some of a pathogen's antigens on their own cell-surface membrane
- body cells invaded by a virus present some of the viral antigens on their own cell-surface membrane.

Study tip

It is always necessary to describe events in detail using the appropriate scientific terms. For example, in immunity questions vague references to 'cells fighting disease' should be avoided.

- transplanted cells from individuals of the same species have different antigens on their cell-surface membrane
- cancer cells are different from normal body cells and present antigens on their cell-surface membranes.

Cells that display foreign antigens on their surface are called **antigen-presenting cells** because they can present antigens of other cells on their own cell-surface membrane.

T lymphocytes will only respond to antigens that are presented on a body cell (rather than to antigens within the body fluids). This type of response is called **cell-mediated immunity** or **the cellular response**. The role of the receptors on T cells is important. The receptors on each T cell respond to a single antigen. It follows that there is a vast number of different types of T cell, each one responding to a different antigen (see Topic 14.1). The stages in the response of T lymphocytes to infection by a pathogen are summarised in Figure 1 and explained below.

1. Pathogens invade body cells or are taken in by phagocytes.
2. The phagocyte places antigens from the pathogen on its cell-surface membrane.
3. Receptors on a specific helper T cell (T_H cell) fit exactly onto these antigens.
4. This attachment activates the T_H cell to divide rapidly by mitosis and form a clone of genetically identical cells.
5. The cloned T_H cells:
   a. develop into memory cells that enable a rapid response to future infections by the same pathogen
   b. stimulate phagocytes to engulf pathogens by phagocytosis
   c. stimulate B cells to divide and secrete their antibody
   d. activate cytotoxic T cells (T_C cells).

Hint

Three terms that are frequently confused are *antigen*, *antibody* and *antibiotic*. When dealing with immunity put *antibiotic* out of your mind – it has nothing to do with immunity.

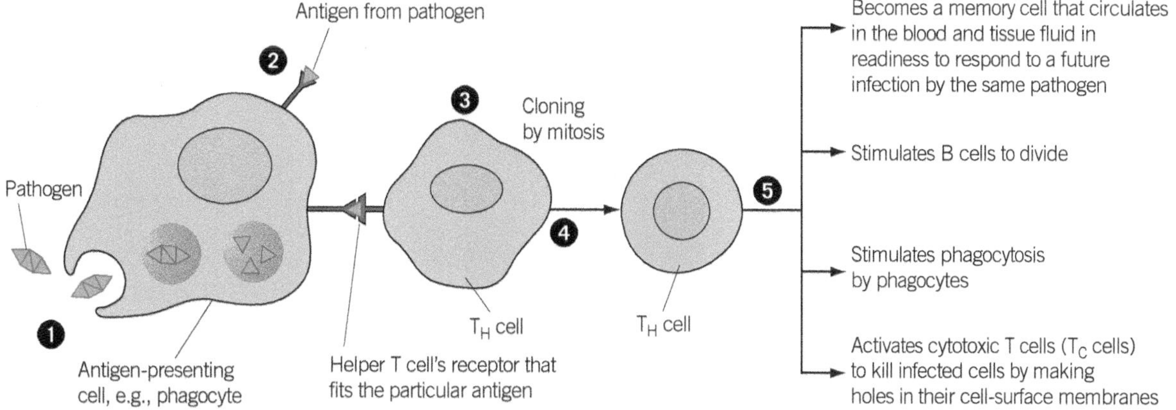

▲ **Figure 1** *Summary of the role of T cells in cell-mediated immunity*

How cytotoxic T cells kill infected cells

Cytotoxic T cells (T_C cells) kill abnormal cells and body cells that are infected by pathogens by producing a protein called perforin that makes holes in the cell-surface membrane. These holes mean the cell membrane becomes freely permeable to all substances and the cell dies as a result. This illustrates the vital importance of cell-surface membranes in maintaining the integrity of cells and hence their survival. The action of T cells is most effective against viruses because viruses replicate inside cells. As viruses use living cells to replicate in, this sacrifice of body cells prevents viruses multiplying and infecting more cells.

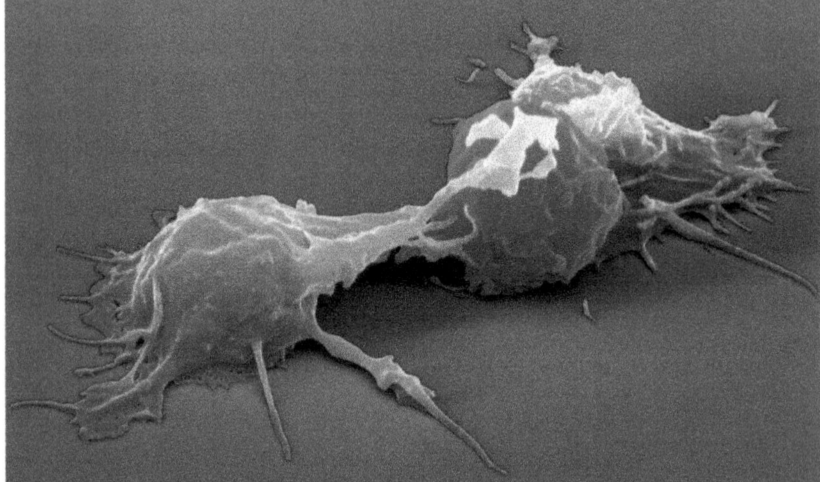

▲ **Figure 2** *False-colour scanning electron micrograph (SEM) of two human cytotoxic T cells (yellow) attacking a cancer cell (red)*

Summary questions

1 Define an antigen.

2 State **two** similarities between T cells and B cells.

3 State **two** differences between T cells and B cells.

Bird flu

Avian (bird) flu is caused by one of many strains of the influenza virus. Although it is adapted primarily to infect birds, the H5N1 strain of the virus can infect other species, including humans. Avian flu affects the lungs and can cause the immune system to go into overdrive. This results in a massive overproduction of T cells.

1 From your knowledge of cell-mediated immunity and lung structure, suggest why humans infected with the H5N1 virus may sometimes die from suffocation.

2 Suggest a reason why any spread of bird flu across the world is likely to be very rapid.

14.4 B lymphocytes, humoral immunity, and antibodies

You saw in Topic 14.3 that the first phase of the specific response to infection is the mitotic division of specific T cells to form a clone of the relevant T cells to build up their numbers. Some of these T cells produce factors that stimulate **B cells** to divide. It is these B cells that are involved in the next phase of the immune response – humoral immunity.

Humoral immunity

Humoral immunity is so called because it involves **antibodies**, and antibodies are soluble in the blood and tissue fluid of the body. An old-fashioned word for body fluids is 'humour'. There are many different types of B cell, possibly as many as 10 million, and each B cell starts to produce a specific antibody that responds to one specific **antigen**. When an antigen (e.g., a protein on the surface of a pathogen, foreign cell, toxin, or damaged or abnormal cell) enters the blood or tissue fluid, there will be one B cell that has an antibody on its surface whose shape exactly fits the antigen, that is, they are complementary. The antibody therefore attaches to this complementary antigen. The antigen enters the B cell by **endocytosis** and gets presented on its surface (processed). T_H cells bind to these processed antigens and stimulate this B cell to divide by mitosis (see Topic 17.1) to form a clone of identical B cells, all of which produce the antibody that is specific to the foreign antigen. This is called **clonal selection** and accounts for the body's ability to respond rapidly to any of a vast number of antigens.

In practice, a typical pathogen has many different proteins on its surface, all of which act as antigens. Some pathogens, such as the bacterium that causes cholera, also produce toxins. Each toxin molecule also acts as an antigen. Therefore many different B cells make clones, each of which produces its own type of antibody. In each clone, the cells produced develop into one of two types of cell:

- **Plasma cells** secrete antibodies, usually into blood plasma. These cells survive for only a few days, but each can make around 2000 antibodies every second during its brief lifespan. These antibodies lead to the destruction of the antigen. The plasma cells are therefore responsible for the immediate defence of the body against infection. The production of antibodies and memory cells (see below) is known as the **primary immune response**.
- **Memory cells** are responsible for the **secondary immune response.** Memory cells live considerably longer than plasma cells, often for decades. These cells do not produce antibodies directly, but circulate in the blood and tissue fluid. When they encounter the same antigen at a later date, they divide rapidly and develop into plasma cells and more memory cells. The plasma cells produce the antibodies needed to destroy the pathogen, whilst the new memory cells circulate in readiness for any future infection. In this way, memory cells provide long-term immunity against the original infection. An increased quantity of antibodies is secreted at a faster rate than in the primary immune response. It ensures that

Learning objectives:

- → Explain the role of B cells (B lymphocytes) in humoral immunity.
- → Explain the roles of plasma cells and antibodies in the primary immune response.
- → Explain the role of memory cells in the secondary immune response.
- → Explain how antigenic variation affects the body's response to infection.
- → Describe the structure and mode of action of antibodies.

Specification reference: 3.2.5.2 and 3.2.5.3

Hint

Remember that B cells with the appropriate antibody to bind to antigens of a pathogen are *not* produced in response to the pathogen. They are present from birth. Being present, they simply *multiply* in response to the pathogen.

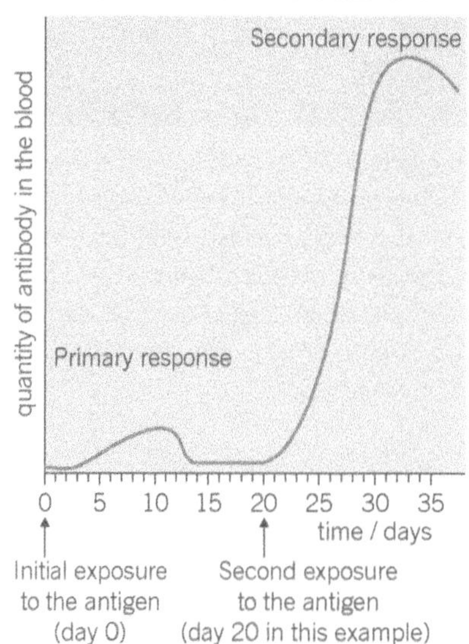

▲ **Figure 1** *Primary and secondary responses to an antigen*

a new infection is destroyed before it can cause any harm – and individuals are often totally unaware that they have been infected. Figure 1 illustrates the relative amounts of antibody produced in the primary and secondary immune responses.

The role of B cells in immunity is explained below and summarised in Figure 2.

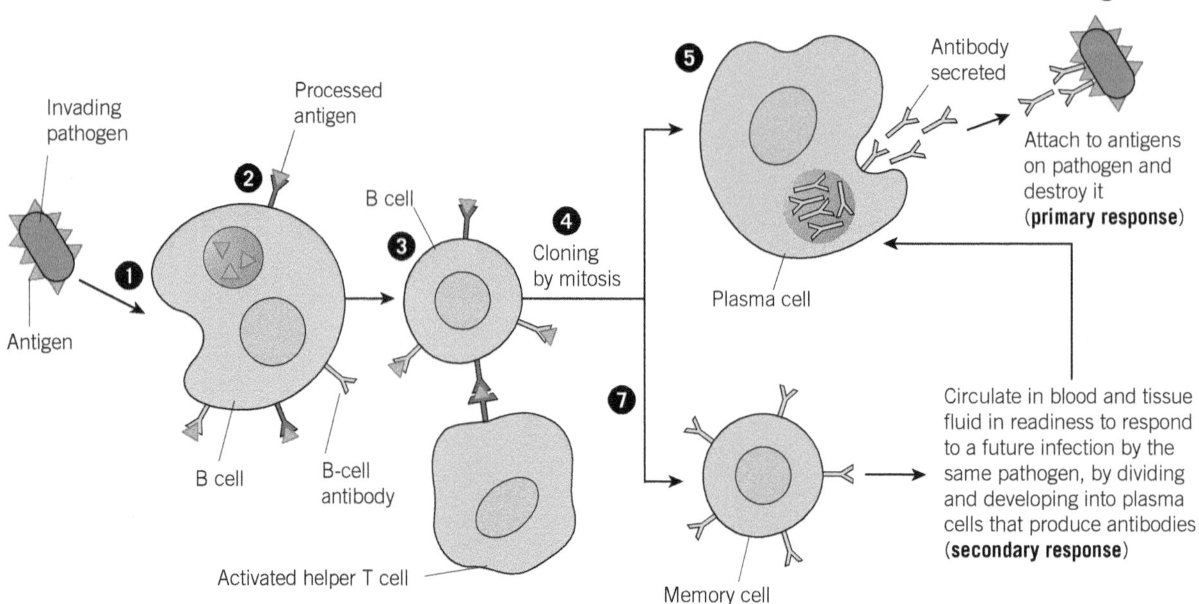

▲ **Figure 2** *Summary of the role of B cells in humoral immunity*

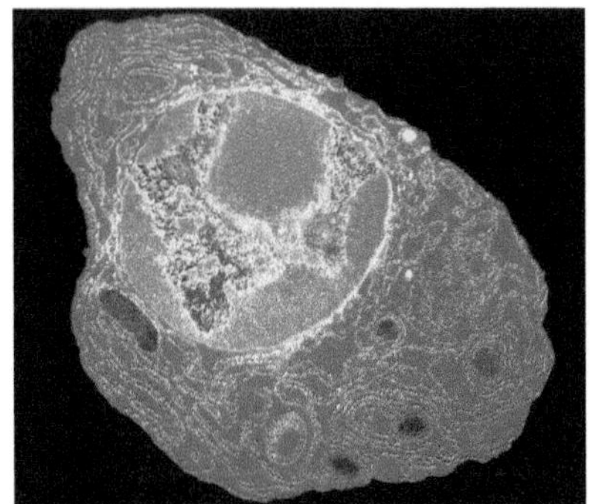

▲ **Figure 3** *False-colour transmission electron micrograph (TEM) of a plasma cell. Plasma cells are mature B lymphocytes that secrete antibodies. Note the well-developed rough endoplasmic reticulum (yellow dotted lines) where the antibodies are synthesised.*

1 The surface antigens of an invading pathogen are taken up by a B cell.
2 The B cell processes the antigens and presents them on its surface.
3 Helper T cells (activated in the process described in Topic 14.3) attach to the processed antigens on the B cell, thereby activating the B cell.
4 The activated B cell divides by **mitosis** to give a clone of plasma cells.
5 The cloned plasma cells produce and secrete the specific antibody that exactly fits the antigen on the pathogen's surface.
6 The antibody attaches to antigens on the pathogen and destroys it.
7 Some B cells develop into memory cells. These can respond to future infections by the same pathogen by dividing rapidly and developing into plasma cells that produce antibodies. This is the secondary immune response.

Antibodies

Antibodies are proteins with specific binding sites, which are synthesised by B cells. When the body is infected by non-self material, a B cell produces a specific antibody. This specific antibody reacts with an antigen on the surface of the non-self material by binding to it. Each antibody has two identical binding sites. The antibody binding sites are complementary to a specific antigen. The massive variety of antibodies is possible because they are made of proteins – molecules that occur in an almost infinite number of forms.

Antibodies are made up of four polypeptide chains. The chains of one pair are long and are called **heavy chains**, whereas the chains of the other pair are shorter and are known as **light chains**. Each antibody has a specific binding site that fits very precisely onto a specific antigen to form what is known as an **antigen–antibody complex**. A specific (type of) antibody has a binding site that binds to a specific antigen. The binding site is part of the **variable region**. Each binding site consists of a sequence of amino acids that form a specific 3-D shape that binds directly to a specific antigen. The rest of the antibody is known as the **constant region**. This binds to receptors on cells such as B cells. The structure of an antibody is illustrated in Figure 4.

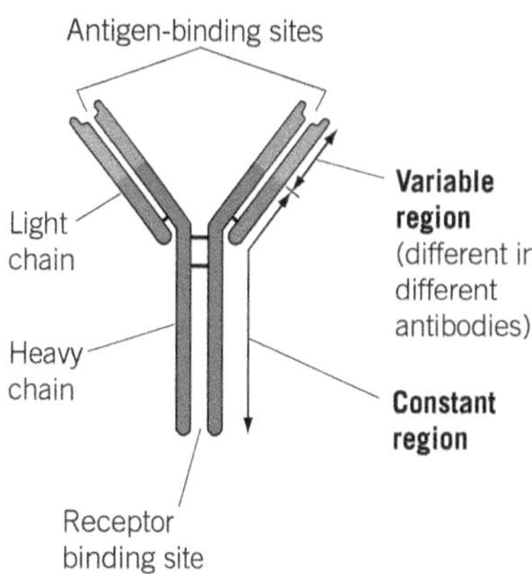

▲ **Figure 4** *Structure of an antibody*

How the antibody leads to the destruction pathogens and foreign cells

It is important to understand that antibodies do not destroy the antigens on foreign cells and microorganisms directly, but rather they prepare the antigen. Different antibodies lead to the destruction of a cell or virus carrying the complementary antigen in a range of ways. Take the example of when the antigen is on a bacterial cell – antibodies assist in the destruction of the cell in two ways:

- They cause agglutination of the bacterial cells (Figure 5). In this way, clumps of bacterial cells are formed, making it easier for the phagocytes to locate them as they are less spread out within the body.
- They then serve as markers that stimulate phagocytes to engulf the bacterial cells to which they are attached.

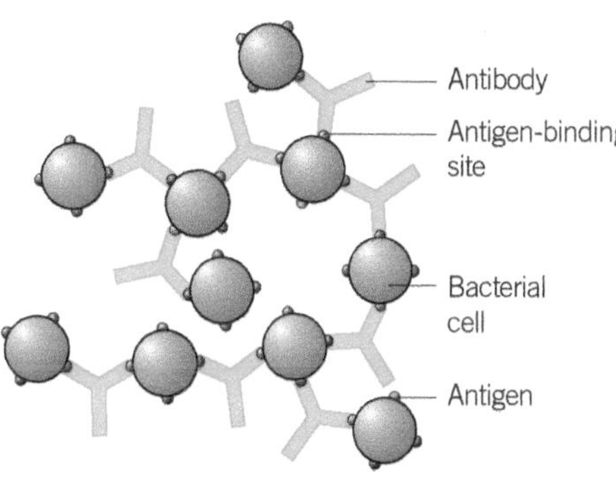

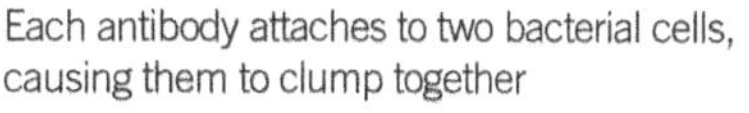

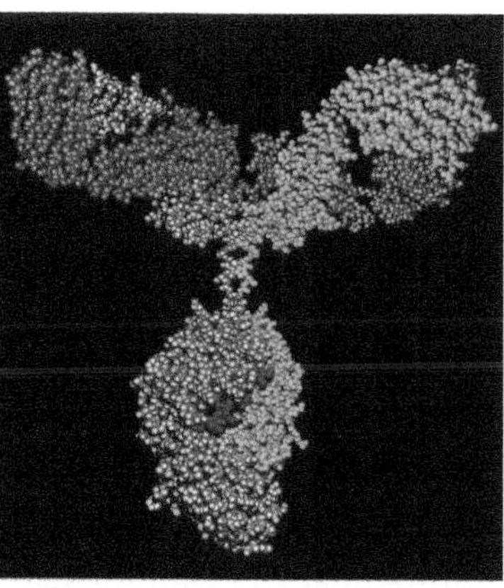

▲ **Figure 5** *Antibodies cause agglutination of bacterial cells (left); molecular model of an antibody (right). This Y-shaped protein is produced by B lymphocytes as part of the immune response.*

Hint

Write about binding sites on antibodies, not active sites. You will not get marks in the exams for saying active sites!

Hint

One molecule fitting neatly with another is a recurring theme throughout biology. We met it with enzymes (see Topic 3.1) and with T cells (see Topic 14.3) and it features again here. Whilst the 'lock and key' image is helpful, remember that, with the induced fit model of enzyme action, the molecules are flexible rather than rigid. This is the same for antibodies. The image of a hand fitting a glove is therefore perhaps a better one when it comes to understanding the process.

Summary questions

1. Explain why the secondary immune response is much more rapid than the primary one.
2. Contrast the cell-mediated and humoral responses to a pathogen.
3. Plasma cells can produce around 2000 protein antibodies each second. Suggest three cell organelles that you might expect to find in large quantities in a plasma cell, and explain why.

Study tip

Agglutination is possible because each antibody has two antigen-binding sites.

14.5 Vaccination

Learning objectives:

- → Describe the nature of vaccines.
- → Describe the features of an effective vaccination programme.
- → Explain why vaccination rarely eliminates a disease.
- → Discuss the ethical issues associated with vaccination programmes.

Specification reference: 3.2.5.3

Immunity

Immunity is the ability of an organism to resist infection. This immunity takes two forms.

- **Passive immunity** is produced by the introduction of antibodies into individuals from an outside source. No direct contact with the pathogen or its **antigen** is necessary to induce immunity. Immunity is acquired immediately. As the antibodies are not being produced by the individuals themselves, the antibodies are not replaced when they are broken down, no memory cells are formed, and so there is no lasting immunity. Examples of passive immunity include anti-venom given to the victims of snake bites and the immunity acquired by the fetus when antibodies pass across the placenta from the mother.
- **Active immunity** is produced by stimulating the production of antibodies by the inividual's own immune system. Direct contact with the pathogen or its antigen is necessary. Immunity takes time to develop. It is generally long lasting and is of two types:
  - **Natural active immunity** results from an individual becoming infected with a disease under normal circumstances. The body produces its own antibodies and may continue to do so for many years.
  - **Artificial active immunity** forms the basis of vaccination (immunisation). It involves inducing an immune response in an individual, without them suffering the symptoms of the disease.

Vaccination is the introduction of the appropriate disease antigens into the body, either by injection or by mouth. The intention is to stimulate an immune response against a particular disease. The material introduced is called a **vaccine** and, in whatever form (see below), it contains one or more types of antigen from the pathogen. These antigens stimulate the immune response as described in Topics 14.3 and 14.4. The response is slight because only a small amount of antigen has been introduced, However, the crucial factor is that memory cells (see Topic 14.4) are produced. These remain in the blood and allow a greater, and more immediate, response to a future infection with the pathogen. The result is that there is a rapid production of antibodies and the new infection is rapidly overcome before it can cause any harm and with few, if any, symptoms.

When carried out on a large scale, vaccination provides protection against disease for individuals, and for whole populations.

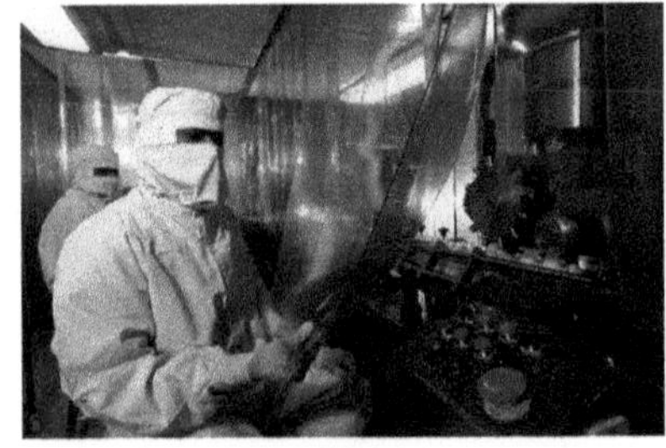

▲ **Figure 1** *The development of new vaccines is a highly technological process requiring sterile conditions*

Features of a successful vaccination programme

It is important to understand that vaccination is used as a precautionary measure to prevent individuals contracting a disease. It is not a means of treating individuals who already have the disease. Some programmes of vaccination against diseases have had considerable success. Yet, in other instances, similar measures have been less successful. The success of a vaccination programme depends on a number of factors:

- A suitable vaccine must be economically available in sufficient quantities to immunise most of the vulnerable population.

- There must be few, if any, side effects from vaccination. Unpleasant side effects may discourage individuals in the population from being vaccinated.
- Means of producing, storing, and transporting the vaccine must be available. This usually involves technologically advanced equipment, hygienic conditions, and refrigerated transport.
- There must be the means of administering the vaccine properly at the appropriate time. This involves training staff with appropriate skills at different centres throughout the population.
- It must be possible to vaccinate the vast majority of the vulnerable population to produce **herd immunity**.

Herd immunity

Herd immunity arises when a sufficiently large proportion of the population has been vaccinated, which makes it difficult for a pathogen to spread within that population. The concept is based on the idea that pathogens are passed from individual to individual when in close contact. Where the vast majority of the population is immune, it is highly improbable that a susceptible individual will come in contact with an infected person. In this way, those individuals who are not immune to the disease are nevertheless protected.

Herd immunity is important because it is never possible to vaccinate everyone in a large population. For example, babies and very young children might not get vaccinated because their immune system is not yet fully functional. It could also be dangerous to vaccinate those who are ill or have compromised immune systems. The percentage of the population that must be vaccinated in order to achieve herd immunity is different for each disease. To achieve herd immunity, vaccination is best carried out at one time. This means that, for a certain period, there are very few individuals in the population with the disease and the transmission of the pathogen is interrupted.

Why vaccination may not eliminate a disease

Even when these criteria for successful vaccination are met, it can still prove extremely difficult to eradicate a disease. The reasons are as follows:

- Vaccination fails to induce immunity in certain individuals, for example, people with defective immune systems.
- Individuals may develop the disease immediately after vaccination but before their immunity levels are high enough to prevent it. These individuals may harbour the pathogen and reinfect others.
- The pathogen may mutate frequently, so that its antigens change suddenly rather than gradually. This means that vaccines suddenly become ineffective because the new antigens on the pathogen are no longer recognised by the immune system. As a result the immune system does not produce the antibodies to destroy the pathogen. This **antigenic variability** happens with the influenza virus, which changes its antigens frequently. Immunity is therefore short lived and individuals may develop repeated bouts of influenza during their lifetime.

▲ **Figure 2** *Vaccination programmes for children have considerably reduced deaths from infectious diseases*

- There may be so many varieties of a particular pathogen that it is almost impossible to develop a vaccine that is effective against them all. For example, there are over 100 varieties of the common cold virus and new ones are constantly evolving.
- Certain pathogens 'hide' from the body's immune system, either by concealing themselves inside cells, or, like the cholera pathogen, by living in places out of reach, such as within the intestines.
- Individuals may have objections to vaccination for religious, ethical, or medical reasons. For example, unfounded concerns over the measles, mumps, and rubella (MMR) triple vaccine has led a number of parents to opt for separate vaccinations for their children, or to avoid vaccination altogether.

The ethics of using vaccines

As vaccinations have saved millions of lives, it is easy to accept vaccination programmes without question. However, they do raise ethical issues that need to be addressed if such programmes are to command widespread support. The production and use of vaccines raises the following questions:

- The production of existing vaccines, and the development of new ones, often involves the use of animals. How acceptable is this?
- Vaccines have side effects that may sometimes cause long-term harm. How can the risk of side effects be balanced against the risk of developing a disease that causes even greater harm?
- On whom should vaccines be tested? How should such trials be carried out? To what extent should individuals be asked to accept risk in the interests of public health?
- Is it acceptable to trial a new vaccine with unknown health risks only in a country where the targeted disease is common, on the basis that the population there has most to gain if it proves successful?
- To be fully effective, the majority, and preferably all, of the population should be vaccinated. Is it right, in the interests of everyone's health, that vaccination should be compulsory? If so, should this be at any time, or just when there is a potential epidemic? Can people opt out? If so, on what grounds: religious belief, medical circumstances, personal belief?
- Should expensive vaccination programmes continue when a disease is almost eradicated, even though this might mean less money for the treatment of other diseases?
- How can any individual health risk from vaccination be balanced against the advantages of controlling a disease for the benefit of the population at large?

Study tip

When discussing ethical issues, always present a balanced view that reflects both sides of the debate, and support your arguments with relevant biological information.

Summary questions

1. Distinguish between active immunity and passive immunity.
2. Explain why vaccinating against influenza is not always effective.

MMR vaccine

In 1988, a combined vaccine for measles, mumps, and rubella (MMR) was introduced into the UK to replace three separate vaccines. All three diseases are potentially disabling. Mumps can lead to orchitis in men, possibly causing sterility, and measles is potentially lethal. Ten years later a study was published in a well-respected medical journal. This suggested that there was a higher incidence of autism among children who had received the triple MMR vaccine than in those who had received separate vaccinations. Autism is a condition in which individuals have impaired social interaction and communication skills.

In the wake of the media furore that followed, many parents decided to have their children vaccinated separately for the three diseases, whereas others opted for no vaccination at all. Parents of autistic children recalled that symptoms of the disorder emerged at around 14 months of age – shortly after the children had been given the MMR vaccination, adding to public concern about the MMR vaccine. The incidence of measles, mumps, and rubella rose.

The vast majority of scientists now think that the vaccine is safe. A number of facts have emerged since the first research linking the MMR vaccine to autism:

- The author of the research had a conflict of interests. He was also being paid by the Legal Aid Board to discover whether parents who claimed their children had been damaged by MMR had a case. Some children were included in both studies.
- Further studies, including one in Japan involving over 30 000 children, have found no link between the MMR vaccine and autism.
- The sample size of the initial research was very small relative to later studies.
- The journal that published the initial research has publicly declared that, had it known all the facts, it would not have published the work.

1 Autism experts point out that many of the symptoms of autism first occur around the age of 14 months. Explain why this information is relevant to the debate on whether the MMR vaccine and autism are linked.

2 Discuss how an organisation funding research might influence the outcome of that research without dishonestly altering the findings.

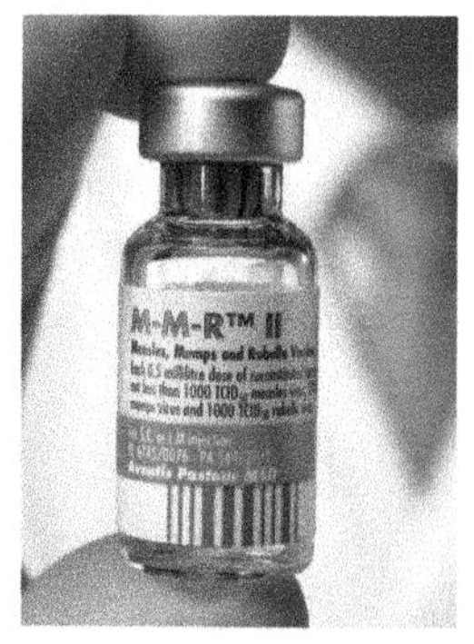

▲ **Figure 3** *MMR vaccination phial*

Even without additional evidence, care has to be exercised when looking at data, especially where there are correlations between two factors. In this example, almost all the population had been vaccinated with the MMR vaccine. There would therefore be a correlation between people who had been vaccinated and almost everything – what they ate, where they lived, etc. For example, data would have shown that that the majority of children who died in road accidents had been given the MMR vaccine. It does not follow that MMR vaccination causes road accidents. It was clearly a difficult choice for parents. Some parents, understandably, opted for separate vaccinations. Others mistrusted vaccinations in general and left their children unprotected. As a result, some children have developed disabilities that could have been avoided. However, had the research proved valid, it would have been those who held faith with the MMR vaccine who would have been putting their children's health at risk. It was a real dilemma.

The public sometimes believe that all such evidence must be true and accept it uncritically. However, all scientific evidence should initially be treated with caution – after all, it is fellow scientists who are often quickest to criticise. There are various reasons for this caution:

- To be universally accepted, a scientific theory must first be critically appraised and confirmed by other scientists in the field. The confirmation of a theory takes time.
- Some scientists may not be acting totally independently but may be funded by other people or organisations who are anticipating a particular outcome from the research.
- Scientists' personal beliefs, views, and opinions may influence the way they approach or represent their research.
- The facts as presented by media headline writers, companies, governments, and other organisations may have been biased or distorted to suit their own interests.
- New knowledge may challenge accepted scientific beliefs – theories are being modified all the time.

Practice questions: Chapter 14

1 (a) What is a pathogen? (*1 mark*)

(b) When a pathogen enters the body it may be destroyed by phagocytosis. Describe how. (*4 marks*)

(c) When a pathogen causes an infection, plasma cells secrete antibodies which destroy this pathogen. Explain why these antibodies are only effective against a specific pathogen. (*2 marks*)

AQA June 2012

2 Figure 1 shows an antibody molecule.

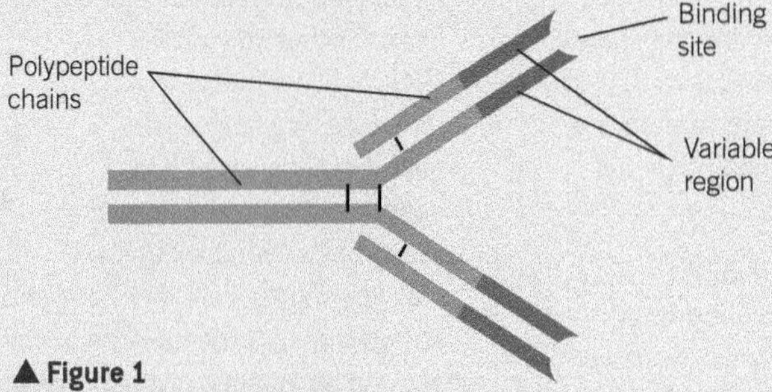

▲ **Figure 1**

(a) What is the evidence from Figure 1 that this antibody has a quaternary structure? (*1 mark*)

(b) Scientists use this antibody to detect an antigen on the bacterium that causes stomach ulcers. Explain why the antibody will only detect this antigen. (*3 marks*)

AQA Jan 2012

3 The table shows the cumulative rise in cases of the infectious disease Ebola over a 5-week period in 2014.

Week	Number of cases	Number of deaths
1	70	20
2	112	40
3	168	95
4	200	119
5	230	134
6	250	148

(a) Plot a graph of the above information with the number of weeks on the X axis. (*1 mark*)

(b) Calculate the rate of increase in number of cases of Ebola in the time period shown on the graph. (*2 marks*)

(c) A scientist suggests that the increase in the number of cases in the following 6 months will be exponential. Explain how plotting the next 6 months' data on a log scale would show whether the increase is exponential. (*1 mark*)

(d) Ebola is a rare disease in the human population, but can be passed on to humans from wild animals. Suggest, using your knowledge of the immune system, why the disease spreads fast once it is present in one human in an urban area. (*4 marks*)

4 Read the following passage.

Microfold cells are found in the epithelium of the small intestine. Unlike other epithelial cells in the small intestine, microfold cells do not have adaptations for the absorption of food.

Microfold cells help to protect against pathogens that enter the intestine. They have receptor proteins on their cell-surface membranes that bind to antigens on the surface of pathogens. The microfold cells take up the antigens and transport them to cells of the immune system. Antibodies are then produced which give protection against the pathogen.

Scientists believe that it may be possible to develop vaccines that make use of microfold cells. These vaccines could be swallowed in tablet form.

Use information from the passage and your own knowledge to answer the following questions.

Microfold cells do not have adaptations for the absorption of food (lines 2–3). Give **two** adaptations that other epithelial cells have for the absorption of food. *(2 marks)*

5 **(a)** Changes to the protein coat of the influenza virus cause antigenic variability. Explain how antigenic variability has caused some people to become infected more than once with influenza viruses. *(2 marks)*

(b) Figure 2 shows the changes in a B lymphocyte after stimulation by specific antigens.

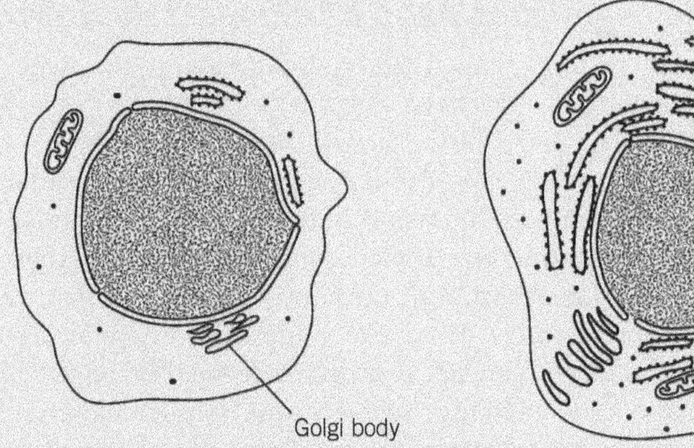

▲ **Figure 2**

Explain how the changes shown in the drawings are related to the function of B lymphocytes. *(4 marks)*

AQA 2004

6 Figure 3 shows one way in which white blood cells protect the body against disease.

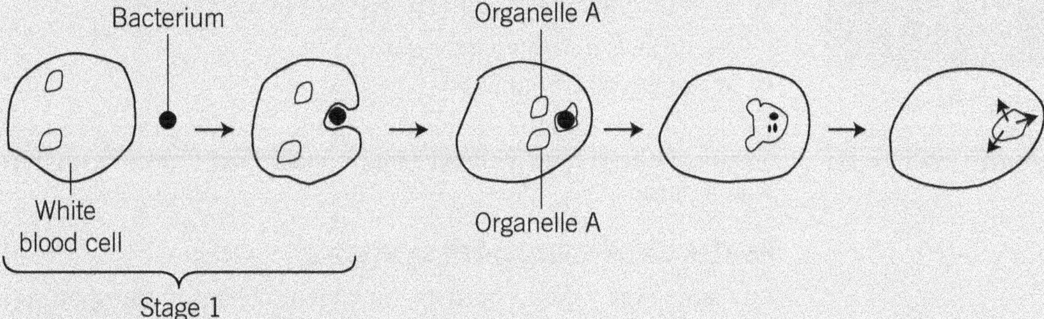

▲ **Figure 3**

(a) Describe what is happening during stage 1. *(2 marks)*

(b) (i) Name organelle **A**.

(ii) Describe the role of organelle **A** in the defence against disease. *(3 marks)*

AQA 2001

7 **(a)** What is vaccination? *(2 marks)*

(b) A test has been developed to find out whether a person has antibodies against the mumps virus. The test is shown in Figure 4.

(i) Explain why this test will detect mumps antibodies, but not other antibodies in the blood.

(ii) Explain why it is important to wash the well at the start of **step 4**.

(iii) Explain why there will be no colour change if mumps antibodies are not present in the blood. *(5 marks)*

AQA 2006

Answers to the Practice Questions are available at
www.oxfordsecondary.com/oxfordaqaexams-alevel-biology

15 Mammalian blood – the circulatory system

15.1 Circulatory system of a mammal

Learning objectives:

- → Describe how large organisms move substances around their bodies.
- → Describe the features of the transport systems of large organisms.

Specification reference: 3.2.6.1

Diffusion is adequate for transport over short distances (see Topic 4.2). The efficient supply of materials over larger distances requires a mass transport system.

Why large organisms need a transport system

All organisms need to exchange materials between themselves and their environment. You have seen that in small organisms this exchange takes place over the surface of the body (see Topic 5.2). However, with increasing size, the surface area to volume ratio decreases to a point where the needs of the organism cannot be met by the body surface alone (see Topic 5.1). A specialist exchange surface is therefore needed to absorb nutrients and respiratory gases, and to remove excretory products. These exchange surfaces are located in specific regions of the organism. A transport system is required to take materials from cells to exchange surfaces and from exchange surfaces to cells. Materials have to be transported between exchange surfaces and the environment. They also need to be transported between different parts of the organism. As organisms have evolved into larger and more complex structures, the tissues and organs of which they are made have become more specialised and dependent upon one another (see Topic 2.4). This makes a transport system all the more essential.

▲ **Figure 1** *Large organisms require a transport system to take materials from exchange surfaces to the cells that need them*

Whether or not there is a specialised transport medium, and whether or not it is circulated by a pump, depends on two factors:

- the surface area to volume ratio
- how active the organism is.

The lower the surface area to volume ratio, and the more active the organism, the greater is the need for a specialised transport system with a pump.

Features of transport systems

Any large organism encounters the same problems in transporting materials within itself. Not surprisingly, the transport systems of many organisms have many common features:

- A suitable medium in which to carry materials, for example, blood. This is normally a liquid based on water, because water readily dissolves substances and can be moved around easily.
- A form of mass transport in which the transport medium is moved around in bulk over large distances.
- A closed system of tubular vessels that contains the transport medium and forms a branching network to distribute it to all parts of the organism.
- A mechanism for moving the transport medium within vessels. This requires a pressure difference between one part of the system and another. It is achieved in two main ways:
    - Animals use muscular contraction either of the body muscles or of a specialised pumping organ, such as the heart (see Topic 15.3).

- Plants do not possess muscles and so often rely on passive natural physical processes such as the evaporation of water (see Topic 16.3).

- A mechanism to maintain the mass flow movement in one direction, for example, valves.
- A means of controlling the flow of the transport medium to suit the changing needs of different parts of the organism.

Transport systems in mammals

Mammals have a closed blood system in which blood is confined to vessels. A muscular pump called the heart circulates the blood around the body. Mammals have a double circulatory system (Figure 2). This refers to the fact that blood passes twice through the heart for each complete circuit of the body. This is because, when blood is passed through the lungs, its pressure is reduced. If it were to pass immediately to the rest of the body, its low pressure would make circulation very slow. Blood is therefore returned to the heart to boost its pressure before being circulated to the rest of the tissues. Substances are then delivered to the rest of the body quickly, which is necessary as mammals have a high body temperature and hence a high rate of **metabolism**. The vessels that make up the circulatory system of a mammal are divided into three types: arteries, veins, and capillaries.

Although a transport system is used to move substances longer distances, the final part of the journey into cells is by diffusion. The final exchange from blood vessels into cells is rapid because it takes place over a large surface area, across short distances, and there is a steep diffusion gradient.

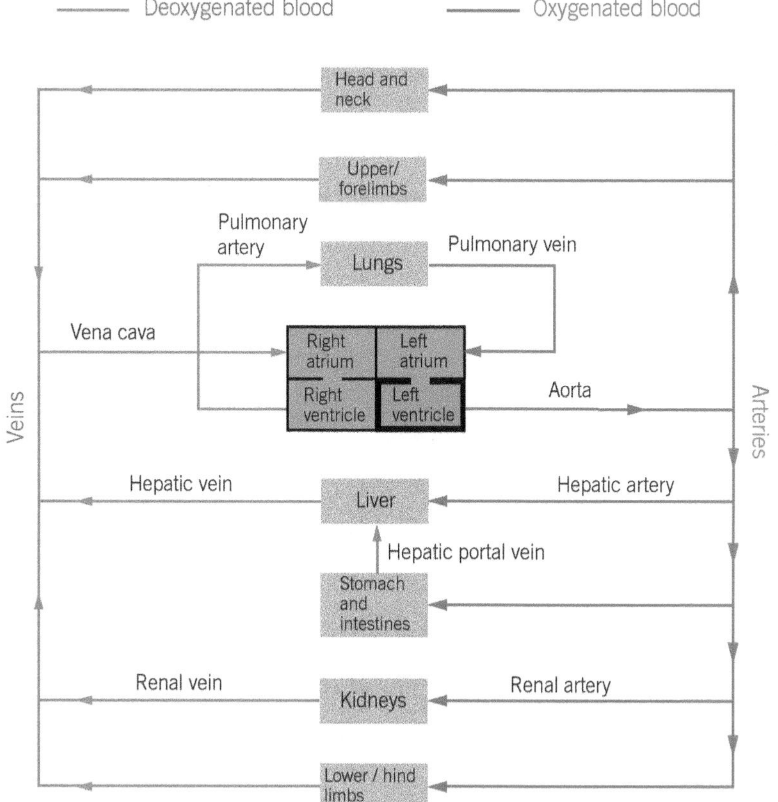

▲ **Figure 3** *Plan of the mammalian circulatory system*

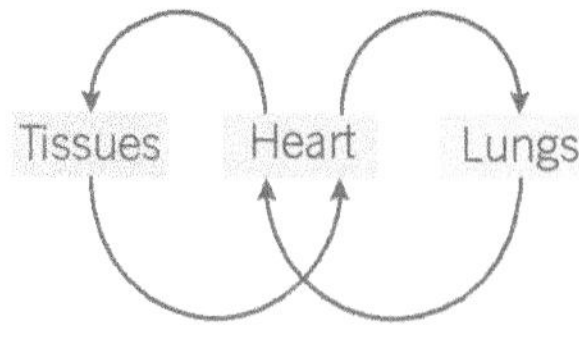

▲ **Figure 2** *Double circulation of a mammal*

Study tip

Almost all cells in the body are within 1 mm of a capillary – a short diffusion path.

Summary questions

1 Using Figure 3, name the blood vessel in each of the following descriptions:
   a joins the right ventricle of the heart to the capillaries of the lungs
   b carries oxygenated blood away from the heart
   c carries deoxygenated blood away from the liver
   d the first main blood vessel that an oxygen molecule reaches after being absorbed from an alveolus
   e has the highest blood pressure.

2 State **two** factors that make it more likely that an organism will have a circulatory pump such as the heart.

3 What is the main advantage of the double circulation found in mammals?

15.2 Blood vessels and their functions

Learning objectives:

- → Describe the structure of arteries, arterioles, and veins.
- → Describe how the structure of each of the above vessels is related to its function.
- → Describe the structure of capillaries and relate this to their function.

Specification reference: 3.2.6.1

Study tip

Arteries, arterioles, and veins carry out transport *not* exchange – only capillaries carry out exchange.

Study tip

The elastic tissue of arteries will stretch and recoil. It is not muscle and will not contract and relax.

In Topic 15.1 you saw that, in larger organisms, materials are transported around the body by the blood. To allow rapid transport of blood and to control its flow, the blood is confined to blood vessels.

Structure of blood vessels

There are different types of blood vessels:

- **Arteries** carry blood away from the heart and into arterioles.
- **Arterioles** are smaller arteries that control blood flow from arteries to capillaries.
- **Capillaries** are tiny vessels that link arterioles to veins.
- **Veins** carry blood from capillaries back to the heart.

Arteries, arterioles, and veins all have the same basic layered structure. From the outside inwards, these layers are:

- a **tough outer layer** that resists pressure changes from both within and outside
- a **muscle layer** that can contract and so control the flow of blood
- an **elastic layer** that helps to maintain blood pressure by stretching and springing back
- a **thin inner lining (endothelium)** that is smooth to prevent friction and thin to allow diffusion
- a lumen that is not actually a layer but the central cavity of the blood vessel through which the blood flows.

What differs between each type of blood vessel is the relative proportions of each layer. These differences are shown in Figure 1. Arterioles are not included because they are similar to arteries. They differ from arteries in being smaller in diameter and having a relatively larger muscle layer and lumen. The differences in structure are related to the differences in the function that each type of vessel performs.

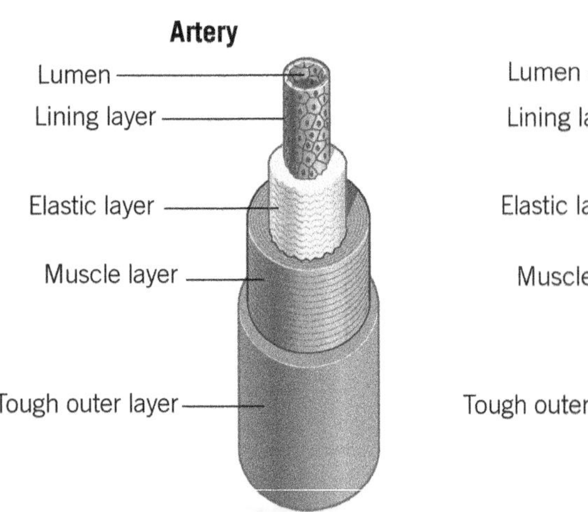

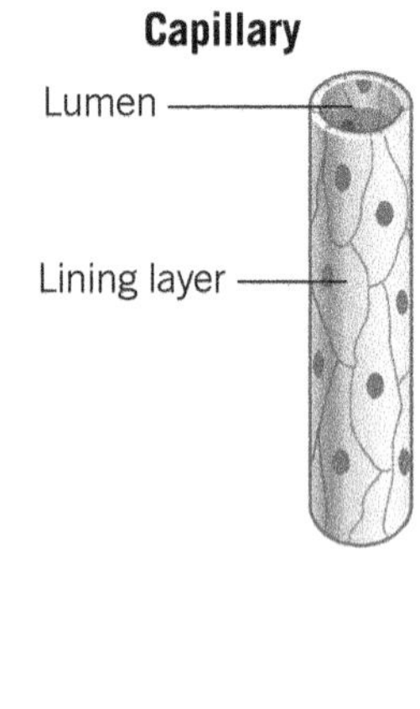

▲ **Figure 1** *Comparison of arteries, veins, and capillaries*

Artery structure related to function

The function of arteries is to transport blood rapidly under high pressure from the heart to the tissues. Their structure is adapted to this function as follows:

- **The muscle layer is thick compared to veins**. This means that smaller arteries can be constricted and dilated in order to control the volume of blood passing through them.
- **The elastic layer is relatively thick compared to veins** because it is important that blood pressure in arteries is kept high if blood is to reach the extremities of the body. The elastic wall is stretched at each beat of the heart (systole). It then springs back when the heart relaxes (diastole) in the same way as a stretched elastic band. This stretch and recoil action helps to maintain high pressure and to smooth pressure surges created by the beating of the heart.
- **The overall thickness of the wall is large**. This also resists the vessel bursting under pressure.
- **There are no valves** (except in the arteries leaving the heart) because blood is under constant high pressure and therefore does not tend to flow backwards.

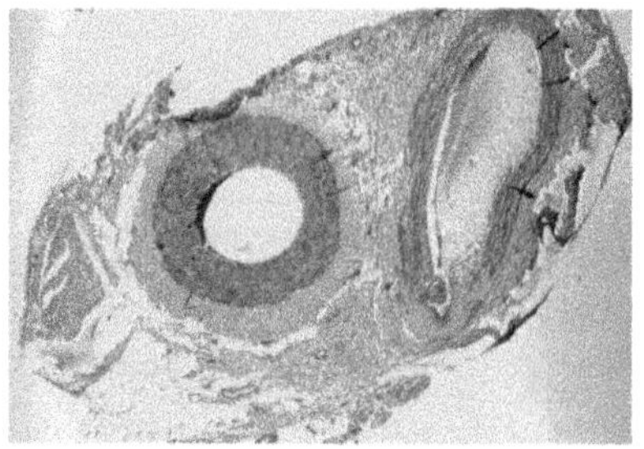

▲ **Figure 2** *Artery (left) and vein (right)*

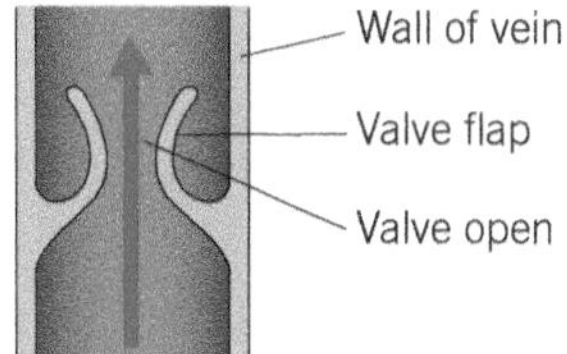

Blood flowing towards the heart passes easily through the valves

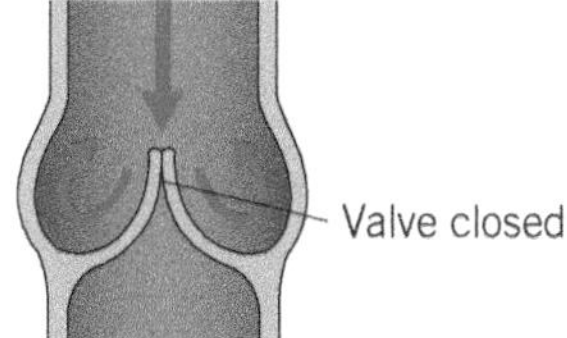

Blood flowing away from the heart pushes valves closed and so blood is prevented from flowing any further in this direction

▲ **Figure 3** *Action of valves in veins in ensuring one-way flow of blood*

Arteriole structure related to function

Arterioles carry blood, under lower pressure than arteries, from arteries to capillaries. They also control the flow of blood between the two. Their structure is related to these functions as follows:

- **The muscle layer is relatively thicker than in arteries**. The contraction of this muscle layer allows constriction of the lumen of the arteriole. This restricts the flow of blood and so controls its movement into the capillaries that supply the tissues with blood.
- **The lumen is relatively larger than in arteries** as the vessel wall is thinner and the lumen forms a larger proportion of the vessel.
- **The elastic layer is relatively thinner than in arteries** because blood pressure is lower.

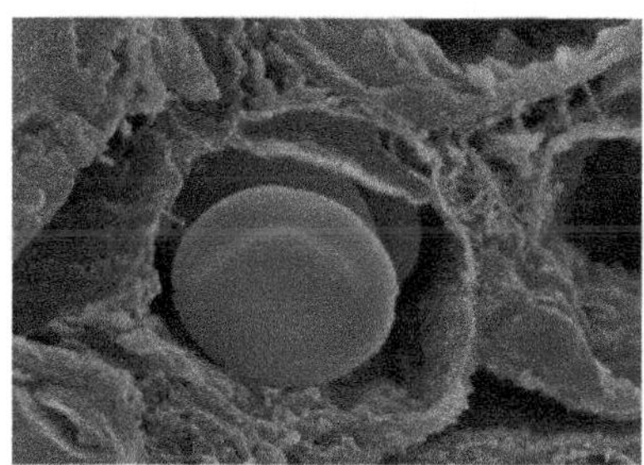

▲ **Figure 4** *False-colour scanning electron micrograph (SEM) of a section through a capillary with red blood cells passing through it*

Vein structure related to function

Veins transport blood slowly, under low pressure, from the tissues to the heart. Their structure is related to this function as follows:

- **The muscle layer is relatively thin** compared to arteries because veins carry blood away from tissues and therefore their constriction and dilation cannot control the flow of blood to the tissues.
- **The elastic layer is relatively thin** compared to arteries because pressure in the veins is too low to create a stretch and recoil action.
- **The overall thickness of the wall is small** because there is no need for a thick wall as the pressure within the veins is too low to create any risk of bursting. It also allows them to be flattened easily, aiding the flow of blood within them (see below).
- **There are valves throughout** to ensure that blood does not flow backwards, which it might otherwise do because the pressure is so low. When body muscles contract, veins are compressed, pressurising the blood within them. The valves ensure that this pressure directs the blood in one direction only – towards the heart (Figure 3).

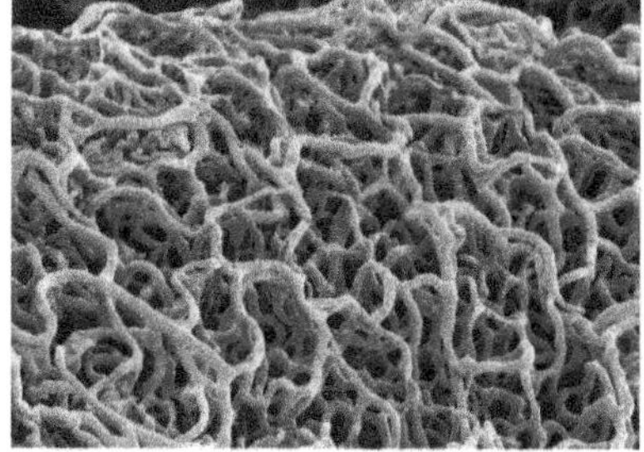

▲ **Figure 5** *Resin cast of a capillary network from the large intestine*

Capillary structure related to function

The function of capillaries (Figures 4 and 5) is to exchange metabolic materials such as oxygen, carbon dioxide, and glucose between the blood and the cells of the body. The flow of blood in capillaries is much slower. This allows more time for the exchange of materials.

The structure of capillaries is related to their function as follows:

- **Their walls consist only of endothelium**, which is a single layer of flattened cells sitting on a basement membrane. This makes them extremely thin, so the distance over which diffusion takes place is short. This allows for rapid diffusion of materials between the blood and the cells.
- **They are numerous and highly branched**, thus providing a large surface area for diffusion.
- **They have a narrow diameter** and so permeate tissues, which means that no cell is far from a capillary.
- **Their lumen is so narrow** that red blood cells are squeezed flat against the side of a capillary. This brings them even closer to the cells to which they supply oxygen. This again reduces the diffusion distance.
- **There are spaces between the lining (endothelial) cells** that allow white blood cells to escape in order to deal with infections within tissues.

Although capillaries are small, they cannot serve every single cell directly. Therefore, the final journey of metabolic materials is made in a liquid solution that bathes the tissues. This liquid is called **tissue fluid**.

Tissue fluid and its formation

Tissue fluid is a watery liquid that contains glucose, amino acids, fatty acids, salts, and oxygen. Tissue fluid supplies all of these substances to the tissues. In return, it receives carbon dioxide and other waste materials from the tissues. Tissue fluid is therefore the means by which materials are exchanged between blood and cells. As such, it bathes all the cells of the body. It is the immediate environment of cells and is, in effect, where they live. Tissue fluid is formed from blood plasma, and the composition of blood plasma is controlled by various homeostatic systems. As a result, tissue fluid provides a mostly constant environment for the cells it surrounds.

Formation of tissue fluid

Blood pumped by the heart passes along arteries, then the narrower arterioles, and, finally, the even narrower capillaries. This creates a pressure, called **hydrostatic pressure**, at the arterial end of the capillaries. This hydrostatic pressure forces tissue fluid out of the blood plasma. The outward pressure is, however, opposed by two other forces:

- hydrostatic pressure of the tissue fluid outside the capillaries, which prevents outward movement of liquid
- the lower **water potential** of the blood, due to the plasma proteins, that pulls water back into the blood within the capillaries.

However, the combined effect of all these forces is to create an overall pressure that pushes tissue fluid out of the capillaries. This pressure is only enough to force small molecules out of the capillaries, leaving all cells and proteins in the blood. This type of filtration under pressure is called **ultrafiltration**.

Return of tissue fluid to the circulatory system

Once tissue fluid has exchanged metabolic materials with the cells it bathes, it is returned to the circulatory system. Most tissue fluid returns to the blood plasma directly via the capillaries. This return occurs as follows:

- The loss of the tissue fluid from the capillaries reduces the hydrostatic pressure inside them.
- As a result, by the time the blood has reached the venous end of the capillary network its hydrostatic pressure is less than that of the tissue fluid outside it.
- Therefore tissue fluid is forced back into the capillaries by the higher hydrostatic pressure outside them.
- In addition, the water potential of the blood plasma is more negative (lower) than the tissue fluid. Therefore, there is net movement of water from the tissue fluid to the blood plasma by osmosis.

The tissue fluid has lost much of its oxygen and nutrients by diffusion into the cells that it bathed, but it has gained carbon dioxide and waste materials in return. These events are summarised in Figure 6.

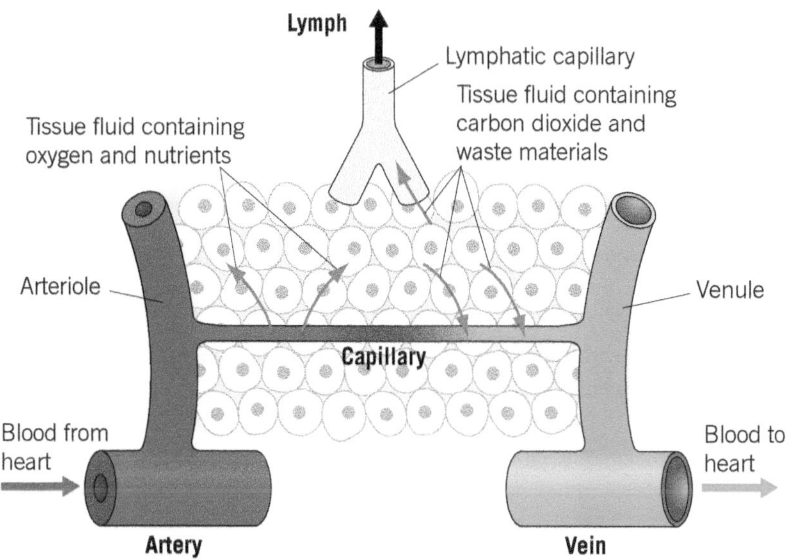

◀ **Figure 6** *Formation and return of tissue fluid*

Not all the tissue fluid can return to the capillaries – the remainder is carried back via the lymphatic system. This is a system of vessels that begin in the tissues. Initially they resemble capillaries, but they are dead ends. This is important, since pressure on them can only produce movement in one direction. They gradually merge into larger vessels that form a network throughout the body. These larger vessels drain their contents back into the bloodstream via two ducts that join veins close to the heart.

The contents of the lymphatic system (lymph) are not moved by the pumping of the heart. Instead they are moved by:

- **hydrostatic pressure** of the tissue fluid that has left the capillaries
- **contraction of body muscles** that squeeze the lymph vessels – valves in the lymph vessels ensure that the fluid inside them moves away from the tissues in the direction of the heart.

A summary of the methods of tissue fluid formation and its return to the bloodstream is shown in Figure 7.

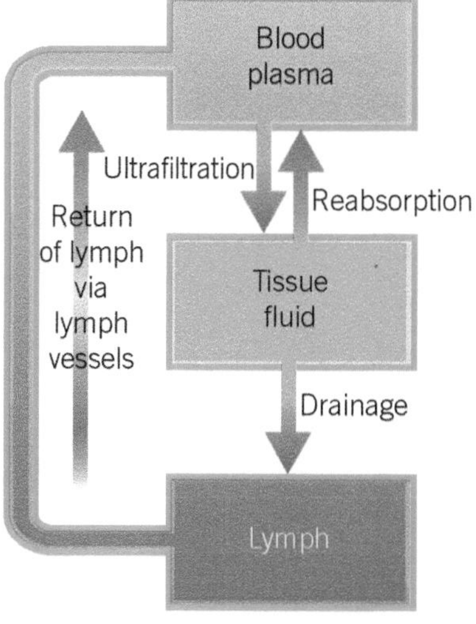

▲ **Figure 7** *Formation and return of tissue fluid to the bloodstream*

Summary questions

1 Give **one** advantage of having:
   a thick elastic tissue in the walls of arteries
   b relatively thick muscle walls in arterioles
   c valves in veins
   d only a lining layer in capillaries.

2 Table 1 shows the mean wall thickness of different blood vessels in a mammal. Suggest the letter that is most likely to refer to **a** the aorta, **b** a capillary, **c** a vein, **d** an arteriole, and **e** the renal artery.

▼ **Table 1**

Blood vessel	Mean wall thickness / mm
A	1.000
B	0.001
C	2.000
D	0.500
E	0.030

3 What forces tissue fluid out of the blood plasma in capillaries and into the surrounding tissues?

4 By which **two** routes does tissue fluid return to the bloodstream?

Blood flow in various blood vessels

The graph in Figure 8 shows certain features of the flow of blood from and to the heart through a variety of blood vessels.

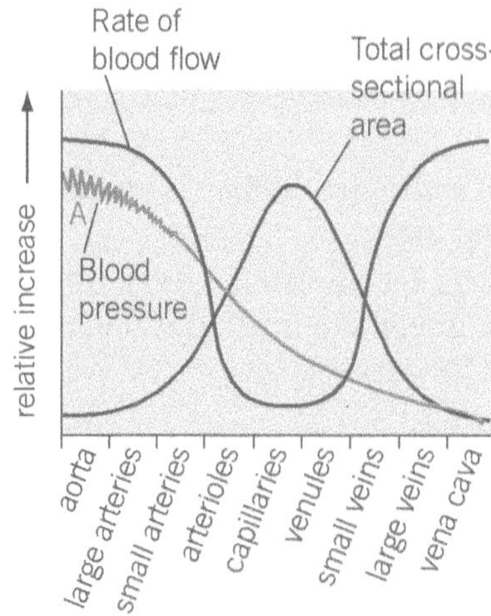

▲ **Figure 8** *Flow of blood to and from the heart*

1 Describe the changes in the rate of blood flow as blood passes from the aorta to the vena cava.
2 Explain why blood pressure in region A fluctuates up and down.
3 Why does the rate of blood flow decrease between the aorta and capillaries?
4 Explain how the rate of blood flow in the capillaries increases the rate of exchange of metabolic materials.
5 How does the structure of capillaries increase the efficiency of the exchange of metabolic substances?

15.3 The structure of the heart

The heart is a muscular organ that lies in the thoracic cavity behind the sternum (breastbone). It operates continuously and tirelessly throughout the life of an organism.

Structure of the human heart

The human heart is really two separate pumps lying side by side. The pump on the left deals with oxygenated blood from the lungs, whilst the one on the right deals with deoxygenated blood from the body. Each pump has two chambers:

- The **atrium** is thin walled and elastic and stretches as it collects blood. It only has to pump blood the short distance to the ventricle and therefore has only a thin muscular wall.
- The **ventricle** has a much thicker muscular wall as it has to pump blood some distance, either to the lungs or to the rest of the body.

Why have two separate pumps? Why not just pump the blood through the lungs to collect oxygen and then straight to the rest of the body before returning it to the heart? The problem with such a system is that the blood has to pass through tiny capillaries in the lungs in order to present a large surface area for the exchange of gases (see Topic 5.6). In doing so, there is a very large drop in pressure and so the blood flow to the rest of the body would be very slow. This drop in pressure is illustrated in Figure 1. Mammals have evolved a system in which the blood is returned to the heart to increase its pressure before it is distributed to the rest of the body. It is essential to keep the oxygenated blood in the pump on the left side separate from the deoxygenated blood in the pump on the right.

Because the right ventricle pumps blood to the lungs, a distance of only a few centimetres, and has a smaller capillary network to service than the left ventricle, it has a thinner muscular wall than the left ventricle. The left ventricle, in contrast, has a thick muscular wall, enabling it to create enough pressure to pump blood to the extremities of the body, a distance of about 1.5 m. Although the two sides of the heart are separate pumps and, after birth, there is no mixing of the blood in each of them, they nevertheless pump the same volume from each ventricle per unit time – both atria contract together and then both ventricles contract together.

Between each atrium and ventricle are valves that prevent the backflow of blood into the atria when the ventricles contract. There are two sets of valves:

- the **left atrioventricular (bicuspid) valves**, formed of two cup-shaped flaps on the left side of the heart
- the **right atrioventricular (tricuspid) valves**, formed of three cup-shaped flaps on the right side of the heart.

Each of the four chambers of the heart is served by large blood vessels that carry blood towards or away from the heart. The ventricles pump blood away from the heart and into the arteries. The atria receive blood from the veins.

Learning objectives:

- → Describe the appearance of the heart and its associated blood vessels.
- → Relate the structure of the heart to its function in a double circulatory system.

Specification reference: 3.2.6.2

Study tip

Although the left ventricle has a thicker wall than the right ventricle, their internal volumes are the same. They have to be, otherwise more blood would be pumped out of one side of the heart than the other.

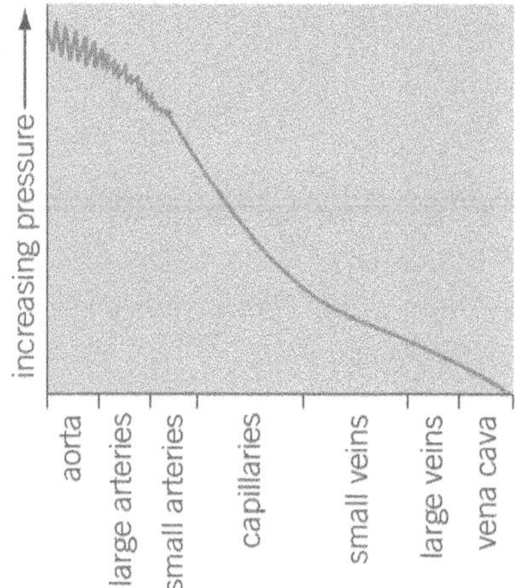

▲ **Figure 1** *Pressure changes in blood vessels*

Study tip

The left and right sides of the heart both contract together.

Hint

An easy way to recall which heart chambers are attached to which type of blood vessel is to remember that A and V always go together. Hence: **A**tria link to **V**eins and **A**rteries link to **V**entricles.

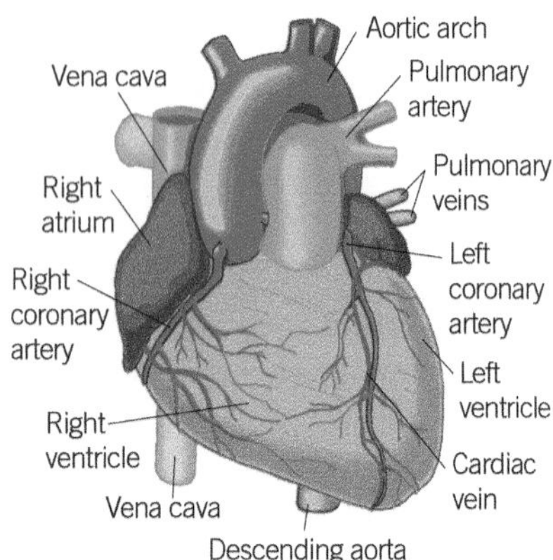

▲ **Figure 3** *External appearance of the human heart showing the blood supply to the heart muscle*

Summary questions

1 What is the name of the blood vessel that supplies the heart muscle with oxygenated blood?

2 State whether the blood in each of the following structures is oxygenated or deoxygenated:

a vena cava

b pulmonary artery

c left atrium.

3 List the correct sequence of the four main blood vessels and four heart chambers that a red blood cell passes through on its journey from the lungs, though the heart and body, and back again to the lungs.

4 Suggest why it is important to prevent mixing of the blood in the two sides of the heart.

Vessels connecting the heart to the lungs are called **pulmonary** vessels. The vessels connected to the four chambers are therefore as follows:

- The **aorta** is connected to the left ventricle and carries oxygenated blood to all parts of the body except the lungs.
- The **vena cava** is connected to the right atrium and brings deoxygenated blood back from the tissues of the body.
- The **pulmonary artery** is connected to the right ventricle and carries deoxygenated blood to the lungs, where its oxygen is replenished and its carbon dioxide is removed. Unusually for an artery, it carries deoxygenated blood.
- The **pulmonary vein** is connected to the left atrium and brings oxygenated blood back from the lungs. Unusually for a vein, it carries oxygenated blood.

The structure of the heart and its associated blood vessels is shown in Figure 2.

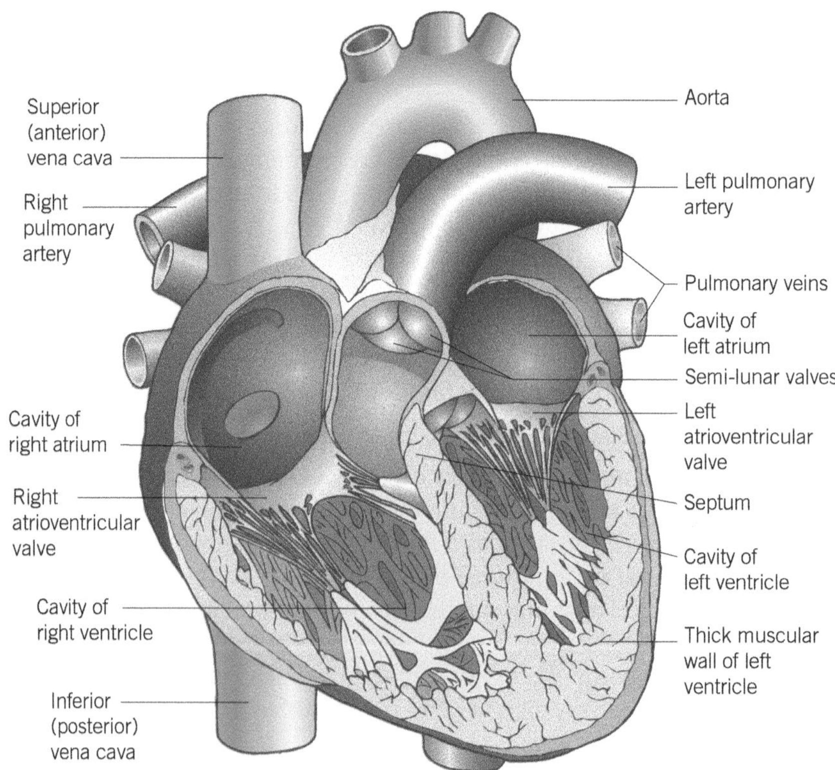

▲ **Figure 2** *Section through the human heart*

Supplying the heart muscle with oxygen

Although oxygenated blood passes through the left side of the heart, the heart does not use this oxygen to meet its own great respiratory needs. Instead, the heart muscle is supplied by its own blood vessels, called the **coronary arteries**, which branch off the aorta shortly after it leaves the heart. Blockage of these arteries, for example, by a blood clot, leads to **myocardial infarction**, or heart attack, because an area of the heart muscle is deprived of oxygen and so dies.

15.4 The cardiac cycle

The heart undergoes a sequence of events that is repeated in humans around 70 times each minute when at rest. This is known as the **cardiac cycle**. There are two phases to the beating of the heart: contraction (systole) and relaxation (diastole). Contraction occurs separately in the ventricles and the atria and is therefore described in two stages. For some of the time, relaxation takes place simultaneously in all chambers of the heart and is therefore treated as a single phase in the account below, which is illustrated in Figure 1. The unidirectional flow of blood is maintained by pressure changes and the action of valves.

Learning objectives:

- → Describe the events in the cardiac cycle.
- → Explain the role of valves in controlling blood flow through the heart.
- → Describe the myogenic nature of cardiac muscle.
- → Explain how the cardiac cycle is coordinated.

Specification reference: 3.2.6.2 and 3.4.8

Relaxation of the heart (diastole)

Blood returns to the atria of the heart through the pulmonary vein (from the lungs) and the vena cava (from the body). As the atria fill, the pressure in them rises, pushing open the atrioventricular valves and allowing the blood to pass into the ventricles. The muscular walls of both the atria and ventricles are relaxed at this stage. The relaxation of the ventricle wall reduces the pressure within the ventricle. This causes the pressure to be lower than that in the aorta and the pulmonary artery, and so the semi-lunar valves in the aorta and the pulmonary artery close, accompanied by the characteristic 'dub' sound of the heart beat.

1 Blood enters atria and ventricles from pulmonary veins and vena cava

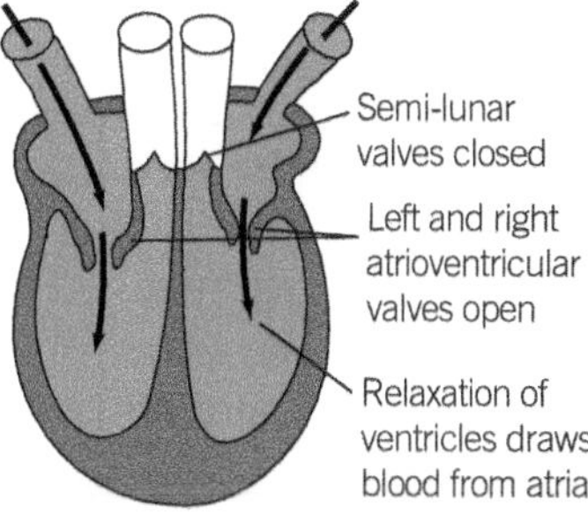

Relaxation of heart (diastole)

Atria are relaxed and fill with blood. Ventricles are also relaxed.

2

Atria contract to push remaining blood into ventricles

Semi-lunar valves closed

Left and right atrioventricular valves open

Blood pumped from atria to ventricles

Contraction of atria (atrial systole)

Atria contract, pushing blood into the ventricles. Ventricles remain relaxed.

3 Blood pumped into pulmonary arteries and the aorta

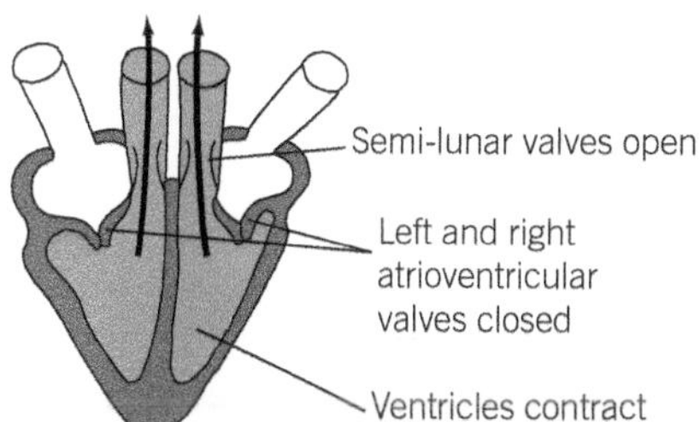

Contraction of ventricles (ventricular systole)

Atria relax. Ventricles contract, pushing blood away from heart through pulmonary arteries and the aorta.

▲ **Figure 1** *The cardiac cycle*

Contraction of the atria (atrial systole)

The muscle of the atrial walls contracts, forcing the remaining blood that they contain (around 20 per cent of the total blood in the heart) into the ventricles. The blood has only to be pushed a very short distance and therefore the muscular walls of the atria are very thin. During this stage, the muscle of the ventricle walls remains relaxed.

Contraction of the ventricles (ventricular systole)

After a short delay to allow the ventricles to fill with blood, their walls contract simultaneously. This increases the blood pressure within them, forcing shut the atrioventricular valves and preventing backflow of blood into the atria. The 'lub' sound of these valves closing is a characteristic of the heart beat. With the atrioventricular valves

Study tip

Don't make the common mistake of thinking that the left ventricle has a thicker wall than the right ventricle because it needs to *resist* the higher pressure.

It really has a thicker wall because it needs to *generate* a higher pressure to send blood around the body.

a *Valve open*

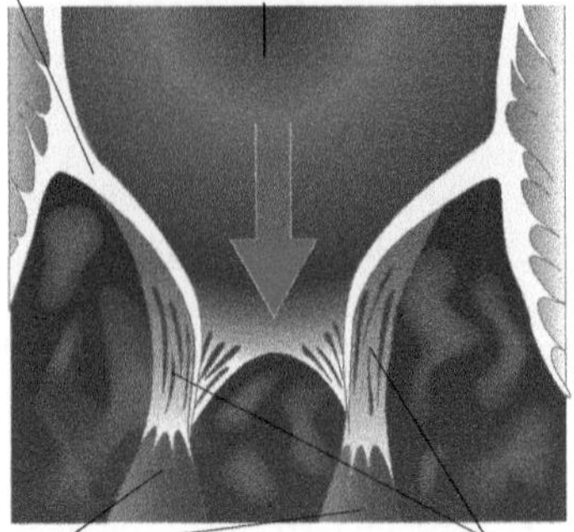

b *Valve closed*

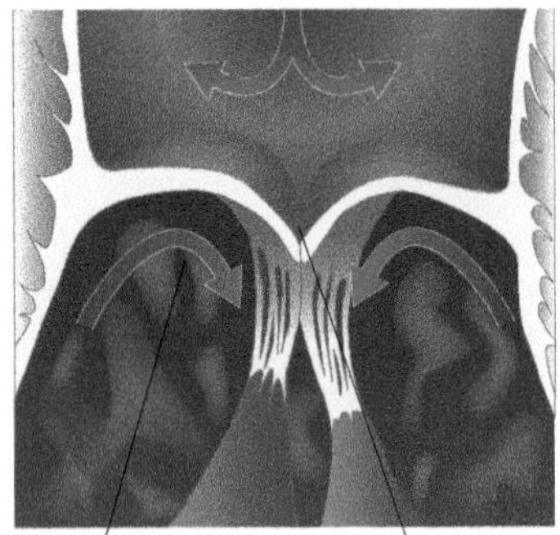

▲ **Figure 2** *Action of the valves*

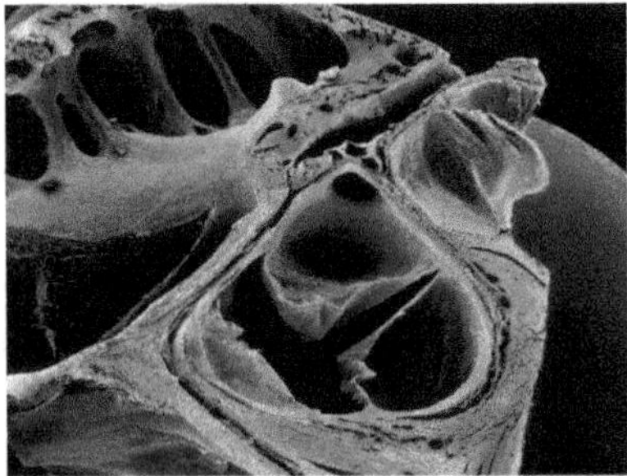

▲ **Figure 3** *False-colour scanning electron micrograph (SEM) of the semi-lunar valve of the aorta*

Study tip

Do not mix up **cardiac output** and **pulmonary ventilation** (see Topic 5.5). Whilst both measure volumes, the first involves blood and the heart (cardiac) and the second involves air and the lungs (pulmonary).

closed, the pressure rises further, forcing open the semi-lunar valves and pushing blood into the pulmonary artery and aorta. The walls of the ventricles are much thicker than those of the atria as they have to pump the blood much further. The wall of the left ventricle has to pump blood to the extremities of the body and so is much thicker than that of the right ventricle, which only has to pump blood as far as the adjacent lungs.

Valves in the control of blood flow

It is important to keep blood flowing in the right direction through the heart and around the body. This is achieved mainly by the pressure created by the heart muscle. Blood, as with all liquids and gases, will always move from a region of higher pressure to one of lower pressure. There are, however, situations within the circulatory system when pressure differences would result in blood flowing in the opposite direction from that which is desirable. In these circumstances, valves prevent any unwanted backflow of blood.

Valves in the cardiovascular system have evolved so that they open whenever the difference in blood pressure either side of them favours the movement of blood in the required direction. When pressure differences are reversed, that is, when blood would tend to flow in the opposite direction to that which is desirable, the valves have evolved to close. Examples of such valves include:

- **atrioventricular valves** between the left atrium and ventricle and the right atrium and ventricle. These prevent backflow of blood when contraction of the ventricles means that ventricular pressure exceeds atrial pressure. Closure of these valves ensures that, when the ventricles contract, blood within them moves to the aorta and pulmonary artery rather than back to the atria.
- **semi-lunar valves** in the aorta and pulmonary artery. These prevent backflow of blood into the ventricles when the recoil action of the elastic walls of these vessels creates a greater pressure in the vessels than in the ventricles.
- **pocket valves** in veins (see Topic 15.2) that occur throughout the venous system. These ensure that when the veins are squeezed, for example, when skeletal muscles contract, blood flows back to the heart rather than away from it.

The structure of all these valves is basically the same. They are made up of a number of flaps of tough, but flexible, fibrous tissue, which are cusp-shaped, that is, like deep saucers or bowls. When pressure is greater on the convex side of these cusps, rather than on the concave side, they move apart to let blood pass between the cusps. However, when pressure is greater on the concave side than on the convex side, blood collects within the 'bowl' of the cusps. This pushes them together to form a tight fit that prevents the passage of blood (Figure 2). So great are the pressures created within the ventricles of the heart that the atrioventricular valves are at risk of becoming inverted. To prevent this, the valves have string-like tendons that are attached to pillars of muscle in the ventricle wall (Figure 2).

Cardiac output

Cardiac output is the volume of blood pumped by one ventricle of the heart in one minute. It is usually measured in $dm^3 min^{-1}$ and depends upon two factors:

- the heart rate (the rate at which the heart beats)
- the stroke volume (volume of blood pumped out at each beat).

Cardiac output = heart rate × stroke volume

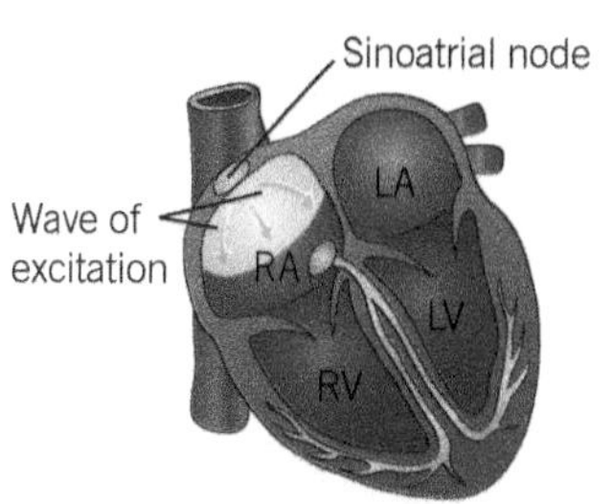

a Wave of electrical activity spreads out from the sinoatrial node

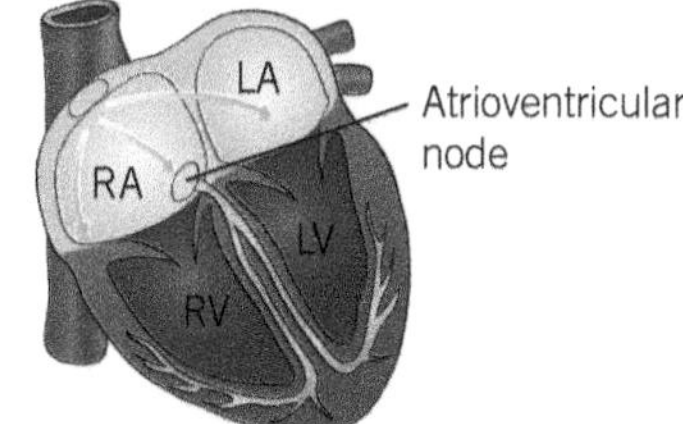

b Wave spreads across both atria, causing them to contract, and reaches the atrioventricular node

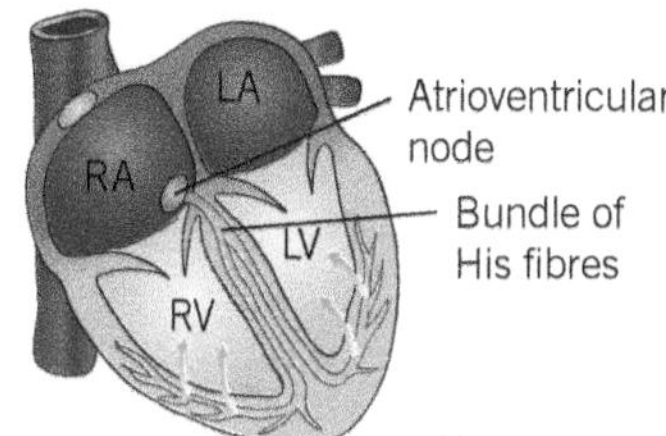

c Atrioventricular node conveys wave of electrical activity between the ventricles along the bundle of His and releases it at the apex, causing the ventricles to contract

▲ **Figure 4** *Control of the cardiac cycle*

How is the cardiac cycle controlled?

The control of heart rate is covered in Topic 27.5. Here we will just consider the coordination of the cardiac cycle. Cardiac muscle is **myogenic**, that is, its contraction is initiated from within the muscle itself, rather than by nervous impulses from outside (neurogenic), as is the case with other muscles. Within the wall of the right atrium of the heart is a distinct group of cells known as the **sinoatrial node (SAN)**. It is from here that the initial stimulus for contraction originates. The SAN has a basic rhythm of stimulation that determines the beat of the heart. For this reason it is often referred to as the **pacemaker**. The sequence of events that controls the cardiac cycle is as follows:

- A wave of electrical activity spreads out from the SAN across both atria, causing them to contract.
- A layer of non-conductive tissue (the atrioventricular septum) prevents the wave crossing to the ventricles.
- The wave of electrical activity is allowed to pass through a second group of cells called the **atrioventricular node (AVN)**, which lies between the atria.
- The AVN, after a short delay, conveys a wave of electrical activity between the ventricles along a series of specialised muscle fibres called the **bundle of His**.
- The bundle of His conducts the wave through the atrioventricular septum to the base of the ventricles, where the bundle branches into smaller fibres.
- The wave of electrical activity is released from these fibres, causing the ventricles to contract quickly at the same time, from the apex of the heart upwards.

These events are summarised in Figure 4.

Pressure and volume changes of the heart

Mammals have a closed circulatory system, that is, the blood is confined to vessels, and this allows the pressure within them to be maintained and regulated. Figure 5 illustrates the pressure and volume changes that take place in the heart during a typical cardiac cycle.

Study tip

Don't simply state that the role of the sinoatrial node (SAN) is to act as a pacemaker. The SAN produces a wave of electrical activity that causes both atria to contract, followed by the ventricles. The SAN initiates the heart beat.

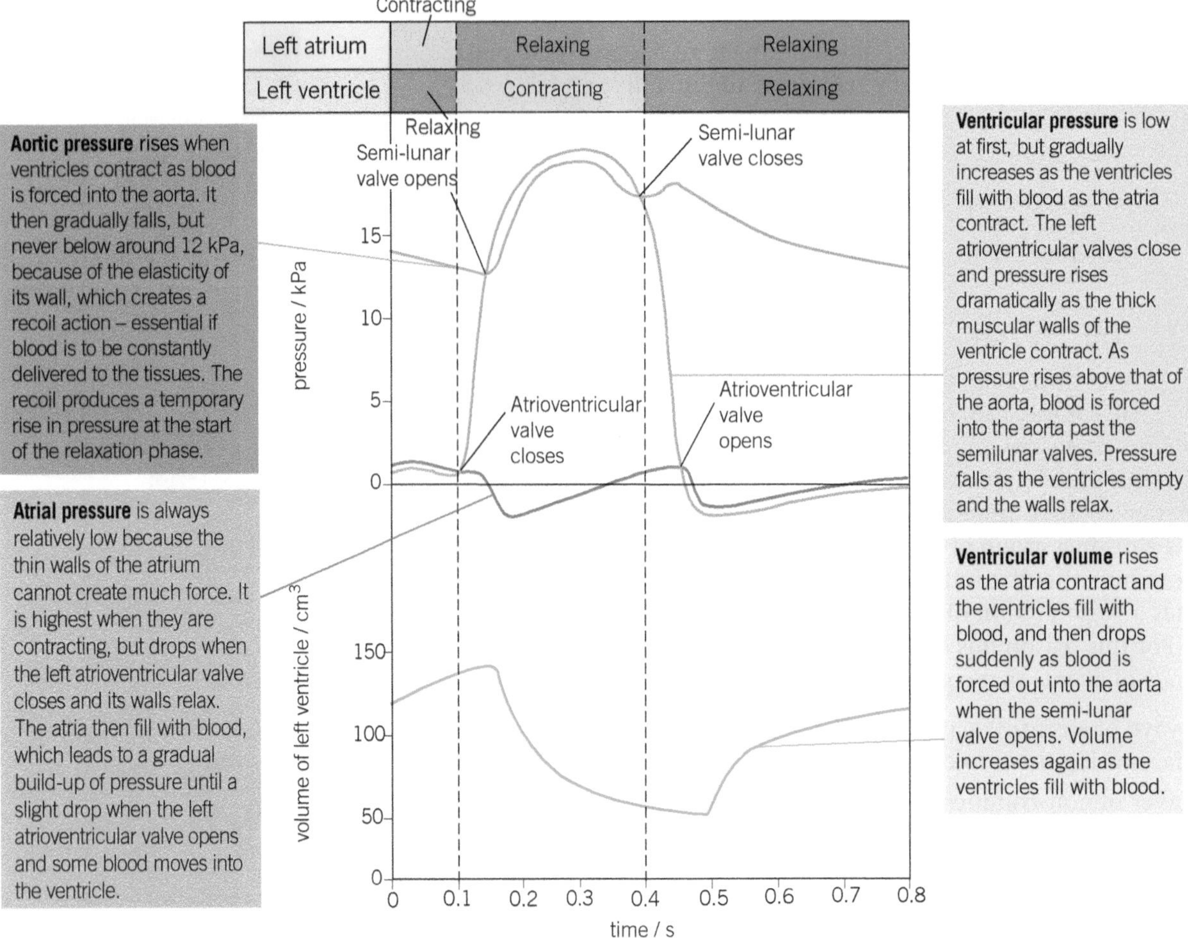

▲ **Figure 5** *Pressure and volume changes in the left side of the heart during the cardiac cycle*

Hint

Two facts will help you to understand the rather complex graph shown in Figure 5.

- Pressure and volume within a closed container are inversely related. When pressure increases, volume decreases, and vice versa.
- Blood, like all fluids, moves from a region where its pressure is greater to one where it is lower, that is, it moves down a pressure gradient.

Summary questions

1. Which chamber of the heart produces the greatest pressure?
2. Indicate whether each of the following statements is true or false.
   a. The left and right ventricles contract together.
   b. Heart muscle is myogenic.
   c. Semi-lunar valves occur between the atria and ventricles.
   d. The wave of electrical activity from the atrioventricular node is conveyed along the bundle of His.
   e. The wave of electrical activity from the sinoatrial node directly causes the ventricles to contract.
3. In each case, name the structure being described.
   a. On contraction it forces blood into the ventricle.
   b. It acts as the heart's pacemaker.
   c. It relays a wave of excitation to the apex of the heart.
4. After a period of training, the heart rate is often decreased when at rest although the cardiac output is unchanged. Suggest an explanation for this.
5. Using Figure 5, calculate the heart rate in beats per minute. Show your working.

Required practical 4

The effect of exercise on heart rate

Changes in heart rate in response to exercise can be a good indicator of physical fitness and this is the basis of any 'step test'. If the heart rate returns to its resting value quickly following a period of exercise (in this case stepping up and down from a bench), this indicates the person has good **aerobic** fitness. Heart rate can be monitored easily by taking a pulse.

Risk assessment

Before carrying out the test you should ensure that your subject is physically well enough. A 'pre-assessment' should be carried out, such as walking up and down a short flight of steps three times. If the subject is out of breath, dizzy, reports any pain, or has a high pulse rate (160 beats per minute or over) then the test should not be attempted. Similarly if the subject reports feeling unwell during the test then they should be stopped immediately.

A step test

The apparatus consists of a stopwatch to measure pulse rate and a step or bench about 25 cm in height. Subjects step up onto the bench with one foot and then bring the second foot up so they are standing on the bench. They then step down with one foot and bring the second foot down so they are standing on the floor. This is one 'cycle' and this cycle is repeated in a regular rhythm.

Pulse rate can be measured at the wrist (the radial artery) or on the neck (the carotid artery). Use fingers to detect the pulse as shown in Figure 6. Alternatively a 'pulse meter' could be used (Figure 7).

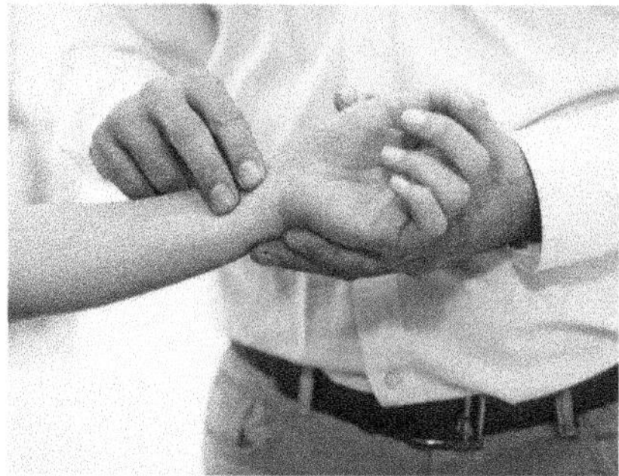

▲ **Figure 6** *Taking a radial pulse. As there is a pulse in the thumb, using fingers gives a more accurate count*

Procedure

- The subject steps up and down at a regular rate for exactly 4 minutes without stopping.
- They then rest for 1 minute, after which the subject's pulse rate is taken for exactly 15 seconds.
- The subject rests for a further 45 seconds, then another pulse rate reading is taken.
- The subject rests for a further 45 seconds and then the final pulse rate reading is taken.

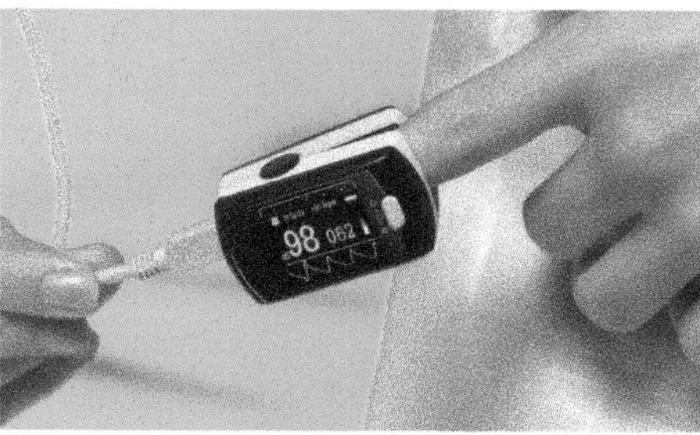

▲ **Figure 7** *An electronic 'pulse meter' can be linked to a data logger to record pulse rates over the exercise period*

You have taken readings at 1, 2, and 3 minutes from the end of exercise. Convert the readings to 'beats per minute' by multiplying each figure by 4 and then add the totals.

Analysis

The fitter you are, the faster your pulse rate will return to normal, and the lower the total number of beats will be. By using a 'constant' (in this case 24 000) and dividing this by the total, it is possible to compare the fitness of different subjects by comparing this fitness score.

Score (24 000 ÷ total beats)	Fitness rating
60 or less	poor
61–70	average
71–80	very good

1. Suggest why the pulse is only taken for 15 seconds.
2. Use your knowledge of the role of the circulatory system to explain why heart rate increases in response to exercise.
3. Explain why a 'pre-assessment' test is necessary before carrying out the step test.

15.5 Heart disease

Learning objectives:

- Describe what is meant by an atheroma.
- Explain the link between atheroma formation and an increased risk of aneurysm and thrombosis.
- Explain what a myocardial infarction is.
- Discuss the risk factors associated with coronary heart disease.

Specification reference: 3.2.7.1 and 3.2.7.2

Heart disease kills more people in the UK than any other disease. Almost half of heart disease deaths are from **coronary heart disease (CHD)**. CHD affects the pair of blood vessels, the coronary arteries, which supply the heart muscle with the glucose and oxygen that it requires for respiration. Blood flow through these vessels may be impaired by the build-up of fatty deposits known as **atheroma**. If blood flow to the heart muscle is interrupted, it can lead to **myocardial infarction**, in other words, a heart attack. Heart disease is not inevitable – most of it can be prevented.

Atheroma

Atheroma is a fatty deposit that forms within the wall of an artery. It begins as fatty streaks that are accumulations of white blood cells that have taken up low-density lipoproteins (LDLs). These streaks enlarge to form an irregular patch, or **atheromatous plaque**. Atheromatous plaques most commonly occur in larger arteries and are made up largely of deposits of cholesterol, fibres, and dead muscle cells. They bulge into the lumen of the artery, causing it to narrow so that the blood flow through it is reduced, as shown in Figure 1. The build-up of atheroma is shown in Figure 2. Atheromas increase the risk of two potentially very dangerous conditions: thrombosis and aneurysm.

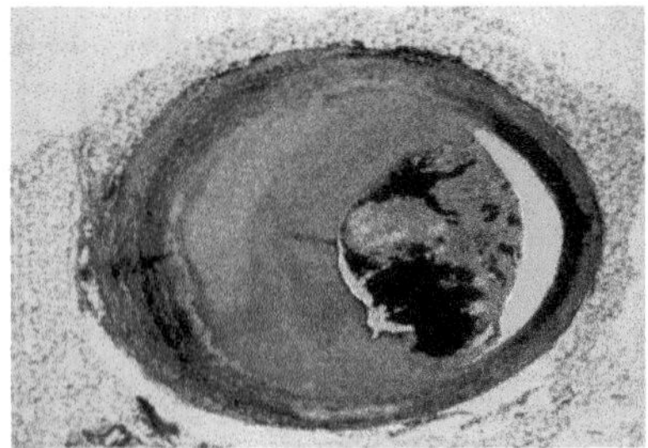

▲ **Figure 1** *Human coronary artery with a fatty atheroma partly blocking the interior*

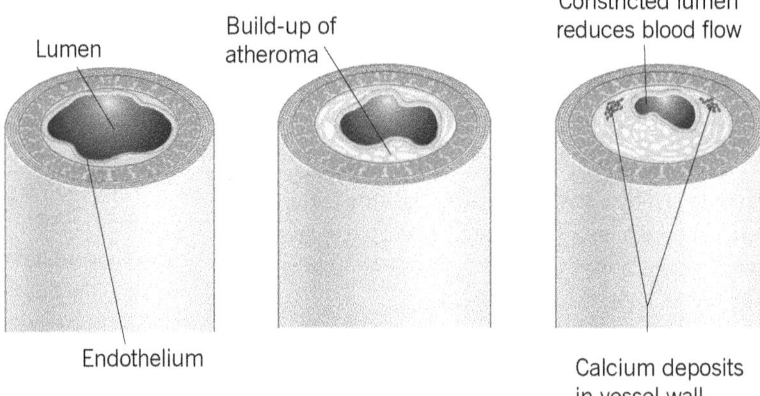

▲ **Figure 2** *Build-up of atheroma*

Hint

Remember that plaques are formed *within* the artery wall. They are *not* deposits on the inner surface of the artery wall.

Study tip

Use the term atheroma to describe the fatty deposits in an artery. Avoid writing furred-up arteries or similar terms.

Thrombosis

If an atheroma breaks through the lining (endothelium) of the blood vessel, it forms a rough surface that interrupts the otherwise smooth flow of blood. This may result in the formation of a blood clot, or **thrombus**, in a condition known as thrombosis. This thrombus may block the blood vessel, reducing or preventing the supply of blood to tissues beyond it. The region of tissue deprived of blood often dies as a result of the lack of oxygen, glucose, and other nutrients that the blood normally provides. Sometimes, the thrombus is carried from its place of origin and lodges in, and blocks, another artery. Figure 3 shows a thrombus in a coronary artery.

Aneurysm

Atheromas that lead to the formation of a thrombus also weaken the artery walls. These weakened points swell to form a balloon-like, blood-filled structure called an **aneurysm**. Aneurysms frequently burst, leading to haemorrhage and therefore loss of blood to the region of the body served by that artery. A brain aneurysm is known as a cerebrovascular accident (CVA), or stroke.

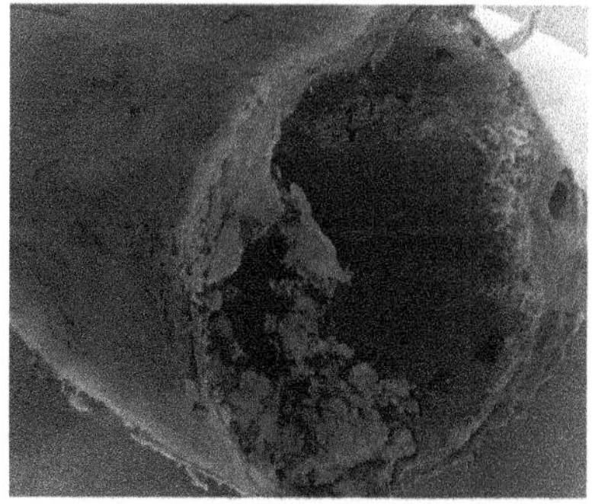

▲ **Figure 3** *Human coronary artery containing a thrombus*

Myocardial infarction

More commonly known as a heart attack, the expression 'myocardial infarction' refers to a reduced supply of oxygen to the muscle (myocardium) of the heart. It results from a blockage in the coronary arteries. If this occurs close to the junction of the coronary artery and the aorta, the heart will stop beating because its blood supply will be completely cut off. If the blockage is further along the coronary artery, the symptoms will be milder, because a smaller area of muscle will suffer oxygen deprivation. In Britain, about half a million people a year have a heart attack, although fewer than one-third of them die as a result. Almost all show signs of atheroma and many have coronary thrombosis (clot formation in the coronary arteries).

Risk factors associated with CHD

There are a number of factors that seperately increase the risk of an individual developing heart disease. When combined together, four, five, or six of these factors produce a disproportionately greater risk (Figure 4). These risk factors include the following.

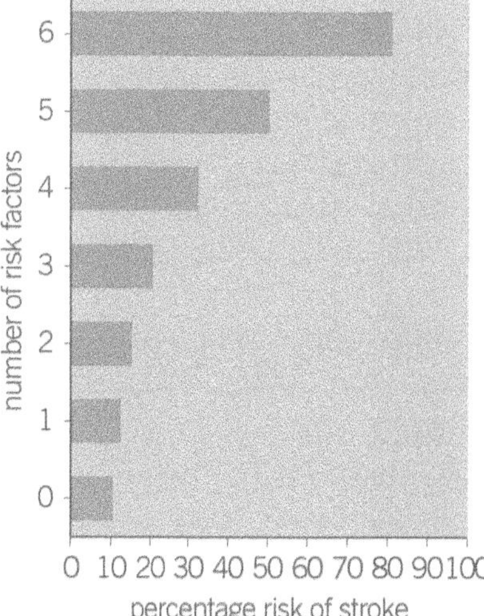

▲ **Figure 4** *The combined impact of six risk factors on the likelihood of a 70-year-old man experiencing a stroke in the next 10 years*

Smoking

Smokers are between two and six times more likely to have heart disease than non-smokers. Giving up smoking is the single most effective way of increasing life expectancy. There are two main constituents of tobacco smoke that increase the likelihood of heart disease:

- **Carbon monoxide** combines easily, but irreversibly, with the haemoglobin in red blood cells to form carboxyhaemoglobin. It thereby reduces the oxygen-carrying capacity of the blood. To supply the equivalent quantity of oxygen to the tissues, the heart must work harder. This can lead to raised blood pressure, which increases the risk of CHD and strokes. In addition, the reduction in the oxygen-carrying capacity of the blood means that it may be insufficient to supply the heart muscle during exercise. This leads to chest pain (angina) or, in severe cases, a myocardial infarction.
- **Nicotine** stimulates the production of the hormone adrenaline, which increases heart rate and raises blood pressure. As a consequence there is a greater risk of smokers having CHD or a stroke. Nicotine also makes the platelets in the blood more 'sticky', and this leads to a higher risk of thrombosis and hence of strokes or myocardial infarction.

High blood pressure

If your genes cause you to have high blood pressure, altering your lifestyle will not change this fact. Lifestyle factors such as excessive prolonged stress, certain diets, and lack of exercise increase the risk of high blood pressure. These are factors over which the individual can exert control. High blood pressure increases the risk of heart disease for the following reasons:

- As there is already a higher pressure in the arteries, the heart must work harder to pump blood into them and is therefore more prone to failure.
- Higher blood pressure within the arteries means that they are more likely to develop an aneurysm and burst, causing haemorrhage.
- In response to the higher pressure within them, the walls of the arteries tend to become thickened and may harden, restricting the flow of blood.

Blood cholesterol

Cholesterol is an essential component of membranes. As such, it is an essential biochemical that must be transported in the blood. It is carried in the plasma as tiny spheres of lipoproteins (lipid and protein). There are two main types:

- **high-density lipoproteins (HDLs)**, which remove cholesterol from tissues and transport it to the liver for excretion. They help protect arteries against heart disease.
- **low-density lipoproteins (LDLs)**, which transport cholesterol from the liver to the tissues, including the artery walls, which they infiltrate, leading to the development of atheroma and hence heart disease.

Diet

There are a number of aspects of diet that increase the risk of heart disease, both directly and indirectly:

- **High levels of salt** raise blood pressure.
- **High levels of saturated fat** increase LDL levels and hence blood cholesterol concentration.

By contrast, foods that act as **antioxidants**, for example, vitamin C, reduce the risk of heart disease, and so does non-starch polysaccharide (dietary fibre).

Hint

Always remember that risk factors increase the *probability* of getting heart disease, but they do not mean that someone will certainly get it. Heavy smokers, with high blood pressure and high blood cholesterol, may never develop heart disease, they are just more likely to (see Figure 5).

Study tip

Use the terms high-density lipoproteins and low-density lipoproteins, *not* good cholesterol and bad cholesterol.

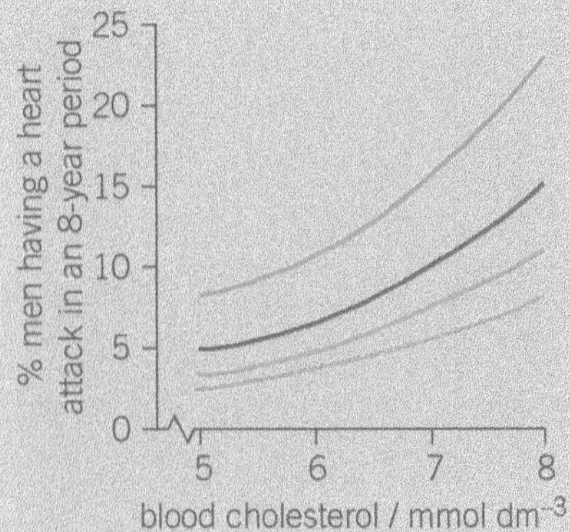

▲ **Figure 5** *Effects of blood pressure, smoking and blood cholesterol on the risk of heart attack in American men*

Summary questions

1 Atheroma, thrombosis, aneurysm and myocardial infarction are four types of heart disease. Link each of the following descriptions to one of these diseases.

   a Commonly known as a heart attack.

   b Build-up of fatty deposits.

   c The formation of a blood clot.

   d Stretched region of an artery wall.

2 State **three** ways in which high blood pressure increases the risk of heart disease.

A calculated risk

Figure 5 shows the effect of three of the above risk factors on the chance of heart attack in American men. Study the data and answer the questions.

1 A smoker with high blood pressure wishes to reduce his risk of heart attack. If he could only alter one factor, would he be better giving up smoking or reducing his blood pressure? Explain your answer.

2 A non-smoker with high blood pressure has a blood cholesterol level of 5 mmol dm^{-3}, but over a period of 3 years this concentration increases to 8 mmol dm^{-3}. How many times greater is his risk of heart disease? Show your working.

3 Two non-smoking men with low blood pressure both have a blood cholesterol level of 5 mmol dm^{-3}. One of them starts to smoke and the blood cholesterol level of the other increases to 7 mmol dm^{-3}. Which man is now at the greater risk from heart disease? Explain your answer.

Electrocardiogram

During the cardiac cycle, the heart undergoes a series of electrical current changes. These are related to the waves of electrical activity created by the sinoatrial node and the heart's response to these. If picked up by a cathode ray oscilloscope, these changes can produce a trace known as an **electrocardiogram** (ECG). Doctors can use this trace to provide a picture of the heart's electrical activity and hence its health. In a normal ECG there is a pattern of large peaks and small troughs that repeat identically at regular intervals. An ECG produced during a heart attack shows less pronounced peaks and larger troughs that are repeated in a similar, but not identical, way. During a condition called fibrillation, the heart muscle contracts in a disorganised way that is reflected in an irregular ECG.

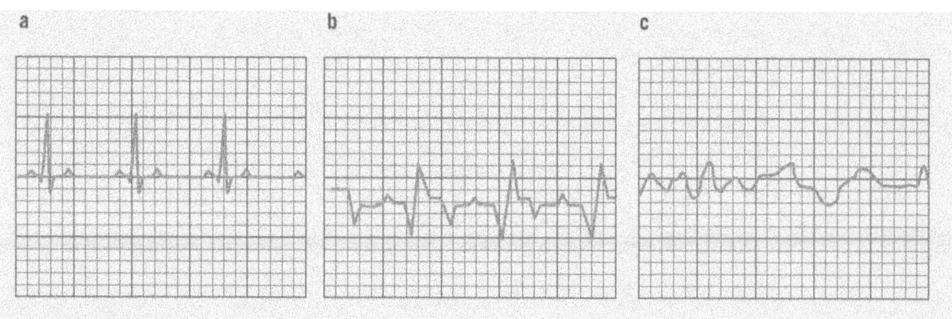

▲ **Figure 6** *Three different electrocardiogram (ECG) traces*

1 The three ECG traces shown in Figure 6 represent an ECG trace for a normal heart, one in fibrillation, and one during a heart attack. Using the letters a, b, and c, suggest which trace corresponds to which heart condition. Give reasons for your answers.

Practice questions: Chapter 15

1 (a) (i) An arteriole is described as an organ. Explain why. *(1 mark)*
(ii) An arteriole contains muscle fibres. Explain how these muscle fibres reduce blood flow to capillaries. *(2 marks)*
(b) (i) A capillary has a thin wall. This leads to rapid exchange of substances between the blood and tissue fluid. Explain why. *(1 mark)*
(ii) Blood flow in capillaries is slow. Give the advantage of this. *(1 mark)*
(c) Kwashiorkor is a disease caused by a lack of protein in the blood. This leads to a swollen abdomen due to a build-up of tissue fluid. Explain why a lack of protein in the blood causes a build-up of tissue fluid. *(3 marks)*
AQA Jan 2013

2 Figure 1 shows tissue fluid and body cells surrounding a capillary in a leg.

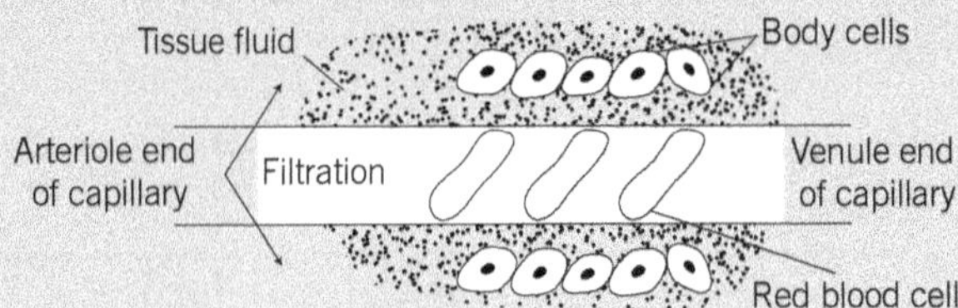

▲ **Figure 1**

(a) Name **two** substances which are at a higher concentration in the blood at the arteriole end of the capillary in a leg than at the venule end. *(1 mark)*
(b) Explain how fluid may be returned to the blood. *(3 marks)*
(c) People with high blood pressure often have swollen ankles and feet. This is the result of an accumulation of tissue fluid.
(i) Suggest an explanation for the link between high blood pressure and the accumulation of tissue fluid.
(ii) Suggest why tissue fluid accumulates more in the ankles and feet than in other parts of the body. *(3 marks)*
AQA 2002

3 Figure 2 shows an external view of a mammalian heart.

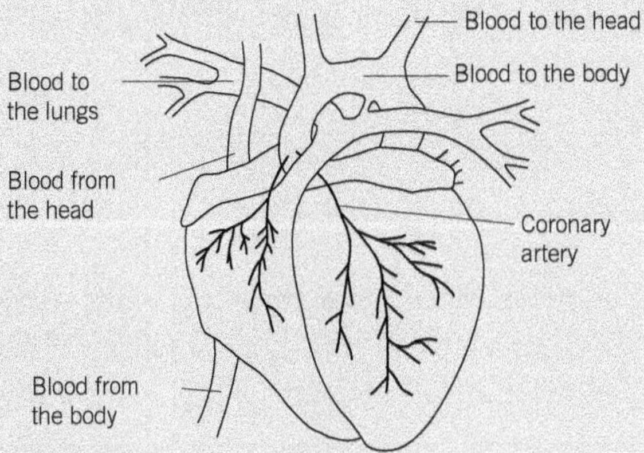

▲ **Figure 2**

(a) Name the blood vessel which:
(i) brings blood from the body *(1 mark)*
(ii) takes blood to the lungs. *(1 mark)*
(b) (i) What is the function of the coronary artery? *(1 mark)*
(ii) From which blood vessel does the coronary artery originate? *(1 mark)*

(c) The information below compares some features of different blood vessels.

		Blood vessel		
		Artery	Capillary	Vein
Property	Mean diameter of vessel	4.0 mm	8.0 μm	5.0 mm
	Mean thickness of wall	1.0 mm	0.5 μm	0.5 mm
		Relative thickness (shown by length of bar)		
Tissues present in wall	Endothelium	▬	▬	▬
	Elastic tissue	▬▬		▬▬
	Muscle	▬▬▬		▬▬

Use the information to explain how the structures of the walls of arteries, veins, and capillaries are related to their functions. *(6 marks)*

AQA 2004

4 Figure 3 shows a cross section of a blood vessel.

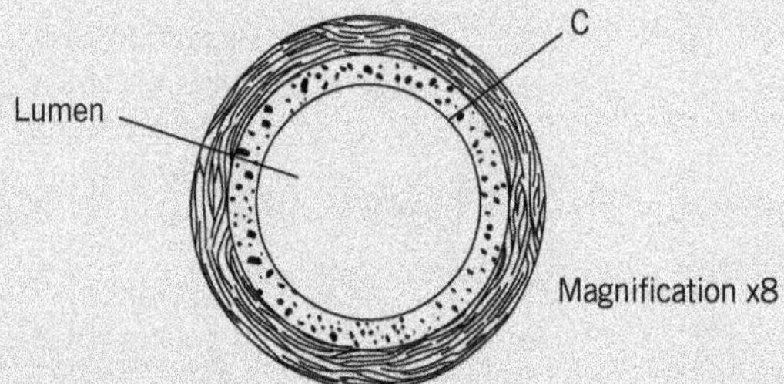

▲ **Figure 3**

(a) Name layer **C**. *(1 mark)*

(b) Calculate the actual diameter of the lumen of this blood vessel in millimetres. Show your working. *(2 marks)*

(c) (i) The aorta has many elastic fibres in its wall. An arteriole has many muscle fibres in its wall. Explain the importance of elastic fibres in the wall of the aorta. *(2 marks)*

(ii) Explain the importance of muscle fibres in the wall of an arteriole. *(2 marks)*

AQA June 2010

Answers to the Practice Questions are available at
www.oxfordsecondary.com/oxfordaqaexams-alevel-biology

16 Mass transport systems in plants

16.1 Movement of water through roots

Learning objectives:

- → Describe how water is taken up by root hairs.
- → Explain the passage of water across the root cortex by apoplastic and symplastic pathways.
- → Explain the role of the endodermis and the concept of root pressure.

Specification reference: 3.2.8.1

Study tip

Always refer to osmosis and to the water potential gradient when explaining how water enters a root-hair cell from the soil.

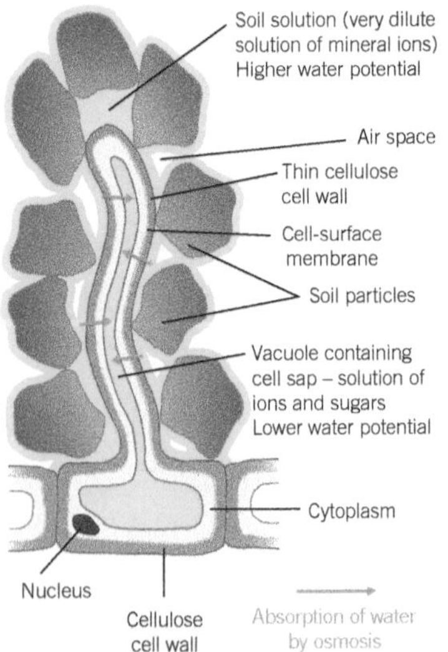

▲ **Figure 2** *Absorption of water by a root-hair cell*

Synoptic link

If you are uncertain about osmosis or water potential it would be worth reading Topic 4.3 again.

The vast majority of plants are terrestrial organisms. As a result they have evolved adaptations to conserve water and are covered by a waterproof layer. Therefore they cannot absorb water over the general body surface. Instead they have a special exchange surface in the soil – the **root hairs**. Before learning how the root hairs absorb water and how water is transported, it is worth looking at the basic structure of a root. The arrangement of tissues in a **dicotyledonous** root is illustrated in Figure 1.

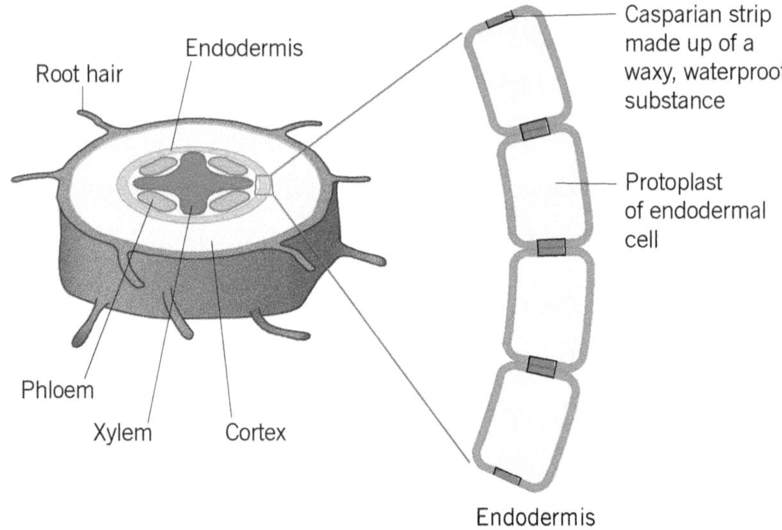

▲ **Figure 1** *Root of a dicotyledonous plant*

Uptake of water by root hairs

Root hairs are the exchange surfaces in plants that are responsible for the absorption of water and mineral ions. Plants constantly lose water by the process of **transpiration** (see Topic 16.3). This loss can amount to up to 700 dm^3 per day in a large tree. All of this water must be replaced by water that is absorbed through the root hairs (Figure 2).

Each root hair is a long, thin extension of a root epidermal cell. These root hairs remain functional for a few weeks before dying back, to be replaced by others nearer the growing tip. They are efficient surfaces for the exchange of water and mineral ions because:

- they provide a large surface area as they are very long extensions and occur in thousands on each of the branches of a root
- they have a thin surface layer (the cell-surface membrane and cellulose cell wall), across which materials can move rapidly.

Root hairs arise from epidermal cells a little way behind the tips of young roots. These hairs grow into the spaces around soil particles.

In damp conditions they are surrounded by a soil solution that contains small quantities of mineral ions. The soil solution is, however, mostly water and therefore has a very high **water potential** – only slightly less than zero. In contrast, the root hairs, and other cells of the root, have sugars, amino acids, and mineral ions dissolved inside them. These cells therefore have a much lower water potential. As a result, water moves by osmosis from the soil solution into the root-hair cells down this water potential gradient.

After being absorbed into the root-hair cell, water continues its journey across the root in two ways:

- the apoplastic pathway (the apoplast)
- the symplastic pathway (the symplast).

The apoplastic pathway

As water enters by osmosis into endodermal cells, it pulls more water along behind it, due to the **cohesive** properties of the water molecules. This creates a tension that draws water along the cell walls of the cells of the root cortex. The mesh-like structure of the cellulose cell walls of these cells has many water-filled spaces and so there is little or no resistance to this pull of water along the cell walls (Figure 3).

The symplastic pathway

This takes place across the cytoplasm of the cells of the cortex as a result of **osmosis**. The water passes through the cell walls along tiny openings called plasmodesmata. Each plasmodesma (singular) is filled with a thin strand of cytoplasm. Therefore there is a continuous column of cytoplasm extending from the root-hair cell to the xylem at the centre of the root. Water moves along this column as follows:

- Water entering by osmosis increases the water potential of the root-hair cell.
- The root-hair cell now has a higher water potential than the first cell in the cortex.
- Water therefore moves from the root-hair cell to the first cell in the cortex by osmosis, down the water potential gradient.
- This first cell now has a higher water potential than its neighbour to the inside of the stem.
- Water therefore moves into this neighbouring cell by osmosis along the water potential gradient.
- This second cell now has a higher water potential than its neighbour to the inside, and so water moves from the second cell to the third cell by osmosis along the water potential gradient.
- At the same time, this loss of water from the first cortical cell lowers its water potential, causing more water to enter it by osmosis from the root-hair cell.
- In this way, a water potential gradient is set up across all the cells of the cortex, which carries water along the cytoplasm from the root-hair cell to the endodermis.

The apoplastic and symplastic pathways are summarised in Figure 4.

Hint

Remember that all water potential values of solutions are negative. Water moves from a higher (less negative) water potential to a lower (more negative) water potential.

Hint

Cohesion is the mutual attraction of molecules for one another. In other words, it is the ability of molecules (in this case water molecules) to stick to one another.

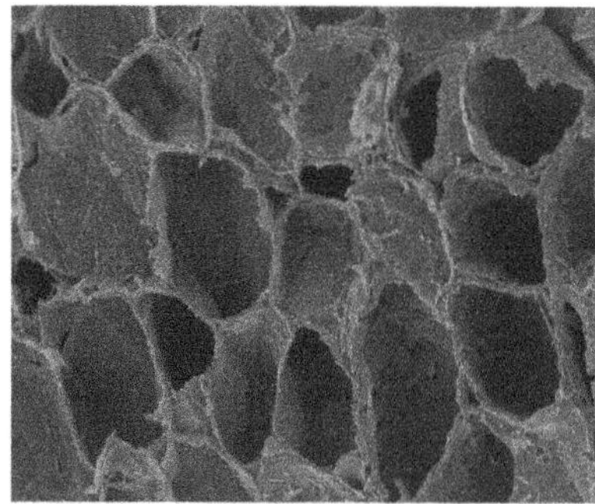

▲ **Figure 3** *Scanning electron micrograph (SEM) of plant cell walls showing the many spaces through which water can pass along the apoplastic pathway*

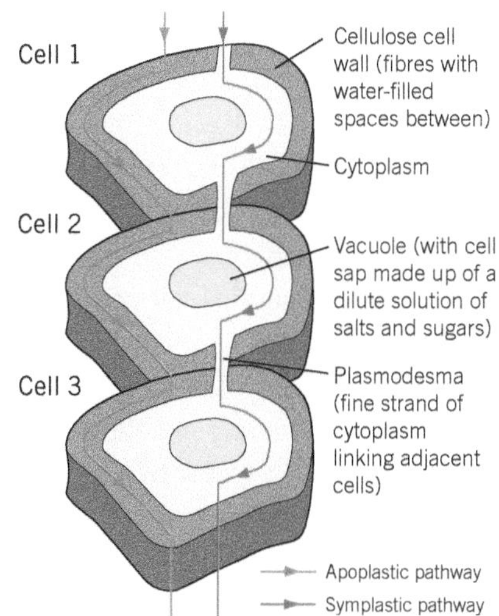

▲ **Figure 4** *Apoplastic and symplastic pathways across root cortex*

Passage of water into the xylem

When water reaches the **endodermis** by the apoplastic pathway, a waterproof band called the Casparian strip in endodermal cells prevents it progressing further along the cell wall (Figure 5). At this point, water enters into the living protoplast of the cell, where it joins water that has arrived there by the symplastic pathway.

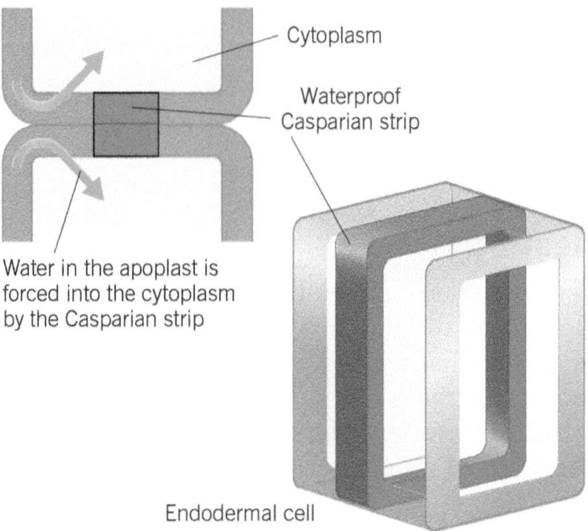

▲ **Figure 5** *Movement of water across the endodermis*

Active transport of salts is the most likely mechanism by which water now gets into the xylem. Endodermal cells actively transport salts into the xylem. This process requires energy and can therefore only occur within living tissue. It takes place through carrier proteins in the cell-surface membrane. If water is to enter the xylem, it must first enter the cytoplasm of endodermal cells.

The active transport of mineral ions into the xylem by the endodermal cells creates a lower water potential in the xylem. Water now moves down a water potential gradient. This creates a force that helps to move water up the plant. This force is called **root pressure**. Whilst its contribution to water movement up a large tree is minimal compared to the transpiration pull (see Topic 16.2), root pressure can make a significant contribution to water movement in small, herbaceous plants. Evidence for the existence of root pressure due to the active pumping of salts into the xylem includes the following:

- The pressure increases with a rise in temperature and decreases at lower temperatures.
- Metabolic inhibitors, such as cyanide, prevent most energy release by respiration and also cause root pressure to cease.
- A decrease in the availability of oxygen or respiratory substrates causes a reduction in root pressure.

Summary questions

1. Explain how water enters a root-hair cell.
2. State **two** differences between the apoplastic and symplastic pathways of water movement.
3. Why is water following the apoplastic pathway unable to cross the endodermis in the cell wall?
4. State whether each of the following processes is active or passive:
   - a uptake of water by root hairs
   - b movement of water from a cortex cell into an endodermal cell
   - c movement of water from an endodermal cell into the xylem.
5. When seedlings are transplanted they sometimes wilt and die. Suggest **two** possible reasons for this.

16.2 Movement of water up stems

The main force that pulls water up the stem of a plant is the evaporation of water from leaves – a process called **transpiration** (see Topic 16.3). It is therefore logical to begin from the point where water molecules evaporate from the leaves through the tiny openings, called **stomata**, on the surface of a leaf. The evaporation takes place from the cell walls of cells lining air spaces in the leaves. Water diffuses through the stomata down a water potential gradient.

Learning objectives:

→ Explain how transpiration pull moves water in the xylem.

Specification reference: 3.2.8.2

Movement of water out through stomata

The humidity of the atmosphere (and hence its water potential) is usually less than that of the air spaces next to the stomata. Provided the stomata are open, water vapour molecules diffuse out of the air spaces down the water potential gradient into the surrounding air. Water lost from the air spaces is replaced by water evaporating from the cell walls of the surrounding mesophyll cells. By changing the size of the stomatal pores, plants can control their rate of transpiration.

Movement of water across the cells of a leaf

Water is lost from mesophyll cells by evaporation from their cell walls to the air spaces of the leaf. This is replaced by water reaching the mesophyll cells from the xylem by either the apoplastic or symplastic pathways (see Topic 16.1). In the case of the symplastic pathway, the water movement occurs because:

- mesophyll cells lose water to the air spaces
- these cells now have a lower water potential and so water enters by osmosis from neighbouring cells
- the loss of water from these neighbouring cells lowers their water potential
- they, in turn, take in water from their neighbours by osmosis.

In this way, a water potential gradient is established that pulls water from the xylem, across the leaf mesophyll, and finally out into the atmosphere. These events are summarised in Figure 1 on the next page.

Movement of water up the stem in the xylem

The two main factors that are responsible for the movement of water up the xylem, from the roots to the leaves, are cohesion–tension and root pressure. We looked at root pressure in Topic 16.1, so let us turn our attention to the cohesion–tension theory. This operates as follows:

- Water evaporates from leaves as a result of transpiration (see Topic 16.3).
- Water molecules form hydrogen bonds between one another and hence tend to stick together. This is known as **cohesion**.
- Water forms a continuous, unbroken pathway across the mesophyll cells and down the xylem.
- As water evaporates from the mesophyll cells in the leaf into the air spaces beneath the stomata, more molecules of water are drawn up behind it as a result of this cohesion.

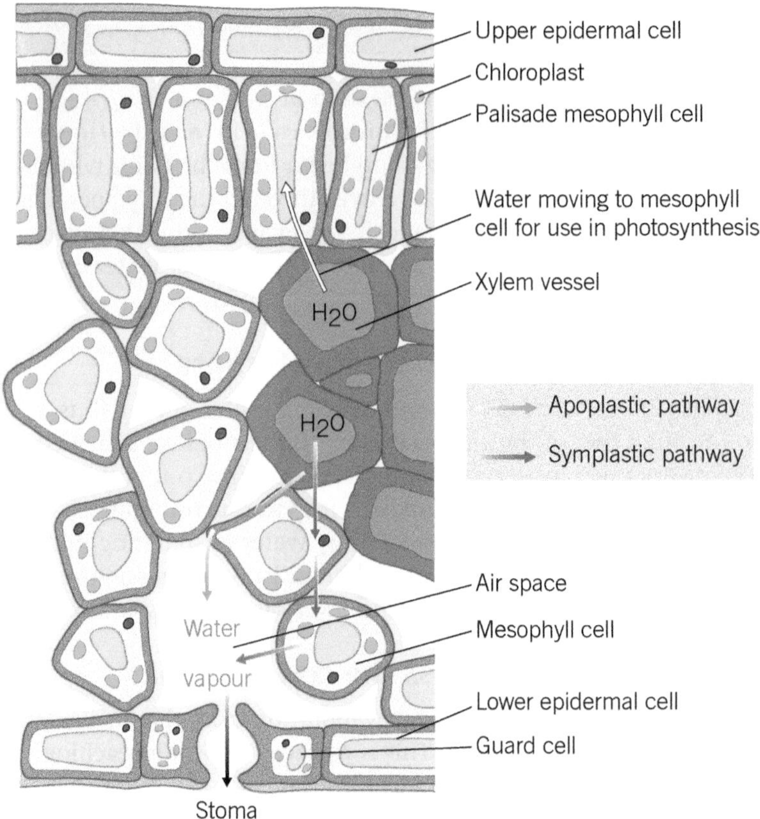

▲ **Figure 1** *Movement of water across a leaf*

- Water is hence pulled up the xylem as a result of transpiration. This is called the **transpiration pull**.
- Transpiration pull puts the xylem under tension, that is, there is a negative pressure within the xylem, hence the name **cohesion–tension theory**.

Such is the force of the transpiration pull that it can easily raise water up the 100 m or more of the tallest trees. There are several pieces of evidence to support the cohesion–tension theory. These include:

- Change in the diameter of tree trunks according to the rate of transpiration. During the day, when transpiration is at its greatest, there is more tension (more negative pressure) in the xylem. This causes the trunk to shrink in diameter. At night, when transpiration is at its lowest, there is less tension in the xylem and so the diameter of the trunk increases.
- If a xylem vessel is broken and air enters it, the tree can no longer draw up water. This is because the continuous column of water is broken and so the water molecules can no longer stick together.
- When a xylem vessel is broken, water does not leak out, as would be the case if it were under pressure. Instead air is drawn in, which is consistent with it being under tension.

Transpiration pull is a passive process and therefore does not require metabolic energy to take place. Indeed, the xylem vessels through which the water passes are dead and so cannot actively move the water. As they are dead, their end walls can break down. This means

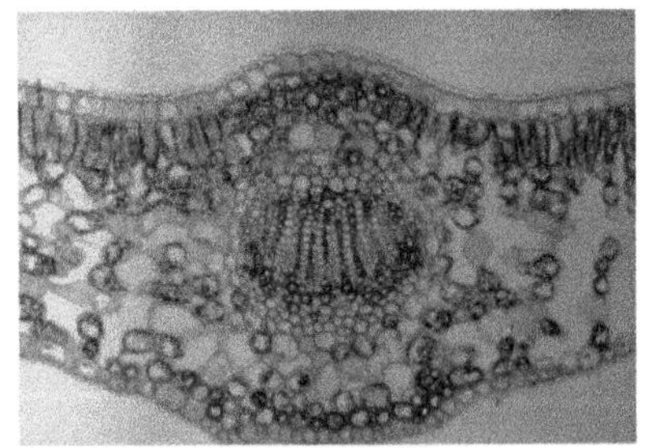

▲ **Figure 2** *Section through a leaf showing the tissues involved in the movement of water*

Study tip

Read the questions carefully. If a question says 'Explain how water in the xylem in the root reaches the leaves', don't describe uptake of water by root hairs and its movement through the endodermis.

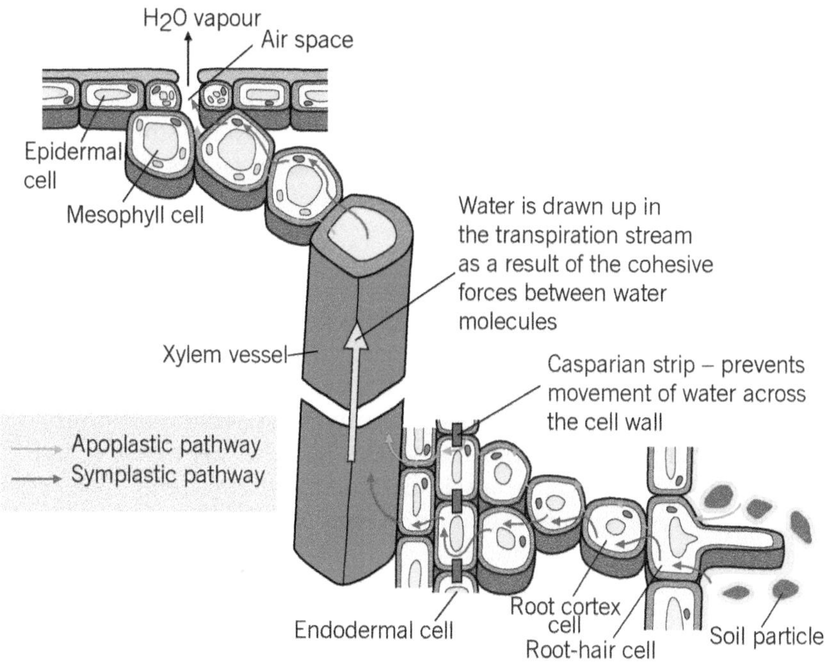

▲ **Figure 3** *Summary of water transport through a plant*

that xylem forms a series of continuous, unbroken tubes from root to leaves, which is essential to the cohesion–tension theory of water flow up the stem. Energy is nevertheless needed to drive the process of transpiration. This energy is in the form of heat that ultimately comes from the Sun.

Figure 3 summarises the movement of water from the soil, through the plant, and into the atmosphere.

Hug a tree

If you put your arms around a suitably sized tree trunk in the middle of the day your fingers will just touch on the far side of the tree. Try to hug the same tree at night and your fingers will probably no longer meet. The graph in Figure 4 shows why. It shows the rate of water flow up a tree and the diameter of the tree trunk over a 24-hour period.

1. At what time of day is transpiration rate greatest? Explain your answer.
2. Describe the changes in the rate of flow of water during the 24-hour period.
3. Explain in terms of the cohesion–tension theory the changes in the rate of flow of water during the 24-hour period.
4. Explain the changes in the diameter of the tree trunk over the 24-hour period.
5. If the tree was sprayed with ammonium sulfamate, a herbicide that kills living cells, the rate of water flow would be unchanged. Explain why.

Summary questions

1. Give the most suitable word, or words, represented by **a–g** in the passage below.

   Water evaporates from the air spaces in a plant by a process called **a**. This evaporation takes place mainly through pores called **b** in the epidermis of the leaf. Water evaporates into the air spaces from mesophyll cells. As a result these cells have a **c** water potential and so draw water by **d** from neighbouring cells. In this way, a water potential gradient is set up that draws water from the xylem. Water is pulled up the xylem because water molecules stick together – a phenomenon called **e**. During the night the diameter of a tree trunk **f**. Other forces helping to move water up the stem include **g**, which is the result of the movement of water into the xylem in the root following the active transport of ions into the xylem.

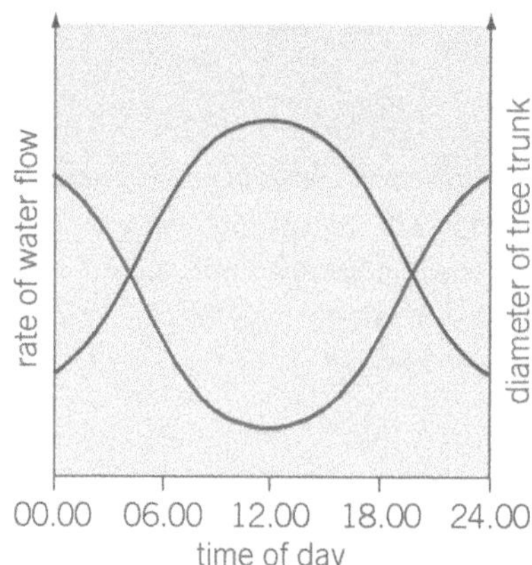

▲ **Figure 4** *Variation of rate of water flow and diameter of a tree trunk*

16.3 Transpiration and factors affecting it

Learning objectives:

- → Explain why transpiration occurs.
- → Describe how different environmental factors affect transpiration rate.

Specification reference: 3.2.8.1

You have seen in Topic 16.2 that transpiration is the evaporation of water vapour from plants. It takes place mostly through stomata in the leaves. Let us now look at why it occurs and what factors affect its rate.

Role of transpiration

Transpiration is sometimes referred to as 'a necessary evil'. This is because, although transpiration is universal in flowering plants, it is the unavoidable result of plants having leaves adapted for photosynthesis. Leaves have a large surface area to absorb light, and stomata to allow inward **diffusion** of carbon dioxide. Both features result in an immense loss of water – up to 700 dm^3 per day in a large tree (Figure 1). Although transpiration helps bring water to the leaves, it is not essential because osmotic processes could achieve this. Less than 1 per cent of water moved in the transpiration stream is used by the plant. So what are the benefits of transpiration?

Materials such as mineral **ions**, sugars, and hormones are moved around the plant dissolved in water. This water is carried up the plant by the transpiration pull. Without transpiration, water would not be so plentiful and the transport of materials would not be as rapid.

▲ **Figure 1** *Transpiration pull is sufficient to transport water up plants such as this giant redwood tree, which is over 80 m tall*

Factors affecting transpiration

A number of factors affect the rate of transpiration. These are discussed below.

Light

Stomata are the openings in leaves through which the carbon dioxide needed for photosynthesis diffuses. Photosynthesis only occurs in the light. The stomata of most plants open in the light and close in the dark. When stomata are open, water moves out of the leaf into the atmosphere. Consequently an increase in light intensity causes an increase in the rate of transpiration.

Temperature

Temperature changes affect two factors that influence the rate of transpiration:

- the speed at which water molecules move
- how much water the air can hold, that is, the **water potential** of air.

A rise in temperature:

- increases the kinetic energy and hence the speed of movement of water molecules. This increased movement of water molecules increases the rate of evaporation of water. This means that water evaporates more rapidly from leaves and so the rate of transpiration increases.
- decreases the humidity of the air outside the leaf, that is, it decreases its water potential.

Both these changes lead to an increase in transpiration rate. A reduction in temperature has the reverse effect – it reduces transpiration rate.

Humidity

Humidity is a measure of the number of water molecules in the air and thus water potential – the drier the air, the lower its water potential. The humidity of the air affects the water potential gradient between the air outside the leaf and the air inside the leaf. When the air outside the leaf has a high humidity, the gradient is reduced and the rate of transpiration is lower. Conversely, low humidity increases the transpiration rate.

Air movement

As water diffuses through stomata, it accumulates as vapour around the stomata on the outside of the leaf. The water potential around the stomata is therefore increased. This reduces the water potential gradient between the moist atmosphere in the air spaces within the leaf and the drier air outside. The transpiration rate is therefore reduced. Any movement of air around the leaf will disperse the humid layer at the leaf surface and so decrease the water potential of the air. This increases the water potential gradient and hence the rate of transpiration. The faster the air movement, the more rapidly the humid air is removed and the greater the rate of transpiration. These effects are shown in Figure 2.

Hint

Remember that the greater the water potential gradient between the inside and the outside of a leaf, the faster water will move out, and therefore the greater the rate of transpiration. Any factor that increases the gradient increases transpiration; any factor that reduces the gradient reduces transpiration.

Study tip

Make sure you understand water potentials, both in terms of solutions and in terms of the humidity of air. The less humid the air, the lower its water potential.

a *Still air*

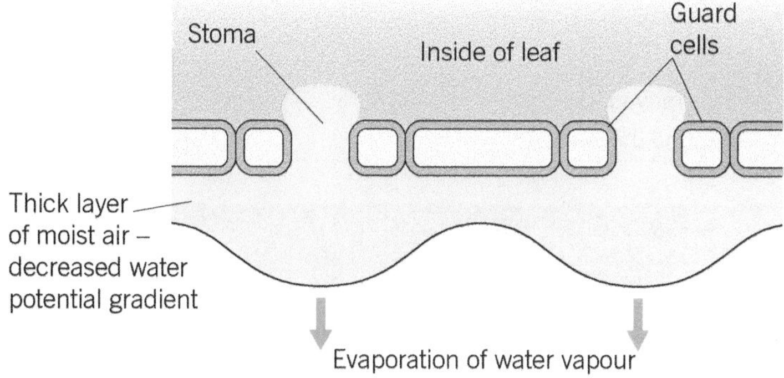

b *Moving air*

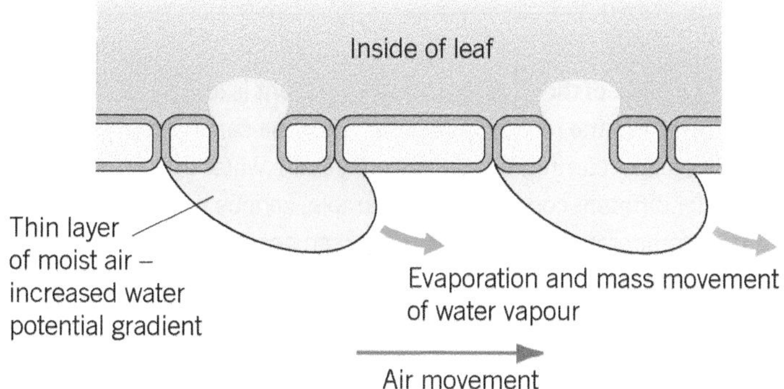

▲ **Figure 2** *Effect of air movement on the rate of transpiration*

The factors affecting transpiration rate are summarised in Table 1 on the next page.

The energy for transpiration comes from the evaporation of water from leaves. You have seen that this evaporation depends on factors such as light, temperature, humidity, and air movement. All these factors are either directly or indirectly the result of the Sun's energy. Therefore it is the Sun that ultimately drives transpiration.

Hint

The conditions inside the leaf are more constant than those outside it. Therefore it is changes to the *external* environment that largely affect transpiration rate.

Summary questions

1 Explain why transpiration is often described as a 'necessary evil'.

2 In each of the following cases, state whether the rate of transpiration increases or decreases:

- **a** the temperature falls
- **b** the wind speed increases
- **c** the humidity decreases
- **d** light intensity increases
- **e** the water potential of the outside air is lowered.

3 When a potted plant is placed inside a black polythene bag, its transpiration rate falls. Give **two** reasons why this happens.

▼ **Table 1** *Summary of factors affecting transpiration rate*

Factor	How factor affects transpiration	Increase in transpiration caused by:	Decrease in transpiration caused by:
Light	Stomata open in the light and close in the dark	Higher light intensity	Lower light intensity
Temperature	Alters the **kinetic energy** of the water molecules and the relative humidity of the air	Higher temperatures	Lower temperatures
Humidity	Affects the water potential gradient between the air spaces in the leaf and the atmosphere	Lower humidity	Higher humidity
Air movement	Changes the water potential gradient by altering the rate at which moist air is removed from around the leaf	More air movement	Less air movement

Required practical 5

Measurement of water uptake using a potometer

It is almost impossible to measure transpiration because it is extremely difficult to condense and collect all the water vapour that leaves all the parts of a plant. What you can easily measure, however, is the amount of water that is taken up in a given time by a part of the plant such as a leafy shoot. About 99 per cent of the water taken up by a plant is lost during transpiration, which means that the rate of uptake is almost the same as the rate at which transpiration is occurring. You can then measure water uptake by the same shoot under different conditions, for example, various humidities, wind speeds, or temperatures. In this way, you can get a reasonably accurate measure of the effects of these conditions on the rate of transpiration.

The rate of water loss in a plant can be measured using a potometer (Figure 3). The experiment is carried out in the following stages:

- A leafy shoot is cut under water to stop air entering xylem and breaking the water column. Care is taken not to get water on the leaves.
- The potometer is filled completely with water, making sure there are no air bubbles.
- Using a rubber tube, the leafy shoot is fitted to the potometer under water.
- The potometer is removed from under the water and all joints are sealed with waterproof jelly.
- An air bubble is introduced into the capillary tube.

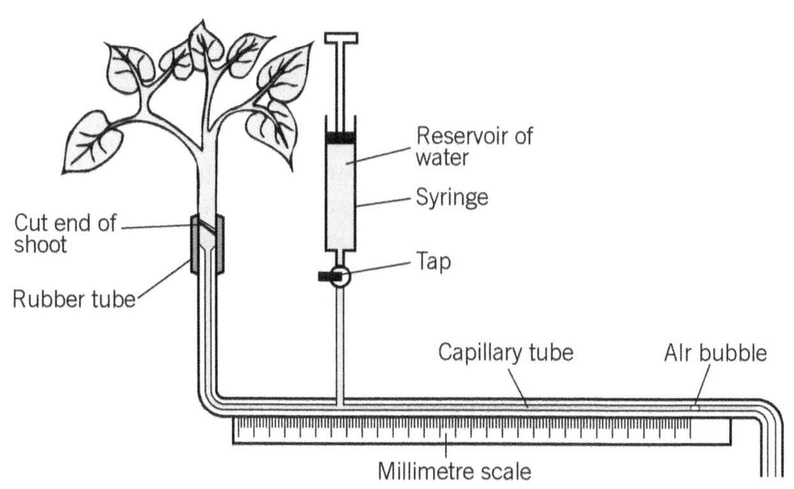

▲ **Figure 3** *A potometer*

- The distance moved by the air bubble in a given time is measured a number of times and the mean is calculated.
- Using this mean value, the volume of water lost is calculated.
- The volume of water lost against the time in minutes can be plotted on a graph.
- Once the air bubble nears the junction of the reservoir tube and the capillary tube, the tap on the reservoir is opened and the syringe is pushed down until the bubble is pushed back to the start of the scale on the capillary tube. Measurements then continue as before.
- The experiment can be repeated to compare the rates of water uptake under different conditions, for example, at different temperatures, humidity, or light intensity, or the differences in water uptake between different species under the same conditions.

1. From your knowledge of how water moves up the stem, suggest a reason why each of the following procedures is carried out:
   a. The leafy shoot is cut under water rather than in the air.
   b. All joints are sealed with waterproof jelly.
2. What assumption must be made if a potometer is used to measure the rate of transpiration?
3. The volume of water taken up in a given time can be calculated using the formula $\pi r^2 l$ (where $\pi = 3.142$, r = radius of the capillary tube, and l = the distance moved by the air bubble). In an experiment the mean distance moved by the air bubble in a capillary tube of radius 0.5 mm during 1 min was 15.28 mm. Calculate the rate of water uptake in $mm^3\ h^{-1}$. Show your working.
4. If a potometer is used to compare the transpiration rates of two different species of plant, suggest one feature of both plant shoots that should, as far as possible, be kept the same.
5. Suggest reasons why the results obtained from a laboratory potometer experiment may not be representative of the transpiration rate of the same plant in the wild.

16.4 Transport of organic substances in the phloem

Learning objectives:

- ➔ Describe the mass flow mechanism for the transport of organic substances in the phloem.
- ➔ Summarise the evidence for and against the mass flow mechanism.

Specification reference: 3.2.8.2

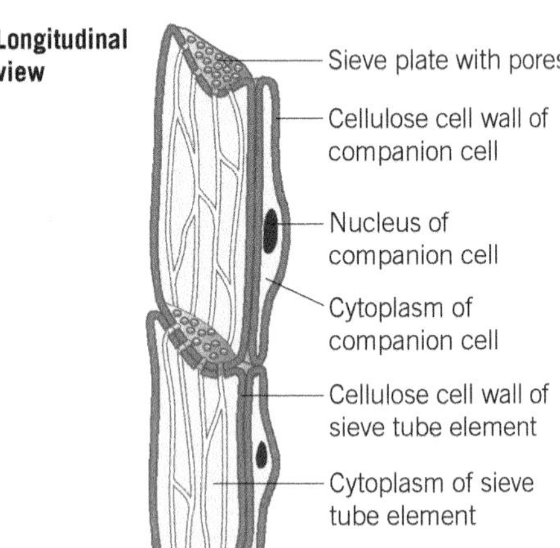

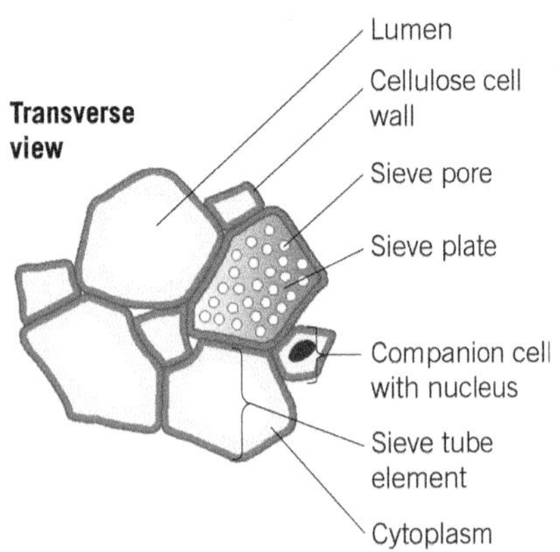

▲ **Figure 1** *Phloem as seen under a light microscope*

The process by which organic molecules and some mineral ions are transported from one part of a plant to another is called **translocation**. In flowering plants, the tissue that transports biological molecules is called phloem. Phloem is made up of sieve tube elements – long thin structures arranged end to end. Their end walls are perforated to form sieve plates. Associated with the sieve tube elements are cells called companion cells. The structure of phloem is shown in Figure 1.

Having produced sugars during photosynthesis, the plant transports them from the sites of production, known as **sources**, to the places where they will be used directly or stored for future use, known as **sinks**. As sinks can be anywhere in a plant – sometimes above and sometimes below the source – it follows that the translocation of molecules in phloem can be in either direction. Organic molecules to be transported include sucrose and amino acids. The phloem also transports inorganic ions such as potassium, chloride, phosphate, and magnesium ions.

Mechanism of translocation

It is accepted that materials are transported in the phloem and that the rate of movement is too fast to be explained by diffusion. What is in doubt is the precise mechanism by which translocation is achieved. Current thinking favours the **mass flow theory**, a theory that can be divided into three phases:

1. Transfer of sucrose into sieve elements from photosynthesising tissue

- Sucrose is manufactured from the products of photosynthesis in cells with chloroplasts.
- The sucrose diffuses down a concentration gradient by facilitated diffusion from the photosynthesising cells into companion cells.
- Hydrogen ions are actively transported from companion cells into the spaces within cell walls using ATP.
- These hydrogen ions then diffuse down a concentration gradient through carrier proteins into the sieve tube elements.
- Sucrose molecules are transported along with the hydrogen ions in a process known as co-transport (Topics 4.4 and 12.3). The protein carriers are therefore also known as **co-transport proteins**.

2. Mass flow of sucrose through sieve tube elements

Mass flow is the bulk movement of a substance through a given channel or area in a specified time. Mass flow of sucrose through sieve tube elements takes place as follows:

- The sucrose produced by photosynthesising cells (source) is actively transported into the sieve tubes as described above.
- This causes the sieve tubes to have a lower (more negative) water potential.
- As the xylem has a much higher (less negative) water potential (see Topic 16.2), water moves from the xylem into the sieve tubes by osmosis, creating a high hydrostatic pressure within them.

- At the respiring cells (sink), sucrose is either used up during respiration or converted to starch for storage.
- These cells therefore have a low sucrose content and so sucrose is actively transported into them from the sieve tubes, lowering their water potential.
- Due to this lowered water potential, water also moves into these respiring cells, from the sieve tubes, by osmosis.
- The hydrostatic pressure of the sieve tubes in this region is therefore lowered.
- As a result of water entering the sieve tube elements at the source and leaving at the sink, there is a high hydrostatic pressure at the source and a low one at the sink.
- There is therefore a mass flow of sucrose solution down this hydrostatic gradient in the sieve tubes.

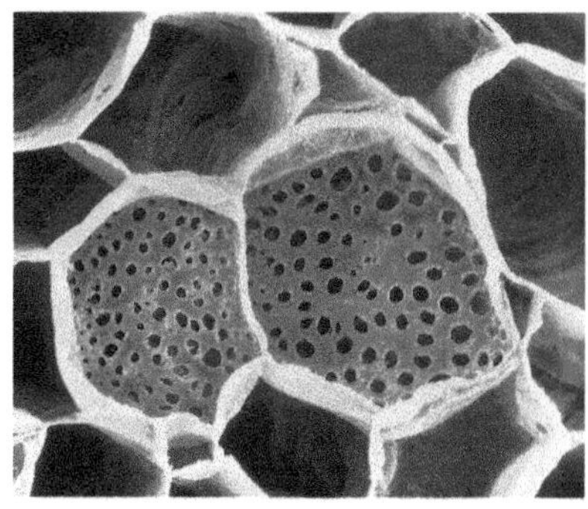

▲ **Figure 2** *False-colour scanning electron micrograph (SEM) of sieve plates*

Although mass flow is a passive process, it occurs as a result of the active transport of sugars. Therefore the process as a whole is active, which is why it is affected by, for example, temperature and metabolic poisons. A model of this theory is shown in Figure 3 and the evidence for and against the mass flow theory is listed in Table 1.

3. Transfer of sucrose from the sieve tube elements into storage or other sink cells

The sucrose is actively transported by companion cells out of the sieve tubes and into the sink cells.

The process of translocation of sucrose in phloem is illustrated in Figure 4.

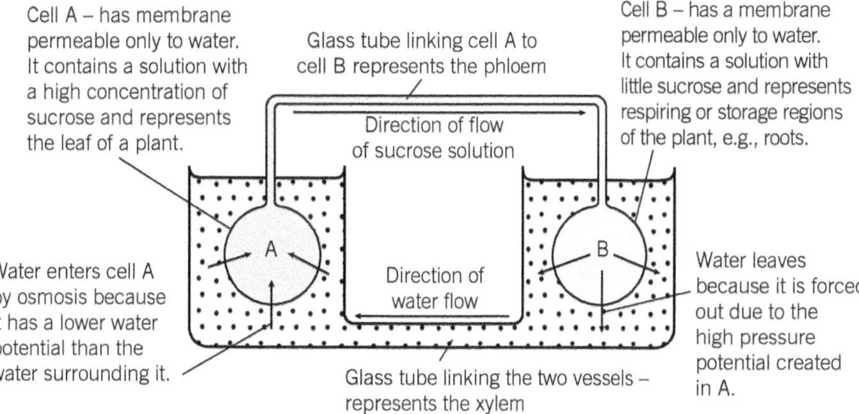

▲ **Figure 3** *Model illustrating the movement of sucrose by mass flow in phloem*

▼ **Table 1** *Evidence for and against the mass flow theory*

Evidence supporting the mass flow hypothesis	Evidence questioning the mass flow hypothesis
• there is a pressure within sieve tubes, as shown by sap being released when they are cut • the concentration of sucrose is higher in leaves (source) than in roots (sink) • downward flow in the phloem occurs in daylight, but ceases when leaves are shaded, or at night • increases in sucrose levels in the leaf are followed by similar increases in sucrose levels in the phloem a little later • metabolic poisons and/or lack of oxygen inhibit translocation of sucrose in the phloem • companion cells possess many mitochondria and readily produce ATP	• the function of the sieve plates is unclear, as they would seem to hinder mass flow (it has been suggested that they may have a structural function, helping to prevent the tubes from bursting under pressure) • not all solutes move at the same speed – they should do so if movement is by mass flow • sucrose is delivered at more or less the same rate to all regions, rather than going more quickly to the ones with the lowest sucrose concentration, which the mass flow theory would suggest

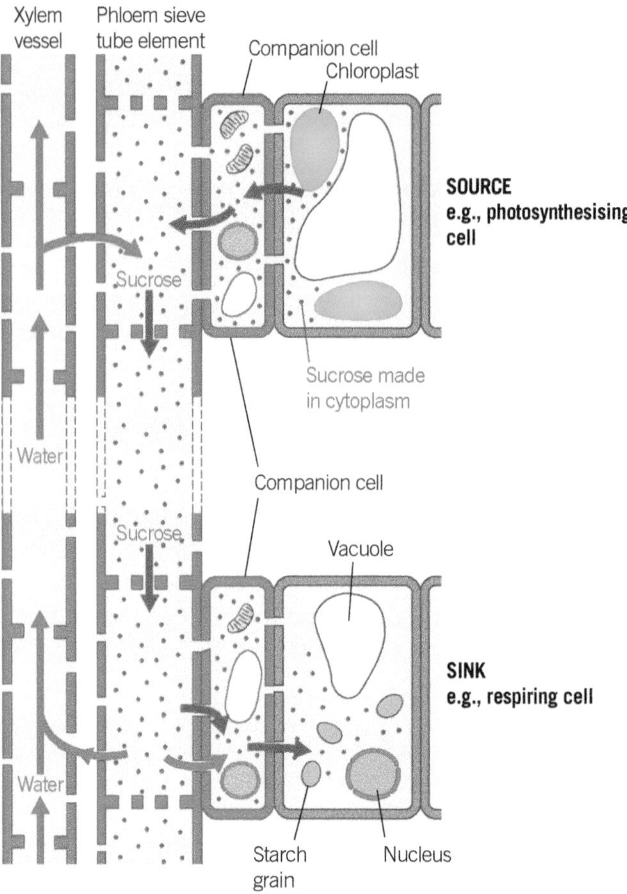

▲ **Figure 4** *Movement of sucrose from source to sink through the phloem of a plant*

Summary question

1 State the most suitable word or words represented by the letters **a**–**k** in the passage below.

Transport of sucrose in plants occurs in the tissue called **a**, from places where it is produced, known as **b**, to places where it is used up or stored, called **c**. One theory of how it is translocated is called the **d** theory. Initially the sucrose is transferred into **e** elements by the process of **f**. The sucrose is produced by **g** cells that therefore have a **h** water potential due to this sucrose. Water therefore moves into them from the nearby **i** tissue that has a **j** water potential. The opposite occurs in those cells (sinks) using up sucrose, and water therefore leaves them by the process of **k**.

16.5 Investigating transport in plants

You have seen that water is carried in xylem whereas sugars and amino acids are carried in phloem. How can we be sure that this is the case? This topic looks at some of the evidence and how it is obtained.

Learning objectives:

- → Describe the use of ringing experiments to investigate transport in plants.
- → Describe the use of tracer experiments to investigate transport in plants.
- → Explain the evidence that translocation of organic molecules occurs in the phloem.

Specification reference: 3.2.8.2 and 3.2.9.2

Ringing experiments

Woody stems have an outer protective layer of bark, on the inside of which is a layer of phloem that extends all round the stem. Inside the phloem layer is xylem (Figure 1).

At the start of a ringing experiment, a section of the outer layers (protective layer and phloem) is removed around the complete circumference of a woody stem whilst it is still attached to the rest of the plant. After a period of time, the region of the stem immediately above the missing ring of tissue is seen to swell (Figure 1). Samples of the liquid that has accumulated in this swollen region are found to be rich in sugars and other dissolved organic substances. Some non-photosynthetic tissues in the region below the ring (towards the roots) are found to wither and die, whereas those above the ring continue to grow.

These observations suggest that removing the phloem around the stem has led to:

- the sugars of the phloem accumulating above the ring, leading to swelling in this region as cells take up more water by osmosis in response to the sugars they contain lowering the water potential.
- the interruption of flow of sugars to the region below the ring and the death of tissues in this region.

The conclusion drawn from this type of ringing experiment is that phloem, rather than xylem, is the tissue responsible for translocating sugars in plants. As the ring of tissue removed did not extend into the xylem, its continuity had not been broken. If xylem were the tissue responsible for translocating sugars, you would not expect sugars to accumulate above the ring nor tissues below it to die.

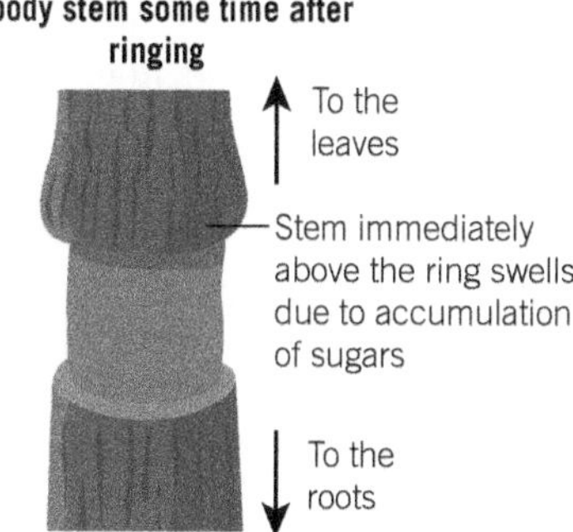

▲ **Figure 1** *Ringing of a woody stem and its results*

Tracer experiments

Radioactive **isotopes** are useful for tracing the movement of substances in plants. For example the isotope ^{14}C can be used to make radioactively labelled carbon dioxide ($^{14}CO_2$). If a plant is then grown in an atmosphere containing $^{14}CO_2$, the ^{14}C isotope will be incorporated into the sugars produced during photosynthesis. These radioactive sugars can then be traced using autoradiography as they move within the plant. In our example, this involves taking thin cross sections of the plant stem and placing them on a piece of X-ray film. The film becomes blackened where it has been exposed to the radiation produced by the ^{14}C in the sugars. The blackened regions are found to correspond to where phloem tissue is in the stem. As the other tissues do not blacken the film, it follows that they do not carry sugars and that phloem alone is responsible for their translocation.

Evidence that translocation of organic molecules occurs in phloem

The techniques described are only two of the pieces of evidence supporting the view that translocation of organic molecules such as sugars takes place in phloem. A more complete list of evidence is given below.

- When phloem is cut, a solution of organic molecules flows out.
- Plants provided with radioactive carbon dioxide can be shown to have radioactively labelled carbon in phloem after a short time.
- Aphids are a type of insect that feed on plants. They have needle-like mouthparts that penetrate the phloem. They can therefore be used to extract the contents of the sieve tubes. These contents show daily variations in the sucrose content of leaves that are mirrored a little later by identical changes in the sucrose content of the phloem (Figure 2). You will learn more about aphids in Topic 16.6.
- The removal of a ring of phloem from around the whole circumference of a stem leads to the accumulation of sugars above the ring and their disappearance from below it.

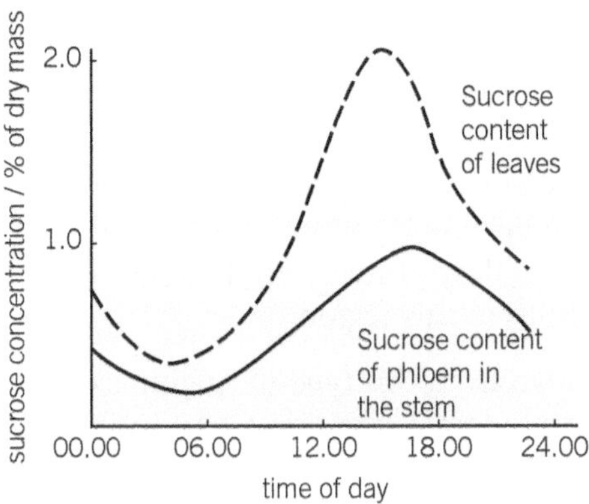

▲ **Figure 2** *Diurnal variation in sucrose content of leaves and phloem*

Summary questions

1 a Suggest what difference there would be between the results of a ringing experiment carried out in the summer and one carried out in the winter.

b Explain the reason for the difference you have suggested.

2 Squirrels sometimes strip sections of bark from around branches. Explain why these branches might die.

3 Suggest how a branch with a complete ring of phloem stripped from it by squirrels might still survive.

4 Explain why squirrels are unlikely to cause the death of a large mature tree by stripping some bark from its trunk.

5 Study Figure 2 and suggest:

a why there is a time lag between the maximum sucrose content in the leaves and in the phloem in the stem

b why the sucrose concentration in the phloem in the stem is lower than that in the leaves.

Using radioactive tracers to find which tissue transports minerals

In an experiment to determine whether minerals are transported in xylem or phloem, a plant was grown in a pot. One branch (Y) on each plant had a 225 mm section of its phloem and xylem separated by inserting strips of impervious wax paper between them, as shown in Figure 3. A 225 mm section of another branch (X) of the same plant that had *not* had its xylem and phloem separated by wax paper was used as a control.

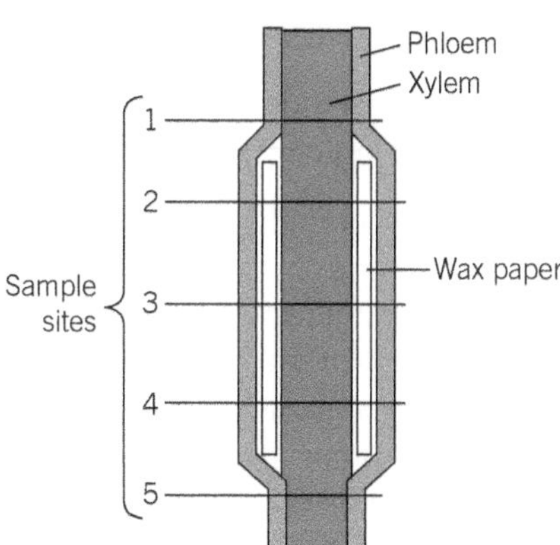

▲ **Figure 3** *Portion of branch of plant showing how xylem and phloem are separated by wax paper and from where samples were taken*

The plant was watered with a solution that contained radioactive potassium (^{42}K). After 5 hours absorbing radioactive ^{42}K, sections of the experimental branch were tested for the quantity of ^{42}K in the xylem and phloem. The sections tested are labelled on Figure 3.

The equivalent positions on the control branch were also tested for ^{42}K.

The results are shown in Table 1.

▼ **Table 1**

Section of stem	Percentage of total ^{42}K			
	Branch X (phloem and xylem together)		Branch Y (phloem and xylem separated)	
	Phloem	Xylem	Phloem	Xylem
1	53	47	53	47
2			09	91
3	56	44	01	99
4			15	85
5	52	48	59	41

1. Draw a conclusion from the data in the table.
2. Justify your conclusion with supporting evidence.
3. Explain the fact that the levels of ^{42}K are similar in the xylem and phloem of branch Y in sections 1 and 5.
4. The control (branch X) was an identical length of a different branch that had not had wax paper placed between the xylem and phloem. Suggest a way in which this control could have been improved. Explain why the change you suggest is an improvement.

16.6 Aphids and plant viruses

Learning objectives:

- → Describe the life cycle of aphids.
- → Describe how aphids feed on plants.
- → Outline the symptoms of some plant viral diseases.
- → Describe the consequences of viral diseases on crop production.

Specification reference: 3.2.9.1 and 3.2.9.2

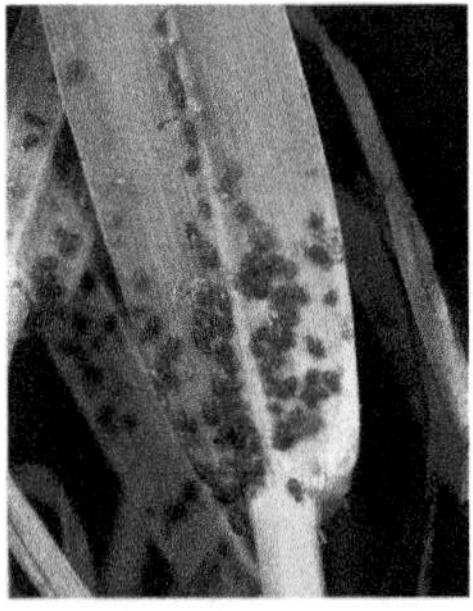

▲ **Figure 1** *Bird-cherry aphids (*Rhopalosiphum padi*) feeding on a wheat leaf. The aphid can be responsible for wheat losses of over 40 per cent.*

Synoptic link

Parthenogenesis is a type of **asexual** reproduction found in some animal species. In asexual reproduction the only type of cell division is mitosis which is covered in Topic 17.1.

As you saw in Topic 16.5, the feeding habits of aphids have been used to investigate translocation in phloem, but what is an aphid?

Aphids are a group of insects belonging to the order Hemiptera. More than 4000 aphid species have been described, classified into 10 families. They feed on phloem sap and are probably the single most important pest of crop plants. Not only do they reduce the quality of the crop by depriving it of the sugars and amino acids in the phloem, but, as the phloem can also contain plant viruses, they are also key vectors in the transmission of viruses from plant to plant. The losses to agriculture varies with crop species but it can be as much as 40 per cent in wheat and 50 per cent in barley.

Aphid life cycle

The generalised life cycle of an aphid is summarised in Figure 2. Aphids overwinter as eggs, which are laid in the autumn. The eggs are highly resistant to extremes of temperature and hatch into wingless females. These females feed on the plant on which they hatch and, once mature, they give birth to live daughter aphids without any fertilisation taking place. This type of reproduction is known as **parthenogenesis** and it means that aphid numbers can increase rapidly. Some of the offspring have wings and can migrate to another plant where, once mature, they can again reproduce wingless females through parthenogenesis. In autumn winged male aphids and female aphids are produced. These migrate to a plant and the females give birth to wingless females that can reproduce sexually. These mate with the males and then lay eggs, which can survive over winter.

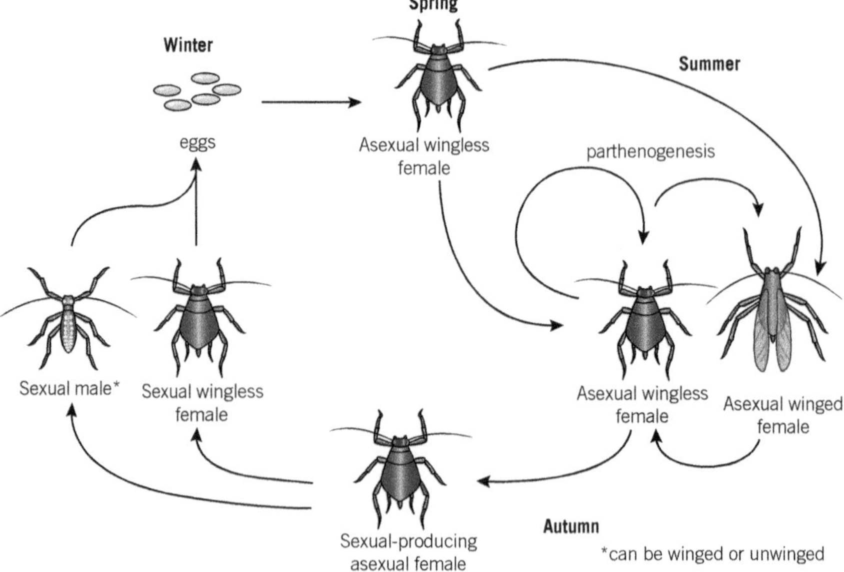

▲ **Figure 2** *Generalised life cycle of an aphid*

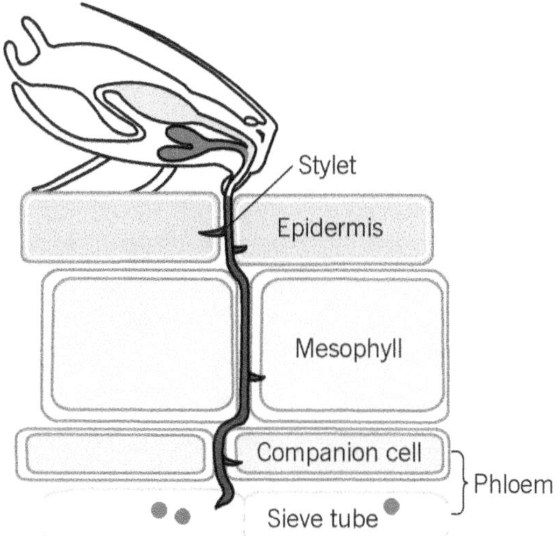

▲ **Figure 3** *The stylet contains a saliva canal and a feeding canal. The saliva gels around the stylet as it penetrates the tissues to reach the phloem. Most cells on the route will be penetrated. The red circles represent virus particles present in the sieve tube.*

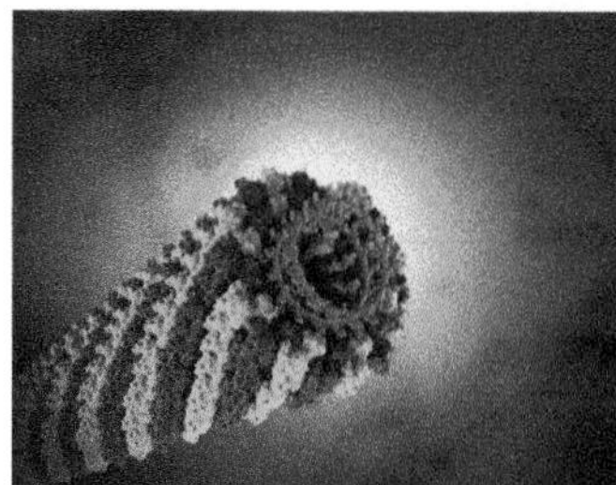

▲ **Figure 4** *Tobacco mosaic virus (TMV) has a single strand of RNA (shown in red) surrounded by a helical arrangement of proteins to form a rod shape. The virus can infect many different species of plant.*

Aphid feeding mechanism

Aphids feed by inserting their mouthparts, or stylet, into the phloem and extracting the phloem sap (Figure 3)

Sap in the sieve tube is under pressure (see Topics 16.4 and 16.5). This results in some sap being exuded from the rear of the aphid. This is sticky and rich is sugars and provides a substrate for moulds to grow on. In a heavy aphid infestation, these black moulds reduce the photosynthetic capacity of the leaves and further reduce crop yields.

Aphids can reduce yields in crop plants in three ways – by removing phloem sap and depriving the plant of sugars and amino acids, by encouraging the growth of moulds on leaves, which reduces photosynthesis and also look unsightly, and by transmitting plant viruses.

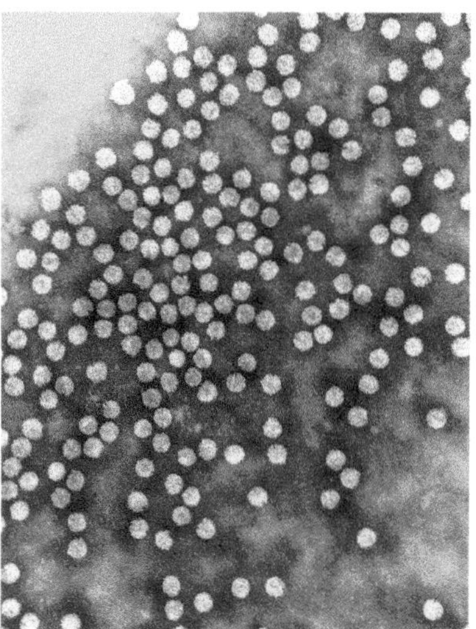

▲ **Figure 5** *False-colour transmission electron micrograph (TEM) image of cucumber mosaic virus*

Plant virus diseases

Like animal viruses such as HIV (Topic 13.3), plant viruses also consist of a nucleic acid and a protein 'coat'. An outer membrane layer can also be present. Plant viruses vary in shape. Some, such as tobacco mosaic virus (TMV), form rigid rods (see Figure 4), whereas other viruses form more spherical shapes (see Figure 5).

Like animal viruses, plant viruses can only reproduce inside a host cell but they cannot penetrate plant cuticles or plant cell walls. They rely on entering plant cells either through damage sites or by using vectors such as aphids, fungi, and nematode worms. Once inside a plant, virus particles are systemic – they can spread throughout the plant – and so can be transmitted to the next generation through seeds and vegetative reproductive structures such as tubers.

▲ **Figure 6** *Young barley plants infected with barley yellow dwarf virus (BYDV). This virus is spread by aphids.*

▲ **Figure 7** *Mosaic pattern and vein clearing in virus-infected plant*

Symptoms of plant viral diseases

Some plant viruses, such as TMV and barley yellow dwarf virus (BYDV) produce yellowing in leaves know as **chlorosis**. This can occur in stripes or mosaic patterns on leaves (Figures 6 and 7) and can also occur in flowers.

Other viruses lead to distortion of parts of the plant. This can include leaf roll (Figure 8) and malformation of flowers and fruits, all of which reduce the market value of the crop as well as the yield.

Yellowing and distortion of plant leaves reduces the photosynthetic efficiency of the crop. In cereals, this can result in a reduction in both the number of grains and in the size of the grain, leading to a loss of yield.

▲ **Figure 8** *A broad bean plant infected by a leaf roll virus*

Global importance of plant viral diseases

Plant viral infections reduce both the quality of the crop (and hence its market value) and the yield. One virus alone, rice tungro virus, accounts for losses in South East Asia that are valued at more than 1.5 billion US dollars annually. Many plant viruses, for example BYDV, can infect a wide range of plants, including most of the staple cereals. Although fungal diseases are probably responsible for greater losses than viral diseases, viral diseases are much more difficult to control. A new strain of a virus infecting cassava emerged in Uganda in the late 1980s. By 1999 the virus had spread to Kenya, Tanzania, Sudan, and

Study tip

A staple food is one that provides the bulk of the energy in the diet of a population. Globally, 60 per cent of the population relies on just three cereal plants: wheat, rice, and maize. Other important non-cereal staples include cassava and sweet potato.

several other African countries. The losses in yield were so great that many farmers stopped growing the crop and the food security of the whole region was badly affected.

Summary questions

1 Complete the following table showing how one species of aphid, *Myzus persicae,* is classified.

Kingdom	
	Arthropoda
Class	
Order	
	Aphididae

2 Give three reasons why the life cycle of aphids makes them successful as plant pests.

3 Discuss the ways that viruses lead to a reduction in both the yield and the economic value of crop plants.

Practice questions: Chapter 16

1 Figure 1 shows some cells from the tissues in a root.

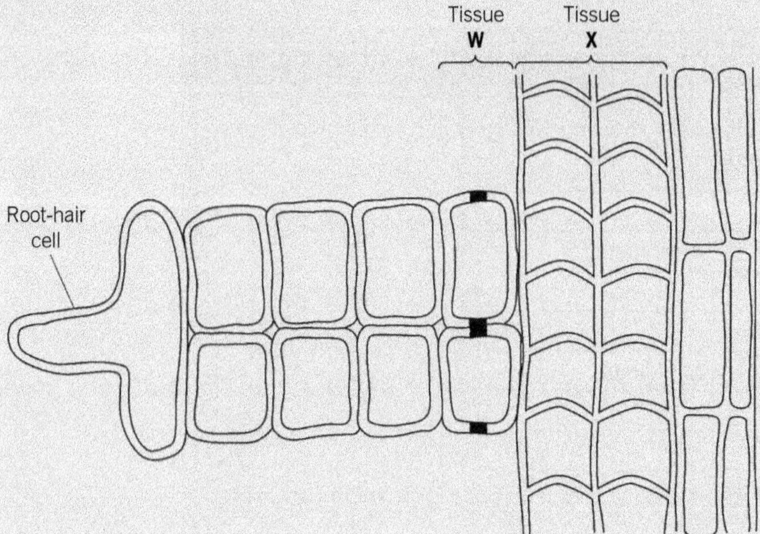

▲ **Figure 1**

(a) Name the tissues labelled **W** and **X**. *(2 marks)*

(b) Explain why water moves from the apoplast pathway to the symplast pathway when it reaches the tissue labelled **W**. *(2 marks)*

AQA 2005

2 A student investigated the rate of transpiration from a leafy shoot. She used a potometer to measure the rate of water uptake by the shoot. Figure 2 shows the potometer used by the student.

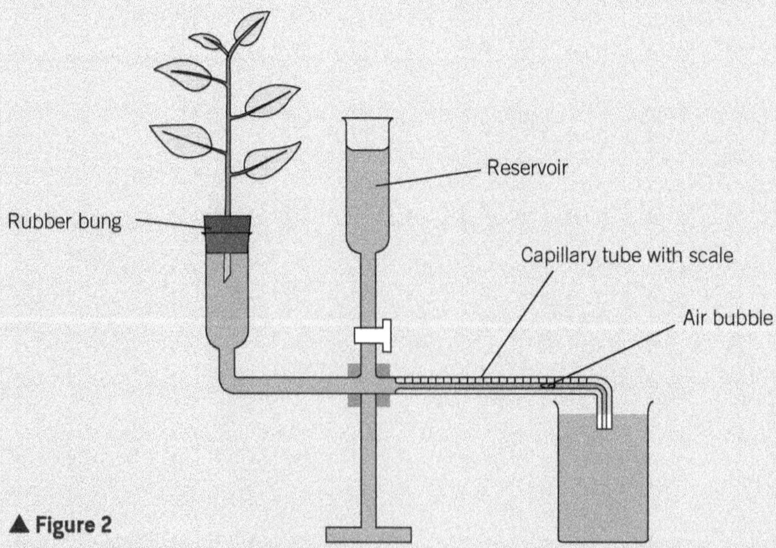

▲ **Figure 2**

(a) Give one environmental factor that the student should have kept constant during this investigation. *(1 mark)*

(b) The student cut the shoot and put it into the potometer under water. Explain why. *(1 mark)*

(c) The student wanted to calculate the rate of water uptake by the shoot in cm^3 per minute. What measurements did she need to make? *(2 marks)*

(d) The student assumed that water uptake was equivalent to the rate of transpiration. Give two reasons why this might not be a valid assumption. *(2 marks)*

(e) (i) The student measured the rate of water uptake three times. Suggest how the reservoir allows repeat measurements to be made. *(1 mark)*
(ii) Suggest why she made repeat measurements. *(1 mark)*
AQA June 2010

3 One leaf on a young plant was supplied with carbon dioxide containing the radioactive isotope of carbon, ^{14}C. The plant was kept in bright light for 1 hour. The amount of radioactivity was then measured at three places in the plant. Figure 3 shows the results.

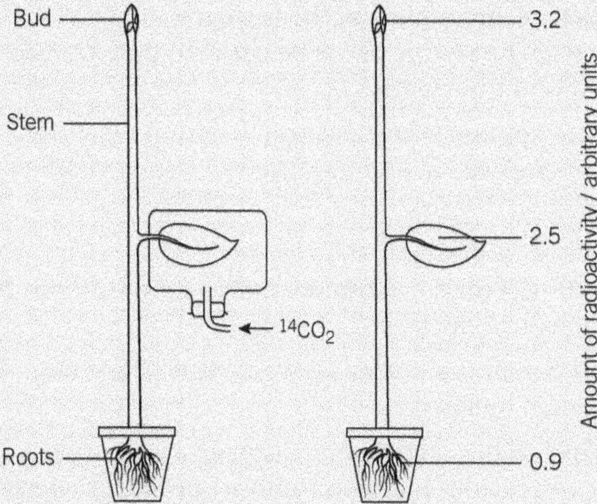

▲ **Figure 3**

Only the treated leaf is shown.

(a) The radioactive carbon is transported as a carbohydrate in the stem. Name this carbohydrate. *(1 mark)*

(b) (i) Suggest one explanation for the difference in the amount of radioactivity in the bud and the roots. *(2 marks)*
(ii) Suggest why some radioactivity remains in the leaf. *(1 mark)*

(c) Describe how a ringing experiment could be carried out to determine which tissue transports the substances containing the radioactive carbon. *(3 marks)*
AQA June 2006

4 Figure 4 is a photomicrograph of two cells found in one of the tissues in a plant stem.

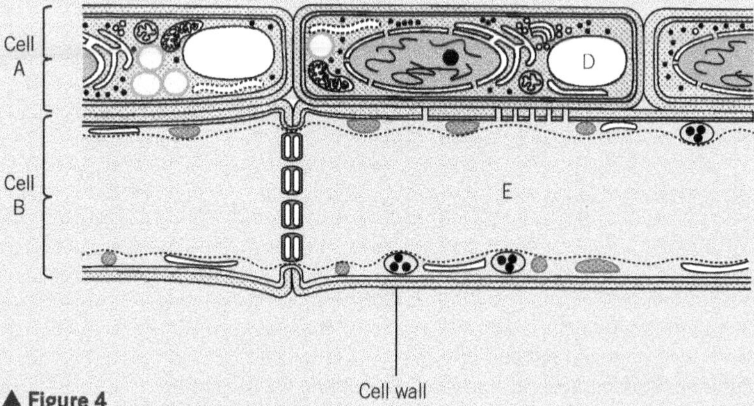

▲ **Figure 4**

(a) Identify cells D and E. *(2 marks)*

Viruses such as tobacco mosaic virus (TMV) consist of single-stranded RNA surrounded by a layer of protein units.

(b) Explain fully why TMV can reproduce in cell A but not cell B. *(2 marks)*

(c) TMV can usually only enter plant cells if they are damaged. Explain why. *(1 mark)*

(d) Once inside a plant cell, virus particles are able to pass between adjacent cells. Explain why. *(1 mark)*

Answers to the Practice Questions are available at
www.oxfordsecondary.com/oxfordaqaexams-alevel-biology

17 Cell division

17.1 The cell cycle and mitosis

Learning objectives:

- → Describe the events of the cell cycle.
- → Describe the events of mitosis.
- → Explain that bacterial cells divide by binary fission.

Specification reference: 3.2.10.1

Eukaryotic cells do not divide continuously, but undergo a regular cycle of division separated by periods of cell growth. This is known as the **cell cycle** and has three stages:

1. **interphase**, which occupies most of the cell cycle, and is sometimes known as the resting phase because no division takes place. It is divided into three parts:
   - **(a)** first growth (G_1) phase, when the proteins from which cell organelles are synthesised and produced so new organelles can be formed
   - **(b)** synthesis (S) phase, when DNA is replicated
   - **(c)** second growth (G_2) phase, when organelles grow and divide and energy stores are increased.
2. **nuclear division**, when the nucleus divides either into two (mitosis) or into four (meiosis)
3. **cell division**, which follows nuclear division and is the process by which the whole cell divides into two (mitosis) or four (meiosis).

Hint

Interphase is sometimes known as the resting phase because no division takes place. In one sense, this description could hardly be further from the truth because interphase is a period of intense chemical activity.

The length of a complete cell cycle varies greatly amongst organisms. Typically, a mammalian cell takes about 24 hours to complete a cell cycle, of which about 90 per cent is interphase.

The various stages of the cell cycle are shown in Figure 1.

Figure 2 shows the variations in mass of the cell and the DNA within it during the cell cycle.

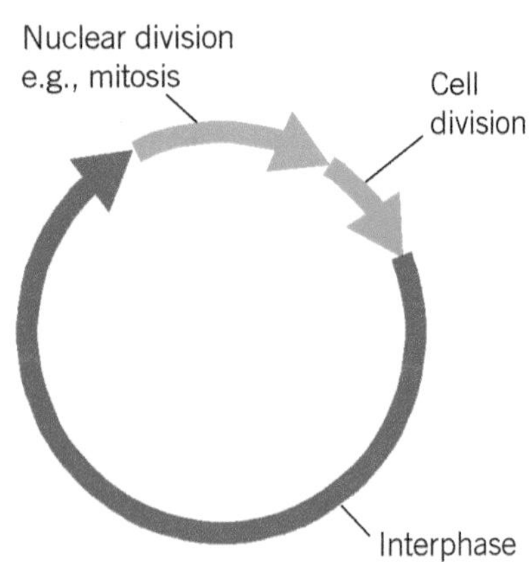

▲ **Figure 1** *The cell cycle*

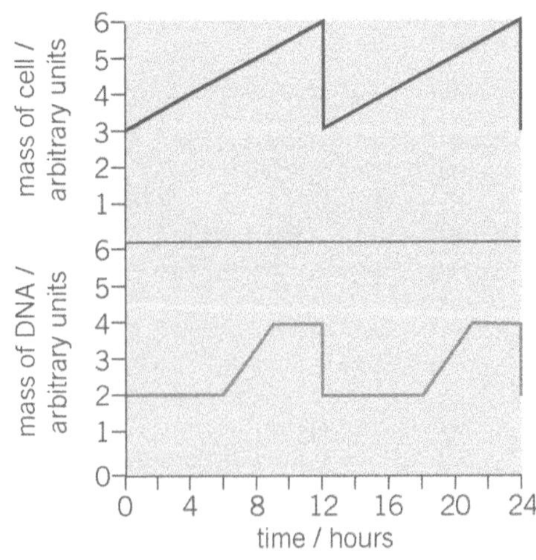

▲ **Figure 2** *Variations in cell and DNA mass during the cell cycle*

Nuclear division can take place by either mitosis or meiosis:

- Mitosis produces two daughter nuclei that have the same number of **chromosomes** as the parent cell and as each other.
- Meiosis produces four daughter nuclei, each with half the number of chromosomes of the parent cell.

After looking at meiosis in Topic 9.1, let us now turn our attention to mitosis.

Synoptic link

Revising your knowledge about the structure of chromosomes (see Topic 7.2) will help you to understand mitosis.

Mitosis

Mitosis is the division of the nucleus of a cell that results in each of the daughter cells having an exact copy of the DNA of the parent cell. Except in the rare event of a **mutation**, the genetic make-up of the two daughter nuclei is therefore identical to that of the parent nucleus. Mitosis is always preceded by a period during which the cell is not dividing. This period is called **interphase**. It is a period of considerable cellular activity that includes a very important event, the replication of DNA (see Topic 7.3). Although mitosis is a continuous process, it can be divided into four stages for convenience:

1 **prophase**, in which the chromosomes become visible and the nuclear envelope disappears
2 **metaphase**, in which the chromosomes arrange themselves at the centre of the cell
3 **anaphase**, in which each of the two strands (or chromatids) of a chromosome migrates to an opposite pole
4 **telophase**, in which the nuclear envelope reforms.

The process is illustrated and explained in Figure 3.

Study tip

It is important to remember that the replication of DNA takes place during interphase before the nucleus and the cell divide.

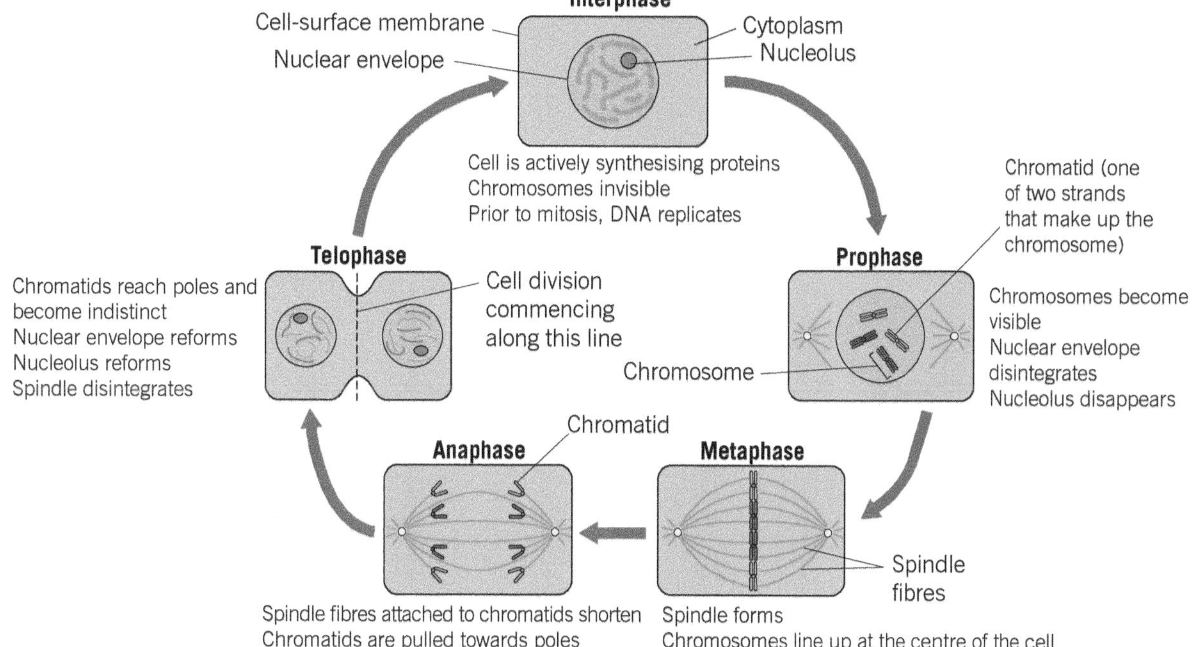

▲ **Figure 3** *The stages of mitosis in an animal cell*

The importance of mitosis

Mitosis is important in organisms because it produces daughter cells that are genetically identical to the parent cells. Why then is it so essential to make exact copies of existing cells? There are three main reasons:

- **growth**. When two **haploid** cells (e.g., a sperm and an ovum) fuse together to form a **diploid** cell, this diploid cell has all the genetic information needed to form the new organism. If the new organism is to resemble its parents, all the cells that grow from this original cell must possess this same set of genetic information. Mitosis ensures that this happens. The cell first divides to give a group of identical cells.

- **differentiation**. These cells change, or differentiate, to give groups of specialised cells, for example, epithelium in animals or xylem in plants (see Topic 2.4). These different cell types each divide by mitosis to give tissues made up of identical cells, which perform a particular function. This is essential as the tissue can only function efficiently if all its cells have the same structure and perform the same function.
- **repair**. If cells are damaged or die it is important that the new cells produced have an identical structure and function to the ones that have been lost. If they were not exact copies the tissue would not function as effectively as before. Mitosis is therefore the means by which new cells replace damaged or dead ones.

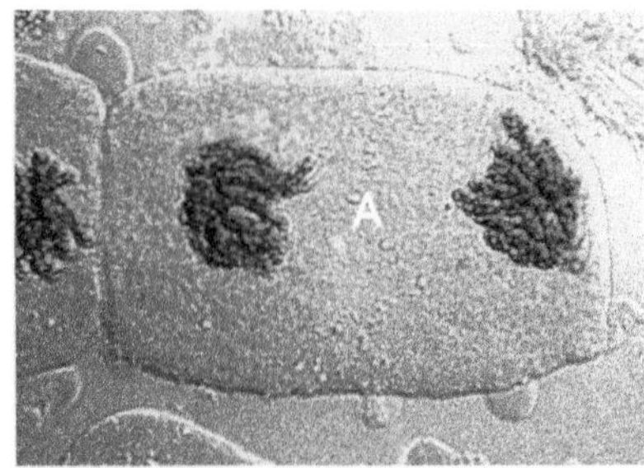

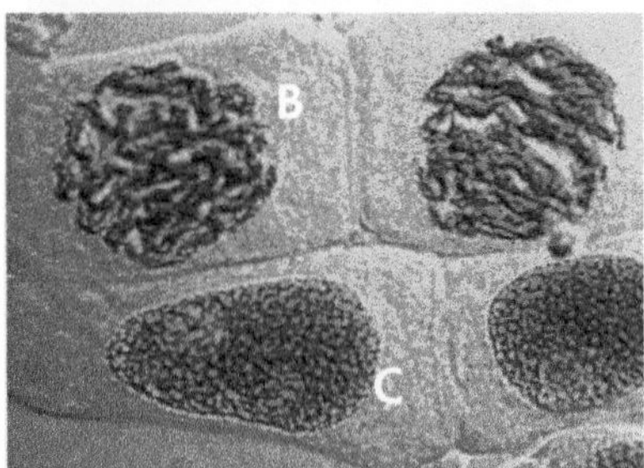

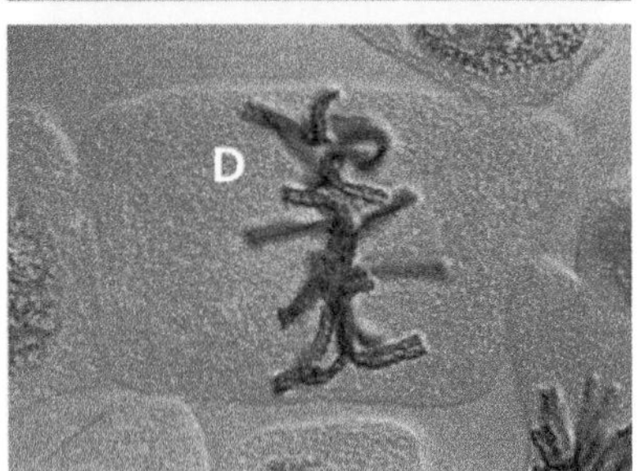

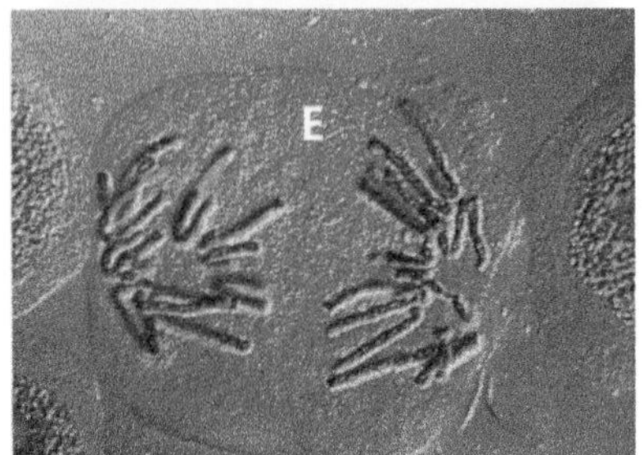

▲ **Figure 4** *Stages of mitosis*

▼ **Table 1**

Stage	Number of cells
Interphase	890
Prophase	73
Metaphase	20
Anaphase	9
Telophase	8

Summary question

1 In the following passage about mitosis, give the most appropriate word that is represented by each of the letters **a** to **i**.

The period when a cell is not dividing is called **a**. The stage of mitosis when the chromosomes are first visible as distinct structures is called **b**. During this stage thin threads develop that span the cell from end to end and together form a structure called the **c**. Towards the end of this stage, the **d** breaks down and the **e** disappears. The stage when the chromosomes arrange themselves across the centre of the cell is called **f**. During the stage called **g**, the chromatids move to opposite ends of the cell. Mitosis is important in **h** and **i** because it produces genetically identical cells.

Recognising the stages of mitosis

The photographs in Figure 4 show cells at various stages of mitosis.

Mitosis is a continuous process. When mitosis is viewed under a microscope, the observer only gets a snapshot of the process at one moment in time. In this snapshot, the number of cells at each stage of mitosis is proportional to the time each cell spends undergoing that stage. Table 1 shows the number of cells at each stage of mitosis during one observation.

1. State the names of the five different stages represented by the letters A–E in Figure 2. In each case give a reason for choosing your answer.
2. From Table 1, if one complete cycle takes 20 hours, how many minutes were spent in metaphase? Show your working.
3. In what percentage of the cells would the chromosomes have been visible? Show your working.

Study tip

You must be able to recognise the stages of mitosis from drawings and photographs and explain the events occurring during each stage.

Required practical 6

Calculating a mitotic index

For tissues in which cells rarely divide, an examination of the tissue under a microscope would show most cells in interphase. However in tissues such as plant meristematic tissue in root tips, where cell division occurs rapidly, many cells will be in one or other of the recognisable stages of mitosis. A mitotic index can be calculated allowing a comparison to be made between different tissues.

$$\textbf{mitotic index} = \frac{\textbf{number of cells in mitosis}}{\textbf{total number of cells}}$$

Meristematic tissue in root tips can be investigated using stained squashes of growing roots of plants such as garlic or onion. The method used is as follows:

- Using a scalpel, carefully cut a 2 cm length of the root tip and place in a watch glass.
- Add a small volume of ethanoic acid and leave for 10 minutes.
- Heat approximately 20 cm^3 of 1M hydrochloric acid to 60 °C in a boiling tube in a water bath.
- Wash the root tips carefully and then transfer them to the acid and leave them for 5 minutes.

1 Suggest what is achieved by soaking the plants in hydrochloric acid.

- Remove the roots, wash them carefully, and then dry them on filter paper and transfer two root sections onto a microscope slide.
- Carefully remove approximately 2 mm of the growing tip, discarding the remainder of the root.

2 Why will mitotic cells only be found in the root tip?

- Add a few drops of aceto-orcein stain and leave the tips for at least 2 minutes.
- Place a coverslip over the roots and carefully apply pressure to squash the tissue and release the cells.
- Examine the cells under a light microscope at high power (×400).

3 Calculate the mitotic index for the tissue shown in Figure 3.

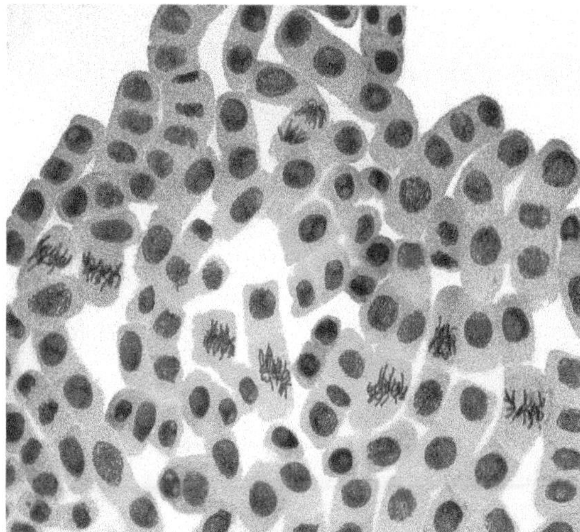

▲ **Figure 3** *Stages of mitosis*

17.2 Binary fisson

Learning objectives:

- → Explain that bacterial cells divide by binary fission.
- → Describe the transfer of DNA horizontally by conjugation.

Specification reference: 3.2.10.3

In Topic 17.1 you learnt about the cell cycle and nuclear division in eukaryotic cells. Prokaryotic cells divide by a process called **binary fission**. As in eukaryotic cells, before binary fission can take place, the circular DNA molecule that is the bacterial genome (sometimes referred to as the bacterial chromosome) must be replicated. The site on the DNA molecule where this begins is called the origin of replication. The stages in binary fission are shown in Figure 1.

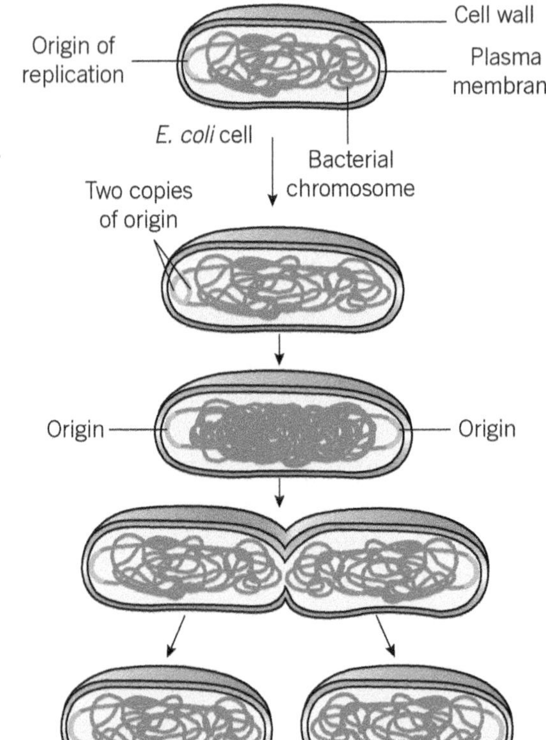

▲ **Figure 1**

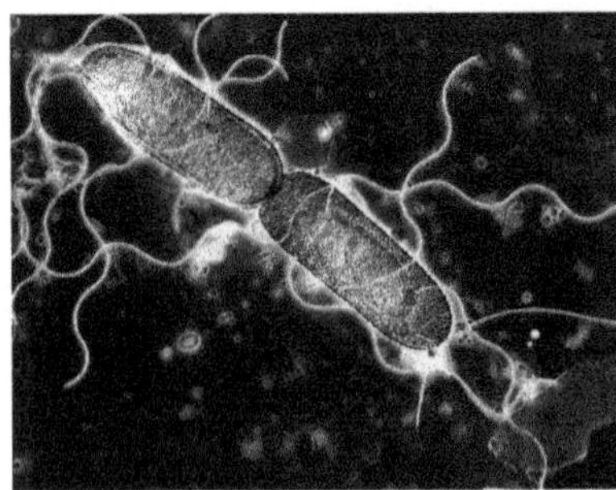

▲ **Figure 2** *False-colour transmission electron micrograph (TEM) image of a bacterial cell dividing. Under optimum conditions, cells can divide every 10 minutes, doubling the size of the population each time.*

Bacterial cells also contain plasmids (Topic 2.5). Plasmids each have their own independent origin of replication. This means that bacterial cells can have many copies of a single plasmid.

Conjugation

DNA can also be transferred from one bacterial cell to another. This is called **conjugation** and it takes place as follows.

- One cell produces a thin projection that meets another cell and forms a thin conjugation tube between the two cells.
- The donor cell replicates one of its small circular pieces of DNA (plasmid).
- The circular DNA is broken to make it linear before it passes along the tube into the recipient cell.

- Contact between the cells is brief, leaving only time for a portion of the donor's DNA to be transferred.
- In this way, the recipient cell acquires new characteristics from the donor cell.

These events are illustrated in Figures 3 and 4.

In conjugation, DNA in the form of genes can be passed from one species to another species. This is known as **horizontal gene transmission**. Where genes are passed down from one generation of a species to the next generation of the same species, the process is known as vertical gene transmission.

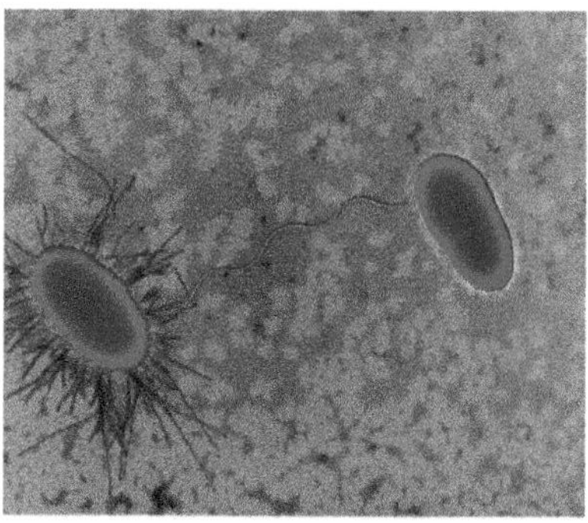

▲ **Figure 3** *False-colour scanning electron micrograph (SEM) of* Escherichia coli *(*E. coli*) bacteria conjugating*

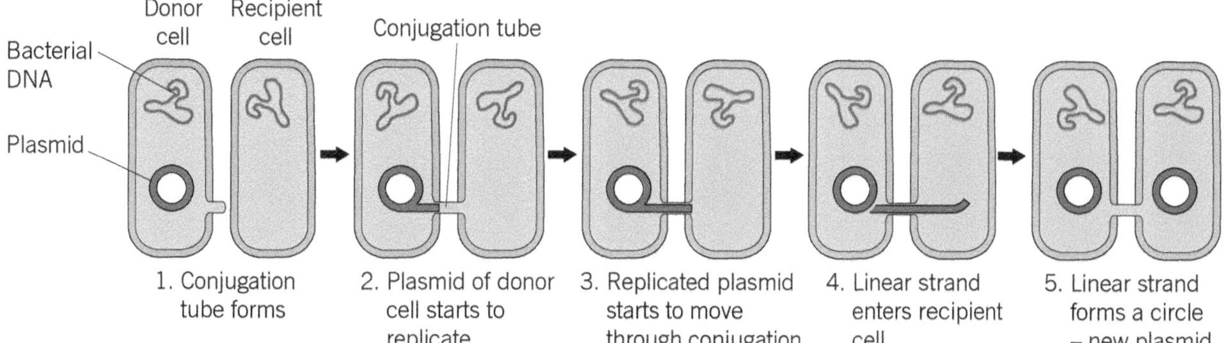

▲ **Figure 4** *DNA transfer by conjugation in bacteria*

Discovering conjugation in bacteria

Joshua Lederberg was a medical student whose observations led him to question the prevailing scientific view that bacteria passed down exact copies of genetic information to the next generation. In other words, that they always produced genetically identical offspring, known as **clones**. In 1946 he teamed up with Edward Tatum to test his hypothesis that there was transfer of genetic information between the bacteria, thereby producing variety by a means other than mutation. They carried out experiments designed to demonstrate that genes could be transferred between different strains of the bacterium *Escherichia coli* (*E. coli*).

Lederberg and Tatum used two different strains of *E. coli* that were grown in dishes of a jelly-like substance (growth medium) that contained certain nutrients. Neither of the two strains could grow on a growth medium that contained only the basic nutrients (minimal medium). This was because each strain was unable to synthesise for itself two different nutrients that were essential for its growth.

- Strain 1 could not synthesise the amino acid methionine and the vitamin biotin. Strain 1 would therefore only grow if these were added to the growth medium.
- Strain 2 could not synthesise the amino acids threonine and leucine. Strain 2 would only grow if these were added to the growth medium.

Summary questions

1. How does the bacterial chromosome differ to those found in eukaryotic cells?
2. Distinguish between horizontal and vertical gene transmission.
3. Discuss the significance of the fact that many antibiotic resistance genes are found on plasmids.

They mixed the two strains of *E. coli* together and grew them for several hours in a medium that contained supplements of all the missing nutrients (methionine, biotin, threonine, and leucine).

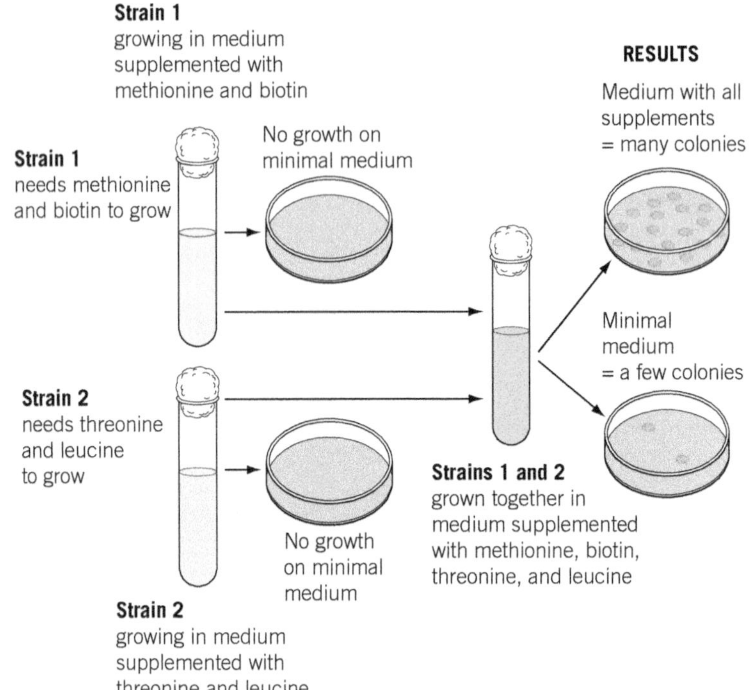

▲ **Figure 5** *Experiment to show transfer of DNA between two strains of* E. coli

1 Which of the strains of *E. coli* should grow on the medium that contained all the missing nutrients?

The bacteria were then removed from the medium by centrifugation (see Topic 2.1) and washed. Next they were transferred either to a medium with all the supplements (methionine, biotin, threonine, and leucine) or to a minimal medium without the supplements.

2 Which strains would you expect to grow on the minimal medium?

Lederberg and Tatum found that, whilst most cells did not grow on unsupplemented medium, around 1 in 10 million did divide and develop into a colony of cells. The experiment and its results are summarised in Figure 5.

In further experiments, they were able to show that DNA had passed from one strain (the donor) to the other (the recipient) by conjugation.

3 What information had the DNA from the donor strain transferred to the recipient strain?

4 These further experiments were carried out to disprove explanations other than conjugation. Suggest one alternative explanation that might account for the growth of a few colonies on the minimal medium.

17.3 Gene mutation

Any change to the quantity or the structure of the DNA of an organism is known as a **mutation**. Mutations arising in body cells are not passed on to the next generation. Mutations occurring during the formation of gametes may be inherited, often producing sudden and distinct differences between individuals. They are therefore the basis of discontinuous variation. Any change to one or more nucleotide bases, or any rearrangement of the bases, in DNA is known as a **gene mutation**.

You have seen that a sequence of triplets on DNA is transcribed into mRNA and is then translated into a sequence of amino acids that make up a polypeptide. It follows that any changes to one or more bases in the DNA triplets could result in a change in the amino acid sequence of the polypeptide. There are a number of ways in which the DNA bases can change, two examples being substitution and deletion of bases.

Substitution of bases

The type of gene mutation in which a nucleotide in a DNA molecule is replaced by another nucleotide that has a different base is known as a substitution. Depending on which new base is substituted for the original base, there are three possible consequences. As an example, let us take the DNA triplet of bases, guanine–thymine–cytosine (GTC) that codes for the amino acid glutamine. A change to a single base could result in one of the following:

- A **nonsense mutation** occurs if the base change results in the formation of one of the three stop codons that mark the end of a polypeptide chain. For example, if the first base, guanine, is replaced by adenine, then GTC becomes ATC. The triplet ATC is transcribed as UAG in mRNA. UAG is one of the three stop codons. As a result the production of the polypeptide would be stopped prematurely. The final protein would almost certainly be significantly different and the protein could not perform its normal function.
- A **missense mutation** arises when the base change results in a different amino acid being coded for. In our example, if the final base, cytosine, is replaced by guanine, then GTC becomes GTG. GTG is one of the DNA triplet codes for the amino acid histidine and this then replaces the original amino acid glutamine. The polypeptide produced will differ in a single amino acid. The significance of this difference will depend upon the precise role of the original amino acid. If it is important in forming bonds that determine the tertiary structure of the final protein, then the replacement amino acid may not form the same bonds. The protein may then be a different shape and therefore not function properly. For example, if the protein is an enzyme, its active site may no longer fit the substrate and it will no longer catalyse the reaction.
- A **silent mutation** occurs when the substituted base, although different, still codes for the same amino acid as before. This is due to the degenerate nature of the genetic code, in which most amino acids have more than one codon. For instance, if the third base in our example is replaced by thymine, then GTC becomes GTT. However, as both DNA triplets code for glutamine, there is no change in the polypeptide produced and so the mutation will have no effect.

Learning objectives:

- → Explain what is meant by a gene mutation.
- → Explain the effects of different types of mutation on the primary structure of a polypeptide.
- → Discuss the causes of gene mutations.
- → Explain the link between mutation and cancers.

Specification reference: 3.2.11.1 and 3.2.11.2

Synoptic link

Discontinuous variation is due to genetic factors and is covered in Topic 6.2.

Hint

The various gene mutations are illustrated by specific examples that name bases and amino acids. These are only to illustrate the points being made and do not need to be remembered.

Hint

Consider the following sentence, which consists only of three-letter words: THE RED HEN ATE HER TEA. If we delete the first letter 'T' but continue to divide the sentence into three-letter words, it becomes HER EDH ENA TEH ERT EA and is incomprehensible. If we delete the final 'T,' this leaves the strange but mostly readable sentence THE RED HEN ATE HER EA.

Examples of all these types of substitution mutation are shown in Table 1.

▼ **Table 1** *Types of substitution mutation*

	Usual triplet of DNA bases	Nonsense mutation	Missense mutation	Silent mutation
Sequence of bases in DNA	GTC	ATC	GTG	GTT
Sequence of bases in mRNA	CAG	UAG	CAC	CAA
Amino acid in polypeptide	Glutamine	None (stop code)	Histidine	Glutamine

Deletion of bases

A gene mutation by deletion arises when a nucleotide is lost from the normal DNA sequence. The loss of a single nucleotide from the thousands in a typical gene may seem a minor change but the consequences can be considerable. Usually the amino acid sequence of the polypeptide is entirely different. This is because the genetic code is read in units of three bases (triplet). One deleted nucleotide creates what is known as a 'frame-shift' because the reading frame that contains each three letters of the code has been shifted to the left by one letter. The gene is now read in the wrong three-base groups and the genetic message is altered. One deleted base at the very start of a sequence could alter every triplet in the sequence. A deleted base near the end of the sequence is likely to have a smaller impact but can still have consequences (see 'Hint'). An example of the effect of a deletion mutation is shown in Figure 1.

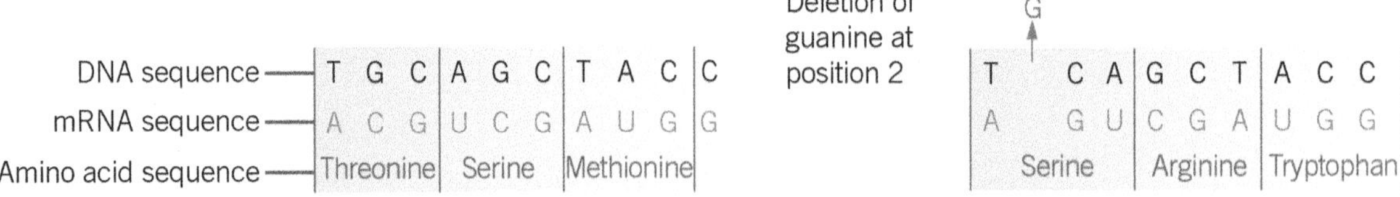

▲ **Figure 1** *Effects of the deletion of a DNA nucleotide on the amino acid sequence in the final polypeptide*

Causes of mutations

Gene mutations can arise spontaneously during DNA replication. Spontaneous mutations are permanent changes in DNA that occur without any outside influence. Despite being random occurrences, mutations occur with a relatively predictable frequency. The natural mutation rate varies from species to species, but is typically around one or two mutations per 100 000 genes per generation. This basic mutation rate is increased by outside factors known as **mutagenic agents** or mutagens. These include the following:

- high-energy radiation that can disrupt the DNA molecule
- chemicals that alter the DNA structure or interfere with transcription.

Mutations have both costs and benefits. On the one hand they produce the genetic diversity necessary for natural selection and speciation (see Topics 23.7 and 23.8). On the other hand they often produce an organism that is less well suited to its environment. Additionally, mutations that occur in body cells rather than in gametes can disrupt normal cellular activities, such as cell division.

▲ **Figure 2** *This albino hedgehog is the result of a mutation that prevents the production of the pigment melanin*

Genetic control of cell division

Cell division is controlled by genes. Most cells divide at a fairly constant rate to ensure that dead or worn-out cells are replaced. In normal cells, this rate is tightly controlled by different types of genes:

- **proto-oncogenes** that stimulate cell division
- **tumour suppressor genes** that slow cell division.

Role of proto-oncogenes

Proto-oncogenes stimulate cell division. In a normal cell, growth factors attach to a receptor protein on the cell-surface membrane and, via relay proteins in the cytoplasm, 'switch on' the genes necessary for DNA replication. This process is shown in Figure 3. A gene mutation can cause proto-oncogenes to mutate into oncogenes. These oncogenes can affect cell division in two ways:

- The receptor protein on the cell-surface membrane can be permanently activated, so that cell division is switched on even in the absence of growth factors.
- The oncogene may code for a growth factor that is then produced in excessive amounts, again stimulating excessive cell division.

The result is that cells divide too rapidly and a tumour, or cancer, develops.

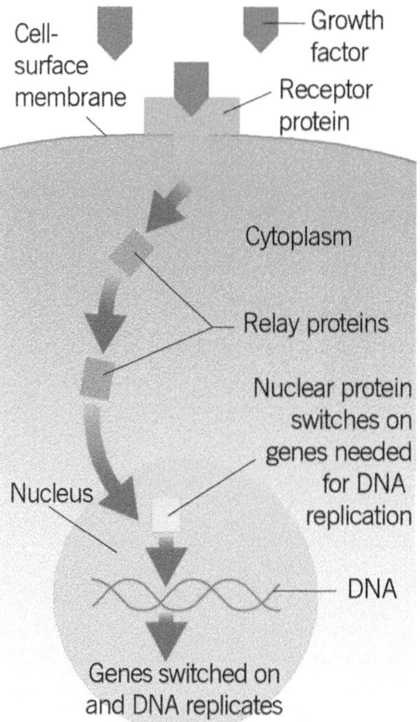

▲ **Figure 3** *Control of cell division in a normal cell*

Benign tumours do not spread to other parts of the body. They can often be removed relatively easily and generally do not come back. In malignant tumours, the cancerous cells can invade nearby tissues and spread to other part of the body where they continue to grow, forming further tumours (metastases).

Role of tumour suppressor genes

Research into hereditary forms of cancer led to the discovery of tumour suppressor genes. These genes have the opposite role to proto-oncogenes in the cell, that is, they inhibit cell division. A normal tumour suppressor gene will therefore maintain normal rates of cell division and prevent the formation of tumours – hence its name. If a tumour suppressor gene becomes mutated it is inactivated. In other words it stops inhibiting cell division, which therefore increases. The mutant cells so formed are usually structurally and functionally different from normal cells. Most mutated cells die. However, any that survive are capable of making clones of themselves and forming tumours. Not all tumours are harmful (malignant) – some are benign.

Hint

The interaction of the two genes that control cell division can be likened to the controls of a car. The proto-oncogene is like the accelerator pedal that increases the speed of the car. The tumour suppressor gene is like the brake pedal that reduces the speed of the car.

Mutagenic agents

Mutations can be induced by external influences called mutagenic agents. These cause damage in a number of ways.

- **Certain chemicals can remove groups from nucleotide bases.** Nitrous acid can remove an—NH_2 group from cytosine in DNA, changing it into uracil.

1 Suggest what might be the result of this change on the codons on an mRNA molecule that is transcribed from a section of DNA with the triplets GCA CTC ATC.

Summary questions

1. The following is a sequence of 12 nucleotides within a much longer mRNA molecule: AUGCAUGUUACU. Following a gene mutation the same 12-nucleotide portion of the mRNA molecule is now AUGCUGUUACUG. What type of gene mutation has occurred? Show your reasoning.
2. Explain why a deletion gene mutation is more likely to result in a change to an organism than a substitution gene mutation.
3. Explain why a mutation that is transcribed on to mRNA may not result in any change to the polypeptide that it codes for.
4. Errors in transcription occur about 100 000 times more often than errors in DNA replication. Explain why errors in DNA replication can be far more damaging than errors in transcription.
5. Which two types of genes control cell division in normal cells. What is the role of each?

▲ **Figure 4** *Ultraviolet radiation from sunbeds has the potential to disrupt DNA replication*

- **Other chemicals can add groups to nucleotides.** Benzopyrene is a chemical found in tobacco smoke. It adds a large group to guanine that makes it unable to pair with cytosine. When DNA polymerase reaches the affected guanine it inserts any of the other bases.

2 What type of mutation is caused by benzopyrene?

- **Ionising radiation**, such as X-rays, can produce highly reactive agents, called free radicals, in cells. These free radicals can alter the shape of bases in DNA so that DNA polymerase can no longer act on them.

3 Explain why DNA polymerase cannot act on DNA that has been damaged by X-rays.

4 State **one** genetic effect of DNA polymerase being unable to act on DNA.

- **Ultraviolet radiation** from the Sun or tanning lamps affects thymine in DNA, causing it to form bonds with the nucleotides on either side of it. This seriously disrupts DNA replication.

Scientific research and experimentation has enabled us to identify potentially dangerous mutagenic agents. The effects of such agents are complex and the amount of harm they cause is often a matter of debate. Commercial organisations such as the tobacco industry, manufacturers of sunbeds, and producers and retailers of sun-block lotions all have an interest in the research that is undertaken. They are more likely to fund research that may benefit their business than research that may harm it. It is therefore important that the results of any research are subjected to the scrutiny of other scientists from a wide variety of backgrounds, views, interests, and organisations, in a process known as peer review.

This is usually achieved by publishing research findings in reputable scientific journals that have an extensive global readership. The conclusions and claims made by researchers and their sponsors can then be debated and the scientific community at large can test the claims by further experimentation. These claims then become accepted, modified, or rejected, depending on the outcome of this further research.

Armed with all this scientific information, decision makers such as governments and heads of business can take appropriate action that benefits society. Governments, for example, can introduce legislation that controls cigarette sales and smoking, and the use of sunbeds, and sets a minimum age at which cigarettes or tanning treatments can be bought. The decisions are often not clear-cut, however. X-rays, for example, can be harmful on one hand but are an invaluable diagnostic tool, with countless health benefits, on the other.

5 Leaders in business and government have to make decisions about the use of scientific discoveries. Who else, apart from research scientists, might influence the advice that these leaders give to the public on the use of sunbeds?

Cancer and its treatment

The treatment of cancer often involves blocking some part of the cell cycle. In this way, the cell cycle is disrupted and cell division, and hence cancer growth, ceases. Drugs used to treat cancer (chemotherapy) disrupt the cell cycle by:

- preventing DNA from replicating, for example, cisplatin
- inhibiting the metaphase stage of mitosis by interfering with spindle formation, for example, vinca alkaloids.

The problem with such drugs is that they also disrupt the cell cycle of normal cells. However, the drugs are more effective against rapidly dividing cells. As cancer cells have a particularly fast rate of division, they are damaged to a greater degree than normal cells. Those normal body cells, such as hair-producing cells, which divide rapidly are also vulnerable to damage. This explains the hair loss that is frequently seen in patients undergoing cancer treatment.

The graph in Figure 5 shows the effect of a chemotherapy drug that kills dividing cells. It was given to a cancer patient once every 3 weeks starting at time zero. The graph plots the changes in the number of healthy cells and cancer cells in a tissue over the treatment period of 12 weeks.

1 How many fewer healthy cells were there after 3 weeks compared to the start of the treatment?
2 What percentage of the original number of healthy cells were still present at 12 weeks?
3 How many times greater is the number of healthy cells compared to the number of cancer cells after 12 weeks?
4 Give a reason for the lower number of cancer cells compared to healthy cells at 12 weeks.
5 Describe two differences between the effect of the drug on cancer cells compared with healthy cells throughout the treatment.
6 Use the graph to explain why chemotherapy drugs have to be given a number of times if they are to be effective in treating cancer.

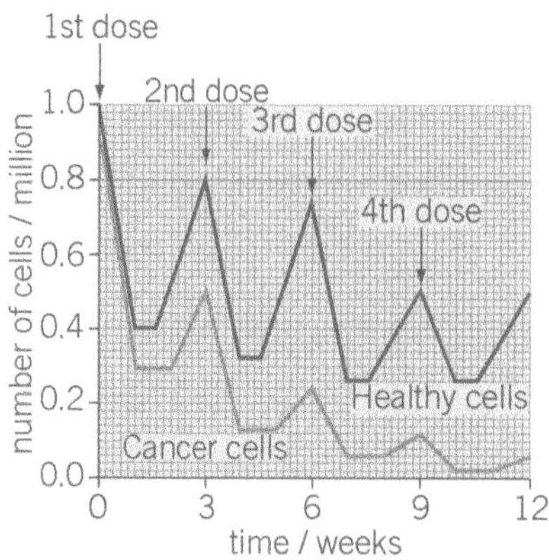

▲ **Figure 5** *Changes in the number of healthy cells and cancer cells in a tissue during a chemotherapy treatment of 12 weeks*

1 (a) The graph shows information about the movement of chromatids in a cell that has just started metaphase of mitosis.

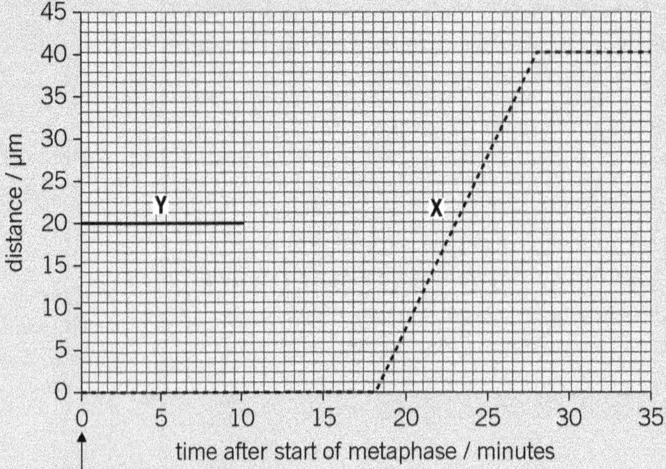

(i) What was the duration of metaphase in this cell in minutes? *(1 mark)*

(ii) Use line **X** to calculate the duration of anaphase in this cell in minutes. *(1 mark)*

(iii) Complete line **Y** on the graph. *(2 marks)*

(b) A doctor investigated the number of cells in different stages of the cell cycle in two tissue samples, **C** and **D**. One tissue sample was taken from a cancerous tumour. The other was taken from non-cancerous tissue. The table shows her results.

Stage of the cell cycle	Percentage of cells in each stage of the cell cycle	
	Tissue sample C	Tissue sample D
Interphase	82	45
Prophase	4	16
Metaphase	5	18
Anaphase	5	12
Telophase	4	9

(i) In tissue sample **C**, one cell cycle took 24 hours. Use the data in the table to calculate the time in which these cells were in interphase during one cell cycle. Show your working. *(2 marks)*

(ii) Explain how the doctor could have recognised which cells were in interphase when looking at the tissue samples. *(1 mark)*

(iii) Which tissue sample, **C** or **D**, was taken from a cancerous tumour? Use information in the table to explain your answer. *(2 marks)*

AQA Jan 2013

2 (a) A to E list some of the events of the cell cycle:

A Chromatids separate
B Nuclear envelope disappears
C Cytoplasm divides
D Chromosomes condense and become visible
E Chromosomes on the equator of the spindle

(i) List these events in the correct order, starting with **D**. *(1 mark)*

(ii) Name the stage described in **E**. *(1 mark)*

(b) Name the phase during which DNA replication occurs. *(1 mark)*

(c) Bone marrow cells divide rapidly. As a result of a mutation during DNA replication, a bone marrow cell may become a cancer cell and start to divide in an uncontrolled way. A chemotherapy drug that kills cells when they are dividing was given to a cancer patient. It was given once every 3 weeks, starting at time 0. The graph shows the changes in the number of healthy bone marrow cells and cancer cells during 12 weeks of treatment.

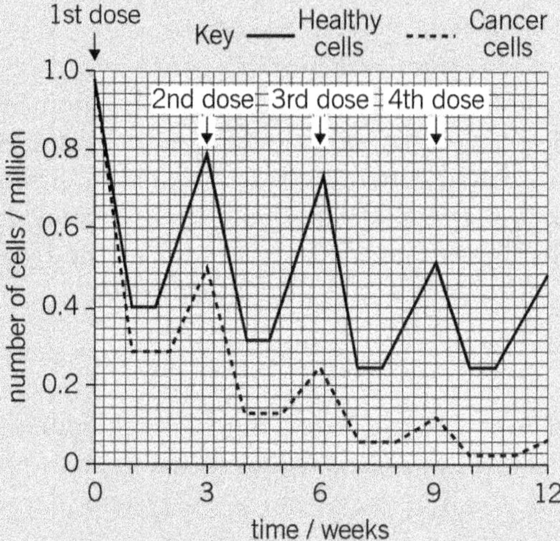

(i) Using the graph calculate the number of cancer cells present at week 12 as a percentage of the original number of cancer cells. Show your working. *(2 marks)*

(ii) Suggest one reason for the lower number of cancer cells compared to healthy cells at the end of the first week. *(1 mark)*

(iii) Describe two differences in the effect of the drug on the cancer cells compared with healthy cells in the following weeks. *(2 marks)*

AQA June 2006

3 **(a)** Gene mutations occur naturally. Give one factor that increases the rate of gene mutations. *(1 mark)*

(b) The table shows the DNA base sequences that code for three amino acids.

DNA base sequence(s) coding for amino acids	Amino acid
CCA CCG CCT CCC	Glycine
TAC	Methionine
TAA TAG TAT	Isoleucine

Some substitution mutations would affect the sequence of amino acids in a polypeptide, and others would not. Using only the information in the table, explain why. *(3 marks)*

AQA January 2007

Answers to the Practice Questions are available at
www.oxfordsecondary.com/oxfordaqaexams-alevel-biology

18 Populations

18.1 Populations and ecosystems

Learning objectives:

- → Understand what is meant by the terms population, community, and ecosystem.
- → Explain that ecosystems have biotic and abiotic components.
- → Describe what is meant by a habitat.
- → Understand the concept of an ecological niche.

Specification reference: 3.3.1.1

Hint

Many students confuse the terms biotic and abiotic. Biotic refers to living things, for example predation or competition between different species or individuals of one species. Abiotic refers to non-living factors, such as temperature, rainfall, or light intensity.

▲ **Figure 1** *Woodland ecosystem*

Study tip

'Population' is a very important concept that needs to be understood. It comes up in many areas of biology, including genetics and evolution.

In this chapter you shall look at how living organisms form communities within ecosystems through which energy is transferred and elements are recycled.

Ecology is the study of the inter-relationships between organisms and their environment. The environment includes both non-living (**abiotic**) components, such as temperature and rainfall, and living (**biotic**) components, such as competition and predation. Ecology is a complex area of study that includes most aspects of biology. It is, in effect, the study of the life-supporting layer of land, air, and water that surrounds the Earth. This layer is called the **biosphere**.

Ecosystems

An ecosystem is made up of all the interacting biotic (living) and abiotic (non-living) features in a specific area. Ecosystems are more or less self-contained functional units. Within an ecosystem there are two major processes to consider:

- the flow of energy through the system
- the cycling of elements within the system.

An example of an ecosystem is a freshwater pond or lake. It has its own community of plants to collect the necessary sunlight energy to supply the organisms within it. Nutrients such as nitrates and phosphates are recycled within the pond or lake. There is little or no loss or gain between it and other ecosystems. Another example of an ecosystem is an oak woodland (Figure 1). Within each ecosystem, there are a number of species. Each species is made up of many individuals that together make up a population.

Populations

A **population** is a group of interbreeding organisms of one species in a habitat. In the different habitats of an oak woodland there are populations of nettles, worms, green woodpeckers, beetles, etc. The boundaries of a population are often difficult to define. In our oak woodland, for example, all the mature green woodpeckers can breed with one another and so form a single population. The woodlice on a decaying log at one side of the wood can, in theory, breed with those on a log a kilometre or more away at the other side of the wood. In practice, the sheer distance makes interbreeding unlikely and therefore they can be considered as separate populations. Where exactly the boundary lies between these two populations is, however, unclear. Populations of different species form a community.

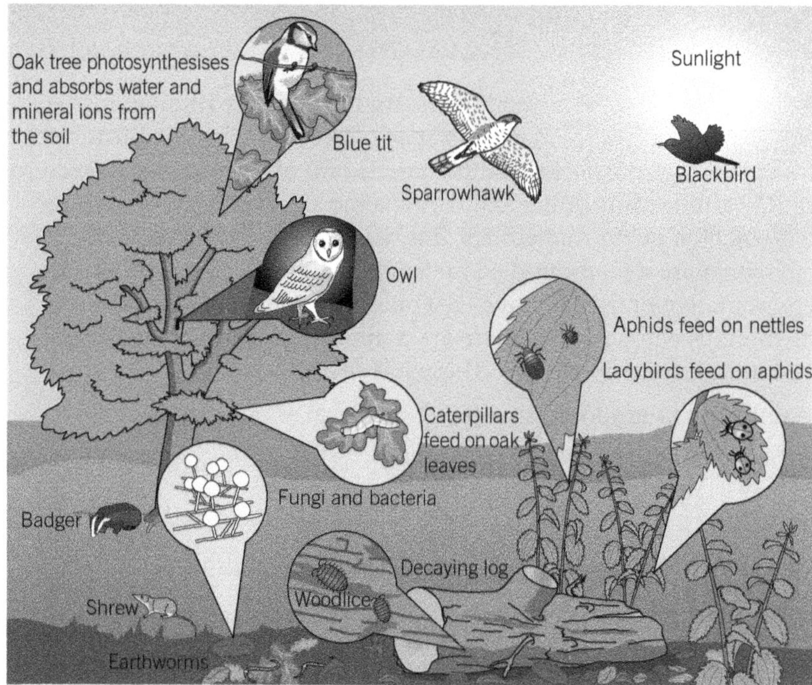

▲ **Figure 2** *Part of an oak woodland ecosystem*

Community

A **community** is defined as all the populations of different species living and interacting in a particular place at the same time. Within an oak woodland, a community may include a large range of organisms, such as oak trees, hazel shrubs, bluebells, nettles, sparrowhawks, blue tits, ladybirds, aphids, woodlice, earthworms, fungi, and bacteria (see Figure 2).

Habitat

A **habitat** is the place where a community of organisms lives. Within an ecosystem there are many habitats. For example, in an oak woodland, the leaf canopy of the trees may be a habitat for blue tits, whereas a decaying log is the habitat for woodlice. A stream flowing through the woodland provides a very different habitat, within which aquatic plants and water beetles live. For a water vole, the stream and its banks are its habitat. Within each habitat there are smaller units, each with their own microclimate. These are called **microhabitats**. The mud at the bottom of the stream may be the microhabitat for a bloodworm whereas a crevice on the bark of an oak tree may be the microhabitat for a lichen.

Ecological niche

A **niche** describes how an organism fits into the environment. A niche refers to where an organism lives and what it does there. It includes all the biotic and abiotic conditions required for an organism to survive, reproduce, and maintain a viable population. Some species may appear very similar, but their nesting habits or other aspects of their behaviour will be different, or they may show different levels of tolerance to environmental factors, such as a pollutant or a shortage of oxygen or nitrates. No two species occupy exactly the same niche.

Hint

Organisms are found in places where the local environmental conditions fall within the range that their adaptations enable them to cope with.

Study tip

Make sure that you can accurately define the basic ecological terms described in this topic.

▲ **Figure 3** *This lake is an example of a habitat*

Summary question

1 In the following passage, state the word that best replaces each of the letters **a** to **g**.

The study of the inter-relationships between organisms and their environment is called **a**. The layer of land, air, and water that surrounds the Earth is called the **b**. An ecosystem is a more or less self-contained functional unit made up of all the living or **c** features and non-living or **d** features in a specific area. Within each ecosystem are groups of different organisms, called a **e**, which live and interact in a particular place at the same time. A group of interbreeding organisms occupying the same place at the same time is called a **f**, and the place where they live is known as a **g**.

18.2 Investigating populations

Learning objectives:

- → Describe the use of quadrats in ecological measurement.
- → Describe how a transect is used to obtain quantitative data about changes in communities along a line.
- → Describe how the abundance of different species is measured.
- → Explain the use of mark-release-recapture methods in measuring the abundance of mobile species.

Specification reference: 3.3.1.1 and 3.3.1.2

To study a **habitat**, it is often necessary to count the number of individuals of a species in a given space. This is known as **abundance**. It is virtually impossible to identify and count every organism. To do so would be time-consuming and would almost certainly cause damage to the habitat being studied. For this reason only small samples of the habitat are usually studied in detail. As long as these samples are representative of the habitat as a whole, any conclusion drawn from the findings will be valid. There are a number of sampling techniques used in the study of habitats. These include:

- sampling at random using frame quadrats or point quadrats
- systematic sampling along transects.

Quadrats

There are three factors to consider when using quadrats:

- **the size of quadrat to use**. This will depend upon the size of the plants or animals being counted and how they are distributed within the area. Larger species require larger quadrats. Where a species occurs in a series of groups rather than being evenly distributed throughout the area, a large number of small quadrats will give more representative results than a small number of large ones.
- **the number of sample quadrats to record within the study area**. The larger the number of sample quadrats the more reliable the results will be. As the recording of species within a quadrat is a time-consuming task, a balance needs to be struck between the validity of the results and the time available. The greater the number of different species present in the area being studied, the greater the number of quadrats required to produce valid results.
- **the position of each quadrat within the study area**. To produce statistically significant results a technique known as sampling at random must be used.

Random sampling

It is important that sampling is random to avoid any bias in collecting data. Avoiding bias ensures that the data obtained are valid. How do you make sampling random?

Suppose you wish to investigate the effects of grazing animals on the species of plants growing in a field. You begin by choosing two fields as close together as possible in order to minimise soil, climatic, and other abiotic differences. One field is regularly grazed by animals such as sheep, whereas the other has not been grazed for many years. You then take sampling at random at many sites in each field by placing the quadrat on the ground and recording the names and numbers of every species found within the area of the quadrat.

But how do you get a truly random sample? You could simply stand in one of the fields and throw the quadrat over your shoulder. However, even with the best of intentions, it is difficult not to introduce an element of personal bias using this method. For example, are you more likely to stand in a dry area than a wet muddy one? Will you deliberately try to avoid an area covered in sheep droppings or full of nettles?

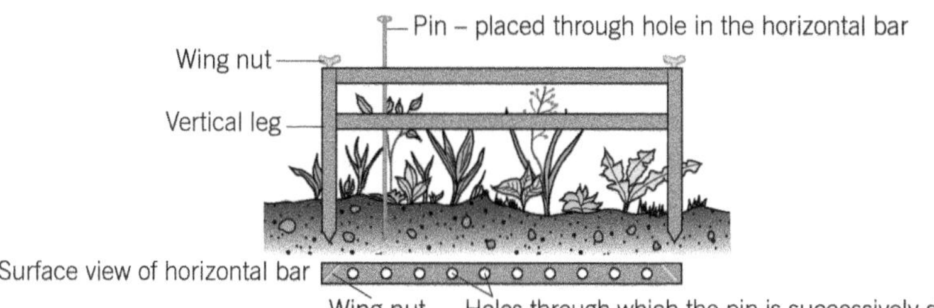

▲ **Figure 1** *A point quadrat*

A better method of sampling at random is to:

1 Lay out two long tape measures at right angles, along two sides of the study area.
2 Obtain a series of coordinates by using random numbers taken from a table or generated by a computer.
3 Place a quadrat at the intersection of each pair of coordinates and record the species within it.

Systematic sampling along transects

It is sometimes more informative to measure the abundance and distribution of a species in a systematic rather than a random manner. This is particularly important where some form of transition in the communities of plants and animals takes place. For example, the distribution of organisms on a tidal seashore is determined by the relative periods of time that they spend under water and exposed to the air, that is, by their vertical height up the shore. The stages of zonation are especially well shown using transects. A line transect comprises a string or tape stretched across the ground in a straight line. Any organism over which the line passes is recorded. A belt transect is a strip, usually a metre wide, marked by putting a second line parallel to the first. The species occurring within the belt between the lines are recorded.

Measuring abundance

Random sampling with quadrats and counting along transects are used to obtain measures of **abundance**. Abundance is the number of individuals of a species within a given space. It can be measured in several ways, depending upon the size of the species being counted and the habitat. Examples include:

- **frequency**, which is the likelihood of a particular species occurring in a quadrat. If, for example, a species occurs in 15 out of 30 quadrats, the frequency of its occurrence is 50 per cent. This method is useful where a species, such as grass, is hard to count. It gives a quick idea of the species present and their general distribution within an area. However, it does not provide information on the density and detailed distribution of a species.
- **percentage cover**, which is an estimate of the area within a quadrat that a particular plant species covers. It is useful where a species is particularly abundant or is difficult to count. The advantages in these situations are that data can be collected rapidly and individual plants do not need to be counted. It is less useful where organisms occur in several overlapping layers (more probably plants).

Study tip

Types of quadrat

A point quadrat consists of a horizontal bar supported by two legs. At set intervals along the horizontal bar are 10 holes, through each of which a long pin may be dropped (Figure 1). Each species that the pin touches is then recorded.

A frame quadrat is a square frame divided by string or wire into equally sized subdivisions (Figure 2). It is often designed so that it can be folded to make it more compact for storage and transport. The quadrat is placed in different locations within the area being studied. The abundance of each species within the quadrat is then recorded.

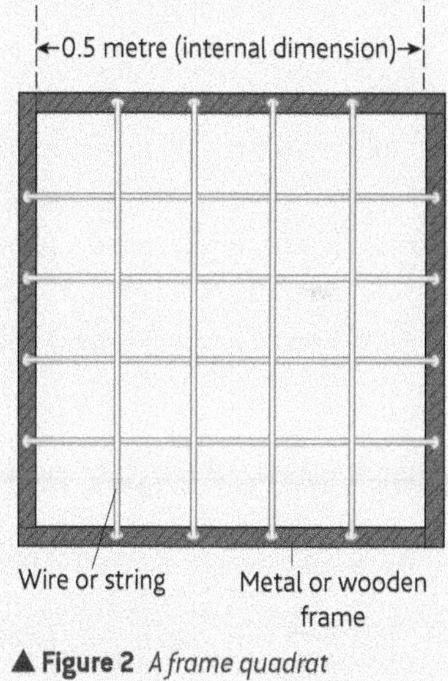

▲ **Figure 2** *A frame quadrat*

▲ **Figure 3** *Ecology student carrying out fieldwork*

To obtain reliable results, it is necessary to ensure that the sample size is large, that is, that many quadrats are used and the mean of all the samples is obtained. The larger the number of samples, the more representative of the community as a whole will be the results.

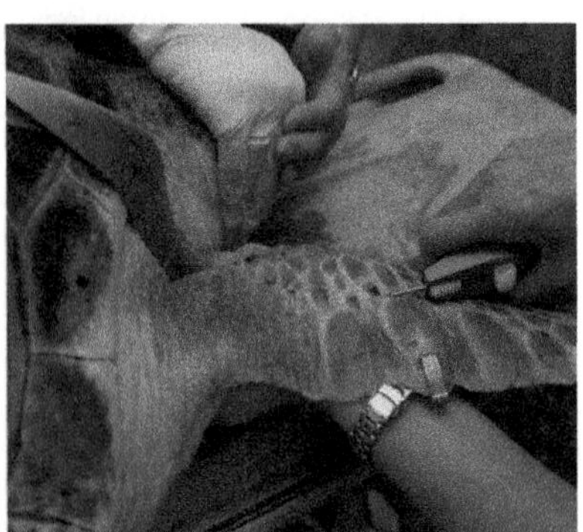

▲ **Figure 4** *A green sea turtle with a flipper tag used in estimating population size by the mark-release-recapture technique*

Mark-release-recapture technique

The methods of measuring abundance described above work well with plant communities but not with most animals. Most animals are mobile and move away when approached. Others are often hidden and are therefore difficult to find and identify. To estimate the abundance of most animals requires an altogether different technique.

In the mark-release-recapture technique, a known number of animals are caught, marked in some way, and then released back into the community. Some time later, a given number of individuals is collected randomly and the number of marked individuals is recorded. The size of the population is then calculated as follows:

$$\text{estimated population size} = \frac{\text{total number of individuals in the first sample} \times \text{total number of individuals in the second sample}}{\text{number of marked individuals recaptured}}$$

This technique relies on a number of assumptions:

- The proportion of marked to unmarked individuals in the second sample is the same as the proportion of marked to unmarked individuals in the population as a whole.
- The marked individuals released from the first sample distribute themselves evenly amongst the remainder of the population and have sufficient time to do so.
- The population has a definite boundary so that there is no immigration into or emigration out of the population.
- There are few, if any, deaths and births within the population.
- The method of marking is not toxic to the individual nor does it make the individual more conspicuous and therefore more liable to predation.
- The mark or label is not lost or rubbed off during the investigation.

Synoptic link

Standard deviations form part of 3.1.6 of the AS specification and are covered in Topic 6.2.

Analysing data

The quantitative data collected from an ecological survey then have to be analysed and interpreted. The first stage is usually to present the data in the form of a table or graph. This makes it easier to compare data, for example, from two different locations. Such comparisons can be made more precisely using statistical analysis of the data. Comparing the mean of two sets of data is helpful but, as you saw in Topic 6.2 that, this tells us nothing of the spread of the data about the mean. It is therefore useful to calculate the standard deviation as well.

Only tentative conclusions may be drawn from comparing two sets of data as there are many factors that can contribute to these differences. One such factor, that chance alone is the reason for the difference, can be checked statistically. This can be done using one of the various methods for testing the significance of differences between two sets of data. The results indicate the probability that these differences are due to some particular factor or are just a matter of chance.

Data can also be analysed for possible correlations and causes. You should remember from Topic 11.2 that two factors are said to be correlated when they vary in relation to each other. A positive correlation is where an increase in the value of one variable is accompanied by an increase in the value of the other variable, for example, the rate of photosynthesis increases with an increase in light intensity. A negative correlation is where an increase in the value of one variable is accompanied by a decrease in the value of the other variable.

Statistical tests can be used to calculate the strength and direction of any correlation between two variables. One example is the Spearman rank correlation. You may be required to use this in your practical work. Showing that two variables correlate statistically does not prove that one causes the other. The numbers of two species in a population may correlate very well but it is possible that both of them are affected by the same environmental factor, for example, a rise in temperature.

Ethics and fieldwork

Understanding the complex inter-relationships between organisms in a community helps us to minimise the effect of human activities on the environment and ensure these communities are conserved. For us to understand communities we need to collect data about them. However, the very collection of such data may be harmful to the communities that we are trying to conserve. To minimise the impact of an ecological investigation on the environment a number of basic procedures should be observed:

- Where possible, the organisms should be studied *in situ*. If it is necessary to remove them, the numbers taken should be kept to the absolute minimum.
- Any organisms removed from a site should be returned to their original habitat as soon as possible. This applies even if they are dead.
- A sufficient period of time should elapse before a site is used for future studies.
- Disturbance and damage to the habitat should be avoided. Trampling, overturning stones, permanently removing organisms, etc. can all adversely affect a habitat.

Field studies often lead to varying degrees of damage to habitats and some of the organisms they contain. The important thing is that there is an appropriate balance between the damage done and the value of the information gained.

1 Suggest a reason why even dead organisms should be returned to the habitat from which they came.
2 Suggest why it is beneficial to a habitat that further investigations are not carried out too soon after an initial study.
3 In the study of a seashore, students turn over large stones to record the numbers of different organisms on their underside. Suggest reasons why it is important that these stones are replaced the same way up as they were originally.
4 In the case of experienced ecologists obtaining data that enables habitats to be conserved, the benefits usually outweigh any damage that they cause to the habitats. This makes their work ethically justifiable. It might be said that the same is not true of school or college students performing field studies. Give reasons for and against A-level students carrying out ecological investigations in the field.

Summary questions

1 An ecologist was estimating the population of sandhoppers on a beach. One hundred sandhoppers were collected, marked, and released again. A week later 80 sandhoppers were collected, of which five were marked. Calculate the estimated size of the sandhopper population on the beach. Show your working.

2 When using the mark-release-recapture technique, explain how each of the following might affect the final estimate of a population.

   a The marks put on the individuals captured in the first sample make them more easily seen by predators and so proportionately more marked individuals are eaten than unmarked individuals.

   b Between the release of marked individuals and the collection of a second sample, an increased birth rate leads to a very large increase in the population.

   c Between the release of marked individuals and the collection of a second sample, disease kills large numbers of all types of individual.

18.3 Variation in population size

Learning objectives:

- → Describe the factors that determine the size of a population.
- → Explain how abiotic factors affect the size of populations.

Specification reference: 3.3.1.2

▲ **Figure 1** *A population of lesser flamingos*

Hint

Humans exist in populations just like other species and therefore the rules also apply to us.

Hint

Algae are a group of mostly aquatic photosynthetic organisms belonging to the kingdom Protoctista. Although many are unicellular, some seaweeds can be up to 45 m long.

A population is a group of interbreeding individuals of the same species in a habitat. The number of individuals in a population is the **population size**. You saw in Topic 18.1 that all the populations of the different organisms that live and interact together are known as a community.

Population growth curves

The usual pattern of growth for a natural population has three phases (see Figure 2):

1. a period of slow growth as the initially small number of individuals reproduce to slowly build up their numbers
2. a period of rapid growth where the ever-increasing number of individuals continue to reproduce. The population size doubles during each interval of time, as seen by the gradient of the curve in Figure 2, which becomes increasingly steep.
3. a period when the population growth declines until its size remains more or less stable. The decline may be due to the food supply limiting numbers or to increased predation. The graph therefore levels out with only cyclic fluctuations due to variations in factors such as food supply or the population size of predators.

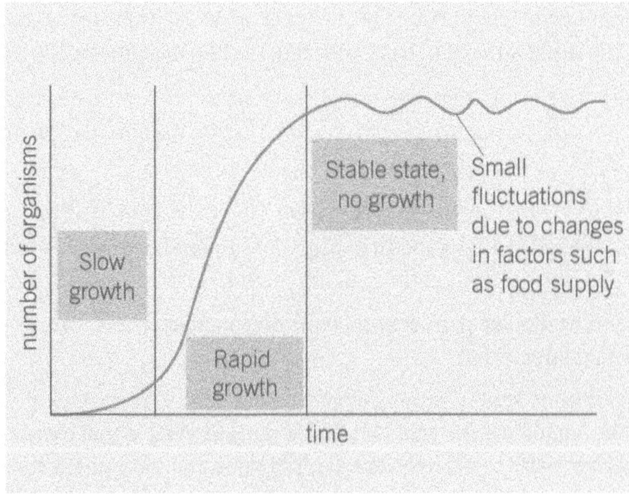

▲ **Figure 2** *Growth curve of most populations*

Population size

Imagine a situation in which a single algal cell, capable of asexual reproduction, is placed in a newly created pond. It is summer and so there is plenty of light and the temperature of the water is around 12 °C. Mineral nutrients have been added to the water. In these circumstances the algal cell divides rapidly because all the factors needed for the growth of the population are present. There are no **limiting factors**. In time, however, things change. For example:

- Mineral ions are used up as the population becomes larger.
- The population becomes so large that the algae at the surface prevent light reaching those at deeper levels.

- Other species are introduced into the pond, carried by animals or the wind, and some of these species may use the algae as food or compete for light or minerals.
- Winter brings much lower temperatures and lower light intensity of shorter duration.

In short, the good life ends and the going gets tough. As a result the growth of the population slows, and possibly ceases altogether, and the population size may even diminish. Ultimately the population is likely to reach a relatively constant size. There are many factors, living (biotic) and non-living (abiotic), which affect this ultimate size. Changes in these factors will influence the rate of growth and the final size of the population.

In summary, no population continues to grow indefinitely because certain factors limit growth, for example, the availability of food, light, water, oxygen, and shelter, and the accumulation of toxic waste, disease, and predators. Each population has a maximum size that can be sustained over a relatively long period and this is determined by these limiting factors.

The various limiting factors that affect the size of a population are of two basic types:

- **Abiotic factors** are concerned with the non-living part of the environment.
- **Biotic factors** are concerned with the activities of living organisms and include, for example, competition and predation.

▲ **Figure 3** *The collared dove only arrived in Britain in the 1950s but its population has increased rapidly since then*

▲ **Figure 4** *A population of migrating birds, like these terns, fluctuates seasonally*

Abiotic factors

The abiotic conditions that influence the size of a population include:

- **temperature**. Each species has a different optimum temperature at which it is best able to survive. The further away from this optimum, the smaller the population that can be supported. In plants and cold-blooded animals, as temperatures fall below the optimum, the enzymes work more slowly and so their metabolic rates are reduced. Populations therefore grow more slowly. At temperatures above the optimum, the enzymes work less efficiently because they gradually undergo **denaturation**. Again, the population grows more slowly.

  Warm-blooded animals, that is, birds and mammals, can maintain a relatively constant body temperature regardless of the external temperature. Therefore you might think that their population growth and size would be unaffected by temperature. However, the further the temperature of the external environment gets from their optimum temperature, the more energy these organisms expend in trying to maintain their normal body temperature. This leaves less energy for individual growth and so they mature more slowly and their reproductive rate slows. The population size therefore gets smaller.
- **light**. As the ultimate source of energy for most **ecosystems**, light is a basic necessity of life. The rate of photosynthesis increases as light intensity increases. The greater the rate of photosynthesis, the faster plants grow and the more spores or seeds they produce. Their population growth and size is therefore potentially greater. In turn, the populations that feed on plants is potentially larger.

Hint

Remember that the growth of any population is eventually slowed by a limiting factor.

Hint

A species can only live within a certain range of abiotic factors and this range differs from species to species.

Synoptic link

To remind yourself of the effects of temperature and pH on enzyme action revisit Topic 3.2.

▲ **Figure 5** *This cactus is adapted to survive in conditions where water is scarce. Its population in dry regions is therefore relatively large as there is little competition from other species, most of which are not adapted to survive in such conditions.*

- **pH**. This affects the action of enzymes. Each enzyme has an optimum pH at which it operates most effectively. A population of organisms is larger where the appropriate pH exists and smaller, or non-existent, where the pH is very different from the optimum. This can be seen, for example, with plants where species such as heather are limited to acidic soils.
- **water and humidity**. Where water is scarce, populations are small and consist only of species that are well adapted to living in dry conditions. Humidity affects the **transpiration** rates in plants and the evaporation of water from the bodies of animals. Again, in dry air conditions, the populations of species adapted to tolerate low humidity will be larger than those with no such adaptations.

The influence of abiotic factors on plant populations

Species X and species Y are two species of flowering plants. Each is able to tolerate different temperatures and different pHs. The chart (Figure 6) below illustrates the way each species is able to tolerate each of these two abiotic factors.

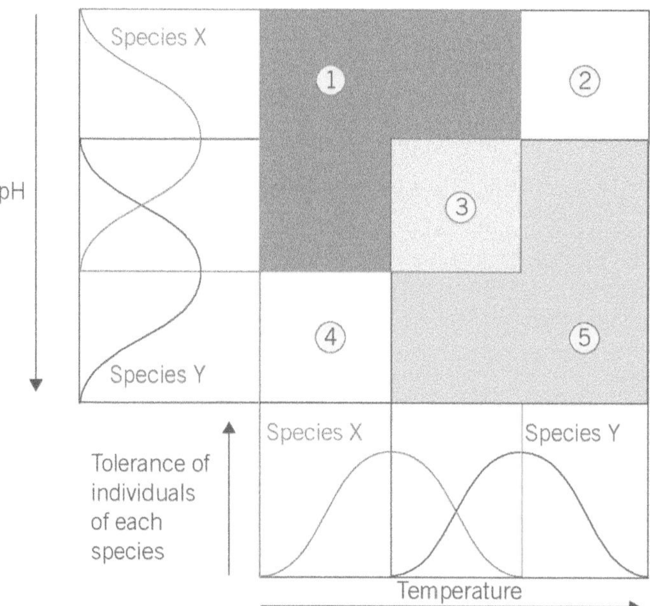

▲ **Figure 6**

1 State the numbered box that best fits each of the descriptions below.
   a Only a population of species X is found.
   b Both temperature and pH allow a population of both species to exist.
   c The temperature is too high for a population of species X and the pH is too low for a population of species Y.
   d There is competition between species X and species Y.
2 Explain why there is no population of either species in box 4.

Summary questions

1 Why do populations never grow indefinitely?
2 Distinguish between biotic and abiotic factors.
3 Suggest the level and type of abiotic factor that is most likely to limit the population size of the organisms and their habitats given below:
   a ground plants on a forest floor
   b hares in a sandy desert
   c bacteria on the summit of a high mountain.

18.4 Interactions between organisms – competition

Where two or more individuals share any resource (e.g., light, food, space, oxygen) that is insufficient to satisfy all their requirements fully, then competition results. Where such competition arises between members of the same species it is called **intraspecific competition**. Where it arises between members of different species it is termed **interspecific competition**.

Intraspecific competition

Intraspecific competition occurs when individuals of the *same* species compete with one another for resources such as food, water, breeding sites, etc. It is the availability of such resources that determines the size of a population. The greater the availability, the larger the population. The lower the availability, the smaller the population. Examples of intraspecific competition include:

- limpets competing for algae, which is their main food. The more algae available, the larger the limpet population becomes.
- oak trees competing for resources. In a large population of small oak trees some will grow larger and restrict the availability of light, water, and minerals to the rest, which then die. In time the population will be reduced to relatively few large, dominant oaks.
- robins competing for breeding territory. Female birds are normally only attracted to males who have established territories. Each territory provides adequate food for one family of birds. When food is scarce, territories become larger to provide enough food. There are therefore fewer territories in a given area and fewer breeding pairs, leading to a smaller population size.

Interspecific competition

Interspecific competition occurs when individuals of *different* species compete for resources such as food, light, water, etc. Where populations of two species initially occupy the same niche, one will normally have a competitive advantage over the other. The population of this species will gradually increase in size whilst the population of the other will diminish. If conditions remain the same, this will lead to the complete removal of one species. This is known as the competitive exclusion principle.

This principle states that where two species are competing for limited resources, the one that uses these resources most effectively will ultimately eliminate the other. In other words, no two species can occupy the same niche indefinitely when resources are limiting. Two species of sea birds, shags and cormorants, appear to occupy the same niche, living and nesting on the same type of cliff face and eating fish from the sea. Analysis of their food, however, shows that shags feed largely on sand eels and herring, whereas cormorants eat mostly flat fish, gobies, and shrimps. They therefore occupy different niches.

Learning objectives:

- → Explain what is meant by intraspecific competition.
- → Explain what is meant by interspecific competition.
- → Explain the effect of competition on population size.

Specification reference: 3.3.1.2

Hint

Which of two species in a niche has the competitive advantage depends upon the conditions at any point in time. If one species can tolerate a higher temperature than another, a rise in environmental temperature will favour it. If however there is a fall in environmental temperature, the other species is more likely to become dominant.

Summary questions

1 Distinguish between intraspecific competition and interspecific competition.

2 Name any two resources that species compete for.

To show how a factor influences the size of a population it is necessary to link it to the birth rate and death rate of individuals in a population. For example, an increase in food supply does not necessarily mean there will be more individuals – it could just result in bigger individuals. It is therefore important to show how a factor, such as a change in food supply, affects the number of individuals in a population. For example, a decrease in food supply could lead to individuals dying of starvation and directly reduce the size of a population. An increase in food supply means that more individuals are likely to survive and so there is an increased probability that they will produce offspring and the population will increase. This effect therefore takes longer to influence population size.

The effects of interspecific competition on population size

The red squirrel is native to the British Isles and exclusively occupied a particular niche until around 130 years ago, when the grey squirrel was introduced from North America. Since then the two species have been competing for food and territory. There are now an estimated 2.5 million grey squirrels and just 160 000 red squirrels in the British Isles. The red squirrel population occurs mostly in Wales and Scotland, with smaller groups in northeastern England and on islands such as Anglesey and the Isle of Wight. Figure 1 illustrates the changes in red and grey squirrel populations in Wales and Scotland between 1970 and 1990.

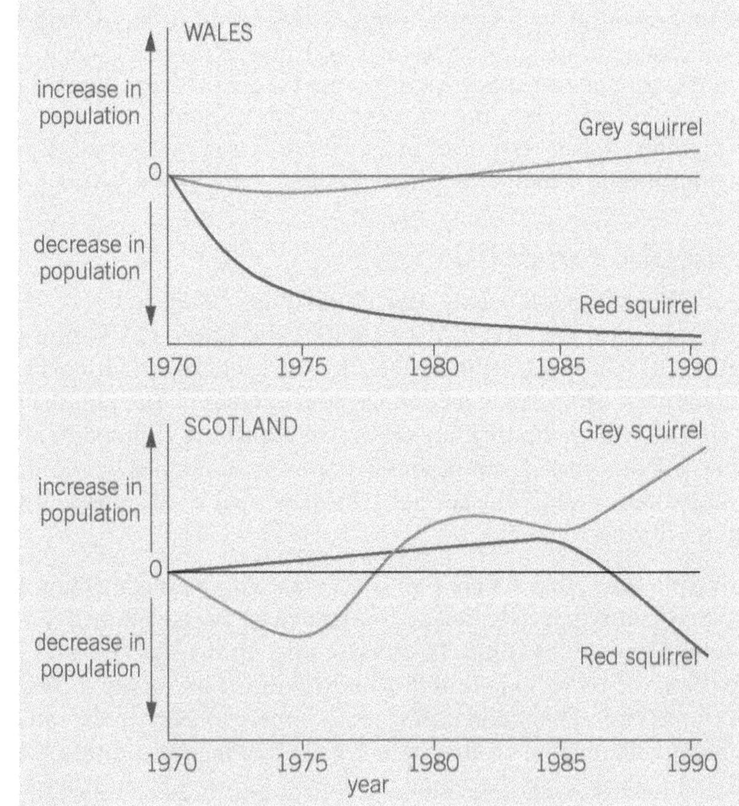

▲ **Figure 1** *Changes in red and grey squirrel populations in Wales and Scotland between 1970 and 1990. The lines show changes in comparison with the 1970 population.*

In many cases we suspect that competition is the reason for variations in population. In practice it is difficult to prove for a number of reasons:

- There are many other factors that influence population size, such as abiotic factors.
- A causal link has to be established to show that competition is the cause of an observed correlation.
- There is a time lag in many cases of competition and so a population change may be due to competition that took place many years earlier.
- Data on natural population sizes are hard to obtain and not always reliable.

Study Figure 1 and answer the following questions.

1. State one piece of evidence from the graph for Scotland which shows that changes in the red squirrel population are due to competition from the grey squirrel.
2. In Wales the populations of both grey and red squirrels declined between 1970 and 1975. Suggest a possible reason for this.
3. Both types of squirrels eat nuts, seeds, and fruit as part of their diet. Grey squirrels spend more time foraging on the forest floor than red squirrels. Suggest how this behaviour might give the grey squirrel a competitive advantage over the red squirrel.
4. Suggest an explanation why islands such as Anglesey and the Isle of Wight still have significant red squirrel populations whilst they have disappeared from much of the rest of England and Wales.

▲ **Figure 2** *Red squirrel*

▲ **Figure 3** *Grey squirrel*

Competing to the death

In an experiment, two species of a genus of unicellular organism called *Paramecium* were grown separately in different test tubes that contained yeast as a source of food. The two species were then grown together in the same test tube – again with yeast as a food source. In each case the populations of both species were measured over a period of 20 days. The results are shown in Figure 4.

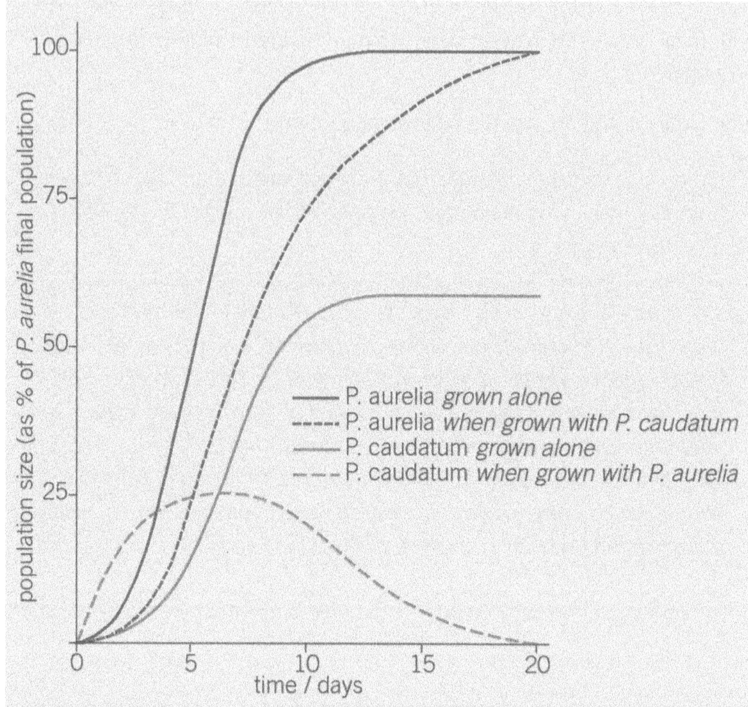

▲ **Figure 4** *Population growth of* Paramecium aurelia *and* P. caudatum *grown separately and together*

1 Describe the population growth curve of *P. caudatum* when grown alone over the 20-day period.
2 Compare the population growth curve of *P. caudatum* when grown with *P. aurelia* to the curve when *P. caudatum* is grown alone.
3 Suggest an explanation for the difference in the final population size of *P. caudatum* when grown with *P. aurelia* compared to when it is grown alone.
4 Suggest why the growth rate of *P. aurelia* is slower in the presence of *P. caudatum* than when grown alone.
5 Suggest why, after 20 days, the population size of *P. aurelia* grown with *P. caudatum* is the same as that when *P. aurelia* is grown alone.

Hint

Although the population of one species may increase as another decreases, this does not prove that this is due to direct competition between them. To be certain, it is necessary to establish a causal link for the observed correlation.

Required practical 7

Investigating competition

The relative effects of interspecific and intraspecific competition on seedling growth can be investigated using a technique called replacement.

The experiment is carried out as follows:
In a series of equally sized pots, seeds from two different species of plants, for example, marigolds (M) and zinnia (Z), are planted as follows:

Pot number	Number of marigold seeds	Number of zinnia seeds	Total number of seeds
1	10	10	20
2	10	100	110
3	100	10	110
4	100	100	200

The pots are left for several weeks and then the plants are harvested and measured.

1 What variables should be kept constant in this experiment?

2 A preliminary test on the germination rates of each species was carried out prior to the main investigation – suggest why?

After several weeks, the pots were harvested by cutting each plant at soil level and quickly weighing each plant.

3 Why was it necessary to weigh the plants quickly?

In order to analyse the relative effects of interspecific competition and intraspecific competition, a graph known as a De Wit replacement plot can be produced (de Wit, C. T. 1961. Space relationship within populations of one or more species. Society of Experimental Biology Symposium 15:314-329).

4 In this experiment, what do you understand by the terms intraspecific and interspecific competition?

Figure 5 shows such a theoretical graph obtained from the experiment for marigold plants.

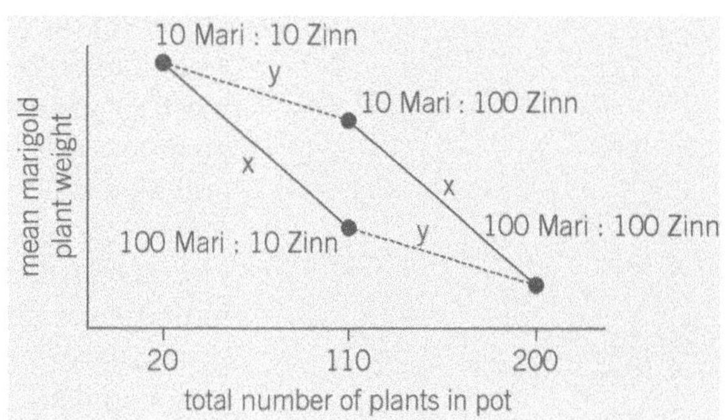

▲ **Figure 5**

Lines labelled X on the diagram represent intraspecific competition, whilst lines labelled Y represent interspecific competition.

5 Which type of competition has most effect on marigold plant weight? Explain your answer.

18.5 Interactions between organisms – predation

Learning objectives:

→ Explain how predator–prey relationships affect the population size of both predators and prey.

Specification reference: 3.3.1.2

In Topic 18.4 you looked at interspecific competition. In this topic, you will look at another type of interspecific relationship, the predator–prey relationship. A **predator** is an organism that feeds on another organism, known as their **prey**.

As predators have evolved they have become better adapted for capturing their prey – faster movement, more effective camouflage, better means of detecting prey. Prey have equally become more adept at avoiding predators – better camouflage, more protective features such as spines, concealment behaviour. In other words the predator and the prey have evolved together. If either of them had not matched the improvements of the other, it would most probably have become extinct.

Predation

Predation occurs when one organism is consumed by another. When a population of a predator and a population of its prey are brought together in a laboratory, the prey is usually exterminated by the predator. This is largely because the range and variety of the habitat provided is normally limited to the confines of the laboratory. In nature the situation is different. The area over which the population can travel is far greater and the environment is much more diverse. In particular, there are many more potential refuges. In these circumstances some of the prey can escape predation, so although the prey population falls to a low level, it rarely becomes extinct.

Evidence collected on predator and prey populations in a laboratory does not necessarily reflect what happens in the wild. At the same time, it is difficult to obtain reliable data on natural populations because it is not possible to count all the individuals in a natural population. Its size can only be estimated from sampling and surveys. These are only as good as the techniques used, none of which guarantee complete accuracy. You must therefore treat all data produced in this way with caution.

Effect of predator–prey relationship on population size

The relationship between predators and their prey and its effect on population size can be summarised as follows:

- Predators eat their prey, thereby reducing the population of prey.
- With fewer prey available, the predators are in greater competition with each other for the prey that are left.
- The predator population is reduced as some individuals are unable to obtain enough prey for their survival.
- With fewer predators left, fewer prey are eaten.

- The prey population therefore increases.
- With more prey now available as food, the predator population in turn increases.

This general predator–prey relationship is illustrated in Figure 1. In natural **ecosystems**, however, organisms eat a range of foods and therefore the fluctuations in population size shown in the graph are often less severe.

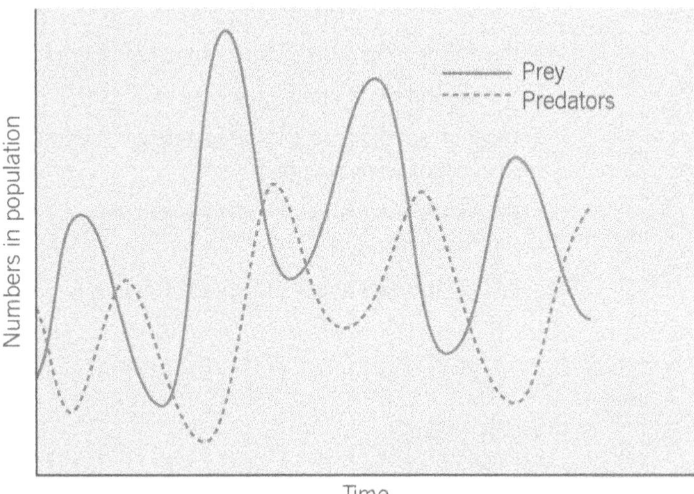

▲ **Figure 1** *Relationships between prey and predator populations*

Although predator–prey relationships are significant reasons for cyclic fluctuations in populations, they are not the only reasons – disease and climatic factors also play a part. These periodic population crashes are important in evolution as they create a **selection pressure** whereby only those individuals who are able to escape predators, or withstand disease or an adverse climate, will survive to reproduce. The population therefore evolves to be better adapted to the prevailing conditions.

Study tip

When asked to describe predator–prey relationships from a graph in examinations, candidates often just write 'one goes up and then the other one does'. To gain credit you should use names to describe precisely the changes taking place.

▲ **Figure 2** *Canadian lynx catching a snowshoe hare*

Summary questions

1. Explain why a predator population often exterminates its prey population in a laboratory but rarely does so in natural habitats.
2. Explain how a fall in the population of a predator can lead to a rise in its prey population.
3. A species of mite (A) is fed on oranges in a laboratory tank until its population is stable. A second mite species (B) that preys on species A is introduced into the tank. Sketch a graph of the likely cycle of population change that the two species will undergo. Explain the changes that the graph illustrates.

The Canadian lynx and the snowshoe hare

The long-term study of the predator–prey relationship between the Canadian lynx and the snowshoe hare was made possible because records exist of the number of furs traded by companies such as the Hudson Bay Company in Canada over 200 years. By analysing these records the relative population size of the Canadian lynx and the snowshoe hare can be determined. The data collected are shown as a graph in Figure 3.

1 What assumption is being made if we use the number of each type of fur traded as a measure of the population size of each species?
2 Describe the changes that occur in the populations of Canadian lynx and snowshoe hare.
3 Explain the changes that you have described.

It has long been observed that the population of snowshoe hares fluctuates in cycles. The question is whether these fluctuations are due mostly to predation by the lynx, mostly to changes in the food supply, or mostly to a combination of both. To find out, ecologists fenced off 1 km^2 areas of coniferous forest in Canada where the hares lived. Separate areas were treated in four different ways:

- In the first set of areas, the hares were given extra food.
- In the second set of areas, lynx were excluded.
- In the third set of areas, the hares were given extra food and lynx were excluded.
- In the fourth set of areas, conditions were left unaltered as a control.

The results of the experiment are shown in Figure 4.

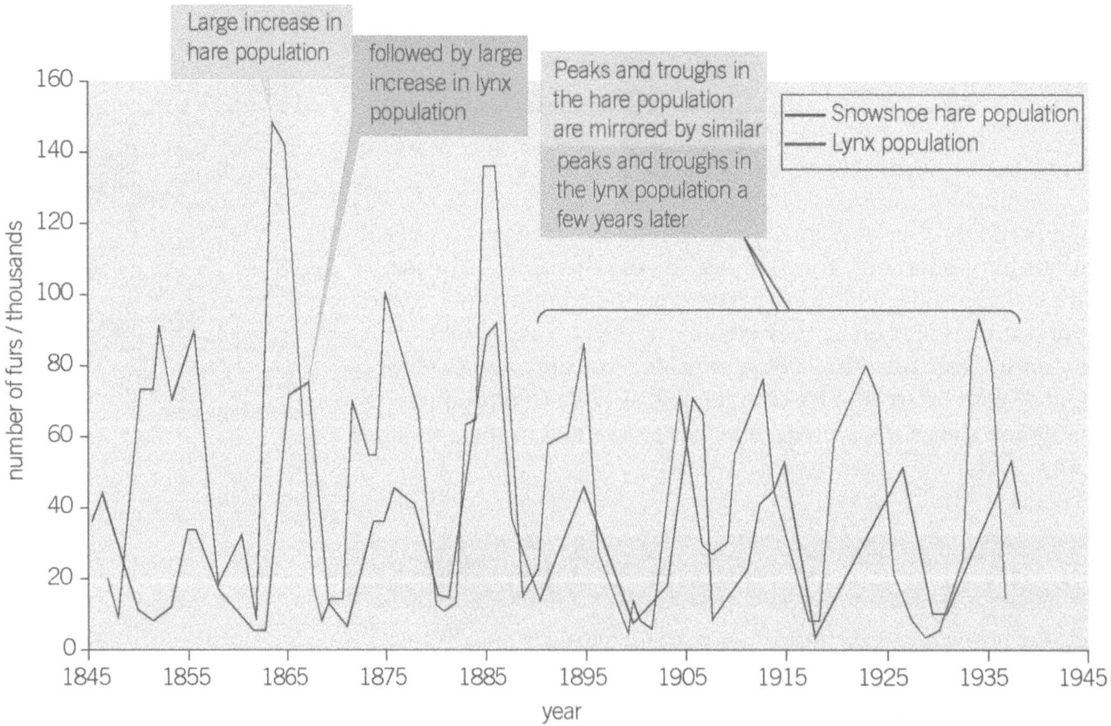

▲ **Figure 3** *The predator–prey relationship illustrated by the number of snowshoe hare and lynx trapped for the Hudson Bay Company between 1845 and 1940*

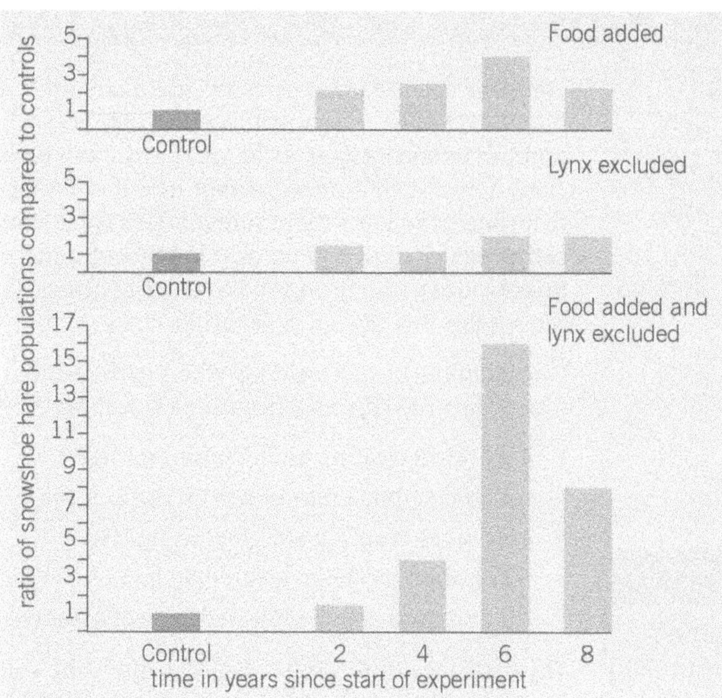

▲ **Figure 4** *Snowshoe hare population experiment*

4 By how many times had the addition of food increased the population after 6 years compared to the control?

5 Which had the greater influence on the population of hares – the addition of food or the exclusion of the lynx? Give a reason for your answer.

6 What conclusions can be drawn from this experiment?

18.6 Succession

Learning objectives:

- → Describe the changes that occur in the variety of species that occupy an area over time.
- → Explain the meaning of the terms succession and climax community.

Specification reference: 3.3.1.3

Hint

Pioneer communities put some organic material into the soil when they die. This allows recycling to start and increases mineral ions in the soil.

▲ **Figure 1** *Lichens, with their ability to withstand dry conditions and to colonise bare rock, are frequently the first pioneer species on barren terrain*

You have seen that ecosystems are made up of all the interacting biotic and abiotic factors in a particular area within which there are a number of communities of organisms. As you look around at natural ecosystems, such as moorland or forest, you may get the impression that they have been there forever. This is far from the case. Ecosystems constantly change, sometimes slowly and sometimes very rapidly. **Succession** is the term used to describe these changes, over time, in the species that occupy a particular area.

One example of succession is when bare rock or other barren land is first colonised. This may occur as a result of:

- a glacier retreating and depositing rock
- sand being piled into dunes by wind or sea
- volcanoes erupting and depositing lava
- lakes or ponds being created by land subsiding
- silt and mud being deposited at river estuaries.

The first stage of this type of succession is the colonisation of an inhospitable environment by organisms called **pioneer species**. Pioneer species often have adaptations that suit them to colonisation. These may include:

- the production of vast quantities of wind-dispersed seeds or spores, so they can easily reach isolated situations such as volcanic islands
- rapid germination of seeds on arrival as they do not require a period of dormancy
- the ability to photosynthesise, as light is normally available but other 'food' is not. They are therefore not dependent on animal species.
- the ability to fix nitrogen from the atmosphere because, even if there is soil, it has few or no nutrients
- **tolerance** to extreme conditions.

Succession takes place in a series of stages. At each stage, certain species can be identified that change the environment, especially the soil, so that it becomes more suitable for other species. These other species may then outcompete the species in the existing community and so a new community is formed.

Imagine an area of bare rock. One of the few kinds of organism capable of surviving on such an inhospitable area is lichens. Lichens are therefore pioneer species. Lichens can survive considerable drying out.

In time, weathering of the base rock produces sand or soil, although this in itself cannot support other plants. However, as the lichens die and decompose they release sufficient nutrients to support a community of small plants. In this way, the lichens change the abiotic environment by creating soil and nutrients for the organisms that follow. Mosses are typically the next stage in succession, followed by ferns. With the continuing erosion of the rock and the increasing amount of organic matter available from the death of these plants, a thicker layer of soil is built up. Again these species change the abiotic environment, making it more suitable for the organisms that follow, for example, small flowering plants such as grasses and, in turn, shrubs and trees. In the UK the ultimate community is most likely to be **deciduous** oak woodland. This stable state comprises a balanced equilibrium of species with few, if any, new species replacing those that have become established. In this state, many species flourish. This is called the **climax community**. This community consists of animals as well as plants.

The animals have undergone a similar series of successional changes, which have been largely determined by the plant types available for food and as **habitats**. The dead lichens provide food for animals such as detritus-feeding mites. The growth of mosses and grasses provides food and habitats for insects, millipedes, and worms. These are followed in turn by secondary consumers, such as centipedes, which feed on these organisms. The development of flowering plants, including trees, helps to support communities of butterflies and moths as well as larger organisms, such as reptiles, mammals, and birds.

Hint

The climax community is determined by the main abiotic factor. For example, trees may not develop on very high mountains because it is too windy or the soil layer is too thin.

Within the climax community there is normally a dominant plant species and a dominant animal species.

During any succession there are a number of common features that emerge:

- **the non-living environment becomes less hostile**, for example, soil forms, nutrients are more plentiful, and plants provide shelter from the wind. This leads to:
- **a greater number and variety of habitats** that in turn produce:
- increased **biodiversity** as different species occupy these habitats. This is especially evident in the early stages, reaching a peak in mid-succession, but decreasing as the climax community is reached. The decrease is due to dominant species outcompeting pioneer and other species, leading to their elimination from the community. With increased biodiversity comes:
- **more complex food webs**, leading to:
- **increased stability of the community**, and
- **increased biomass**, especially during mid-succession.

▲ **Figure 2** *Deciduous woodland is normally the climax community in lowland Britain*

Climax communities are in a stable equilibrium with the prevailing climate. It is abiotic factors such as climate that determine the dominant species of the community. In the lowlands of the UK, the climax community is deciduous woodland. In other climates of the world it may be tundra, steppe, or rainforest.

Another type of succession occurs when land that has already sustained life is suddenly altered. This may be the result of land clearance for agriculture or a forest fire. The process by which the ecosystem returns to its climax community is the same as described above, except that it normally occurs more rapidly. This is because spores and seeds often remain alive in the soil, and there is an influx of animals and plants through dispersal and migration from the surrounding area. This type of succession therefore does not begin with pioneer species, but with organisms from subsequent successional stages. Because the land has been altered in some way, for example, by fire, some of the species in the climax community will be different.

Figure 3 summarises the events of ecological succession on land.

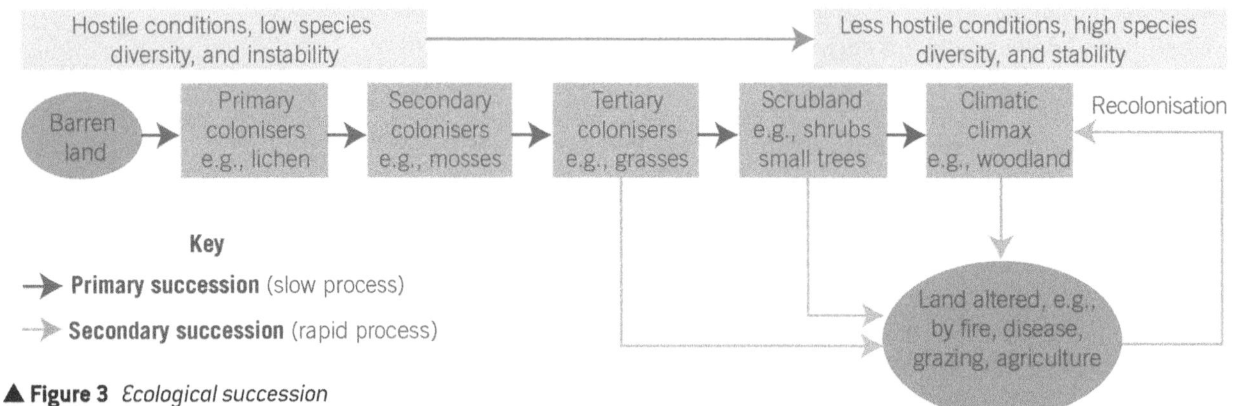

▲ **Figure 3** *Ecological succession*

▲ **Figure 4** *The grassland in the foreground is grazed by sheep and so is prevented from reaching its natural climax. The land behind the fence has not been grazed for many years and has reverted to the climax community of woodland. This is therefore an example of secondary succession.*

Warming to succession

Many glaciers in the northern hemisphere have been melting over the past 200 years. This retreat is, in part, the result of the additional global warming that has taken place since the industrial revolution and the burning of fossil fuels that has accompanied it. When glaciers melt and retreat they leave behind gravel deposits known as moraines. The retreat of the glaciers in Glacier Bay, Alaska, has been measured since 1794 and so the age of the moraines in this region is recorded.

Although no ecologist has been present to watch the succession that has taken place on these moraines, they can infer the changes that have occurred by examining the plant and animal communities on the moraines of different ages. The 'youngest' moraines (those nearest the retreating glacier) have the earliest colonisers (pioneer species), whereas those successively further away from the glacier show a time sequence of later communities.

Each stage of a succession has its own distinctive community of plants and animals that alters the environment in a way that allows the next stage and its community to develop. The stages that follow the retreat of an Arctic glacier are:

- **pioneer stage.** In the early years after the ice has retreated, photosynthetic bacteria and lichens colonise patches of land. Both of these pioneers fix nitrogen. This is essential because nitrogen is virtually absent from glacial moraines. They also form tough mats that help to stabilise the loose surface of the moraines. When these pioneer species die, they decompose to form humus. Humus provides the nutrients that enable mosses to colonise. The pioneer stage occurs when the land has been ice free for 10–20 years.
- ***Dryas* stage**. Some 30 years after the ice has retreated, the ground is an almost continuous mat of the herbaceous plant *Dryas* (see Figure 5). Its roots stabilise the thin and fragile soil layer formed from the erosion of the rocks that make up the moraine. *Dryas* also fixes nitrogen, further adding nitrogenous nutrients to this poor-quality soil. Other plants found at this stage are the Arctic poppy (Figure 6) and moss campion.
- **alder stage.** This arises about 50–70 years after the ice has retreated. Alder is a small shrub-like tree that has nitrogen-fixing nodules on its roots, enabling it to grow on nitrogen-poor soil. Alder sheds its leaves, which decompose into nitrogen-rich humus that further enriches the soil.

Summary questions

1. What name is given to the first organisms to colonise bare land?
2. Describe how changes in the environment lead to increased biodiversity during succession.
3. What name is given to the stable, final stage of any succession?

- **spruce stage.** About 100 years after the ice has retreated, spruce trees develop amongst the alder. A period of transition takes place and during the next 50 years or so the taller spruce outcompete the alder and ultimately displace it altogether (see Figure 7).

Figure 8 summarises changes in soil nitrogen, plant diversity, and biomass following the retreat of an Arctic glacier.

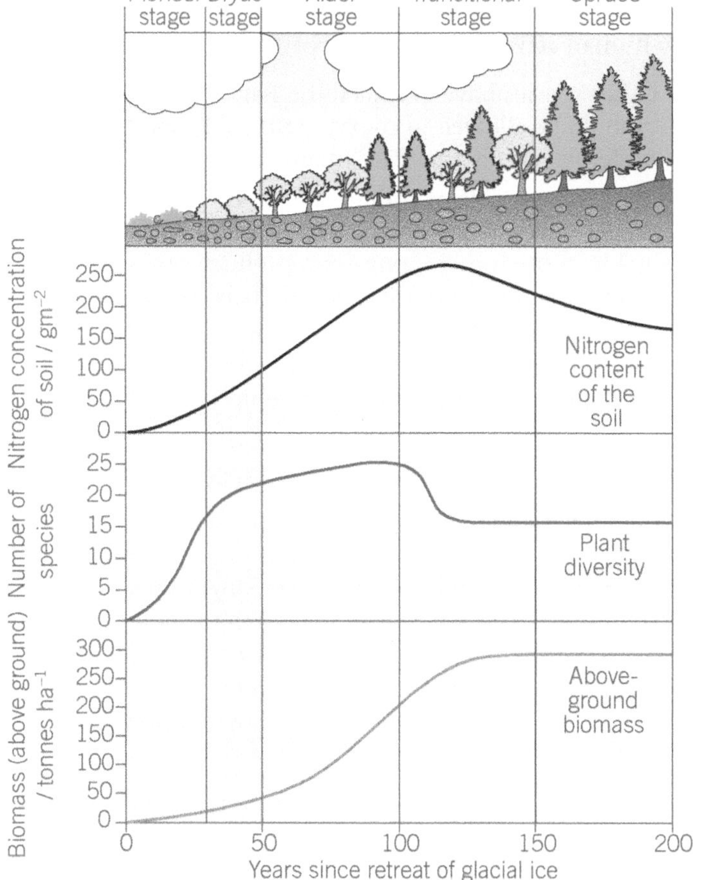

▲ **Figure 8**

1. Using the information on the graphs, describe and explain the changes in above-ground biomass over the 200-year period.
2. a. Using your knowledge of the nitrogen cycle, explain how nitrogen from the atmosphere becomes incorporated into the soil, causing its level to increase during the first 100 years after the glacier retreats. (see Topic 22.3)
   b. Suggest two reasons for the fall in soil nitrogen levels after 150 years.
3. Suggest a reason for:
   a. the rapid increase in plant species during the first 30 years after the retreat of the glacier
   b. the fall in the number of plant species 100 years after the retreat of the glacier.
4. Explain why it would be more appropriate to use a transect rather than random quadrats when investigating this succession.

▲ **Figure 5** Dryas *(mountain avens) is the most common pioneer species in Glacier Bay, Alaska. It is able to fix nitrogen and forms dense mats and therefore enriches and stabilises the thin fragile soil.*

▲ **Figure 6** *Arctic poppy (yellow flower) and moss campion (pink flowers) are early flowering pioneer species on Arctic moraines*

▲ **Figure 7** *Spruce trees are the final succession stage following the retreat of glacial ice in the Arctic. They begin to grow around 100 years after the ice has retreated and persist as the dominant vegetation for centuries.*

18.7 Conservation of habitats

Learning objectives:

- → Describe what is meant by conservation.
- → Explain how the managing succession can help to conserve habitats.

Specification reference: 3.3.1.3

▲ **Figure 1** *Moorland is an example of the conservation of a habitat by managing succession. Burning of heather and grazing by sheep has prevented shrubs and trees from developing.*

What is conservation?

Conservation is the management of the Earth's natural resources in such a way that maximum use of them can be made in the future. This involves active intervention by humans to maintain **ecosystems** and **biodiversity**. It is therefore a dynamic process that entails careful management of existing resources and reclamation of those already damaged by human activities. The main reasons for conservation are:

- **ethical**. Other species have occupied the Earth far longer than we have and should be allowed to coexist with us. Respect for living things is preferable to disregard for them.
- **economic**. Living organisms contain a gigantic pool of genes with the capacity to make millions of substances, many of which may prove valuable in the future. Long-term productivity is greater if ecosystems are maintained in their natural balanced state.
- **cultural and aesthetic**. Habitats and organisms enrich our lives. Their variety adds interest to everyday life and inspires writers, poets, artists, composers, and others who entertain and fulfill us.

Conserving habitats by managing succession

You saw in Topic 18.6 that any **climax community** has undergone a series of successional changes to reach its current state. Many of the species that existed in the earlier stages are no longer present as part of the climax community. This is because their habitats have disappeared as a result of succession, or they have been outcompeted by other species. One way of conserving these habitats, and hence the species they contain, is by managing succession in a way that prevents a change to the next stage.

One example is the moorland that exists over much of the higher ground in the UK. The burning of heather and grazing by sheep has prevented this land from reaching its climax community. The burning and grazing destroy the young tree saplings and so prevent the natural succession into deciduous woodland.

Around 4000 years ago, much of lowland UK was a climax community of oak woodland, but most of this forest was cleared to allow grazing and cultivation. The many heaths and grasslands that we now refer to as 'natural' are the result of this clearance and subsequent grazing by animals.

If the factor that is preventing further succession is removed, then the ecosystem develops naturally into its climatic climax (secondary succession). For example, if grasslands are no longer grazed or mowed, or if farmland is abandoned, shrubs initially take over, followed by deciduous woodland.

Summary questions

Fenland is an area of waterlogged marsh and peat land. It supports a rich and unique community of plants and animals. If left alone, reeds initially dominate and the area gradually dries out as dead vegetation accumulates. Grasses, shrubs, and trees in turn replace the fenland species.

1. Give reasons for conserving habitats such as fenland.
2. Suggest practical measures that may be taken to prevent succession by grasses, shrubs, and trees in fenland.

Conflicting interests

One challenging conservation issue in the UK is the conflict between the conservation of hen harriers and the commercial hunting of red grouse.

One scientific survey investigated the effect of predation by hen harriers on the breeding success of red grouse on managed moorland in Scotland. Some of the results included:

- On moorland where hen harriers were present there were, on average, 17 per cent fewer young grouse than on moorlands without hen harriers.
- Over a 3-year period grouse nests were intensively observed during the 6 weeks following the hatching of chicks. In this period, predation by harriers accounted for 91 per cent of grouse chick losses.
- Prey remains found around harrier nests were examined. Of the 300 items identified, 32 per cent were grouse chicks.

1 How many of the items of prey identified around harrier nests were grouse chicks?
2 Harriers also feed on voles and meadow pipits. Explain how a rise in the population of these organisms might affect the population of grouse.

Moorland is considered one of the most attractive landscapes in the UK. Many of the national parks are made up of moorland and are visited by millions of people each year. To rear grouse, moorland has to be carefully managed. Controlled grazing by sheep and the periodic burning of vegetation are used to maintain the low-growing plant populations of heather, bilberry, and crowberry that grouse feed on and nest within. The money to support this management comes largely from charges made to those who shoot grouse.

3 Explain what might happen to moorland if sheep-grazing and burning of the vegetation ceased.

The population of grouse in the UK is in decline due mainly to disease. Currently there are around 250 000 breeding pairs. The hen harrier was persecuted to such an extent that, by 1900, it was only found on a few Scottish islands. It recolonised the UK mainland in the 1970s and there are now around 750 breeding pairs. Both harriers and grouse normally produce one clutch of eggs each year. Hen harriers are protected by law and it is illegal to kill them, collect their eggs, or destroy their nests. Conservationists want to retain this protection so that the population of hen harriers can increase. Grouse managers want to be allowed to control hen harrier populations to prevent them threatening the declining grouse populations.

4 Outline the arguments for and against continued protection of hen harrier populations.

Practice questions: Chapter 18

1 An ecosystem can be described as a dynamic system involving the interaction of biotic and abiotic components. Within an ecosystem, populations of organisms occupy ecological niches.

(a) Explain what is meant by the following terms:

(i) population *(1 mark)*

(ii) ecological niche. *(2 marks)*

(b) Figure 1 shows the ranges of mean annual temperatures and precipitation (water falling as rain or snow) for six types of ecosystem.

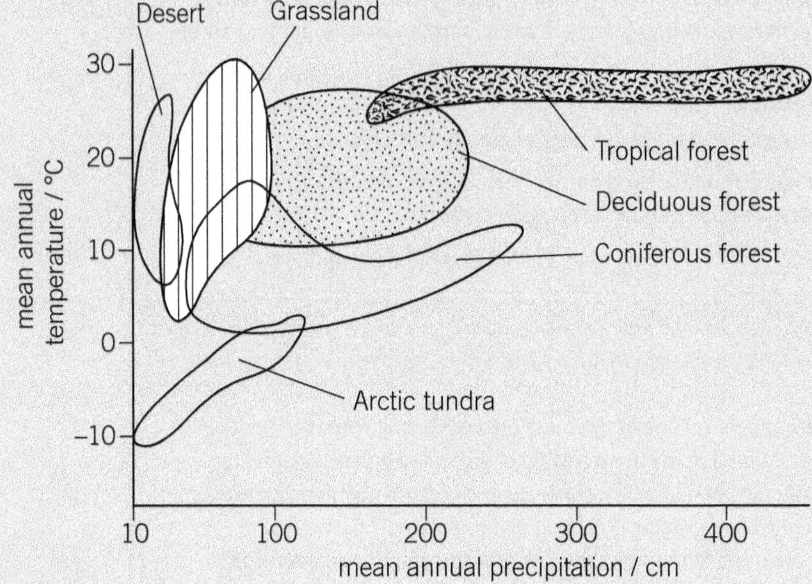

▲ **Figure 1**

Arctic tundra is considered to be an extreme environment whereas tropical forest is physically less hostile to living organisms.

(i) Explain how the information in the diagram supports this view. *(1 mark)*

(ii) Describe and explain the difference in the effect of abiotic factors on the diversity of organisms in the tundra and in tropical forest. *(2 marks)*

AQA 1998

2 The mark-release-recapture technique may be used to estimate population size.

(a) Give **two** assumptions that must be made when using this technique. *(2 marks)*

(b) In estimating the size of a ladybird population, 70 ladybirds were trapped, marked, and released. A week later, a second sample was captured. In this second sample, 27 were marked and 13 were not marked.

(i) Calculate the estimated size of the ladybird population. Show your working.

(ii) Explain why it is important that the samples contain as many ladybirds as possible. *(3 marks)*

AQA 2003

3 Scientists used a line transect to find the distribution of three species of *Ranunculus* (buttercup) in a field. The field consists of a series of ridges and furrows. Figure 2 shows the distribution of the species of *Ranunculus* along the line transect.

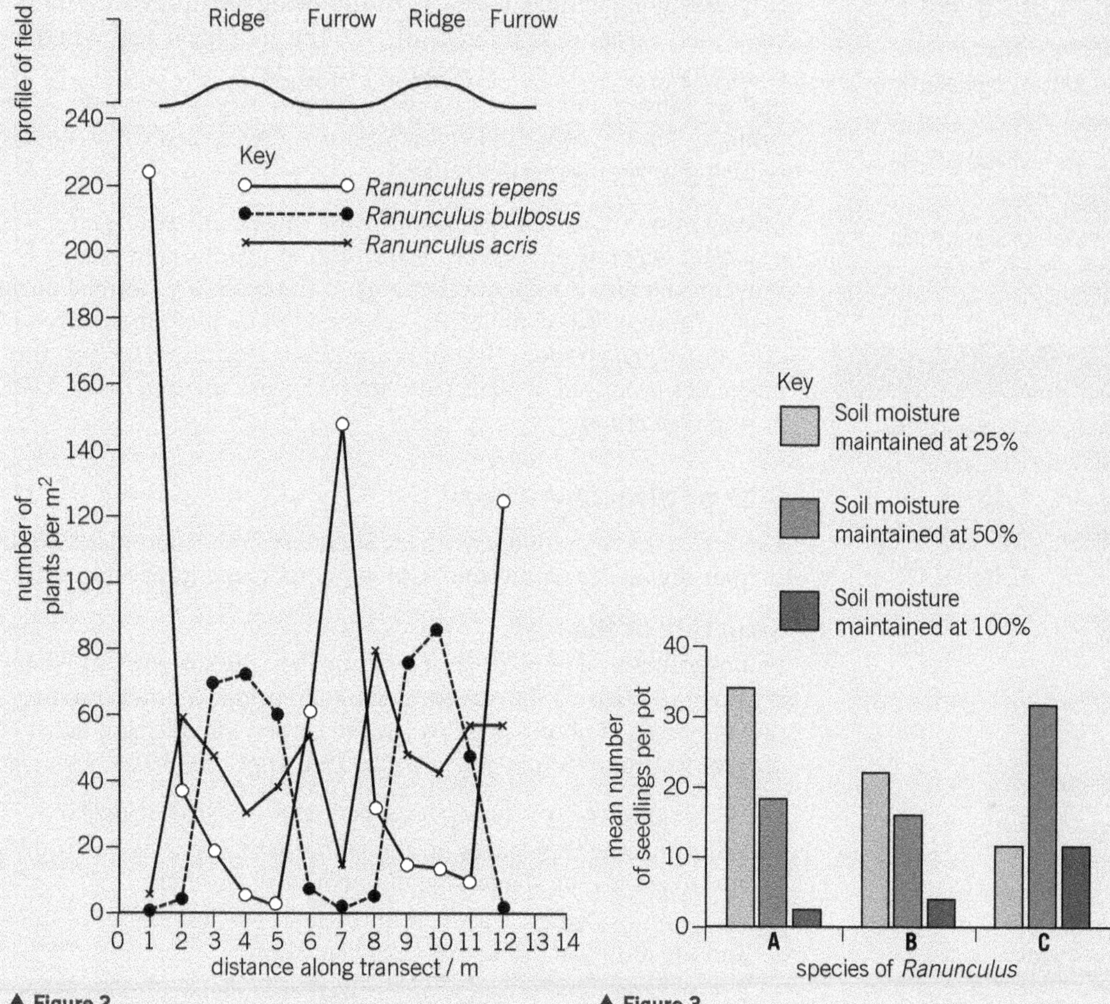

▲ **Figure 2**

▲ **Figure 3**

(a) Describe how you would use a transect to obtain data on the distribution of the *Ranunculus* species as shown in Figure 2. *(2 marks)*

(b) Other than soil moisture, give **one** abiotic factor and explain how it could lead to the abundance of *Ranunculus repens* in the furrows of this field. *(2 marks)*

(c) The scientists then investigated the effect of soil moisture on seed germination of the three species of *Ranunculus*. They planted seeds of species **A** in three sets of pots. The soil in one set of pots was maintained at 25% water content, the soil in the second set was maintained at 50% water content, and the soil in the third set was maintained at 100% water content. They repeated this with seeds of species **B** and species **C**. Four weeks later, the scientists recorded the mean number of seedlings in the pots in each set. Their results are shown in Figure 3.

(i) Suggest **two** factors which should be controlled during this investigation.

(ii) Use information from Figure 2 and Figure 3 to identify each species of *Ranunculus* **A–C**.

(iii) Describe how you could determine whether two *Ranunculus* plants belong to the same species. *(5 marks)*

AQA 2007

Answers to the Practice Questions are available at
www.oxfordsecondary.com/oxfordaqaexams-alevel-biology

19 Photosynthesis

19.1 Overview of photosynthesis

Learning objectives:

- → Describe how plant leaves are adapted to carry out photosynthesis.
- → Outline the main stages of photosynthesis.

Specification reference: 3.3.2.1 and 3.3.2.2

Synoptic link

The structure of a leaf and of the chloroplast were considered in Topics 5.3 and 2.3. Remember plants are eukaryotic organisms.

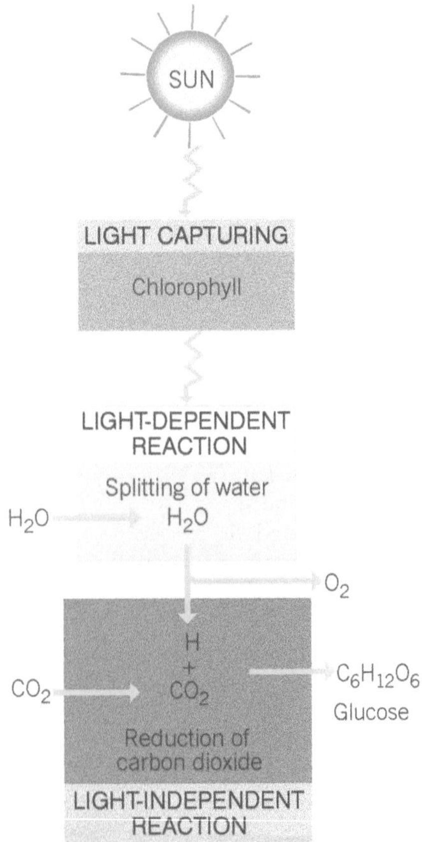

Humans, along with almost every other living organism, owe their continued existence to photosynthesis. The energy we use, whether it comes from food when we respire or from the wood, coal, oil, or gas that we burn in our homes, has been captured by photosynthesis from sunlight. Photosynthesis likewise produces the oxygen we breathe by releasing it from water molecules.

Energy flows though all organisms. How this energy enters an organism depends on its type of nutrition. In plants, light energy is transformed into the chemical energy of the molecules formed during photosynthesis. These molecules are used by the plant to produce ATP during respiration. Non-photosynthetic organisms feed on the molecules produced by plants and then also use them to make ATP during respiration.

Site of photosynthesis

The leaf is the main photosynthetic structure. The chloroplasts are the cellular organelles within the leaf where photosynthesis takes place.

Structure of the leaf

Photosynthesis takes place largely in the leaf, the structure of which is shown in Figure 2. Leaves are adapted to bring together the three raw materials of photosynthesis (water, carbon dioxide, and light) and remove its products (oxygen and glucose). These adaptations include:

- a large surface area that collects as much sunlight as possible
- an arrangement of leaves on the plant that minimises overlapping and so avoids the shadowing of one leaf by another
- thin, as most light is absorbed in the first few millimetres of the leaf and the diffusion distance is thus kept short
- a transparent cuticle and epidermis that let light through to the photosynthetic mesophyll cells beneath
- long, narrow upper mesophyll cells packed with chloroplasts that collect sunlight
- numerous stomata for gaseous exchange
- stomata that open and close in response to changes in light intensity
- many air spaces in the lower mesophyll layer to allow diffusion of carbon dioxide and oxygen
- a network of xylem that brings water to the leaf cells, and phloem that carries away the sugars produced in photosynthesis.

▲ **Figure 1** *Overview of photosynthesis*

An outline of photosynthesis

The overall equation for photosynthesis is:

$$\underset{\text{carbon dioxide}}{6CO_2} + \underset{\text{water}}{6H_2O} \xrightarrow{\text{light}} \underset{\text{glucose}}{C_6H_{12}O_6} + \underset{\text{oxygen}}{6O_2}$$

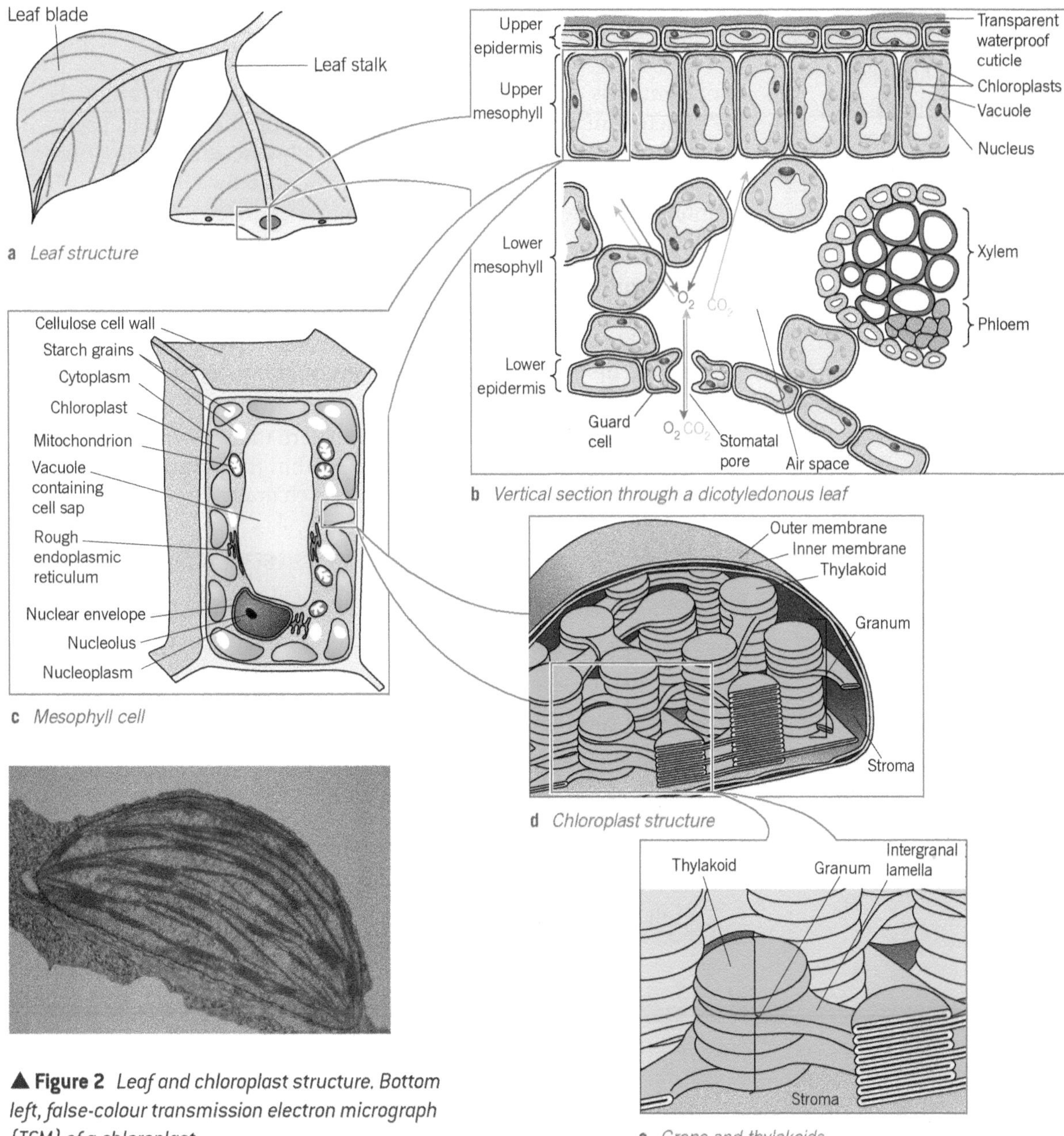

▲ **Figure 2** *Leaf and chloroplast structure. Bottom left, false-colour transmission electron micrograph (TEM) of a chloroplast.*

The equation shown is highly simplified. Photosynthesis is a complex metabolic pathway involving many intermediate reactions. It is a process of energy transformation in which light energy is first changed into electrical energy and then into chemical energy. There are three main stages to photosynthesis (see Figure 1 on the previous page):

1. **capturing of light energy** by chloroplast pigments such as chlorophyll
2. **the light-dependent reaction**, in which light energy is converted into chemical energy. During the process an electron flow is created by the effect of light on chlorophyll and this causes water to split (**photolysis**) into protons, electrons, and oxygen. The products are reduced NADP, ATP, and oxygen.
3. **the light-independent reaction**, in which these protons (hydrogen ions) are used to reduce carbon dioxide to produce sugars and other organic molecules.

Study tip

Remember that all plant cells respire all the time, whereas only those plant cells with chloroplasts carry out photosynthesis – and then only in the light.

▲ **Figure 3** *Photomicrograph of a moss leaf showing cells that contain chloroplasts (green) around their margins*

Structure and role of chloroplasts in photosynthesis

Photosynthesis takes place within cell organelles called chloroplasts, the structure of which is shown in Figure 2d on the previous page. These vary in shape and size but are typically disc-shaped, 2–10 μm long and 1 μm in diameter. They are surrounded by a double membrane. Inside the chloroplast membranes are two distinct regions:

- **The grana** are stacks of up to 100 disc-like structures called **thylakoids** where the light-dependent stage of photosynthesis takes place. Within the thylakoids is the photosynthetic pigment called chlorophyll. Some thylakoids have tubular extensions that join up with thylakoids in adjacent grana. These are called intergranal lamellae.
- **The stroma** is a fluid-filled matrix where the light-independent stage of photosynthesis takes place. Within the stroma are a number of other structures such as starch grains.

Summary questions

1 Which **two** molecules are the raw materials of photosynthesis?

2 Which **two** molecules are the products of photosynthesis?

3 In which parts of the chloroplast does each of the following occur?

 a the light-dependent reaction

 b the light-independent reaction.

4 What are the products of each of the following?

 a the light-dependent reaction

 b the light-independent reaction.

19.2 The light-dependent reaction

The light-dependent reaction of photosynthesis involves the capture of light whose energy is used for two purposes:

- to add an inorganic phosphate (Pi) molecule to ADP, thereby making ATP (see Topic 4.4)
- to split water into H^+ ions (protons) and OH^- ions. As the splitting is caused by light, it is known as **photolysis**.

Learning objectives:

- → Explain what is meant by oxidation and reduction.
- → Explain how ATP is synthesised in the light-dependent reaction.
- → Explain what is meant by photolysis.
- → Describe how chloroplasts are adapted to carry out the light-dependent reaction.

Specification reference: 3.3.2.1

Oxidation and reduction

Before you look at what happens in the light-dependent reaction, it is necessary to understand what oxidation and reduction are.

When a substance combines with oxygen the process is called **oxidation**. The substance to which oxygen has been added is said to be oxidised. When one substance gains oxygen from another, the one losing the oxygen is said to be reduced, the process being known as **reduction**. In practice, when a substance is oxidised it loses electrons and when it is reduced it gains electrons. This is the more usual way to define oxidation and reduction. Oxidation results in energy being given out, whereas reduction results in it being taken in. Oxidation and reduction always take place together.

Hint

Oxidation and reduction can each be described in three ways:

Oxidation – loss of electrons or loss of hydrogen or gain of oxygen.

Reduction – gain of electrons or gain of hydrogen or loss of oxygen.

The making of ATP

When a chlorophyll molecule absorbs light.energy, it boosts the energy of a pair of electrons within this chlorophyll molecule, raising them to a higher energy level. These electrons are said to be in an excited state. In fact the electrons become so energetic that they leave the chlorophyll molecule altogether. The electrons that leave the chlorophyll are taken up by a molecule called an **electron carrier**. Having lost a pair of electrons, the chlorophyll molecule has been oxidised. The electron carrier, which has gained electrons, has been reduced.

Study tip

Make sure you know the three ways in which something can be oxidised or reduced.

The electrons are now passed along a number of electron carriers in a series of oxidation-reduction reactions. These electron carriers form a transfer chain that is located in the membranes of the thylakoids. Each new carrier is at a slightly lower energy level than the previous one in the chain, and so the electrons lose energy at each stage. This energy is used to combine an inorganic phosphate molecule with an ADP molecule in order to make ATP. This process is summarised in Figure 1.

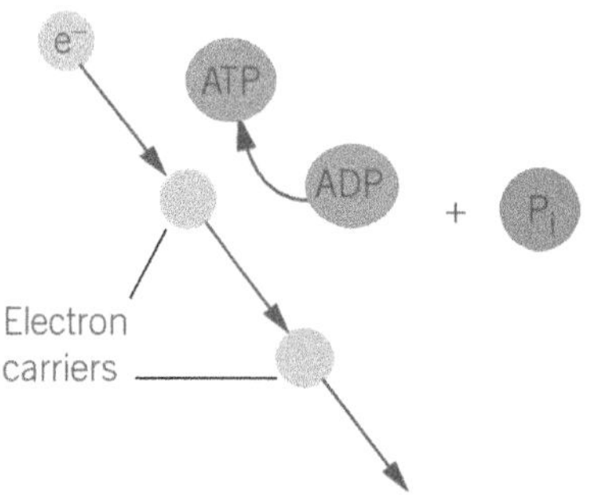

▲ **Figure 1** *Electron transfer chain and the making of ATP*

Photolysis of water

The loss of electrons when light strikes a chlorophyll molecule leaves it short of electrons. If the chlorophyll molecule is to continue absorbing light energy, these electrons must be replaced. The replacement electrons are provided from water molecules that are split using light energy. This photolysis of water also yields hydrogen ions (protons). The equation for this process is:

$$\underset{\text{water}}{2H_2O} \longrightarrow \underset{\text{protons}}{4H^+} + \underset{\text{electrons}}{4e^-} + \underset{\text{oxygen}}{O_2}$$

Hint

Reduced NADP is the most important product of the light-dependent reaction.

Hint

To picture how thylakoids are arranged in the grana, think of a thylakoid as a coin and the grana as a stack of many such coins, one on top of the other.

These hydrogen ions (protons) are taken up by an electron carrier called NADP. On taking up the hydrogen ions (protons) the NADP becomes reduced. The reduced NADP then enters the light-independent reaction (see Topic 19.3) along with the electrons from the chlorophyll molecules. The reduced NADP is important because it is a further potential source of chemical energy to the plant. The oxygen by-product from the photolysis of water is either used in respiration or diffuses out of the leaf as a waste product of photosynthesis.

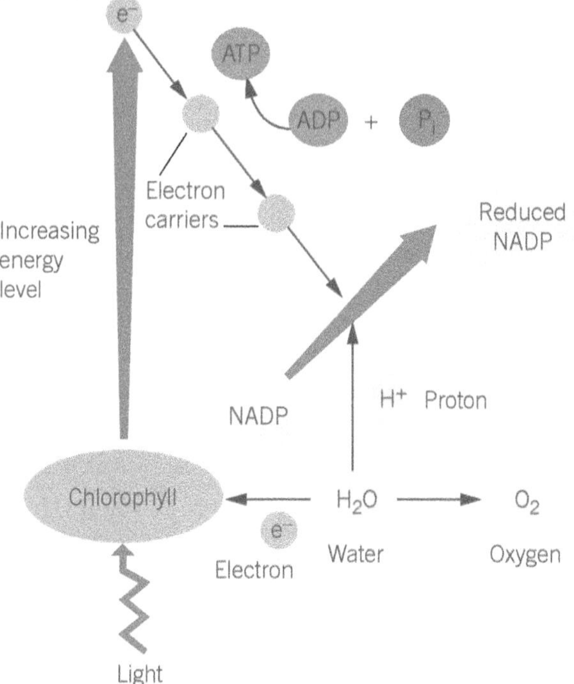

▲ **Figure 2** *Summary of the light-dependent reaction of photosynthesis*

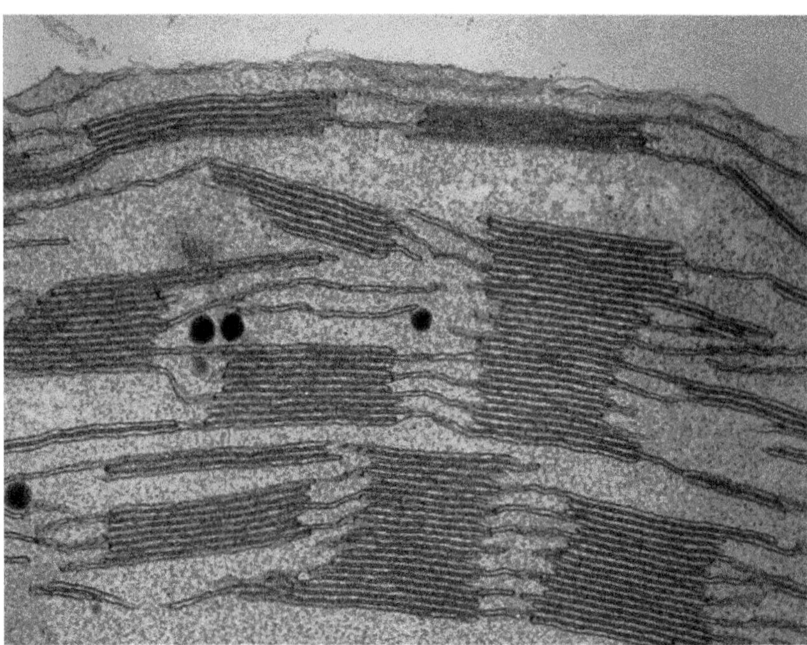

▲ **Figure 3** *False-colour transition electron micrograph (TEM) of grana in a chloroplast from a leaf of maize. The grana are made up of disc-like thylakoids, where the light-dependent reaction of photosynthesis takes place.*

Site of the light-dependent reaction

The light-dependent reaction of photosynthesis (see Figure 2) takes place in the thylakoids of chloroplasts. The thylakoids are disc-like structures that are stacked together in groups called grana.

Chloroplasts are structurally adapted to their function of capturing sunlight and carrying out the light-dependent reaction of photosynthesis in the following ways:

- The thylakoid membranes provide a large surface area for the attachment of chlorophyll, electron carriers, and enzymes that carry out the light-dependent reaction.
- A network of proteins in the grana hold the chlorophyll in a very precise manner that allows maximum absorption of light.
- The granal membranes have enzymes attached to them, which help to manufacture ATP.
- Chloroplasts contain both DNA and ribosomes so that they can quickly and easily manufacture some of the proteins needed for the light-dependent reaction.

Chloroplasts and the light-dependent reaction

Figure 4 shows the structure of a chloroplast.

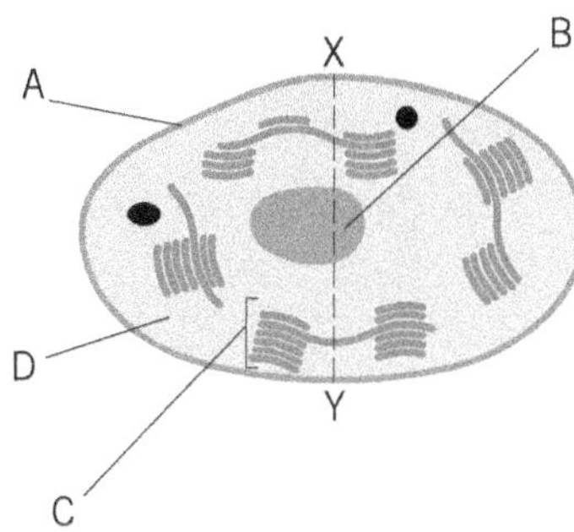

▲ **Figure 4**

1 Name the parts labelled **A**, **C**, and **D**.
2 In which of these labelled parts does the light-dependent reaction take place?
3 Structure **B** is used for storage. Suggest the name of the substance likely to be stored in **B**.
4 ATP is produced in the light-dependent reaction of photosynthesis. Suggest **two** reasons why plants cannot use this as their only source of ATP.
5 The actual length of X–Y in this chloroplast is 2 μm. What is the magnification used in Figure 4? Show your working.

Summary questions

1 Where precisely within a plant cell are the electron carriers involved in the light-dependent reaction found?

2 Describe what happens in the photolysis of water.

3 In each of the following, state whether the process involves oxidation or reduction of the molecule named.

  a An unsaturated fat molecule gains a hydrogen atom.

  b Oxygen is lost from a carbon dioxide molecule.

  c Light causes an electron to leave a chlorophyll molecule.

19.3 The light-independent reaction

Learning objectives:

- → Describe how carbon dioxide is absorbed by plants and incorporated into organic molecules.
- → Explain the roles of ATP and reduced NADP in the light-independent reaction.
- → Outline the events in the Calvin cycle.

Specification reference: 3.3.2.2

The products of the light-dependent reaction of photosynthesis, namely ATP and reduced NADP, are used to reduce carbon dioxide in the second stage of photosynthesis. Unlike the first stage, this stage does not require light directly and, in theory, occurs whether or not light is available. It is therefore called the light-independent reaction. In practice, it requires the products of the light-dependent stage and so rapidly ceases when light is absent. The light-independent reaction takes place in the stroma of the chloroplasts. The details of this stage were worked out by Melvin Calvin and his co-workers and so it is often referred to as the Calvin cycle.

The Calvin cycle

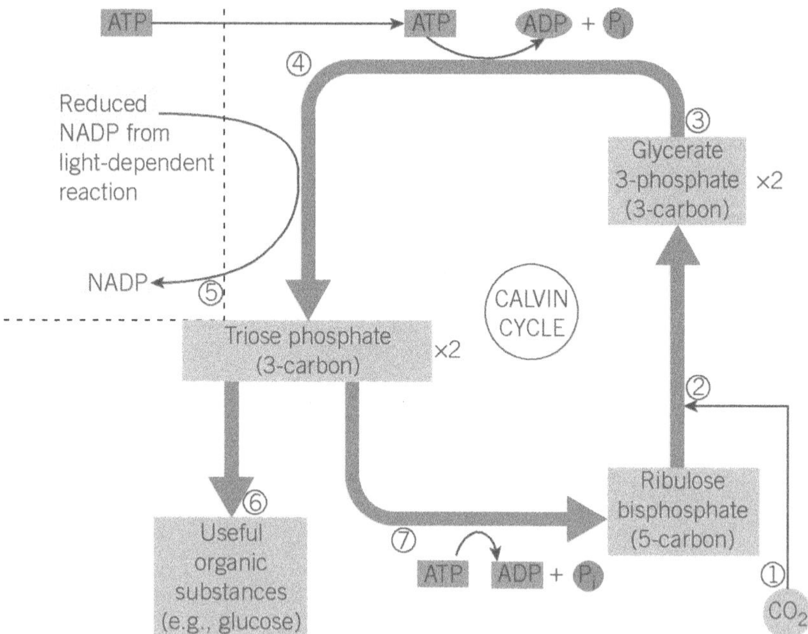

▲ **Figure 1** *Summary of the light-independent reaction of photosynthesis (or Calvin cycle)*

In the following account of the Calvin cycle, the numbered stages are illustrated in Figure 1.

1. Carbon dioxide from the atmosphere diffuses into the leaf through stomata and dissolves in water around the walls of the mesophyll cells. It then diffuses through the plasma membrane, cytoplasm, and chloroplast membranes into the stroma of the chloroplast.
2. In the stroma, the carbon dioxide combines with the 5-carbon compound **ribulose bisphosphate (RuBP)** using an enzyme.
3. The combination of carbon dioxide and RuBP produces two molecules of the 3-carbon **glycerate 3-phosphate (GP)**.
4. ATP and reduced NADP from the light-dependent reaction are used to reduce the activated GP to **triose phosphate (TP)**.
5. The NADP is re-formed and goes back to the light-dependent reaction to be reduced again by accepting more hydrogen.

6 Some TP molecules are converted to useful organic substances, such as glucose.

7 Most TP molecules are used to regenerate RuBP using ATP from the light-dependent reaction.

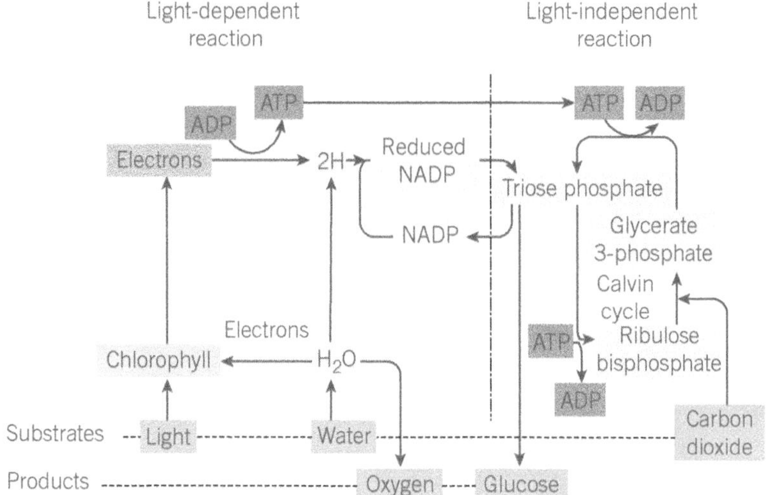

▲ **Figure 2** *Summary of photosynthesis*

Hint

Any substance whose name ends in 'ose' is a sugar. The ending 'ate' usually means that the substance is an acid (in solution).

Summary questions

1 What is the role of ribulose bisphosphate (RuBP) in the Calvin cycle?

2 How is the reduced NADP from the light-dependent reaction used in the light-independent reaction?

3 Apart from reduced NADP, which other product of the light-dependent reaction is used in the light-independent reaction?

4 Where precisely in a plant cell are the enzymes involved in the Calvin cycle found?

5 Light is not required for the Calvin cycle to take place. Explain therefore why the Calvin cycle cannot take place for long in the absence of light.

Site of the light-independent reaction

The light-independent reaction of photosynthesis takes place in the stroma of the chloroplasts.

The chloroplast is adapted to carrying out the light-independent reaction of photosynthesis in the following ways:

- The fluid of the stroma contains all the enzymes needed to carry out the light-independent reaction (reduction of carbon dioxide).
- The stroma fluid surrounds the grana and so the products of the light-dependent reaction in the grana can readily diffuse into the stroma.
- It contains both DNA and ribosomes so it can quickly and easily manufacture some of the proteins needed for the light-independent reaction.

Using a lollipop to work out the light-independent reaction

The details of the light-independent reaction were worked out by Melvin Calvin and his co-workers using his 'lollipop' experiment. It was so called because the apparatus, shown in Figure 3, resembled a lollipop.

In the experiment, single-celled algae are grown under light in a thin transparent 'lollipop'. Radioactive hydrogencarbonate is injected into the 'lollipop'. This supplies radioactive carbon dioxide to the algae. At 5-second intervals, samples of the photosynthesising algae are dropped into hot methanol to stop chemical reactions instantly. The compounds in the algae are then separated out and those that are radioactive are identified. The results are given in Table 1.

▼ **Table 1**

Time / s	Substances found to be radioactive
0	Carbon dioxide
5	Glycerate 3-phosphate
10	Glycerate 3-phosphate + triose phosphate
15	Glycerate 3-phosphate + triose phosphate + glucose
20	Glycerate 3-phosphate + triose phosphate + glucose + ribulose bisphosphate

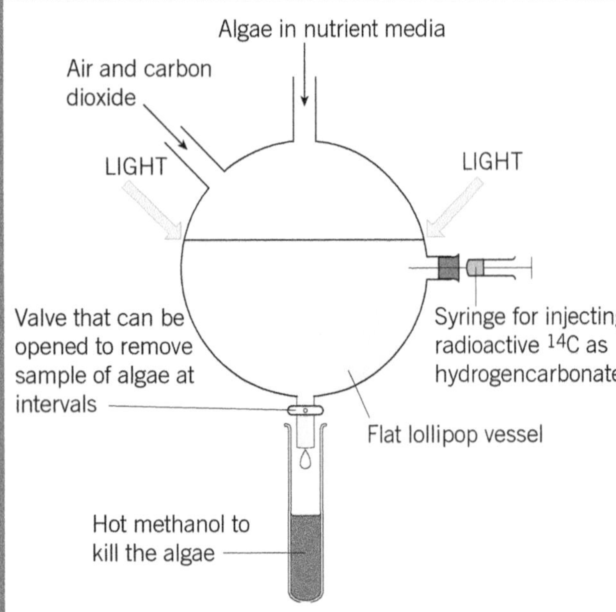

▲ **Figure 3** *The 'lollipop' apparatus used by Melvin Calvin*

- Algae are grown under light in the thin transparent lollipop.
- Radioactive ^{14}C in the form of hydrogencarbonate is injected.
- At intervals (seconds to minutes) samples of the photosynthesising algae are dropped into the hot methanol to stop chemical reactions instantly.
- The compounds in the algae are separated by two-way chromatography.
- The radioactive compounds are identified and the pathway determined by the time at which each first appeared. The first to appear is the first in the pathway, etc.

In a further experiment, samples of algae were collected at 1-minute intervals over a period of 5 minutes. The quantities of glycerate 3-phosphate (GP) and ribulose bisphosphate (RuBP) were measured. At the beginning of the experiment, the concentration of carbon dioxide supplied was high. After 2 minutes the concentration of carbon dioxide was reduced. The graph in Figure 4 shows the results of this experiment.

1. Suggest a reason why the carbon dioxide supplied to the algae was radioactively labelled.
2. Explain how information in Table 1 provides evidence that GP is converted into TP.
3. Suggest an explanation for how the hot methanol might stop further chemical reactions taking place.
4. Referring to Figure 4, describe the effects on the quantities of GP and RuBP of the decrease in carbon dioxide after 2 minutes.
5. Suggest explanations for these changes to the levels of GP and RuBP.

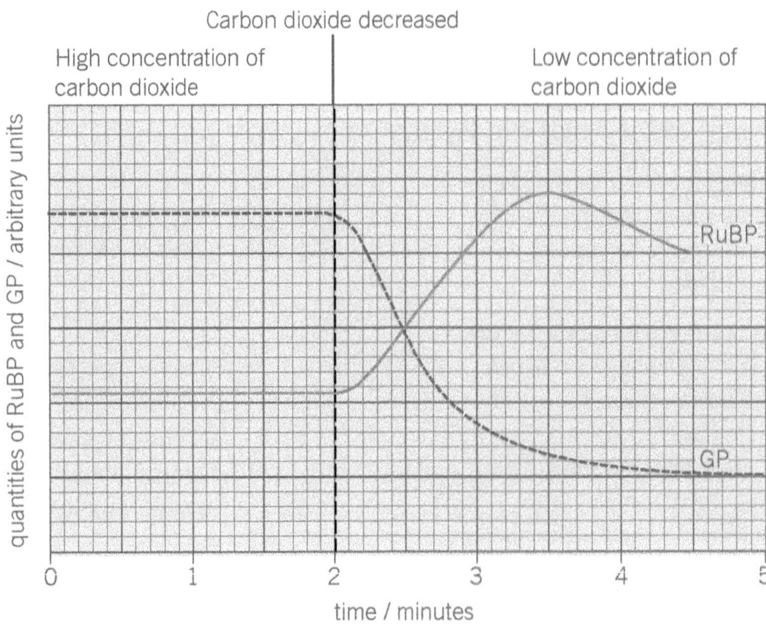

◀ **Figure 4**

19.4 Factors affecting photosynthesis

You have seen that photosynthesis is the process that captures light energy and thereby makes life on Earth possible. It follows that an understanding of those factors which influence the rate of photosynthesis can help to ensure that an adequate supply of energy (food) is available, not only to ourselves, but also to those other organisms with which we share this planet. Before we consider some of these factors, it is necessary to understand the concept of limiting factors.

Learning objectives:

- → Explain what is meant by limiting factors in the context of photosynthesis.
- → Describe how photosynthesis can be measured experimentally.
- → Describe the effects of temperature, carbon dioxide concentration, and light intensity on the rate of photosynthesis.

Specification reference: 3.3.2.3

Limiting factors

In any complex process such as photosynthesis, the factors that affect its rate all operate together. However, the rate of the process at any given moment is not affected by all the factors, but rather by the one whose level is at the least favourable value. This factor is called a **limiting factor** because it limits the rate at which the process can take place. Changing the levels of the other factors will not alter the rate of the process.

To take the example of light intensity limiting the rate of photosynthesis:

- In complete darkness, it is the absence of light alone that prevents photosynthesis occurring. No matter how much we raise or lower the temperature or change the concentration of carbon dioxide, there will be no photosynthesis. Light, or rather the absence of it, is the factor determining the rate of photosynthesis at that moment.
- If we provide light, however, the rate of photosynthesis will increase.
- As we add more light, the rate increases further. This does not continue indefinitely, however, because there comes a point at which further increases in light intensity have no effect on the rate of photosynthesis.
- At this point some other factor, such as the concentration of carbon dioxide, is in short supply and so limits the process. Carbon dioxide is now the limiting factor and only an increase in its level will increase the rate of photosynthesis.
- As with light, providing more carbon dioxide will lead to more photosynthesis.
- Further increases in carbon dioxide level will have no effect on the rate of photosynthesis.
- At this point a different factor, such as temperature, is the limiting factor and only an alteration in its level will affect the rate of photosynthesis.

These events are illustrated in Figure 1.

Processes such as photosynthesis are made up of a series of small reactions. It is the slowest of these reactions that determines the overall rate of photosynthesis. In turn, it is the level of factors such as temperature and the supply of raw materials that determines the speed of each step.

The law of limiting factors can therefore be expressed as:

At any given moment, the rate of a physiological process is limited by the factor that is at its least favourable value.

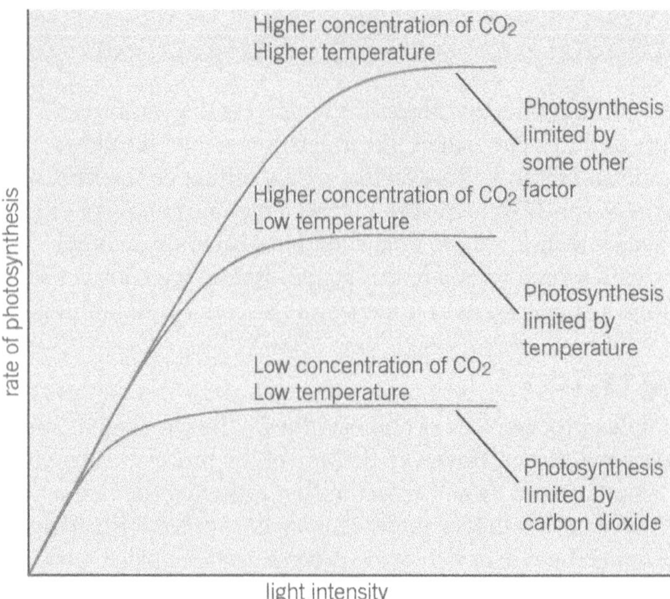

▲ **Figure 1** *Concept of limiting factors as illustrated by the effects of the levels of different conditions on the rate of photosynthesis*

Effect of light intensity on the rate of photosynthesis

The rate of photosynthesis is usually measured in one of two ways:

- the volume of oxygen released by a plant
- the volume of carbon dioxide taken up by a plant.

When light is the limiting factor, the rate of photosynthesis is directly proportional to light intensity. As light intensity is increased, the volume of oxygen produced and carbon dioxide absorbed due to photosynthesis will increase to a point at which it is exactly balanced by the oxygen absorbed and the carbon dioxide produced by cellular respiration. At this point there will be no net exchange of gases into or out of the plant. This is known as the light **compensation point**. Further increases in light intensity will cause a proportional increase in the rate of photosynthesis and increasing volumes of oxygen will be given off and carbon dioxide taken up. A point will be reached at which further increases in light intensity will have no effect on photosynthesis. At this point some other factor, such as carbon dioxide concentration or temperature, is limiting the reaction. These events are illustrated in Figure 2.

Effect of carbon dioxide concentration on the rate of photosynthesis

Carbon dioxide is present in the atmosphere at a concentration of around 0.04 per cent. This level continues to increase as the result of human activities such as burning fossil fuels and the clearing of rainforests. It is still one of the rarest gases present and is often the factor that limits the rate of photosynthesis under normal conditions. The optimum concentration of carbon dioxide for a consistently high rate of photosynthesis is 0.1 per cent and growers of some greenhouse crops, such as tomatoes, enrich the air in the greenhouses with more carbon

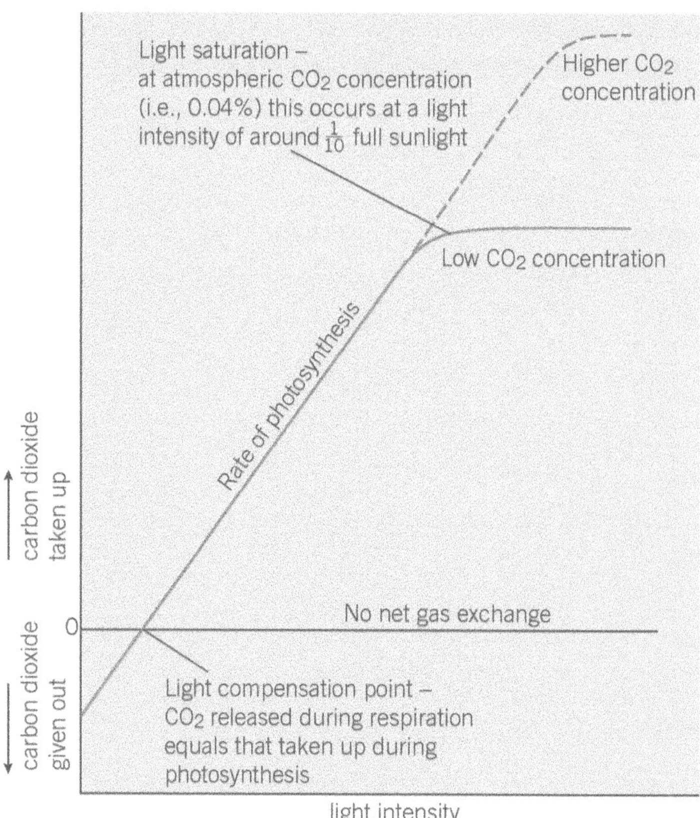

▲ **Figure 2** *Graph showing the effect of light intensity on the rate of photosynthesis as measured by the amount of CO_2 exchange*

dioxide to provide higher yields. Carbon dioxide concentration affects enzyme activity, in particular the enzyme that catalyses the combination of ribulose bisphosphate with carbon dioxide in the light-independent reaction (see Topic 19.2). Figures 1 and 2 illustrate the effect of different carbon dioxide levels on photosynthesis.

Effect of temperature on the rate of photosynthesis

Provided that other factors are not limiting, the rate of photosynthesis increases in direct proportion to the temperature. Between the temperatures of 0 °C and 25 °C the rate of photosynthesis is approximately doubled for each 10 °C rise in temperature. In many plants, the optimum temperature is 25 °C, above which the rate levels off and then declines – largely as a result of enzyme denaturation. Purely photochemical reactions are not usually affected by temperature, and so the fact that photosynthesis is temperature sensitive suggested to early researchers that there was also a totally chemical process involved as well as a photochemical one. We now know that this chemical process is the light-independent reaction (see Topic 19.3).

Summary questions

Figure 3 illustrates the influence of light intensity, carbon dioxide levels, and temperature on the rate of photosynthesis.

1 0.1% carbon dioxide at 25 °C
2 0.04% carbon dioxide at 35 °C
3 0.04% carbon dioxide at 25 °C
4 0.04% carbon dioxide at 15 °C

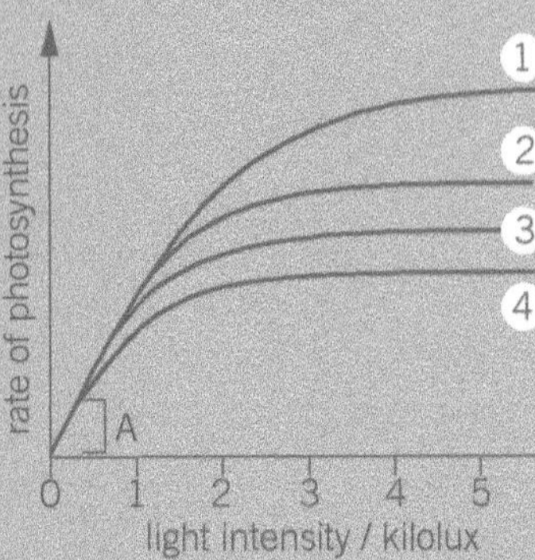

▲ **Figure 3**

1 State one measurement that could be taken to determine the rate of photosynthesis in this experiment?

2 Name the factor that is limiting the rate of photosynthesis over the region marked A on the graph. Explain your answer.

3 In the spring a commercial grower of tomatoes keeps his greenhouses at 25 °C and at a carbon dioxide concentration of 0.04 per cent. The light intensity is 4 kilolux at this time of year. Using the graph, predict whether the tomato plants would grow more if the carbon dioxide level was raised to 0.1 per cent or if the temperature was increased to 35 °C. Explain your answer.

4 Why is there no point in the grower heating his greenhouses on a dull day?

5 Using your knowledge of the light-independent reaction, explain why, at 25 °C, raising the level of carbon dioxide from 0.04 to 0.1 per cent increases the amount of glucose produced.

Hint

The light-independent reaction is enzyme controlled. Temperature affects enzyme activity and so must affect the light-independent reaction.

Required practical 8

Measuring photosynthesis

The rate of photosynthesis in an aquatic plant such as Canadian pondweed (*Elodea*) can be found by measuring the volume of oxygen produced using the apparatus (called a photosynthometer) illustrated in Figures 4 and 5.

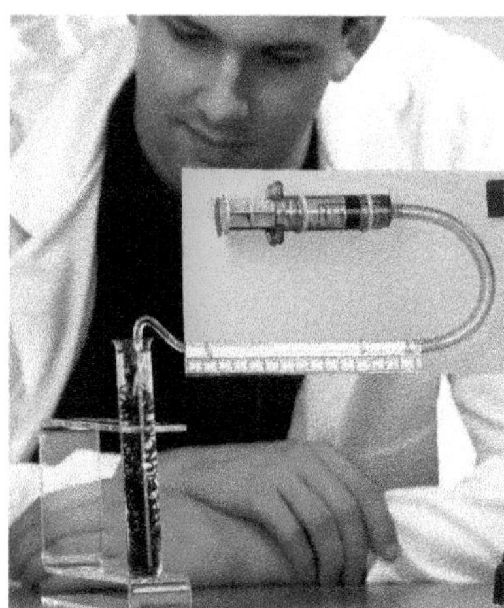

▲ **Figure 4** *Student using a photosynthometer to measure the rate of photosynthesis*

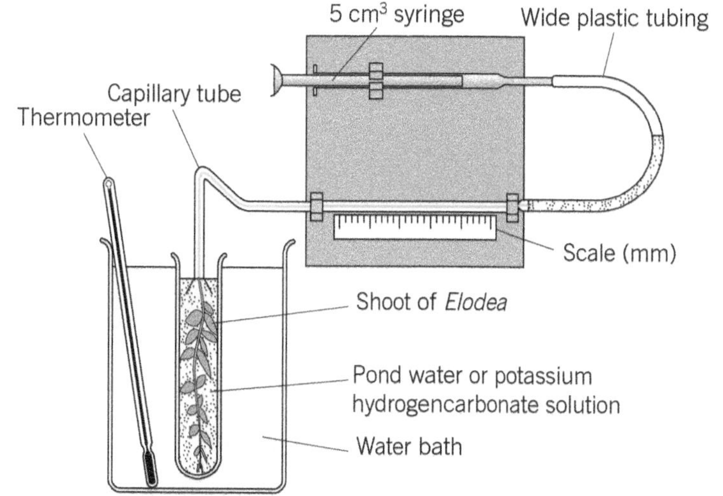

▲ **Figure 5** *Apparatus used to measure the rate of photosynthesis under various conditions*

- The apparatus is set up as in Figure 5, taking care not to introduce any air bubbles into it and that the apparatus is completely airtight.
- The water bath is used to maintain a constant temperature throughout the experiment and can be adjusted as necessary.
- Potassium hydrogencarbonate solution is used around the plant to provide a source of carbon dioxide.
- A source of light, whose intensity can be adjusted, is arranged close to the apparatus, which is kept in an otherwise dark room.
- The apparatus is kept in the dark for 2 hours before the experiment begins.
- The light source is switched on and timing is begun.
- Oxygen produced by the plant during photosynthesis collects in the funnel end of the capillary tube above the plant.
- After 30 minutes this oxygen is drawn up the capillary tube by gently withdrawing the syringe until its volume can be measured on the scale, which is calibrated in mm^3.
- The gas is drawn up into the syringe, which is then depressed again before the process is repeated at the same light intensity four or five times and the average volume of oxygen produced per hour is calculated.
- The apparatus is left in the dark for 2 hours before the procedure is repeated with the light source set at a different light intensity.

1. Why does the apparatus need to be airtight?
2. Why does the temperature of the water bath need to be kept constant?
3. Suggest an advantage of providing an additional source of carbon dioxide.
4. Suggest a reason for carrying out the experiment in a room that is dark except for the light source.
5. Suggest why the plant is kept in the dark before the experiment begins.
6. Suggest a reason why measuring the volume of oxygen produced by the plant in this experiment may not be an accurate measure of photosynthesis.

Practice questions: Chapter 19

1 (a) The diagram shows the light-dependent reaction of photosynthesis. Using the diagram, explain the part played by water in the production of ATP during photosynthesis. (*2 marks*)

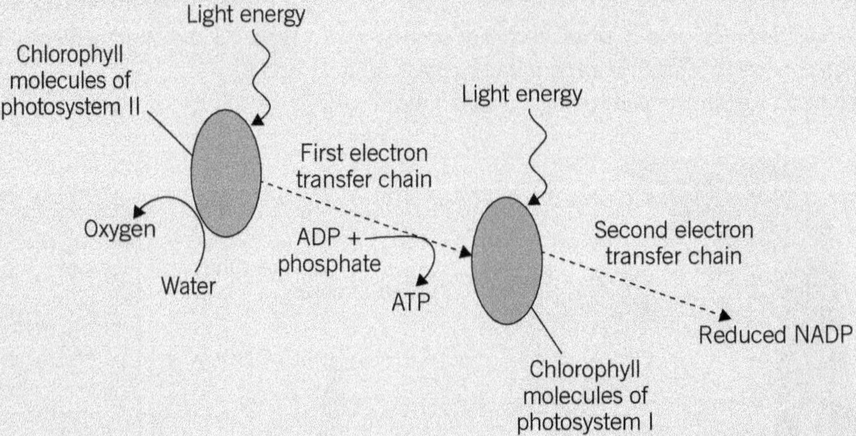

(b) Scientists investigated the effect of a herbicide on the light-dependent reaction. They measured the effects of different concentrations of the herbicide on the production of oxygen and on both electron transfer chains. The graph shows the results. Using all of the information, suggest how the herbicide causes each of the following:

(i) the reduction in the production of oxygen (*3 marks*)

(ii) the death of plants. (*3 marks*)

AQA Jan 2009

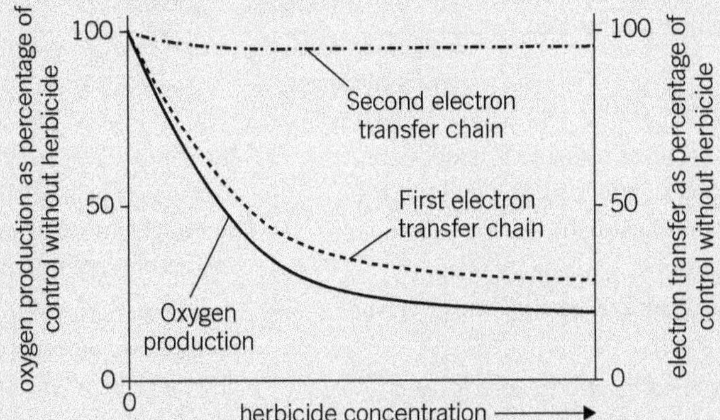

2 (a) Name two products of the light-dependent reaction of photosynthesis which are used in the light-independent reaction. (*2 marks*)

(b) In early experiments on the biochemistry of photosynthesis, scientists discovered that the photosynthetic reactions occurred very rapidly. They measured photosynthesis by green protoctists when exposed to flashes of light. Suggest how photosynthesis could have been measured. (*1 mark*)

(c) The experiments were carried out at both a high light intensity and a high carbon dioxide concentration. Suggest why. (*1 mark*)

AQA June 2008

3 (a) The diagram summarises the pathways involved in photosynthesis. Name molecule **Z**. (*1 mark*)

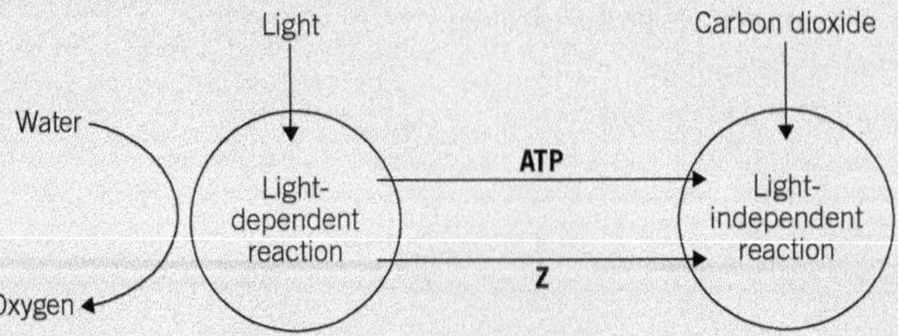

(b) Under some conditions oxygen reacts with ribulose bisphosphate to give glycerate 3-phosphate and phosphoglycolate. This reaction is summarised in the equation

RuBP + oxygen ⟶ glycerate 3-phosphate + phosphoglycolate

Phosphoglycolate takes no part in the light-independent reaction.

(i) Give the number of carbon atoms in one molecule of phosphoglycolate. *(1 mark)*

(ii) The production of phosphoglycolate could lead to a reduction in the rate of photosynthesis. Explain how. *(3 marks)*

(iii) An investigation was carried out on the effect of temperature and oxygen concentration on the rate of photosynthesis in leaves. The results are shown in the graph. Describe and explain the effect of oxygen concentration on the rate of photosynthesis. *(2 marks)*

AQA June 2007

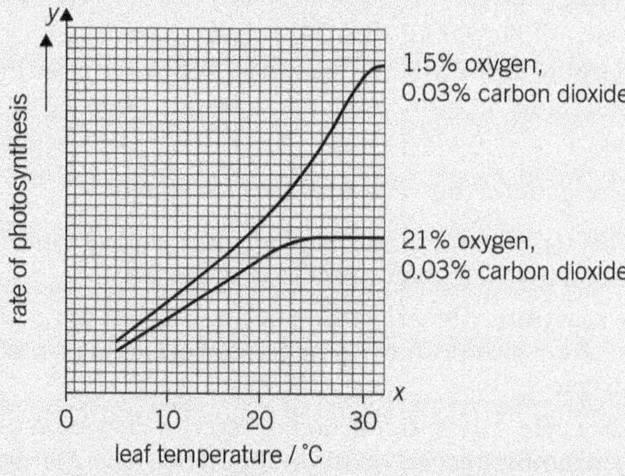

4 Different wavelengths of light are used to illuminate a tube containing a suspension of photosynthetic algal cells. The percentage of light absorbed and the rate of photosynthesis are measured using the apparatus shown. The light meter is calibrated to read 100 per cent using a glass tube containing water but no algal cells.

(a) (i) Explain how the percentage of light absorbed by the algal cells is calculated. *(1 mark)*

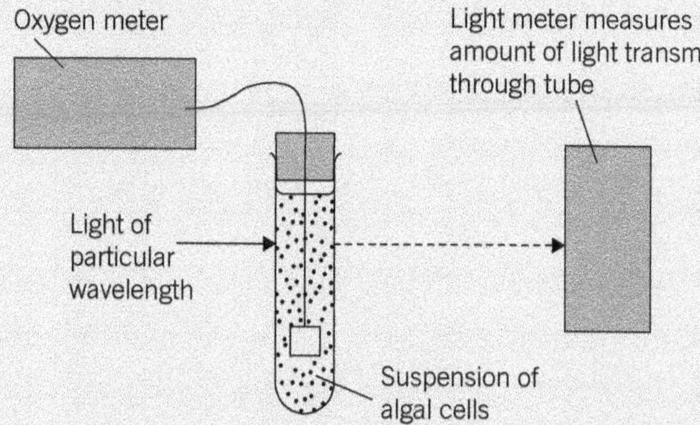

(ii) What measurements should be taken to determine the rate of photosynthesis? *(1 mark)*

(b) (i) Apart from temperature and pH, give one factor that should be kept constant during the investigation. *(1 mark)*

(ii) A buffer is used to maintain a constant pH. Explain why the pH of the suspension would increase during photosynthesis in the absence of a buffer. *(1 mark)*

(c) Explain why temperature has little influence on the absorption of light by photosynthetic organisms. *(2 marks)*

AQA Jan 2007

Answers to the Practice Questions are available at
www.oxfordsecondary.com/oxfordaqaexams-alevel-biology

20 Respiration
20.1 Glycolysis

Learning objectives:
- → Describe the process of glycolysis and its role in cellular respiration.
- → Describe the products of glycolysis.

Specification reference: 3.3.3.1 and 3.3.3.2

You have seen in Chapter 19 that photosynthesis converts energy in the form of sunlight into the chemical energy of carbohydrates such as glucose. You also saw in Topic 4.4 that it is ATP, rather than glucose, which cells use directly as an energy source. The conversion of glucose into ATP takes place during the process of cellular respiration. There are two different forms of cellular respiration, dependent upon whether oxygen is available or not:

- **Aerobic respiration** requires oxygen and produces carbon dioxide, water, and much ATP.
- **Anaerobic respiration (fermentation)** takes place in the absence of oxygen and produces lactate (in animals) or ethanol and carbon dioxide (in plants) but only a little ATP in both cases.

Aerobic respiration can be divided into four stages:

1 **glycolysis** – the splitting of the 6-carbon glucose molecule into two 3-carbon pyruvate molecules
2 **link reaction** – the conversion of the 3-carbon pyruvate molecule into carbon dioxide and a 2-carbon molecule called acetyl coenzyme A
3 **Krebs cycle** – the introduction of acetyl coenzyme A into a cycle of oxidation-reduction reactions that yield some ATP and a large number of electrons
4 **electron transfer chain** – the use of the electrons produced in the Krebs cycle to synthesise ATP with water produced as a by-product.

The main respiratory pathways are summarised in Figure 1.

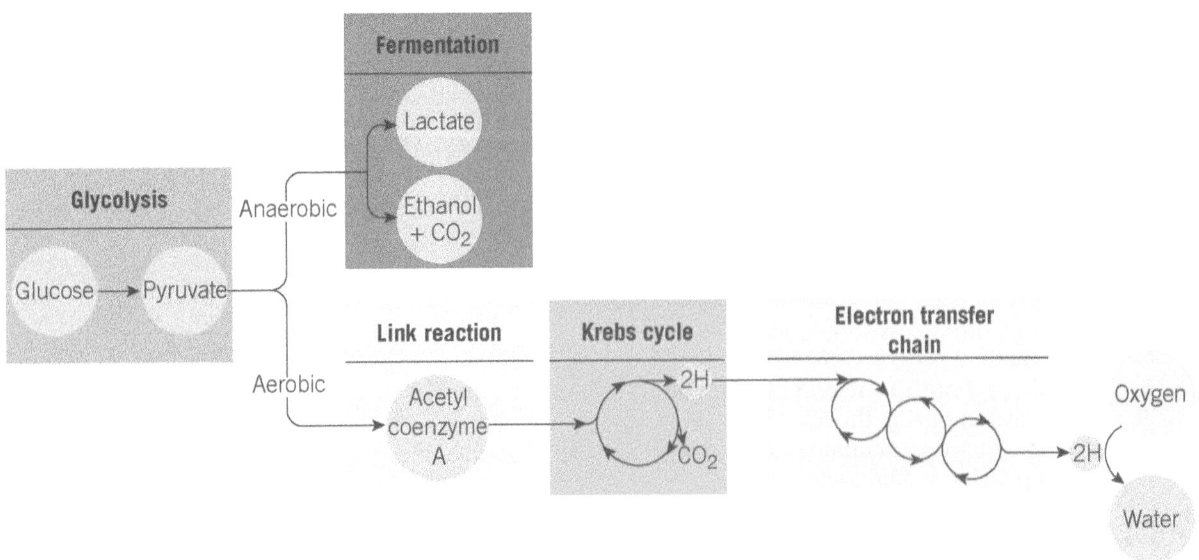

▲ **Figure 1** *Summary of respiratory pathways*

Glycolysis is the initial stage of both aerobic and anaerobic respiration. It occurs in the cytoplasm of all living cells and is the process by which a hexose (6-carbon) sugar, usually glucose, is split into two molecules of the 3-carbon molecule pyruvate. Although there are a number of smaller enzyme-controlled reactions in glycolysis, these can be conveniently grouped into four stages:

1 **activation of glucose by phosphorylation**. Before it can be split into two, glucose must first be made more reactive by the addition of two phosphate molecules (phosphorylation). The phosphate molecules come from the **hydrolysis** of two ATP molecules to ADP. This provides the energy to activate glucose (lowers the **activation energy** for the enzyme-controlled reactions that follow).
2 **splitting of the phosphorylated glucose**. Each glucose molecule is split into two 3-carbon molecules known as triose phosphate.
3 **oxidation of triose phosphate**. Hydrogen is removed from each of the two triose phosphate molecules and transferred to a hydrogen-carrier molecule known as NAD to form reduced NAD.
4 **the production of ATP**. Enzyme-controlled reactions convert each triose phosphate into another 3-carbon molecule called pyruvate. In the process, two molecules of ATP are regenerated from ADP.

The events of glycolysis are summarised in Figure 2.

Study tip

NAD (and FAD and NADP) are coenzymes that can be reduced and then oxidised repeatedly.

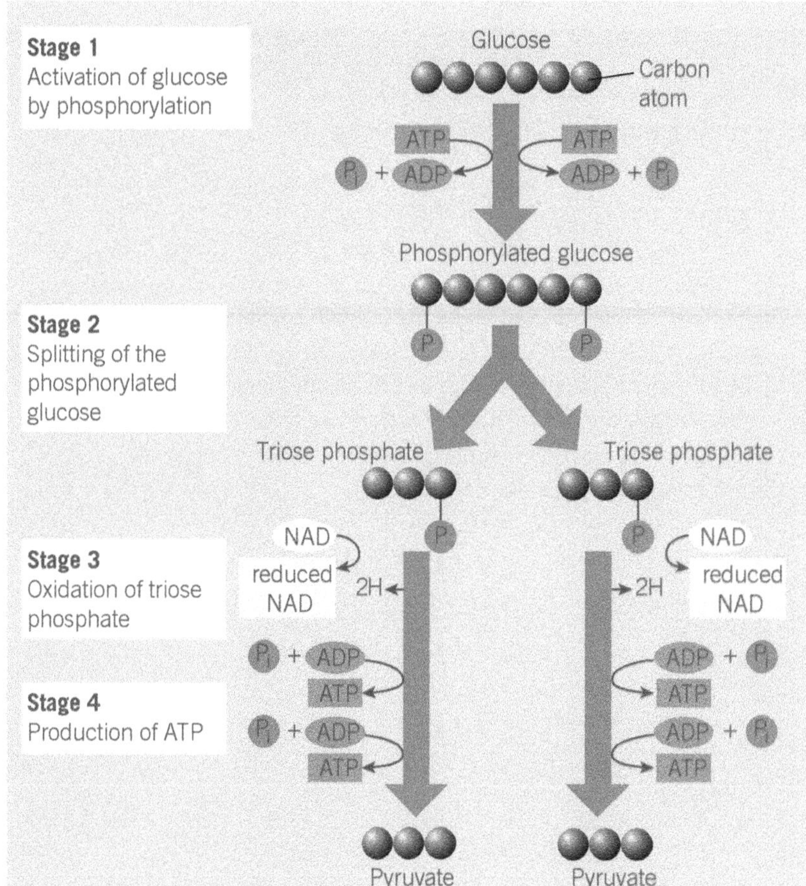

▲ **Figure 2** *Summary of glycolysis*

Hint

It must be remembered that for each molecule of glucose at the start of the process two molecules of triose phosphate are produced. Therefore the yields must be doubled, that is, four molecules of ATP and two molecules of reduced NAD.

Hint

Gluc**ose**, a sugar, is oxidised to pyru**vate**, an acid.

Energy yields from glycolysis

The overall yield from one glucose molecule undergoing glycolysis is therefore:

- two molecules of ATP (four molecules of ATP are produced, but two were used up in the initial phosphorylation of glucose and so the net increase is two molecules)
- two molecules of reduced NAD (these have the potential to produce more ATP as you shall see in Topic 20.4)
- two molecules of pyruvate.

Glycolysis is a universal feature of every living organism. The enzymes for the glycolytic pathway are found in the cytoplasm of cells and so glycolysis does not require any organelle or membrane for it to take place. It does not require oxygen and therefore it can take place whether or not oxygen is present. In the absence of oxygen the pyruvate produced by glycolysis can be converted into either lactate or ethanol by a process called anaerobic respiration. This is necessary in order to re-oxidise NAD so that glycolysis can continue. This is explained, along with details of the reactions, in Topic 20.2. Anaerobic respiration, however, yields only a small fraction of the potential energy stored in the pyruvate molecule. In order to release the remainder of this energy, most organisms use oxygen to break down pyruvate further in a process called the Krebs cycle.

Summary question

1 In the following passage, state the most suitable word to replace each of the letters **a** to **j**.

Glycolysis takes place in the **a** of cells and begins with the activation of the main respiratory substrate, namely the hexose sugar called **b**. This activation involves the addition of two **c** molecules provided by two molecules of **d**. The resultant activated molecule is known as **e** and in the next stage of glycolysis it is split into two molecules called **f**. The third stage entails the oxidation of these molecules by the removal of **g**, which is transferred to a carrier called **h**. The final stage is the production of the 3-carbon molecule **i**, which also results in the formation of two molecules of **j**.

20.2 Anaerobic respiration

You will see in Topic 20.3 that the hydrogen atoms released in glycolysis and other respiratory pathways are converted to water, and this process requires oxygen. What happens if oxygen is temporarily or permanently unavailable to a tissue or a whole organism?

In the absence of oxygen, neither the Krebs cycle nor the electron transfer chain can take place (See Topics 20.3 and 20.4). Instead only the anaerobic process of glycolysis is available to produce ATP. For glycolysis to continue, its products of pyruvate and hydrogen must be constantly removed. In particular, the hydrogen must be released from the reduced NAD in order to regenerate NAD. Without this, the already tiny supply of NAD in cells will be entirely converted to reduced NAD, leaving no NAD to take up the hydrogen newly produced from glycolysis. Glycolysis will then grind to a halt. The replenishment of NAD is achieved by the pyruvate molecule from glycolysis accepting the hydrogen from reduced NAD.

In eukaryotic cells, only two types of anaerobic respiration occur with any regularity:

- In plants, and in microorganisms such as yeast, the pyruvate is converted to ethanol and carbon dioxide.
- In animals, the pyruvate is converted to lactate.

Learning objectives:

- → Describe how ATP is produced in the absence of oxygen.
- → Explain how ethanol is produced in anaerobic respiration.
- → Explain how lactate is produced in anaerobic respiration.

Specification reference: 3.3.3.2

Production of ethanol in plants and some microorganisms

Anaerobic respiration leading to the production of ethanol occurs in organisms such as certain bacteria and fungi (e.g., yeast) as well as in some cells of higher plants, for example, root cells under waterlogged conditions.

The pyruvate molecule formed at the end of glycolysis loses a molecule of carbon dioxide and accepts hydrogen from reduced NAD to produce ethanol. The summary equation for this is:

pyruvate + reduced NAD ⟶ ethanol + carbon dioxide + NAD

This form of anaerobic respiration in yeast has been exploited by humans for thousands of years in the brewing industry. In brewing, ethanol is the important product. Yeast is grown in anaerobic conditions in which it ferments natural carbohydrates in plant products, such as grapes (wine production) or barley seeds (beer production), into ethanol.

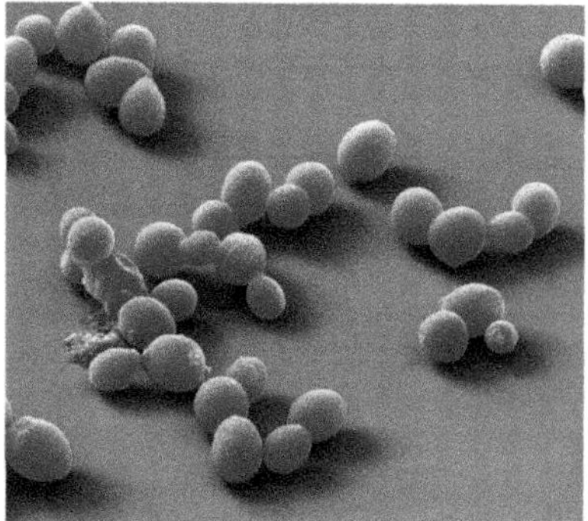

▲ **Figure 1** *False-colour scanning electron micrograph (SEM) of yeast cells. Yeast produces ethanol and carbon dioxide during anaerobic respiration, making it useful in fermentation and baking.*

Production of lactate in animals

Anaerobic respiration leading to the production of lactate occurs in animal cells as a means of overcoming a temporary shortage of oxygen. Clearly, such a mechanism has considerable survival value, for example, in a baby mammal in the period immediately after birth, and in an animal living in water where the amount of oxygen may sometimes be very low.

However, lactate production occurs most commonly in muscles as a result of strenuous exercise. In these conditions oxygen may be used up

Hint

During strenuous exercise, muscles carry out aerobic respiration. If this cannot supply ATP fast enough, they also carry out some anaerobic respiration as well. It is a case not of one or the other but of both together.

▲ **Figure 2** *During strenuous exercise, muscle tissue may temporarily respire anaerobically*

more rapidly than it can be supplied and therefore an oxygen debt occurs. It is often essential, however, that the muscles continue to work despite the lack of oxygen, for example, if the organism is fleeing from a predator. In the absence of oxygen, glycolysis would normally cease as reduced NAD accumulates. If glycolysis is to continue and release some energy, the reduced NAD must be removed. To achieve this, each pyruvate molecule produced takes up the two hydrogen atoms from the reduced NAD produced in glycolysis to form lactate as shown below:

pyruvate + reduced NAD ⟶ lactate + NAD

At some point the lactate produced needs to be oxidised back to pyruvate. This can then be either further oxidised to release energy or converted into glycogen. This happens when oxygen is once again available. In any case, lactate will cause cramp and muscle fatigue if it is allowed to accumulate in the muscle tissue. Although muscle has a certain tolerance to lactate, it is nevertheless important that it is removed by the blood and taken to the liver to be converted to glycogen. Figure 3 shows how the NAD needed for glycolysis to continue is regenerated in both common forms of anaerobic respiration.

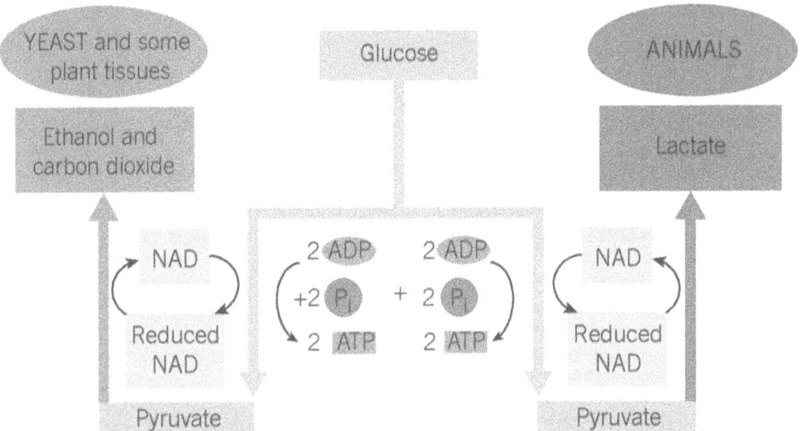

▲ **Figure 3** *How the NAD needed for glycolysis is regenerated in various organisms*

Energy yields from anaerobic and aerobic respiration

Energy from cellular respiration is derived in two ways:

- **substrate-level phosphorylation** in glycolysis and the Krebs cycle. This is the direct linking of inorganic phosphate (P_i) to ADP to produce ATP.
- **oxidative phosphorylation** in the electron transfer chain (Topic 20.4). This is the indirect linking of inorganic phosphate to ADP to produce ATP using the hydrogen atoms carried on NAD and FAD from glycolysis and the Krebs cycle. Cells produce most of their ATP in this way.

In anaerobic respiration, pyruvate is converted to either ethanol or to lactate. Consequently it is not available for the Krebs cycle. Therefore in anaerobic respiration neither the Krebs cycle nor the electron transfer chain can take place. The only ATP that can be produced by anaerobic respiration is that formed by glycolysis. This amount is very small per molecule of respiratory substrate – although a large amount of ATP may be produced this way – when compared to the much greater quantity produced during aerobic respiration.

Synoptic link

To help you follow the experiment described in the application and to answer the questions, you need to understand cell fractionation and enzyme inhibition. You should therefore review Topics 2.1 and 3.3.

Summary question

The diagram below shows the relationship between some respiratory pathways.

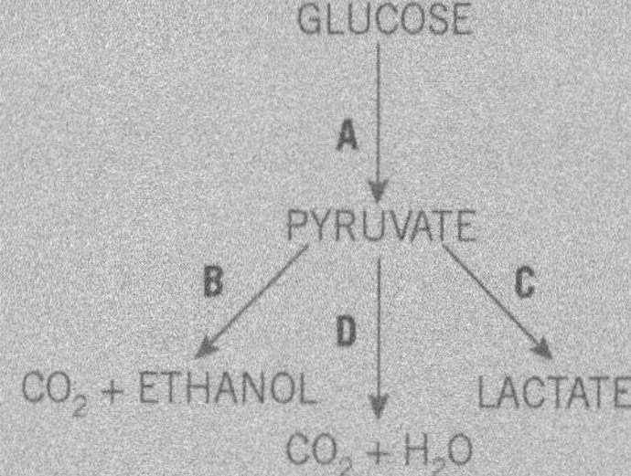

1 State which of the pathways, A, B, C, or D, apply to each of the following statements. There may be more than one answer in each case.

- **a** Only occurs in the presence of oxygen.
- **b** Takes place in animals.
- **c** Produces ATP.
- **d** Is carried out by yeast in the absence of oxygen
- **e** Produces reduced NAD.
- **f** Regenerates NAD from reduced NAD.
- **g** Is known as glycolysis.

Investigating where certain respiratory pathways take place in cells

Cyanide is a non-competitive inhibitor of an enzyme in the electron transfer chain. It therefore prevents the transfer of electrons along this chain.

To determine where in the cell some of the respiratory pathways take place, scientists carried out the following experiment involving cyanide. Given the poisonous nature of cyanide, the scientists carried out risk assessments and imposed appropriate safety precautions to ensure their own safety and that of others. These precautions included the wearing of safety glasses and gloves, working in well-ventilated areas, disposing of the chemical appropriately, and having a cyanide antidote kit available at all times.

- Mammalian liver cells were broken up (homogenised) and the resulting homogenate was centrifuged.
- Portions containing only nuclei, ribosomes, mitochondria, and the remaining cytoplasm were separated out.
- Samples of each portion, and of the complete homogenate, were incubated as follows:

– with glucose	– with glucose and cyanide
– with pyruvate and cyanide	– with pyruvate

After incubation the presence or absence of carbon dioxide and lactate in each sample was recorded. The results are shown in Table 1, in which ✓ = present and ✗ = absent.

▼ **Table 1**

Incubated with	Complete homogenate		Nuclei only		Ribosomes only		Mitochondria only		Remaining cytoplasm only	
	Carbon dioxide	Lactate	Carbon dioxide	Lactate	Carbon dioxide	Lactate	Carbon dioxide	Lactate	Carbon dioxide	Lactate
Glucose	✓	✓	✗	✗	✗	✗	✗	✗	✗	✓
Pyruvate	✓	✓	✗	✗	✗	✗	✓	✗	✗	✓
Glucose + cyanide	✗	✓	✗	✗	✗	✗	✗	✗	✗	✓
Pyruvate + cyanide	✗	✓	✗	✗	✗	✗	✗	✗	✗	✓

1. Briefly describe how the different portions of the homogenate may have been separated out by centrifuging.
2. From the results of this experiment, name **two** organelles that appear not to be involved in respiration. Explain your answer.
3. a. In which cell organelle would you expect to find the enzymes of the Krebs cycle?
   b. Explain how the results in the table support your answer.
4. Which portion of the homogenate contains the enzymes that convert pyruvate into lactate?
5. Explain why lactate is produced in the presence of cyanide but carbon dioxide is not.
6. Explain why carbon dioxide can be produced by the complete homogenate when none of the separate portions can do so.
7. If glucose were incubated with cytoplasm from yeast cells, which two products would be formed?
8. Which **three** of the following structures might you expect to be rich in mitochondria: xylem vessel, liver cell, red blood cell, epithelial cell, muscle cell?

20.3 Link reaction and Krebs cycle

The pyruvate molecules produced during glycolysis possess potential energy that can only be released using oxygen in a process called the Krebs cycle. Before they can enter the Krebs cycle, these pyruvate molecules must first be oxidised in a procedure known as the **link reaction**. In eukaryotic cells both the Krebs cycle and the link reaction take place exclusively inside mitochondria.

Learning objectives:

→ Describe the link reaction and the Krebs cycle.

→ Explain the role of hydrogen carriers in the link reaction and Krebs cycle.

Specification reference: 3.3.3.3

The link reaction

The pyruvate molecules produced in the cytoplasm during glycolysis are actively transported into the matrix of mitochondria. Here pyruvate undergoes a series of reactions during which the following changes take place:

- The pyruvate is oxidised by removing hydrogen. This hydrogen is accepted by NAD to form reduced NAD, which is later used to produce ATP (see Topic 20.4).
- The 2-carbon molecule that is formed, and which is called an acetyl group, combines with a molecule called coenzyme A (CoA) to produce a compound called **acetylcoenzyme A**.
- A carbon dioxide molecule is formed from each pyruvate.

The overall equation can be summarised as:

pyruvate + NAD + CoA ⟶ acetyl CoA + reduced NAD + CO_2

Synoptic link

It will be helpful in this section to remind yourself of the structure of mitochondria. Details can be found in Topic 2.3.

The Krebs cycle

The Krebs cycle was named after the British biochemist, Hans Krebs, who worked out its sequence. The Krebs cycle involves a series of oxidation-reduction reactions that take place in the matrix of mitochondria. Its events are illustrated in Figure 1 and can be summarised as follows:

- The 2-carbon acetylcoenzyme A from the link reaction combines with a 4-carbon molecule to produce a 6-carbon molecule.
- This 6-carbon molecule loses carbon dioxide and hydrogens to give a 4-carbon molecule and a single molecule of ATP produced as a result of substrate-level phosphorylation (see Topic 20.2).
- The 4-carbon molecule can now combine with a new molecule of acetylcoenzyme A to begin the cycle again.

For each molecule of pyruvate, the link reaction and the Krebs cycle therefore produce:

- reduced **coenzymes** such as NAD and FAD (see next page). These have the potential to provide the energy to produce ATP molecules (see Topic 20.4) and are therefore the important products of the Krebs cycle.
- one molecule of ATP
- three molecules of carbon dioxide.

As two pyruvate molecules are produced for each original glucose molecule, the yield from a single glucose molecule is double the quantities above.

A summary of the link reaction and the Krebs cycle is shown in Figure 1.

Study tip

Remember that the carbon dioxide produced in respiration is formed directly from molecules involved in the link reaction and the Krebs cycle.

Hint

Only a small amount of ATP is formed directly by the Krebs cycle. The vast majority of energy is carried away from the Krebs cycle by reduced NAD and reduced FAD and only later used to make ATP.

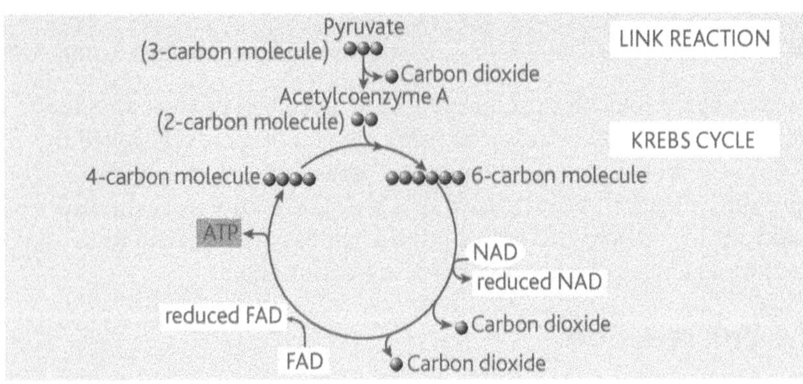

▲ **Figure 1** *Summary of the link reaction and the Krebs cycle*

Coenzymes

Coenzymes are molecules that some enzymes require in order to function. Coenzymes play a major role in photosynthesis and respiration where they carry hydrogen atoms from one molecule to another. Examples include:

- **NAD**, which is important throughout respiration
- **FAD**, which is important in the Krebs cycle
- **NADP**, which is important in photosynthesis (see Topics 19.2 and 19.3).

In respiration, NAD is the most important carrier. It works with dehydrogenase enzymes that catalyse the removal of hydrogen ions from substrates and transfer them to other molecules such as the hydrogen carriers involved in oxidative phosphorylation (see Topic 20.4).

The significance of the Krebs cycle

The Krebs cycle performs an important role in the cells of organisms for four reasons:

- It breaks down or oxidises macromolecules into smaller ones – pyruvate is broken down into carbon dioxide.
- It produces hydrogen atoms that are carried by NAD to the electron transfer chain for oxidative phosphorylation. This leads to the production of ATP that provides metabolic energy for the cell.
- It regenerates the 4-carbon molecule that combines with acetylcoenzyme A, which would otherwise accumulate.
- It is a source of intermediate compounds used by cells in the manufacture of other important substances such as fatty acids, amino acids, and chlorophyll.

Summary questions

1 How many carbon molecules are there in a single molecule of pyruvate?

2 What 2-carbon molecule is pyruvate converted to during the link reaction?

3 State precisely in which part of the cell the Krebs cycle takes place.

4 Table 1 lists statements about some biochemical processes in a plant cell. State whether each of the numbers 1–18 represents 'true' or 'false'.

▼ **Table 1**

Statement	Glycolysis	Krebs cycle	Light-dependent reaction of photosynthesis
ATP is produced	1	2	3
ATP is needed	4	5	6
NAD is reduced	7	8	9
NADP is reduced	10	11	12
CO_2 is produced	13	14	15
CO_2 is needed	16	17	18

Coenzymes in respiration

Coenzymes such as NAD are important in respiration. They help enzymes to function by carrying hydrogen atoms from one molecule to another. Scientists can model the way coenzymes work in cells using a blue dye called methylene blue. It can accept hydrogen atoms and so become reduced. Reduced methylene blue is colourless.

methylene blue (blue colour) + hydrogen ⟶ reduced methylene blue (colourless)

In an investigation into respiration in yeast, three test tubes were set up as shown in Table 2.

▼ **Table 2**

Tube A	Tube B	Tube C
2 cm^3 yeast suspension	2 cm^3 distilled water	2 cm^3 yeast suspension
2 cm^3 glucose solution	2 cm^3 glucose solution	2 cm^3 distilled water
1 cm^3 methylene blue	1 cm^3 methylene blue	1 cm^3 methylene blue

All three tubes were incubated at a temperature of 30 °C. The colour of each tube was recorded at the start of the experiment and after 5 minutes and 15 minutes. The results are shown in the table below:

▼ **Table 3**

Time / min	Colour of tube contents		
	Tube A	Tube B	Tube C
0	blue	blue	blue
5	colourless	blue	blue
15	colourless	blue	pale blue

1 Tube B acts as a control. Explain why this control was necessary in this investigation.

2 Using your knowledge of respiration, suggest an explanation for the colour change after 15 minutes in:

   a tube A

   b tube C.

3 How might the results in tube A after 15 minutes have been different if the experiment had been carried out at 70 °C? Explain your answer.

4 After 20 minutes the contents of tube A were mixed with air by shaking it vigorously, turning the methylene blue back to a blue colour. Suggest a reason for this colour change.

5 Suggest why conclusions made only on the basis of the results of this experiment may not be valid.

20.4 Electron transfer chain

Learning objectives:

- → Describe the location of the electron transfer chain.
- → Explain how ATP is synthesised in the electron transfer chain.
- → Describe the role of oxygen in aerobic respiration.

Specification reference: 3.3.3.3

So far in the process of aerobic respiration, you have seen how hexose sugars such as glucose are split (glycolysis) and how the 3-carbon pyruvate that results is fed into the Krebs cycle to yield carbon dioxide and hydrogen atoms. The carbon dioxide is a waste product and is removed during the process of gaseous exchange. The hydrogen atoms (or more particularly the electrons they possess) are valuable as a potential source of energy. These hydrogen atoms are carried by the coenzymes NAD and FAD into the next stage of the process – the **electron transfer chain**. This is the mechanism by which the energy of the electrons within the hydrogen atoms is converted into a form that cells can use, namely adenosine triphosphate (ATP). The synthesis of ATP by this mechanism is known as oxidative phosphorylation.

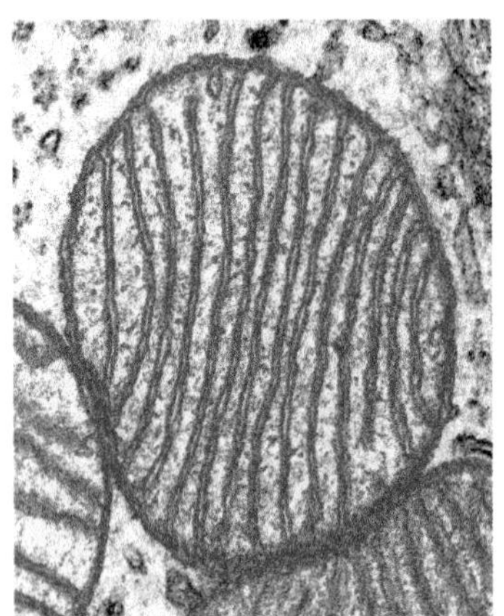

▲ **Figure 1** *False-colour transmission electron micrograph (TEM) of a sectioned mitochondrion (red and yellow). It has two membranes – an outer surrounding membrane and an inner membrane that forms folds called* cristae, *seen here as red lines. The cristae are the sites of the electron transfer chain.*

The electron transfer chain and mitochondria

Mitochondria are rod-shaped organelles that are found in eukaryotic cells. You saw in Topic 2.3 that each mitochondrion is bounded by a smooth outer membrane and an inner one that is folded into extensions called cristae. The inner space, or matrix, of the mitochondrion is made up of a semi-rigid material of protein, lipids, and traces of DNA.

Mitochondria are the sites of the electron transfer chain. Attached to the inner folded membrane (cristae) are the enzymes and other proteins involved in the electron transfer chain and hence ATP synthesis.

As mitochondria play such a vital role in respiration and the release of energy, it is hardly surprising that they occur in greater numbers in metabolically active cells, such as those of the muscles, liver, and epithelial cells, which carry out active transport. The mitochondria in these cells also have more densely packed cristae, which provide a greater surface area for the attachment of enzymes and other proteins involved in electron transport.

Synoptic link

Re-reading about the structure of mitochondria in Topic 2.3 will help you follow the processes of the electron transfer chain.

Hint

Remember that a single hydrogen atom is made up of one proton (H^+) and one electron (e^-).

The electron transfer chain and the synthesis of ATP

ATP is synthesised using the electron transfer chain as follows:

- The hydrogen atoms produced during glycolysis and the Krebs cycle combine with the coenzymes NAD and FAD that are attached to the cristae of the mitochondria.
- The reduced NAD and FAD donate the electrons of the hydrogen atoms they are carrying to the first molecule in the electron transfer chain.
- This releases the protons (H^+) from the hydrogen atoms and these protons are actively transported across the inner mitochondrial membrane.
- The electrons, meanwhile, pass along a chain of electron transport carrier molecules in a series of oxidation-reduction reactions. The electrons lose energy as they pass down the chain and some of this is used to combine ADP and inorganic phosphate to make ATP. The remaining energy is released in the form of heat.
- The protons accumulate in the space between the two mitochondrial membranes before they diffuse back into the mitochondrial matrix

through special protein channels. They diffuse back through a complex including an ATP synthase. This diffusion, down a diffusion electrochemical gradient, is what powers the ATP synthesis.

- At the end of the chain the electrons combine with these protons and oxygen to form water. Oxygen is therefore the final acceptor of electrons in the electron transfer chain.

These events are summarised in Figure 2.

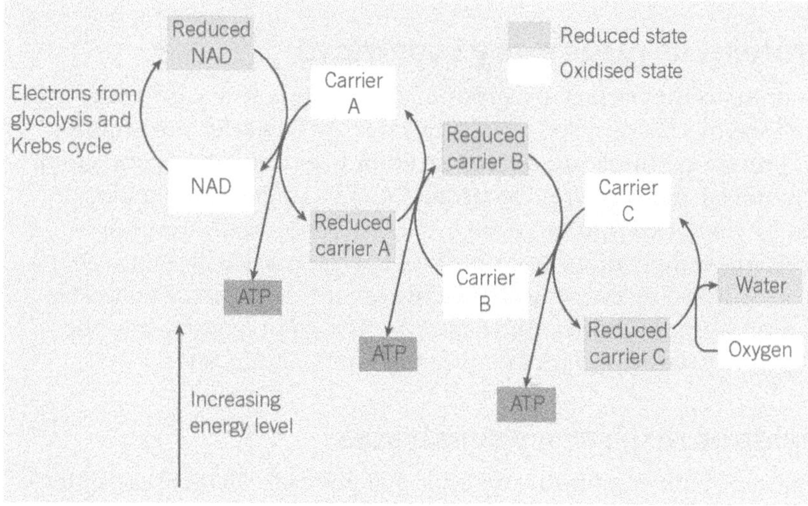

▲ **Figure 2** *Summary of electron transfer chain*

The importance of oxygen in respiration is to act as the final acceptor for the electrons produced from the hydrogen atoms removed in glycolysis and the Krebs cycle. Without its role in removing electrons at the end of the chain, the hydrogen ions (protons) and electrons (the components of hydrogen atoms) would 'back up' along the chain and the process of respiration would come to a halt. This point is illustrated by the effect of cyanide on respiration. Most people are aware that cyanide is a very potent poison that causes death rapidly. It is lethal because it is a non-competitive inhibitor of the final enzyme in the electron transfer chain. This enzyme catalyses the addition of the hydrogen ions and electrons to oxygen to form water. Its inhibition causes hydrogen ions and electrons to accumulate on the carriers, bringing cellular respiration to a halt.

Study tip

Oxygen is used as the final acceptor of hydrogen atoms at the end of the electron transfer chain. It is therefore used to form water and *not* carbon dioxide, as you may think.

Summary questions

1. The processes that occur in the electron transfer chain are also known as oxidative phosphorylation. Suggest why this term is used.
2. The surface of the inner mitochondrial membrane is highly folded to form cristae. State one advantage of this arrangement to the electron transfer chain.
3. The oxygen taken up by organisms has an important role in aerobic respiration. Explain this role.
4. As part of which molecule does the oxygen taken in leave an organism after being respired?

Sequencing the chain

The order in which the carrier molecules of the electron transfer chain are arranged can be determined experimentally. The experiments rely on the fact that each transfer of electrons between one molecule and the next is catalysed by a specific enzyme. In a series of experiments, three different inhibitors, 1, 2, and 3, are added to four electron transport molecules, A, B, C, and D. Table 1 shows whether the molecules A–D are oxidised or reduced after the inhibitor is added.

▼ **Table 1**

Inhibitor added	Electron transport molecules			
	A	B	C	D
1	reduced	oxidised	reduced	oxidised
2	oxidised	oxidised	reduced	oxidised
3	reduced	oxidised	reduced	reduced

1. Using the information in the table, state the order of the electron transport molecules in this chain. Explain your answer.

20.5 Respiratory substrates

Learning objectives:

- → Describe how energy can be released from molecules other than glucose.
- → Explain what is meant by a respiratory quotient.

Specification reference: 3.3.3.4

Synoptic link

When blood glucose concentration is low, amino acids can be converted to glucose in liver cells. This is regulated by the hormone glucagon and is known as gluconeogenesis (Topic 27.4).

Study tip

Remind yourself of the molecular structure of amino acids, glycerol, and fatty acids in Chapter 1.

So far in this chapter it has been assumed that it is always glucose which is broken down by cells in respiration. This is true for some cells such as the neurones in the brain or red blood cells. However other cells and tissues can break down different molecules through the respiratory pathways and release energy for ATP synthesis.

Proteins as respiratory substrates

As proteins themselves are macromolecules, it is amino acids that are used by cells as respiratory substrates. Amino acids not required for protein synthesis are **deaminated** in liver cells (the amine group is removed and converted to **urea**). The remainder of the molecule can be converted into glycogen or fat. However, in the absence of sufficient carbohydrate as a source of energy, muscle proteins can be hydrolysed to amino acids. These are then deaminated and used as respiratory substrates (see Figure 1). The point of entry into the respiratory pathway varies between different amino acids.

Lipids as respiratory substrates

Lipids are converted into fatty acids and glycerol. Glycerol can enter glycolysis (see Figure 1) and fatty acids are broken down by a process known as **ß oxidation** to produce acetylcoenzyme A, which then enters the Krebs cycle (see Figure 1). Since large amounts of reduced FAD and reduced NAD are produced from ß oxidation and the Krebs cycle, fatty acids can only be broken down aerobically since oxygen will be required as the final electron acceptor in order to recycle the two coenzymes.

Table 1 shows the relative energy content of lipids and proteins compared to carbohydrate.

▼ **Table 1**

Respiratory substrate	Energy content / kJ g^{-1}
Carbohydrate	15.8
Lipid	39.4
Protein	17.0

The differences in the energy content values reflects the number of hydrogen atoms released during the breakdown of the molecule. This is because most energy is released through the electron transfer chain and the final oxidation of hydrogen to water.

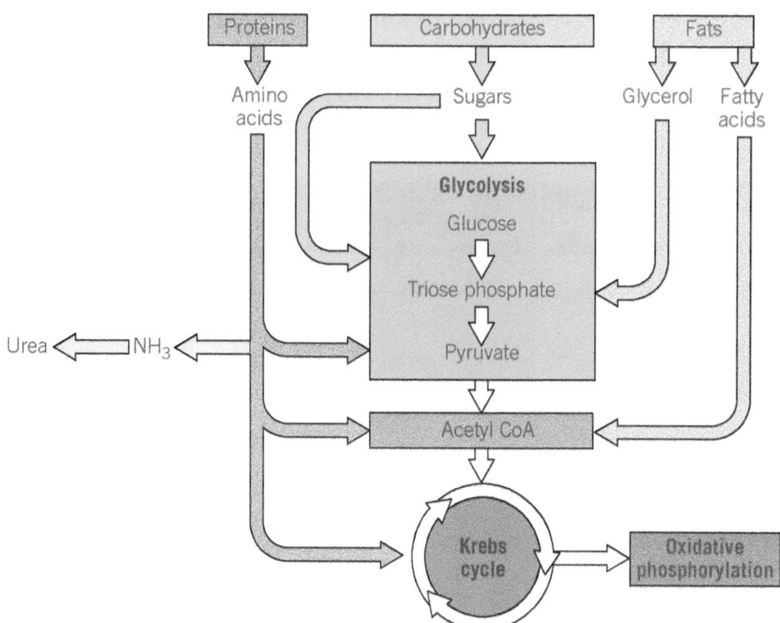

Figure 1 *A summary of the pathways used in the breakdown of different respiratory substrates*

The respiratory quotient

In the breakdown of molecules in aerobic respiration, carbon dioxide is produced from the breakdown of the carbon 'framework' of the molecule. Water is produced from the removal of hydrogen atoms and their subsequent transfer to oxygen in the electron transfer chain. Look at the simplified equation for aerobic respiration below:

$$\mathbf{C_6H_{12}O_6 + 6O_2 \rightarrow 6CO_2 + 6H_2O + energy}$$

When glucose is respired the ratio of oxygen taken in and carbon dioxide released is 1:1. As different substrates are broken down in aerobic respiration, the ratio of carbon dioxide to oxygen may vary as the ratio of carbon atoms to hydrogen atoms varies. The ratio of carbon dioxide to oxygen is called the **respiratory quotient (RQ)**.

RQ values can be calculated from the equation for the aerobic breakdown of the substrate. The formula for the aerobic breakdown of glucose would be:

$$\mathbf{RQ = \frac{moles\ of\ carbon\ dioxide\ given\ out}{moles\ of\ oxygen\ taken\ in} = \frac{6CO_2}{6O_2} = 1}$$

The RQ for a fatty acid such as oleic acid ($C_{18}H_{34}O_2$) would be calculated as follows:

$$\mathbf{C_{18}H_{34}O_2 + 25.5\ O_2 \rightarrow 18CO_2 + 17H_2O + energy}$$

The RQ value would be $\frac{18CO_2}{25.5O_2} = 0.7$

When respiration is anaerobic, no oxygen is consumed hence the denominator in the equation for the RQ will be zero, giving a value of infinity!

RQ values can also be determined experimentally by measuring oxygen consumption and carbon dioxide production by organisms using a **respirometer**. It is possible that by measuring this and calculating the RQ value, the substrates being used in respiration by an organism and the type of respiration being carried out (aerobic or anaerobic) can be deduced.

$$RQ = \frac{\text{volume of carbon dioxide produced min}^{-1}}{\text{volume of oxygen consumed min}^{-1}}$$

Summary questions

1 Calculate the RQ value for the aerobic respiration of a fatty acid with the formula $C_{18}H_{36}O_2$.

2 Explain why mature human red blood cells can only respire glucose.

Required practical 9

Calculating RQ values in germinating seeds

A **respirometer** is set up as shown in Figure 2. The potassium hydroxide will absorb any carbon dioxide being produced by the seeds in tube B. Tube A acts a control and compensates for any atmospheric changes that would cause the air in the apparatus to expand or contract. This means that any changes in the position of the fluid in the capillary U-tube reflect the changes brought about by the respiring seeds.

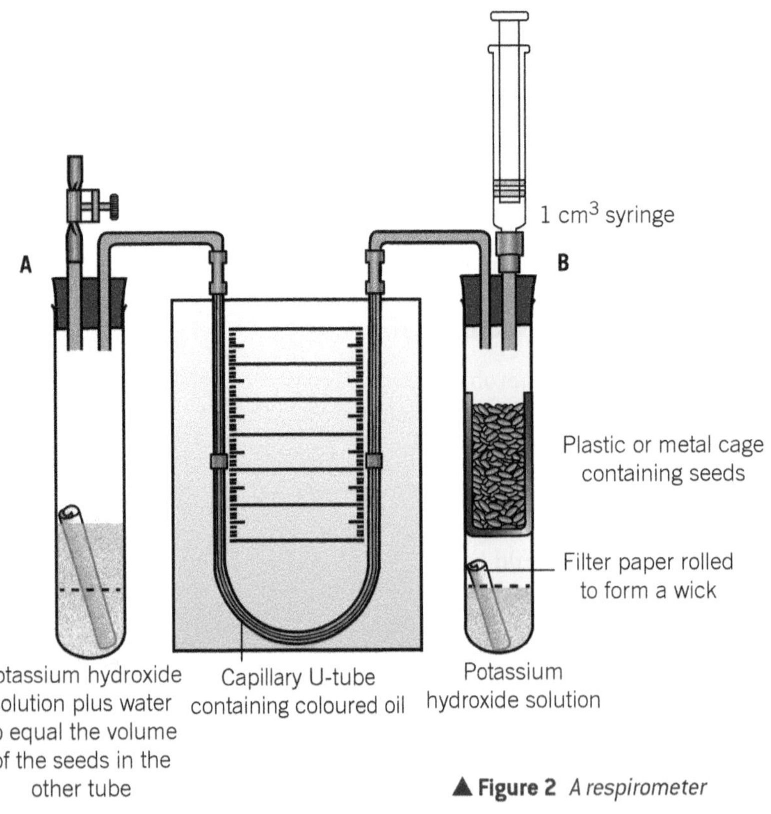

▲ **Figure 2** *A respirometer*

1 State one further variable that would need to be controlled and explain how that could be achieved.

Following a period of equilibration, the screw clip is closed and the syringe is used to adjust the levels in the U-tube so they are equal. After a fixed time (for example, 30 minutes) the manometer fluid will have moved.

2 Which direction will the manometer liquid have moved? Explain your answer.

The volume in the syringe is adjusted until the fluid in the U-tube is back at the starting position. This volume (volume X) corresponds to the oxygen consumed by the seeds during the 30-minute period.

The potassium hydroxide is then removed from both tubes A and B and the procedure above is repeated. A second volume (volume Y) is recorded. The volume of carbon dioxide produced is (volume X + volume Y). A value for RQ can now be calculated.

3 Germinating seeds were found to have an RQ value of 0.8. What could you deduce about their respiratory substrate?

4 Explain why an RQ value of 1 would *not necessarily* indicate that carbohydrates were being respired aerobically.

5 How would you modify the procedure to investigate the effect of temperature on respiration rate?

Practice questions: Chapter 20

1 (a) Figure 1 shows glycolysis and the Krebs cycle. Name
(i) molecule X *(1 mark)*
(ii) molecule Y. *(1 mark)*

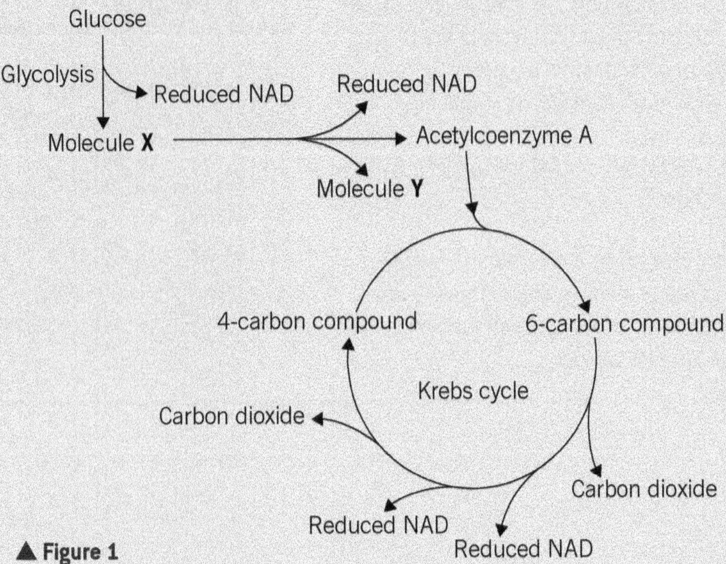

▲ **Figure 1**

(b) Where in a cell does glycolysis occur? *(1 mark)*

(c) (i) High concentrations of ATP inhibit an enzyme involved in glycolysis. Describe how inhibition of glycolysis will affect the production of ATP by the electron transfer chain. *(1 mark)*
(ii) Explain this effect. *(3 marks)*

AQA June 2007

2 (a) Table 1 contains some statements relating to biochemical processes in a plant cell. Complete the table with a tick if the statement is true or a cross if it is not true for each biochemical process. *(4 marks)*

▼ **Table 1**

Statement	Glycolysis	Krebs cycle	Light-dependent reaction of photosynthesis
NAD is reduced			
NADP is reduced			
ATP is produced			
ATP is required			

(b) An investigation was carried out into the production of ATP by mitochondria. ADP, phosphate, excess substrate, and oxygen were added to a suspension of isolated mitochondria.
(i) Suggest the substrate used for this investigation. *(1 mark)*
(ii) Explain why the concentration of oxygen and amount of ADP fell during the investigation. *(2 marks)*
(iii) A further investigation was carried out into the effect of three inhibitors, A, B, and C, on the electron transfer chain in these mitochondria. In each of three experiments, a different inhibitor was added. Table 2 shows the state of the electron carriers, W–Z, after the addition of inhibitor. Give the order of the electron carriers in this electron transfer chain. Explain your answer. *(2 marks)*

▼ Table 2

Inhibitor added	Electron carrier			
	W	X	Y	Z
A	oxidised	reduced	reduced	oxidised
B	oxidised	oxidised	reduced	oxidised
C	reduced	reduced	reduced	oxidised

AQA June 2006

3 (a) Figure 2 shows the structure of a mitochondrion. In which part of the mitochondrion does the Krebs cycle take place? *(1 mark)*

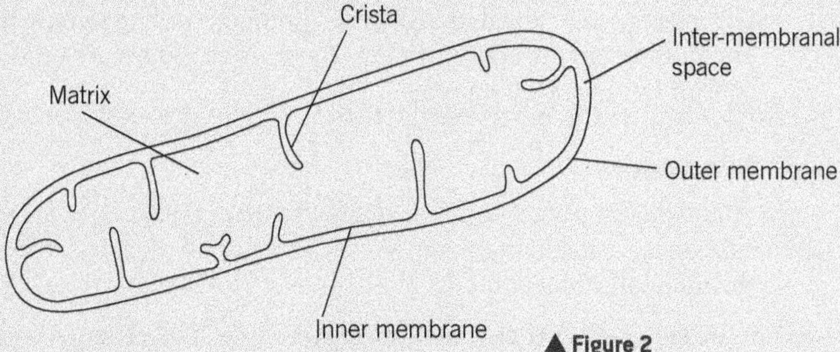

▲ Figure 2

(b) Name two substances for which there would be net movement into the mitochondrion. *(2 marks)*

(c) The mitochondria in muscles contain many cristae. Explain the advantage of this. *(2 marks)*

AQA Jan 2006

4 (a) RQ values can be calculated experimentally from respiring organisms. Write an equation for calculating an RQ value experimentally. *(1 mark)*

(b) RQ values were measured for an individual over a period of 400 minutes. The individual had been fasting prior to the experiment. During the experiment, two periods of high intensity training (HIT) were carried out. Figure 3 is a graph showing the changes in RQ values over that period of time. What evidence in Figure 3 indicates that the subject had been fasting. Explain your answer. *(3 marks)*

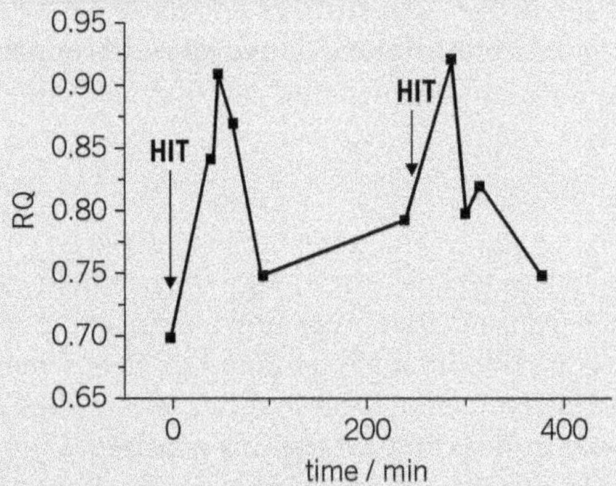

▲ Figure 3

(c) Explain why the RQ value increased during the periods of intensive exercise. *(2 marks)*

Answers to the Practice Questions are available at
www.oxfordsecondary.com/oxfordaqaexams-alevel-biology

21 Energy and ecosystems

21.1 Food chains and food webs

Learning objectives:

- → Describe how energy enters an ecosystem.
- → Describe how energy is transferred between organisms in the ecosystem.
- → Define the terms trophic level, food web, producer, consumer, and decomposer.
- → Explain how energy is lost from ecosystems.

Specification reference: 3.3.4.1

The organisms found in any ecosystem rely on a source of energy to carry out all their activities. The ultimate source of this energy is sunlight, which is converted to chemical energy by photosynthesising organisms and is then passed as food between other organisms. In this chapter we shall look at how this energy transfer takes place. We shall also compare natural ecosystems with those based on modern intensive farming and consider how farming practices increase the efficiency of energy conversion.

Organisms can be divided into three groups according to how they obtain their energy and nutrients. These three groups are producers, consumers, and decomposers.

Producers

Producers are photosynthetic organisms that manufacture organic substances using light energy, water, and carbon dioxide according to the following equation:

$6CO_2$	+	$6H_2O$	+	energy	$\longrightarrow$	$C_6H_{12}O_6$	+	$6O_2$
carbon dioxide		water		light		glucose		oxygen

Green plants are producers.

Hint

You may be familiar with the following terms:

- Herbivore – an animal that eats plants (producers) and is therefore a primary consumer.
- Carnivore – an animal that eats animals and may therefore be a secondary or a tertiary consumer.
- Omnivore – an animal that eats both plants and animals and is therefore a primary consumer and also a secondary or a tertiary consumer.

As you see, these terms are not very precise and you should avoid them when writing about food chains and food webs.

Consumers

Consumers are organisms that obtain their energy by feeding on (consuming) other organisms rather than by using the energy of sunlight directly. Animals are consumers. Those that directly eat producers (green plants) are called **primary consumers** because they are the first in the chain of consumers. Those animals eating primary consumers are called **secondary consumers** and those eating secondary consumers are called **tertiary consumers**. Secondary and tertiary consumers are usually predators but they may also be scavengers or parasites.

Decomposers

When producers and consumers die, the energy they contain can be used by a group of organisms that break down these complex materials into simple components again. In doing so, they release valuable minerals and elements in a form that can be absorbed by plants and so contribute to recycling. Organisms which feed on dead organic material in this way are referred to as saprophytes. The majority of this work is carried out by fungi and bacteria, called **decomposers**, and to a lesser extent by certain animals such as earthworms, called **detritivores**.

Food chains

The term food chain describes a feeding relationship in which the producers are eaten by primary consumers. These in turn are eaten by secondary consumers, who are then eaten by tertiary consumers. In a long food chain the tertiary consumers may in turn be eaten by further consumers called quaternary consumers. Each stage in this chain is referred to as a trophic level. The arrows on food chain diagrams represent the direction of energy flow. Shorter food chains may have three levels:

grass (producer) → **sheep** (primary consumer) → **human** (secondary consumer)

Longer food chains may have five trophic levels:

nettle (producer) → **aphid** (primary consumer) → **ladybird** (secondary consumer) → **blue tit** (tertiary consumer) → **sparrowhawk** (quaternary consumer)

Table 1 describes three further food chains, each from a different habitat.

▼ **Table 1** *Examples of food chains*

Trophic level	Grassland	Pond	Seashore
Quaternary consumer	Stoat	Pike	Seagull
Tertiary consumer	Grass snake	Stickleback	Crab
Secondary consumer	Toad	Leech	Whelk
Primary consumer	Caterpillar	Water snail	Limpet
Producer	Grass	Pondweed	Seaweed

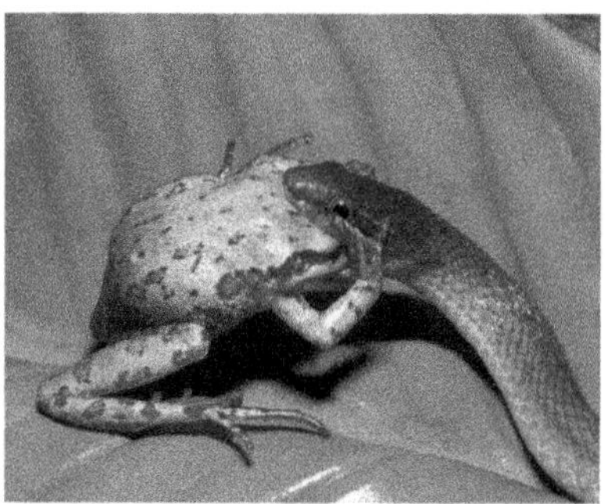

▲ **Figure 1** *The snake (tertiary consumer) is swallowing an insect-eating frog (secondary consumer) on a plant leaf (producer)*

Study tip

Make sure you learn simple definitions of ecological terms and use them appropriately.

Food webs

In reality, most animals do not rely upon a single food source and within a single **habitat** many food chains will be linked together to form a food web. For example, on the edge of an oak woodland, the food chain shown above (that starts with the nettle and ends with the sparrowhawk) can be combined with others to form the web shown in Figure 2.

The problem with food webs is their complexity. In the example shown in Figure 2, the nettle is shown as being eaten by aphids only. In reality it may be eaten by at least 20 different organisms. The same is true of the oak leaves and some of the consumers in the food web. Indeed, it is likely that all organisms within a habitat, even within an ecosystem, will be linked to others in the food web. Charting the feeding inter-relationships of thousands of species is not feasible. In any case, these relationships are not fixed but change depending on the time of year, age, and population size of the organisms. Although not an exact science, describing food webs is valuable in helping us to understand populations.

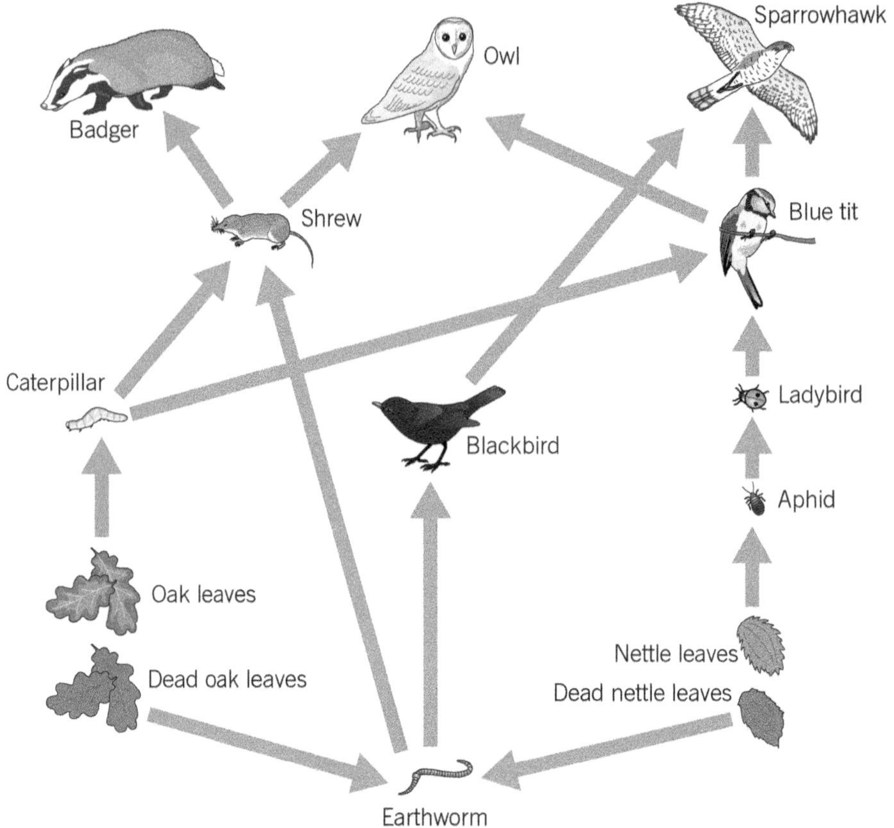

▲ **Figure 2** *Part of a woodland food web*

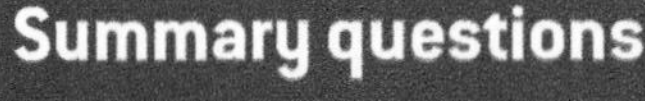

Summary questions

The diagram below shows a simplified food web within an aquatic ecosystem.

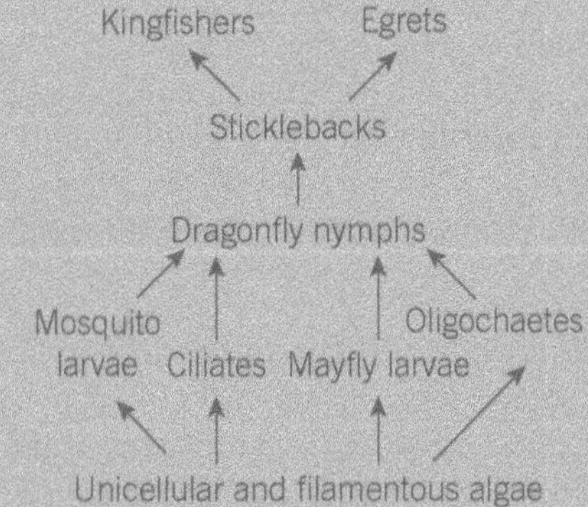

1. Which organisms are secondary consumers?
2. Which organisms carry out photosynthesis?
3. Which organisms are at the fourth trophic level?
4. What do the arrows in the diagram show?
5. When the organisms in this web die they will be broken down by bacteria and fungi. What general term is used to describe these bacteria and fungi?

21.2 Energy transfer between trophic levels

Learning objectives:

→ Explain why energy is lost along a food chain.

→ Explain why there are no more than five trophic levels in a food chain.

→ Calculate the transfer of energy between trophic levels.

Specification reference: 3.3.4.1 and 3.3.4.2

The Sun is the source of energy for ecosystems. However, as little as one per cent of this light energy may be captured by green plants and so made available to organisms in the food chain. These organisms in turn pass on only a small fraction of the energy that they receive to each successive stage in the chain. How then is so much energy lost?

Energy losses in food chains

Plants normally convert between one per cent and three per cent of the Sun's energy available to them into organic matter. Most of the Sun's energy is not converted to organic matter by photosynthesis because:

- over 90 per cent of the Sun's energy is reflected back into space by clouds and dust or absorbed by the atmosphere
- not all wavelengths of light can be absorbed and used for photosynthesis
- light may not fall on a chlorophyll molecule
- a factor, such as low carbon dioxide levels, may limit the rate of photosynthesis (see Topic 19.4).

The total quantity of energy that the plants in a community convert to organic matter is called the **gross production**. However, plants use 20–50 per cent of this energy in respiration, leaving little to be stored. The rate at which they store energy is called the **net production**.

net production = gross production − respiratory losses

Even then, only about 10 per cent of this food stored in plants is used by primary consumers for growth. Secondary and tertiary consumers are slightly more efficient, transferring about 20 per cent of the energy available from their prey into their own bodies. The low percentage of energy transferred at each stage is the result of the following:

- Some of the organism is not eaten.
- Some parts are eaten but cannot be digested and are therefore lost in faeces.
- Some of the energy is lost in excretory materials, such as urine.
- Some energy losses occur as heat from respiration and directly from the body to the environment. These losses are high in mammals and birds because of their high body temperature. Much energy is needed to maintain their body temperature when heat is constantly being lost to the environment.

Energy flow along food chains, showing the percentage transferred at each trophic level, is summarised in Figure 1.

It is the relative inefficiency of energy transfer between trophic levels that explains why:

- most food chains have only four or five trophic levels because there is insufficient energy available to support a large enough breeding population at trophic levels higher than these

- the total mass of organisms in a particular place (**biomass**) is less at higher trophic levels
- the total amount of energy stored is less at each level as one moves up a food chain.

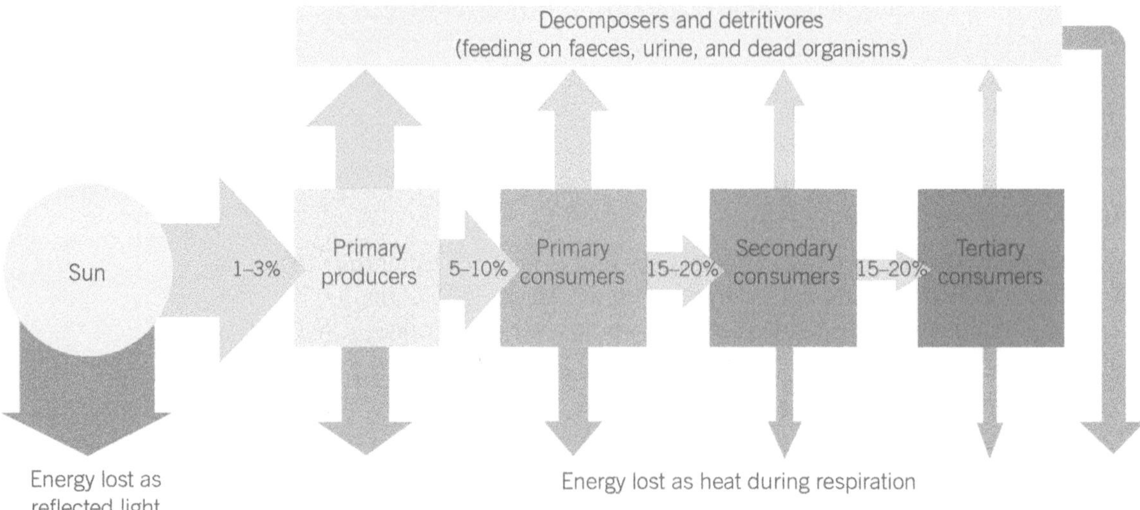

▲ **Figure 1** *Energy flow through different trophic levels of a food chain. The arrows are not to scale and give only an idea of the proportion of energy transferred at each stage. Likewise, the figures for percentage energy transfer between trophic levels are only a rough average as they vary considerably between different plants, animals, and habitats.*

Calculating the efficiency of energy transfers

Data are often presented showing the amount of energy available at each trophic level of a food chain. The energy available is usually measured in kilojoules per square metre per year (kJ m^{-2} $year^{-1}$). It is often useful to calculate the efficiency of the energy transfer between each trophic level of these food chains. This is calculated as follows:

$$\textbf{energy transfer} = \frac{\textbf{energy available after the transfer}}{\textbf{energy available before the transfer}} \times 100$$

Let us take an example. Look at Figure 3, which shows the amount of energy available at different trophic levels in a lake in the USA. Suppose you wanted to calculate the percentage efficiency of the transfer of energy from trout to humans. You would make the calculation as follows.

Energy available after the transfer (i.e., energy available to humans) = 50 kJ m^{-2} $year^{-1}$

Energy available before the transfer (i.e., energy available to trout) = 250 kJ m^{-2} $year^{-1}$

$$\textbf{percentage efficiency} = \frac{50}{250} \times 100 = \frac{5000}{250} = 20\%$$

Hint

If you were ever in any doubt about the considerable loss of energy from organisms, just think about how much food you have eaten in your whole life – and all there is to show for it is what you are now.

▲ **Figure 2** *Only about 10 per cent of the energy in the plant being eaten by this swallowtail butterfly larva will be used for its growth*

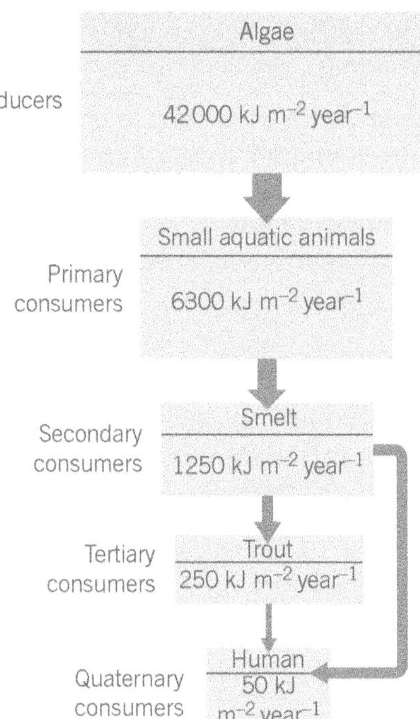

◀ **Figure 3** *Food chain in Cayuga Lake, New York State. Figures illustrate the relative amount of energy available at each trophic level in the food chain.*

Hint

When making calculations involving energy transfers, always remember that energy cannot be created or destroyed. In this type of question, this means that the total amount of energy entering a box must equal the amount of energy in the box plus the amount leaving the box.

Summary questions

1. State **three** reasons for the small percentage of energy transferred at each trophic level.
2. Explain why most food chains rarely have more than four trophic levels.
3. Using Figure 3, calculate the percentage efficiency of energy transfer between:
   a. primary consumers and secondary consumers
   b. algae and humans.

   Show your working in both cases.

Adding up the totals

Figure 4 shows the flow of energy through a terrestrial ecosystem each year. All the values are in kJ m^{-2} year^{-1}.

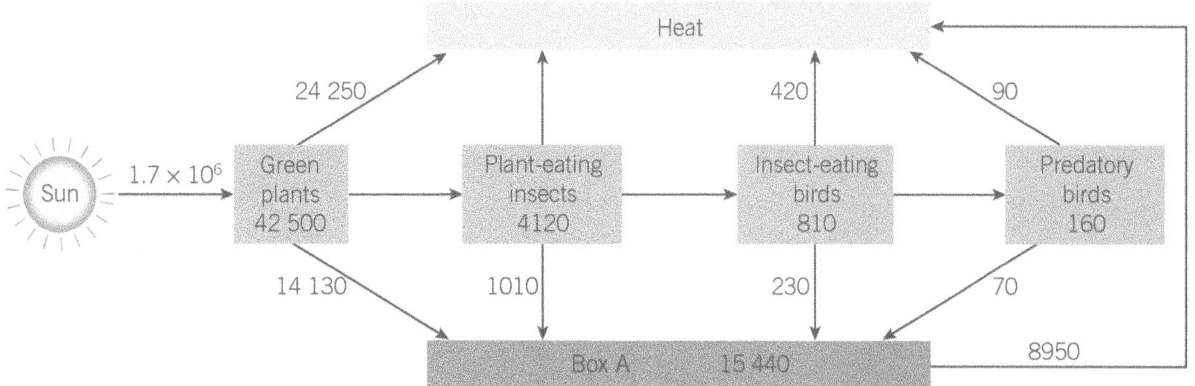

▲ **Figure 4**

1. Give the name of the group of organisms represented by box A.
2. Which group of organisms are secondary consumers?
3. Calculate the percentage efficiency with which light energy is transferred to energy in green plants. Show your working.
4. State **three** reasons why so little of the solar energy is transferred to energy in green plants.
5. Calculate the amount of energy that is lost as heat from plant-eating insects. Show your working.

21.3 Ecological pyramids

Learning objectives:

- → Describe the different types of ecological pyramid and compare their advantages and disadvantages.

Specification reference: 3.3.4.1

Diagrams of food chains and food webs are a useful means of showing what different organisms eat and therefore how energy flows between them. They do not, however, provide any quantitative information. Sometimes it is useful to know the number, mass, or amount of energy stored by organisms at each **trophic level**. To do this we construct ecological pyramids.

Pyramids of numbers

Usually the numbers of organisms at lower trophic levels are greater than the numbers at higher levels. This can be shown by drawing bars with lengths proportional to the numbers present at each trophic level. Figure 1a shows the typical number pyramid as illustrated by the food chain:

grass ⟶ rabbits ⟶ foxes

There are considerably more grass plants than rabbits and considerably more rabbits than foxes.

There can be significant drawbacks to using a number pyramid to describe a food chain:

- No account is taken of size – one giant tree is treated the same as one tiny aphid and each parasite has the same numerical value as its larger host. This means that sometimes the pyramid is not a pyramid at all (Figure 1b) or it is inverted (Figure 1c).
- The number of individuals can be so great that it is impossible to represent them accurately on the same scale as other species in the food chain. For example, one tree may have millions of greenfly living off it.

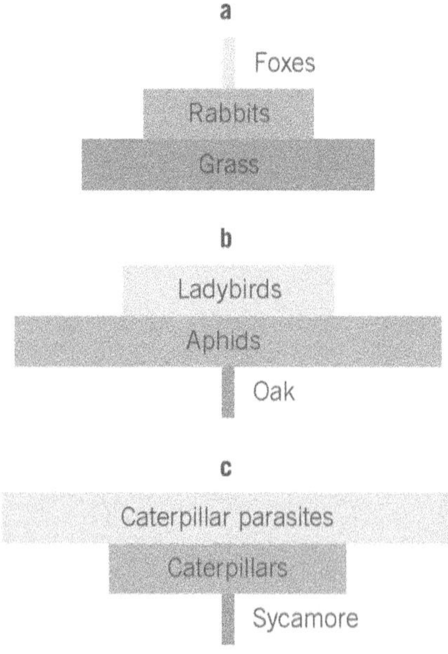

▲ **Figure 1** *Pyramids of numbers*

Pyramids of biomass

A more reliable, quantitative description of a food chain is provided when, instead of counting the organisms at each level, their biomass is measured. **Biomass** is the total mass of the plants and/or animals in a particular place. The fresh mass is quite easy to assess, but the presence of varying amounts of water makes it unreliable. The use of dry mass measurement overcomes this problem but, because the organisms must be killed, it is usually only made on a small sample, and this sample may not be representative. Biomass is measured in grams per square metre ($g\,m^{-2}$) when an area is being sampled, for example, on grassland or a seashore. Where a volume is being sampled, for example, in a pond or an ocean, it is measured in grams per cubic metre ($g\,m^{-3}$).

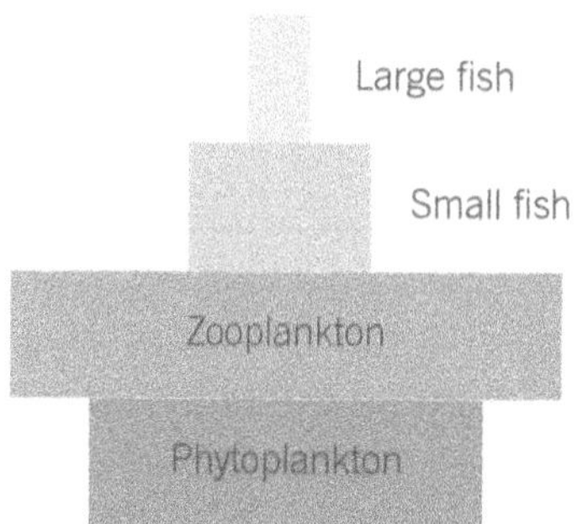

▲ **Figure 2** *Pyramid of biomass for a marine ecosystem*

In both pyramids of numbers and pyramids of biomass, only the organisms present at a particular time are shown – seasonal differences are not apparent. This is particularly significant when the biomass of some marine ecosystems is measured. Over the course of a whole year, the mass of phytoplankton (plants) must exceed that of zooplankton (animals), but at certain times of the year this is not seen. For example, in early spring around the British Isles, zooplankton consume phytoplankton so rapidly that the biomass of zooplankton is greater than that of phytoplankton (Figures 2 and 4).

Pyramids of energy

The most accurate representation of the energy flow through a food chain is to measure the energy stored in organisms. However, collecting the data for pyramids of energy (Figure 3) can be difficult and complex. Data are collected in a given area (e.g., one square metre) for a set period of time, usually a year. The results are much more reliable than those for biomass, because two organisms of the same dry mass may store different amounts of energy. For example, one gram of fat stores twice as much energy as one gram of carbohydrate. An organism with more fat will therefore have more stored energy than one with less fat, even though their biomasses are equal. The energy flow in these pyramids is usually measured in $kJ\,m^{-2}\,year^{-1}$.

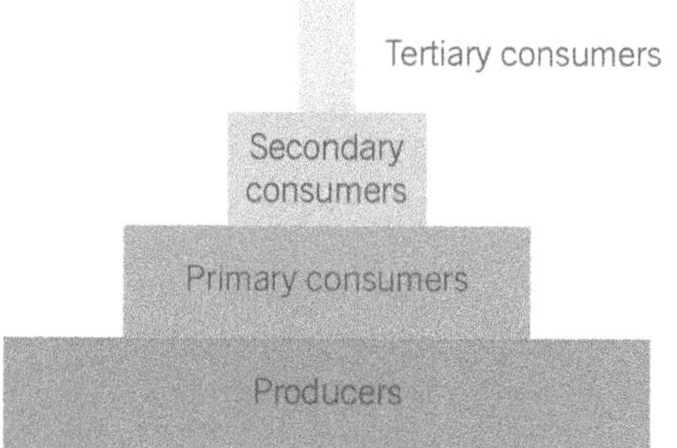

▲ **Figure 3** *Pyramid of energy based on oak trees*

Study tip

Make sure you are clear what units are used for each type of pyramid.

Summary questions

1 State **two** advantages of using a pyramid of biomass rather than a pyramid of numbers when representing quantitative information on a food chain.

2 Explain how a pyramid of biomass for a marine ecosystem may sometimes show producers (phytoplankton) with a smaller biomass than primary consumers (zooplankton).

3 Name suitable units for the measurement of biomass.

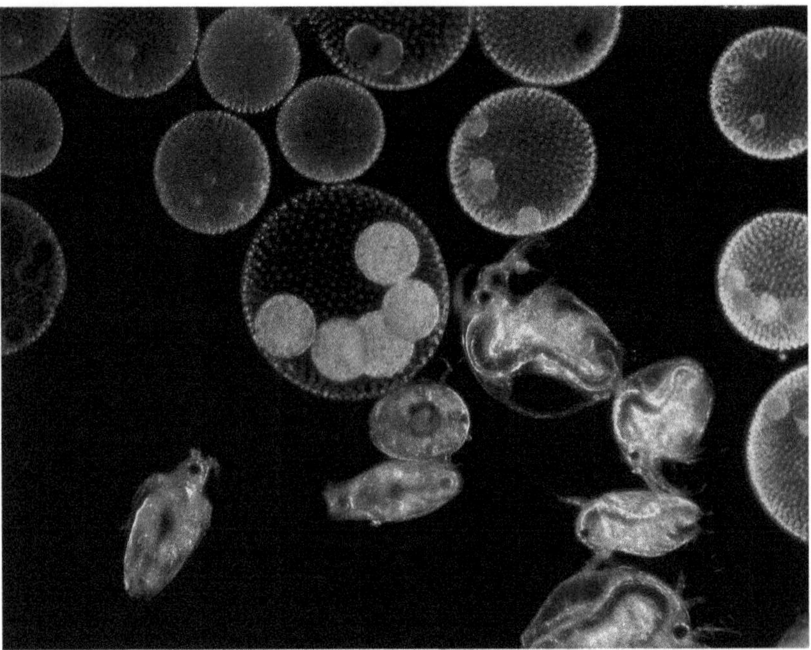

▲ **Figure 4** *Dark field photomicrograph of phytoplankton (green algae in upper region) and zooplankton (water fleas in lower region). At certain times of year the biomass of the consumer (zooplankton) may temporarily exceed the biomass of the producer (phytoplankton).*

A woodland food chain

A study of a woodland food chain produced the ecological pyramids shown in Figure 5.

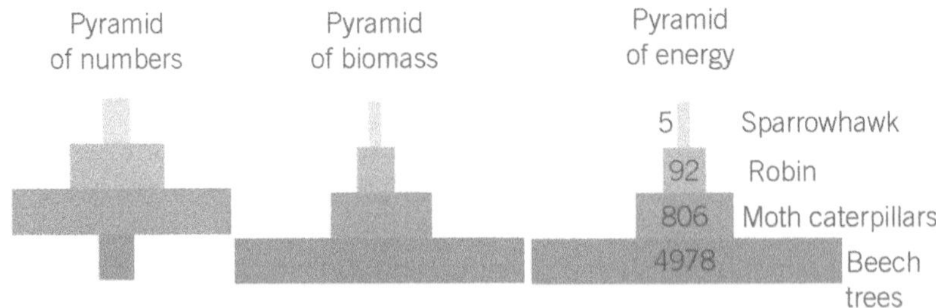

▲ **Figure 5**

1 Which organisms are the primary consumers?
2 Calculate the percentage efficiency with which energy is transferred from moth caterpillars to robins. Show your working.
3 Suggest suitable units for the figures shown in the pyramid of energy.
4 In the pyramid of numbers, the block representing beech trees is smaller than that of moth caterpillars. In the other pyramids it is larger. Explain this difference.
5 State **two** ways in which energy is lost between the robin and the sparrowhawk.
6 Parasitic fleas obtain their energy from the sparrowhawks on which they live. Redraw each of the pyramids to show how they might appear if parasitic fleas were included.

21.4 Energy transfer in agricultural ecosystems

Learning objectives:

- → Compare natural and agricultural ecosystems.
- → Explain how agricultural ecosystems are manipulated to improve productivity.

Specification reference: 3.3.4.2

You have looked at how energy is transferred through natural ecosystems. When we look around us, however, much of the landscape that we see is not a natural ecosystem but an agricultural one that has been created by humans. How then do these agricultural ecosystems differ from natural ones?

What is an agricultural ecosystem?

Agricultural ecosystems are made up largely of domesticated animals and plants used to produce food for mankind. You have seen in Topic 21.2 that there are considerable energy losses at each trophic level of a food chain. Humans are often at the third, or even fourth, trophic level of a food chain. This means that the energy we receive from the food we eat is often only a tiny proportion of that available from the Sun at the start of the food chain. Agriculture tries to ensure that as much of the available energy from the Sun as possible is transferred to humans. In effect, it is channelling the energy flowing through a food web into the human food chain and away from other food chains. This increases the productivity of the human food chain.

Hint

Agriculture is about simplifying food webs, allowing more of the available energy to be transferred to humans. This inevitably has a bad effect on the other organisms in the food web!

What is productivity?

Productivity is the rate at which something is produced. Plants are called producers because they 'produce' chemical energy (food) by converting light energy into it during photosynthesis. The rate at which plants assimilate this chemical energy is called **gross primary productivity**. It is measured for a given area over a given period of time, usually in units of $kJ\,m^{-2}\,year^{-1}$.

Some of this chemical energy is utilised by the plant for its respiration, typically 20 per cent. The remainder is known as the **net primary productivity**. This is available to the next organism in the food chain – the primary consumer. Net productivity can therefore be expressed as:

$$\textbf{net primary productivity} = \textbf{gross primary productivity} - \textbf{respiratory losses}$$

Net primary productivity is important in agricultural ecosystems and is affected by two main factors:

- the efficiency of the crop at carrying out photosynthesis. This is improved if all the necessary conditions for photosynthesis are supplied.
- the area of ground covered by the leaves of the crop.

▲ **Figure 1** *Natural ecosystems, such as this oak woodland, have lower productivity*

▲ **Figure 2** *Agricultural ecosystems, such as this soya bean field, have higher productivity*

Comparison of natural and agricultural ecosystems

Although governed by the same basic ecological principles as a natural ecosystem, agricultural ecosystems differ in a number of ways. Some of these differences are shown in Table 1. The two basic differences, energy input and productivity, are also discussed overleaf.

Energy input

In natural ecosystems the only source of energy is the Sun. Most land in Britain would be covered by forest if left to develop naturally. This is known as a climax community (see Topic 18.6). To maintain an agricultural ecosystem we have to prevent this climax community developing. We do this by excluding most of the species in that community, leaving only the particular crop that we are trying to grow.

To remove or suppress the unwanted species and to maximise growth requires an additional input of energy. This energy is used to plough fields, sow crops, remove weeds, suppress pests and diseases, feed and house animals, transport materials, and the many other tasks carried out by farmers. This additional energy comes in two forms:

- **food**. Farmers and other people that work on farms expend energy as they work. This energy comes from the food they eat.
- **fossil fuels**. As farms have become more mechanised, energy has increasingly come from the fuel used to plough, harvest, and transport crops; to produce and apply fertilisers and pesticides; and to house, feed, and transport livestock.

Hint

Remember that the energy from fossil fuels required in an agricultural ecosystem is *in addition* to solar energy, not instead of it.

Productivity

In natural ecosystems productivity is relatively low. The additional energy input to agricultural ecosystems is used to increase the productivity of a crop by reducing the effect of limiting factors on its growth. The energy used to exclude other species means that the crop has little competition for light, carbon dioxide, water, and the minerals needed for photosynthesis. The ground is therefore covered almost exclusively by the crop. Fertilisers are added to provide essential ions, and pesticides are used to destroy pests and prevent disease. Together these factors mean that productivity is much higher in an agricultural ecosystem than in a natural one.

▼ **Table 1** *Comparison of natural and agricultural ecosystems*

Natural ecosystem	Agricultural ecosystem
Solar energy only – no additional energy input	Solar energy plus energy from food (labour) and fossil fuels (machinery and transport)
Lower productivity	Higher productivity
More species diversity	Less species diversity
More genetic diversity within a species	Less genetic diversity within a species
Nutrients are recycled naturally within the ecosystem with little addition from outside	Natural recycling is more limited and supplemented by the addition of artificial fertilisers
Populations are controlled by natural means, such as competition and climate	Populations are controlled by both natural means and by use of pesticides and cultivation
Is a natural climax community	Is an artificial community prevented from reaching its natural climax

Summary questions

1. What is meant by the term 'net primary productivity'?
2. In what units is net productivity usually measured?
3. Explain why the productivity of an agricultural ecosystem is greater than that of a natural ecosystem.
4. What are the differences between the ways that energy is provided in a natural ecosystem and in an agricultural ecosystem?

Increasing productivity

Food production depends upon photosynthesis. As the rate of photosynthesis is determined by the factor that is in shortest supply (the limiting factor), it follows that there is a commercial value in determining which factor is limiting photosynthesis at any one time. By supplying more of this factor, photosynthesis, and hence food production, can be increased.

It is not feasible to control the environment of crops in natural conditions. Plants grown in greenhouses are a different matter, however. In the enclosed environment of a greenhouse it is possible to regulate temperature, humidity, light intensity, and carbon dioxide concentration. Scientists are able to predict the effects of changing these factors on the rate of photosynthesis. They can then advise commercial growers on the beneficial applications of their findings in order to increase the rate of photosynthesis and hence the growth of their crops.

It may seem logical to simply increase the level of all factors, to ensure maximum yield from photosynthesis. In practice different plants have different optimum conditions and too high a level of a particular factor may reduce yield or kill the plant altogether. For example, high temperatures may increase the yield of one species but **denature** the enzymes of another, thereby killing it. It is also uneconomic and wasteful to expend energy raising temperature and other levels beyond what is necessary. Precise control of the environment is therefore essential. This can be brought about in ways ranging from total manual control to the use of sophisticated computerised systems.

Let us take the example of carbon dioxide concentration. The average concentration in the atmosphere is around 400 parts per million (ppm). It has been shown that by raising this level to 1000 ppm the yields from tomato plants can be increased by 20 per cent or more. The actual level to which carbon dioxide concentration should be enhanced depends on many factors. Table 2 shows some of the recommended levels of carbon dioxide for maximum yield depending on certain factors.

▼ **Table 2** *Recommended levels of carbon dioxide for maximum yield*

Conditions	CO_2 / ppm
Bright sunny weather (short duration)	5000
Bright sunny weather (long duration)	1000
Cloudy weather	750
Young plants	700
Greenhouse open for ventilation	400

1 Why is the suggested level of carbon dioxide set lower on a cloudy day than a sunny one?
2 Suggest a reason for the level of 400 ppm when the greenhouse is open to the outside air.
3 Prolonged exposure to high levels of carbon dioxide causes stomata to close. Suggest how this fact may have influenced the recommended levels of carbon dioxide on bright, sunny days.

▲ **Figure 3** *Commercial greenhouse*

21.5 Chemical and biological control of agricultural pests

Learning objectives:

- → Explain what is meant by biological pests.
- → Discuss the use of pesticides in controlling pests.
- → Explain the use of biological agents in pest control.
- → Explain the use of integrated pest management in agricultural ecosystems.

Specification reference: 3.3.4.2

Hint

A pest is just an organism growing or living where we do not want it.

Hint

Details of the predator–prey relationship and the relative sizes of their populations are covered in Topic 18.5.

You saw in Topic 21.4 that agricultural ecosystems attempt to channel as much available energy from the Sun as possible along human food chains. You also know that each food chain is part of a much more complex food web. This means that many other organisms are competing for the energy in our food chains. To us, these competing organisms are pests. They may be controlled using chemicals, biological agents, or a combination of both (integrated system).

What are pests and pesticides?

Although it is difficult to define what is meant by a **pest**, it is generally taken to be an organism that competes with humans for food or space, or it could be a danger to health.

Pesticides are poisonous chemicals that kill pests. They are named after the pests they destroy – herbicides kill plants (herbs), fungicides kill fungi, and insecticides kill insects.

An effective pesticide should:

- **be specific**, so that it is only toxic to the organisms at which it is directed. It should be harmless to humans and other organisms, especially the natural predators of the pest, to earthworms, and to pollinating insects such as bees.
- **biodegrade**, so that, once applied, it will break down into harmless substances in the soil. At the same time, it needs to be chemically stable, so that it has a long shelf-life.
- **be cost-effective**, because development costs are high and new pesticides remain useful only for a limited time. This is because pests can develop genetic resistance, making the pesticide useless.
- **not accumulate**, so that it does not build up, either in specific parts of an organism or as it passes along food chains.

Biological control

It is possible to control pests by using organisms that are either predators or parasites of the pest organism. The aim is to control the pest, not to eradicate it, which might be counterproductive. If the pest was reduced to such an extent that there was insufficient food for its predators, the predators would die. The surviving pests would then be able to multiply unchecked. Ideally, the control agent and the pest should exist in balance with one another, at a level where the pest has little, or no, adverse effect.

Using biological control instead of chemical pesticides can have some advantages. These are shown in Table 1. Biological control methods also have a number of disadvantages:

- They do not act as quickly, so there is often some interval of time between introducing the control organism and a significant reduction in the pest population.
- A control organism may itself become a pest. For example, its population may increase, especially where there are few natural

predators. As the pest population is reduced, the control organism may use alternative sources of food, such as crops.

▼ **Table 1** *Comparison of biological and chemical control of organisms*

Biological control	Chemical pesticides
Very specific	Always have some effect on non-target species
Once introduced, the control organism reproduces itself so re-introduction rarely necessary	Must be reapplied at intervals, making them very expensive
Pests do not become resistant	Pests develop genetic resistance, and new pesticides have to be developed

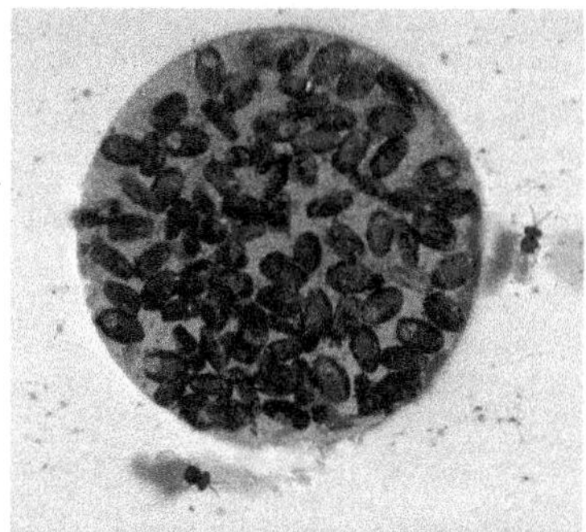

▲ **Figure 1** *Adult* Encarsia formosa *wasps hatched on card for commercial use in protecting crops from infestations of whitefly*

Integrated pest-control systems

Integrated pest-control systems aim to integrate all forms of pest control rather than being reliant on one type. The emphasis is on deciding the acceptable level of the pest rather than trying to eradicate it altogether. Eradication is, in any case, costly to carry out, counterproductive, and almost impossible to achieve. Integrated control involves:

- choosing animal or plant varieties that suit the local area and are as pest-resistant as possible
- managing the environment to provide suitable habitats, close to the crops, for natural predators
- regularly monitoring the crop for signs of pests so that early action can be taken
- removing the pests mechanically (hand-picking, vacuuming, erecting barriers) if the pest exceeds an acceptable population level
- using biological agents if necessary and available
- using pesticides as a last resort if pest populations start to get out of control.

Such systems can be effective with minimum impact on the environment.

Hint

Remember that pests are part of food webs. They provide food for other organisms. Their removal can disrupt these food webs and have long-term consequences for us all.

How controlling pests affects productivity

Pests reduce productivity in agricultural ecosystems. Weeds compete with crop plants for water, mineral ions, carbon dioxide, space, and light. As these are often in limited supply, any amount taken by the pest means less is available for the crop plant. One or more of them may become the limiting factor in photosynthesis, thus reducing the rate of photosynthesis, and hence productivity. Insect pests may damage the leaves of crops, limiting their ability to photosynthesise and thus reducing their productivity. Alternatively, they may be in direct competition with humans, eating the crop itself. Many crops are now grown in **monoculture**, and this enables insect and fungal pests to spread rapidly. Pests of domesticated animals may cause disease. The animals may not grow as rapidly, be unfit for human consumption, or die – all of which lead to reduced productivity.

▲ **Figure 2** *Monocultures, such as this area of rape seed, enable insect and fungal pests to spread rapidly*

The aim of pest control is to limit the effect of pests on productivity to a commercially acceptable level. In other words, to balance the cost of pest control with the benefits it brings. The problem is that at least two different interests are involved – the farmer who has to satisfy our demand for cheap food whilst still making a living, and the **conservation**

Summary questions

1 Pesticides are used to increase productivity. Explain how their use might sometimes reduce productivity.

2 State **two** advantages and **two** disadvantages of biological pest control.

3 Weeds are growing amongst wheat in a field. Explain how these weeds might reduce the productivity of a crop.

of natural resources, which will enable us to continue to have food in the future. The trick is to balance these two, often conflicting, interests.

To weed or not to weed?

The graph in Figure 3 shows the effects of weeds on the productivity of two crops, wheat and soya bean.

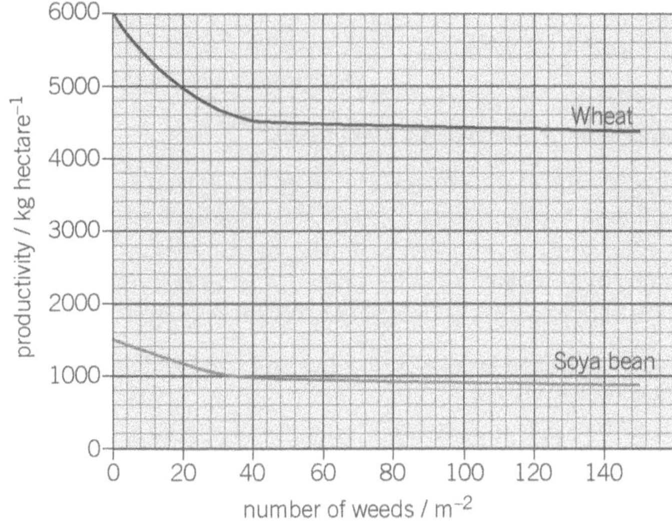

▲ **Figure 3**

1 Describe the effects of weeds on the productivity of wheat.

2 A herbicide that reduces the number of weeds from 40 m^{-2} to 0 m^{-2} is applied to both crops. Which crop would show the greatest percentage change in productivity?

3 It will cost a farmer £100 to treat each hectare of his wheat crop with a herbicide. The herbicide will reduce the number of weeds from 40 m^{-2} to 20 m^{-2}. He can sell his wheat at £150 a tonne (one tonne = 1000 kg). Is it economically worthwhile for the farmer to apply the herbicide to his crop? Explain your reasoning.

A mighty problem

The two-spotted spider mite, *Tetranychus urticae*, is an important pest of crops, especially those in greenhouses. Control is mostly achieved using chemicals. However, the spider mite has increasingly developed resistance to these chemicals and they are therefore less effective in controlling its populations.

Studies have been carried out to investigate the use of biological control to combat spider mites. One such study investigated the effectiveness of the predatory mite *Phytoseiulus persimilis* against the two-spotted spider mite. This predatory mite feeds on the spider mite. Mites were introduced into two separate groups of 100 bean plants as follows:

- Experiment 1 – spider mites only

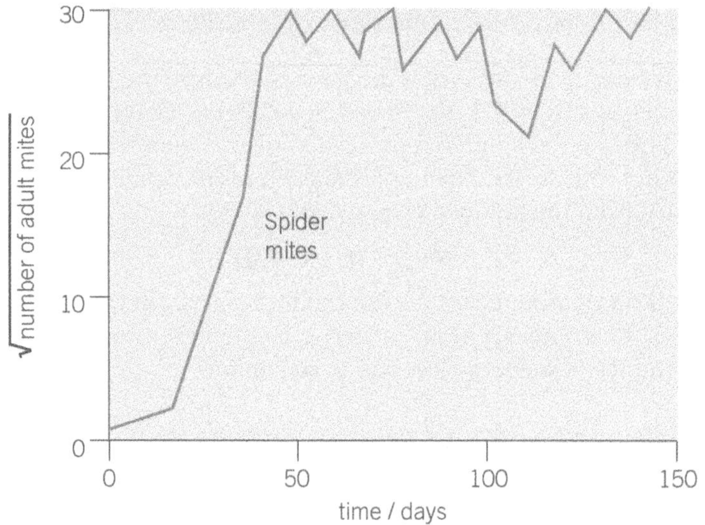

- Experiment 2 – spider mites and predatory mites

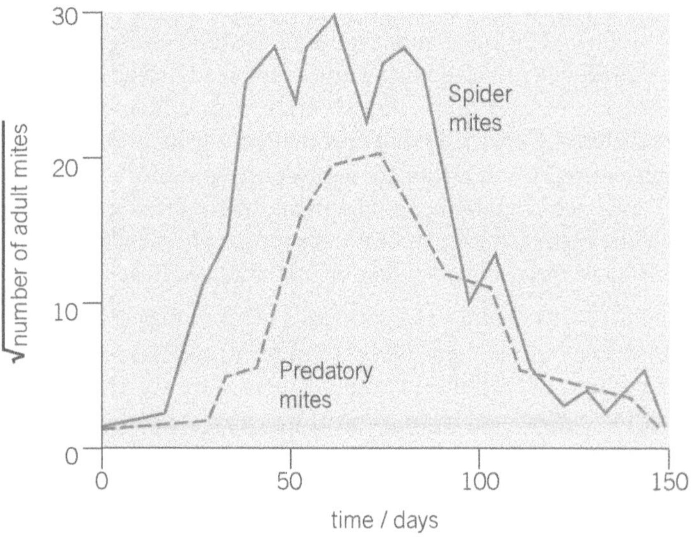

▲ **Figure 4**

1 Describe and explain the differences between the spider mite populations in experiment 1 and experiment 2.
2 Comment on the effectiveness of predatory mites in controlling populations of spider mites.
3 Suggest what the levels of the two populations in experiment 2 might be over a period of 150–300 days if the experiment was continued. Explain the reasons for the levels you suggest.

▲ **Figure 5** *Biological pest control: orange predatory mite (*Phytoseiulus persimilis*) attacking the red spider mite (*Tetranychus urticae*)*

21.6 Intensive rearing of domestic livestock

Learning objectives:

→ Describe how rearing animals intensively increases the efficiency of energy conversion.

Specification reference: 3.3.4.2

Intensive rearing of livestock is designed to produce the maximum yield of meat, eggs, and milk at the lowest possible cost. Cows, pigs, chickens, and turkeys are the animals most commonly reared intensively. These animals are consumers. The net production for consumers can be calculated from the following equation:

$$\mathbf{N = I - (F + U + R)}$$

where **N** = net production, **I** is the chemical energy in the ingested food, **F** is the chemical energy in faeces, **U** is the chemical energy in urine, and **R** is the energy loss due to respiration.

Intensive rearing and energy conversion

As energy passes along a food chain only a small percentage passes from one organism in the chain to the next (see Topic 21.2). This is because much of the energy is lost as heat during respiration. Intensive rearing of domestic livestock is about converting the smallest possible amount of food energy into the greatest quantity of animal mass. One way to achieve this is to minimise the energy losses from domestic animals during their lifetime. This means that more of the food energy taken in by the animals will be converted into body mass, ready to be passed on to the next link in the food chain, namely us. Energy conversion can be made more efficient by ensuring that as much energy from respiration as possible goes into growth rather than other activities or other organisms. This is achieved by keeping animals in confined spaces, such as small enclosures, barns, or cages, a practice often called 'factory farming'. This increases the energy-conversion rate because:

- movement is restricted and so less energy is used in muscle contraction
- the environment can be kept warm in order to reduce heat loss from the body (most intensively reared species are warm blooded)
- feeding can be controlled so that the animals receive the optimum amount and type of food for maximum growth, with no wastage
- predators are excluded so that there is no loss to other organisms in the food web.

Other means of improving the energy-conversion rate include:

- selective breeding of animals to produce varieties that are more efficient at converting the food they eat into body mass
- using hormones to increase growth rates.

Hint

It is worth remembering that intensively reared animals still require large amounts of plant material (producers) as a source of food and this needs to be grown somewhere.

Summary questions

1 How does rearing animals intensively in small covered enclosures increase their energy-conversion rate?

2 Suggest a reason why keeping animals in the dark for longer periods might improve the energy-conversion rate.

Features of intensive rearing of livestock

Food is essential for life. With an ever-expanding human population, there is pressure to produce more and more food intensively. What then are the main features of intensive rearing? These include:

- **efficient energy conversion**. By restricting wasteful loss of energy, more energy is passed to humans along the food chain.

- **low cost**. Foods such as meat, eggs, and milk can be produced more cheaply than by other methods.
- **quality of food**. It is often argued that the taste of the foods produced by intensive rearing is inferior to foods produced less intensively.
- **use of space**. Intensive rearing uses less land whilst efficient production means that less of the countryside is required for agriculture, leaving more as natural habitats.
- **safety**. Smaller, concentrated units are easier to control and regulate. The high density animal populations are more vulnerable to the rapid spread of disease but it is easier to prevent infections being introduced from the outside and to isolate the animals if this happens.
- **disease**. Large numbers of animals living in close proximity means that infections can spread easily amongst them. To control this, the animals are regularly given antibiotics.
- **use of drugs**. Overuse of antibiotics to prevent disease in animals has lead to the evolution of antibiotic resistance. This resistance can be transferred to bacteria that cause human diseases, making their treatment with certain antibiotics ineffective. Other drugs may be given to animals to improve their growth or to reduce aggressive behaviour. These may alter the flavour of the meat or pass into the meat and then into humans, affecting human health.
- **animal welfare**. The larger intensive farms have the resources to maintain a high level of animal welfare and are more easily regulated. However, animals are kept unnaturally and this may cause stress, resulting in aggressive behaviour. This may cause them to harm each other or themselves, which is why battery chickens are de-beaked. Restricted movement can lead to osteoporosis and joint pain. The well-being of animals may be sacrificed for financial gain.
- **pollution**. Intensively reared animals produce large concentrations of waste in a small area. Rivers and ground water may become polluted. Pollutant gases may be dangerous and smell. Large intensive farms may have their own disposal facilities that enable them to treat waste more effectively than smaller farms.
- **reduced genetic diversity**. Selective breeding is used to develop animals with high energy-conversion rates and a tolerance of confined conditions. This reduces the genetic diversity of domestic animals, resulting in the loss of genes that might later prove to have been beneficial.
- **use of fossil fuels**. High energy-conversion rates are possible because fossil fuels are used to heat the buildings that house the animals, in the production of the materials in the buildings (especially cement), and in the production and transportation of animal feeds. The carbon dioxide emitted increases global warming.

▲ **Figure 1** *Rows of intensively reared battery hens*

1. Explain how antibiotic resistance in bacteria causing animal diseases can be transferred to bacteria causing human disease.
2. Egg production often involves hens being kept in battery cages. Give reasons for and against this form of egg production.
3. Discuss some of the ethical issues that arise from the features of intensive farming listed above.

Practice questions: Chapter 21

1 Potato plants originate from the Andes mountains in South America. They are adapted for survival in a cool climate. The potatoes we eat are food storage organs, called tubers, and are produced on underground stems.

Figure 1 shows the rate of photosynthesis and respiration for one variety of potato plant.

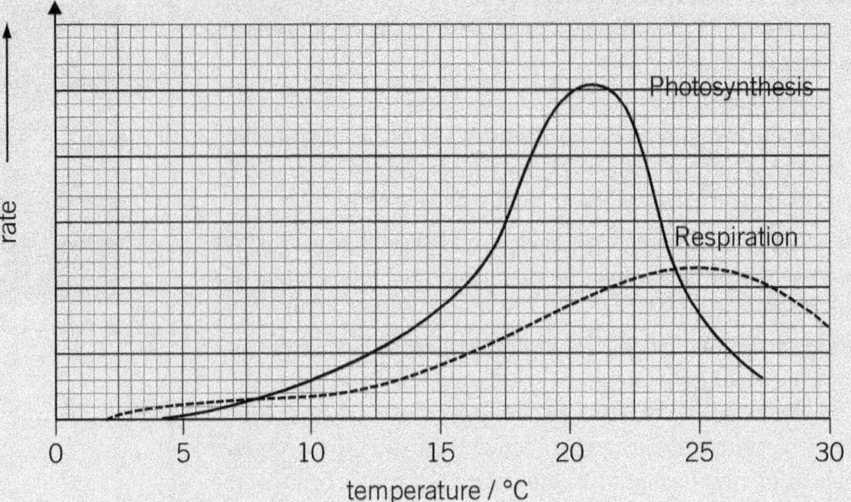

▲ **Figure 1**

(a) Between which temperatures is there a net gain in energy by the potato plant? *(1 mark)*

(b) When this variety was grown in a hot climate, with a mean daytime temperature of 23.5 °C, it failed to produce tubers. Use information in Figure 1 to explain why no tubers were produced. *(2 marks)*

(c) Suggest what causes the rate of photosynthesis to decrease at temperatures above 21 °C. *(2 marks)*

AQA 2003

2 Purple loosestrife is a plant which grows in Europe. It was introduced into the USA where it became a pest.

(a) Suggest why purple loosestrife became a pest when it was introduced into the USA, but is not a pest in Europe. *(2 marks)*

(b) A European beetle was tested to see whether it could be used for the biological control of purple loosestrife in the USA. In an investigation beetles were released in an area where purple loosestrife was a pest. The table shows some of the results.

Time after releasing beetles / years	Mean number of purple loosestrife stems per square metre	Mean number of beetles per square metre
1	22	5
2	8	40
3	6	68
4	7	62

Are the beetles effective in controlling purple loosestrife? Give evidence from the table to support your answer. *(2 marks)*

(c) Fire-ants are a serious pest in parts of the USA. An investigation was carried out to find the best way to control the fire-ant population. Figure 2 shows the results of this investigation.

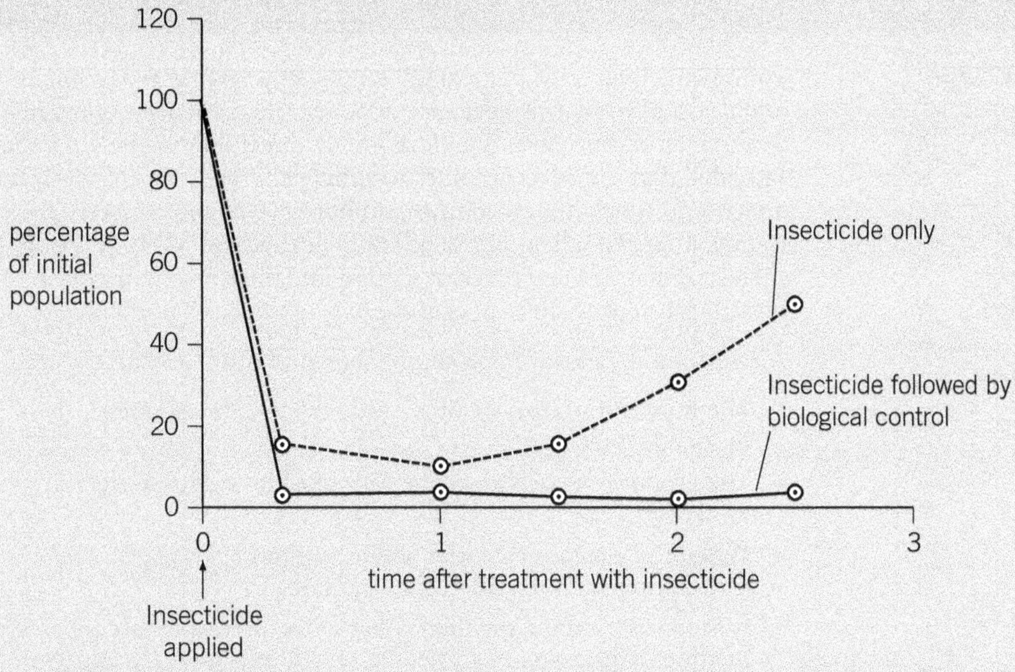

▲ **Figure 2**

(i) Describe the effect of using insecticide followed by biological control.

(ii) Explain the change in fire-ant population over the period when they were treated with insecticide alone. *(5 marks)*

(d) Give the advantages and disadvantages of using biological control. *(6 marks)*

AQA 2006

3 Figure 3 shows the transfer of energy through a cow. The figures are in kJ × 10^6 $year^{-1}$.

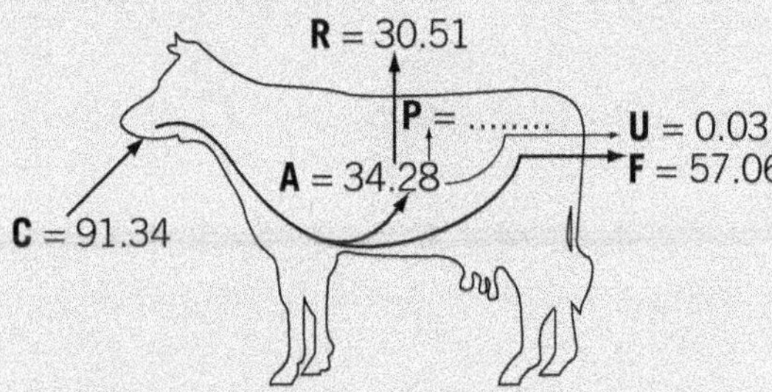

Key: **A** = energy absorbed from the gut
C = energy consumed in food
F = energy lost in faeces
P = energy used in production of new tissue
R = energy lost by respiration
U = energy lost in urine

▲ **Figure 3**

(a) (i) Complete the following equation for the energy used in the production of new tissue. Use only the letters **C**, **F**, **R**, and **U**.

P = ... *(1 mark)*

(ii) Calculate the value of **P** in kJ × 10^6 $year^{-1}$. *(1 mark)*

(b) It has been estimated that an area of 8100 m^2 of grassland is needed to keep one cow. The productivity of grass is 21 135 kJ m^{-2} $year^{-1}$. What percentage of the energy in the grass is used in the production of new tissue in one cow? Show your working. *(2 marks)*

(c) Keeping cattle indoors, in barns, leads to a higher efficiency of energy transfer. Explain why. *(1 mark)*

AQA 2004

Answers to the Practice Questions are available at
www.oxfordsecondary.com/oxfordaqaexams-alevel-biology

22 Nutrient cycles

22.1 The carbon cycle

Learning objectives:

- → Explain that nutrients are cycled within ecosystems.
- → Describe how carbon is cycled in ecosystems.
- → Explain the roles of decomposers and saprophytic microorganisms in the carbon cycle.

Specification reference: 3.3.5.1 and 3.3.5.2

You saw in Topic 21.2 that energy enters an **ecosystem** as sunlight and is lost as heat. This heat cannot be recycled. The flow of energy through an ecosystem is therefore in one direction, that is, it is linear. Provided that the Sun continues to supply energy to Earth, this is not a problem. Nutrients, by contrast, do not have an extraterrestrial source. There is only a certain quantity of them on Earth. It is essential therefore that elements such as carbon and nitrogen are recycled. The flow of nutrients within an ecosystem is not linear, but cyclic.

All nutrient cycles have one simple sequence at their heart.

- The nutrient is taken up by **producers** (plants) as simple, inorganic molecules.
- The producer incorporates the nutrient into complex organic molecules.
- When the producer is eaten or decomposes after death, the nutrient passes into **consumers** (animals).
- It then passes along the food chain when these animals are eaten by other consumers.
- When the producers and consumers die, their complex molecules are broken down by **saprophytic microorganisms** (decomposers) that release the nutrient in its original, simple form. The cycle is then complete.

Although other processes and non-living sources are also involved, it is this sequence, illustrated in Figure 1, that forms the basis of all nutrient cycles.

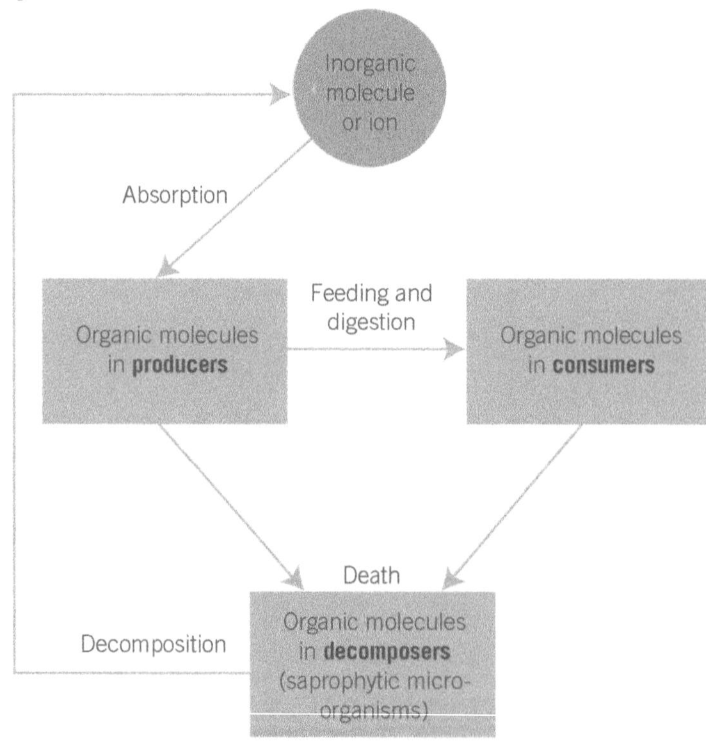

▲ **Figure 1** *Basic sequence of all nutrient cycles*

Carbon is a component of all the major macromolecules in living organisms. It is the basic building block for life itself. The main source of carbon for terrestrial organisms is the carbon dioxide in the atmosphere and yet it makes up a mere 0.04 per cent of the air around us. It is not surprising therefore that the turnover of atmospheric carbon dioxide is considerable. Photosynthetic organisms remove it from the air to build it up into macromolecules such as carbohydrates, fats, and proteins. All organisms return it to the air through the process of respiration.

Variations in the rates of respiration and photosynthesis give rise to short-term fluctuations in the proportions of oxygen and carbon dioxide in the atmosphere. For example, the concentration of carbon dioxide at night is greater than during the day. This is because the absence of light means that no photosynthesis can take place. Respiration is still occurring, however, although its rate may be slightly reduced as the temperature is normally lower at night. Similarly, the daytime concentration of carbon dioxide on a summer's day, when photosynthesis is at its greatest, is lower than on a winter's day.

Globally the concentration of carbon dioxide in the atmosphere has increased over the past few hundred years. The main reasons for this are two human activities:

- **The combustion of fossil fuels**, such as coal, oil and peat, has released carbon dioxide that was previously locked up within these fuels.
- **Deforestation**, especially of the rainforests, has removed enormous amounts of photosynthesising biomass and so less carbon dioxide is being removed from the atmosphere.

The additional production of carbon dioxide through these human activities is threatening to upset the delicate balance of the carbon cycle. As carbon dioxide is a greenhouse gas this additional production is contributing to global warming. This issue is discussed in Topic 22.2.

The oceans contain a massive reserve of carbon dioxide. This store is some 50 times greater than that in the atmosphere. It helps to keep the concentration of atmospheric carbon dioxide more or less constant. Some of any excess carbon dioxide in the atmosphere dissolves in the waters of the oceans. When atmospheric concentration is low, the reverse occurs. Aquatic photosynthetic organisms (phytoplankton) use this dissolved carbon dioxide to form the macromolecules that make up their bodies.

The carbon in photosynthetic organisms passes along food chains to animals. On their death, both plants and animals are usually broken down by saprophytic microorganisms known collectively as **decomposers**. Saprophytic microorganisms secrete enzymes on to the dead organisms. These enzymes break down complex molecules into smaller, soluble molecules that the saprophytic microorganisms absorb by diffusion. The carbon in the dead organisms is then released as carbon dioxide during respiration by the decomposer.

If decay is prevented for any reason, then the organisms may become fossilised into coal, oil, or peat. Not all parts of organisms decompose. The shells and bones of aquatic organisms sink to the bottom of the oceans and, over millions of years, form carbon-containing sedimentary rocks such as chalk and limestone. This carbon eventually returns to the atmosphere as these rocks are weathered.

Study tip

It is important to remember the key reactions in photosynthesis, in which carbon dioxide is fixed into biological molecules, and in respiration, in which carbon dioxide is released from these molecules.

Synoptic link

The processes of photosynthesis and respiration are dealt with in Chapters 19 and 20 respectively.

▲ **Figure 2** *Burning rainforest contributes to global warming not only by releasing carbon dioxide but also by removing enormous amounts of photosynthesising biomass that would have absorbed carbon dioxide from the atmosphere*

Hint

All organisms die and so decomposers feed on all trophic levels in food chains and food webs.

▲ **Figure 3** *Fungi such as this bracket fungus are decomposers that help to recycle carbon*

The carbon cycle is summarised in Figure 4.

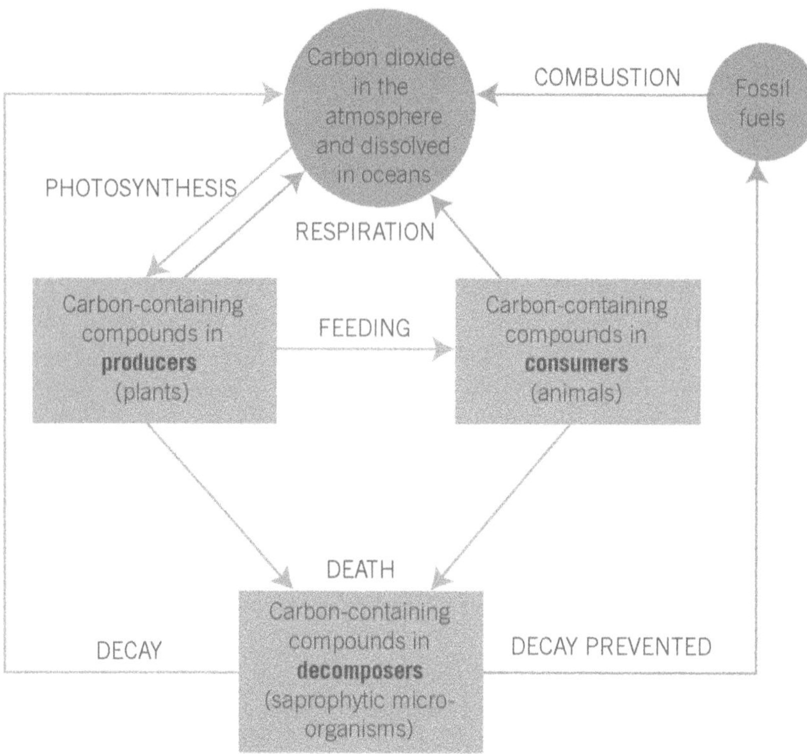

▲ **Figure 4** *The carbon cycle*

Summary questions

1. Explain why the carbon dioxide concentration of the atmosphere is often less on a summer's day than on a winter's day.
2. Figure 5 is an illustration of the carbon cycle. Each box represents a process. Name the process in each of the boxes A, B, C, and D.

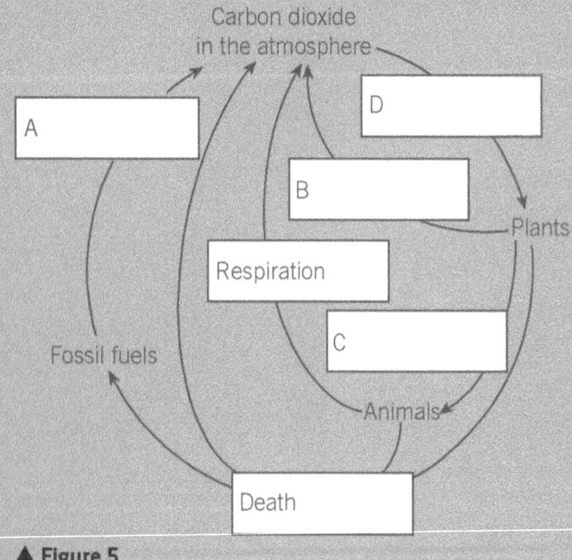

▲ **Figure 5**

3. Suggest which of the sequences below best represents the flow of carbon (C) through the carbon cycle:
   - **a** atmospheric C ⟶ respiration ⟶ animal C ⟶ plant C ⟶ decay ⟶ atmospheric C
   - **b** atmospheric C ⟶ plant C ⟶ photosynthesis ⟶ atmospheric C
   - **c** atmospheric C → animal C ⟶ respiration and decay ⟶ atmospheric C
   - **d** atmospheric C ⟶ plant C ⟶ animal C ⟶ respiration and decay ⟶ atmospheric C
   - **e** atmospheric C ⟶ plant C ⟶ photosynthesis ⟶ animal C ⟶ atmospheric C

22.2 The greenhouse effect and global warming

It was mentioned in Topic 22.1 that carbon dioxide is a greenhouse gas. A greenhouse gas is any gas that has a greenhouse effect.

Learning objectives:

- → Explain what is meant by the greenhouse effect.
- → Identify the major greenhouse gases and explain why they are increasing.
- → Discuss the consequences of global warming and the contribution made by greenhouse gases.

Specification reference: 3.3.5.2

The greenhouse effect

The greenhouse effect is a natural process that occurs all the time and keeps average global temperatures at around 17 °C. Without it, the average temperature at the surface of the Earth would be about minus 18 °C. This effect is the result of the heat and light of the Sun (solar radiation) that reaches our planet. Some solar radiation is reflected back into space, some is absorbed by the atmosphere, and, fortunately, some reaches the Earth's surface. Some of this radiation reaching the Earth's surface is reflected back as heat and is lost into space. However, some is radiated back to Earth by clouds and the 'greenhouse gases' that form part of the atmosphere. The gases trap this heat close to the Earth's surface, keeping it warm. The greenhouse gases act like the glass in a greenhouse by trapping heat beneath them, hence the name 'greenhouse effect'. The process is illustrated in Figure 1.

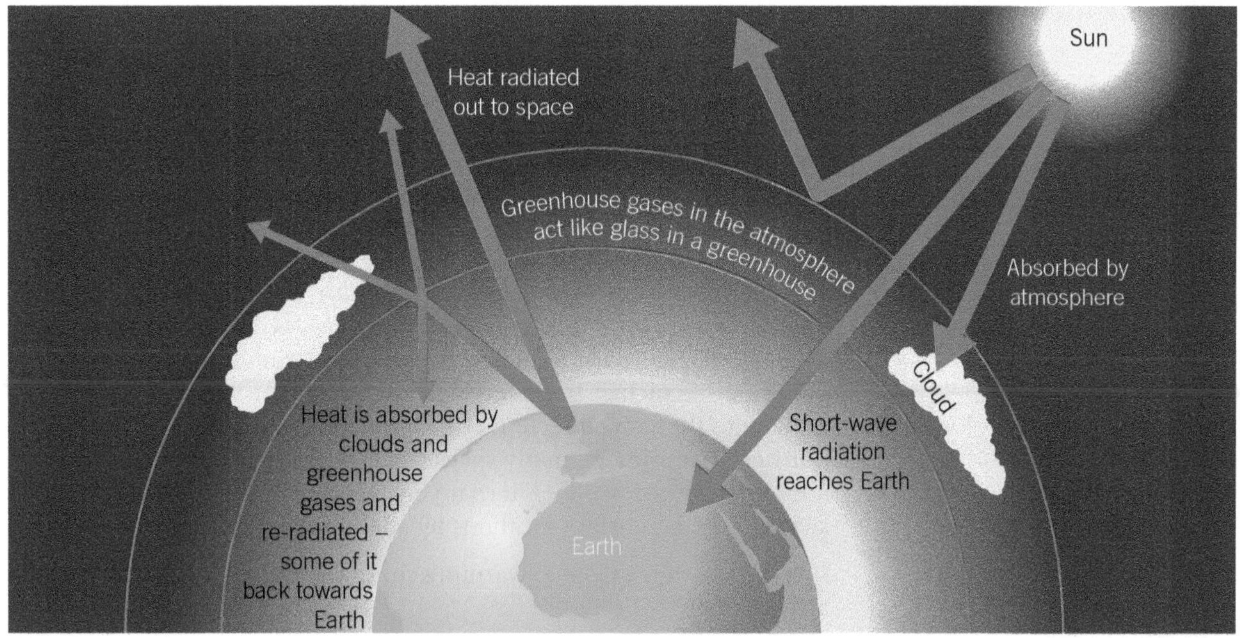

▲ **Figure 1** *The greenhouse effect*

Greenhouse gases

The most important greenhouse gas is carbon dioxide, partly because there is so much of it and partly because it remains in the atmosphere for so much longer than other greenhouse gases (100 years, compared with 10 years for methane). It has been estimated that 50–70 per cent of global warming is due to carbon dioxide in the atmosphere. It is mainly as a result of human activities that the concentration of carbon dioxide is increasing, enhancing the greenhouse effect and causing environmental concerns. Another natural greenhouse gas is methane. Methane is

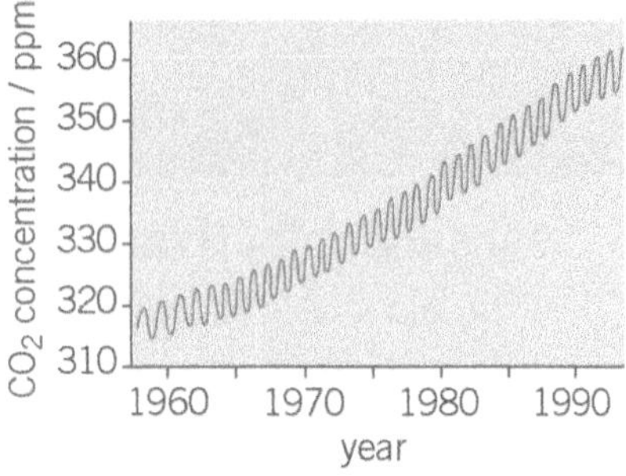

▲ **Figure 2** *Trends in atmospheric carbon dioxide concentration since 1958 (recorded at Mauna Loa, Hawaii)*

produced when microorganisms break down the organic molecules of which organisms are made. This occurs mostly in two situations:

- when decomposers break down the dead remains of organisms
- when microorganisms in the intestines of primary consumers, such as cattle, digest the food that has been eaten.

Global warming

The mean global temperature has increased by around 0.6 °C since 1900, a change known as global warming. The Earth has always shown periodic fluctuations in temperature. We cannot therefore say for certain that this recent temperature increase is due to human activities, such as the burning of fossil fuels and deforestation, generating additional carbon dioxide. What we can say is that the concentration of carbon dioxide has risen from 270 parts per million (ppm) before the industrial revolution to 370 ppm today. This rise in carbon dioxide concentration over recent years is shown in Figure 2. We also know that carbon dioxide is a greenhouse gas. Most scientists therefore think that these human activities have contributed to global warming.

Consequences of global warming

Even if humans continue to release greenhouse gases at the present rate, it is by no means certain what the effects would be. Carbon dioxide concentration could even be reduced naturally, as phytoplankton and other plants increase their rates of photosynthesis.

Global warming is expected to bring about changes in temperature and precipitation, the timing of the seasons, and the frequency of extreme events such as storms, floods, and drought. Climate change will affect the **niches** that are available in a **community**. As each organism is adapted to a particular niche the distribution of species will alter. If the rate of climate change is slow, species may have time to gradually migrate to new areas, where they will compete for the available niches. This could lead to the loss of native species that occupy those niches. It is already being noted that many species are moving in the same direction as the climate. For example, in the northern hemisphere, many species of butterflies, lichens, and birds are moving northwards. Alternatively, some species may be able to survive in their current locations by adapting to different food sources.

If present trends in global warming continue, the following changes are possible:

- Melting of polar ice caps could cause the extinction of some wild plants and animals, for example, the polar bear, and cause sea levels to rise.
- A rise in sea level due to the thermal expansion of oceans could flood low-lying land, including much of Bangladesh. It could also flood many major cities, as well as fertile land such as the Nile delta. Salt water would extend further up rivers and make cultivation of crop plants difficult.
- Higher temperatures and less rainfall could lead to the failure of the present crops in some areas. More drought-resistant species would have to be grown and, in severe cases, the land might become desert and so not sustain any crops at all. The distribution of wild plants in

Synoptic link

To help you to understand the terms in this section, it is worth revising Topic 18.1.

Hint

Global warming changes the niches of many organisms and so affects the composition of communities.

▲ **Figure 3** *Carbon dioxide from domestic and industrial burning of fossil fuels may contribute to global warming*

these areas would naturally change, with only **xerophytes** being able to survive. As animals are ultimately dependent on plants for food, their distribution would also be affected. Species that can feed on xerophytes and can withstand hot dry conditions would move in.

- Greater rainfall and intense storms would occur in some areas, due to the disturbance of climate patterns. Again, the distribution of plants and animals would change in favour of those adapted to withstand such conditions.
- The life cycles and populations of insect pests would alter as they adapt to the changed conditions. As insects carry many human and crop **pathogens**, tropical diseases could spread towards the poles. Species that damage crops could move towards the poles into areas that they have not previously inhabited. For example, the green shield bug, a common pest of vegetable crops in Mediterranean countries, has recently been found in southern England for the first time.

However, it should be remembered that there could also be benefits to some parts of the world. The increased rainfall would fill reservoirs, the warmer temperatures would allow crops to be grown where it is presently too cold, the rate of photosynthesis (and hence productivity) could increase, and it might be possible to harvest twice a year instead of once.

▲ **Figure 4** *Global warming is contributing to the melting of polar ice*

Digging into the past

To determine whether there is a correlation between the concentration of carbon dioxide in the atmosphere and global temperature requires measurements of both these factors over many thousands of years. But no measurements were taken thousands of years ago. How then can we look into the past? The answer lies in the ingenuity of scientists.

We can determine the age of fossils by measuring the amount of radioactive decay of an **isotope** of carbon (radiocarbon dating). Fossil shells from beneath the ocean bed contain oxygen isotopes as well as carbon ones. The proportion of these oxygen isotopes reflects the proportion of them in the sea water at the time the organism took them into its shell when it was alive. The proportion of each oxygen isotope in the sea water varied according to the coming and going of each ice age, which in turn reflected the global temperature. By measuring the isotopes of carbon and oxygen in fossil shells we get both an estimate of its age and of the temperature of the Earth at that time.

What about the carbon dioxide concentration? When ice forms, air bubbles become trapped in it. If we drill down into the Antarctic glaciers, the deeper we go the older is the ice. The air bubbles in the ice will reflect the composition of the atmosphere at the time when the ice was formed. Scientists have drilled 3.6 km down through an Antarctic glacier. At this depth the ice is 420 000 years old. By measuring the carbon dioxide concentration of air bubbles in the ice at different depths, we can find out how its concentration in the atmosphere has changed over the past 420 000 years.

The results of these two different studies are shown in the graphs in Figure 5.

1. Giving reasons from the information in Figure 5, state whether you think there is a correlation between the concentration of carbon dioxide in the atmosphere and global temperature.
2. Do the data indicate that changes in carbon dioxide concentration cause changes in global temperature. Explain your answer.

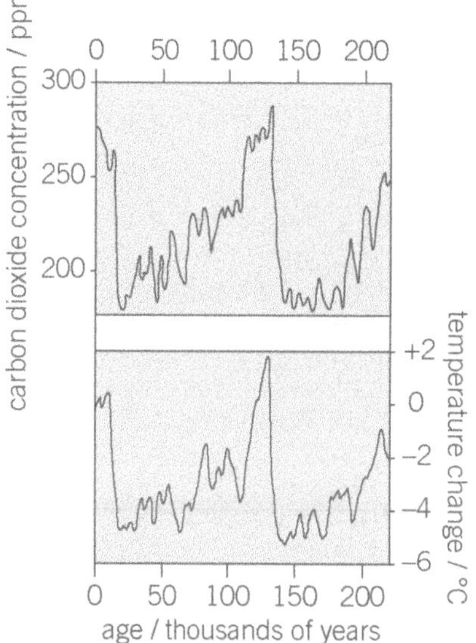

▲ **Figure 5** *Comparison of changes in atmospheric carbon dioxide (from analysis of bubbles in glacial ice) and temperature changes (from oxygen isotope studies)*

Summary questions

1 In Figure 2 the line on the graph fluctuates. Suggest an explanation for this regular up-and-down pattern.

2 Tropical rainforest around the world is being burnt so that the land can be used for other purposes. Explain all the possible effects on carbon dioxide levels in the atmosphere and global warming if the cleared land was used to produce palm oil for the manufacture of biofuels for use in vehicles.

3 State how the effect on global warming might be different if the land was used to raise beef cattle.

Global warming and crop yields

Most studies of the influence of global warming on crop yields have concentrated on the effect of an increase in daytime temperature. A study carried out on the yield of rice, however, looked at the effects of an increase in both the maximum (day) temperature and the minimum (night) temperature.

The study found that there was no correlation between the maximum (day) temperature and crop yields. The results for the effects of the minimum (night) temperature on the yield of rice are shown in the graphs in Figure 6.

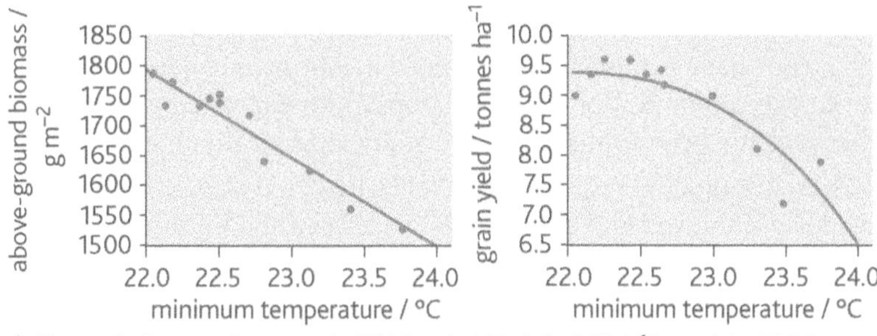

▲ **Figure 6** *Source: Peug* et al., *PNAS, vol. 101, July 2004 (Copyright 2004, National Academy of Science, USA)*

1 Describe the effect of minimum (night) temperature on the grain yield of rice.
2 Calculate the percentage decrease in above-ground biomass that occurs when the minimum temperature is increased from 22.0 °C to 24.0 °C. Show your working.
3 Suggest a possible explanation for some of the reduction in yield when the minimum temperature increases.

Global warming and insect pests

Global warming can affect the numbers and life cycles of insect pests. Rice crops can be infected by the rice stem-borer, *Chilo suppressalis*, a moth whose larvae (caterpillars) over-winter in the paddy fields. They emerge in spring and enter the stems of rice plants, limiting growth and sometimes resulting in the death of the plant.

A study in Japan looked at the mean winter (November to April) temperature over the past 50 years, how this temperature affected the number of larvae in rice plants, and how, in turn, this affected the yield of rice. The results are summarised in the graphs in Figure 7.

1 Describe and explain the relationship shown in graph B.
2 From the graphs, suggest an explanation of how global warming can lead to a loss of yield in rice crops.
3 From the graphs, calculate the likely % loss in rice yield in 1980 as a result of rice stem-borer larvae. Show your working.
4 Suggest an explanation for why infection with stem-borer larvae might kill rice plants.

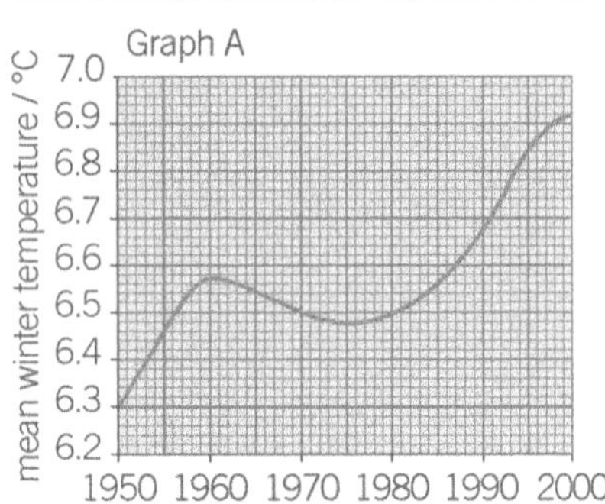

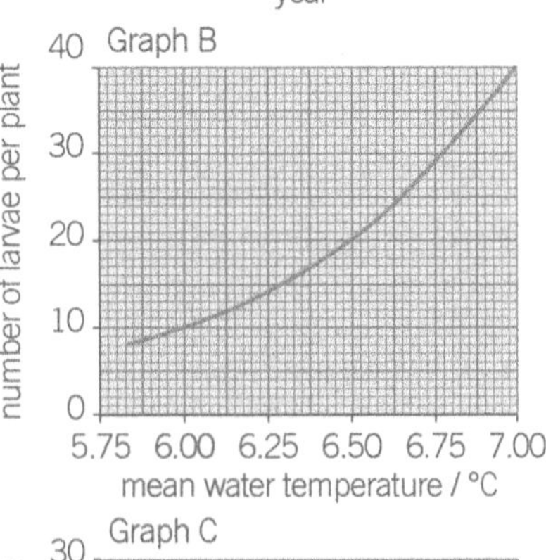

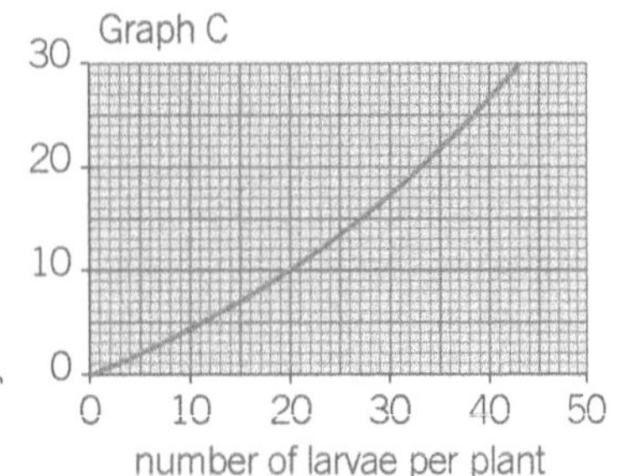

▲ **Figure 7**

22.3 The nitrogen cycle

All living organisms require a source of nitrogen from which to manufacture proteins, nucleic acids, and other nitrogen-containing compounds. Although 78 per cent of the atmosphere is nitrogen, there are very few organisms that can use nitrogen gas directly. Plants take up most of the nitrogen they need in the form of nitrate ions (NO_3^-) from the soil. These ions are absorbed, using **active transport**, by the root hairs. This is where nitrogen enters the living component of the ecosystem. Animals obtain nitrogen-containing compounds by eating and digesting plants.

Nitrate ions are very soluble and easily leach through the soil, beyond the reach of plant roots. In natural ecosystems, the nitrate concentration in the soil water is restored through the recycling of nitrogen-containing compounds. In agricultural ecosystems, the concentration of soil nitrate can be further increased by the addition of fertilisers. When plants and animals die, the process of decomposition begins, in a series of steps by which microorganisms replenish the nitrate concentration in the soil. This release from decomposition is most important because, in natural ecosystems, there is very little nitrate available from other sources.

There are four main stages in the nitrogen cycle (Figure 1), **ammonification, nitrification, nitrogen fixation**, and **denitrification**, each of which involves **saprophytic microorganisms**.

Learning objectives:

- → Describe the role of microorganisms in the cycling of nitrogen in ecosystems.
- → Explain the terms ammonification, nitrification, nitrogen fixing, and denitrification.

Specification reference: 3.3.5.3

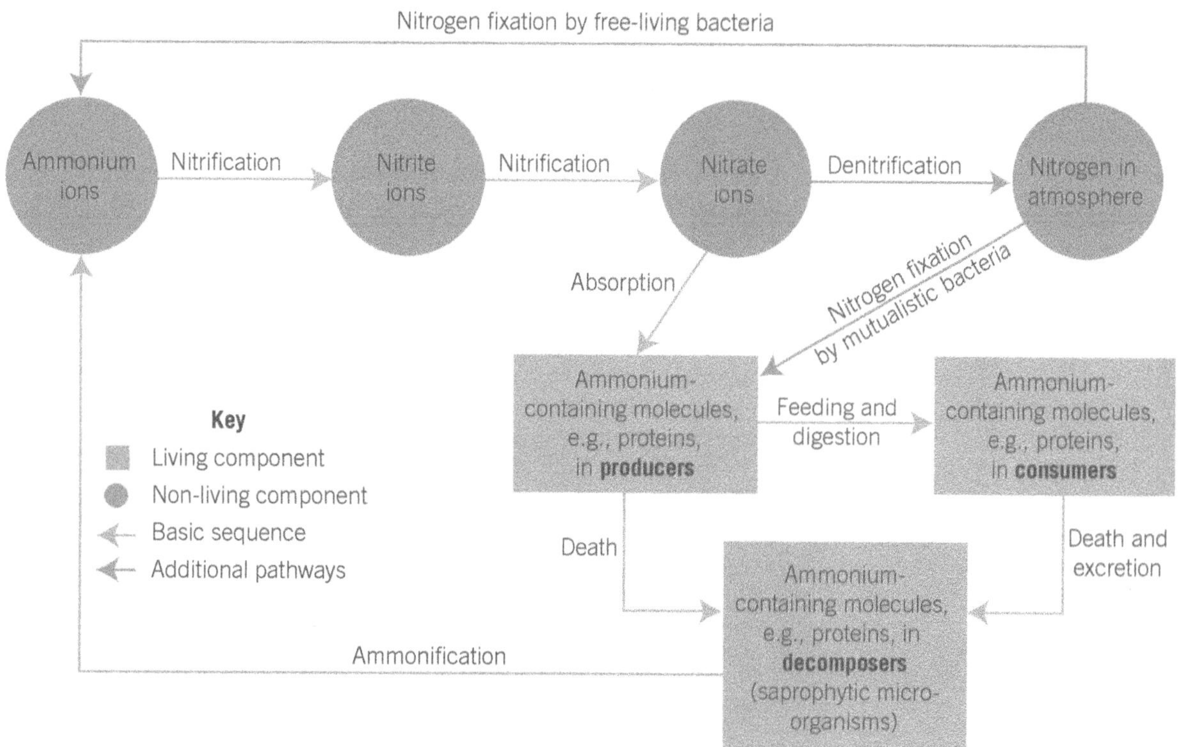

▲ **Figure 1** *The nitrogen cycle*

Synoptic link

The processes of absorption by root hairs and active transport are dealt with in Topics 16.1 and 4.4 respectively.

Ammonification

Ammonification is the production of ammonia from organic ammonium-containing compounds. In nature, these compounds include urea (from the breakdown of excess amino acids), and proteins, nucleic acids, and vitamins (found in faeces and dead organisms). Saprophytic microorganisms, mainly fungi and bacteria, feed on these materials, releasing ammonia, which then forms ammonium ions in the soil. This is where nitrogen returns to the non-living component of the ecosystem.

Hint

In many ecosystems, the availability of nitrates is the factor that limits plant growth. As plants are the primary producers, this means that nitrate availability affects the whole ecosystem.

Nitrification

Plants use light energy to produce organic compounds. Some bacteria, however, obtain their energy from chemical reactions involving inorganic ions. One such reaction is the conversion of ammonium ions to nitrate ions. This is an **oxidation** reaction and so releases energy. It is carried out by free-living soil microorganisms called nitrifying bacteria. This conversion occurs in two stages:

1 oxidation of ammonium ions to nitrite ions (NO_2^-)

2 oxidation of nitrite ions to nitrate ions (NO_3^-).

Nitrifying bacteria require oxygen to carry out these conversions and so they need a soil that has many air spaces. To raise productivity, it is important for farmers to keep soil structure light and well aerated by ploughing. Good drainage also prevents the air spaces being filled with water and so prevents air being forced out of the soil.

Study tip

The word nitrogen is often misused. Nitrogen is an element that forms a part of ions, such as nitrites and nitrates, as well as a part of complex molecules, such as proteins and nucleic acids. Do not use the term nitrogen when referring to these substances. Instead, use nitrogen-containing ions or nitrogen-containing molecules.

Nitrogen fixation

This is a process by which nitrogen gas is converted into nitrogen-containing compounds. It can be carried out industrially and also occurs naturally when lightning passes through the atmosphere. By far the most important form of nitrogen fixation is carried out by microorganisms, of which there are two main types:

- **free-living nitrogen-fixing bacteria**. These bacteria reduce gaseous nitrogen to ammonia, which they then use to manufacture amino acids. Nitrogen-rich compounds are released from them when they die and decay.
- **mutualistic nitrogen-fixing bacteria**. These bacteria live in nodules on the roots of plants such as peas and beans (Figure 2). They obtain carbohydrates from the plant and the plant acquires amino acids from the bacteria.

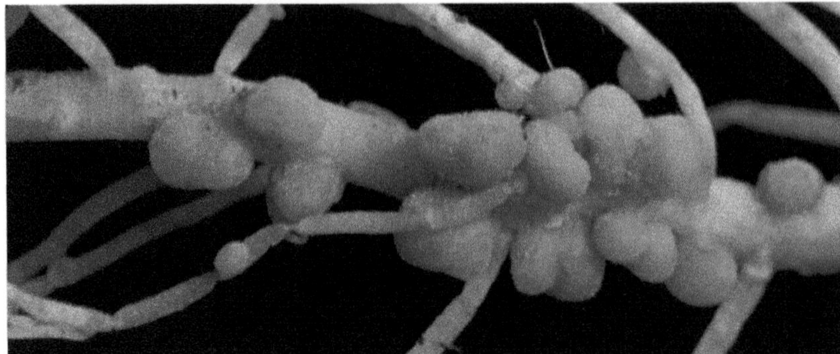

▲ **Figure 2** *Nitrogen-fixing nodules on the roots of a pea plant allow the plant to use free nitrogen in the atmosphere and soil. Mutualistic bacteria in the nodules fix the nitrogen, transforming it into a form useable by the plant.*

Denitrification

When soils become waterlogged, and therefore short of oxygen, the type of microorganisms present changes. Fewer aerobic nitrifying and nitrogen-fixing bacteria are found, and there is an increase in anaerobic **denitrifying bacteria**. These convert soil nitrates into gaseous nitrogen. This reduces the availability of nitrogen-containing compounds for plants. For land to be productive, the soils on which crops grow must therefore be kept well aerated to prevent the build-up of denitrifying bacteria.

As with any nutrient cycle, the delicate balance can be easily upset by human activities. Some of the effects of these activities are considered in Topic 22.5.

▲ **Figure 3** *Ploughing helps to aerate the soil and so prevents the build-up of denitrifying bacteria that can reduce the level of soil nitrates*

Summary question

1 In the following passage, suggest the most appropriate word to replace each of the letters **a** to **j**:

A few organisms can convert nitrogen gas into compounds that are useful to other organisms, in a process known as **a**. These organisms can be free living or live in a relationship with certain **b**. Most plants obtain their nitrogen by absorbing **c** from the soil through their **d** by active transport. They then convert this to **e**, which is passed to animals when they eat the plants. On death, **f** break down these organisms, releasing **g**, which can then be oxidised to form nitrites by **h** bacteria. Further oxidation by the same type of bacteria forms **i** ions. These ions may be converted back to atmospheric nitrogen by the activities of **j** bacteria.

22.4 Use of natural and artificial fertilisers

Learning objectives:

- → Explain why agricultural ecosystems need fertilisers.
- → Compare natural and artificial fertilisers.
- → Explain how fertilisers increase productivity.

Specification reference: 3.3.5.3

In Topic 21.4, you saw how agricultural ecosystems increase the efficiency of energy conversion along human food chains. They do so by improving productivity. One farming practice that contributes to this improved productivity is the use of fertilisers. Let us see how this is achieved.

The need for fertilisers

All plants need mineral ions, especially those containing nitrogen or nitrates, from the soil. Much food production in the developed world is intensive, that is, it is concentrated on specific areas of land that are used repeatedly to achieve maximum yield from the crops and animals grown on them. Intensive food production makes large demands on the soil because mineral ions are continually taken up by the crops being grown on it. These crops are either used directly as food or as fodder for animals that are then eaten. Either way, the mineral ions that the crops have absorbed from the soil are removed.

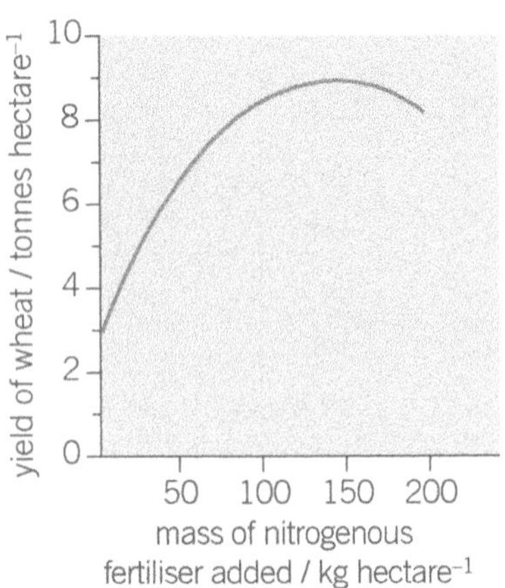

▲ **Figure 1** *The effect of different quantities of nitrogenous fertiliser on the yield of wheat*

In natural ecosystems the minerals that are removed from the soil by plants are returned when the plant is broken down by microorganisms on its death. In agricultural systems the crop is harvested and then transported from its point of origin for consumption. The urine, faeces, and dead remains of the consumer are rarely returned to the same area of land. Under these conditions the concentration of the mineral ions in agricultural land will fall. It is therefore necessary to replenish these mineral ions because, otherwise, their reduced concentration will become the main limiting factor to plant growth. Productivity will consequently be reduced. To offset this loss of mineral ions, fertilisers need to be added to the soil. These fertilisers are of two types:

- **natural (organic) fertilisers**, which consist of the dead and decaying remains of plants and animals, including bone meal, as well as animal wastes such as manure
- **artificial (inorganic) fertilisers**, which are mined from rocks and deposits and then converted into different forms and blended together to give the appropriate balance of minerals for a particular crop. Compounds containing the three elements nitrogen, phosphorus, and potassium are almost always present.

Research suggests that a combination of natural and artificial fertilisers gives the greatest long-term increase in productivity. However, it is important that minerals are added in appropriate quantities as there is a point at which further increases in the quantity of fertiliser no longer results in increased productivity. This is illustrated in Figure 1.

▲ **Figure 2** *Cattle slurry, a natural fertiliser, being spread on to a crop of wheat*

How fertilisers increase productivity

Plants require minerals for their growth. Let us look at nitrogen as an example. Nitrogen is an essential component of proteins and DNA. Both are needed for plant growth. Where nitrates are readily available, plants are likely to develop earlier, grow taller, and have a greater leaf area. This increases the rate of photosynthesis and improves crop productivity. There can be no doubt that nitrogen fertilisers have been of considerable benefit in providing us with cheaper food. It is estimated that the use of fertilisers has increased agricultural food production in the UK by around 100 per cent since 1955.

Different forms of nitrogen fertiliser

Nitrogen fertiliser can be applied to crops in a number of different forms. These include ammonium salts, animal manure, the ground-up bones of animals (bone meal), and urea (a waste product found in the urine of mammals).
An investigation was carried out in which the same crop was grown on six separate plots of land each of the same area. No nitrogen fertiliser was added to the first plot. To each of the remaining five plots, a different form of nitrogen fertiliser was added at the rate of 140 kg total nitrogen per hectare. The graph in Figure 3 shows the results of the investigation.

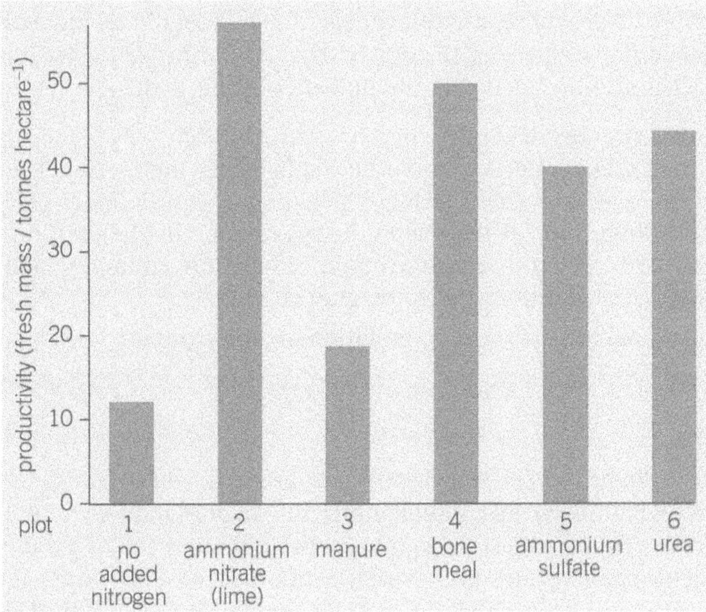

▲ **Figure 3**

1 Which forms of nitrogen used in the investigation are natural fertilisers?
2 Why did the investigation include a plot to which no nitrogen fertiliser was added?
3 Suggest how the addition of nitrogen fertiliser, in whatever form, increased productivity.
4 The mass of each fertiliser used was different in each case. Suggest why this was necessary.
5 It is sometimes claimed that nitrogen fertilisers in the form of ammonium salts increase productivity of crops better than other forms of nitrogen fertilisers. State, with your reasons, whether or not you think the results of this experiment support this view.
6 The increase in productivity when manure was applied was lower than for other forms of nitrogen fertiliser. This is because the manure has to break down before its nitrogen is released and this process takes a few months. How might farmers who spread manure on their crops use this information in order to improve productivity?

Summary questions

1 Explain why fertilisers are needed in an agricultural ecosystem.
2 Using Figure 1, state what concentration of fertiliser you would advise a farmer to apply to a field of wheat.
3 Suggest a reason why, after a certain point, the addition of more fertiliser no longer improves the productivity of a crop.
4 Distinguish between natural and artificial fertilisers.

22.5 Environmental consequences of using nitrogen fertilisers

Learning objectives:

- → Discuss the environmental effects of using fertiliser.
- → Explain how leaching and eutrophication affect the environment.

Specification reference: 3.3.5.3

In natural ecosystems minerals such as nitrate, which are removed from the soil by plants, are returned when the plant is broken down. However, in agricultural systems, the crop is removed and so the nitrate is not returned and has to be replaced. This is done by the addition of natural or artificial fertilisers.

Effects of nitrogen fertilisers

Nitrogen is an essential component of proteins and is needed for growth and, therefore, an increase in the area of leaves. This increases the rate of photosynthesis and improves crop productivity. There can be no doubt that nitrogen-containing fertilisers have benefited us considerably by providing us with cheaper food. Most of this increase is due to additional nitrogen (Figure 1). The use of nitrogen-containing fertilisers has also had some detrimental effects. These include:

- **reduced species diversity**, because nitrogen-rich soils favour the growth of grasses, nettles, and other rapidly growing species. These outcompete many other species, which die as a result. Species-rich hay meadows, such as the one in the photograph in Figure 3, only survive when soil nitrogen concentrations are low enough to allow other species to compete with the grasses.
- **leaching**, which may lead to pollution of watercourses
- **eutrophication**, caused by leaching of fertiliser into watercourses.

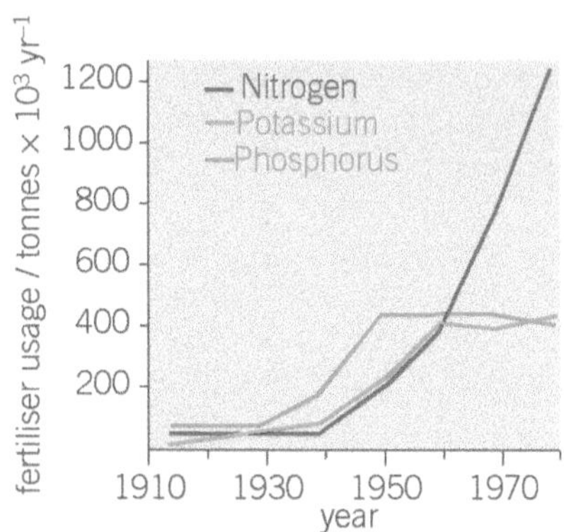

▲ **Figure 1** *Use of fertilisers in the UK*

▲ **Figure 2** *Low species diversity in a field grown for silage that has had nitrogen-containing fertiliser added*

▲ **Figure 3** *High species diversity in a meadow grown for hay without the addition of nitrogen-containing fertiliser*

Leaching

Leaching is the process by which nutrients are removed from the soil. Rainwater will dissolve any soluble nutrients, such as nitrates, and carry them deep into the soil, eventually beyond the reach of plant roots. The leached nitrates find their way into watercourses, such as streams and rivers, that in turn may drain into freshwater lakes. Here they may have a harmful effect on humans if the river or lake is a source of drinking water. Very high nitrate concentration in drinking water can prevent efficient oxygen transport in babies and a link to stomach cancer in humans has been suggested. The leached nitrates are also harmful to the environment as they can cause eutrophication.

Eutrophication

Eutrophication is the process by which nutrients build up in bodies of water. It is a natural process that occurs mostly in freshwater lakes and the lower reaches of rivers. Eutrophication consists of the following sequence of events:

1. In most lakes and rivers there is naturally very little nitrate and so nitrate is a limiting factor for plant and algal growth.
2. As the nitrate concentration increases as a result of leaching, it ceases to be a limiting factor for the growth of plants and algae and both grow exponentially.
3. As algae mostly grow at the surface, the upper layers of water become densely populated with algae. This is called an 'algal bloom'.

4 This dense surface layer of algae absorbs light and prevents it from penetrating to lower depths.
5 Light then becomes the limiting factor for the growth of plants and algae at lower depths and so they eventually die.
6 The lack of dead plants and algae is no longer a limiting factor for the growth of saprophytic algae and so these, too, grow exponentially, using the dead organisms as food.
7 The saprophytic bacteria require oxygen for their respiration, creating an increased demand for oxygen.
8 The concentration of oxygen in the water is reduced and nitrates are released from the decaying organisms.
9 Oxygen then becomes the limiting factor for the population of **aerobic** organisms, such as fish. These organisms ultimately die as the oxygen is used up altogether.
10 Without the aerobic organisms, there is less competition for the anaerobic organisms, whose populations now rise exponentially.
11 The anaerobic organisms further decompose dead material, releasing more nitrates and some toxic wastes, such as hydrogen sulfide, which make the water putrid.

Organic manures, animal slurry, human sewage, ploughing old grassland, and natural leaching can all contribute to eutrophication, but the leaching of artificial fertilisers is the main cause.

▲ **Figure 4** *Algal bloom in a canal as a result of eutrophication caused by nitrogen fertiliser run-off*

Summary questions

1 What is eutrophication?
2 How may an increase in algal growth at the surface lead to the death of plants growing beneath them?
3 Explain how the death of these plants can result in the death of animals such as fish.

Troubled waters

A farmer applied a large quantity of fertiliser to fields next to a small lake. A period of heavy rain followed. After 10 days, scientists monitoring the lake noticed changes to the algal population, the clarity of the water, and the levels of dissolved oxygen. These changes are shown in the three graphs in Figure 5. Secchi depth is a measure of the clarity of water. Measurements are taken by lowering a black-and-white disc (called a Secchi disc) into the water and recording the depth at which it is no longer visible.

1 Suggest a reason why the changes in the lake do not occur until 10 days after the application of the fertiliser to the fields.
2 Explain a possible cause of the increase in the density of algae after 10 days.
3 Describe and explain the relationship between the density of algae and water clarity in the lake.
4 Describe and explain changes to the levels of dissolved oxygen over the 100-day period.

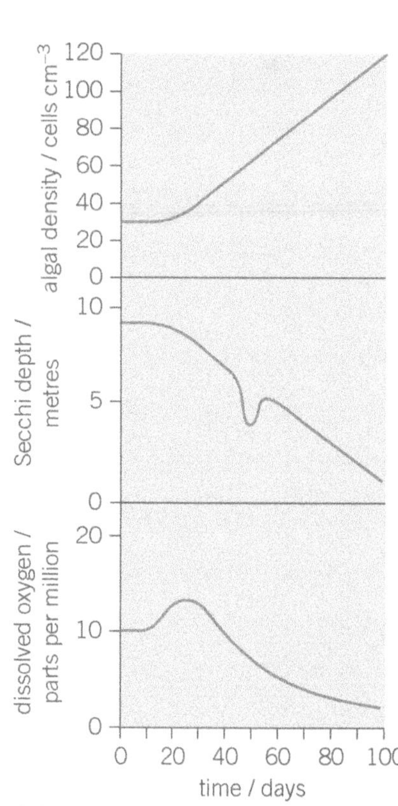

▲ **Figure 5**

Practice questions: Chapter 22

1 Arctic tundra is an ecosystem found in very cold climates. Figure 1 shows some parts of the carbon and nitrogen cycles in arctic tundra.

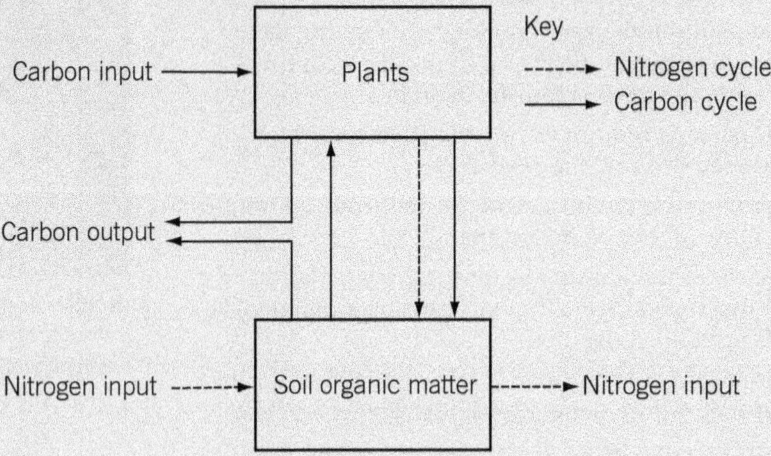

▲ **Figure 1**

(a) Name the process represented by:
  (i) carbon output *(1 mark)*
  (ii) nitrogen input. *(1 mark)*

(b) An increase in temperature causes an increase in carbon input. Explain why. *(2 marks)*

(c) The nitrogen compounds in the organic matter in soil are converted to nitrates. Explain how nitrogen output can occur from these nitrates. *(3 marks)*

AQA 2007

2 Figure 2 shows a river system in an area of farmland. The numbers show the nitrate concentration in parts per million (ppm) in water samples taken at various locations along the river. Concentrations above 250 ppm encourage eutrophication in the river.

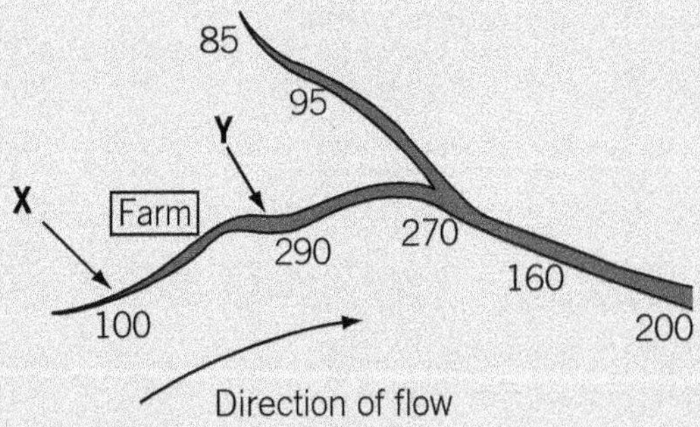

▲ **Figure 2**

(a) Explain how farming practices might be responsible for the change in nitrate concentration in the water between point **X** and point **Y**. *(2 marks)*

(b) Describe the effect the nitrate concentration may have in the river at point **Y**. *(5 marks)*

AQA 2005

3 Figure 3 shows the cumulative mass of carbon removed from the atmosphere by a pine forest in the 20 years after planting.

(a) Explain how the growth of the forest results in a decrease in the carbon content of the atmosphere. *(2 marks)*

(b) A new power station is to be built which will emit a total of 3800 tonnes of carbon over 20 years. In order to balance the carbon emissions a pine forest will be planted to remove an equivalent amount over 20 years. Use the graph to work out the smallest area of forest that would be needed. Show your working. *(2 marks)*

(c) Explain how carbon-containing compounds present in the pine leaves that fall from the trees are used for growth by microorganisms that live in the soil. *(3 marks)*

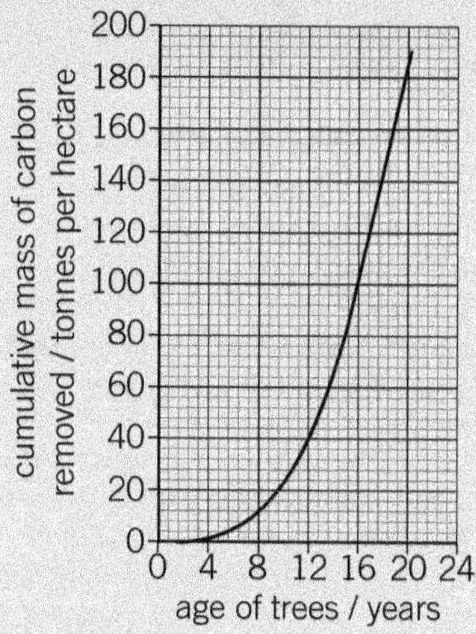

▲ **Figure 3**

(d) Give **one** reason to explain why the rate of recycling of carbon would be greater in summer than in winter. *(1 mark)*

AQA 2004

4 Two fields, **A** and **B**, were used to grow the same crop. In the previous year field **A** was used for grazing cattle and field **B** was used for the same crop. The fields were divided into plots. Different masses of fertiliser containing sodium nitrate were applied to these plots. After 6 weeks, samples of crop plants from each plot were collected and their mass determined. The results are shown in the table.

Mass of fertiliser added / kg ha^{-1}	Mass of crop / kg m^{-2}	
	Field A	Field B
0	14.5	6.4
10	16.7	9.8
20	17.4	12.9
30	17.5	16.2
40	17.5	17.1
50	17.5	17.1
60	17.5	17.1

(a) (i) Describe the pattern shown by the data for field **B**.

(ii) Explain the change in the mass of crop produced from field **B** when the mass of fertiliser added increases from 0 to 20 kg ha^{-1}.

(iii) Explain why the mass of crop produced stays the same in both fields when more than 40 kg of fertiliser is added. *(5 marks)*

(b) When no fertiliser was added, the mass of crop from field **A** was higher than from field **B**. Explain this difference. *(2 marks)*

(c) Explain **two** advantages and **one** disadvantage of an inorganic fertiliser such as sodium nitrate compared with an organic fertiliser such as manure. *(3 marks)*

AQA 2004

23 Inheritance and selection

23.1 Studying inheritance

Learning objectives:

- → Explain what is meant by the terms genotype and phenotype.
- → Explain what is meant by the terms dominant, recessive, and co-dominant.
- → Explain the concept of multiple alleles.

Specification reference: 3.3.6.1

Synoptic link

An understanding of inheritance depends upon an understanding of the way chromosomes behave during meiosis. It is therefore advisable to study Topic 9.1 before starting this chapter.

Hint

All individuals of the same species have the same genes, but not necessarily the same alleles of these genes.

In this chapter you shall look at the way characteristics are passed from one generation to the next and how this can produce genetic variety within a population. You shall see that, when populations are geographically isolated, there is no interbreeding between members of each population and so their genes are also isolated. In time, the genes within each population will change and this can lead to the formation of new species. Let us begin by looking at some of the terms and conventions that are used in studying inheritance.

The fact that children resemble both their parents to a greater or lesser degree and yet are identical to neither has long been recognised. However, it took the rediscovery at the beginning of the last century of the work of a scientist and monk, Gregor Mendel, to establish the basic laws by which characteristics are inherited.

Genotype and phenotype

Genotype is the genetic constitution (make-up) of an organism. It describes all the alleles that an organism contains. The genotype sets the limits within which the characteristics of an individual may vary. It may determine that a human baby could grow to be 1.8 m tall, but the actual height that this individual reaches is affected by other factors, such as diet. A lack of an element such as calcium (for the growth of bone) at a particular stage of development could mean that the individual never reaches his/her potential maximum height. Any change to the genotype as a result of a change to the DNA is called a mutation and it may be inherited if it occurs in the formation of gametes or spores.

Phenotype is the observable characteristics of an organism. It is the result of the interaction between the expression of the genotype and the environment. The environment can alter an organism's appearance. Any change to the phenotype that does not affect the genotype is not inherited and is called a modification.

Genes and alleles

A **gene** is a section of DNA, that is, a sequence of nucleotide bases, which usually determines a single characteristic of an organism (for example, eye colour). It does this by coding for particular polypeptides. These make up the enzymes that are needed in the biochemical pathway that leads to the production of the characteristic (for example, a gene could code for a brown pigment in the iris of the eye). Genes exist in two, or more, different forms called alleles. The position of a gene on a chromosome is known as the locus.

An **allele** is one of the different forms of a gene. In pea plants, for example, there is a gene for the colour of the seed pod. This gene has two different forms, or alleles – an allele for a green pod and another allele for a yellow pod.

Only one allele of a gene can occur at the locus position of any one chromosome. However, in diploid organisms the chromosomes occur in pairs called homologous chromosomes. There are therefore two loci that can each carry one allele of a gene. If the allele on each of the chromosomes is the same (for example, both alleles for green pods are present) then the organism is said to be **homozygous** for the characteristic. If the two alleles are different (for example, one chromosome has an allele for green pods and the other chromosome has an allele for yellow pods) then the organism is said to be **heterozygous** for the characteristic.

In most cases where two different alleles are present in the genotype (heterozygous state), only one of them shows itself in the phenotype.

For instance, in our example where the alleles for green pods and yellow pods are present in the genotype, the phenotype is always green pods. The allele of the heterozygote that expresses itself in the phenotype is said to be **dominant**, and the one that is not expressed is said to be **recessive**. A homozygous organism with two dominant alleles is called **homozygous dominant**, whereas one with two recessive alleles is called **homozygous recessive**. The effect of a recessive allele is apparent in the phenotype of a diploid organism only when it occurs in the presence of another identical allele, that is, when it is in the homozygous state.

These different genetic types are shown in Figure 1.

Study tip

Many people do not know the difference between an allele and a gene. Make certain you are not one of them.

Study tip

Remember that, in diploid cells, there are *two* copies of each allele – one copy inherited from the mother and the other copy from the father.

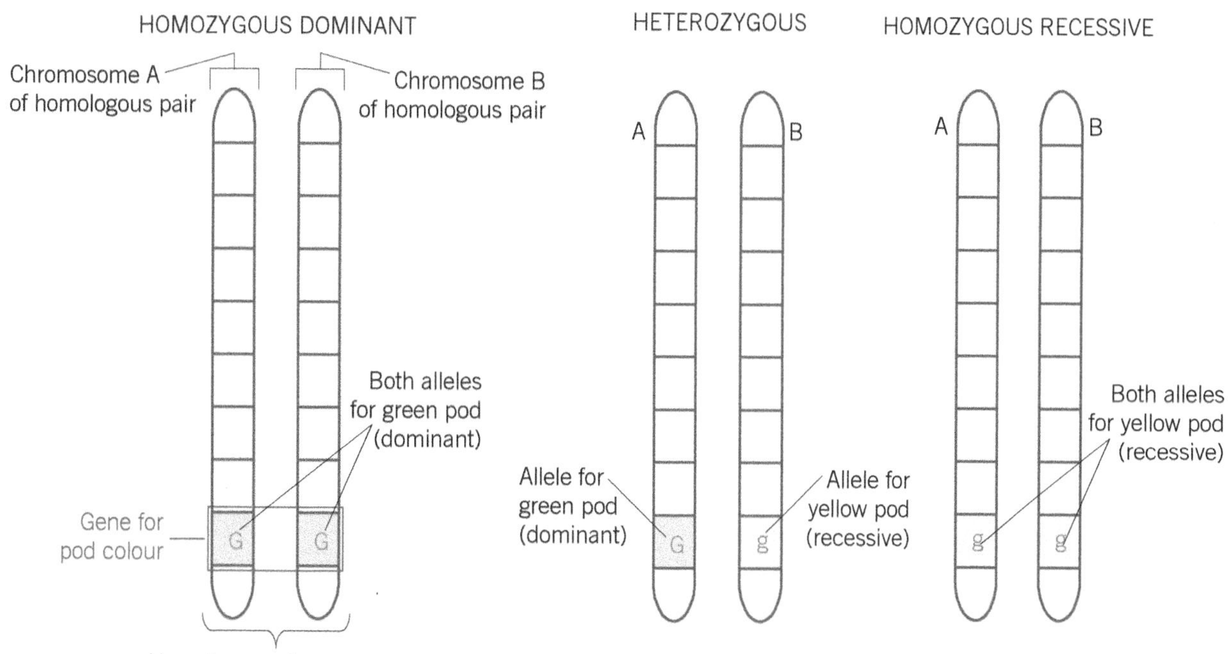

▲ **Figure 1** *Pair of homologous chromosomes showing different possible pairings of dominant and recessive alleles*

In some cases, two alleles both contribute to the phenotype, in which case they are referred to as **co-dominant**. In this situation, when both alleles occur together the phenotype is either a blend of both features (for example, snapdragons with pink flowers resulting from an allele for red-coloured flowers and an allele for white-coloured flowers) or both features are represented (for example, the presence of both A and B antigens in blood group AB). You will learn more about co-dominance in Topic 23.4.

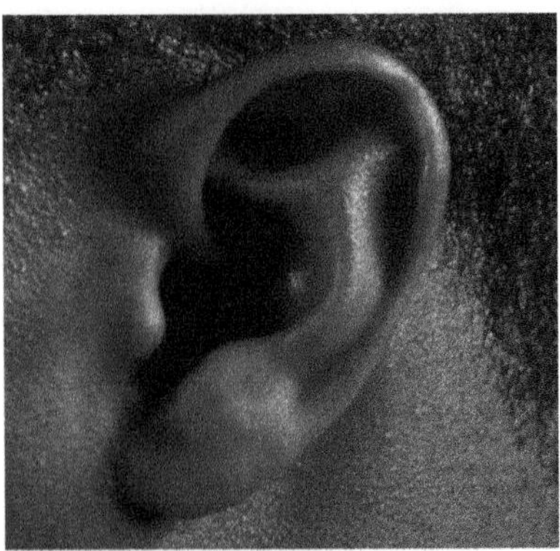

▲ **Figure 2** *'Attached' earlobe. As with other inherited physical body characteristics, the earlobe is subject to genetic differences. Here, the earlobe is firmly attached to the facial skin rather than hanging freely.*

Sometimes a gene has more than two allelic forms. In this case, the organism is said to have **multiple alleles** for the character. However, as there are always only two chromosomes in a homologous pair, it follows that only two of the three or more alleles in existence can be present in a single organism. Multiple alleles occur in the human ABO blood grouping system. Again, you shall learn more about multiple alleles in Topic 23.4.

Figure 3 summarises the different terms used in genetics.

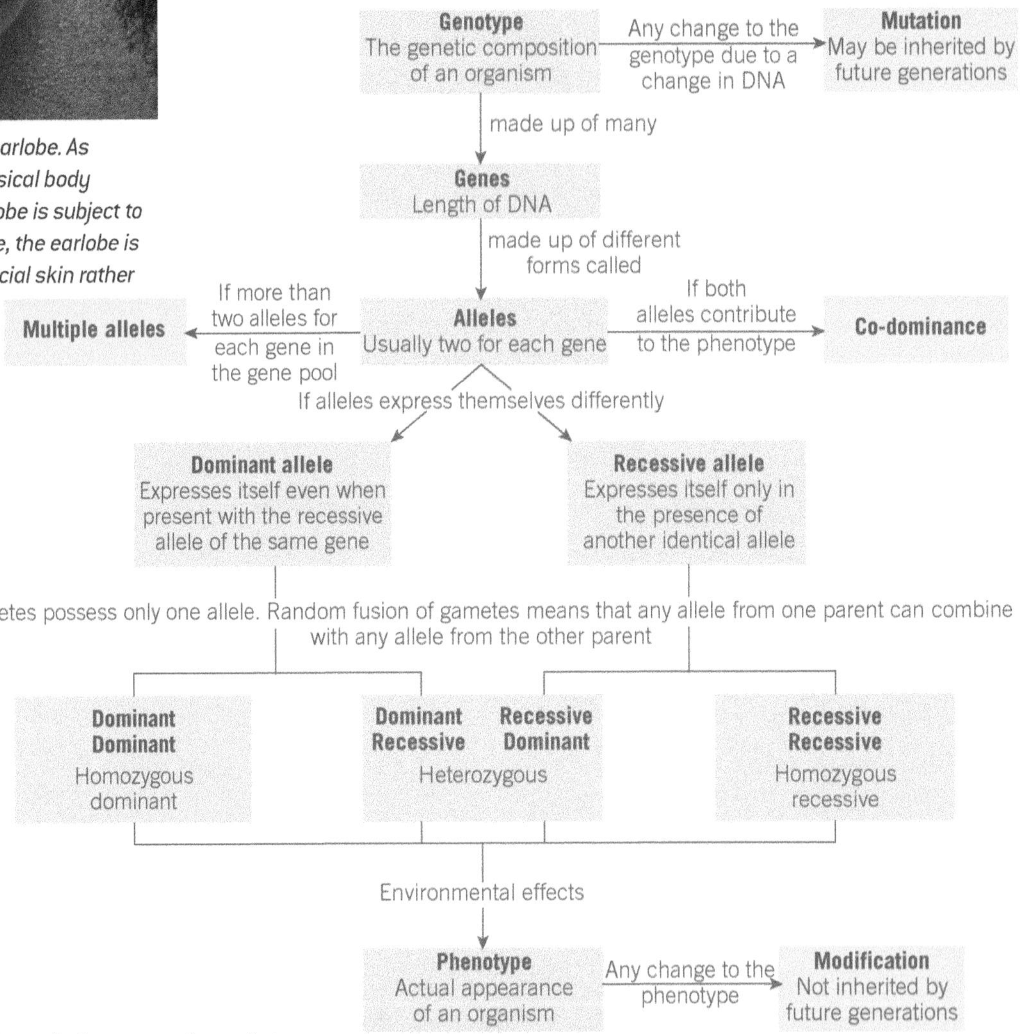

Genotype
The genetic composition of an organism

Any change to the genotype due to a change in DNA →

Mutation
May be inherited by future generations

made up of many ↓

Genes
Length of DNA

made up of different forms called ↓

Alleles
Usually two for each gene

If more than two alleles for each gene in the gene pool → **Multiple alleles**

If both alleles contribute to the phenotype → **Co-dominance**

If alleles express themselves differently

Dominant allele
Expresses itself even when present with the recessive allele of the same gene

Recessive allele
Expresses itself only in the presence of another identical allele

Gametes possess only one allele. Random fusion of gametes means that any allele from one parent can combine with any allele from the other parent

Dominant
Dominant
Homozygous dominant

Dominant **Recessive**
Recessive **Dominant**
Heterozygous

Recessive
Recessive
Homozygous recessive

Environmental effects ↓

Phenotype
Actual appearance of an organism

Any change to the phenotype →

Modification
Not inherited by future generations

▲ **Figure 3** *Summary of genetic terms*

Summary question

1 In the following passage, give the word that best replaces each letter **a** to **j**.

The genetic composition of an organism is called the **a** and any change to it is called a **b** and may be inherited by future generations. The actual appearance of an organism is called the **c**. A gene is a sequence of **d** along a piece of DNA that determines a single characteristic of an organism. It does this by coding for particular **e** that make up proteins including the enzymes needed in a biochemical pathway. The position of a gene on a chromosome is called the **f**. Each gene has two or more different forms called alleles. If the two alleles on a homologous pair of chromosomes are the same they are said to be **g**, but if they are different, they are said to be **h**. An allele that is not apparent in the phenotype when paired with a dominant allele is said to be **i**. Two alleles are called **j** where they contribute equally to the appearance of a characteristic.

23.2 Monohybrid inheritance

Learning objectives:

→ Explain the pattern of inheritance of a single gene using genetic diagrams.

Specification reference: 3.3.6.2

Representing genetic crosses

Genetic crosses are usually represented in a standard form of shorthand. This shorthand form is described in Table 1. Although you may occasionally come across variations to this scheme, that outlined in Table 1 is the one normally used. Once you have practised a number of genetic crosses, you may be tempted to miss out stages or explanations. Not only is this likely to lead to errors, it often makes your explanations difficult for others to follow.

▼ **Table 1** *Representing genetic crosses*

Instruction	Reason/notes	Example [green pod and yellow pod]
Questions in exam usually give the symbols to be used in which case always use the ones provided. Choose a single letter to represent each characteristic.	An easy form of shorthand.	–
Choose the first letter of one of the contrasting features.	When more than one character is considered at one time such a logical choice means it is easy to identify which letter refers to which character.	Choose **G** (green) or **Y** (yellow).
If possible, choose the letter in which the higher and lower case forms differ in shape as well as size.	If the higher and lower case forms differ it is almost impossible to confuse them, regardless of their size.	Choose **G** because the higher case form (**G**) differs in shape from the lower case form (**g**), whereas **Y** and **y** are very similar and are likely to be confused.
Let the higher case letter represent the dominant feature and the lower case letter the recessive one. Never use two different letters where one character is dominant.	The dominant and recessive feature can easily be identified. Do not use two different letters as this indicates co-dominance.	Let **G** = green and **g** = yellow. Do not use **G** for green and **Y** for yellow.
Represent the parents with the appropriate pairs of letters. Label them clearly as 'parents' and state their phenotypes.	This makes it clear to the reader what the symbols refer to.	Green pod, Yellow pod Parents **GG** × **gg**
State the gametes produced by each parent. Label them clearly, and encircle them. Indicate that meiosis has occurred.	This explains why the gametes only possess one of the two parental factors. Encircling them reinforces the idea that they are separate.	Meiosis Meiosis Gametes (G) (G) (g) (g)
Use a type of chequerboard or matrix, called a Punnett square, to show the results of the random crossing of the gametes. Label male and female gametes even though this may not affect the results.	This method is less liable to error than drawing lines between the gametes and the offspring. Labelling the sexes is a good habit to acquire – it has considerable relevance in certain types of crosses, for example, sex-linked crosses.	♂ GAMETES: (G) (G) ♀ GAMETES (g): Gg Gg ♀ GAMETES (g): Gg Gg
State the phenotypes of each different genotype and indicate the numbers of each type. Always put the higher case (dominant) letter first when writing out the genotype.	Always putting the dominant feature first can reduce errors in cases where it is not possible to avoid using symbols with the higher and lower case letters of the same shape.	All offspring are plants producing green pods (Gg).

Study tip

It is important to always use the accepted method of representing genetic crosses and to do so in full. *You* may understand what you are doing, but if your teacher cannot follow it, you are most unlikely to get full credit for your efforts.

Hint

The term F_1 should be used only for the offspring of crosses between the original parents, whereas the term F_2 should be used only for the offspring resulting from crossing the F_1 individuals.

Summary questions

1. In humans, Huntington's disease is caused by a dominant, mutant gene. Draw a genetic diagram to show the possible genotypes and phenotypes of the offspring produced by a man with one allele for the disease and a woman who does not have the disease.
2. In cocker spaniels, black coat colour is the result of a dominant allele and red coat colour is the result of a corresponding recessive allele.
   a. Draw a genetic diagram to show a cross between a pure-breeding bitch with a black coat and a pure-breeding dog with a red coat.
   b. If a dog and a bitch from this first cross are mated, what is the probability that any one of the offspring will have a red coat? Use a genetic diagram to show your working.

Inheritance of pod colour in peas

Monohybrid inheritance is the inheritance of a single gene. To take a simple example we will look at one of the features Gregor Mendel studied – the colour of the pods of pea plants. Pea pods come in two basic colours – green and yellow.

If pea plants with green pods are bred repeatedly with each other so that they consistently give rise to plants with green pods, they are said to be **pure breeding** for the character of green pods. Pure-breeding strains can be bred for almost any character. This means that the organisms are homozygous (that is, they have two **alleles** that are the same) for that particular gene.

If these pure-breeding green-pod plants are then crossed with pure-breeding yellow-pod plants, all the offspring, known as the **first filial**, or **F_1, generation**, produce green pods. This means that the allele for green pods is **dominant** to the allele for yellow pods, which is therefore **recessive**. This cross is shown in Figure 1.

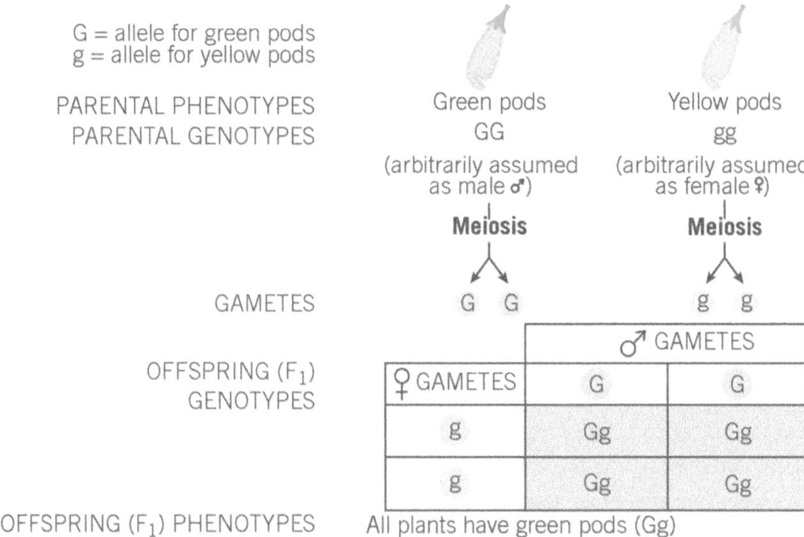

▲ **Figure 1** *Cross between a pea plant that is pure breeding for green pods and one that is pure breeding for yellow pods*

When the heterozygous plants (Gg) of the F_1 generation are crossed with one another (the F_1 intercross), the offspring (known as the second filial, or F_2, generation) are always in an approximate ratio of three plants with green pods to one plant with yellow pods. This cross is shown in Figure 2.

These observed facts led to the formation of a basic law of genetics (the law of segregation). This states:

In diploid organisms, characteristics are determined by alleles that occur in pairs. Only one of each pair of alleles can be present in a single gamete.

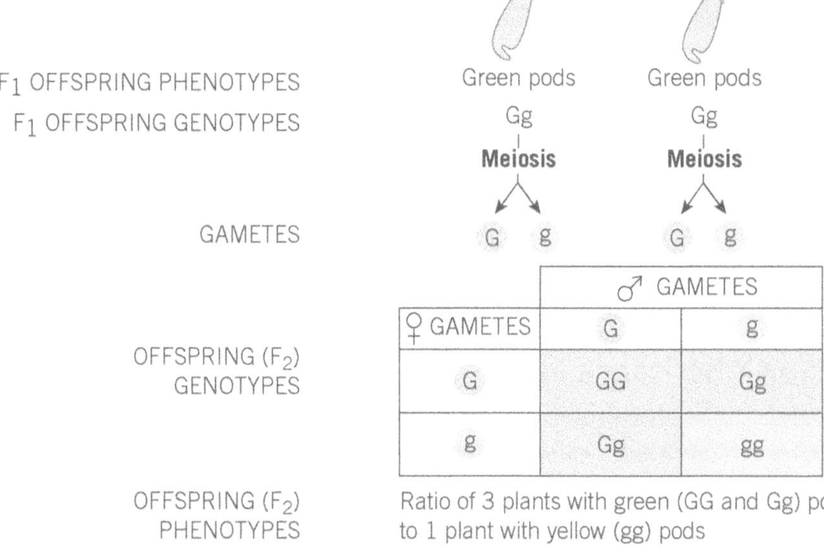

	♂ GAMETES	
♀ GAMETES	G	g
G	GG	Gg
g	Gg	gg

▲ **Figure 2** *F_1 intercross between pea plants that are heterozygous for green pods*

Determining genotypes

One common problem that arises when studying inheritance is that an organism whose **phenotype** displays a dominant characteristic may possess either of two **genotypes**:

- two dominant alleles (homozygous dominant)
- one dominant allele and one recessive allele (heterozygous).

It is not possible to tell which genotype an organism has from outward appearances. It is, however, possible to determine the actual genotype by carrying out a specific genetic cross. It is a very common type of cross known as a test cross because it tests whether an unknown genotype is homozygous dominant or heterozygous.

To see how we carry out this cross, let us use our example of pea plants with different seed-pod colours. Suppose we have a plant that produces green seed pods. This plant has two possible genotypes with respect to pod colour:

- homozygous dominant (GG)
- heterozygous (Gg).

To discover its actual genotype, we cross the plant with an organism displaying the recessive phenotype of the same character, that is, in our case with a pea plant producing yellow pods (gg).

1 Draw genetic diagrams to show the results of a cross between a pea plant with yellow pods and:
   a a pea plant that has a homozygous dominant genotype for green pods
   b a pea plant that has a heterozygous genotype for green pods.
2 A cross was carried out between a pea plant producing green pods and one producing yellow pods. The seeds from this cross were germinated and, of the 63 plants grown, all produced green pods.
   a What is the probable genotype of the parent plant with green pods?
   b Explain why we cannot be absolutely certain of the parent plant's genotype.
3 In a cross between a different pea plant with green pods and a pea plant with yellow pods, 96 plants were produced. Of these, 89 had green pods and seven had yellow pods.
   a What is the probable genotype of the parent plant with green pods?
   b How certain can we be of the genotype of the parent plant with green pods?

23.3 Sex inheritance and sex linkage

Learning objectives:

- → Describe how sex is determined.
- → Described what is meant by sex linkage and explain using genetic diagrams how sex-linked conditions are inherited.

Specification reference: 3.3.6.2

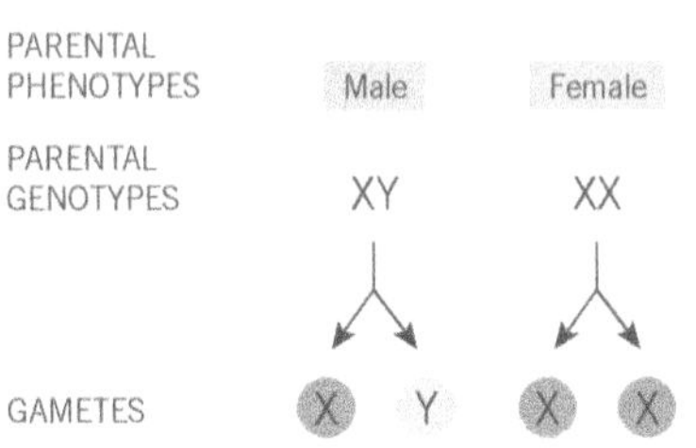

OFFSPRING GENOTYPES

	♂ GAMETES	
♀ GAMETES	X	Y
X	XX	XY
X	XX	XY

OFFSPRING PHENOTYPES

50% Male (XY)
50% Female (XX)

▲ **Figure 1** *Sex inheritance in humans*

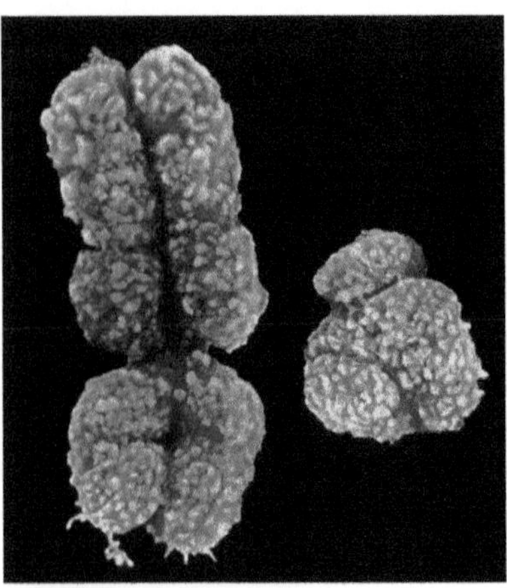

▲ **Figure 2** *Scanning electron micrograph (SEM) of human X (left) and Y chromosomes as found in a male*

Humans have 23 pairs of chromosomes. Twenty two of these pairs have partners that are identical in appearance, whether in a male or a female. The remaining pair are the sex chromosomes. In human females, the two sex chromosomes appear the same and are called the **X chromosomes**. In the human male there is a single X chromosome like that in the female, but the second one of the pair is smaller in size and shaped differently. This is the **Y chromosome**. However, some species have systems other than XY male and XX female.

Sex inheritance in humans

Unlike other features of an organism, sex is determined by chromosomes rather than by genes. In humans:

- as females have two X chromosomes, all the gametes are the same in that they contain a single X chromosome
- as males have one X chromosome and one Y chromosome, they produce two different types of gamete – half have an X chromosome and half have a Y chromosome.

The inheritance of sex is shown in Figure 1.

Sex linkage – haemophilia

Any gene that is carried on either the X or the Y chromosome is said to be sex linked. However, the X chromosome is much longer than the Y chromosome. This means that, for most of the length of the X chromosome, there is no equivalent homologous portion of the Y chromosome. Those characteristics that are controlled by recessive alleles on this non-homologous portion of the X chromosome will appear more frequently in the male. This is because there is no homologous portion on the Y chromosome that might have the dominant allele, in the presence of which the recessive allele does not express itself.

The recessive allele of one of these genes causes the disease haemophilia in humans, in which the blood clots only slowly and there may be slow and persistent internal bleeding, especially in the joints. As such it is potentially lethal if not treated. This has resulted in some selective removal of the gene from the population, making its occurrence relatively rare (about 1 person in 20000 in Europe) (See Topic 23.7). Although haemophiliac females are known, the condition is almost entirely confined to males.

One of a number of causes of haemophilia is a recessive allele with altered DNA nucleotides that therefore do not code for the required protein. This results in the individual being unable to produce a protein that is required in the clotting process. The extraction of this protein from donated blood means that it can now be given to people with haemophilia, allowing them to lead near-normal lives. Figure 3 shows the usual way in which a male inherits haemophilia. Note that the alleles are shown in the usual way (H = dominant allele for production of the clotting protein, and h = recessive allele for the non-production of clotting protein). However, as they are linked to the X chromosome, they are not shown separately, but always attached to the X chromosome, that is, as X^H and X^h respectively. There is no equivalent allele on the Y chromosome as it does not carry the gene for producing clotting protein.

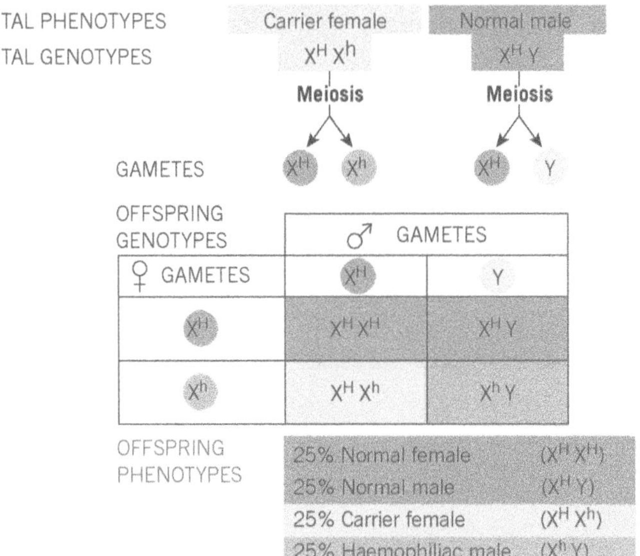

▲ **Figure 3** *Inheritance of haemophilia from a carrier female*

As males can *only* obtain their Y chromosome from their father, it follows that their X chromosome comes from their mother. As the defective allele that does not code for the clotting protein is linked to the X chromosome, males always inherit the disease from their mother. If their mother does not have the disease, she may be **heterozygous** for the character ($X^H X^h$). Such females are called carriers because they carry the allele without showing any signs of the character in their phenotype.

As males pass the Y chromosome on to their sons, they cannot pass haemophilia to them. However, they can pass the recessive allele, via the X chromosome, to their daughters, who would then become carriers of the disease (Figure 4).

Pedigree charts

One useful way to trace the inheritance of sex-linked characteristics such as haemophilia is to use a pedigree chart (see Figure 5). In these:

- a male is represented by a square
- a female is represented by a circle
- shading within either shape indicates the presence of a characteristic, such as haemophilia, in the phenotype
- a dot within a circle signifies a woman with a normal phenotype but who carries the defective allele.

Hint

Remember that males have a Y chromosome and this could only have come from their fathers. Their X chromosome must therefore have come from their mothers.

Study tip

Sometimes people think that the X chromosome passes from father to son. If this were so, the mother must have provided the son's Y chromosome, in which case she would be male! Clearly this is nonsense.

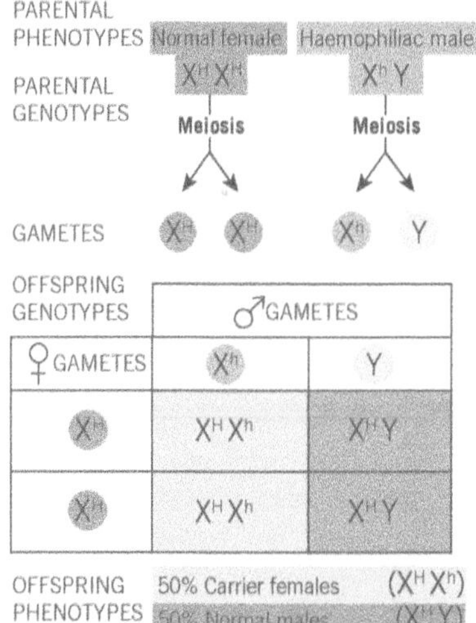

▲ **Figure 4** *Inheritance of the haemophiliac allele from a male with haemophilia*

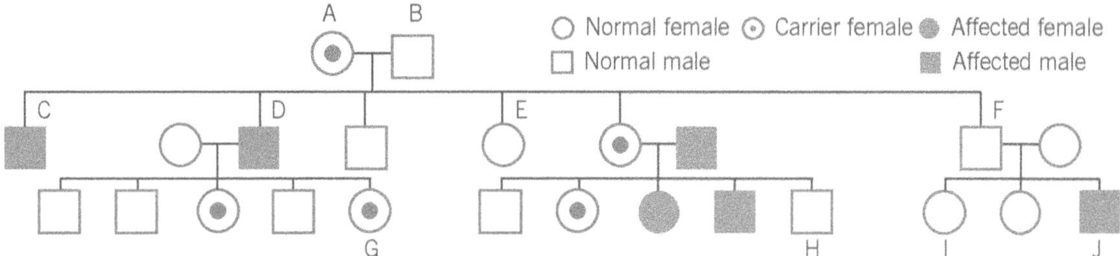

▲ **Figure 5** *A pedigree chart showing the inheritance of red-green colour blindness*

Summary questions

Red-green colour blindness is linked to the X chromosome. The allele (r) for red-green colour blindness is recessive to the normal allele (R). Figure 5 on the previous page shows the inheritance of this characteristic in a family.

1. What sex chromosomes are present in individuals labelled E and F?
2. In terms of colour blindness, what are the phenotypes of each of the individuals labelled A, B, and D?
3. In terms of colour blindness, what are the genotypes of each of the individuals labelled G, H, I, and J?
4. If individual C were to have children with a normal female (one who does not have any r alleles), what would the probability be of any sons having colour blindness?
5. Individual J is colour blind. From the family tree, suggest how this might have occurred.

A right royal disease

The royal families of Europe have been affected by haemophilia for the last few centuries. The origins of the disease stretch back to Queen Victoria. A pedigree chart showing the inheritance of haemophilia from Queen Victoria in members of various European royal families is shown in Figure 6.

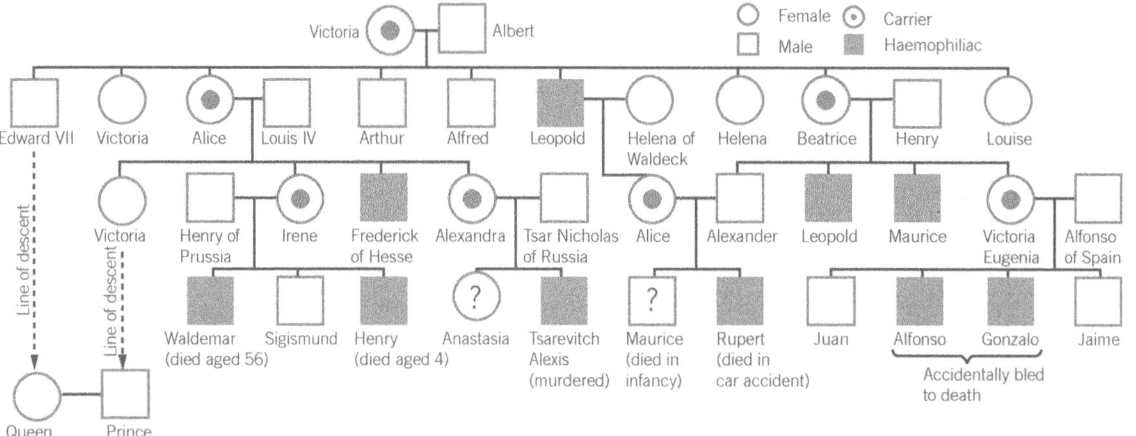

▲ **Figure 6** *Pedigree chart showing the transmission of haemophilia from Queen Victoria*

1. Explain why haemophilia is not present in the current British royal family of Queen Elizabeth II and Prince Philip, and their children.
2. Give evidence from the chart which shows that haemophilia is:
   a. sex linked
   b. recessive.
3. Using the symbols X^H for the chromosome carrying an allele that produces a clotting protein and X^h for a chromosome carrying an allele that does not produce a clotting protein, list the possible genotypes of the following people:
   a. Queen Elizabeth II
   b. Gonzalo
   c. Irene
4. Suppose Waldemar and Anastasia had married and produced children. Using the same symbols, list all the possible genotypes of their:
   a. sons
   b. daughters.

   Explain your answers.

23.4 Co-dominance and multiple alleles

Learning objectives:

- → Explain how co-dominance affects the inheritance of characteristics.
- → Explain how multiple alleles affect inheritance.
- → Explain the inheritance of the ABO blood groups in humans.

Specification reference: 3.3.6.2

In Topics 23.2 and 23.3 you saw straightforward situations in which there were two possible alleles at each locus on a chromosome, one of which was dominant and the other recessive. You shall now look at two different situations:

- **co-dominance**, in which both alleles are equally dominant
- **multiple alleles**, where there are more than two alleles, of which only two may be present at the loci of an individual's homologous chromosomes.

Co-dominance

Co-dominance occurs where, instead of one allele being dominant and the other recessive, both alleles are dominant to some extent. This means that both alleles of a gene are expressed in the phenotype.

▲ **Figure 1** *Snapdragons*

One example occurs in the snapdragon plant, in which one allele codes for an enzyme that catalyses the formation of a red pigment in flowers. The other allele codes for an altered enzyme that lacks this catalytic activity and so does not produce the pigment. If these alleles showed the usual pattern of one dominant and one recessive, the flowers would have just two colours red and white. As they are co-dominant, however, three colours of flower are found:

- In plants that are **homozygous** for the first allele, both alleles code for the enzyme and, hence, pigment is produced. These plants have red flowers.
- In plants that are **homozygous** for the other allele, no enzyme and hence no pigment is produced. These plants have white flowers.
- **Heterozygous** plants, with their single allele for the functional enzyme, produce just sufficient red pigment to produce pink flowers.

Hint

Remember to use different letters, such as R and W, when writing about co-dominance and to put these as superscripts attached to the letter of the gene (C), e.g. C^R and C^W.

If a snapdragon with red flowers is crossed with one with white flowers, the resulting seeds give rise to plants with pink flowers. Note that you cannot use upper and lower case letters for the alleles, as this would imply that one (the upper case) was dominant to the other (the lower case). You therefore use different letters – in this case R for red and W for white – and put them as superscripts on a letter that represents the gene, in this case C for colour. Hence the allele for red pigment is written as C^R and the allele for no pigment as C^W. Figure 2 shows a cross between a red and a white snapdragon, and Figure 3 shows a cross between the resultant pink-flowered plants.

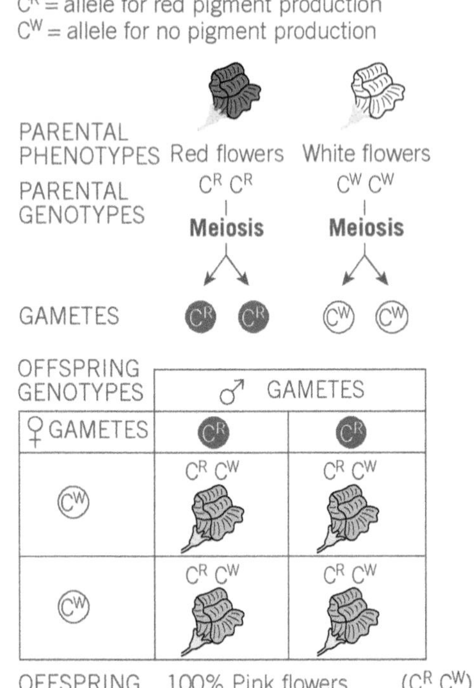

▲ **Figure 2** *Cross between a snapdragon with red flowers and one with white flowers*

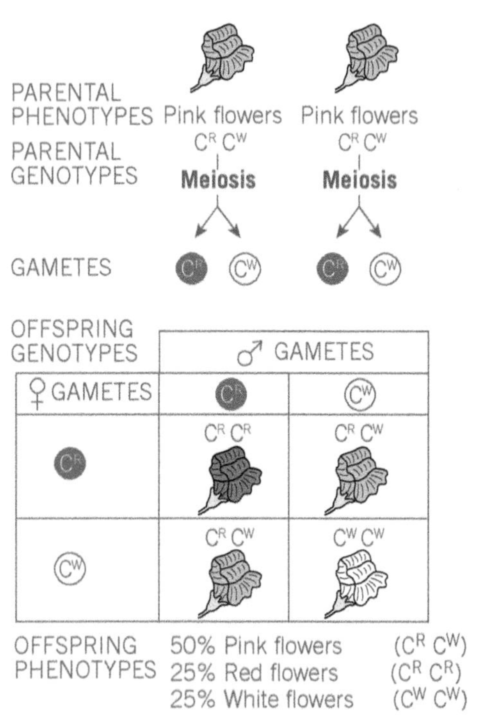

▲ **Figure 3** *Cross between two snapdragons with pink flowers*

Multiple alleles

Sometimes a gene has more than two alleles, that is, it has multiple alleles. The inheritance of the human ABO blood groups is an example. There are three alleles associated with the gene I (immunoglobulin gene), which lead to the production of different **antigens** on the surface membrane of red blood cells:

- allele I^A, which leads to the production of antigen A
- allele I^B which leads to the production of antigen B
- allele I^O, which does not lead to the production of either antigen.

Although there are three alleles, only two can be present in an individual at any one time, as there are only two homologous chromosomes and therefore only two gene loci. The alleles I^A and I^B are co-dominant, whereas the allele I^O is recessive to both. The possible genotypes for the four blood groups are shown in Table 1. There are obviously many different possible crosses between different blood groups, but two of the most interesting are:

▼ **Table 1** *Possible genotypes of blood groups in the ABO system*

Blood group	Possible genotypes
A	I^AI^A or I^AI^O
B	I^BI^B or I^BI^O
AB	I^AI^B
O	I^OI^O

1 A cross between an individual of blood group O and one of blood group AB, rather than producing individuals of either of the parental blood groups, produces only individuals of the other two groups, A and B (see Figure 4).

2 When certain individuals of blood group A are crossed with certain individuals of blood group B, their children may have any of the four blood groups (see Figure 5).

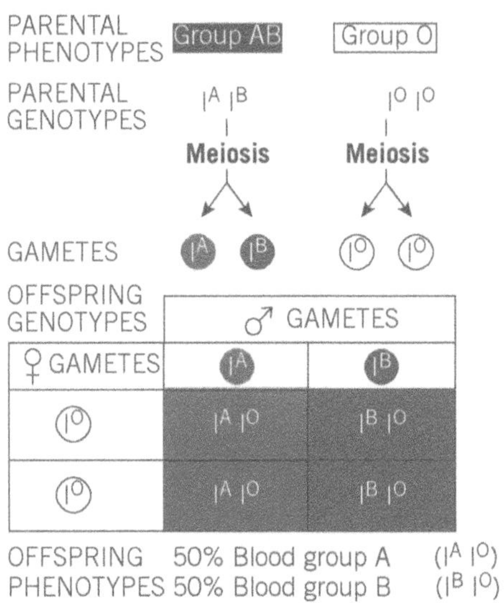

▲ **Figure 4** *Cross between an individual of blood group AB and one of blood group O*

PARENTAL PHENOTYPES Group A Group B

PARENTAL GENOTYPES $I^A I^O$ $I^B I^O$

Meiosis Meiosis

GAMETES I^A I^O I^B I^O

OFFSPRING GENOTYPES

♀ GAMETES	♂ GAMETES I^A	I^O
I^B	$I^A I^B$	$I^B I^O$
I^O	$I^A I^O$	$I^O I^O$

OFFSPRING PHENOTYPES 25% Blood group A ($I^A I^O$)
25% Blood group B ($I^B I^O$)
25% Blood group AB ($I^A I^B$)
25% Blood group O ($I^O I^O$)

▲ **Figure 5** *Cross between an individual of blood group A and one of blood group B*

Multiple alleles and a dominance hierarchy

In human blood groups, alleles I^A and I^B are co-dominant and I^O is recessive to both. Sometimes, however, there may be more than three alleles, each of which is arranged in a hierarchy with each allele being dominant to those below it and recessive to those above it. One example is coat colour in rabbits. The gene for coat colour (C) has four alleles. In order of dominance they are (most dominant first):

Agouti coat (C^A) Chinchilla coat (C^{Ch}) Himalayan coat (C^H) Albino coat (C^a)

Table 2 shows the possible genotypes of rabbits with each of these coat colours.

▼ **Table 2** *Coat colour in rabbits*

Coat colour	Possible genotypes
Full colour (agouti)	C^AC^A, C^AC^{Ch}, C^AC^H, C^AC^a
Chinchilla	$C^{Ch}C^{Ch}$, $C^{Ch}C^H$, $C^{Ch}C^a$
Himalayan	C^HC^H, C^HC^a
Albino	C^aC^a

Summary questions

1 In the example of coat colour in rabbits above, list all the possible genotypes of a rabbit with an agouti coat.

2 A man claims not to be the father of a child. The man is blood group O whilst the mother of the child is blood group A and the child is blood group AB. State, with your reasons, whether you think that the man could be the father of the child.

3 In some breeds of domestic fowl, the gene controlling feather shape has two alleles that are co-dominant. When homozygous, the allele A^S produces straight feathers and the allele A^F produces frizzled feathers. A heterozygote for feather shape has mildly frizzled feathers. Draw a genetic diagram to show the genotypes and phenotypes resulting from a cross between a mildly frizzled cockerel and a frizzled hen. The gene for feather shape is **not** sex-linked.

23.5 Dihybrid crosses and gene interactions

Learning objectives:

- → Describe the pattern of inheritance of two unlinked alleles.
- → Explain what is meant by autosomal linkage.
- → Analyse the outcomes from genetic crosses using the chi^2 test.
- → Describe the interaction between genes controlling the same characteristic.

Specification reference: 3.3.6.2.

In Topic 9.1 you learnt how homologous chromosomes segregate independently during meiosis and how crossing over could occur between pairs of homologous chromosomes. In a **dihybrid cross**, two different genes are either located on different chromosomes or on the same chromosome but at some considerable distance apart, such that there is a likelihood of chiasma formation and crossing over occurring between them.

In Topic 23.2 you looked at the inheritance of pod colour in peas. The gene had two **alleles**, G and g, where G was the allele for green pods and g the allele for yellow pods. The possible genotypes and phenotypes are shown in Table 1:

▼ **Table 1** *Pod colour in pea plants*

Genotype	Phenotype
GG	Green pods
Gg	Green pods
gg	Yellow pods

A second gene controls the shape of the seeds inside the pod. This gene also has two alleles – R is dominant and produces round seeds whereas r produces wrinkled seeds. This gene is on a different chromosome to the gene that controls pod colour so these genes are described as 'not linked'. The possible genotypes and phenotypes would be as shown in Table 2:

▼ **Table 2** *Seed shape in pea plants*

Genotype	Phenotype
RR	Round seeds
Rr	Round seeds
rr	Wrinkled seeds

If you look at the possible genotypes and phenotypes for both genes there are nine possible genotypes and four possible phenotypes, as shown in Table 3:

▼ **Table 3** *Inheritance of pea pod colour and seed shape*

Genotype	Phenotype
GGRR	Green round
GGRr	Green round
GGrr	Green wrinkled
GgRR	Green round
GgRr	Green round
Ggrr	Green wrinkled
ggRR	Yellow round
ggRr	Yellow round
ggrr	Yellow wrinkled

Study tip

When using genetic diagrams involving dihybrid crosses, always keep the two alleles together – so GgRr not GRgr.

A diagram of a cross between plants that are heterozygous for both colour and seed shape is shown in Figure 1.

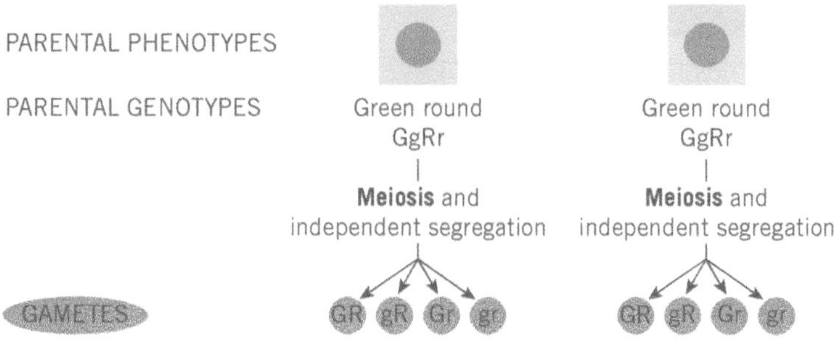

GAMETES	GR	gR	Gr	gr
GR	GGRR	GgRR	GGRr	GgRr
gR	GrRR	ggRR	GrRr	ggRr
Gr	GGRr	GgRr	GGrr	Ggrr
gr	GgRr	ggRr	Ggrr	ggrr

▲ **Figure 1** *A cross between two pea plants that are heterozygous for both pod colour and seed shape*

Looking at the phenotypes that correspond to the genotypes for the F_1 generation in Figure 1, we see a phenotypic ration of 9:3:3:1 – nine green and round, three yellow and round, three green and wrinkled, and one yellow and wrinkled. This is the phenotypic ratio that would be expected from a cross involving two parents who were both heterozygous for two unlinked genes.

Autosomal linkage

In Topic 23.3 you looked at sex linkage and patterns of inheritance for genes carried on the X chromosome. Chromosomes other than the sex chromosomes are known as **autosomes**. Genes that are located close to each other on an **autosomal** chromosome are said to show autosomal **linkage**. As the distance between the gene loci on the chromosome is small, there is less chance of a cross over event occurring between them. This means that the alleles that occupy these gene loci are more likely to be inherited together.

The chi² test

The cross shown in Figure 1 between two plants heterozygous for two unlinked genes should give the **expected** phenotypic ratio of 9:3:3:1. What does this mean experimentally? Our hypothesis is that we have parent plants who are both heterozygous for both genes. This means that, for example, from 128 offspring, we should obtain the following numbers for each phenotype (Table 4):

Study tip

As the two genes are on different chromosomes, and as there are two alleles for each gene, one allele of each gene will be found in each gamete giving four possible gametes.

Study tip

A gene **locus** (plural = loci) is the position occupied by a gene on a chromosome. Where a gene has two or more alleles, that position on the chromosome will be occupied by one of those alleles.

▼ **Table 4** *Expected numbers of offspring with combined characteristics*

Phenotype	Number of plants expected
Green round	72
Yellow round	24
Green wrinkled	24
Yellow wrinkled	8
Total number of plants	128

However, due to the random nature of fertilisation, the outcome of an experimental cross may not result in these exact numbers. So, by how much can the **observed** numbers differ from the **expected** numbers before a hypothesis is rejected? The **chi²** (χ^2) **test** is the statistical test that would be carried out to determine if there is a **significant difference** between the observed and the expected results.

Let us assume that, in our experiment, we have obtained the numbers of offspring from our cross of two plants both with the phenotype green (pods) and round (seeds) that are shown in Table 5.

▼ **Table 5** *Actual numbers of offspring with combined characteristics*

Phenotype	Number of plants obtained
Green round	57
Yellow round	30
Green wrinkled	29
Yellow wrinkled	12
Total number of plants	128

The formula for calculating chi² is as follows, where O = the observed results, E = the expected results and Σ refers to the sum.

$$\chi^2 = \Sigma \frac{(O - E)^2}{E}$$

The calculation can be set out as shown in Table 6:

▼ **Table 6** *How to set out a chi² calculation*

Observed	Expected	$O-E$	$(O-E)^2$	$(O-E)^2/E$
57	72	−15	225	3.13
30	24	6	36	1.5
29	24	5	25	1.04
12	8	4	16	2.00
			$\chi^2 = \Sigma =$	**7.67**

The next step is to compare the calculated value of χ^2 to a table that relates the value obtained to the probability that these results could occur by chance. A table of χ^2 values is given in Table 7.

▼ **Table 7** *Table of χ^2 values*

Degrees of freedom	Probability greater than: 0.1	0.05	0.01	0.001
1	2.71	3.84	6.64	10.83
2	4.60	5.99	9.21	13.82
3	6.25	7.82	11.34	16.27
4	7.78	9.49	13.28	18.46

The calculated value of χ^2 will be compared with the **critical value** selected from the table. In order to select the critical value we take into account the **degrees of freedom**. This is the number of categories (in this case, phenotypes) we are dealing with minus 1. So in this case, the **degrees of freedom** is 3 (4 – 1). Biologists generally work to probabilities that are less than 0.05, meaning that we would expect these results to occur five times for every 100 times the experiment is repeated. That gives us a critical value for χ^2 of 7.82. Our **calculated value** (7.67) is **less than** this critical value. This means that there is **no significant difference** between our observed and our expected results. We can conclude from this that our hypothesis – that we have parent plants who are both heterozygous for both genes – is supported at a probability of $p = 0.05$.

Gene interactions

Some phenotypic characteristics are affected by more than one gene. **Epistasis** is when a gene at one locus affects or inhibits the effect of a gene at another locus. The inheritance of banding pattern in a species of snail called *Cepaea nemoralis* is an example of **dominant epistasis**. The banding pattern is controlled by two unlinked genes.

Gene A has 2 alleles **A** = unbanded and **a** = banded

Gene B has 2 alleles **B** = single band and **b** = five bands

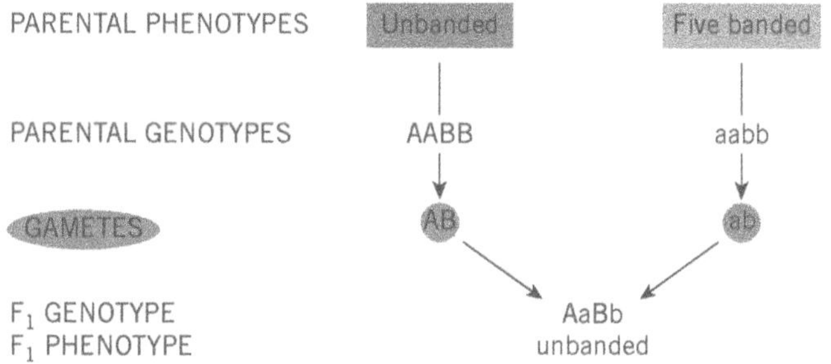

A cross between F_1 snails would lead to the F_2 genotypes and phenotypes shown in Table 8:

▼ **Table 8** *Resulting genotypes and phenotypes for F_2 generation*

GAMETES...	AB	Ab	aB	ab
AB	AABB unbanded	AABb unbanded	AaBb unbanded	AaBb unbanded
Ab	AABb unbanded	AAbb unbanded	AaBb unbanded	Aabb unbanded
aB	AaBB unbanded	AaBb unbanded	aaBB single banded	aaBb single band
ab	AaBB unbanded	Aabb unbanded	aaBb single banded	Aabb five banded

Study tip

In your exams a statement that 'the results are not significant' is wrong and you would not receive any marks. It is the *difference* between the observed and expected results that is significant or not.

▲ **Figure 2** *Banded and unbanded* C. nemoralis

The phenotypic ratio would be 12 unbanded : 3 single banded : 1 five banded. This is different to the 9:3:3:1 ratio we would expect for a normal dihybrid cross. When the epistatic gene A is either homozygous or heterozygous, which means that at least one dominant A allele is present, gene B cannot be expressed in the phenotype.

The inheritance of colour in *Salvia* flowers shows **recessive epistasis**. Two unlinked gene loci A/a and B/b interact as shown in Table 9:

▼ **Table 9** *Inheritance of colour in* Salvia *flowers*

Genotype	Flower colour
A-B- (where - can be either allele of the gene)	purple
AAbb or Aabb	pink
aaBB or aaBb	white
aabb	white

In the absence of at least one dominant allele for gene A, gene B has no effect on the phenotype.

Summary questions

1 Draw genetic diagrams to show the genotypes of the offspring from the following crosses:

a aaBB × AAbb

b GgHH × GgHh

2 The allele for black fur in a species of animal is dominant to white and the allele for long tail is dominant to short tail. What are the possible genotypes for an animal with black fur and a long tail?

3 The metabolic pathway that controls the production of flower colour in a species of plant is shown below:

The genes for the production of enzymes 1 and 2 are on separate chromosomes. Each gene has two alleles, A/a for enzyme 1 and B/b for enzyme 2. The recessive allele prevents the production of a functioning enzyme in each case.

Complete the table to give the colour of the flowers with the following genotypes:

Genotype	Flower colour
aaBB	
AaBB	
Aabb	

Genetics in tomato plants

Some varieties of tomato plants have different coloured stems, which can be green or purple. The stems can also be hairy or hairless.

Homozygous plants with green hairless stems were crossed with homozygous plants with purple hairy stems. All the offspring had purple hairy stems.

1 Choosing appropriate symbols, complete Table 10 to show the genotypes of the two parent plants and the F_1 plants.

▼ **Table 10**

Phenotype	Genotype
Parent 1 - green hairless stem	
Parent 2 - purple hairy stem	
F_1 - purple hairy stem	

The F_1 plants were interbred, and the phenotypes and numbers of offspring in the F_2 generation are given in Table 11.

▼ **Table 11**

Phenotype	Number of plants
Purple hairy stem	293
Purple hairless stem	15
Green hairy stem	12
Green hairless stem	98

2 Assuming that this was a dihybrid cross, what would the expected ratio of the phenotypes in the F_2 generation be?

The parental phenotypes (purple hairy stem and green hairless stem) are more common than expected. The other phenotypes (purple hairless stem and green hairy stem) are known as **recombinant** phenotypes and are present in lower numbers than expected.

3 How do you explain the high numbers of parental phenotypes and low numbers of recombinant phenotypes?

23.6 Allele frequency and population genetics

Learning objectives:

- → Explain what is meant by the terms gene pool and allelic frequency.
- → Outline the Hardy–Weinberg principle.
- → Use the Hardy–Weinberg principle to calculate allele, genotype, and phenotype frequencies.

Specification reference: 3.3.7.1 and 3.3.7.2

Synoptic link

The influence of genetic bottle necks on allele frequencies and the influence of artificial selection of domesticated animals and plants was discussed in Topic 10.1. Genetic drift will be covered in Topic 23.7 and all these topics are relevant here.

Hint

Whether an allele is recessive or dominant has nothing to do with it being harmful or beneficial. People with type O blood group have two recessive alleles for the gene but as it is the most common blood group it can hardly be harmful. Also, Huntington's disease is a fatal condition due to a dominant allele.

We have so far looked at how genes and their alleles are passed between individuals in a population. Let us now consider the genes and alleles of an entire population.

All the alleles of all the genes of all the individuals in a population at any one time are known as the **gene pool**. Sometimes the term is used to refer to all the alleles of one particular gene in a population, rather than all the genes. The number of times an allele occurs within the gene pool is referred to as the **allele frequency**.

Let us look at this more closely by considering just one gene that has two alleles, one of which is dominant and the other recessive. An example is the gene responsible for cystic fibrosis, a human disease in which the mucus produced by affected individuals is thicker than normal. The gene has a dominant allele (F) that leads to normal mucus production, and a recessive allele (f) that leads to the production of thicker mucus and hence cystic fibrosis. Any individual human has two of these alleles in every one of their cells, one on each of the pair of homologous chromosomes on which the gene is found. As these alleles are the same in every cell, we only count one pair of alleles per gene per individual when considering a gene pool. If there are 10 000 people in a population, there will be twice as many (20 000) alleles in the gene pool *of this gene*.

The pair of alleles of the cystic fibrosis gene has three different possible combinations, namely homozygous dominant (FF), homozygous recessive (ff), and heterozygous (Ff). When we look at allele frequencies, however, it is important to appreciate that the heterozygous combination can exist in two different arrangements, namely Ff and fF. (It is just conventional to put the dominant allele first in all cases.)

In any population the total frequency of alleles is taken to be 1.0. In our population of 10 000 people, if everyone had the genotype FF, then the frequency of the dominant allele (F) would be 1.0 and the frequency of the recessive allele (f) would be 0.0. If everyone was heterozygous (Ff), the frequency of the dominant allele (F) would be 0.5 and the frequency of the recessive allele (f) would be 0.5. Of course, in practice, the population is not made up of one genotype but of a mixture of all three, the proportions of which vary from population to population. How then can we work out the allele frequency of these mixed populations?

The Hardy–Weinberg principle

The Hardy–Weinberg principle provides a mathematical equation that can be used to calculate the frequencies of the alleles of a particular gene in a population. The principle predicts that the proportion of dominant and recessive alleles of any gene in a population remains the same from one generation to the next provided that five conditions are met:

- No mutations arise.
- The population is isolated, that is, there is no flow of alleles into or out of the population.

- There is no selection, that is, all alleles are equally likely to be passed to the next generation.
- The population is large.
- Mating within the population is random.

Although these conditions are probably never totally met in a natural population, the Hardy–Weinberg principle is still useful when studying gene frequencies.

To help us understand the principle let us consider a gene that has two alleles: a dominant allele (A) and a recessive allele (a).

Let the frequency of allele A = p

and the frequency of allele a = q

The first equation we can write is:

$$p + q = 1.0$$

because there are only two alleles and so the frequency of one plus the other must be 1.0 (100%).

As there are only four possible arrangements of the two alleles, it follows that the frequency of all four added together must equal 1.0. Therefore we can state that:

AA + Aa + aA + aa = 1.0

or, expressing this as a frequency:

$$p^2 + 2pq + q^2 = 1.0$$

We can now use these equations to determine the frequency of any allele in a population. For example, suppose that a particular characteristic is the result of the recessive allele a, and we know that one person in 25 000 displays the character.

- The character, being recessive, will only be observed in individuals who have two recessive alleles aa.
- The frequency of aa must be 1/25 000 or 0.00004.
- The frequency of aa is q^2.
- If $q^2 = 0.00004$, then $q = \sqrt{0.00004}$ or 0.00063, approximately.
- We know that the frequency of both alleles A and a is $p + q$ and is equal to 1.0.
- If $p + q = 1.0$, and $q = 0.00063$ then:
- $p = 1.0 - 0.00063 = 0.9937$, that is, the frequency of allele A = 0.9937.
- We can now calculate the frequency of the heterozygous individuals in the population.
- From the Hardy–Weinberg equation we know that the frequency of the heterozygotes is $2pq$.
- In this case, $2pq = (2 \times 0.9937 \times 0.0063) = 0.0125$.
- In other words, 125 individuals in 10 000 carry the allele for the character. This is the equivalent of 313 in our population of 25 000.
- These individuals act as a reservoir of recessive alleles in the population, although they themselves do not express the allele in their phenotype.

Summary questions

1 Define the terms:
 a gene pool
 b allelic frequency.

2 What does the Hardy–Weinberg principle predict?

3 What **five** conditions need to be met for this prediction to hold true?

4 The frequency p of a dominant allele is 0.942. Calculate the frequency of the heterozygous genotype in the population. Show your working and express your answer as a percentage of the population.

Hint

Unless an allele leads to a phenotype with an advantage or a disadvantage compared with other phenotypes, its allele frequency in a population will stay the same from one generation to the next.

Not as black and white as it seems

A gene that controls wing colour in the peppered moth has two **alleles**. The expression of the dominant allele produces moths with light-coloured wings and the expression of the recessive allele produces moths with dark-coloured wings. Scientists sampled a population of moths by catching them in a trap and recording their sex and wing colour. The numbers in are Table 1.

▲ **Figure 1** *Dark- and light-coloured wing forms of the peppered moth*

▼ **Table 1**

	Light-coloured wings	Dark-coloured wings
Male	836	269
Female	817	293
Total	**1653**	**562**

1 State, with your reasons, whether you think the gene for wing colour is sex linked.

2 What proportion of the total sample has two recessive alleles?

3 In the Hardy–Weinberg equation $(p^2 + 2pq + q^2 = 1.0)$, p = the frequency of the dominant allele and q = the frequency of the recessive allele. For the population of moths that were caught, use this equation to calculate:

   a the frequency (q) of the recessive allele

   b the frequency (p) of the dominant allele

   c the percentage of heterozygotes.

4 Some scientists wanted to estimate the size of the total moth population. Describe how they might do this.

23.7 Selection

You saw in Topic 23.6 that the Hardy–Weinberg principle predicts that the proportion of dominant and recessive alleles of any gene in a population will remain the same from one generation to the next providing certain conditions are met. One such condition is that 'there is no selection', that is, all alleles are equally likely to be passed to the next generation. In practice, not all alleles of a population are equally likely to be passed to the next generation. This is because only certain individuals are reproductively successful and so pass on their alleles.

Learning objectives:

- → Explain what is meant by genetic drift.
- → Describe how reproductive success affects the allele frequency within a gene pool.
- → Explain what is meant by selection.
- → Describe the environmental factors that exert selection pressures.
- → Explain what is meant by stabilising and directional selection.

Specification reference: 3.3.7.1, 3.3.8.1, and 3.3.8.2

Genetic drift

Allele frequencies can also vary between populations due to genetic bottlenecks and the founder effect (Topic 10.1). **Genetic drift** can also contribute to changes in allele frequency. Purely by chance, some individuals might leave more descendents than others. The alleles of these individuals will therefore slightly increase as a proportion of the population. It is not because these alleles confer any advantage – the process is entirely random. Genetic drift is the random fluctuation of alleles within populations.

Reproductive success and allele frequency

Differences between the reproductive success of individuals affects **allele frequency** in populations. The process works like this:

- Organisms tend to produce more offspring than can be supported by the supply of food, light, space, etc.
- Despite overproduction of offspring, most populations remain relatively constant in size.
- This means that there is competition between members of a species to be the ones that survive.
- Within any population of a species there will be a gene pool containing a wide variety of alleles.
- Some individuals will possess combinations of alleles that make them better able (fitter) to survive in their competition with others. In other words, there is **differential survival**.
- These individuals are more likely to obtain the available resources and so grow more rapidly and live longer. As a result, they will have a better chance of successfully breeding and producing more offspring. In other words, there is **differential reproduction**.
- Only those individuals that successfully reproduce will pass on their alleles to the next generation.
- Therefore it is the alleles that gave the parents an advantage in the competition for survival that are most likely to be passed on to the next generation.
- As these new individuals have 'advantageous' alleles, they in turn are more likely to survive, and so reproduce successfully.
- Over many generations, the number of individuals with the advantageous alleles will increase at the expense of the individuals with the less advantageous alleles.
- Over time, the frequency of the advantageous alleles in the population increases whilst that of the non-advantageous ones decreases.

Synoptic link

Competition between members of the same species is called intraspecific competition and is covered in Topic 18.4.

It must be stressed that what is 'advantageous' depends upon the environmental conditions at any one time. For example, alleles for black body colour may be advantageous as camouflage against a smoke-blackened wall, but non-advantageous against a snowy landscape.

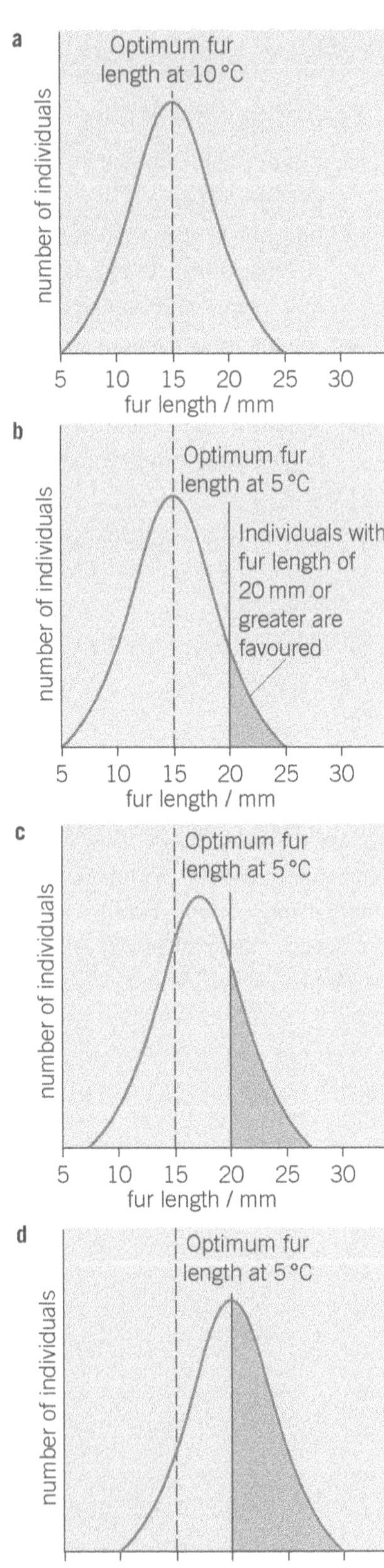

▲ **Figure 1** *Directional selection*

To illustrate the process let us look at the example of the peppered moth. This normally has a light colour that camouflages it against the light background of the lichen-covered trees on which it rests. From time to time black mutant forms of the moth arise. These mutants are highly conspicuous against their light background. As a result, the black mutants are subjected to greater predation from insect-eating birds than the better-camouflaged, normal light forms.

When a dark form of the peppered moth arose in Manchester around 1848, most buildings, walls, and trees were blackened by the soot from 50 years of industrial development. Against this black background the dark form was less, not more, conspicuous than the light natural form. As a result, the light form was eaten by birds more frequently than the dark form. More black-coloured moths than light-coloured moths survived and successfully reproduced. Over many generations, the frequency of the advantageous dark-colour allele increased at the expense of the less advantageous light-colour allele. By 1895, 98 per cent of Manchester's population of the moth was of the black type.

Let us now look in more detail at how selection affects a population.

Types of selection

Selection is the process by which, depending on their phenotypes, organisms that are better adapted to their environment survive and breed, whereas those that are less well adapted fail to do so. Every organism is subjected to a process of selection, based on its suitability for surviving the conditions that exist at the time. Different environmental conditions favour different phenotypic characteristics in the population. Depending on which characteristics are favoured, selection will produce a number of different results.

- Selection may favour individuals that vary in one direction from the mean of the population. This is called **directional selection** and changes the characteristics of the population.
- Selection may favour average individuals. This is called **stabilising selection** and preserves the characteristics of a population.

In this chapter, we have been considering characteristics that are influenced by a single gene. In reality, most characteristics are influenced by more than one gene (**polygenes**). You have already learnt that this type of characteristic is usually influenced by the environment. The effect of the environment on polygenes produces individuals in a population that vary about the mean. When we plot this variation on a graph we get a **normal distribution curve**. Let us look at how these two types of selection affect this curve.

Directional selection

If the environmental conditions change, so will the **phenotypes** needed for survival. Some individuals, which fall to either the left or right of the mean, will possess a phenotype more suited to the new conditions. These will be more likely to survive and breed. They will therefore contribute more offspring (and the alleles these offspring possess) to the next generation than other individuals. Over time, the mean will then move in the direction of these individuals.

To explain, let us imagine a population of a mammal in which there is a range of fur lengths.

- At an average environmental temperature of 10 °C, the optimum fur length for survival is 15 mm. This is the mean fur length of the population, with a number of individuals distributed either side of it (Figure 1a).
- If the average environmental temperature falls to 5 °C, individuals with longer fur (say 20 mm or more) will be better insulated from the cold. These individuals are more likely to survive and so produce more offspring. Those with shorter fur are less likely to survive and so produce fewer offspring (Figure 1b).
- Over many generations, the mean fur length of the population increases as more individuals with longer fur survive, and more individuals with shorter fur die from the cold (Figure 1c). The proportion of alleles for longer fur in the population is increasing at the expense of the alleles for shorter fur.
- Over further generations, the shift in mean fur length continues until it reaches 20 mm – the optimum for the new average environmental temperature of 5 °C (Figure 1d).

Directional selection therefore results in phenotypes at one extreme of the population being selected for and those at the other extreme being selected against.

Stabilising selection

If environmental conditions remain stable, it is the individuals with phenotypes closest to the mean that are favoured. These individuals are more likely to pass their alleles on to the next generation. Those individuals with phenotypes at the extremes are less likely to pass on their alleles. Stabilising selection therefore tends to eliminate the phenotypes at the extremes.

To take the example of mammalian fur length again:

- In years when the average environmental temperature is hotter than usual, individuals with shorter fur will be favoured because they can lose body heat more rapidly.
- In colder years, the opposite is true and individuals with longer fur will be favoured because they are better insulated from the cold.
- Therefore if the temperature fluctuates from year to year (that is, it is unstable), individuals at both extremes will survive. This is because there are some years in which each can thrive at the expense of the other.
- If, however, the environmental temperature is constant, say at 10 °C, individuals at the extremes will never be at an advantage. They will be selected against in favour of those with the mean fur length.
- The mean will remain the same but there will be fewer individuals with fur length at either extreme.

Stabilising selection therefore results in phenotypes around the mean of the population being selected for and those at both extremes being selected against. These events are summarised in Figure 2.

Early selection

The body mass at birth of babies born at a hospital was measured over a 12-year period. In the graph in Figure 3 on the next page the percentage of births in the population (*y*-axis on the left) is plotted against birth mass of the infants as a histogram.

Hint

Selection acts on phenotypes and this has an indirect effect on the inheritance of alleles from one generation to the next.

Synoptic link

Normal distribution curves are covered in Topic 6.2. Knowledge of this will help your understanding of what is discussed here.

Hint

If the environmental change is great enough, there may be no phenotype suited to the new conditions, in which case the population will die out.

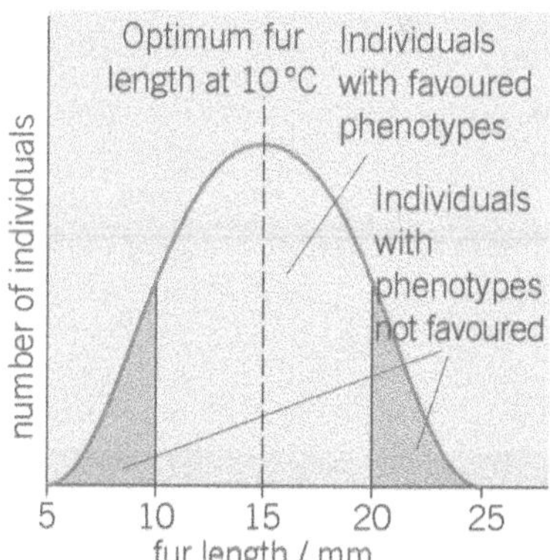

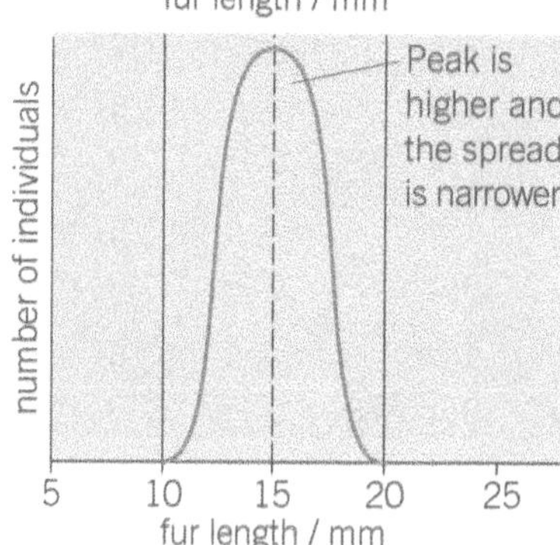

▲ **Figure 2** *Stabilising selection*

Summary questions

1 What is selection?
2 Distinguish between directional selection and stabilising selection.
3 A severe cold spell in 1996 killed over 50% of swallows living on cliffs in Nebraska. Biologists collected nearly 2000 dead swallows from beneath the cliffs and captured around 1000 living ones. By measuring the body mass of the birds, they found that birds with a larger than average body mass survived the cold spell better than ones with a smaller than average body mass. State, giving your reasons, which type of selection was taking place here.

Over the same period, the infant mortality (death) rate was also recorded. The infant mortality rate is measured on a logarithmic scale (*y*-axis on the right) and plotted against infant body mass at birth as a line graph.

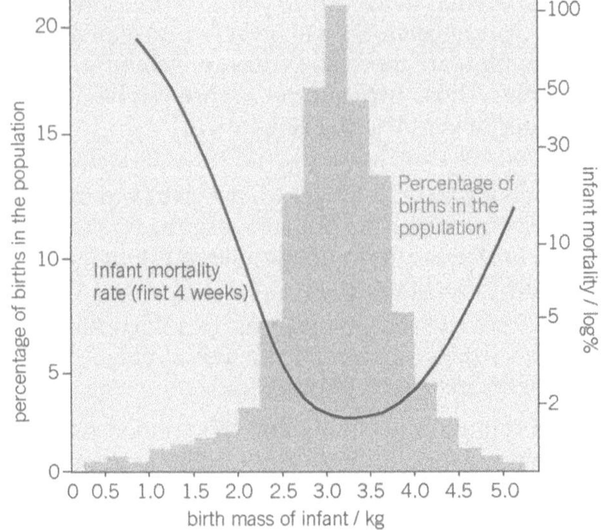

▲ **Figure 3**

1 Describe the relationship between infant birth mass and
   a the percentage of births in the population
   b infant mortality.
2 Which type of selection is shown by the data? Give reasons for your answer.

They must be cuckoo!

Cuckoos lay their eggs in the nests of other birds. The host birds will often raise these parasite chicks alongside their own.

In many valleys in southern Spain, great cuckoos and common magpies have lived together for hundreds of years. In some valleys, however, magpies have been around for centuries but cuckoos have only recently arrived.

Scientists placed artificial cuckoo eggs into magpie nests in both types of valley. Where cuckoos and magpies had lived together for a long period, 78 per cent of the magpies removed the cuckoo eggs from their nests. Where cuckoos had only recently colonised the valleys, only 14 per cent of the magpies removed the cuckoo eggs.

It would appear that, in the valleys where cuckoos are well established, selection has favoured those magpies that removed the cuckoo eggs.

1 Suggest one advantage to the magpies of removing cuckoo eggs from their nest.
2 Explain how removing cuckoo eggs increases the probability of the alleles for this type of behaviour being passed on to subsequent generations.
3 Suggest why this form of behaviour is not shown by magpies in those valleys where cuckoos have only recently arrived.
4 State, with your reasons, which type of selection is taking place here.

▲ **Figure 4** *Cuckoos lay their eggs (bottom row) in the nests of magpies, whose eggs are shown on the upper row*

23.8 Speciation

Speciation is the evolution of new species from existing species. A species is a group of individuals that share similar genes and are capable of breeding with one another to produce fertile offspring. In other words they belong to the same gene pool.

Any species consists of one or more populations. Within each population of a species, individuals breed with one another. Although it is possible for them to breed with individuals from other populations, they breed with each other most of the time. Therefore a single gene pool still exists.

If two populations become separated in some way, the flow of alleles between them may cease. The environmental factors that each group encounters may differ. Selection will affect the two populations in different ways and so the type and frequency of the alleles in each will change. Each population will evolve along separate lines. In time, the gene pools of the two populations may become so different that, even if reunited, they will be incapable of successfully breeding with each other. They would have become separate species, each with its own gene pool. Speciation has taken place. Speciation depends on groups within a population becoming isolated in some way. One such way is geographical isolation.

Learning objectives:

- → Explain what is meant by speciation.
- → Explain how geographical isolation can result in speciation.
- → Distinguish between allopatric and sympatric speciation.

Specification reference: 3.3.8.2

Geographical isolation and allopatric speciation

Geographical isolation occurs when a physical barrier prevents two populations from breeding with one another. Such barriers include oceans, rivers, mountain ranges, and deserts. What proves a barrier to one species may be no problem for another. Whilst an ocean may isolate populations of hedgehogs it can easily be crossed by species of marine fish. Even the smallest stream may separate two groups of woodlice, whereas the whole Pacific Ocean may fail to isolate certain bird populations. To see how geographical separation of populations can lead to the formation of new species by allopatric speciation, let us imagine a species X living in an area of forest.

- The individuals of species X form a single gene pool and freely interbreed.
- Climate changes over the centuries lead to drier conditions that reduce the area of forest and separate it into two regions that are many hundreds of kilometres apart.
- Further climate changes cause one forest region (A) to become much colder and wetter and the other forest region (B) to become warmer and drier.
- In region A, phenotypes are selected that are better able to survive in colder, wetter conditions.
- In region B, different phenotypes are selected – ones that are better able to survive in warmer, drier conditions.
- The frequency of alleles therefore change due to selection for suitable phenotypes, before mutation leads to further changes.
- The type and frequency of the alleles in the gene pools of each group of species X become increasingly different due to mutation.
- In time, the differences between the two gene pools become so great that they are, in effect, separate species. This is known as **allopatric** speciation.
- Further climate change and regrowth of the forest may lead to the two species being reunited geographically. However, they will not be able to interbreed as they are now different species.

These events are summarised in Figure 1.

Synoptic link

The concept of a species is covered in Topic 9.2.

1 Species X occupies a forest area. Individuals within the forest form a single gene pool and freely interbreed.

2 Climatic changes to drier conditions reduce the size of the forest to two isolated regions. The distance between the two regions is too great for the two groups of species X to cross to each other.

3 Further climatic changes result in one region (Forest A) becoming colder and wetter. Group X_1 adapts to these new conditions. The other region (Forest B) becomes warmer and drier. Group X_2 adapts to these conditions.

4 Continued adaptation leads to evolution of new species – Y and Z.

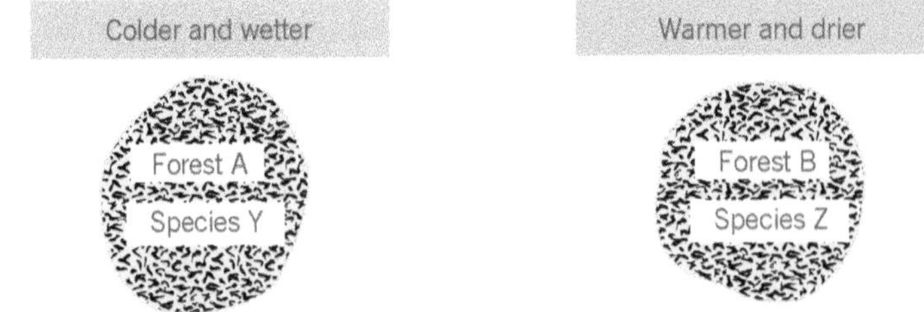

5 A return to the original climatic conditions results in regrowth of forest. Forests A and B merge and the two groups of species are reunited. The two groups are no longer capable of interbreeding. They are now two species, Y and Z, each with its own gene pool.

▲ **Figure 1** *Speciation due to geographical isolation*

Sympatric speciation

Speciation can occur in the absence of a geographical barrier. A change in the population leads to some sections of the population no longer being able to breed successfully with other members of the same population even though they still occupy the same physical area. This is called **sympatric** speciation. One well-documented example is found in salt marshes in the UK.

The original salt marsh species of cord grass was *Spartina maritima*. This grass had a diploid chromosome number of 60 ($2n = 60$). Towards the end of the 19th century another species, *Spartina alterniflora*, was imported from the east coast of America to the UK. This has a diploid chromosome number ($2n$) of 62. The two species hybridised to produce a new species, *Spartina townsendii*. This plant was sterile and could not interbreed with its parent plants but successfully reproduced asexually and was a new species. Within a few years, as a result of a mutation that doubled the chromosome number, a tetraploid grass developed from *S. townsendii*, which has been called *Spartina angelica*. This grass is fertile ($2n = 122$) and again will not interbreed with other species of *Spartina* even though they inhabit the same area. Over time, the size and vigorous growth of *S. angelica* has led to it becoming the dominant species on salt marshes within the UK.

Synoptic link

All the *Spartina* species occupied the same niche. The principle of competitive exclusion which you covered in Topic 18.4 means that there was competition between the species with *S. angelica* outcompeting the other *Spartina* species.

Summary questions

1 What is a species?
2 What is speciation?
3 What is meant by geographical isolation?
4 Distinguish between allopatric and sympatric speciation.
5 Explain how geographical isolation of two populations of a species can result in the accumulation of differences in their gene pools.

1 One form of baldness in humans is controlled by two alleles, B and b, of a single gene. This gene is not on the X chromosome but the expression of the gene is affected by the sex of a person. Men who are BB or Bb will become bald. Men who are bb will not become bald. Women who are BB will become bald. Women who are Bb or bb will not become bald. One type of colour blindness is controlled by a sex-linked gene, found on the X chromosome. The dominant allele X^A leads to normal colour vision and the recessive allele X^a leads to colour blindness.

(a) (i) Give all the possible genotypes of a bald man who has normal colour vision. *(1 mark)*

(ii) Give all the possible genotypes of a woman who will not become bald and who carries one allele for colour blindness. *(1 mark)*

(b) A mother and a father are both heterozygous for the gene for baldness. The father has normal colour vision and the mother is heterozygous for the gene for colour blindness. Complete the genetic diagram to show the probability of a son of this couple being colour blind but not becoming bald.

	Father	Mother
Genotypes of parents		
Gametes		
Genotypes of sons		

2 The ground finch, *Geospiza fortis,* is a species of bird which lives on a small isolated island. These finches feed on seeds of different sizes from different species of plants. The finches show variation in the size of their beaks. Birds with larger beaks can eat large and small seeds. Birds with smaller beaks are only able to eat small seeds.

In 1977 there was a severe drought on the island. This killed many species of plants that the finches fed on. One species of food plant did survive and this produced large seeds. The graphs show the distribution of beak sizes of the finch population before and after the drought. Beak size was measured by the depth of the beak, as shown in the diagram.

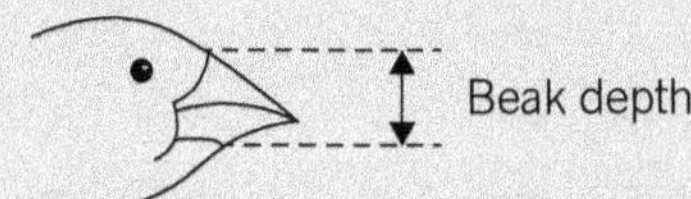

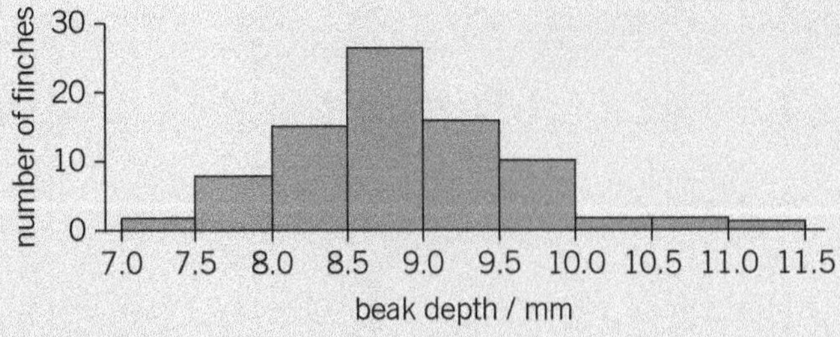

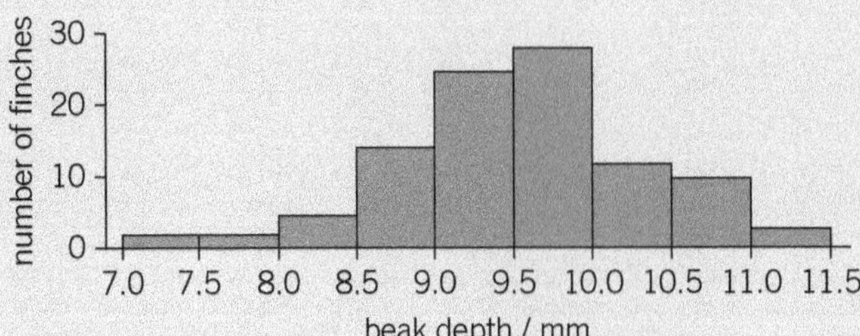

(a) (i) What type of variation is shown in the graphs? (1 mark)
(ii) How is this type of variation genetically controlled? (1 mark)

(b) The evidence that beak size is determined by genetic factors was obtained by comparing beak sizes of parents and their offspring. Explain how this comparison provided evidence for the role of genetic factors. (1 mark)

(c) Explain the changes in beak size from 1976 to 1978. (4 marks)

3 Chickens have a structure called a comb on their heads. The drawings show two types of comb. The shape of the comb is controlled by two alleles of one gene. The allele for pea comb, A, is dominant to the allele for single comb, a.

The colour of chicken eggs is controlled by two alleles of a different gene. The allele for blue eggs, B, is dominant to the allele for white eggs, b.

The genes for comb shape and egg colour are situated on the same chromosome.

A farmer crossed a male chicken with the genotype AaBb with a female chicken that had a single comb and produced white eggs.

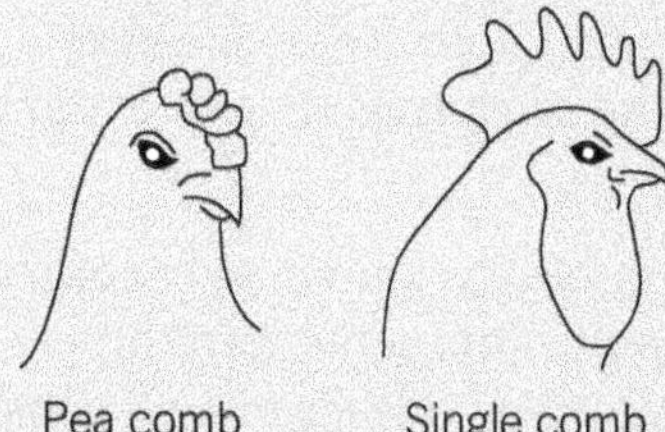

(a) What was the genotype of the female parent? (1 mark)

The diagram shows how the alleles of the genes were arranged on the chromosomes of the male parent.

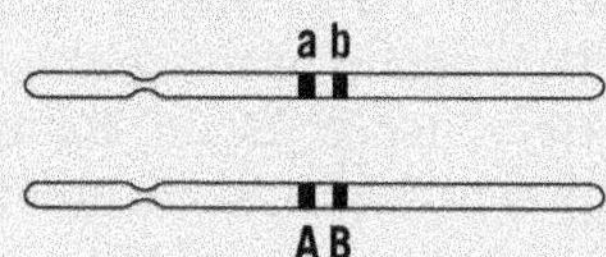

(b) Which two genotypes will be most frequent in the offspring? (1 mark)

(c) The farmer could identify which of the female offspring from this cross would eventually produce blue eggs. Explain how. (2 marks)

(d) Genes A and B are close together on the chromosome. This is important when trying to identify which of the female offspring would produce blue eggs. Explain why. (2 marks)

(e) Suggest two environmental factors which are likely to affect egg production. (2 marks)

4 Phenylketonuria (PKU) is a hereditary disease in humans. The allele responsible for PKU is recessive and carried on chromosome 12. The frequency of children born with PKU is one in 10 000 live births.

(a) In a population of 1000, how many people would be expected to be heterozygous for PKU? (2 marks)

(b) A couple, neither of whom have PKU, have a child who has been diagnosed with the condition. What is the probability that their second child will be heterozygous for PKU? Explain your anwer. (2 marks)

Answers to the Practice Questions are available at
www.oxfordsecondary.com/oxfordaqaexams-alevel-biology

24 Response to stimuli

24.1 Sensory reception

Learning objectives:

- → Explain what is meant by a stimulus and a response.
- → Explain the advantage to organisms of being able to respond to stimuli.
- → Explain what are taxes and kineses.
- → Explain how each type of response increases an organism's chances of survival.

Specification reference: 3.4.1

In this chapter you shall consider how internal and external stimuli are detected by organisms and how they lead to a response.

Stimulus and response

A **stimulus** is a detectable change in the internal or external environment of an organism that produces a **response** in the organism.

The ability to respond to stimuli increases the chances of survival for an organism. For example, to be able to detect and move away from harmful stimuli, such as predators and extremes of temperature, or to detect and move towards a source of food clearly aid survival. Those organisms that survive have a greater chance of raising offspring and of passing their alleles to the next generation. There is always, therefore, a selection pressure favouring organisms that have more appropriate responses.

Stimuli are detected by cells, molecules, or organs known as **receptors**. Receptors transform the energy of a stimulus into some form of energy that can be processed by the organism and leads to a response. The response is carried out by one or more of a range of different cells, molecules, tissues, organs, and systems. These are known as **effectors**. Receptors and effectors are often some distance apart and therefore some form of communication between the two is needed if the organism is to respond effectively. One means of communication occurs via chemicals called hormones (see Chapter 26 and Topic 27.4), which is a relatively slow process found in both plants and animals.

Animals have another, more rapid, means of communication – the nervous system. Their nervous system usually has many different receptors and effectors. Each receptor and effector is linked to a central **coordinator** of some type. The coordinator acts like a switchboard, connecting information from each receptor with the appropriate effector. The sequence of events from stimulus to response can therefore involve either chemical control or nerve cells and may be summarised as:

stimulus → receptor → coordinator → effector → response

Let us look first at the simplest forms of response to stimuli involving neurones in animals, and how they can increase an organism's chances of survival.

Hint

Plants respond to stimuli, but their receptors produce chemicals and not nerve impulses, and their effectors usually respond by growing and not by muscle contraction. This means that plants usually respond more slowly than animals.

Reflex arcs

Figure 1 shows the three neurones involved in a simple reflex arc. The 'circuit' between receptor and effector is relatively short so the speed of response to the stimulus is rapid. A simple reflex response is innate, meaning that it is a fixed pattern of response that is common to all members of that species of organism. Reflexes are protective and prevent damage to the organism as shown in Figure 2. Examples of simple reflexes in humans include the blink reflex.

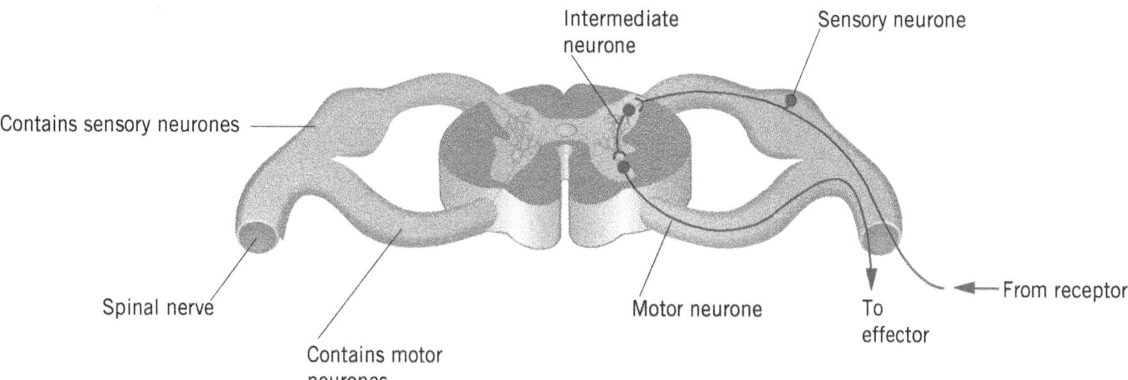

▲ **Figure 1** *Section through spinal cord showing the neurones of a reflex arc*

Taxes

A **taxis** is a simple response whose direction is determined by the direction of the stimulus. As a result, a motile organism (or a part that detaches) responds directly to environmental changes by moving its whole body either towards a favourable stimulus or away from an unfavourable one. Taxes are classified according to whether the movement is towards the stimulus (positive taxis) or away from the stimulus (negative taxis) and also by the nature of the stimulus.

- Single-celled algae will move towards light (positive phototaxis). This increases their chances of survival since, being photosynthetic, they need light to manufacture their food.
- Earthworms will move away from light (negative phototaxis). This increases their chances of survival because it takes them into the soil, where they are better able to conserve water, find food, and avoid predators.
- Some species of bacteria will move towards a region where glucose is more highly concentrated (positive chemotaxis). This increases their chances of survival because they use glucose as a source of food.

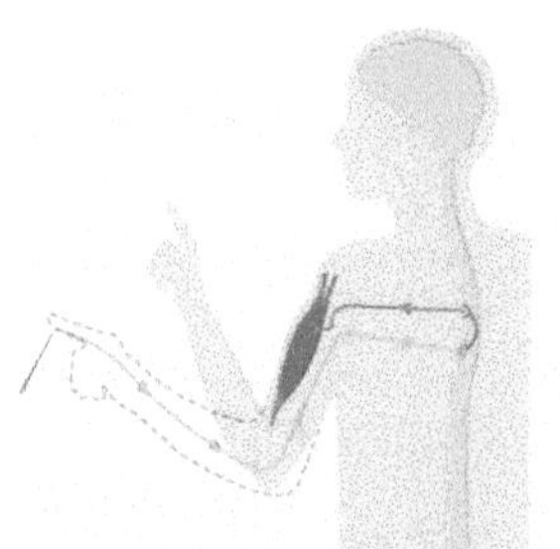

▲ **Figure 2** *Contact with a sharp object triggers pain receptors resulting in a withdrawal response that avoids damaging tissues*

Kineses

A **kinesis** is a form of response in which the organism does not move towards or away from a stimulus. Instead, the more unpleasant the stimulus, the more rapidly it moves and the more rapidly it changes direction. A kinesis therefore results in an increase in random movements. This type of response is designed to keep it moving and changing direction until it happens to find itself in favourable conditions. It is important when a stimulus is less directional. Humidity and temperature, for example, do not always produce a clear gradient from one extreme to another.

An example of a kinesis occurs in woodlice. Woodlice lose water from their bodies in dry conditions. When they are in a dry area they move more rapidly and change direction more often. This increases their chances of moving into a different area. If this different area happens to be moist, they slow down and change direction less often. This means that they are likely to stay where they are. In this way, the

▲ **Figure 3** *Woodlice exhibit a behaviour called kinesis, which ensures that they spend most of their time in the dark moist conditions that prevent them from drying out, and hence aid their survival*

woodlice spend more time in favourable moist conditions than in less favourable drier ones. This prevents them drying out and so increases their chances of survival.

Summary questions

For each of the following statements, name the type of response described and the survival value of the response.

1. Some species of bacteria move away from the waste products that they produce.
2. The sperm cells of a moss plant are attracted towards a chemical produced by the female reproductive organ of another moss plant.
3. When nutrient concentrations are low, the bacterium *E. coli* increases its turning activity.

24.2 Role of receptors

The central nervous system receives sensory information from its internal and external environment through a variety of sense cells and organs called receptors, each type responding to a different type of stimulus. Sensory reception is the function of these sense organs, whereas sensory perception involves making sense of the information from the receptors. This is largely a function of the brain. The concepts of stimulus and response were covered in Topic 24.1. We shall now look in detail at one receptor – the Pacinian corpuscle.

Learning objectives:

- → Explain the main features of sensory reception.
- → Describe how the Pacinian corpuscle works.
- → Explain how receptors in the eye work.

Specification reference: 3.4.2.1 and 3.4.2.2

Features of sensory reception as illustrated by the Pacinian corpuscle

Pacinian corpuscles respond to changes in mechanical pressure. As with all sensory receptors, a Pacinian corpuscle:

- **is specific to a single type of stimulus**. In this case, it responds only to mechanical pressure. It will not respond to other stimuli, such as heat, light, or sound.
- **produces a generator potential by acting as a transducer**. All stimuli are forms of information, but unfortunately not forms that the body can understand. It is the role of the transducer to convert the information provided by the stimulus into a form that can be understood by the body, namely nerve impulses. The stimulus is always some form of energy, for example, heat, light, sound, or mechanical energy. The nerve impulse is also a form of energy. Receptors therefore convert, or transduce, one form of energy into another. All receptors convert the energy of the stimulus into a nervous impulse known as a **generator potential**. For example, the Pacinian corpuscle, whose action is described below, transduces the mechanical energy of the stimulus into a generator potential.

▲ **Figure 1** *Pacinian corpuscle*

Structure and function of a Pacinian corpuscle

Pacinian corpuscles respond to mechanical stimuli such as pressure. They occur deep in the skin and are most abundant on the fingers, the soles of the feet, and the external genitalia. They also occur in joints, ligaments, and tendons, where they enable the organism to know which joints are changing direction. The single sensory neurone of a Pacinian corpuscle is at the centre of layers of tissue, each separated by a gel. This gives it the appearance of an onion when cut vertically (Figure 2). How does this structure transduce the mechanical energy of the stimulus into a generator potential?

You learnt in Topic 4.1 that plasma membranes contain proteins that span them. These form channel along which ions can be transported. Some channels carry one specific ion. Sodium channels, for example, carry only sodium ions.

The sensory neurone ending at the centre of the Pacinian corpuscle has a special type of sodium channel in its plasma membrane. This is called a **stretch-mediated sodium ion channel**. These channels are so called because their permeability to sodium ions changes when they change shape, for example, by stretching. The Pacinian corpuscle functions as follows:

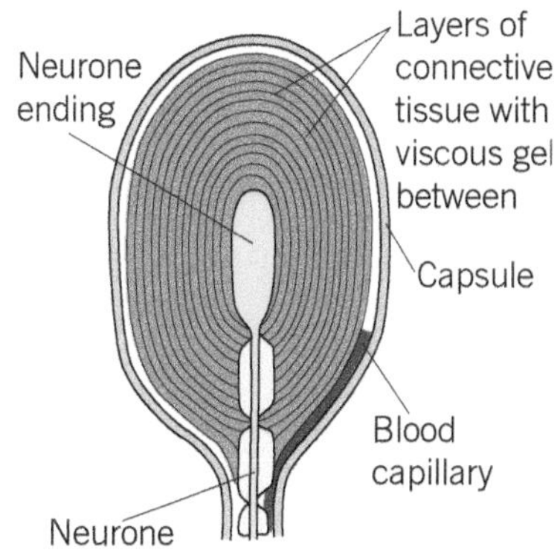

▲ **Figure 2** *Structure of a Pacinian corpuscle*

Synoptic link

Details of carrier proteins in membranes and how they function to transport ions through them are covered in Topic 4.4. Revision of this material will help you to fully understand what follows here and in the next chapter.

- In its normal (resting) state, the stretch-mediated sodium channels of the membrane around the neurone of a Pacinian corpuscle are too narrow to allow sodium ions to pass along them. In this state, the neurone of the Pacinian corpuscle has a resting potential.
- When pressure is applied to the Pacinian corpuscle, it changes shape and the membrane around its neurone becomes stretched (see Figure 3).
- This stretching widens the sodium channels in the membrane and sodium ions diffuse into the neurone.
- The influx of sodium ions changes the potential of the membrane (i.e., it becomes depolarised), thereby producing a generator potential.
- The generator potential in turn creates an action potential (nerve impulse) that passes along the neurone and then, via other neurones, to the central nervous system.

These events are illustrated in Figure 3.

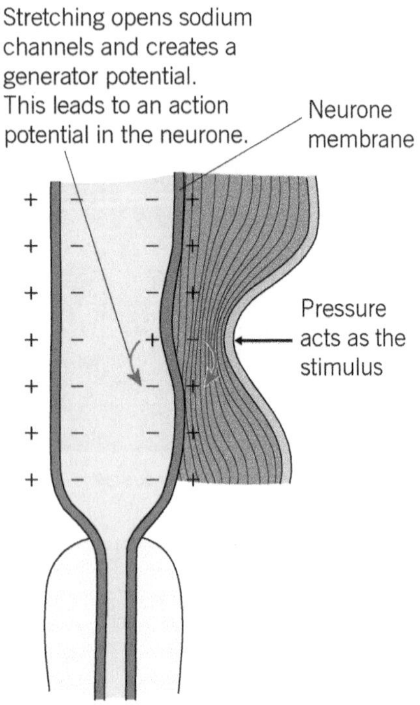

▲ **Figure 3** *Creation of a generator potential in a Pacinian corpuscle*

Receptors working together in the eye

You have seen that an individual receptor responds to only one type of stimulus. It also only responds to a certain intensity of stimulus. This means that, if the body is to be able to distinguish between different intensities of a stimulus (e.g., different light intensities), it must have a range of receptors, each responding to a different intensity of stimulus. To illustrate this, let us consider how the eye works.

The light receptor cells of the mammalian eye are found on its innermost layer – the retina. The millions of light receptors found in the retina are of two main types – rod cells and cone cells. Both rod and cone cells act as transducers by converting light energy into the electrical energy of a nerve impulse.

Rod cells

Rod cells cannot distinguish different wavelengths of light and therefore produce images only in black and white. Rod cells are more numerous than cone cells – there are around 120 million in each eye.

Many rod cells share a single sensory neurone (see Figure 4, overleaf). Rod cells can therefore respond to light of very low intensity. This is because a certain threshold value has to be exceeded before a generator potential is created in the bipolar cells to which they are attached. As a number of rod cells are attached to a single bipolar cell (known as retinal convergence), there is a much greater chance that the threshold value will be exceeded than if only a single rod cell were attached to each bipolar cell. As a result, rod cells allow us to see in low light intensity (i.e., at night), although only in black and white.

In order to create a generator potential, the pigment in the rod cells (rhodopsin) must be broken down. Low-intensity light is sufficient to cause this breakdown. This also helps to explain why rod cells respond to low-intensity light.

A consequence of many rod cells linking to a single bipolar cell is that light received by rod cells sharing the same neurone will only generate a single impulse regardless of how many of the neurones are stimulated. This means that they cannot distinguish between the separate sources of light that stimulated them. Two dots close together will appear as a single blob. Rod cells therefore have low **visual acuity**.

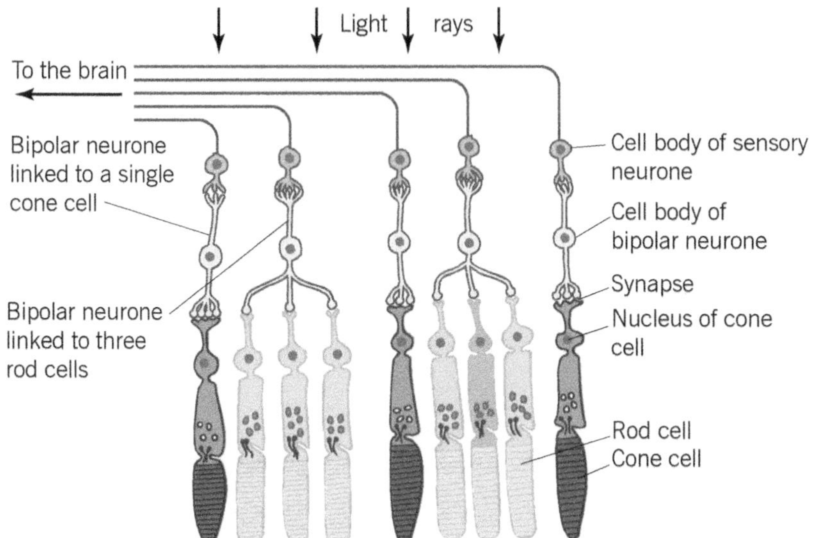

▲ **Figure 4** *Microscopic structure of the retina*

▼ **Table 1** *Differences between rod and cone cells*

Rod cells	Cone cells
Rod-shaped	Cone-shaped
Greater numbers than cone cells	Fewer numbers than rod cells
More at the periphery of the retina, absent at the fovea	Fewer at the periphery of the retina, concentrated at the fovea
Give poor visual acuity	Give good visual acuity
Sensitive to low-intensity light	Not sensitive to low-intensity light

Cone cells

Cone cells are of three different types, each responding to a different wavelength of light. Depending upon the proportion of each type that is stimulated, we can perceive images in full colour.

In each human eye, there are around 6 million cone cells, often with their own separate bipolar cell connected to a sensory neurone (see Figure 4). This means that the stimulation of a number of cone cells cannot be combined to help exceed the threshold value and so create a generator potential. As a result, cone cells only respond to high light intensity and not to low light intensity.

In addition, cone cells each contain one of three different pigments. The pigments in cone cells (each a form of iodopsin) require a higher light intensity for their breakdown. Only light of high intensity will therefore break down the pigment and create a generator potential.

As it is only the cone cells that respond to different wavelengths (colours) of light, this also explains why we cannot see colours in low light intensity (i.e., at night).

Each cone cell has its own connection to a single bipolar cell, which means that, if two adjacent cone cells are stimulated, the brain receives two separate impulses. The brain can therefore distinguish between the two separate sources of light that stimulated the two cone cells. This means that two dots close together will appear as two dots. Therefore cone cells give very accurate vision, that is, they have good visual acuity.

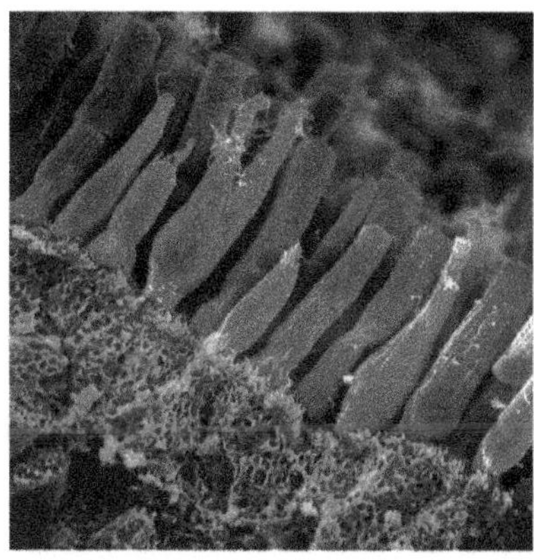

▲ **Figure 5** *False-colour scanning electron micrograph (SEM) of rod and cone cells in the retina of the eye. Rod cells (brown) are long nerve cells responding to dim light whereas cone cells (green) detect colour.*

Table 1 summarises the differences between rod and cone cells.

The distribution of rod and cone cells on the retina is uneven. Light is focused by the lens on the part of the retina opposite the pupil. This point is known as the fovea. The fovea therefore receives the highest intensity of light. Therefore cone cells, but not rod cells, are found at the fovea. The concentration of cone cells diminishes further away from the fovea. At the peripheries of the retina, where light intensity is at its lowest, only rod cells are found.

All this shows how the distribution of rod and cone cells, and the connections they make in the optic nerve, can explain the differences in sensitivity and visual acuity in mammals. By having different types of light receptor, each responding to different stimuli, mammals can benefit from good all-round vision both day and night.

Summary questions

1 What is a stretch-mediated sodium channel?

2 Describe the sequence of events by which pressure on a Pacinian corpuscle results in the creation of a generator potential.

3 Explain why brightly coloured objects often appear grey in dim light.

4 At night, it is often easier to see a star in the sky by looking slightly to the side of it rather than directly at it. Suggest why this is so.

24.3 The nervous system and neurones

Learning objectives:

- → Describe some of the different types of neurone found in the nervous system.
- → Describe the organisation of the nervous system.

Specification reference: 3.4.3.1

The simplest type of nervous response to a stimulus is a reflex arc. Before considering how a reflex arc works, it is helpful to understand how the millions of **neurones** in a mammalian body are organised and to be familiar with the structure of the spinal cord.

Neurones (nerve cells) are specialised cells adapted to rapidly carrying electrochemical charges called **nerve impulses** from one part of the body to another. Neurones are the key cells that make up the nervous system.

Organisation of the nervous system

The nervous system has two major divisions:

- the **central nervous system (CNS)**, which is made up of the brain and spinal cord
- the **peripheral nervous system (PNS)**, which is made up of pairs of nerves that originate from either the brain or the spinal cord.

The peripheral nervous system is divided into:

- **sensory neurones**, which carry nerve impulses from receptors towards the central nervous system
- **motor neurones**, which carry nerve impulses away from the central nervous system to **effectors**.

The motor nervous system can be further subdivided as follows:

- the **voluntary nervous system**, which carries nerve impulses to body muscles and is under voluntary (conscious) control
- the **autonomic nervous system**, which carries nerve impulses to glands, smooth muscle, and cardiac muscle and is not under voluntary control, that is, it is involuntary (subconscious).

A summary of nervous organisation is given in Figure 1.

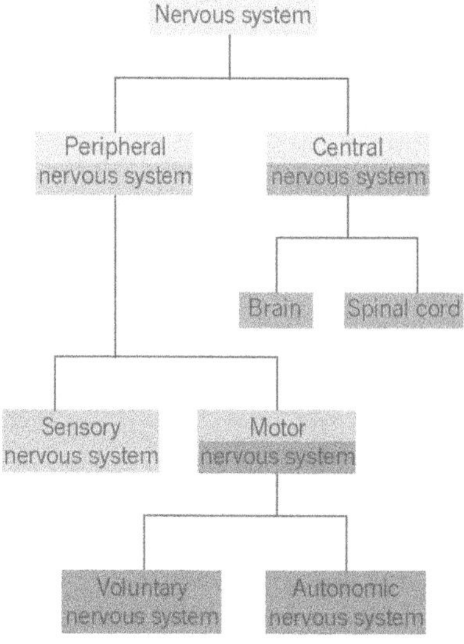

▲ **Figure 1** *Nervous organisation*

The spinal cord

The spinal cord is a column of nervous tissue that runs along the back and lies inside the vertebral column for protection. Emerging at intervals along the spinal cord are pairs of nerves.

The structure of neurones

A mammalian neurone is made up of:

- a **cell body**, which contains a nucleus and large amounts of rough endoplasmic reticulum. This is associated with the production of proteins and **neurotransmitters**.
- **dendrons**, which are small extensions of the cell body that subdivide into smaller branched fibres, called **dendrites**, which carry nerve impulses towards the cell body
- an **axon**, which is a single long fibre that carries nerve impulses away from the cell body
- a **myelin sheath**, which forms a covering to the axon and is made up of the membranes of the Schwann cells. These membranes are rich in a lipid known as **myelin**. Neurones with a myelin sheath

are called myelinated neurones. Some neurones lack a myelin sheath and are called unmyelinated neurones. Myelinated neurones transmit nerve impulses faster than unmyelinated neurones

- **Schwann cells**, which surround the axon, protecting it and providing electrical insulation. They also carry out **phagocytosis** (the removal of cell debris) and play a part in nerve regeneration. Schwann cells wrap themselves around the axon many times, so that layers of their membranes build up around it.
- **nodes of Ranvier**, which are gaps between adjacent Schwann cells where there is no myelin sheath. The gaps are 2–3 μm long and occur every 1–3 mm in humans (see Figure 2).

The structure of a myelinated motor neurone is illustrated in Figure 3 overleaf.

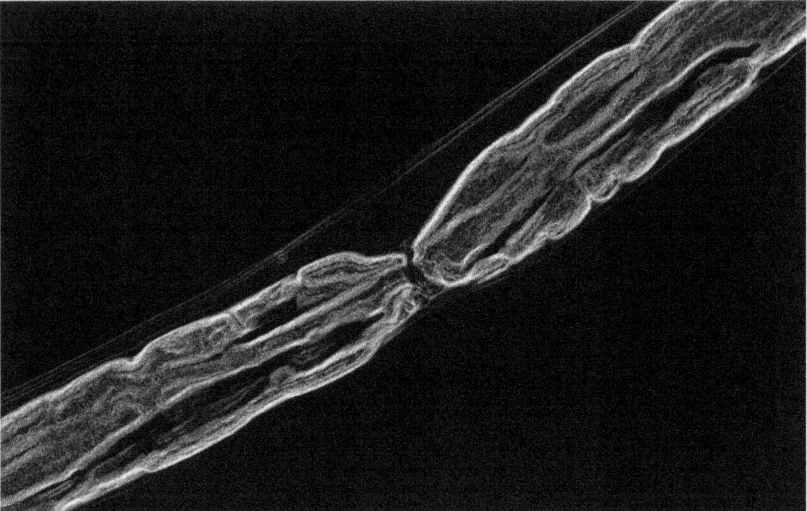

▲ **Figure 2** *Light micrograph of a node of Ranvier in a neurone. The node is the gap in the centre. The gap is a small area without myelin in an otherwise myelinated nerve fibre.*

Neurones can be classified according to their function:

- **Sensory neurones** transmit nerve impulses from a **receptor** to an intermediate or motor neurone. They have one dendron that carries the impulse towards the cell body and one axon that carries it away from the cell body.
- **Motor neurones** transmit nerve impulses from an intermediate or sensory neurone to an **effector**, such as a gland or a muscle. They have a long axon and many short dendrites.
- **Intermediate neurones** transmit impulses between neurones, for example, from sensory to motor neurones. They have numerous short processes.

Figure 4 overleaf shows the structure of all three types of neurone.

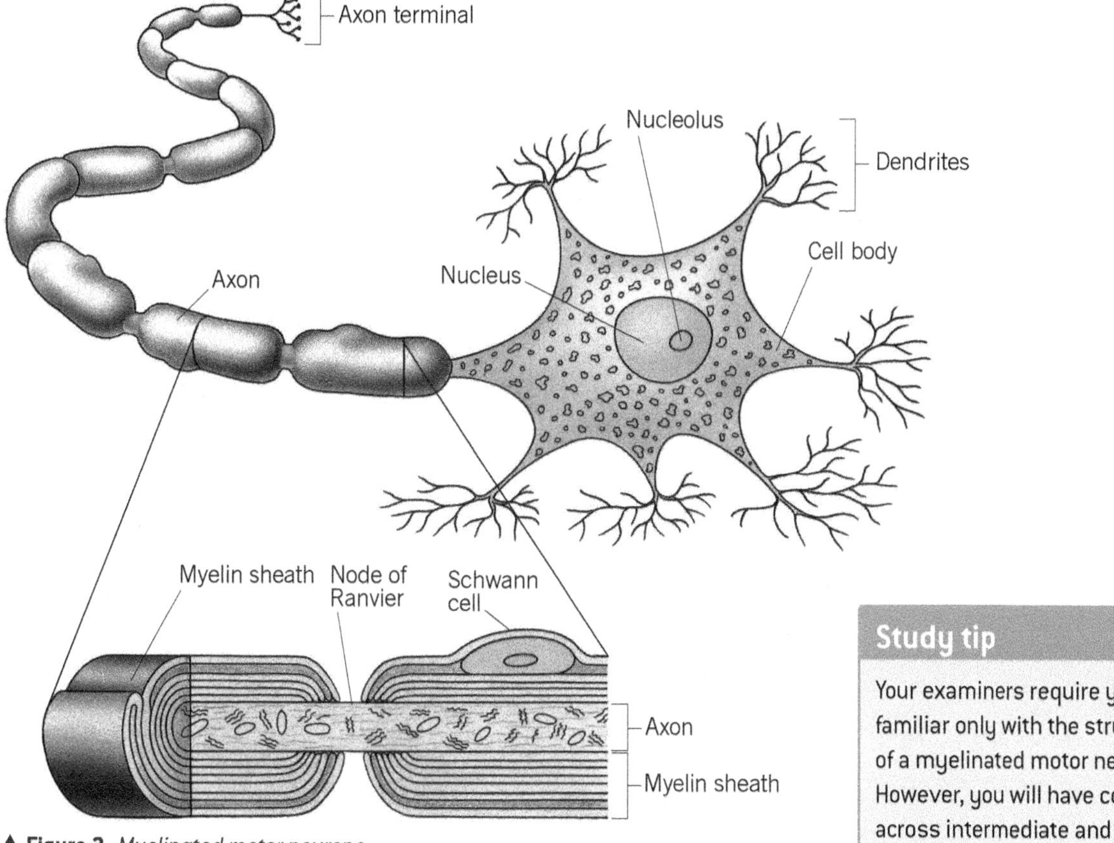

▲ **Figure 3** *Myelinated motor neurone*

Study tip

Your examiners require you to be familiar only with the structure of a myelinated motor neurone. However, you will have come across intermediate and sensory neurones in Topic 24.1 as part of the reflex arc.

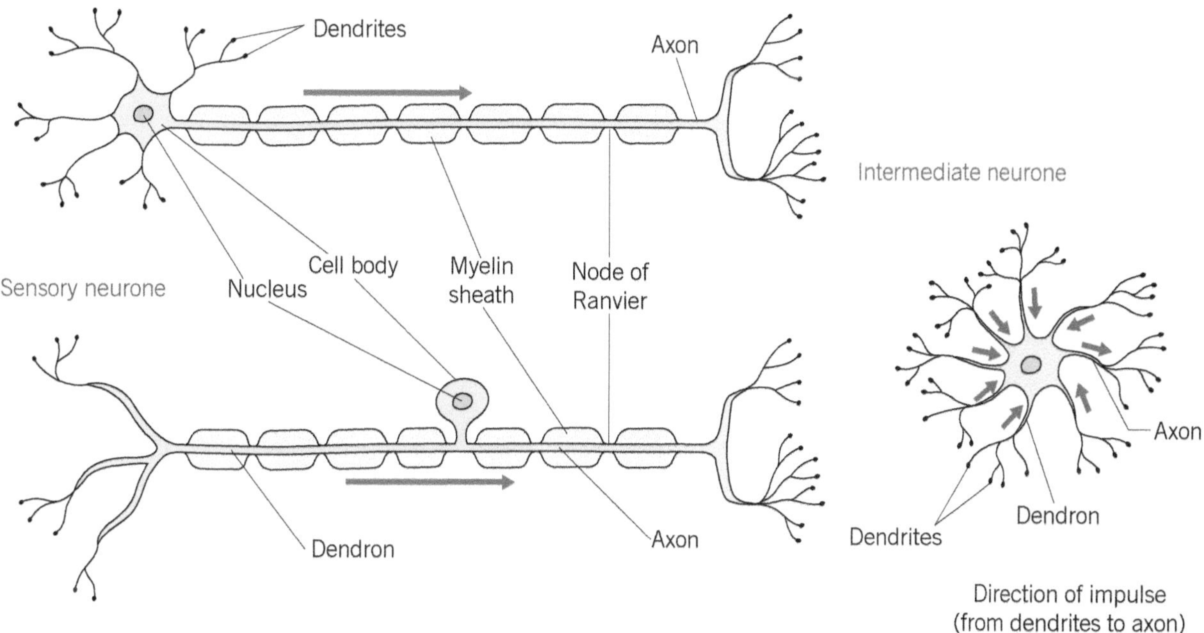

▲ **Figure 4** *Types of neurone*

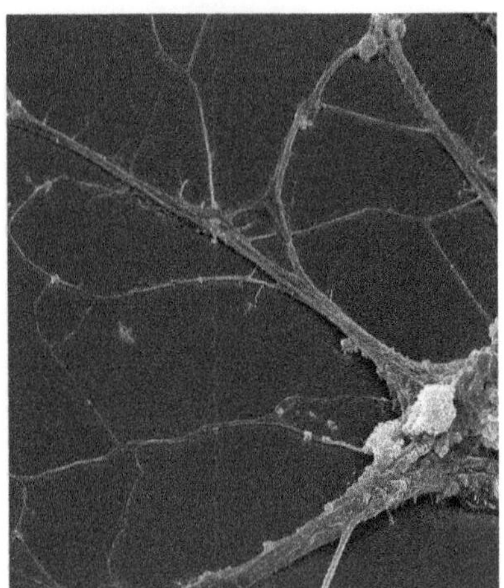

▲ **Figure 5** *Scanning electron micrograph (SEM) of a neurone with the cell body at its centre and dendrites radiating from it*

Summary question

1 In the following passage give the word that best replaces each of the letters **a** to **j**.

Neurones are adapted to carry electrochemical charges called **a**. Each neurone comprises a cell body that contains a **b** and large amounts of **c**, which is used in the production of proteins and neurotransmitters. Extending from the cell body is a single long fibre called an axon and smaller branched fibres called **d**. Axons are surrounded by **e** cells, which protect and provide **f** because their membranes are rich in a lipid known as **g**. There are three main types of neurone. Those that carry nerve impulses to an effector are called **h** neurones. Those that carry impulses from a receptor are called **i** neurones, and those that link the other two types are called **j** neurones.

24.4 The nerve impulse

A nerve impulse may be defined as a self-propagating wave of depolarisation that travels along the surface of the axon membrane. It is not, however, an electrical current, but a temporary reversal of the electrical potential difference across the axon membrane. This reversal is between two states, called the **resting potential** and the **action potential**.

Learning objectives:

→ Describe how the resting potential is established.

→ Describe the events in an action potential.

Specification reference: 3.4.3.1

Resting potential

The movement of ions, such as sodium ions (Na^+) and potassium ions (K^+), across the axon membrane is controlled in a number of ways:

- The phospholipid bilayer of the axon plasma membrane prevents sodium and potassium ions diffusing across it.
- Molecules of proteins, known as intrinsic proteins, span this phospholipid bilayer. These proteins contain channels, called ion channels, which pass through them. Some of these channels have 'gates', which can be opened or closed in order to allow sodium or potassium ions to move through them at any one time, but prevent their movement on other occasions. There are different gated channels for sodium and potassium ions. Some channels, however, remain open all the time, allowing the sodium and potassium ions to diffuse through them unhindered.
- Some carrier proteins actively transport potassium ions into the axon and sodium ions out of the axon. The pump is the carrier protein called a **sodium–potassium pump**.

As a result of these various controls, the inside of an axon is negatively charged relative to the outside. This is known as the **resting potential** and ranges from 50 to 90 millivolts (mV), but is usually around –65 mV. In this condition the axon is said to be **polarised**. The establishment of this potential difference (the difference in charge between the inside and outside of the axon) is due to the following events:

- Sodium ions are actively transported *out* of the axon by the sodium–potassium pumps.
- Potassium ions are actively transported *into* the axon by the sodium–potassium pumps.
- The active transport of sodium ions is greater than that of potassium ions, so three sodium ions move out for every two potassium ions that move in.
- Although both sodium and potassium ions are positive, the outward movement of sodium ions is greater than the inward movement of potassium ions. As a result, there are more sodium ions in the tissue fluid surrounding the axon than in the cytoplasm, and more potassium ions in the cytoplasm than in the tissue fluid, thus creating a chemical gradient.
- The sodium ions begin to diffuse back naturally into the axon whilst the potassium ions begin to diffuse back out of the axon.
- However, most of the gates in the channels that allow the potassium ions to move through are open, whilst most of the gates in the channels that allow the sodium ions to move through are closed.

Synoptic link

To understand the nerve impulse requires a thorough knowledge and understanding of plasma membranes, particularly the structure of plasma membranes and the role of their intrinsic proteins in the sodium–potassium pump. It would be useful to revise Topics 4.1 and 4.4 as a starting point for this section.

Hint

As the phospholipid bilayer does not allow diffusion of sodium and potassium ions, they diffuse back through those sodium and potassium gates that are permanently open.

- As a result the axon membrane is 100 times more permeable to potassium ions, which therefore diffuse back out of the axon faster than the sodium ions diffuse back in. This further increases the potential difference (difference in charge) between the negative inside and the positive outside of the axon.
- Apart from the chemical gradient that causes the movement of the potassium and sodium ions, there is also an electrical gradient. As more and more potassium ions diffuse out of the axon, so the outside of the axon becomes more and more positive. Further outward movement of potassium ions therefore becomes difficult because, being positively charged, they are attracted by the overall negative state inside the axon, which compels them to move into the axon, and repelled by the overall positive state of the surrounding tissue fluid, which prevents them from moving out of the axon.
- An equilibrium is established in which the chemical and electrical gradients are balanced and there is no net movement of ions.

These events are summarised in Figure 1.

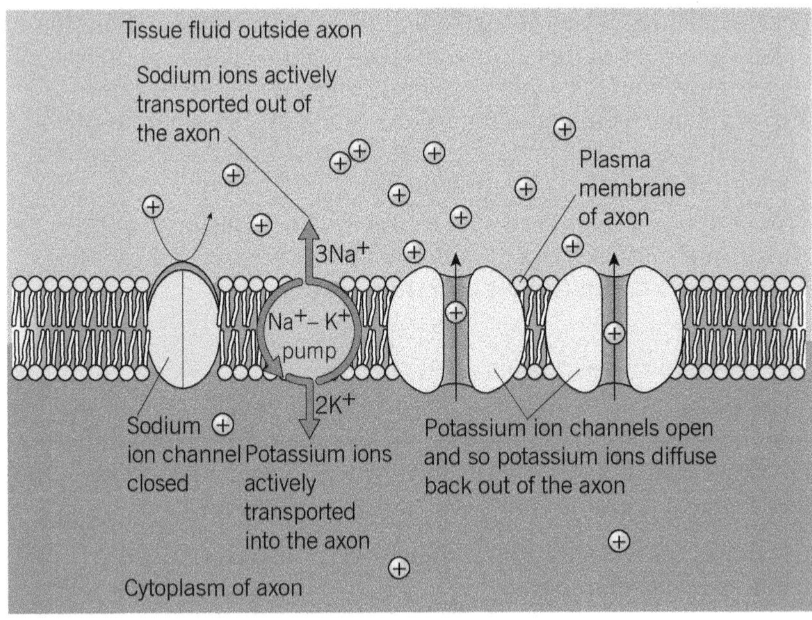

▲ **Figure 1** *Distribution of ions at resting potential*

The action potential

When a stimulus is received by a receptor or nerve ending, its energy causes a temporary reversal of the charges on the axon membrane. As a result, the negative charge of −65 mV inside the membrane becomes a positive charge of around +40 mV. This is known as the **action potential**, and in this condition the membrane is said to be **depolarised**. This depolarisation occurs because the channels in the axon membrane change shape, and hence open or close, depending on the voltage across the membrane. They are therefore called voltage-gated channels. The sequence of events is described below (the numbers relate to the stages illustrated in Figure 2):

Hint

Make sure that you understand the sequence of events during an action potential in terms of the intrinsic (transmembrane) proteins that transport sodium and potassium ions, that is, in terms of sodium ion channel proteins, potassium ion channel proteins, and sodium–potassium pumps (carrier proteins).

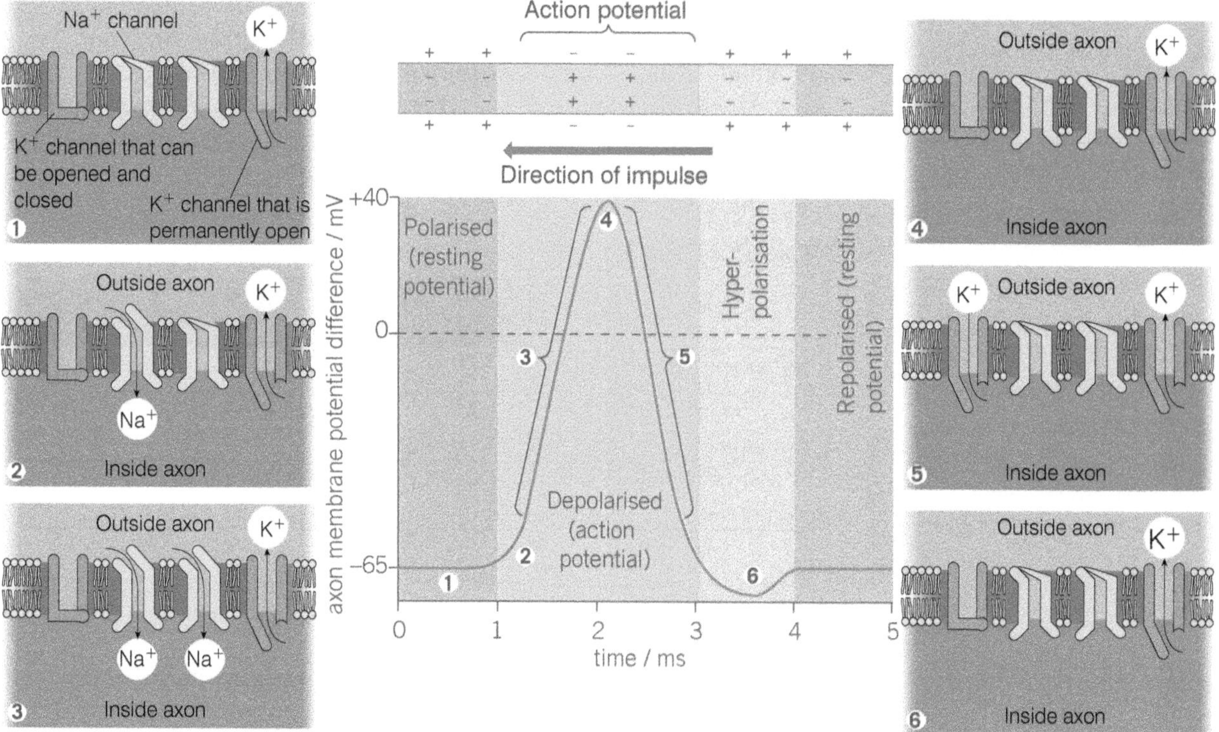

▲ **Figure 2** *The action potential*

1 At resting potential some potassium voltage-gated channels are open (namely those that are permanently open) but the sodium voltage-gated channels are closed.

2 The energy of the stimulus causes some sodium voltage-gated channels in the axon membrane to open and therefore sodium ions diffuse into the axon through these channels along their electrochemical gradient. Being positively charged, they trigger a reversal in the potential difference across the membrane.

3 As the sodium ions diffuse into the axon, so more sodium channels open, causing an even greater influx of sodium ions by diffusion.

4 Once the action potential of around +40 mV has been established, the voltage gates on the sodium ion channels close (thus preventing further influx of sodium ions) and the voltage gates on the potassium ion channels begin to open.

5 With some potassium voltage-gated channels now open, the electrical gradient that was preventing further outward movement of potassium ions is now reversed, causing more potassium ion channels to open. This means that yet more potassium ions diffuse out, causing repolarisation of the axon.

6 The outward diffusion of these potassium ions causes a temporary overshoot of the electrical gradient, with the inside of the axon being more negative (relative to the outside) than usual (hyperpolarisation). The gates on the potassium ion channels now close and the activities of the sodium–potassium pumps once again cause sodium ions to be pumped out and potassium ions in. The resting potential of −65 mV is re-established and the axon is said to be **repolarised**.

Hint

The unit of time given on the y-axis of the graph (see Figure 2) is the millisecond (ms). A millisecond is 0.001 of a second. There are therefore 1000 milliseconds in a second. At 2 ms each, action potentials are very short-lived!

Summary questions

1 Describe how the movement of ions establishes the resting potential in an axon.

2 Table 1 shows the membrane potential of an axon at different stages of an action potential. The table refers to those channels that can be opened and closed, not those that remain permanently open. For each of the letters A–F, indicate the state of the relevant channels, that is, open or closed.

The terms action potential and resting potential can be misleading because the movement of sodium ions inwards during the action potential is purely due to diffusion – which is a passive process – whereas the resting potential is maintained by active transport – which is an active process. The term action potential simply means that the axon membrane is transmitting a nerve impulse, whereas resting potential means that it is not.

▼ **Table 1**

	Resting	Beginning to depolarise	Repolarising
Membrane potential / mV	−70	−50	−20
Na^+ channels in axon membrane	A	B	C
K^+ channels in axon membrane	D	E	F

Measuring action potentials

The plasma membrane of an axon will transmit an action potential when stimulated to do so. The action potential involves changes in the electrical potential across the membrane due to the movement of positive ions.

1 Which **two** positive ions are responsible for this change in electrical potential?

Figure 3 shows two action potentials that were recorded using an instrument called an oscilloscope.

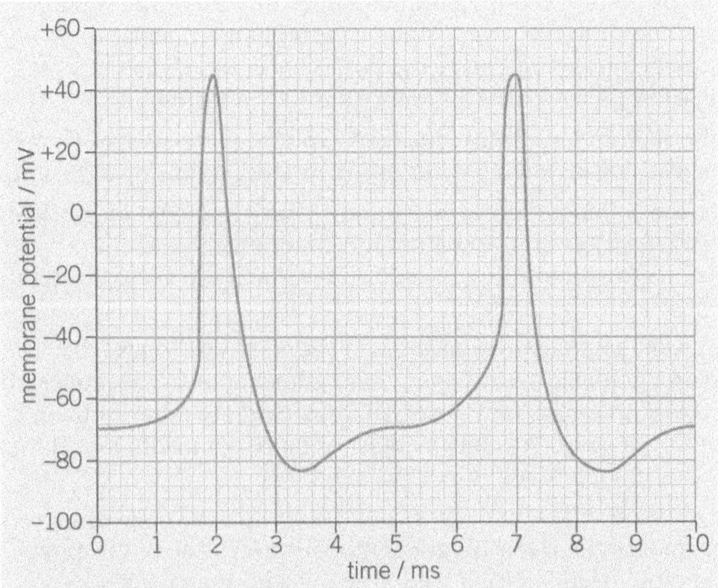

▲ **Figure 3**

2 Between 0.5 ms and 2.0 ms there is a considerable change in membrane potential. Explain how this change is brought about.

3 How many action potentials will occur in 1 second if the frequency shown on the graph is maintained for this period? Show your working.

24.5 Passage of an action potential

Once it has been created, an **action potential** moves rapidly along an **axon**. The size of the action potential remains the same from one end of the axon to the other. Strictly speaking, nothing physically 'moves' from place to place along the axon of the neurone, but rather the reversal of electrical charge is reproduced at different points along the axon membrane. As one region of the axon produces an action potential and becomes depolarised, it acts as a stimulus for the **depolarisation** of the next region of the axon. In this manner, action potentials are regenerated along each small region of the axon membrane. In the meantime, the previous region of the membrane returns to its resting potential, that is, it undergoes **repolarisation**.

Learning objectives:

- → Describe how an action potential passes along an unmyelinated axon.
- → Explain how an action potential passes along a myelinated axon.

Specification reference: 3.4.3.1

Passage of an action potential along an unmyelinated axon

It is easier to understand how a nerve impulse is propagated in a myelinated axon if we first look at how it is propagated in an unmyelinated one. This process is described and illustrated in Figure 2.

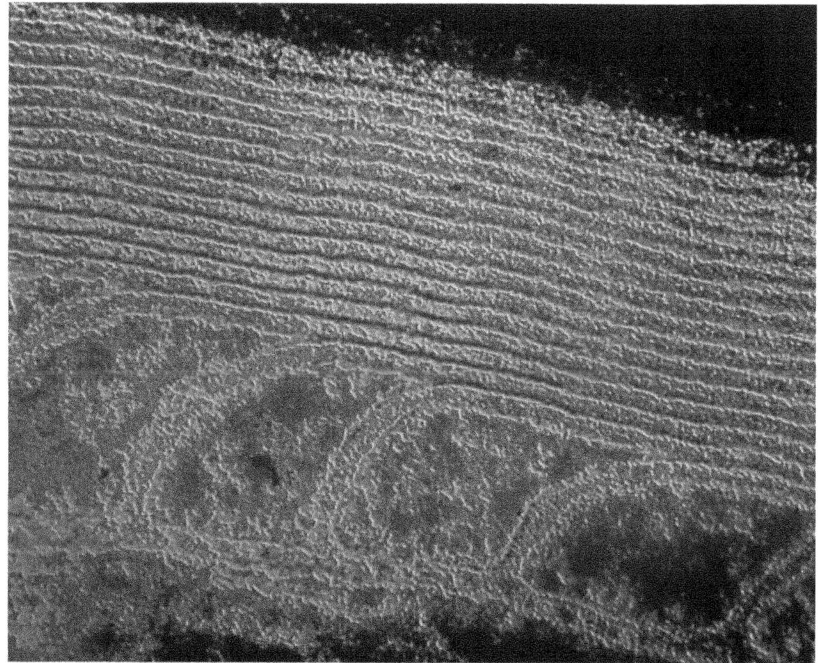

▲ **Figure 1** *False-colour transmission electron micrograph (TEM) of the myelin sheath (orange bands at top) around the axon (bottom)*

Polarised

Depolarised

Repolarised

1
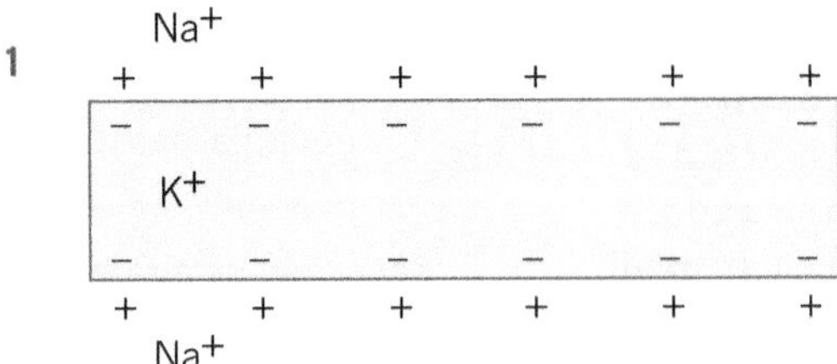

1 At resting potential the concentration of sodium ions outside the axon membrane is high relative to the inside, whereas that of the potassium ions is high inside the membrane relative to the outside. The overall concentration of positive ions is, however, greater on the outside, making this positive compared with the inside. The axon membrane is polarised.

2
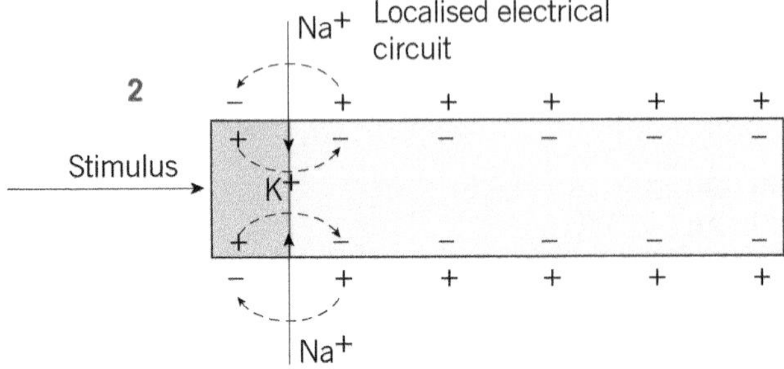

2 A stimulus causes a sudden influx of sodium ions and hence a reversal of charge on the axon membrane. This is the action potential and the membrane is depolarised.

3
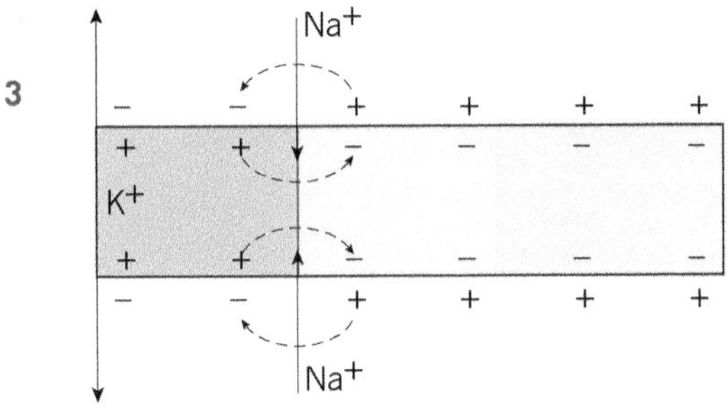

3 The localised electrical circuits established by the influx of sodium ions cause the opening of sodium voltage-gated channels a little further along the axon. The resulting influx of sodium ions in this region causes depolarisation. Behind this new region of depolarisation, the sodium voltage-gated channels close and the potassium ones open. Potassium ions begin to leave the axon along their electrochemical gradient.

4
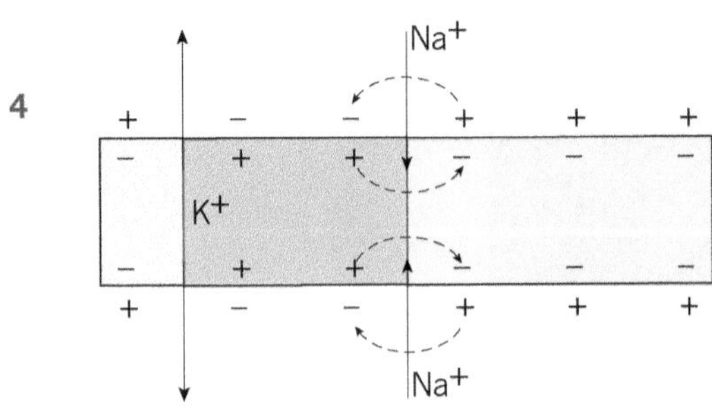

4 The action potential (depolarisation) is propagated in the same way further along the axon. The outward movement of the potassium ions has continued to the extent that the axon membrane behind the action potential has returned to its original charged state (positive outside, negative inside), that is, it has been repolarised.

5
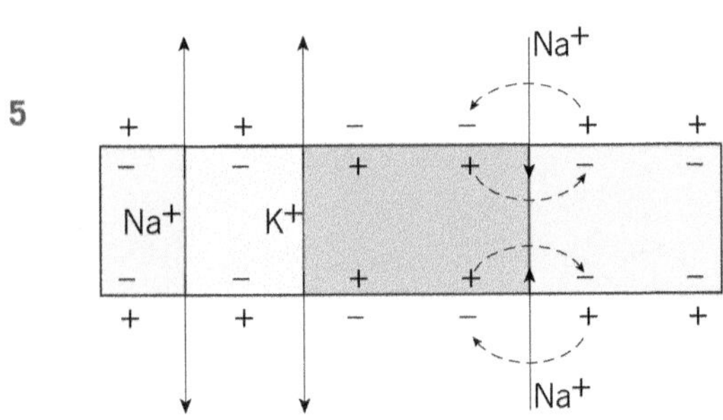

5 Repolarisation of the axon allows sodium ions to be actively transported out, once again returning the axon to its resting potential in readiness for a new stimulus if it comes.

▲ **Figure 2** *Passage of an impulse along the axon of an unmyelinated neurone*

Passage of an action potential along a myelinated axon

In myelinated axons, the fatty sheath of myelin around the axon acts as an electrical insulator, preventing action potentials from forming. At intervals of 1–3 mm there are breaks in this myelin insulation, called nodes of Ranvier (see Topic 24.3). Action potentials can occur at these points. The localised circuits therefore arise between adjacent nodes of Ranvier and the action potentials in effect 'jump' from node to node in a process known as saltatory conduction (Figure 3). As a result, an action potential passes along a myelinated neurone faster than along the axon of an unmyelinated one.

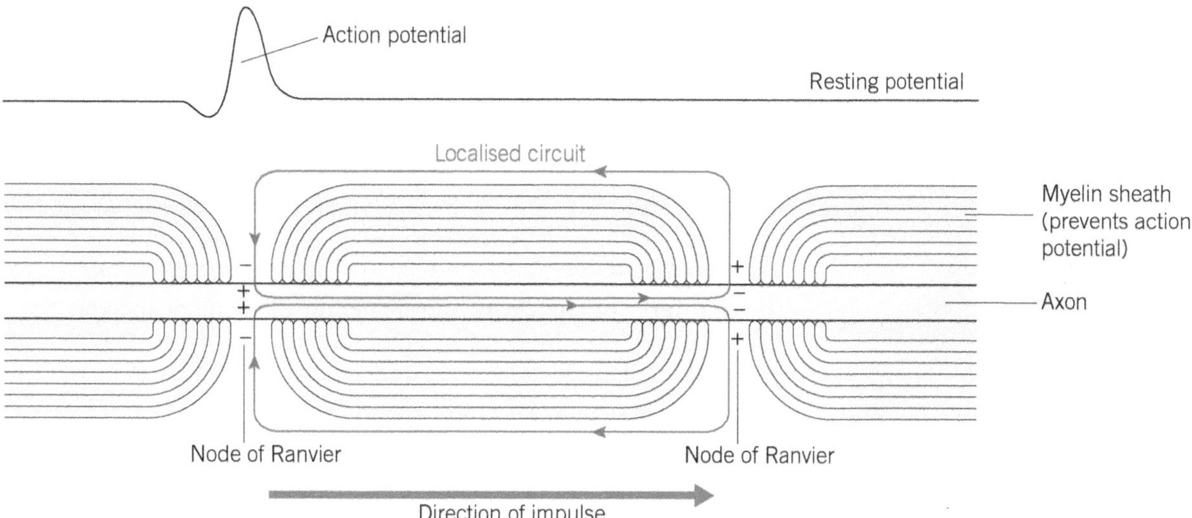

▲ **Figure 3** *Passage of an action potential along a myelinated axon. Action potentials are produced only at nodes of Ranvier. Depolarisation therefore skips from node to node (saltatory conduction).*

Summary questions

1 In a myelinated axon, sodium and potassium ions can only be exchanged at certain points along it.

- **a** What is the name given to these points?
- **b** Explain why ions can only be exchanged at these points.
- **c** What effect does this have on the way an action potential is conducted along the axon?
- **d** What name is given to this type of conduction?
- **e** How does it affect the speed with which the action potential is transmitted compared to the speed along an unmyelinated axon?

2 What happens to the size of an action potential as it moves along an axon?

Hint

The term saltatory in saltatory conduction comes from the Latin word *saltare*, meaning to jump.

24.6 Speed of the nerve impulse

Learning objectives:

- → Describe the factors that affect the speed of conductance of an action potential.
- → Explain the importance of the refractory period in the transmission of a nerve impulse.
- → Describe what is meant by the all-or-nothing principle.

Specification reference: 3.4.3.1

▲ **Figure 1** *False-coloured transmission electron micrograph (TEM) of a section through a myelinated neurone and Schwann cell. Myelin (black) surrounds the axon (purple), increasing the speed at which nerve impulses travel. It is formed when Schwann cells (green) wrap around the axon, depositing layers of myelin between each coil.*

Once an action potential has been set up, it moves rapidly from one end of the axon to the other without any decrease in size. In other words, the final action potential at the end of the axon is the same size as the first action potential. This transmission of the action potential along the axon of a neurone is the **nerve impulse**.

Factors affecting the speed at which an action potential travels

A number of factors affect the speed at which the action potential passes along the axon. Depending upon these factors, an action potential may travel at a speed of as low as $0.5\,m\,s^{-1}$ or as high as $120\,m\,s^{-1}$. These factors include:

- **the myelin sheath**. You saw in Topic 24.5 that the myelin sheath acts as an electrical insulator, preventing an action potential forming in the part of the axon covered in myelin. The action potential does, however, jump from one node of Ranvier to another (saltatory conduction). This increases the speed of conductance from $30\,m\,s^{-1}$ in an unmyelinated neurone to $90\,m\,s^{-1}$ in a similar myelinated one.
- **the diameter of the axon**. The greater the diameter of an axon, the faster the speed of conductance. This is due to less leakage of ions from a large axon (leakage makes membrane potentials harder to maintain).
- **temperature**. This affects the rate of diffusion of ions and therefore the higher the temperature the faster the nerve impulse. The energy for active transport comes from respiration. Respiration, like the sodium–potassium pump, is controlled by enzymes. Enzymes function more rapidly at higher temperatures up to a point. Above a certain temperature, enzymes and the plasma membrane proteins are denatured and impulses fail to be conducted at all. Temperature is clearly an important factor in response times in cold-blooded (ectothermic) animals, whose body temperature varies in accordance with the environment.

The refractory period

Once an action potential has been created in any region of an axon, there is a period afterwards when inward movement of sodium ions is prevented because the sodium voltage-gated channels are closed. During this time it is impossible for a further action potential to be generated. This is known as the **refractory period** (Figure 2, overleaf).

The refractory period serves three purposes:

- **It ensures that an action potential is propagated in one direction only**. An action potential can only pass from an active region to a resting region. This is because an action potential cannot be propagated in a region that is refractory, which means that it can only move in a forward direction. This prevents the action potential from spreading out in both directions, which it would otherwise do.
- **It produces discrete impulses**. Due to the refractory period, a new action potential cannot be formed immediately behind the first one. This ensures that action potentials are separated from one another.

- **It limits the number of action potentials**, which also limits the strength of stimulus that can be detected. As action potentials are separated from one another this limits the number of action potentials that can pass along an axon in a given time.

All-or-nothing principle

Nerve impulses are described as **all-or-nothing** responses. There is a certain level of stimulus, called the **threshold value**, which triggers an action potential. Below the threshold value, no action potential, and therefore no impulse, is generated. Any stimulus, of whatever strength, that is below the threshold value will fail to generate an action potential – this is the 'nothing' part. Any stimulus above the threshold value will succeed in generating an action potential.

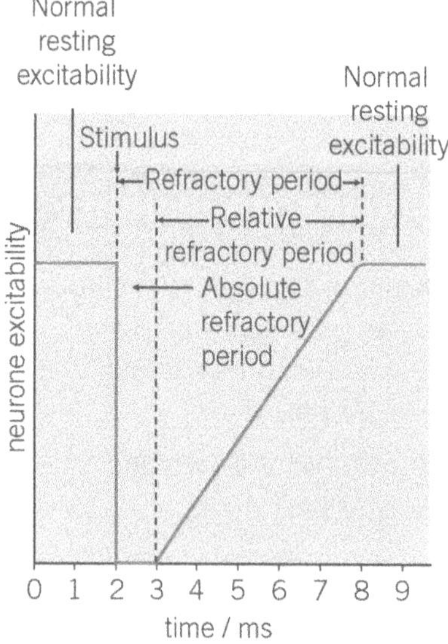

▲ **Figure 2** *Graph illustrating neurone excitability before and after a nerve impulse*

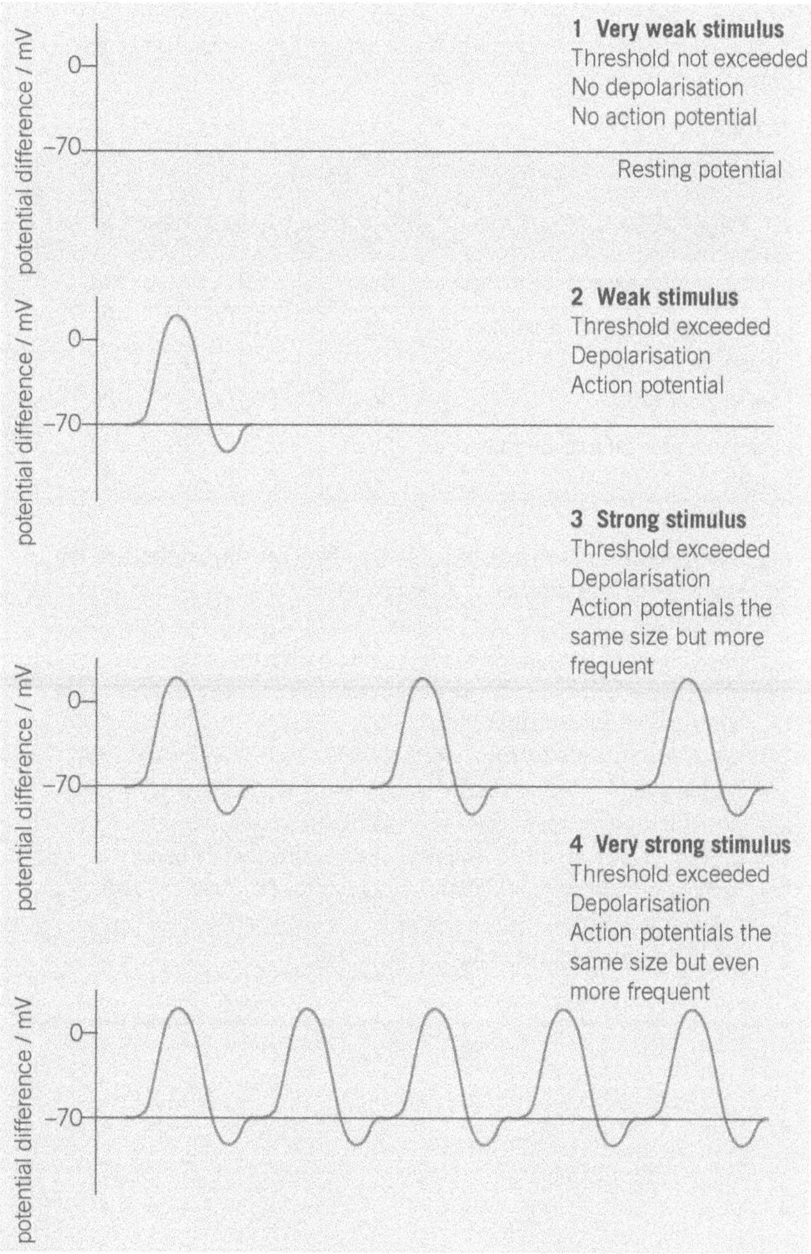

potential difference / mV
0
−70
1 Very weak stimulus
Threshold not exceeded
No depolarisation
No action potential
Resting potential

2 Weak stimulus
Threshold exceeded
Depolarisation
Action potential

3 Strong stimulus
Threshold exceeded
Depolarisation
Action potentials the same size but more frequent

4 Very strong stimulus
Threshold exceeded
Depolarisation
Action potentials the same size but even more frequent

▲ **Figure 3** *Effect of stimulus intensity on impulse frequency*

Hint

The brain would be overloaded with information if it became aware of every little stimulus. The all-or-nothing nature of the action potential acts as a filter, preventing minor stimuli from setting up nerve impulses, and thus preventing the brain becoming overloaded.

It does not matter how much above the threshold a stimulus is, it will still only generate one action potential – this is the 'all' part. How, then, can an organism perceive the size of a stimulus? This is achieved in two ways:

- by the number of impulses passing in a given time. The larger the stimulus, the more impulses that are generated in a given time (Figure 3).
- by having different neurones with different threshold values. The brain interprets the number and type of neurones that pass impulses as a result of a given stimulus and thereby determines its size.

Summary questions

1. Explain how the refractory period ensures that nerve impulses are kept separate from one another.
2. What is the all-or-nothing principle?

Different axons, different speeds

Table 1 below shows the speeds at which different axons conduct action potentials.

▼ Table 1

Axon	Myelin	Axon diameter / μm	Transmission speed / m s^{-1}
Human motor axon to leg muscle	Yes	20	120
Human sensory axon from skin pressure receptor	Yes	10	50
Squid giant axon	No	500	25
Human motor axon to internal organ	No	1	2

1. Using data from the table, describe the effect of axon diameter on the speed of conductance of an action potential.
2. The data show that a myelinated axon conducts an action potential faster than an unmyelinated axon. Explain why this is so.
3. What is the name of the cells whose membranes make up the myelin sheath around some types of axon?
4. State which has the greater effect on the speed of conductance of an action potential – the presence of myelin or the diameter of the axon. Use information from Table 1 to explain your answer.
5. The squid is an ectothermic animal. This means that its body temperature fluctuates with the temperature of the water in which it lives. Suggest how this might affect the speed at which action potentials are conducted along a squid axon.

24.7 Structure and function of synapses

A synapse is the point where the axon of one neurone communicates with the dendrite of another or with an effector, for example, at neuromuscular junctions. They are important in allowing different neurones to interact and, therefore, in achieving coordination.

Learning objectives:

- → Describe the structure of a synapse.
- → Outline the functions of synapses in the nervous system.

Specification reference: 3.4.3.2

Structure of a synapse

Synapses transmit impulses from one neurone to another by means of chemicals known as **neurotransmitters**. Neurones are separated by a small gap, called the **synaptic cleft**, which is 20–30 nm wide. The neurone that releases the neurotransmitter is called the **presynaptic neurone**. The axon of this neurone ends in a swollen portion known as the **synaptic knob**. This possesses many mitochondria and large amounts of endoplasmic reticulum. These are required in the manufacture of the neurotransmitter. Once made, the neurotransmitter is stored in the **synaptic vesicles**. Once the neurotransmitter is released from the vesicles it diffuses across to the postsynaptic neurone, which possesses complementary molecules on its membrane to which the neurotransmitter binds. The structure of a chemical synapse is illustrated in Figure 1.

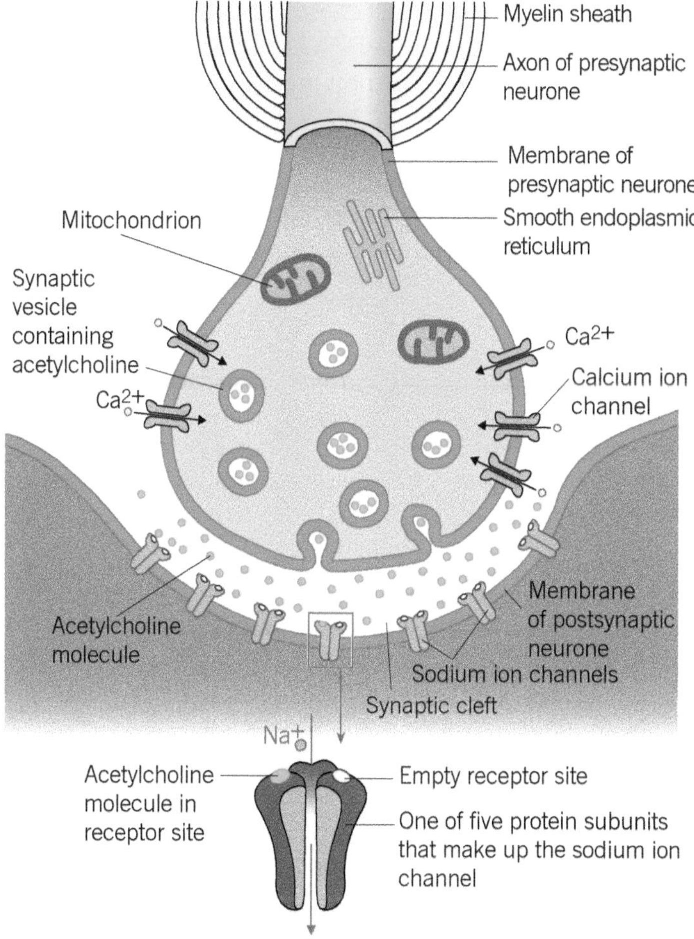

▲ **Figure 1** *Structure of a synapse*

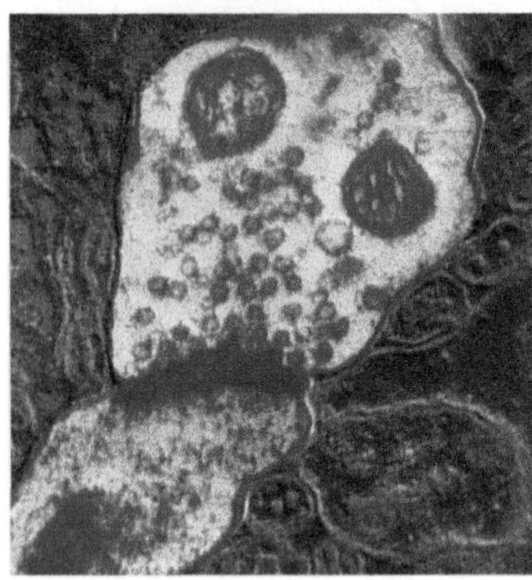

▲ **Figure 2** *False-colour transmission electron micrograph (TEM) of a synapse. The synaptic cleft between the two neurones (centre) appears deep red. The cell above the cleft has many small vesicles (red-yellow spheres) containing neurotransmitter, whereas the two larger spheres above the vesicles are mitochondria.*

Functions of synapses

Synapses transmit information in the form of a release of neurotransmitter. In so doing, they act as junctions, allowing:

- impulses which are travelling along a single neurone to a synapse to initiate an action potential in a number of different neurones that share the same synapse
- impulses travelling along several different neurones to collectively initiate an action potential in a single neurone which shares the same synapse.

You shall look in more detail at transmission in synapses in Topic 24.8. To understand the basic functioning of synapses described here, it is sufficient to appreciate the following:

- A chemical (the neurotransmitter) is made *only* in the presynaptic neurone and not in the postsynaptic neurone.
- The neurotransmitter is stored in synaptic vesicles and released into the synapse when an **action potential** reaches the synaptic knob.
- When released, the neurotransmitter diffuses across the synapse to receptor molecules on the postsynaptic neurone.
- The neurotransmitter binds with the receptor molecules and sets up a new action potential in the postsynaptic neurone.

Features of synapses

The basic way in which synapses function means that they have a number of different features.

Unidirectionality

Synapses only transmit information (in the form of a chemical neurotransmitter) in one direction – from the presynaptic neurone to the postsynaptic neurone. In this way, synapses act like valves.

Summation

Low-frequency action potentials often produce insufficient amounts of neurotransmitter to trigger a new action potential in the postsynaptic neurone. They can, however, be made to do so by a process called summation. This entails a build-up of neurotransmitter in the synapse by one of two methods:

- **spatial summation**, in which a number of different presynaptic neurones together release enough neurotransmitter to exceed the threshold value of the postsynaptic neurone. Together they therefore trigger a new action potential.
- **temporal summation**, in which a single presynaptic neurone releases neurotransmitter many times over a short period. If the total amount of neuro transmitter exceeds the threshold value of the postsynaptic neurone and causes sufficient depolarisation, then a new action potential is triggered.

Spatial and temporal summation are illustrated in Figure 3, overleaf.

Spatial summation

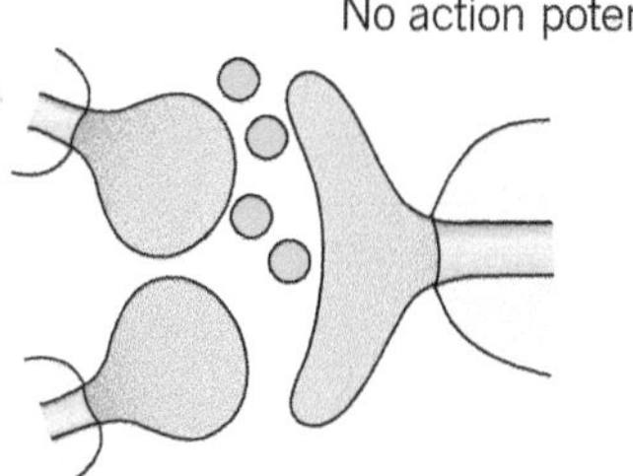

Neurone A releases neurotransmitter but quantity is insufficient to allow the entry of enough sodium ions. The threshold is not reached and no action potential occurs.

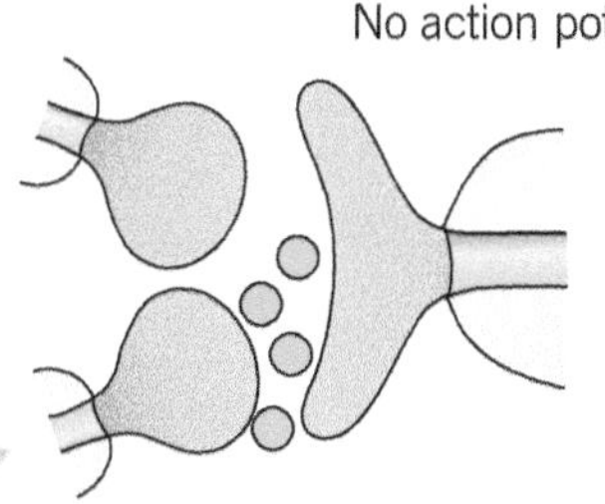

Neurone B releases neurotransmitter but quantity is insufficient to allow the entry of enough sodium ions. The threshold is not reached and no action potential occurs.

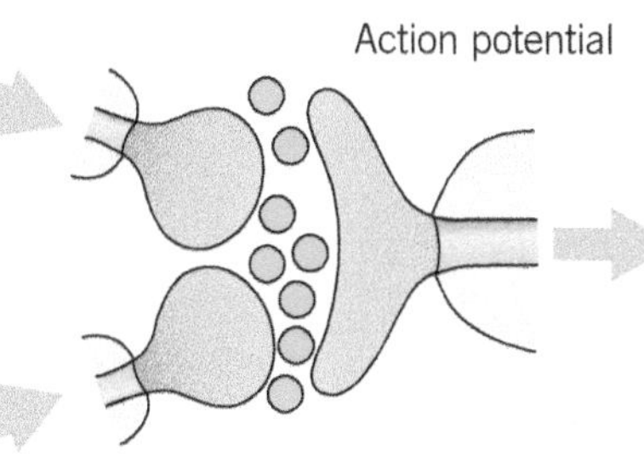

Neurone A and B release neurotransmitter. The binding of the neurotransmitter allows enough sodium ions to enter to exceed the threshold and the membrane is depolarised. An action potential is triggered in the postsynaptic neurone.

Temporal summation

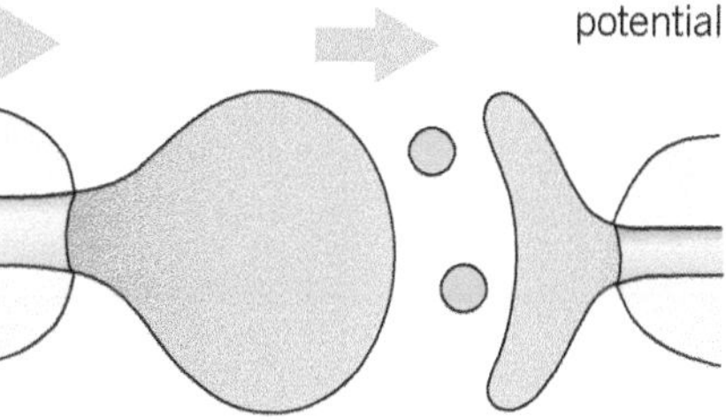

Low-frequency action potentials lead to release of small amount of neurotransmitter. Quantity is below the threshold to trigger an action potential in the postsynaptic neurone.

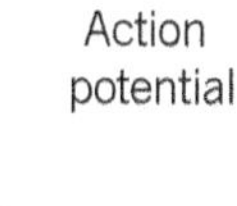
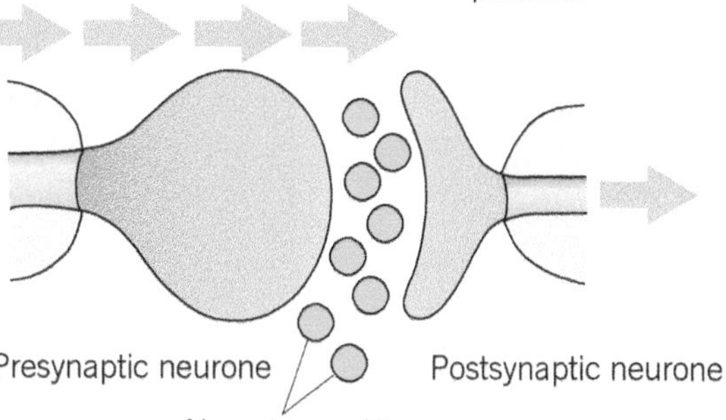

High-frequency action potentials lead to release of large amount of neurotransmitter. Quantity is above the threshold to trigger an action potential in the postsynaptic neurone.

▲ **Figure 3** *Spatial and temporal summation*

Inhibition

On the postsynaptic membrane of some synapses, the protein channels carrying chloride ions (Cl^-) can be made to open. This leads to an inward diffusion of Cl^- ions, making the inside of the postsynaptic membrane even more negative than when it is at resting potential. This is called hyperpolarisation and makes it less likely that a new action potential will be created. For this reason these synapses are called inhibitory synapses.

Summary questions

1. How is a presynaptic neurone adapted for the manufacture of neurotransmitter?
2. How is the postsynaptic neurone adapted to receive the neurotransmitter?
3. Describe the basic events in the transmission of a nerve impulse from one neurone to another.
4. If a neurone is stimulated in the middle of its axon, an action potential will pass both ways along it to the synapses at each end of the neurone. However, the action potential will only pass across the synapse at one end. Explain why.
5. When walking along a street we barely notice the background noise of traffic. However, we often respond to louder traffic noises, such as the sound of a horn.
   - **a** From your knowledge of summation, explain this difference.
   - **b** Suggest an advantage in responding to high-level stimuli but not to low-level ones.
6. Explain why hyperpolarisation reduces the likelihood of a new action potential being created.

24.8 Transmission across a synapse

In Topic 24.7 you learnt how information is passed between neurones in the form of discrete bursts of neurotransmitter. Let us now consider this in more detail by looking at a cholinergic synapse.

A **cholinergic** synapse is one in which the neurotransmitter is a chemical called **acetylcholine**. Acetylcholine is made up of two parts, acetyl (more precisely ethanoic acid) and choline. Cholinergic synapses are common in vertebrates, where they occur in the central nervous system and at neuromuscular junctions (junctions between neurones and muscles). Details of the neuromuscular junction are given in Topic 25.1.

The process of transmission across a cholinergic synapse is described in the series of diagrams in Figure 2 on page 401. To simplify matters, only the relevant structures are shown on each diagram.

Effects of drugs on synapses

There are many different neurotransmitters responsible for the exchange of information across a synapse. There are also many different types of complementary molecules on the postsynaptic membrane to which the neurotransmitter can bind. Some of these neurotransmitters and receptors are excitatory, that is, they create a new action potential in the postsynaptic neurone. Others are inhibitory, that is, they make it less likely that a new action potential will be created in the postsynaptic neurone. Overall, the action of a neurotransmitter depends on the receptor to which it binds.

Given that our perception of the world is through information received by receptors and transferred to the brain by neurones that connect via synapses, it is not surprising that the effects of many medicinal and recreational drugs are due to their actions on synapses. Drugs act on synapses in two main ways:

- **They stimulate the nervous system by creating more action potentials in postsynaptic neurones.** A drug may do this by mimicking a neurotransmitter, stimulating the release of more neurotransmitter or inhibiting the enzyme that breaks down the neurotransmitter. The outcome is to enhance the body's responses to impulses passed along the postsynaptic neurone. For example, if the neurone transmits impulses from sound receptors, a person will perceive the sound as being louder.
- **They inhibit the nervous system by creating fewer action potentials in postsynaptic neurones.** A drug may do this by inhibiting the release of neurotransmitter or blocking the receptors on sodium or potassium ion channels on the postsynaptic neurone. The outcome is to reduce the body's responses to impulses passed along the postsynaptic neurone. In this case, if the neurone transmits impulses from sound receptors, a person will perceive the sound as being quieter.

The effects of a drug on a neurotransmitter depend on the type of neurotransmitter. For example, a drug that inhibits an excitatory neurotransmitter will reduce a particular effect, but a drug that inhibits an

Learning objectives:

→ Describe how information is transmitted across a synapse.

Specification reference: 3.4.3.2

Hint

You need to think in terms of separate 'bursts' of neurotransmitter release from the presynaptic knob. Each one relates to the arrival of an action potential along the neurone.

Study tip

In this topic, there are many possibilities for synoptic questions that bring together other topics. These include membrane structure, enzyme action, mitochondria and ATP production, diffusion (across the synaptic cleft and through channel proteins), and how molecular shapes fit one another (e.g., a neurotransmitter fitting into receptors on the postsynaptic neurone).

Summary questions

1 For each of the following, state as accurately as possible the name of the substance described.

a They diffuse into the postsynaptic neurone where they generate an action potential.

b A neurotransmitter found in a cholinergic synapse.

c It is released by mitochondria to enable the neurotransmitter to be re-formed.

d Their influx into the presynaptic neurone causes synaptic vesicles to release their neurotransmitter.

2 Why is it necessary for acetylcholine to be hydrolysed by acetylcholinesterase?

inhibitory neurotransmitter will enhance a particular effect. Let us look at some examples of the effects of drugs on synapses.

Endorphins are neurotransmitters used by certain sensory nerve pathways, especially pain pathways. Endorphins block the sensation of pain by binding to pain receptor sites. Drugs such as **morphine**, **codeine**, and **heroin** bind to the specific receptors used by endorphins.

1 Suggest the likely effect of drugs like morphine and codeine on the body.
2 Explain how the effect you suggest might be brought about.

Serotonin is a neurotransmitter involved in the regulation of sleep and certain emotional states. Reduced activity of the neurones that release serotonins is thought to be one cause of clinical depression. **Prozac** (fluoxetine) is an antidepressant drug that affects serotonin within synaptic clefts.

3 Suggest a way that the drug Prozac might affect serotonin within synaptic clefts.
4 Explain how the effect you suggest makes Prozac an effective antidepressant.

GABA is a neurotransmitter that inhibits the formation of action potentials when it binds to postsynaptic neurones. **Valium** (diazepam) is a drug that enhances the binding of GABA to its receptors.

5 Suggest the likely effect of Valium on the nerve pathways that cause muscle contractions.
6 Explain the reasoning for your answer.
7 Epilepsy can be the result of an increase in the activity of neurones in the brain due to insufficient GABA. An enzyme breaks down GABA on the postsynaptic membrane. A drug called vigabatrin has a molecular structure similar to GABA and is used to treat epilepsy. Suggest a way in which vigabatrin might be effective in treating epilepsy.

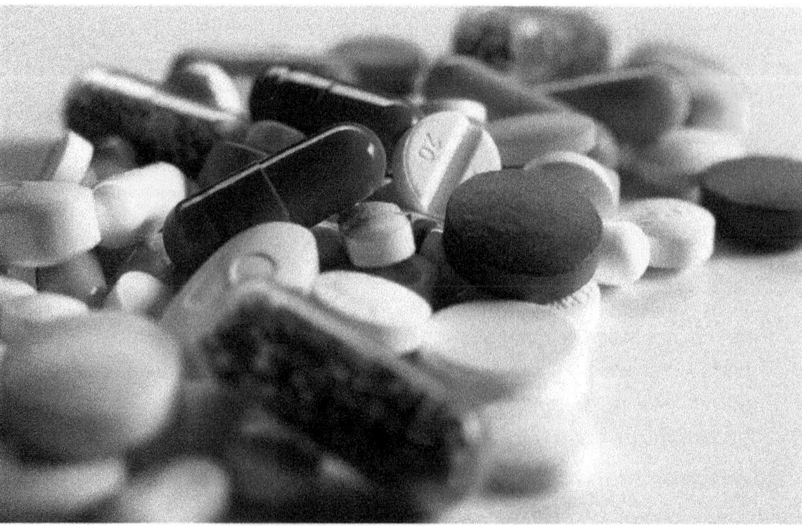

▲ **Figure 1** *Many drugs function by acting on synapses*

1 The arrival of an action potential at the end of the presynaptic neurone causes calcium ion channels to open and calcium ions (Ca^{2+}) enter the synaptic knob.

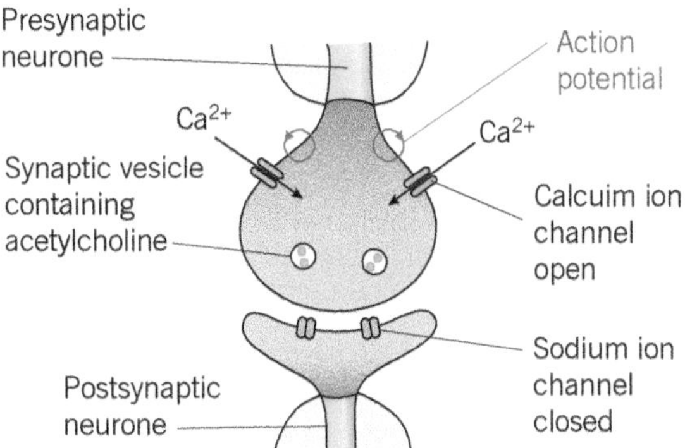

2 The influx of calcium ions into the presynaptic neurone causes synaptic vesicles to fuse with the presynaptic membrane, so releasing acetylcholine into the synaptic cleft.

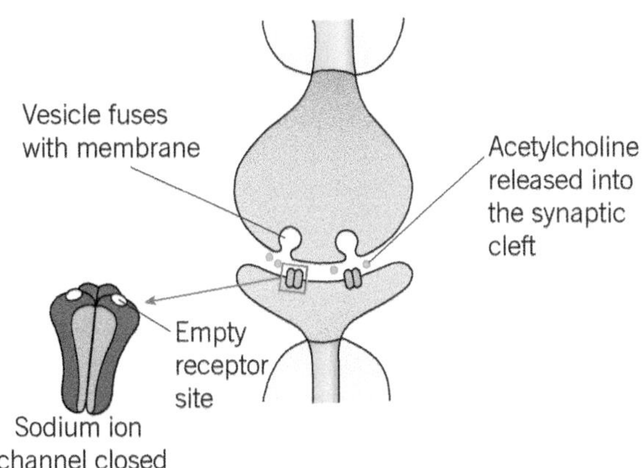

3 Acetylcholine molecules diffuse rapidly across the narrow synaptic cleft and fuse with complementary binding sites on the sodium ion channel in the membrane of the postsynaptic neurone. This causes the sodium ion channels to open, allowing sodium ions (Na^+) to diffuse in rapidly along a concentration gradient.

4 The influx of sodium ions generates a new action potential in the postsynaptic neurone.

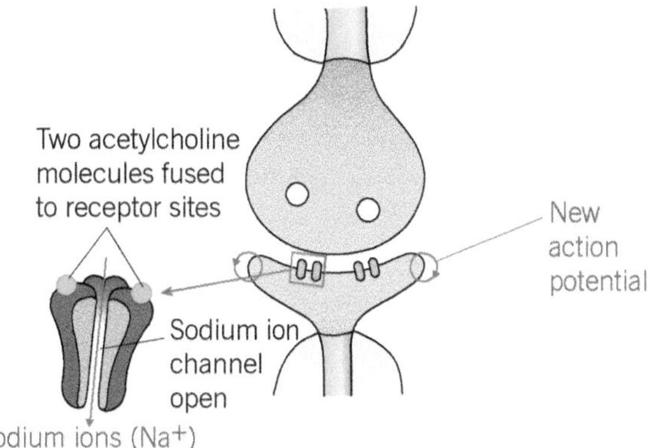

5 Acetylcholinesterase hydrolyses acetylcholine into choline and ethanoic acid (acetyl), which diffuse back across the synaptic cleft into the presynaptic neurone (recycling). In addition to recycling the choline and ethanoic acid, the breakdown of acetylcholine also prevents it from continuously generating a new action potential in the postsynaptic neurone.

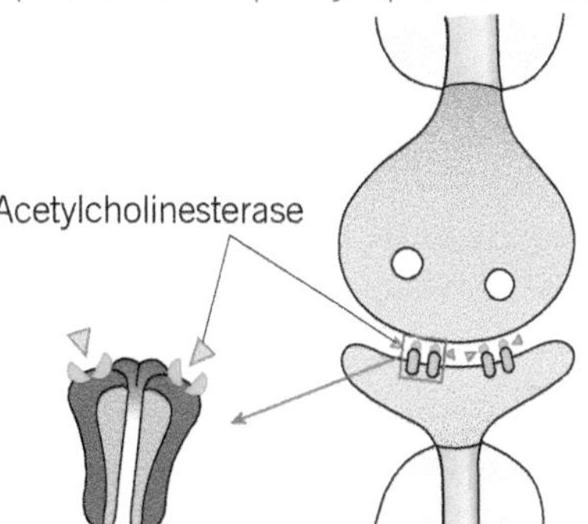

6 ATP released by mitochondria is used to recombine choline and ethanoic acid into acetylcholine. This is stored in synaptic vesicles for future use. Sodium ion channels close in the absence of acetylcholine in the receptor sites.

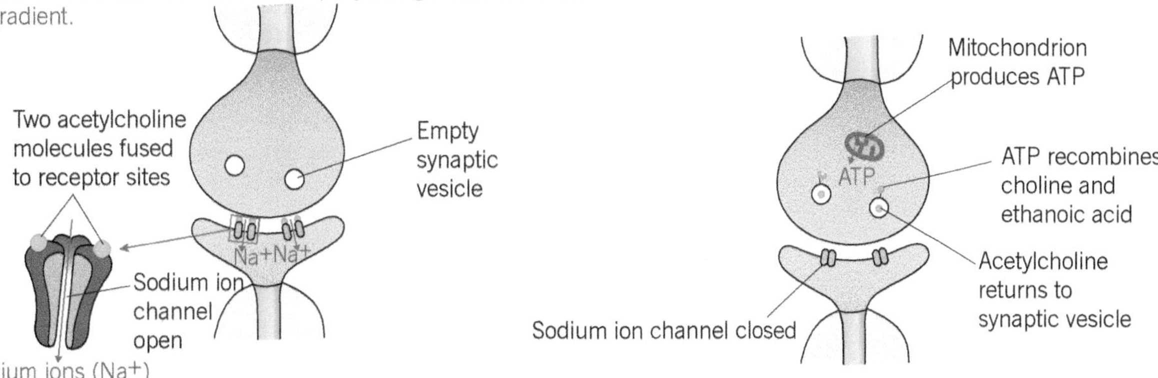

▲ **Figure 2** *Mechanism of transmission across a cholinergic synapse*

Practice questions: Chapter 24

1 **(a)** (i) Describe how a resting potential is maintained in a neurone. *(2 marks)*

(ii) The potential across the membrane is reversed when an action potential is produced. Describe how. *(2 marks)*

(b) The graph shows the relationship between the diameter of the axon and the speed of conduction of nerve impulses in myelinated axons of a cat. As the diameter of the axon increases, the length of myelination between the nodes increases. This could explain the increase in speed of conduction shown in the graph. Suggest how. *(2 marks)*

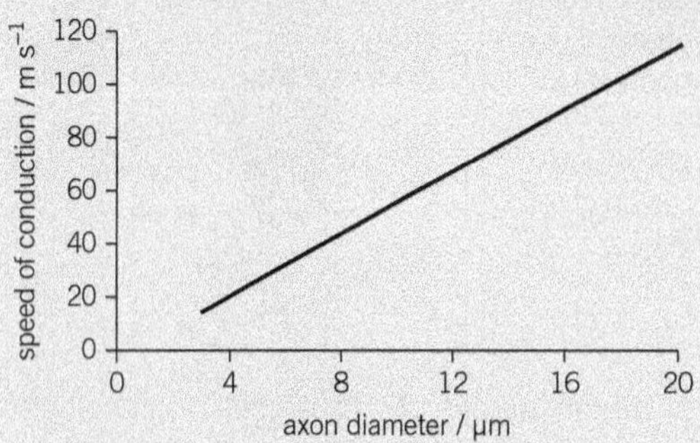

(c) A myelinated axon uses less ATP to transmit a nerve impulse than an unmyelinated axon of the same diameter. Explain why. *(2 marks)*

AQA June 2008

2 **(a)** What is a reflex? *(2 marks)*

(b) The diagram shows the neurones in a reflex arc. Name the types of neurone labelled A, B, and C. *(1 mark)*

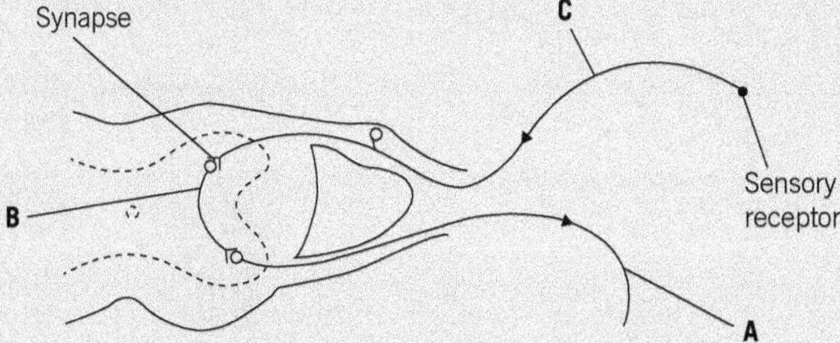

(c) Nervous transmission is delayed at synapses. Explain why. *(2 marks)*

(d) The axon of neurone A is myelinated. The axon of neurone B is non-myelinated. Explain why impulses travel faster along the axon of neurone A. *(2 marks)*

AQA Jan 2008

3 Some people have red-green colour blindness. This may be caused by a mutant allele that results in the failure to produce a light-sensitive pigment in one type of cone cell. The graph shows the absorption spectra of the pigments from the cone cells of a person with this form of colour blindness.

(a) A person with red-green colour blindness finds it difficult to distinguish between orange and green, but can distinguish between blue and green. Use the graph to explain why. *(2 marks)*

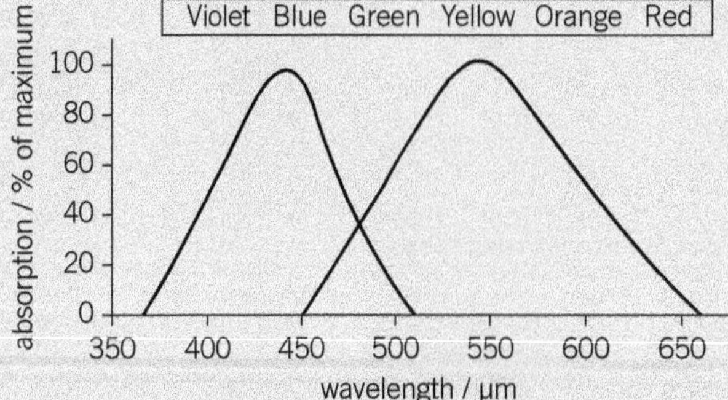

(b) Some people are completely colour blind. This is because they have no cone cells. People with complete colour blindness have difficulty in seeing detail in bright daylight. Suggest why. *(4 marks)*

AQA Jan 2008

4 (a) Heat receptors in the skin are stimulated when a finger touches a hot object. A reflex causes the finger to be pulled away. The diagram shows the reflex arc associated with this response. Draw an arrow on the motor neurone to show the direction in which an impulse travels. *(1 mark)*

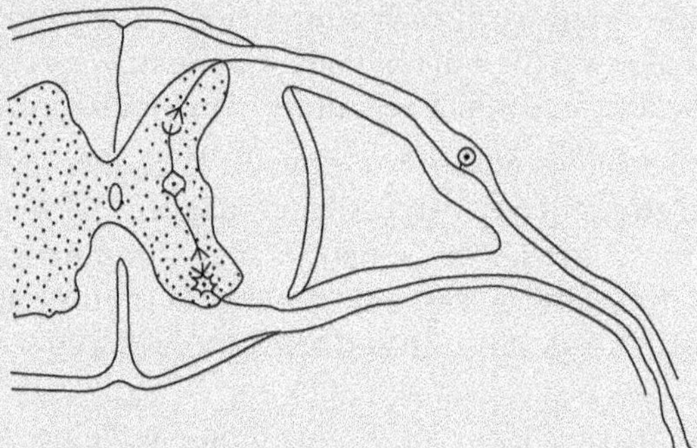

(b) Pain receptors in the skin are also stimulated when a hot object is touched. These receptors send nerve impulses to the brain. Name the area of the brain receiving these impulses. *(1 mark)*

(c) Enkephalins are neurotransmitters released by the brain and spinal cord in response to harmful stimuli. Enkephalin molecules are similar in shape to acetylcholine. Enkephalin molecules act as pain killers by inhibiting synaptic transmission. Explain how this inhibition occurs. *(4 marks)*

AQA June 2007

5 During an action potential, the permeability of the cell-surface membrane of an axon changes. The graph shows changes in permeability of the membrane to sodium ions (Na^+) and to potassium ions (K^+) during a single action potential.

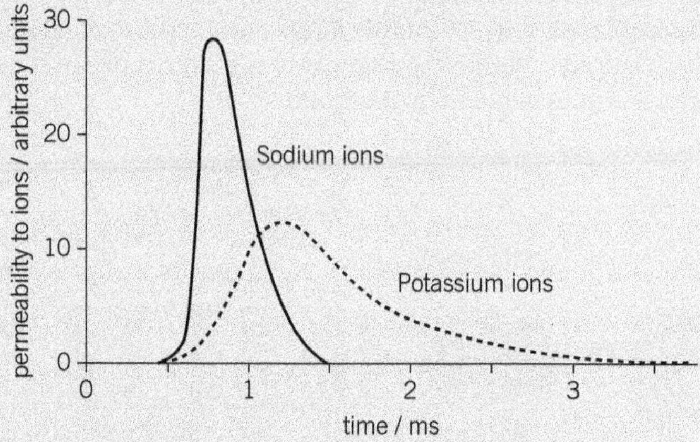

(a) Explain the shape of the curve for sodium ions between 0.5 ms and 0.7 ms. *(3 marks)*

(b) During an action potential, the membrane potential rises to +40 mV and then falls. Use information from the graph to explain the fall in membrane potential. *(3 marks)*

(c) After exercise, some ATP is used to re-establish the resting potential in axons. Explain how the resting potential is re-established. *(2 marks)*

AQA June 2010

Answers to the Practice Questions are available at

www.oxfordsecondary.com/oxfordaqaexams-alevel-biology

25 Skeletal muscles as effectors

25.1 Structure of skeletal muscle

Learning objectives:

- → Describe the gross and microscopic structure of a skeletal muscle.
- → Describe the ultrastructure of a myofibril.
- → Describe actin and myosin arranged within a myofibril.

Specification reference: 3.4.4.1

Muscles are **effector** organs that respond to nervous stimulation by contracting and so bring about movement. There are three types of muscle in the body. **Cardiac muscle** is found exclusively in the heart whereas **smooth muscle** is found in the walls of blood vessels and the gut. Neither of these types of muscle is under conscious control and we remain largely unaware of their contractions. The third type, **skeletal muscle**, makes up the bulk of body muscle in vertebrates. It is attached to bone and acts under voluntary, conscious control.

A rope is made up of millions of separate threads. Each thread has very little individual strength and can easily be snapped. Yet grouped together in a rope, these threads can support a mass running into hundreds of tonnes. In a similar way, a single muscle is made up of millions of tiny muscle fibres called **myofibrils**. Individually they produce almost no force, but collectively they can be extremely powerful. Just as the threads in a rope are lined up parallel to each other in order to maximise its strength, so the myofibrils are arranged in order to give maximum force. And just as the threads of a rope are grouped into strings, the strings are grouped into small ropes, and small ropes are grouped into bigger ropes, so muscle is composed of smaller units bundled into progressively larger ones (see Figure 1).

If muscle were made up of individual cells joined end to end it would not contract efficiently. This is partly because the junction between adjacent cells would be a point of weakness that would reduce the overall strength of the muscle. To overcome this, muscles have a different structure. The separate cells have become fused together into muscle fibres. These muscle fibres share nuclei and also cytoplasm, called **sarcoplasm**, which is mostly found around the circumference of the fibre (Figure 1). Within the sarcoplasm is a large concentration of mitochondria and endoplasmic reticulum.

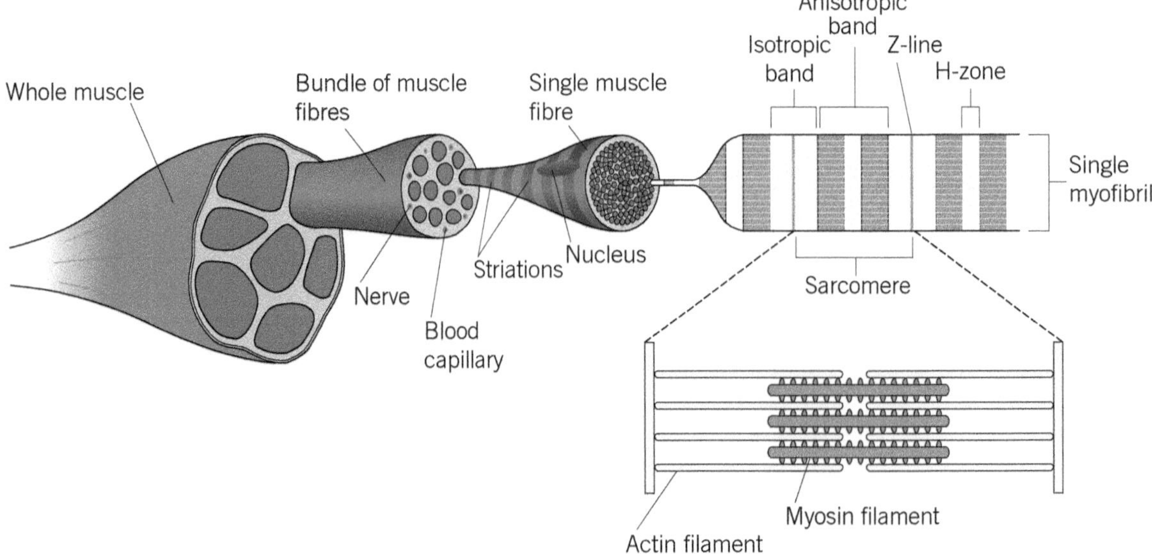

▲ **Figure 1** *The gross and microscopic structure of skeletal muscle*

Microscopic structure of skeletal muscle

You can see from Figure 1 that each muscle fibre is made up of myofibrils. Myofibrils contain very large numbers of two types of protein filaments:

- **Thin filaments** consist of two chains twisted together. Each chain is made of molecules of the protein actin.
- **Thick filaments** are made of molecules of the protein myosin. Each molecule has a tail and a bulbous head.

The arrangement of these filaments is shown in Figures 2 and 3.

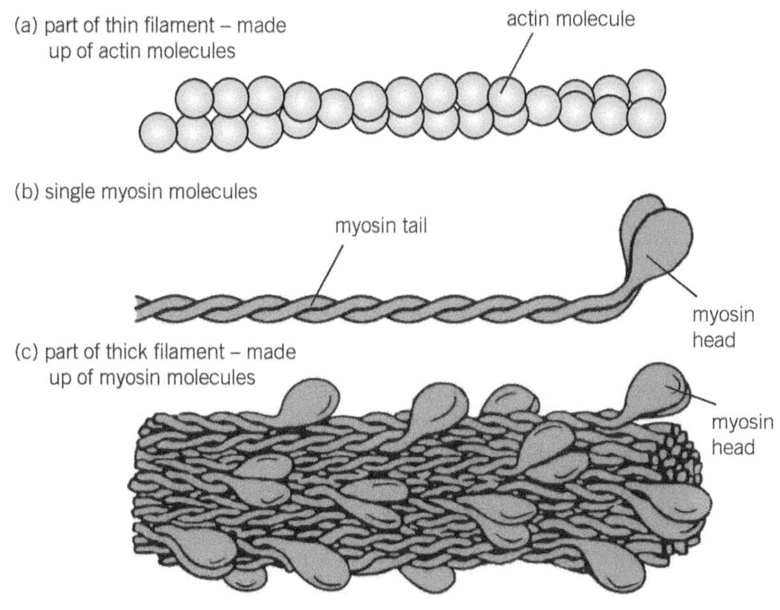

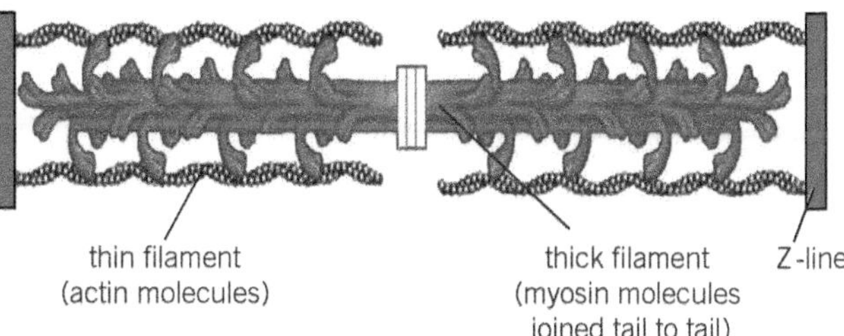

▲ **Figure 2** *Structure of actin and myosin molecules and their arrangement into thick and thin filaments*

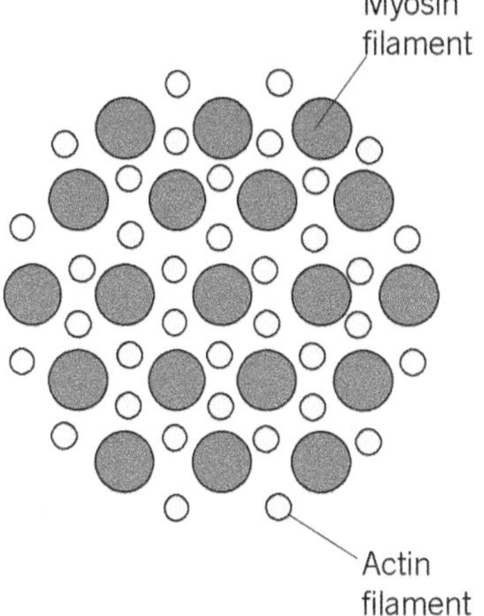

▲ **Figure 3** *Transverse section through part of a muscle fibre showing the arrangement of actin and myosin filaments*

Myofibrils appear striped due to their alternating light-coloured and dark-coloured bands. The light bands are called **isotropic bands** (or **I-bands**). They appear lighter because the actin and myosin filaments do not overlap in this region. The dark bands are called **anisotropic bands** (or **A-bands**). They appear darker because the actin and myosin filaments overlap in this region.

At the centre of each anisotropic band is a lighter-coloured region called the **H-zone**. At the centre of each isotropic band is a line called the **Z-line**. The distance between adjacent Z-lines is called a **sarcomere** (Figure 1). When a muscle contracts, these sarcomeres shorten and the pattern of light and dark bands changes (see Topic 25.2).

Two other important proteins are found in muscle:

- **tropomyosin**, which forms a fibrous strand around the actin filament
- a globular protein (**troponin**) involved in muscle contraction.

You will learn more about these in Topic 25.2.

Types of muscle fibre

There are two types of muscle fibre, the proportions of which vary from muscle to muscle and person to person. The two types are:

- **slow-twitch fibres**. These contract more slowly and provide less powerful contractions over a longer period. They are therefore adapted to endurance work, such as running a marathon. In humans they are more common in muscles like the calf muscle, which need to contract constantly to maintain the body in an upright position. They are suited to this role by being adapted for aerobic respiration in order to avoid a build-up of lactic acid, which would cause them to function less effectively. These adaptations include having:
    - a large store of myoglobin (a bright red molecule that stores oxygen, which accounts for the red colour of slow-twitch fibres)
    - a rich supply of blood vessels to deliver oxygen and glucose
    - numerous mitochondria to produce ATP.
- **fast-twitch fibres**. These contract more rapidly and produce powerful contractions but only for a short period. They are therefore adapted to intense exercise, such as weight-lifting. As a result they are more common in muscles that need to do short bursts of intense activity, like the biceps muscle of the upper arm. Fast-twitch fibres are adapted to their role by having:
    - a high concentration of glycogen
    - thicker and more numerous myosin filaments
    - a high concentration of enzymes involved in anaerobic respiration, which provides ATP rapidly
    - a store of phosphocreatine, a molecule that can rapidly generate ATP from ADP in anaerobic conditions and so provide energy for muscle contraction.

Hint

To help you to remember which band is the dark band and which is the light band, look at the vowels in 'light' and 'dark'. This vowel is the first letter of the relevant band. Therefore the d**a**rk band is the **A**-band and the l**i**ght band is the **I**-band.

Hint

The arrangement of sarcomeres into a long line means that, when one sarcomere contracts a little, the line as a whole contracts a lot! In addition, having the lines of sarcomeres running parallel to each other means that all the force is generated in one direction.

Neuromuscular junctions

A neuromuscular junction is the point where a motor neurone meets a skeletal muscle fibre. There are many such junctions along the muscle. If there were only one junction of this type it would take time for a wave of contraction to travel across the muscle, in which case not all the fibres would contract simultaneously and the movement would be slow. As rapid muscle contraction is frequently essential for survival there are many neuromuscular junctions spread throughout the muscle. This ensures that contraction of a muscle is rapid and powerful when it is simultaneously stimulated by action potentials. All muscle fibres supplied by a single motor neurone act together as a single functional unit and are known as a motor unit. This arrangement gives control over the force that the muscle exerts. If only slight force is needed, only a few units are stimulated. If a greater force is required, a larger number of units are stimulated.

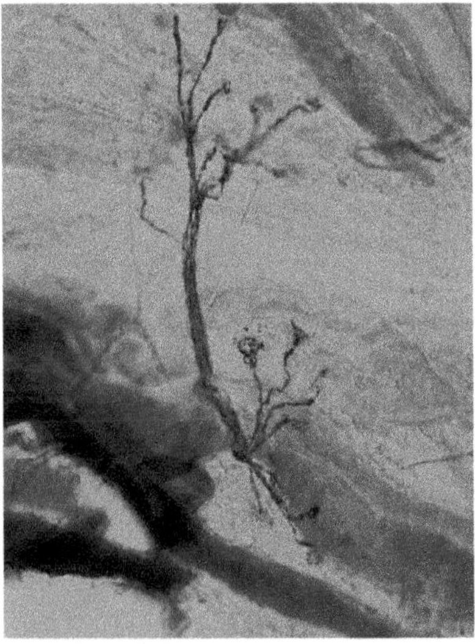

▲ **Figure 4** *Light micrograph of a neuromuscular junction*

When a nerve impulse is received at the neuromuscular junction, the synaptic vesicles fuse with the presynaptic membrane and release their acetylcholine. The acetylcholine diffuses to the postsynaptic membrane, altering its permeability to sodium ions (Na^+), which enter rapidly, depolarising the membrane. A description of how this leads to the contraction of the muscle is given in Topic 25.2.

The acetylcholine is broken down by acetylcholinesterase to ensure that the muscle is not overstimulated. The resulting choline and ethanoic acid (acetyl) diffuse back into the neurone, where they are recombined to form acetylcholine using energy provided by the mitochondria found there.

The structure of a neuromuscular junction is shown in Figure 5 and an account of how it functions is provided in Topic 25.2.

Synoptic link

The sequence of events which occurs in transmission at neuromuscular junctions is very similar to that which occurs at synapses as described in Topic 24.8

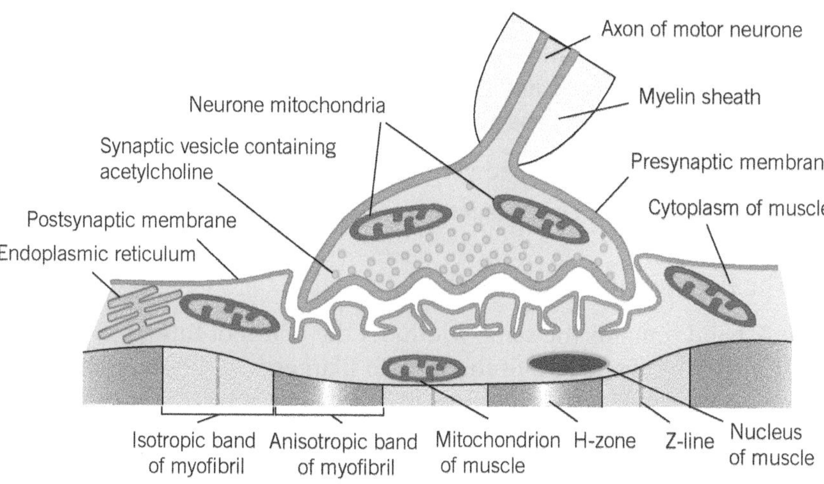

▲ **Figure 5** *The neuromuscular junction*

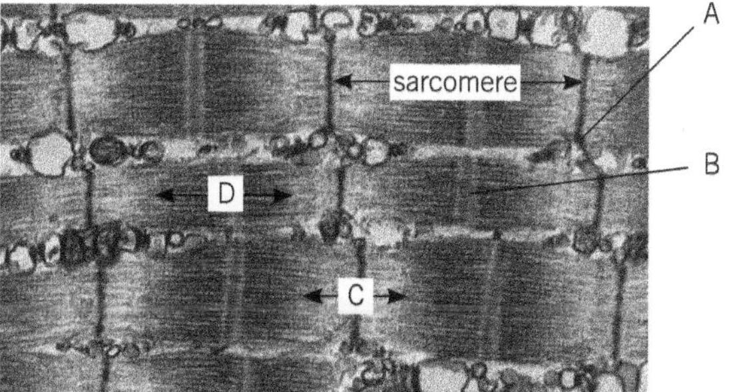

▲ **Figure 6** *False-colour transmission electron micrograph (TEM) of skeletal muscle*

Summary questions

1 Suggest a reason why there are numerous mitochondria in the sarcoplasm.

2 Study Figure 6 and name the structures labelled A–D.

3 If we cut across a myofibril at certain points, we see only thick myosin filaments. Cut at a different point we see only thin actin filaments. At yet other points we see both types of filament. Explain why.

4 How do slow-twitch fibres differ from fast-twitch fibres in the way they function?

5 Describe how each type of fibre is adapted to its functions.

25.2 Contraction of skeletal muscle

Learning objectives

- → Describe the evidence that supports the sliding filament mechanism of muscle contraction.
- → Describe how the sliding filament mechanism causes a muscle to contract and relax.
- → Describe where the energy for muscle contraction comes from.

Specification reference: 3.4.4.2

Now we have looked at the structure of skeletal muscle in Topic 25.1, let us turn our attention to how exactly the arrangement of the various proteins brings about contraction of the muscle fibre. The process involves the actin and myosin filaments sliding past one another and is therefore called the **sliding filament mechanism**.

Evidence for the sliding filament mechanism

In Topic 25.1 you saw that myofibrils appear darker in colour where the actin and myosin filaments overlap and lighter where they do not. If the sliding filament mechanism is correct, then there will be more overlap of actin and myosin in a contracted muscle than in a relaxed one. If you look at Figure 1, you will see that, when a muscle contracts, the following changes occur to a sarcomere:

- The I-band becomes narrower.
- The Z-lines move closer together or, in other words, the sarcomere shortens.
- The H-zone becomes narrower.

The A-band remains the same width. As the width of this band is determined by the length of the myosin filaments, it follows that the myosin filaments have not become shorter. This discounts the theory that muscle contraction is due to the filaments themselves shortening.

Synoptic link

The functioning of muscle depends on the molecular shapes of the four main proteins involved. The importance of shape on the functioning of proteins is covered in Topic 1.5.

Study tip

These diagrams of the two filaments are simplified. Remind yourself about the molecular structure as shown in Figure 2 in Topic 25.1.

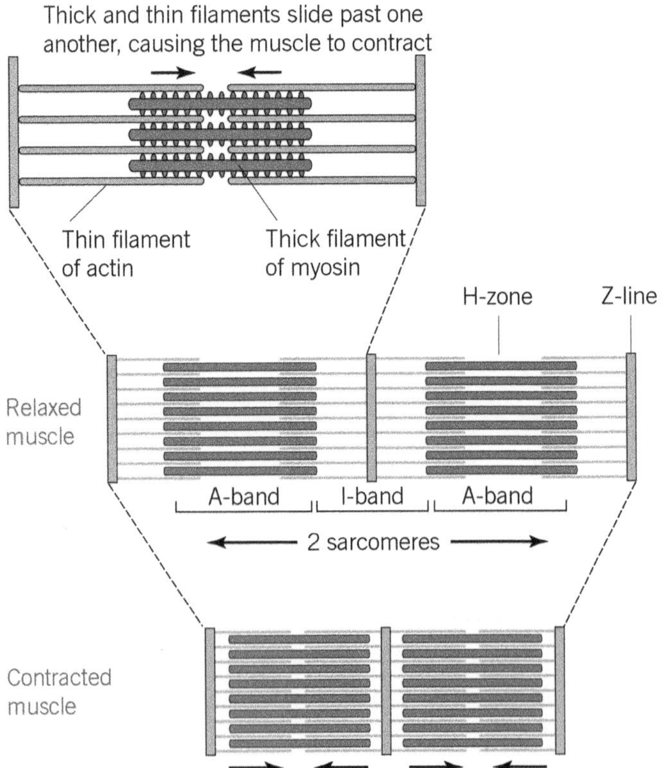

▲ **Figure 1** *Comparison of two sarcomeres in a relaxed and a contracted muscle*

Before we look at how the sliding filament mechanism works, let us take a closer look at the three main proteins involved in the process:

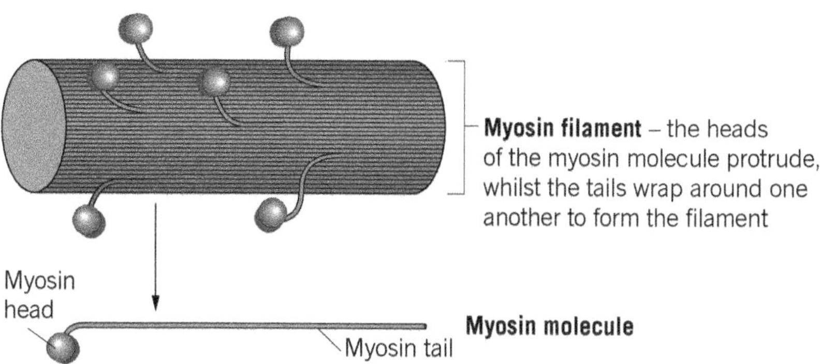

▲ **Figure 2** *Structure of myosin*

- Myosin (see Figure 2) is made up of two types of protein:
  - a fibrous protein arranged into a filament made up of several hundred molecules (the tail)
  - a globular protein formed into two bulbous structures at one end (the head).
- Actin is a globular protein whose molecules are arranged into long chains that are twisted around one another to form a helical strand.
- Tropomyosin forms long thin threads that are wound around actin filaments.

The arrangement of the molecules of actin and tropomyosin is shown in Figure 3.

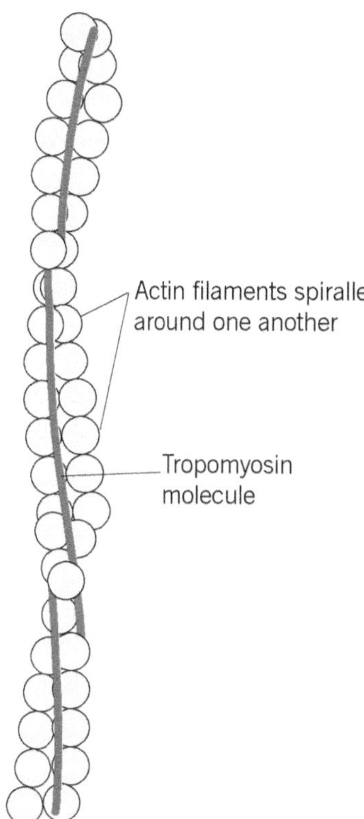

▲ **Figure 3** *The relationship of tropomyosin to an actin filament*

The sliding filament mechanism of muscle contraction

The hypothesis that actin and myosin filaments slide past one another during muscle contraction is supported by the changes seen in the band pattern on myofibrils. The next question for the scientists was, by what mechanism do the filaments slide past one another? Clues to the answer lie in the shape of the various proteins involved.

To summarise, the bulbous heads of the myosin filaments form cross-bridges with the actin filaments. They do this by attaching themselves to binding sites on the actin filaments, and then flexing in unison, pulling the actin filaments along the myosin filaments. They then become detached and, using ATP as a source of energy, return to their original angle and re-attach themselves further along the actin filaments. This process is repeated up to 100 times a second. The action is similar to the way a ratchet operates.

The following account describes the sliding filament mechanism of muscle contraction in detail. The process is continuous but, for ease of understanding, has been divided into stimulation, contraction, and relaxation.

Hint

The action of the myosin heads is similar to the rowing action of oarsmen in a boat. The oars (myosin heads) are dipped into the water, flexed as the oarsmen pull on them, removed from the water, and then dipped back into the water further along. The oarsmen work in unison and the boat and water move relative to one another.

Hint

An action potential is the result of depolarisation of part of a membrane. The spread of the action potential across the muscle is therefore often referred to as a 'wave of depolarisation'.

Muscle stimulation

- An **action potential** reaches many **neuromuscular junctions** simultaneously, causing calcium ion channels to open and calcium **ions** to move into the synaptic knob.
- The calcium ions cause the synaptic vesicles to fuse with the presynaptic membrane and release their **acetylcholine** into the synaptic cleft.
- Acetylcholine diffuses across the synaptic cleft and binds with receptors on the postsynaptic membrane, causing it to depolarise.

Muscle contraction

- The action potential travels deep into the fibre through a system of tubules (T-tubules) that branch throughout the cytoplasm (sarcoplasm) of the muscle.
- The tubules are in contact with the endoplasmic reticulum of the muscle (sarcoplasmic reticulum), which has actively absorbed calcium ions from the cytoplasm of the muscle.
- The action potential opens the calcium ion channels on the endoplasmic reticulum and calcium ions flood into the muscle cytoplasm down a diffusion gradient.
- The calcium ions cause the tropomyosin molecules that were blocking the binding sites on the actin filament to pull away (Figure 4, stages 1 and 2).
- The ADP molecule attached to the myosin heads means that they are now in a state to bind to the actin filament and form a cross-bridge (Figure 4, stage 3).
- Once attached to the actin filament, the myosin heads change their angle, pulling the actin filament along as they do so and releasing a molecule of ADP (Figure 4, stage 4).
- An ATP molecule attaches to each myosin head, causing it to become detached from the actin filament (Figure 4, stage 5).
- The calcium ions then activate the enzyme ATPase, which hydrolyses the ATP to ADP. The hydrolysis of ATP to ADP provides the energy for the myosin head to return to its original position (Figure 4, stage 6).
- The myosin head, once more with an attached ADP molecule, then reattaches itself further along the actin filament and the cycle is repeated as long as nervous stimulation of the muscle continues (Figure 4, stage 7).

Hint

The hydrolysis of ATP releases energy and produces ADP and inorganic phosphate (Pi). For simplicity, this account just refers to ADP as the product.

Muscle relaxation

- When nervous stimulation ceases, calcium ions are actively transported back into the endoplasmic reticulum using energy from the **hydrolysis** of ATP.
- This reabsorption of the calcium ions allows tropomyosin to block the actin filament again.
- Myosin heads are now unable to bind to actin filaments and contraction ceases, that is, the muscle relaxes.

1 Tropomyosin molecule prevents myosin head from attaching to the binding site on the actin molecule.

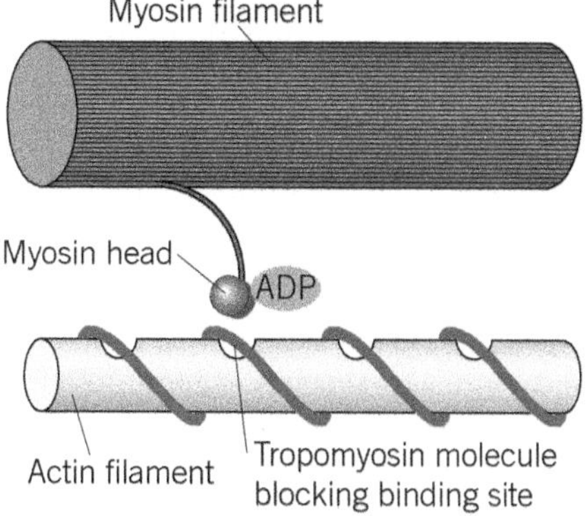

2 Calcium ions released from the endoplasmic reticulum cause the tropomyosin molecule to pull away from the binding sites on the actin molecule.

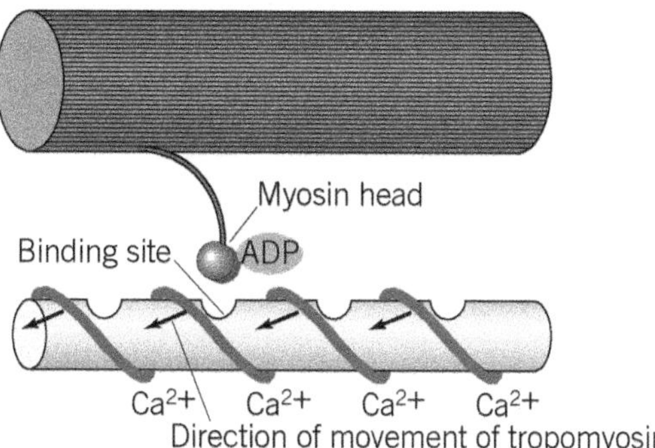

3 Myosin head now attaches to the binding site on the actin filament.

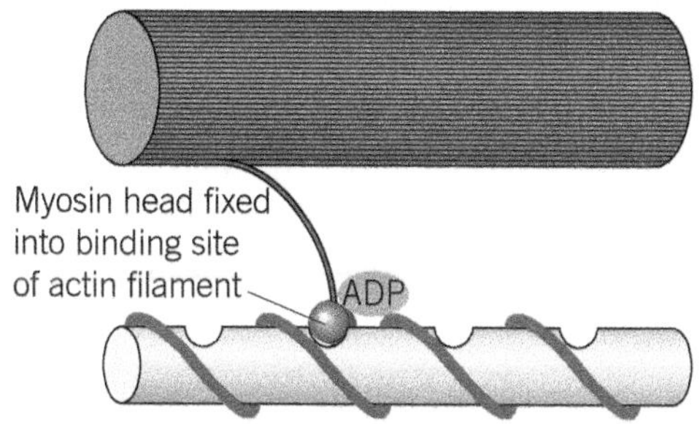

4 Head of myosin changes angle, moving the actin filament along as it does so. The ADP molecule is released.

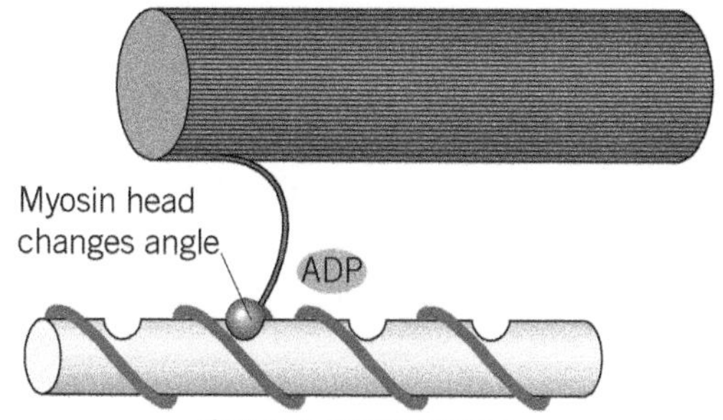

5 ATP molecule fixes to myosin head, causing it to detach from the actin filament.

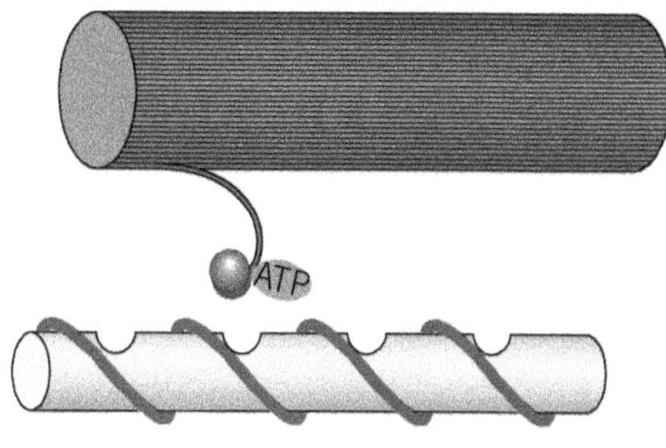

6 Hydrolysis of ATP to ADP by ATPase provides the energy for the myosin head to resume its normal position.

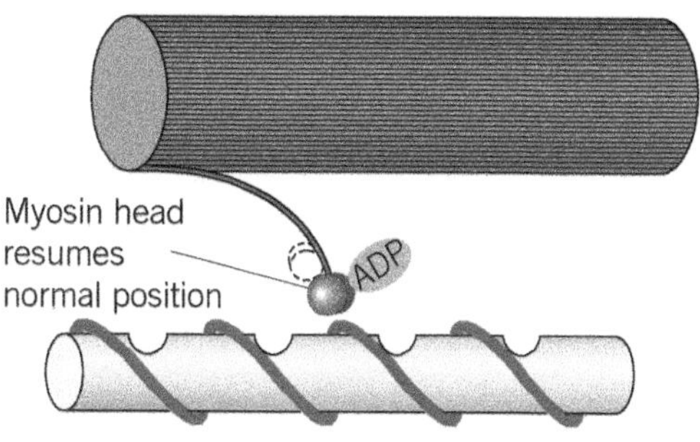

7 Head of myosin reattaches to a binding site further along the actin filament and the cycle is repeated.

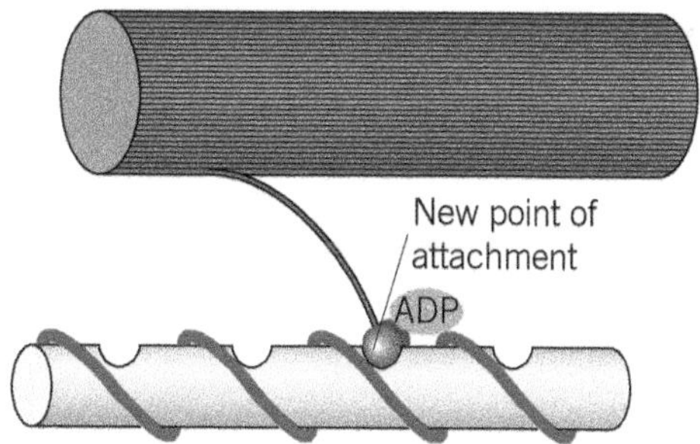

▲ **Figure 4** *Sliding filament mechanism of muscle contraction (showing only one myosin head throughout)*

▲ **Figure 5** *Marathon runners undergoing strenuous exercise*

Energy supply during muscle contraction

Muscle contraction requires considerable energy. This is supplied by the hydrolysis of ATP to ADP and inorganic phosphate (Pi) (Topic 4.4). The energy released is needed for:

- the movement of the myosin heads
- the reabsorption of calcium ions into the endoplasmic reticulum by active transport.

In an active muscle, there is clearly a great demand for ATP. In some circumstances, for example, escaping from danger, the ability of muscles to work intensely can be life saving. Most ATP is regenerated from ADP during the respiration of pyruvate in the mitochondria, which are particularly plentiful in the muscle. However, this process requires oxygen. In a very active muscle oxygen is rapidly used up and it takes time for the blood supply to replenish it. Therefore a means of rapidly generating ATP anaerobically is also required. This is achieved by more glycolysis and also by using a chemical called **phosphocreatine**.

Phosphocreatine cannot supply energy directly to the muscle, so instead it regenerates ATP, which can. Phosphocreatine is stored in muscle and acts as a reserve supply of phosphate, which is available immediately to combine with ADP and so re-form ATP. The phosphocreatine store is replenished using phosphate from ATP when the muscle is relaxed.

Muscles work in pairs

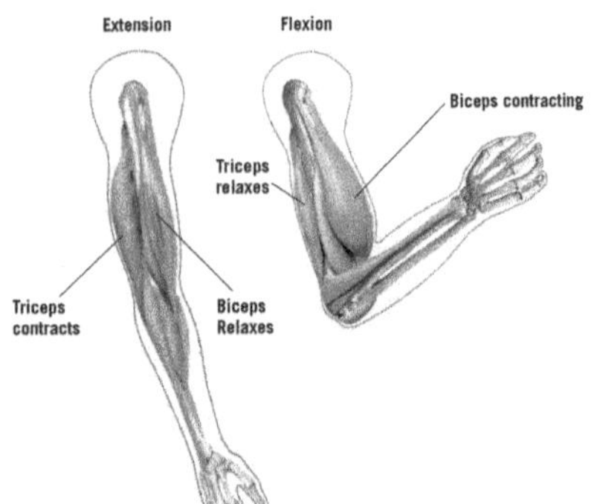

▲ **Figure 6** *Two muscles responsible for moving the elbow joint working as an antagonistic pair*

Skeletal muscles pull on tendons when they contract. Tendons consist of connective tissue that contains closely packed collagen fibres. Tendons are securely attached to the bones of the skeleton, which operates as a system of levers. Contracting muscles pull on inelastic tendons, which in turn pull on bones producing movement. However, muscles can only pull when they contract – they do not push. To bend or extend a limb at a joint, for example, to flex or straighten the arm at the elbow, two muscles operate as an **antagonistic pair**.

In the movement of the elbow joint, the biceps acts as the **flexor** and contracts to bend the arm. The triceps acts as the **extensor** and contracts to straighten the arm. Whilst one member of the pair is contracting, the other is relaxing, as shown in Figure 6.

Summary questions

1. How is the shape of the myosin molecule adapted to its role in muscle contraction?
2. Trained sprinters have high levels of phosphocreatine in the muscles. Explain the advantage of this.
3. During the contraction of a muscle sarcomere, a single actin filament moves 0.8 μm. If the hydrolysis of a single ATP molecule provides enough energy to move an actin filament 40 nm, how many ATP molecules are needed to move the actin filament 0.8 μm? Show your working.
4. Dead cells can no longer produce ATP. Soon after death, muscles contract, making the body stiff – a state known as *rigor mortis*. From your knowledge of muscle contraction, explain the reasons why *rigor mortis* occurs after death.

Practice questions: Chapter 25

1 (a) Figure 1 shows some of the muscles involved in moving the wing bone of a bird. Explain how these muscles are able to move the wing bone up and down. *(2 marks)*

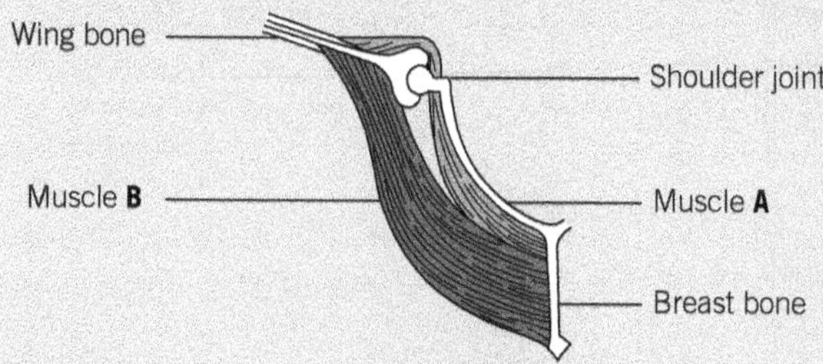

▲ **Figure 1**

(b) Figure 2 shows a sarcomere in a relaxed muscle. Draw the appearance of this sarcomere in a contracted muscle. *(1 mark)*

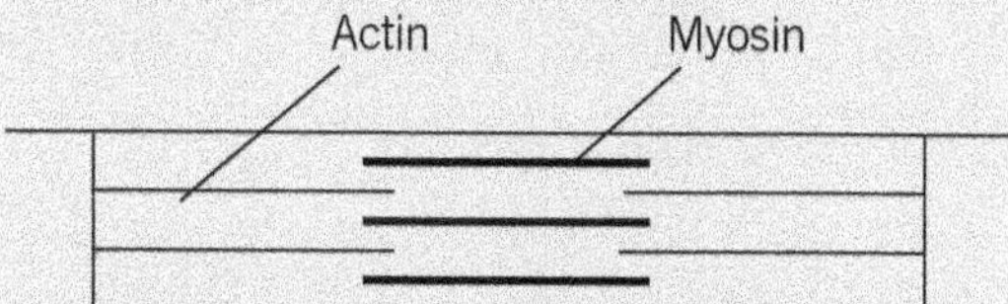

▲ **Figure 2**

(c) Explain the role of each of the following in muscle contraction:

(i) tropomyosin *(2 marks)*

(ii) myosin. *(2 marks)*

AQA Jan 2009

2 Figure 3 shows a neuromuscular junction.

(a) (i) On the diagram, label the myelin sheath. *(1 mark)*

(ii) The myelin sheath is not formed in new-born babies. Explain how this leads to slower reflexes in babies. *(2 marks)*

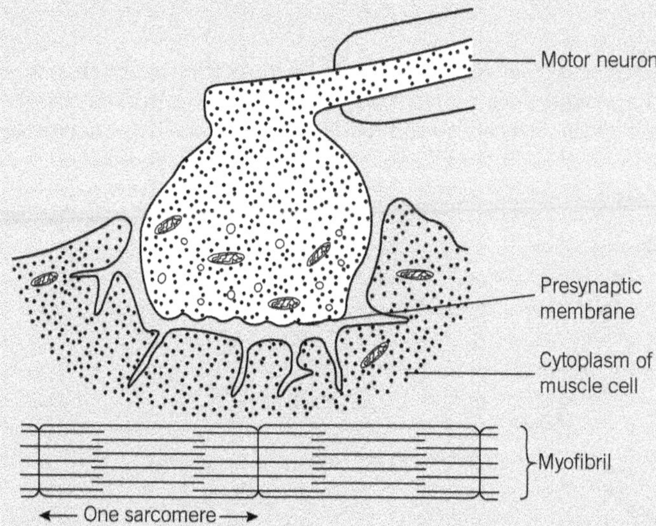

▲ **Figure 3**

(b) Nerve impulses arriving at the presynaptic membrane at the neuromuscular junction result in shortening of sarcomeres. Describe how. *(7 marks)*

AQA June 2007

3 Insects do not have bones, instead their hard outer covering acts as an exoskeleton to which muscles are attached. Figure 4 shows two of the main flight muscles of an insect.

(a) Use your knowledge of antagonistic muscle action to explain how muscles A and B bring about movement of the wing. *(2 marks)*

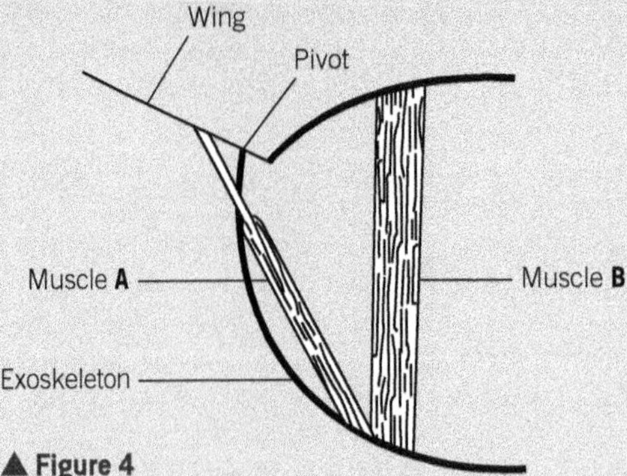

▲ Figure 4

(b) Choline is a chemical needed to synthesise the neurotransmitter acetylcholine. Hemicholinium blocks the absorption of choline into the presynaptic neurone at a neuromuscular junction. Describe and explain how exposure to hemicholinium will affect the muscle action of an insect. *(3 marks)*

(c) Describe the function of calcium ions in muscle contraction. *(3 marks)*

AQA Jan 2007

4 Figure 5 shows the banding pattern observed in part of a relaxed muscle fibril.

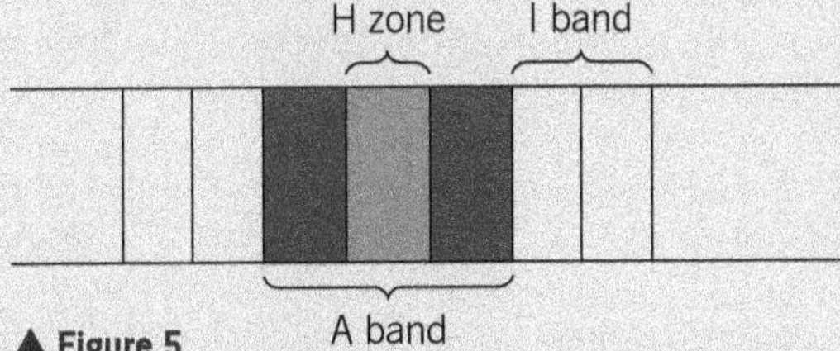

▲ Figure 5

(a) Describe what causes the different bands seen in the muscle fibril. *(2 marks)*

(b) Describe how the banding pattern will be different when the muscle fibril is contracted. *(2 marks)*

AQA Jan 2006

5 Figure 6 is a tracing of a cross section through skeletal muscle tissue. This muscle contains fast-twitch muscle fibres and slow-twitch muscle fibres. The section has been stained to show the distribution of the enzyme succinate dehydrogenase. This enzyme is found in mitochondria.

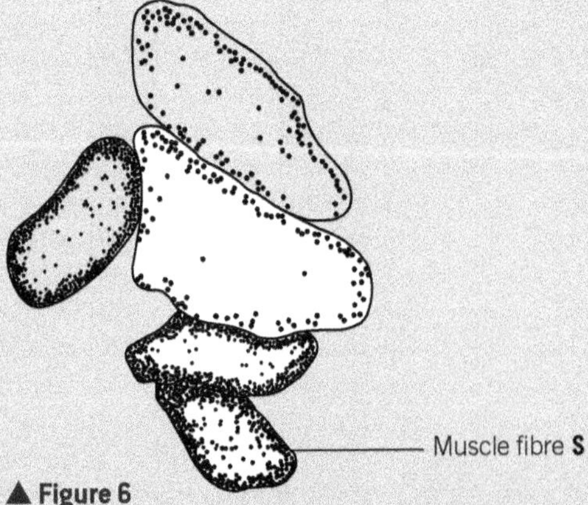

▲ Figure 6

(a) Succinate dehydrogenase catalyses one of the reactions in the Krebs cycle. What is the evidence from the drawing that muscle fibre S is a slow-twitch muscle fibre? Explain your answer. *(2 marks)*

(b) Use evidence from the diagram to describe the distribution of mitochondria inside the slow-twitch muscle fibres. Explain the importance of this distribution. *(3 marks)*

Answers to the Practice Questions are available at
www.oxfordsecondary.com/oxfordaqaexams-alevel-biology

26 Control systems in plants

26.1 Tropisms and auxins

Unlike animals, plants have no nervous system. Nevertheless, in order to survive, plants respond to changes in both their external and internal environments. A **tropism** is a growth movement of part of a plant in response to a directional stimulus. For example, plants respond to:

- **light**. Stems grow towards light (i.e., are positively phototropic) because light is needed for photosynthesis.
- **gravity**. Plants need to be firmly anchored in the soil. Roots are sensitive to gravity and grow in the direction of its pull (i.e., they are positively geotropic).
- **water**. Almost all plant roots grow towards water (i.e., are positively hydrotropic) in order to absorb it for use in photosynthesis and other metabolic processes, as well as for support.

Plants respond to external stimuli by means of plant hormones or, more correctly, **plant growth factors**. This term is better and more descriptive because:

- they exert their influence by affecting growth
- unlike animal hormones, they are made by cells located throughout the plant rather than in particular organs
- unlike animal hormones, some plant growth factors affect the tissues that release them rather than acting on a distant target organ.

Plant growth factors are produced in small quantities. They may have their effect close to the tissues which produce them or act on tissues at some distance. An example of a plant growth factor is **indoleacetic acid (IAA)**, which, amongst other things, causes plant cells to elongate.

Control of tropisms by IAA

You learnt that a tropism is a directional growth movement of a plant in response to a directional stimulus. In the case of light, you can observe that a young shoot will bend towards light that is directed at it from one side. This response is due to the following sequence of events:

1 Cells in the tip of the shoot produce IAA, which is then transported down the shoot.

2 The IAA is initially transported to all sides as it begins to move down the shoot.

3 Light causes the movement of IAA from the light side to the shaded side of the shoot.

4 A greater concentration of IAA builds up on the shaded side of the shoot than on the light side.

5 As IAA causes elongation of cells and there is a greater concentration of IAA on the shaded side of the shoot, the cells on this side elongate more.

6 The shaded side of the shoot grows faster, causing the shoot to bend towards the light.

Learning objectives:

→ Explain what is meant by tropisms in plants.

→ Describe the role of auxins in plant tropisms.

Specification reference: 3.4.5.1 and 3.4.5.2

▲ **Figure 1** *Geotropism is a plant response to the Earth's gravitational field. This bean plant shows a turn in its stem, which occurred after the pot was tipped over. The response also occurs in the dark, showing that it is not phototropism.*

Study tip

Although you may come across the term plant hormone in your wider reading, the term you will be expected to use and the one you will meet in exam questions is plant growth factor or substance.

IAA also controls the bending of roots in the direction of gravity. However, whereas a high concentration of IAA increases growth in stem cells, it decreases growth in root cells. For example, an IAA concentration of 10 parts per million increases stem growth by 200 per cent but decreases root growth by 100 per cent.

IAA and root responses to gravity

Cells at the root tips contain organelles know as statoliths. These starch-filled structures settle to the lowest part of the cell. In Figure 2a, the location of the statoliths is such that IAA is distributed equally back up the root so the direction of growth is straight down. In Figure 2b, the root position has been moved to the horizontal. The statoliths relocate, triggering a redistribution of IAA back along the lower part of the root. As high concentrations of IAA inhibit cell elongation in roots, the lower surface grows more slowly than the upper surface. Consequently the root starts to bend so it can continue to grow downwards in the direction of gravity.

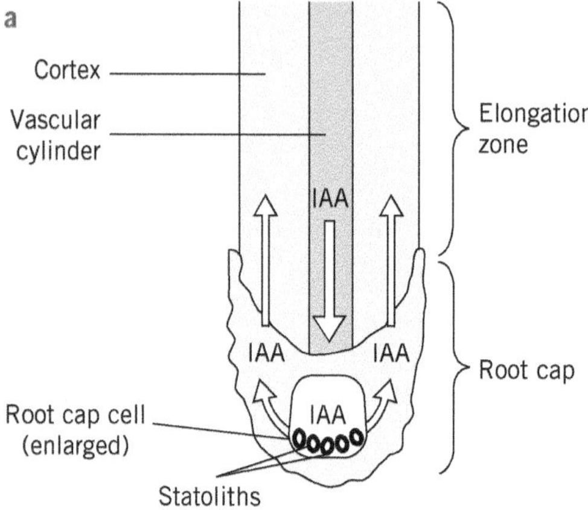

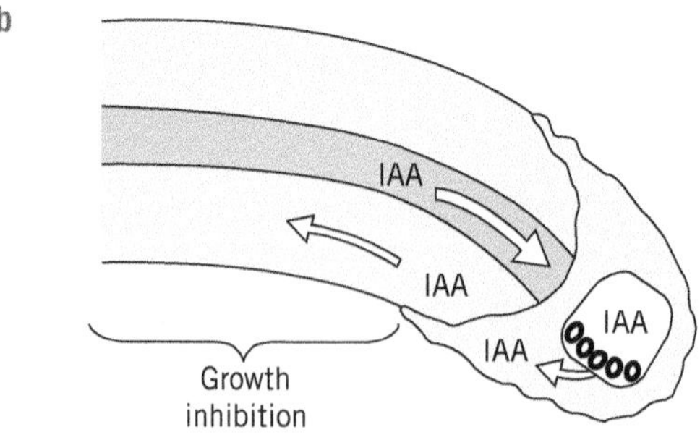

▲ **Figure 2** *The control of geotropism by IAA*

Summary questions

1 Suggest **two** advantages to a plant of having roots that respond to gravity by growing in the direction of its pull.

2 State **two** differences between animal hormones and plant growth factors.

Discovering the role of IAA in tropisms

The mechanism by which IAA causes shoots to bend towards the light was unravelled in a series of experiments stretching over almost a century.

- Scientists use their knowledge and understanding when observing events, in defining a problem, and when questioning the explanations of other scientists.
- Scientists use their observations to form a hypothesis that they test experimentally.
- Evidence from experiments is constantly questioned by scientists and is a stimulus for further investigation involving experimental refinements and new hypotheses.

No less a person than Charles Darwin was one of the earliest scientists to investigate the response of plant shoots to light. He kept birds and so grew grasses that he later fed to them. He observed that young grass shoots bent towards the window. (i.e., they were positively phototropic). Being curious, he proposed the hypothesis that the stimulus of the light was detected by the tip of the shoot, which was therefore the source of the response. He tested his hypothesis in a series of experiments in which he removed the tips of shoots or covered the tips with lightproof covers.

These experiments and the results are summarised in Figure 3 (experiments 1–3).

1 Which of Darwin's three experiments acted as a control?

EXPT NO.	METHOD	RESULT	EXPLANATION
1	Unilateral light; Plant shoot	Shoot bends towards light	The shoot is positively phototropic. Bending occurs behind the tip.
2	Light; Shoot tip removed; Tip discarded	No response	The tip must either detect the stimulus or produce the messenger (or both) as its removal prevents any response.
3	Light; Lightproof cover is placed over intact tip of shoot	No response	The light stimulus must be detected by the tip.

▲ **Figure 3** *Darwin's experiments to show that it is the tips of shoots that are the source of the phototropic response*

Once it had been shown that the tip is the light-sensitive region of the shoot but that the response (bending) occurs lower down the shoot, some scientists proposed another hypothesis, that a chemical substance was being produced in the tip and transported down the stem, where it caused a response. Others disagreed and put forward an alternative theory, that it was an electrical signal that was passing from the tip and causing the response. One scientist, Peter Boysen-Jensen, carried out a further set of experiments designed to prove the hypothesis that the 'messenger' was a chemical. In these experiments, he used mica, which conducts electricity but not chemicals, and gelatin, which conducts chemicals but not electricity. His experiments and the results are summarised in Figure 4 (experiments 4–6).

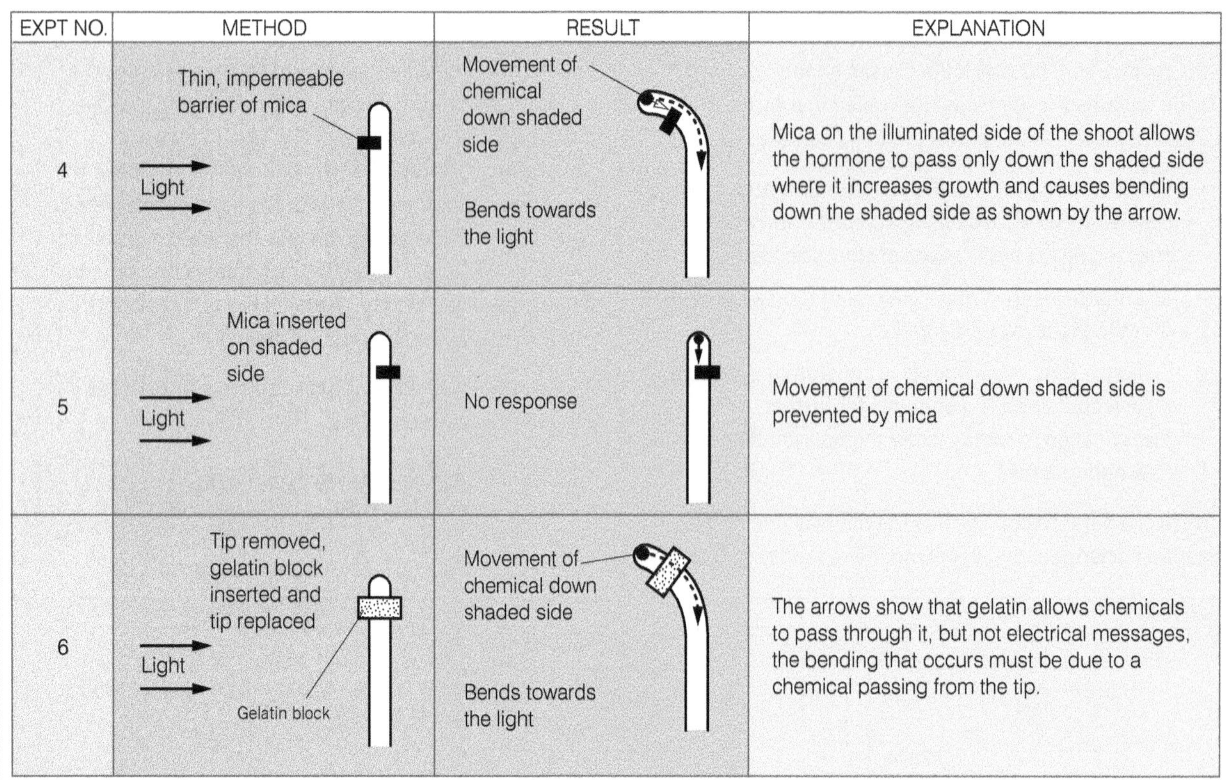

▲ **Figure 4** *Boysen-Jensen's experiments to show the nature of the 'messenger' in the phototropic response*

> 2 Suggest an explanation for the results in experiment 5.

Boysen-Jenson's experiments stimulated another scientist, Arpad Paál, to investigate how the chemical messenger worked. He removed the tips of shoots and placed them on one side of the cut surface. He kept the shoots in total darkness throughout the experiment. His experiment and its results are shown in Figure 5 (experiment 7).

EXPT NO.	METHOD	RESULT
7	Darkness Tips removed and then replaced but displaced to one side	Shoots bend towards side where no tip is present

▲ **Figure 5** *Paal's experiment on the action of the 'messenger'*

> 3 Suggest an explanation for the results in experiment 7.

So far, it had been established that bending was due to a chemical that was produced in the tip and caused growth on the shaded side of the shoot. This chemical was later shown to be indoleacetic acid (IAA). The next question was how did light cause the uneven distribution of IAA? Different theories were put forward, including:

1 Light inhibits IAA production in the tip and so it is only produced on the shaded side.
2 Light destroys the IAA as it passes down the light side of the shoot.
3 IAA is transported from the light side to the shaded side of the shoot.

This prompted Winslow Briggs and his associates to test these hypotheses. They set up experiments as shown in Figure 6 (experiments 8–10).

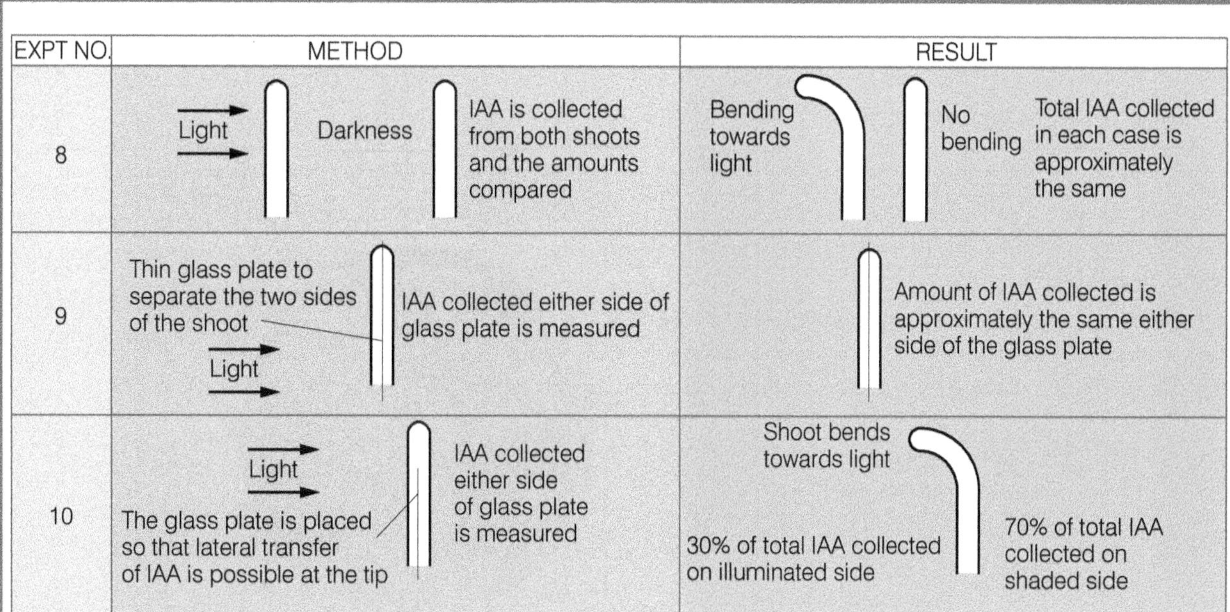

EXPT NO.	METHOD	RESULT
8	Light → Darkness. IAA is collected from both shoots and the amounts compared	Bending towards light. No bending. Total IAA collected in each case is approximately the same
9	Thin glass plate to separate the two sides of the shoot. Light → IAA collected either side of glass plate is measured	Amount of IAA collected is approximately the same either side of the glass plate
10	Light → The glass plate is placed so that lateral transfer of IAA is possible at the tip. IAA collected either side of glass plate is measured	Shoot bends towards light. 30% of total IAA collected on illuminated side. 70% of total IAA collected on shaded side

▲ **Figure 6** *Briggs's experiments to determine how IAA becomes unevenly distributed*

Study experiments 8, 9, and 10.

4 Suggest reasons for using a glass plate in experiments 9 and 10.

5 State which of the three theories the results tend to support. Give reasons for your answer.

Required practical 10

Investigating tropisms in seedlings

Germinating seedlings require water in order to mobilise the food reserves stored in the seed. This means it is essential that emerging roots grow downwards – show **positive geotropism** – to access soil water.

Seeds of the white mustard plant *Sinapis alba* are placed in a solution of dilute bleach for 5–8 minutes to surface sterilise them. The seeds are then rinsed several times in sterile distilled water and then transferred using sterile forceps onto a petri dish containing sterile 0.1% agar. The seeds are pressed gently onto the agar in a row across the top of the agar plate as shown in Figure 1.

▲ **Figure 1** *Seeds placed on the agar plate*

1 Why is it important to work with sterilised seeds and sterilised agar?
2 Why is it important that the seeds are not pushed too deep into the agar?

The agar plates are wrapped in cling film and placed in an upright, vertical position in a well-lit area such as a window sill. The seeds are monitored and after approximately 48–72 hours, the roots should have emerged and grown as shown in Figure 2.

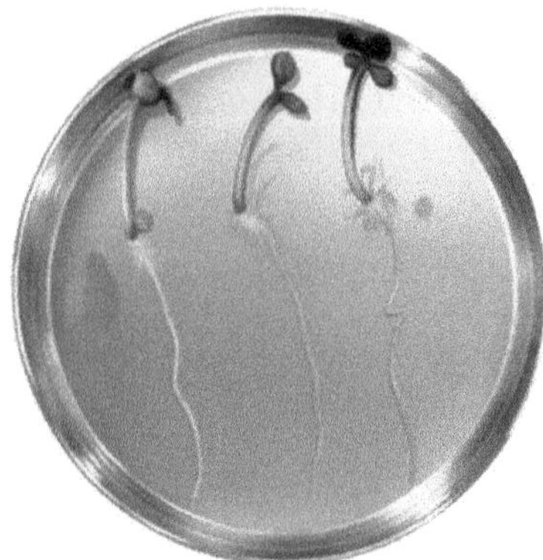

▲ **Figure 2** *Three seedlings on an agar plate after 72 hours*

3 Suggest why growth rate might vary:
  a for agar plates left in different locations
  b for seedlings on the same plate.

4 Suggest what has been done to the agar plate during the 72-hour period to produce the growth pattern shown in Figure 3.

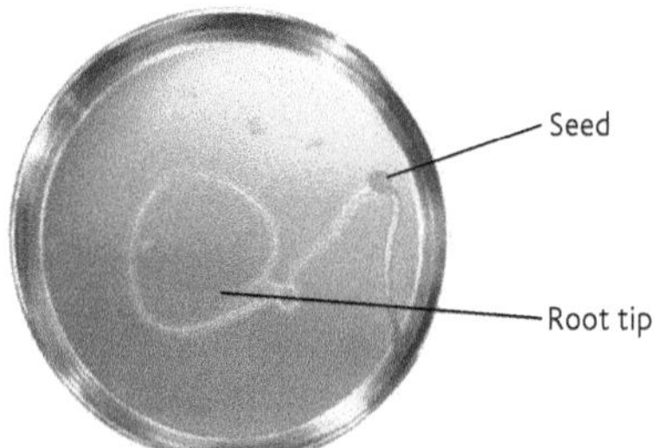

▲ **Figure 3** *An agar plate containing a single seedling after 72 hours*

26.2 Ethene and abscisic acid (ABA)

Learning objectives:

- → Describe the role of ethene in fruit ripening.
- → Explain how ripening can be manipulated in commercial fruit.
- → Describe the role of ABA in plant responses to drought.

Specification reference: 3.4.5.3

Plant growth substances are not a chemically uniform group. Ethene is a hydrocarbon gas that functions as a plant growth regulator. Ethene is not exclusively a plant product. Its effects were first noticed in the 19th century when leakage from gas street lamps caused deformities in nearby plants. Since then it has been shown to be particularly associated with ripening in some fruits and also with ageing in plants and the associated deterioration of plant tissues (senescence).

Climacteric and non-climacteric fruits

Fruits have been traditionally classified into one of two groups on the basis of their ripening behaviour. **Climacteric** fruits include species such as bananas, apples, tomatoes and avocados. In these and other climacteric species, fruits ripen after harvest. This means that they can be picked unripe and then ripened artificially by exposure to ethene. Non-climacteric fruits such as grapes and strawberries do not ripen once picked from the plant. Climacteric plants exhibit an increase in respiration rate linked to ethene production after they have been harvested. So what is the ethene doing, why is an increase in respiration necessary, and why does it happen?

The use of ethene in ripening fruit

Fruits that are unripe have a number of recognisable characteristics in that they are largely green, hard, and sour or acidic to taste. The ripening process in climacteric fruits involves ethene upregulating (switching on) genes for the enzymes responsible for changing the colour (hydrolases that convert chlorophyll to different pigments), for softening fruit (pectinases), and for increasing the sugar content (amylases). The synthesis of these enzymes requires ATP and hence the link to increased respiration rate. Commercial fruit growers frequently harvest climacteric fruits when they are mature but not yet ripe. This enables fruit to be transported considerable distances and stored at low temperatures for some time. Ethene is then used in a controlled way to bring the fruit to ripeness as and when it is required.

▲ **Figure 1** *Banana ripening room*

To trigger ripening, fruits such as bananas are placed in specially constructed ripening rooms and the temperature and humidity are raised to optimum levels. An example of such a room is shown in Figure 1. Ethene concentration in the room is then raised using a catalytic generator such as the one shown in Figure 2. Fans ensure an even temperature and distribution of ethene. The rooms are airtight to prevent gases escaping.

Fruit in the room will ripen at a uniform rate and over a predictable time span, enabling the supplier to meet the demands of the market.

▲ **Figure 2** *Catalytic ethene generator*

Abscisic acid and stomatal closure

Abscisic acid (ABA) is another plant growth substance that plays a key role in a plant's response to environmental stress. One way in which ABA exerts its effect is through the closing of stomata in

Synoptic link

It would be useful to review Topic 16.3. In the event of water loss through transpiration exceeding water uptake through roots, the plant will wilt. Leaves will no longer be photosynthesising effectively as wilted leaves reduce their light harvesting surface area and less water is available for photosynthesis (Topic 19.1). The plant will stop growing and may die.

response to low soil water or high salinity. When soil water is low, ABA is transported from the roots where it is synthesised up to shoots via the xylem. In Topic 5.3 you were told that it is the guard cells that are responsible for opening and closing stomata. The sequence of events that open the stomata is described below:

- Hydrogen ions (H^+) are actively transported out of the guard cells, lowering the concentration of H^+ in the cells and causing the inside of the cell to become negatively charged.
- The negative charge causes *inward* K^+ channels in the membranes to open and K^+ diffuses into the guard cells.
- The K^+ ions lower the water potential of the guard cells and water moves in by **osmosis**.
- The guard cells become turgid and open the stomata.

ABA acts on receptors on the cell-surface membrane of guard cells causing stomata to close. The sequence of events is illustrated in Figure 3 and outlined below:

- ABA binds to receptors leading to the production of hydrogen peroxide (H_2O_2) and other **reactive oxygen species (ROS)**.
- **ROS** in turn triggers an increase in calcium ions (Ca^{2+}) in the cytoplasm from both outside the cell and from the vacuole. ROS also inhibit the H^+ pumps.
- The rise in Ca^{2+} opens anion channels and Cl^- and other negatively charged ions leave the cell, which depolarises the membrane.
- This causes *inward* K^+ channels to close and *outward* K^+ channels to open, and K^+ ions leave the cell.
- Water follows the ions by osmosis and the guard cells become flaccid and close the stomata.

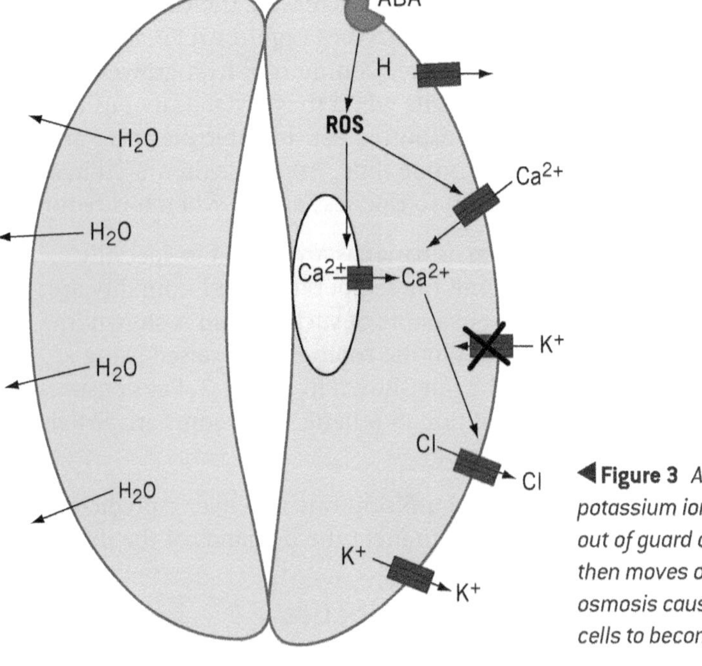

Figure 3 *ABA causes potassium ions to move out of guard cells. Water then moves out by osmosis causing the cells to become flaccid.*

The closure of stomata induced by ABA can be very rapid and is a short-term response to water stress in the plant.

Summary questions

1. Keeping ripe fruit close to flowers shortens the shelf life of the flowers – suggest why.
2. Silver thiosulfate is often supplied with cut flowers to prolong their life in vases. Suggest how silver thiosulfate might function.
3. Why do plants respond to salt stress in the same way as they do to water stress?

1 Scientists investigated the response of lateral roots to gravity. Lateral roots grow from the side of main roots. Figure 1 shows four stages, **A** to **D**, in the growth of a lateral root and typical cells from the tip of the lateral root in each stage. All of the cells are drawn with the bottom of the cell towards the bottom of the page.

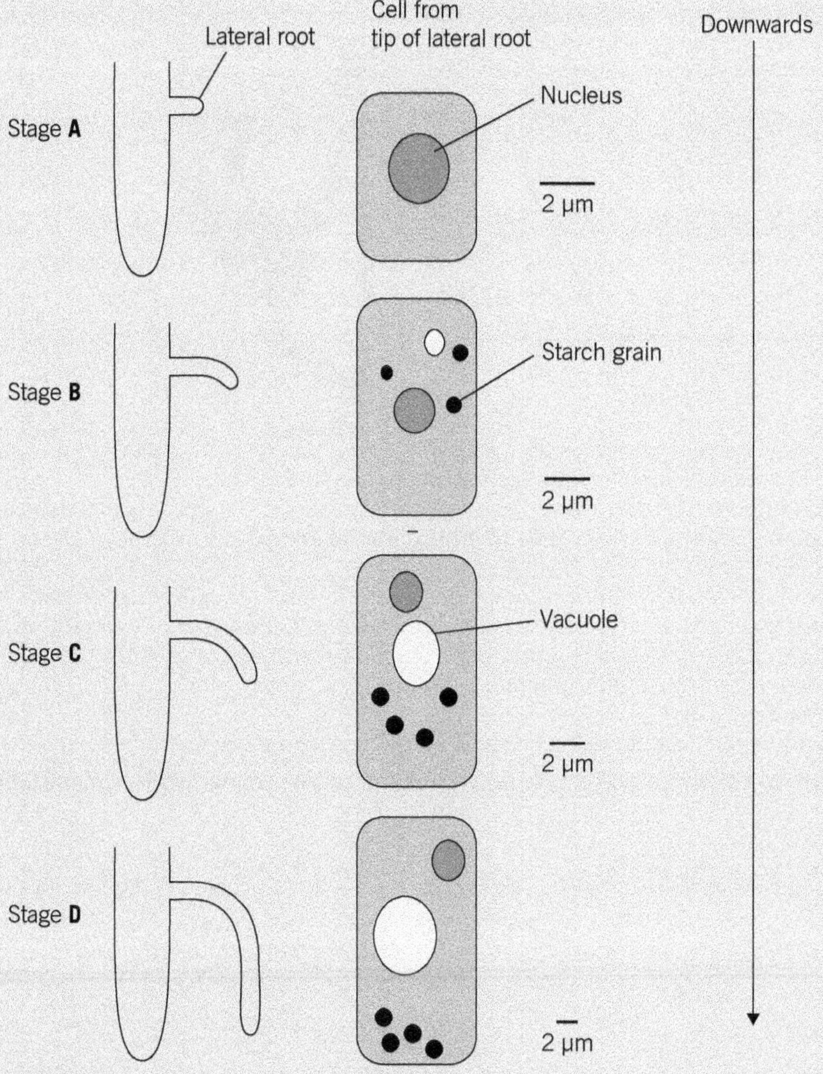

▲ **Figure 1**

(a) Describe three changes in the root tip cells between stages **A** and **D**. *(3 marks)*

(b) The scientists' hypothesis was that there was a relationship between the starch grains in the root tip cells and the bending and direction of growth of lateral roots. Does the information in the diagram support this hypothesis? Give reasons for your answer. *(3 marks)*

(c) Figure 2 shows the distribution of indoleacetic acid (IAA) in the lateral root at stage B. Explain how this distribution of IAA causes the root to bend. *(2 marks)*

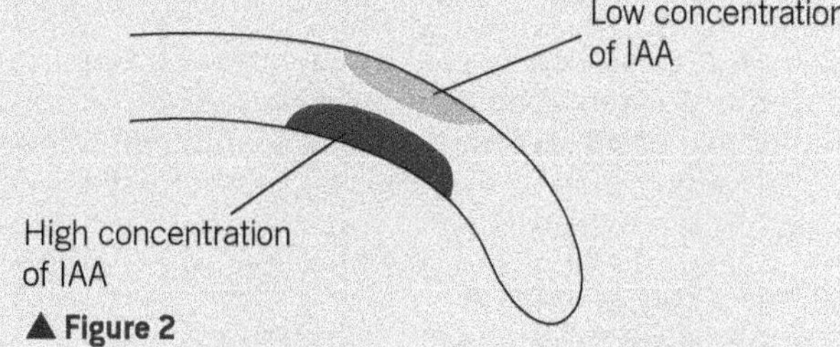

▲ **Figure 2**

AQA June 2013

2 Plant growth substances such as abscisic acid (ABA) are important in events in the life cycles of plants. The concentration of plant growth substances can change, which in turn regulates the expression of specific genes. The concentration of ABA was measured during the development of cotton fruits.
Figure 3 shows how the concentration of ABA changes as the cotton fruit develops.
Figure 4 shows a fully mature cotton bud.

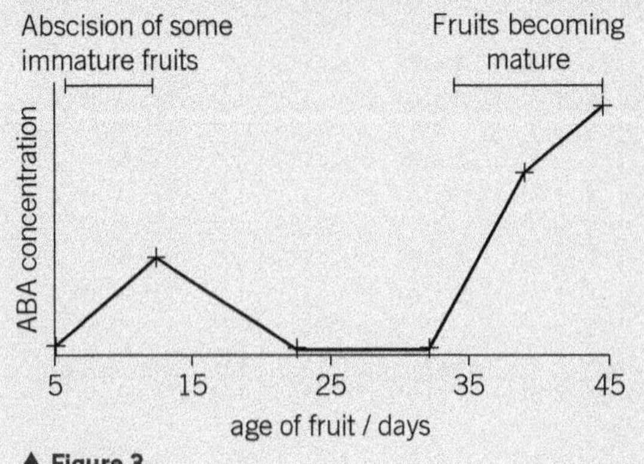

▲ Figure 3

▲ Figure 4

(a) Describe the relationship between fruit development and ABA concentration. *(3 marks)*

(b) Suggest how ABA acts in plant cells to bring about changes such as fruit development.

(c) An experiment was carried out using plants grown in outdoor experimental plots. Concentrations of ABA were measured in the leaves under a variety of conditions. The concentration of ABA was found to increase on warm, dry, and windy days.

(i) Explain why these environmental conditions lead to an increase in ABA. *(2 marks)*

(ii) Describe the effect of raising concentrations of ABA on the leaf. *(2 marks)*

3 Figure 5 shows how the respiration rate in two different fruits changes following picking.

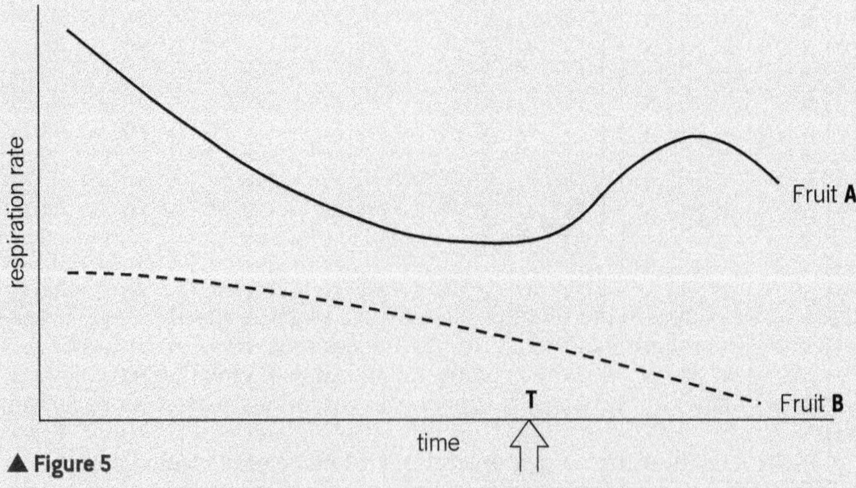

▲ Figure 5

(a) With reference to Figure 5, which of the two fruits, **A** or **B**, is a climacteric fruit? Explain your answer. *(1 mark)*

(b) Both fruits were harvested at time **T** and kept in cold conditions for transport to markets around the world. The fruits were then prepared for sale.
How would the preparation for sale of the two types of fruit be different?

(c) Ripe fruits are normally much sweeter than unripe fruit. Suggest how the sugar content in the cells of ripening fruit is increased. *(2 marks)*

Answers to the Practice Questions are available at
www.oxfordsecondary.com/oxfordaqaexams-alevel-biology

27 Homeostasis and negative feedback

27.1 Principles of homeostasis

In previous chapters you looked at how organisms develop systems to control and coordinate their activities as they become more complex. In particular you have considered the way in which organisms respond rapidly to environmental changes using their nervous system. Another feature of an increase in complexity is the ability of organisms to control their internal environment. By maintaining a relatively constant internal environment for their cells, organisms can limit the external changes these cells experience. This maintenance of a constant internal environment is called **homeostasis**. In this chapter you shall learn about homeostasis and the role of the other coordination system, hormonal coordination, in an organism's physiological control.

The internal environment is made up of tissue fluids that bathe each cell, supplying nutrients and removing wastes. Maintaining the features of this fluid at the optimum levels protects the cells from changes in the external environment, thereby giving the organism a degree of independence.

Learning objectives:

- → Explain what is meant by homeostasis and why it is important.
- → Describe how control mechanisms work.
- → Describe how control mechanisms are coordinated.

Specification reference: 3.4.6.1

What is homeostasis?

Homeostasis is the maintenance of a constant internal environment in organisms. It involves maintaining the chemical make-up, volume, and other features of blood and tissue fluid within restricted limits. Homeostasis ensures that the cells of the body are in an environment that meets their needs and allows them to function normally despite external changes. This does not mean that there are no changes. On the contrary, there are continuous fluctuations brought about by variations in internal and external conditions, such as changes in temperature, pH, and water potential. These changes, however, occur around a set point. Homeostasis is the ability to return to that set point and so maintain organisms in a balanced equilibrium.

Synoptic link

You learnt about the formation of tissue fluid in Topic 15.2. As tissue fluid in mammals is derived from blood plasma, maintaining parameters such as water potential in blood plasma will mean that they are also maintained in tissue fluid.

Water potential and its importance to cells is covered in Topic 4.3.

The importance of homeostasis

Homeostasis is essential for the proper functioning of organisms for the following reasons:

- The enzymes that control the biochemical reactions within cells, and other proteins, such as channel proteins, are sensitive to changes in pH and temperature. Any change to these factors reduces the efficiency of enzymes or may even prevent them working altogether, for example, by denaturing them. Even small fluctuations in temperature or pH can impair the ability of enzymes to carry out their roles effectively. Maintaining a constant internal environment means that reactions take place at a constant and predictable rate.
- Changes to the water potential of the blood and tissue fluids may cause cells to shrink or expand (even to bursting point) as a result of water leaving or entering by osmosis. In both instances the cells cannot operate normally. The maintenance of a constant blood glucose concentration is essential in ensuring a constant water

Synoptic link

The importance of temperature and pH in relation to enzyme activity is covered in Topic 3.2.

▲ **Figure 1** *Homeostasis allows animals such as these camels in the desert (top) and these penguins in the Antarctic (bottom) to survive in extreme environments*

potential. A constant blood glucose concentration also ensures a reliable source of glucose for respiration by cells.

- Organisms with the ability to maintain a constant internal environment are more independent of the external environment. They have a wider geographical range and therefore have a greater chance of finding food, shelter, etc. Mammals, for example, with their ability to maintain a constant temperature, are found in most habitats, ranging from hot arid deserts to frozen polar regions.

Control mechanisms

The control of any self-regulating system involves a series of stages that feature:

- the **set point**, which is the desired level, or norm, at which the system operates. This is monitored by a ...
- **receptor**, which detects any deviation from the set point and informs the ...
- **controller**, which coordinates information from various receptors and sends instructions to an appropriate ...
- **effector**, which brings about the changes needed to return the system to the set point. This return to normality creates a ...
- **feedback loop**, which informs the receptor of the changes to the system brought about by the effector.

Figure 2 illustrates the relationship between these stages using the everyday example of controlling a central heating system.

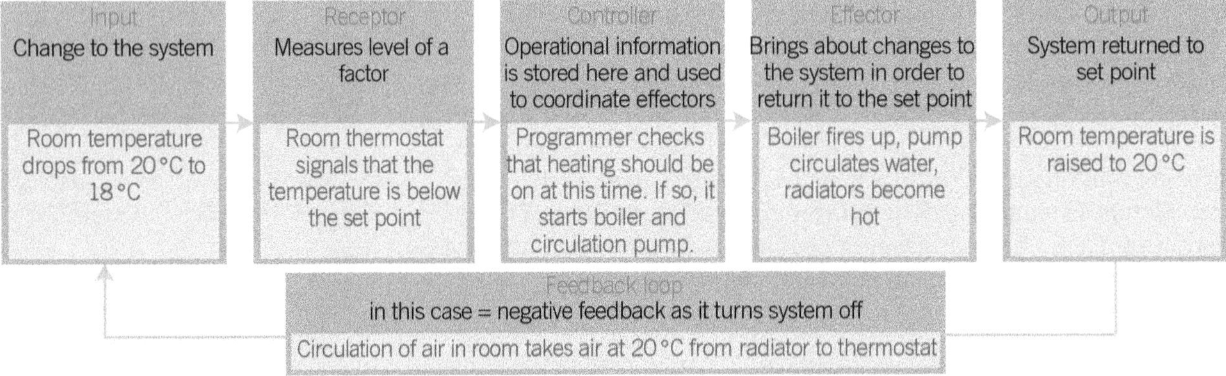

▲ **Figure 2** *Components of a typical control system*

Coordination of control mechanisms

Systems normally have many receptors and effectors. It is important to ensure that the information provided by receptors is analysed by the control centre before action is taken. Receiving information from a number of sources allows a better degree of control. For example, temperature receptors in the skin may signal that the skin itself is cold and that body temperature should be raised. However, information from the temperature centre in the brain may indicate that blood temperature is already above normal. This situation could arise during strenuous exercise when blood temperature rises but sweating cools the skin. By analysing the information from all the detectors, the brain can decide the best course of action – in this case not to raise the body temperature further. In the same way, the control centre must coordinate the action of the effectors so that they operate harmoniously. For example, sweating would be less effective in cooling the body if it were not accompanied by **vasodilation**.

Summary questions

1. What is homeostasis?
2. Explain why maintaining a constant internal temperature is important in mammals.
3. Explain why maintaining a constant blood glucose concentration is important in mammals.

27.2 The principles of feedback mechanisms

Learning objectives:

→ Explain how negative feedback controls homeostatic processes.

→ Compare negative and positive feedback.

Specification reference: 3.4.6.2

You saw in Topic 27.1 that the **homeostatic** control of any system involves a series of stages featuring:

- the **set point**, or desired level (norm), at which the system operates
- a **receptor**, which detects any deviation from the set point (norm)
- a **controller**, which coordinates information from various sources
- an **effector**, which brings about the corrective measures needed to return the system to the set point (norm)
- a **feedback loop**, which informs the receptor of the changes to the system brought about by the effector.

Let us now look in more detail at the last stage in the list – the feedback loop. When an effector has corrected any deviation and returned the system to the set point, it is important that this information is fed back to the receptor. If the information is not fed back, the receptor will continue to stimulate the effector, leading to an over correction and causing a deviation in the opposite direction. There are two types of feedback, negative feedback and positive feedback.

Negative feedback

Negative feedback occurs when the feedback causes the corrective measures to be turned off. In doing so it returns the system to its original (normal) level.

Let us take the example of temperature regulation that you looked at in Topic 27.1. If the temperature of the blood increases, thermoreceptors in the **hypothalamus** send nerve impulses to the heat loss centre, which is also in the hypothalamus. This in turn sends impulses to the skin (effector organ). **Vasodilation**, sweating, and lowering of body hairs all lead to a reduction in blood temperature. If the fact that blood temperature has returned to normal is not fed back to the hypothalamus, it will continue to stimulate the skin to lose body heat. Blood temperature will then fall below normal and may continue to do so, causing **hypothermia** and the death of the organism.

What happens in practice is that the cooler blood returning from the skin passes through the hypothalamus. The thermoreceptors detect that blood temperature is at its normal set point again and so they cease to send impulses to the heat loss centre. This in turn stops sending impulses to the skin and so vasodilation, sweating, etc. cease and blood temperature remains at its normal level rather than continuing to fall. The blood, having been cooled to its normal temperature, has turned *off* the effector (the skin) that was correcting the rise in temperature. This is therefore negative feedback and is illustrated in Figure 1.

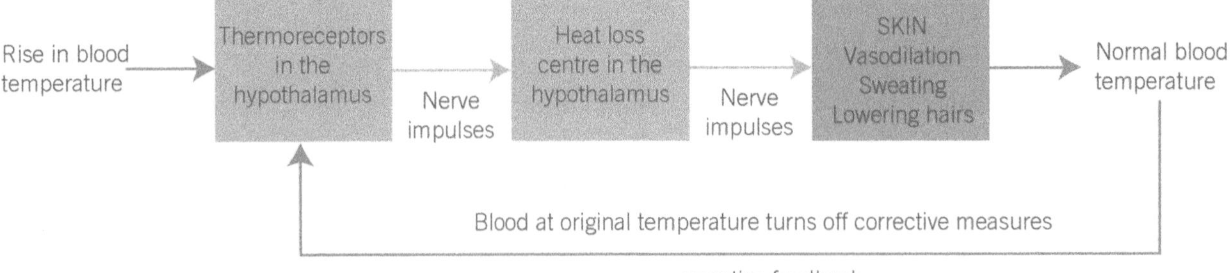

▲ **Figure 1** *Negative feedback in the control of body temperature*

▲ **Figure 2** *Control of body temperature involves negative feedback mechanisms*

There are separate negative feedback mechanisms to regulate departures from the norm in each direction. In temperature regulation, for example, if the blood temperature falls, rather than rises, the heat gain centre will cause **vasoconstriction**, the raising of hairs, and reduced sweating. There will again be negative feedback to turn off these mechanisms when the blood returns to its normal temperature.

Blood pH in mammals is controlled by the combined efforts of the lungs and kidneys acting as effectors. You saw in Topic 5.8 that high concentrations of carbon dioxide lower pH. Chemoreceptors present in the walls of certain blood vessels (see Topic 27.5) detect the fall in pH and impulses are sent to the medulla oblongata. This region of the brain controls, among other things, breathing rate and depth. By increasing the rate and depth of breathing, more carbon dioxide is exhaled and so the pH is raised. The kidneys contribute to pH control by adjusting the concentration of basic or acidic ions in blood plasma. Again, if pH is too low, more basic hydrogen carbonate ions (HCO_3^-) are retained in the blood, which effectively raises the pH. In this way, the pH of blood plasma is maintained with a range of 7.35–7.45.

Another example is in the control of blood glucose that you will learn about in Topic 27.4. If there is a fall in the concentration of glucose in the blood, the α cells in the **islets of Langerhans** in the pancreas produce the hormone **glucagon**. Glucagon causes the conversion of glycogen to glucose and **gluconeogenesis** in the liver. As a result the blood glucose concentration rises to normal. As this blood circulates back to the pancreas, the α cells detect the change and stop producing glucagon (negative feedback). These events are illustrated in Figure 3.

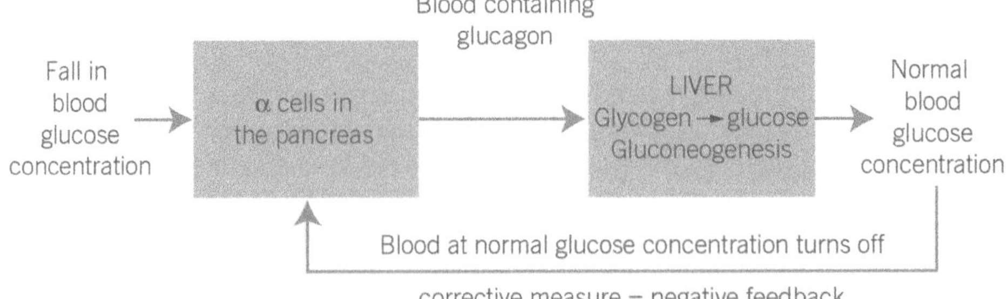

▲ **Figure 3** *Negative feedback in the control of blood glucose concentration*

In a similar way, if the blood glucose concentration rises, rather than falls, **insulin** will be produced from the α cells in the pancreas. Insulin increases the uptake of glucose by cells and its conversion to glycogen and fat. The fall in blood glucose concentration that results will turn off insulin production once blood glucose concentration return to normal (negative feedback).

Having separate negative feedback mechanisms that control departures from the norm in either direction gives a greater degree of homeostatic control.

Positive feedback

Positive feedback occurs when the feedback causes the corrective measures to remain turned on. In doing so it causes the system to deviate even more from the original (normal) level. Examples are rare, but one occurs in neurones when a stimulus causes a small influx of sodium ions (Topic 24.4). This influx increases the permeability of the neurone to sodium ions so more ions enter, causing a further increase in permeability and even more rapid entry of ions. This results in a very rapid build-up of an action potential that allows an equally rapid response to a stimulus.

Positive feedback occurs more often when there is a breakdown of control systems. In certain diseases, for example, typhoid fever, there is a breakdown of temperature regulation resulting in a rise in body temperature leading to **hyperthermia**. In the same way, when the body gets too cold (hypothermia), the temperature control system tends to break down, leading to positive feedback, resulting in the body temperature dropping even lower.

Summary questions

1. Distinguish between positive and negative feedback.
2. Why is negative feedback important in maintaining a system at a set point?
3. What is the advantage of having separate negative feedback mechanisms to control deviations away from normal?

Control of blood water potential

Figure 4 shows some of the changes that occur as a result of water being lost from the blood due to sweating.

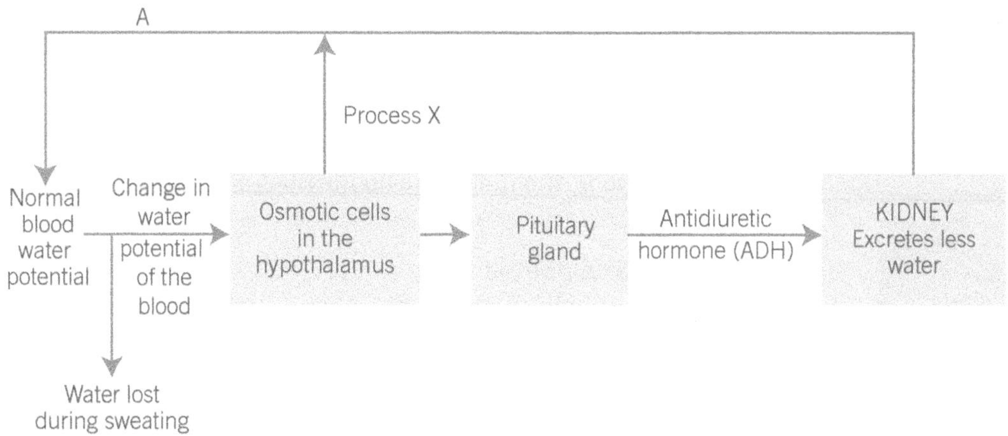

▲ **Figure 4**

1. Describe the change in water potential that occurs in the blood as a result of sweating.
2. Which of the structures shown in Figure 4 acts as:
   a. a receptor
   b. an effector?
3. Describe how ADH gets from the pituitary gland to the kidney.
4. The kidney conserves the water that is already in the blood. Given that the water potential of the blood returns to its normal level prior to sweating, suggest what is happening in process **X**.
5. State as concisely as possible what mechanism is shown by the line labelled A.

27.3 The need to control blood glucose concentration

Learning objectives:

- → Explain why glucose concentrations in blood have to be maintained within set limits.
- → Describe the factors that affect glucose concentration.
- → Outline the role of the liver in glucose homeostasis.
- → Describe diabetes as an example of a disease where blood glucose is not controlled.
- → Explain the difference between type I and type II diabetes.

Specification reference: 3.4.7.1

Hint

Symptoms of diabetes:

- high blood glucose concentration
- presence of glucose in the urine
- increased thirst and hunger
- need to urinate excessively
- genital itching or regular episodes of thrush
- tiredness
- weight loss
- blurred vision.

▲ **Figure 1** *A diabetic person injecting insulin*

Glucose is the main substrate for respiration, providing the source of energy for almost all organisms. It is broken down during glycolysis (see Topic 20.1) and, if oxygen is present, the Krebs cycle (see Topic 20.3) and electron transfer system (see Topic 20.4), to provide **ATP** – the energy currency of cells (see Topic 4.4). It is therefore essential that the blood of mammals contains a relatively constant concentration of glucose for respiration. If the concentration falls too low, cells will be deprived of energy and die – brain cells are especially sensitive in this respect because they can only respire glucose. If the concentration rises too high, it lowers the **water potential** of the blood and creates osmotic problems that can cause dehydration and be equally dangerous. Homeostatic control (see Topic 27.4) of blood glucose is therefore essential.

Blood glucose and variations in its level

The normal concentration of blood glucose is 90 mg in each 100 cm^3 of blood. Blood glucose comes from two sources:

- **directly from the diet** in the form of glucose resulting from the breakdown of other carbohydrates such as starch, maltose, lactose, and sucrose
- **from the breakdown of glycogen (glycogenolysis)** stored in the liver and muscle cells. A normal liver contains 75–100 g of glycogen, produced by converting excess glucose from the diet in a process called glycogenesis.

Diabetes and its control

Diabetes is a chronic disease in which a person is unable to control their blood glucose concentration. There are over 100 million people worldwide with diabetes, and the World Health Organisation predicts that this will rise to 366 million by 2030! There are two distinct forms, of which diabetes mellitus, or 'sugar diabetes', is much more common.

Types of sugar diabetes

Diabetes is a metabolic disorder caused by an inability to control blood glucose concentration due to a lack of the hormone insulin or a loss of responsiveness to insulin.

There are two forms of diabetes:

- **Type I (insulin dependent)** is due to the body being unable to produce insulin. It normally begins in childhood. It may be the result of an autoimmune response whereby the body's immune system attacks its own cells, in this case the β cells of the **islets of Langerhans**. Type I diabetes develops quickly, usually over a few weeks, and the symptoms (see 'Hint') are normally obvious.
- **Type II (insulin independent)** is normally due to the **glycoprotein** receptors on the body cells losing their responsiveness to insulin. However, it may also be due to an inadequate supply of insulin from the pancreas. Type II diabetes usually develops in

people over the age of 40 years. There is, however, an increasing number of cases of obesity and poor diet leading to type II diabetes in adolescents. It develops slowly, and the symptoms are normally less severe and may go unnoticed. People who are overweight are particularly likely to develop type II diabetes. Over 75 per cent of people with diabetes have type II disease.

Figure 2 illustrates the differences in blood glucose concentration between diabetic and non-diabetic individuals who have swallowed a glucose solution.

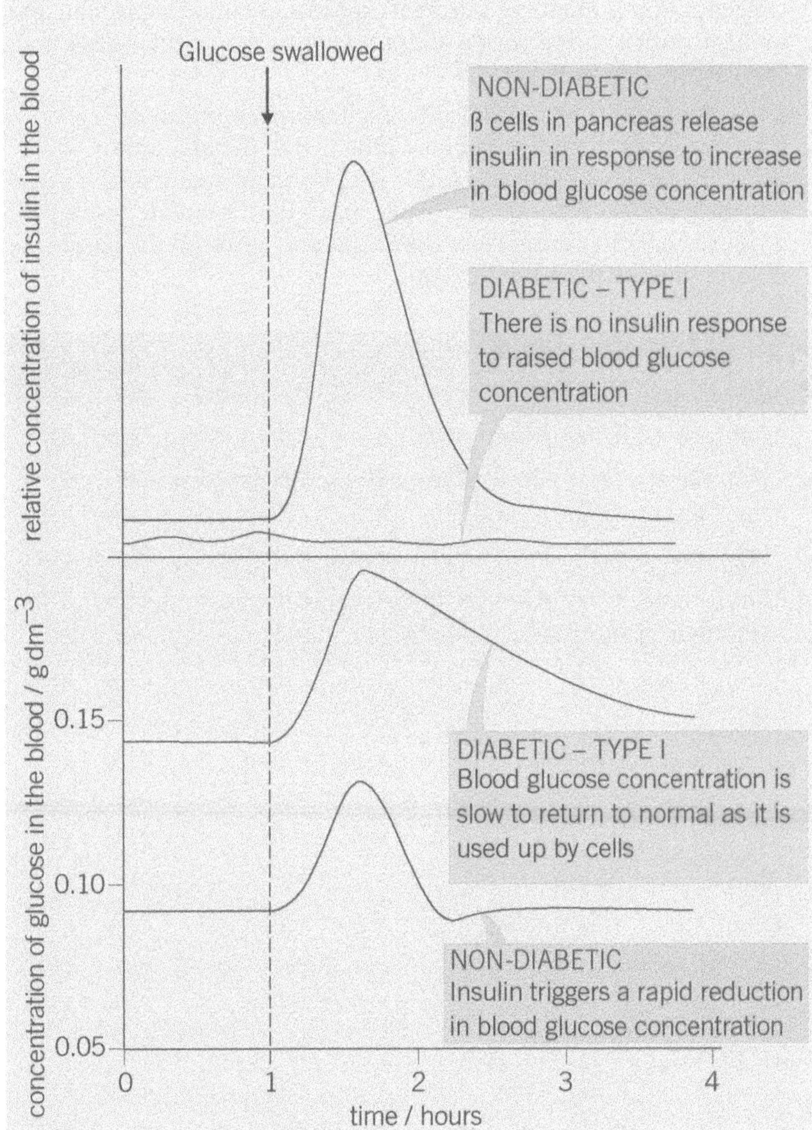

▲ **Figure 2** *Comparison of blood glucose and insulin concentrations in a person with type I diabetes and a non-diabetic person after each has swallowed a glucose solution*

Hint

Blood glucose concentration can be controlled by changing the uptake of glucose from the gut (diet) and by changing the rate at which glucose is removed from the blood (exercise and insulin).

Control of diabetes

Although diabetes cannot be cured, it can be treated very successfully. Treatment varies depending on the type of diabetes.

- **Type I diabetes** is controlled by injections of insulin. This cannot be taken by mouth because, being a protein, it would be digested in the alimentary canal. It is therefore injected, typically either two or four times a day. The dose of insulin must be matched exactly to the glucose intake. If a diabetic takes too much insulin, he or she will experience a low blood glucose concentration that can result in unconsciousness. To ensure the correct dose, blood glucose concentration is monitored using biosensors. By injecting insulin and managing their carbohydrate intake and exercise carefully, diabetics can lead normal lives.
- **Type II diabetes** is controlled by regulating the intake of carbohydrate in the diet and matching this to the amount of exercise taken. In some cases, this may be supplemented by injections of insulin or by the use of drugs that stimulate insulin production. Other drugs slow down the rate at which the body absorbs glucose from the intestine.

Summary questions

1. State **one** difference between the causes of type I and type II diabetes.
2. State **one** difference between the main ways of controlling type I and type II diabetes.
3. Suggest an explanation for why tiredness is a symptom of diabetes.
4. What lifestyle advice might you give someone in order to help them avoid developing type II diabetes?

Effects of diabetes on substance concentrations in the blood

An experiment was carried out with two groups of people. Group X had type I diabetes whereas group Y did not (control group). Every 15 minutes blood samples were taken from all members of both groups and the mean concentrations of insulin, glucagon, and glucose were calculated. After an hour, every person was given a glucose drink. The results are shown in the graphs below.

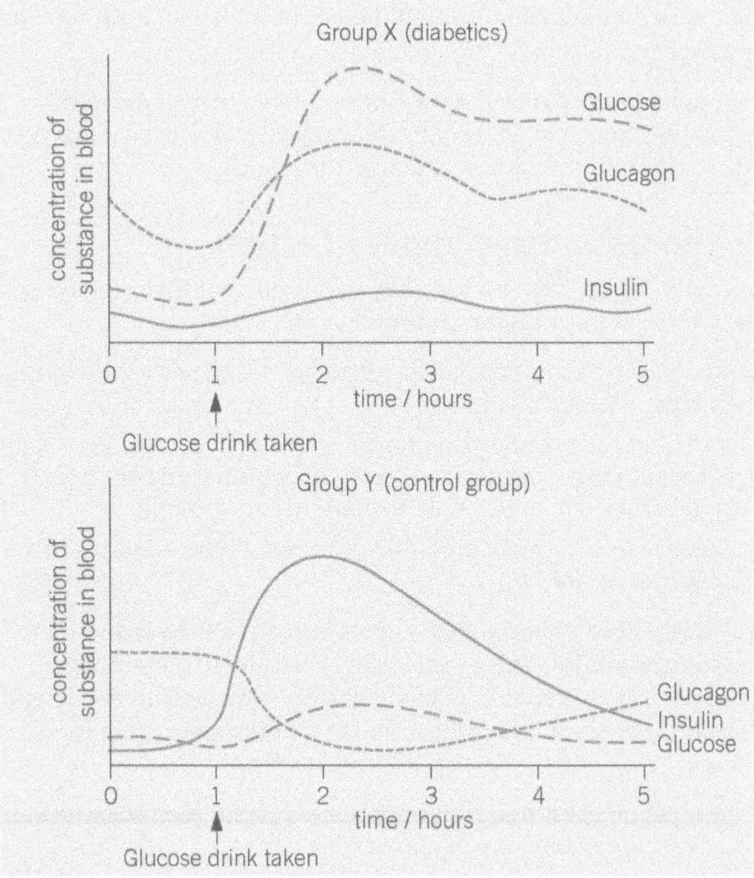

▲ **Figure 3**

1. Name a hormone other than insulin and glucagon that is involved in regulating blood glucose concentration.
2. State **two** differences between groups X and Y in the way insulin secretion responds to the drinking of glucose.
3. Suggest a reason why the glucose level falls in both groups during the first hour.
4. Using information from the graphs, explain the changes in the blood glucose concentration in group Y after drinking the glucose.
5. Explain the differences in blood glucose concentration of group X compared to group Y.
6. Suggest what might happen to the blood glucose concentration of group X if they have no food over the next 24 hours.

27.4 Hormones and the regulation of blood glucose

Learning objectives:

- → Explain how hormones work.
- → Describe the role of the pancreas in regulating blood glucose.
- → Describe the roles of insulin, glucagon, and adrenaline in regulating blood glucose.

Specification reference: 3.4.7.2, 3.4.7.3, and 3.4.7.4

Animals possess two principal coordinating systems – the nervous system, which communicates rapidly, and the hormonal system, which usually communicates more slowly. Both systems interact in order to maintain the constancy of the internal environment whilst also being responsive to changes in the external environment. Both systems also use chemical messengers – the hormonal system exclusively so, and the nervous system through the use of neurotransmitters in chemical synapses.

The regulation of blood glucose is an example of how different hormones interact in achieving homeostasis. However, let us first look at what hormones are and how they work.

Hormones and their mode of action

Hormones differ from one another chemically but they all have certain characteristics in common. Hormones are:

- produced by glands (endocrine glands), which secrete the hormone directly into the blood
- carried in the blood plasma to the cells on which they act – known as **target cells** – which have receptors on their cell-surface membranes that are complementary to the hormone
- effective in very small quantities, but often have widespread and long-lasting effects.

One mechanism of hormone action is known as the **second messenger model**. This mechanism is used by two hormones involved in the regulation of blood glucose, namely adrenaline and glucagon. The second messenger model of hormone action works as follows:

- Adrenaline binds to a transmembrane protein receptor within the cell-surface membrane of a liver cell.
- The binding of adrenaline causes the protein to change shape on the inside of the membrane.
- This change of protein shape leads to the activation of an enzyme called adenyl cyclase. The activated adenyl cyclase converts ATP to cyclic AMP (cAMP).
- The cAMP acts as a second messenger that binds to protein kinase enzyme, changing its shape and therefore activating it.
- The active protein kinase enzyme catalyses the conversion of glycogen to glucose which moves out of the liver cell by facilitated diffusion and into the blood, through channel proteins.

The mechanism for the action of adrenaline is illustrated in Figure 1.

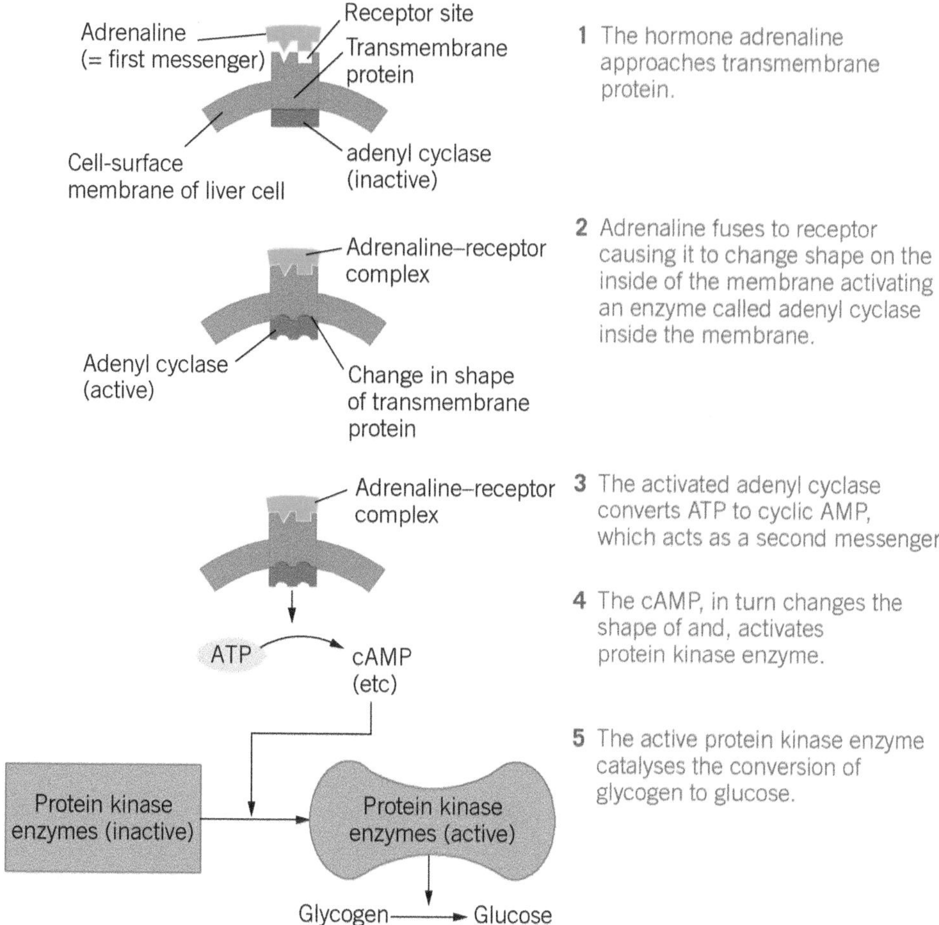

▲ **Figure 1** *Second messenger model of hormone action as illustrated by the action of adrenaline in regulating blood sugar*

The role of the pancreas in regulating blood glucose

The pancreas is a large, pale-coloured gland that is situated in the upper abdomen, behind the stomach. It produces enzymes (protease, amylase and lipase) for digestion and hormones (insulin and glucagon) for regulating blood glucose.

When examined microscopically, the pancreas is made up largely of the cells that produce its digestive enzymes. Scattered throughout these cells are groups of hormone-producing cells known as **islets of Langerhans**. The cells of the islets of Langerhans are of two types:

- α **cells**, which are the larger and produce the hormone **glucagon**
- β **cells**, which are smaller and produce the hormone **insulin**.

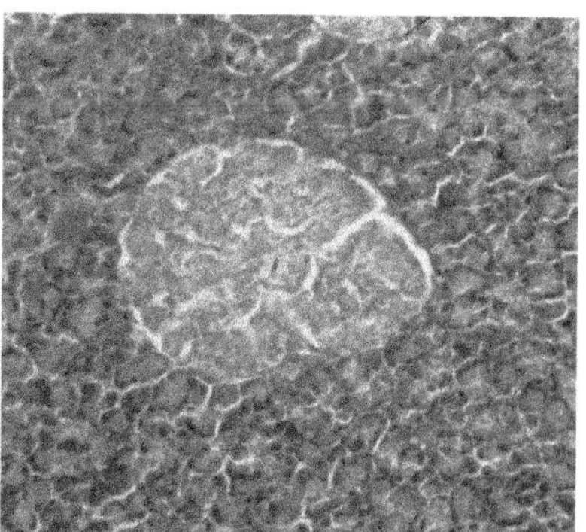

▲ **Figure 2** *Light micrograph of the pancreas showing an islet of Langerhans (centre) containing α cells and β cells. The meshworks of blue and white in the islet are blood capillaries. Around the islet are the enzyme-producing pancreatic cells.*

The role of the liver in regulating blood sugar

The liver is located immediately below the diaphragm, has a mass of up to 1.5 kg and is made up of cells called hepatocytes. It serves a large variety of roles including regulating blood glucose concentration. While the pancreas produces the hormones insulin and glucagon, it is in the liver where they have their effects. There are three important processes associated with regulating blood sugar which take place in the liver.

Hint

You will find it easier to understand the terms used in this topic if you remember the following:

gluco / glyco = glucose

neo = new

lysis = splitting

genesis = birth / origin

Therefore:

glycogen – o – lysis = splitting of glycogen

gluco – neo – genesis = formation of new glucose

- **Glycogenesis** is the conversion of glucose into glycogen. When blood glucose concentration is higher than normal the liver removes glucose from the blood and converts it to glycogen. It can store 75–100 g of glycogen, which is sufficient to maintain a human's blood glucose concentration for about 12 hours when at rest, in the absence of other sources.
- **Glycogenolysis** is the breakdown of glycogen to glucose. When blood glucose concentration is lower than normal, the liver can convert stored glycogen back into glucose which diffuses into the blood to restore the normal blood glucose concentration.
- **Gluconeogenesis** is the production of glucose from sources other than carbohydrate. When its supply of glycogen is exhausted, the liver can produce glucose from non-carbohydrate sources such as glycerol and amino acids.

Factors that influence blood glucose concentration

The normal concentration of blood glucose is 5 mmol dm^{-3}. Blood glucose comes from three sources:

- **directly from the diet** in the form of glucose absorbed following hydrolysis of other carbohydrates such as starch, maltose, lactose, and sucrose
- **from the hydrolysis in the small intestine of glycogen = glycogenolysis** stored in the liver and muscle cells
- **from gluconeogenesis**, which is the production of glucose from sources other than carbohydrate.

As animals do not eat continuously and their diet varies, their intake of glucose fluctuates. Likewise, glucose is used during respiration at different rates depending on the level of mental and physical activity. It is against these changes in supply and demand that the three main hormones, **insulin**, **glucagon** and **adrenaline**, operate to maintain a constant blood glucose concentration.

Insulin and the β cells of the pancreas

The β cells of the islets of Langerhans in the pancreas detect a rise in blood glucose concentration and respond by secreting the hormone insulin directly into the blood plasma. Insulin is a globular protein made up of 51 amino acids.

Almost all body cells (red blood cells being a notable exception) have glycoprotein receptors on their cell-surface membranes that bind with insulin molecules. When it combines with the receptors, insulin brings about:

- a change in the tertiary structure of the glucose transport protein channels, causing them to change shape and open, allowing more glucose into the cells
- an increase in the number of carrier molecules in the cell-surface membrane
- activation of the enzymes that convert glucose to glycogen and fat.

As a result, the blood glucose concentration is lowered in one or more of the following ways:

- by increasing the rate of absorption of glucose into the cells, especially in muscle cells
- by increasing the respiratory rate of the cells, which therefore use up more glucose, thus increasing their uptake of glucose from the blood
- by increasing the rate of conversion of glucose into glycogen (glycogenesis) in the cells of the liver and muscles
- by increasing the rate of conversion of glucose to fat.

The effect of these processes is to remove glucose from the blood and so return its level to normal. This lowering of the blood glucose concentration causes the β cells to reduce their secretion of insulin (negative feedback).

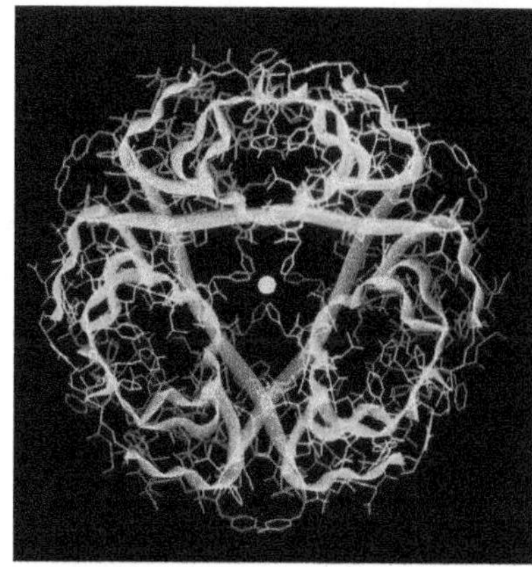

▲ **Figure 3** *Molecular graphic of an insulin molecule. Insulin is made up of 51 amino acids arranged in two chains (shown here as yellow and green ribbons).*

Glucagon and the α cells of the pancreas

The α cells of the islets of Langerhans detect a fall in blood glucose and respond by secreting the hormone glucagon directly into the blood plasma. Only the cells of the liver have receptors that bind to glucagon, so only liver cells respond. They do this by:

- activating an enzyme that converts glycogen to glucose
- increasing the conversion of amino acids and glycerol into glucose (gluconeogenesis).

The overall effect is therefore to increase the amount of glucose in the blood and return it to its normal level. This raising of the blood glucose concentration causes the α cells to reduce the secretion of glucagon (negative feedback).

Role of adrenaline in regulating blood glucose concentration

There are at least four other hormones apart from glucagon that can increase blood glucose concentration. The best known of these is adrenaline. At times of excitement or stress, adrenaline is produced by the adrenal glands that lie above the kidneys. Adrenaline raises the blood glucose concentration by:

- activating an enzyme that causes the breakdown of glycogen to glucose in the liver
- inactivating an enzyme that synthesises glycogen from glucose.

Hormone interaction in regulating blood glucose

The two hormones insulin and glucagon act in opposite directions. Insulin lowers the blood glucose concentration, whereas glucagon increases it. The two hormones are said to act antagonistically. The system is self-regulating in that it is the level of glucose in the blood that determines the quantity of insulin and glucagon produced. In this way, the interaction of these two hormones allows highly sensitive control of the blood glucose concentration. The level of glucose is not, however, constant, but fluctuates around a set point. This is because of the way negative feedback mechanisms work. Only when blood glucose is above the set point is insulin released and only when blood glucose concentration is below the set point is glucagon released.

The control of blood glucose concentration is summarised in Figure 4.

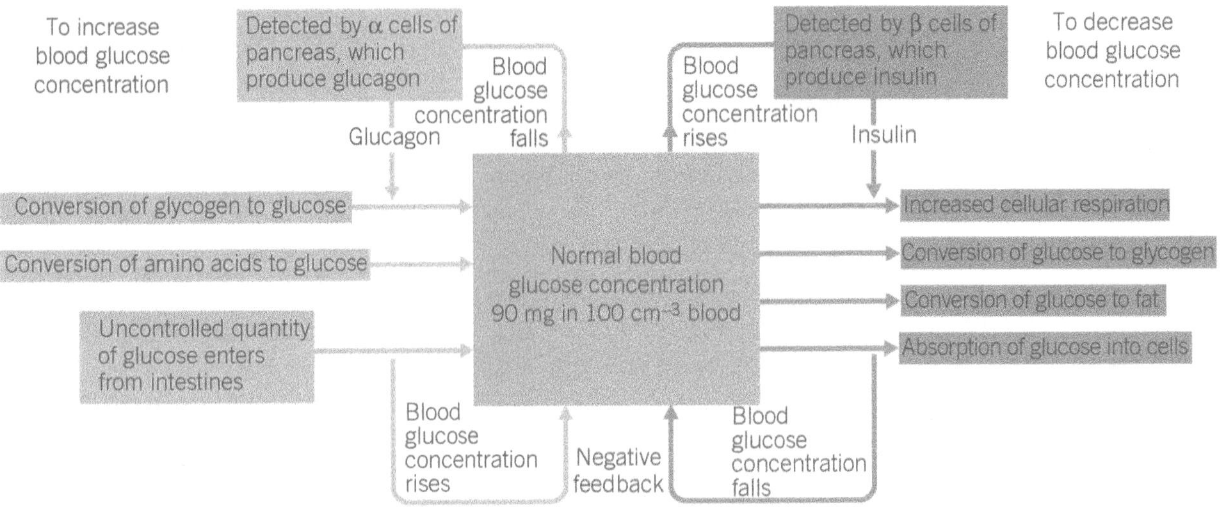

▲ **Figure 4** *Summary of regulation of blood glucose*

Summary question

1 In the following passage, state the most suitable word to replace the letters **a** to **n**.

The chemical energy in glucose is released by cells during the process known as **a**. It is therefore important that the blood glucose concentration is maintained at a constant level because if it falls too low cells are deprived of energy, and **b** cells are especially sensitive in this respect. If it gets too high **c** problems occur that may cause dehydration. Blood glucose is formed directly from **d** in the diet or from the breakdown of **e**, which is stored in the cells of the liver and **f**. The liver can also increase blood glucose concentration by making glucose from other sources, such as glycerol and **g**, in a process known as **h**. Blood glucose is used up when it is absorbed into cells, converted into fat or **i** for storage, or used up during **j** by cells. In order to maintain a constant level of blood glucose the pancreas produces two hormones from clusters of cells within it called **k**. The β cells produce the hormone **l**, which causes the blood glucose concentration to fall. The α cells produce the hormone **m**, which has the opposite effect. Another hormone, called **n**, can also raise blood glucose concentration.

27.5 Control of heart rate

Although we are not aware of it, much of the sensory information reaching our central nervous system comes from receptors within our bodies. All the internal systems of our bodies need to operate efficiently and be ready to adapt to meet the changing demands made upon them. This requires the coordination of a vast amount of information. This information comes from the monitoring of all our internal systems – a process that takes place continuously. Before investigating one example, how heart rate is controlled, let us first look at the part of the nervous system that is responsible for this type of control, the autonomic nervous system.

Learning objectives:

- → Describe what is meant by the autonomic nervous system.
- → Describe how the autonomic nervous system regulates heart rate.
- → Describe the role of chemical and pressure receptors in the regulation of heart rate.

Specification reference: 3.4.8

The autonomic nervous system

In Topic 24.3, you looked at how the motor neurones of the nervous system are organised into voluntary and autonomic nervous systems. Autonomic means self-governing. The autonomic nervous system controls the involuntary (subconscious) activities of internal muscles and glands. It has two divisions:

- **the sympathetic nervous system**. In general, this stimulates effectors and so speeds up any activity. It acts rather like an emergency controller. It stimulates effectors when we exercise strenuously or experience powerful emotions. In other words, it helps us to cope with stressful situations by heightening our awareness and preparing us for activity (the fight or flight response).
- **the parasympathetic nervous system**. In general, this inhibits effectors and so slows down any activity. It controls activities under normal resting conditions. It is concerned with conserving energy and replenishing the body's reserves.

The actions of the sympathetic and parasympathetic nervous systems normally oppose one another. In other words they are **antagonistic**. If one system contracts a muscle, then the other relaxes it. The activities of internal glands and muscles are therefore regulated by a balance of the two systems. Let us look at one such example, the control of heart rate.

Control of heart rate

In Topic 15.4 you saw how the **sinoatrial node (SAN)** and atrioventricular node (AVN) act to coordinate the beating of the atria and ventricles. Remember that cardiac muscle is myogenic – it does not actually need a nerve impulse to make it contract. The resting heart rate of a typical adult human is around 70 beats per minute. However, it is essential that this rate can be altered to meet varying demands for oxygen. During exercise, for example, the resting heart rate may need to more than double.

Changes to the heart rate are controlled by a region of the brain called the **medulla oblongata**. This has two centres:

- a centre that **increases heart rate**, which is linked to the SAN by the sympathetic nervous system
- a centre that **decreases heart rate**, which is linked to the SAN by the parasympathetic nervous system.

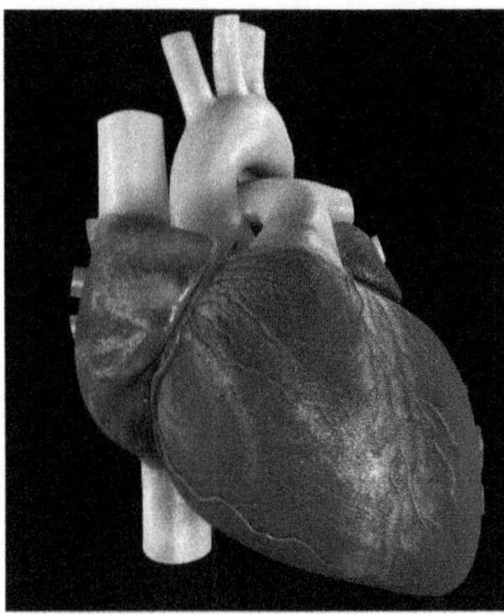

▲ **Figure 1** *The rate at which the human heart beats is controlled by receptors that measure the level of carbon dioxide and pressure of the blood*

Increased muscular/metabolic activity
↓
More carbon dioxide produced by tissues from increased respiration
↓
Blood pH is lowered
↓
Chemical receptors in the carotid arteries increase frequency of impulses to the medulla oblongata
↓
Centre in medulla oblongata that speeds heart rate increases frequency of impulses to SAN via the sympathetic nervous system
↓
SAN increases heart rate
↓
Increased blood flow removes carbon dioxide faster
↓
Carbon dioxide level returns to normal

▲ **Figure 2** *Effects of exercise on cardiac output*

Which of these centres is stimulated depends upon the information they receive from two types of receptor, which respond to one of the following:

- chemical changes in the blood
- pressure changes in the blood.

Control by chemoreceptors

Chemoreceptors are found in the wall of the carotid arteries (the arteries that serve the brain). They are sensitive to changes in the pH of the blood that result from changes in carbon dioxide concentration. In solution, carbon dioxide forms an acid and therefore lowers pH. The process of control works as follows:

- When the blood has a higher than normal concentration of carbon dioxide, its pH is lowered.
- The chemoreceptors in the wall of the carotid arteries and the aorta detect this and increase the frequency of nervous impulses to the centre in the medulla oblongata that increases heart rate.
- This centre increases the frequency of impulses via the sympathetic nervous system to the SAN which, in turn, increases the heart rate.
- The increased blood flow that this causes leads to more carbon dioxide being removed by the lungs and so the carbon dioxide level of the blood returns to normal.
- As a consequence the pH of the blood rises to normal and the chemoreceptors in the wall of the carotid arteries and aorta reduce the frequency of nerve impulses to the medulla oblongata.
- The medulla oblongata reduces the frequency of impulses to the SAN, which therefore decreases the heart rate to normal.

This process is summarised in Figure 2, which shows the sequence of events that follows changes in activity levels.

Control by pressure receptors

Pressure receptors occur within the walls of the carotid arteries and the aorta. They operate as follows:

- **When blood pressure is higher than normal,** they transmit a nervous impulse to the centre in the medulla oblongata that decreases heart rate. This centre sends impulses via the parasympathetic nervous system to the SAN of the heart, which decreases the rate at which the heart beats.
- **When blood pressure is lower than normal**, they transmit a nervous impulse to the centre in the medulla oblongata that increases heart rate. This centre sends impulses via the sympathetic nervous system to the SAN, which increases the rate at which the heart beats.

Figure 3 summarises the control of heart rate

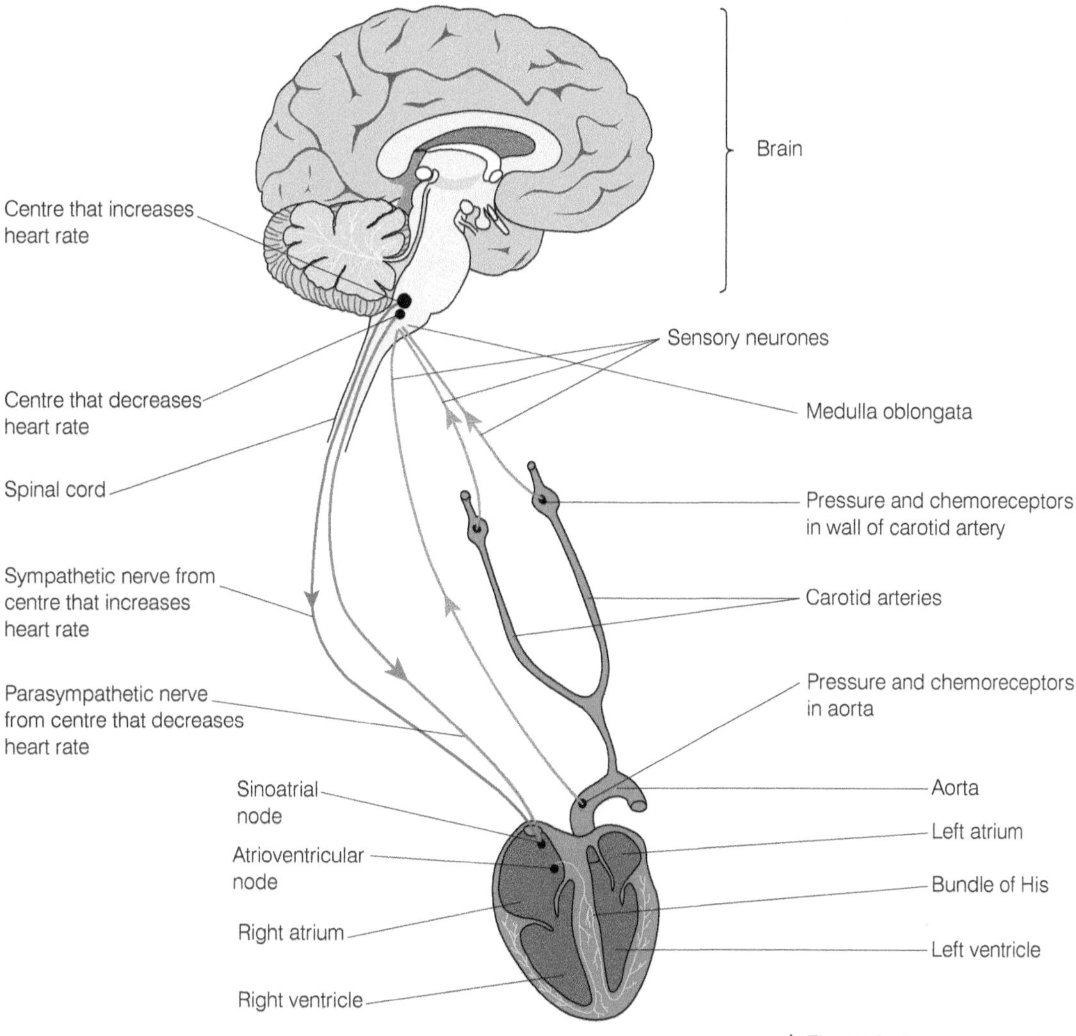

▲ **Figure 3** *Control of heart rate*

Summary questions

1. What is the function of the autonomic nervous system?
2. Distinguish between the functions of the sympathetic and parasympathetic nervous systems.
3. Suppose the parasympathetic nerve connections from the medulla oblongata to the sinoatrial node were cut. Suggest what might happen if a person's blood pressure increases above normal.
4. The nerve connecting the carotid artery to the medulla oblongata of a person is cut. This person then undertakes some strenuous exercise. Suggest what might happen to the person's:
   a heart rate
   b blood carbon dioxide concentration.

   Explain your answers.

1 The diagram shows a section through a mammalian heart as seen from the front.

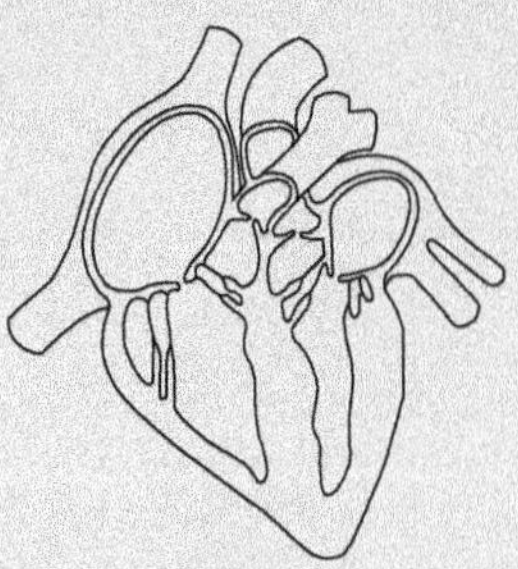

(a) Write the letter S on the diagram to show the position of the sinoatrial node. *(1 mark)*

(b) (i) Write the letter C on the diagram to show where ventricular contraction begins. *(1 mark)*

(ii) Explain the importance of the ventricles beginning to contract where you have indicated. *(2 marks)*

(c) Describe how the heart rate is increased during exercise. *(4 marks)*

AQA Jan 2007

2 Technicians in a hospital laboratory tested urine and blood samples from a girl with diabetes at intervals over a 1-year period. Each time the technicians tested her urine, they also measured her blood glucose concentration. Their results are shown in the graph.

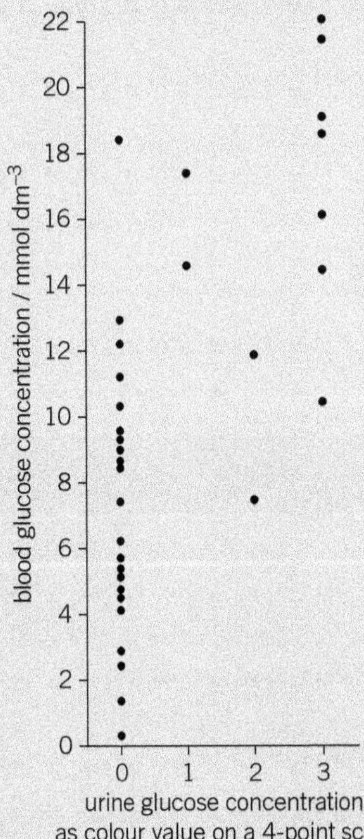

(a) (i) The girl who took part in this investigation was being successfully treated with insulin. The graph shows that on some occasions, the concentration of glucose in her blood was very high. Suggest why. *(2 marks)*

(ii) Use the graph to evaluate the use of the urine test as a measure of blood glucose concentration. *(3 marks)*

(b) Diabetic people who do not control their blood glucose concentration may become unconscious and go into a coma. A doctor may inject a diabetic person who is in a coma with glucagon. Explain how the glucagon would affect the person's blood glucose concentration. *(2 marks)*

AQA June 2010

3 **(a)** (i) What is meant by homeostasis? *(1 mark)*

(ii) Giving one example, explain why homeostasis is important in mammals. *(2 marks)*

(b) A person swimming in cold water may not be able to maintain their core body temperature and begins to suffer from hypothermia. Explain why a tall, thin swimmer is more likely to suffer from hypothermia than a short, stout swimmer of the same body mass. *(2 marks)*

(c) Cross-channel swimmers may suffer from muscle fatigue during which the contraction mechanism is disrupted. One factor thought to contribute to muscle fatigue is a decrease in the availability of calcium ions within muscle fibres. Explain how a decrease in the availability of calcium ions could disrupt the contraction mechanism in muscles. *(3 marks)*

AQA June 2006

4 The release of a substance called dopamine in some areas of the brain increases the desire to eat. Scientists measured increases in the release of dopamine in the brains of rats given different concentrations of sucrose solution to drink. Sucrose stimulates taste receptors on the tongue. The graph shows their results. Each point is the result for one rat.

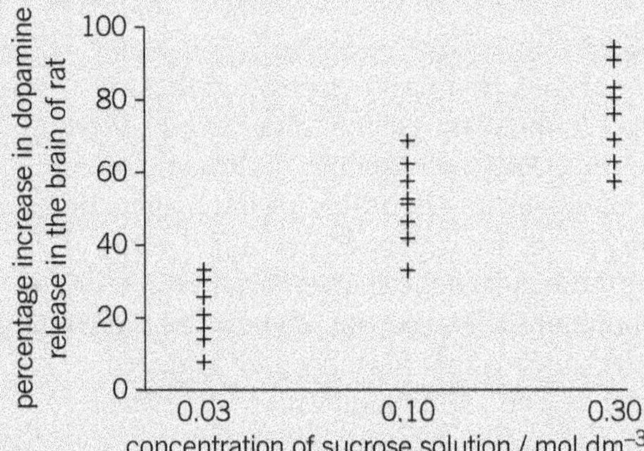

(a) The scientists concluded that drinking a sucrose solution had a positive feedback effect on the rats' desire to eat. How do these data support this conclusion? *(3 marks)*

(b) In this investigation, the higher the concentration of sucrose in a rat's mouth, the higher the frequency of nerve impulses from each taste receptor to the brain. If rats are given very high concentrations of sucrose solution to drink, the refractory period makes it impossible for information about the differences in concentration to reach the brain. Explain why. *(2 marks)*

(c) In humans, when the stomach starts to become full of food, receptors in the wall of the stomach are stimulated. This leads to negative feedback on the desire to eat. Suggest why this negative feedback is important. *(3 marks)*

AQA June 2013

28 Control of transcription and translation

28.1 Epigenetic control of gene expression

Learning objectives:

→ Explain what is meant by the term epigenetics.

→ Explain the role of DNA methylation in gene expression.

→ Explain the role of histone acetylation in gene expression.

→ Describe the effect of methylating genes involved in cancer development.

Specification reference: 3.4.9.1

As all DNA is replicated before a cell divides, then all cells contain copies of every gene present in the organisms genome. However, not every gene is expressed in every cell. You will learn more about this in Topic 28.3. But how are genes switched on and off? In Chapter 8 you learnt about the genetic code and how this is transcribed and then translated in cells to synthesise polypeptides. The enzyme RNA polymerase is responsible for transcription but not every gene present in the genome can necessarily be accessed by this enzyme. Frequently additional **transcription factors** are required and these too may be prevented from accessing the gene or just not present (see Topic 28.2). This means that not every gene is **expressed**.

Epigenetic changes

Epigenetics is the study of changes in gene expression that are not the result of changes to the gene itself. In other words, there has been no change to the DNA base sequence of the gene but in some cells the gene is expressed and in others it isn't. Epigenetic changes modify the way genes are expressed over the lifetime of the organism. Many of these epigenetic changes occur as a result of changes in the environment of the organism. So how are these changes brought about? Two mechanisms that have been investigated extensively are DNA methylation and modification of histones.

Study tip

It would be useful to remind yourself about DNA organisation in cells and the role of histones, which you covered in Topic 8.3.

Methylation of DNA

It is possible to add **methyl** groups (—CH_3) to molecules of DNA. Most commonly these are attached to cytosine bases, particularly when these are adjacent to guanine bases. This is known as a **CpG site**. Methylation tends to prevent the binding of transcription factors and RNA polymerases and hence methylation tends to prevent gene expression. Figure 1 shows chromosomes from identical twins aged 3 years and aged 50 years. At 50 years there are far more differences in the methylation pattern between the twins. Environmental differences such as diet, smoking, and exercise have led, over time, to epigenetic differences between the twins, even though they both have identical copies of the genomic DNA.

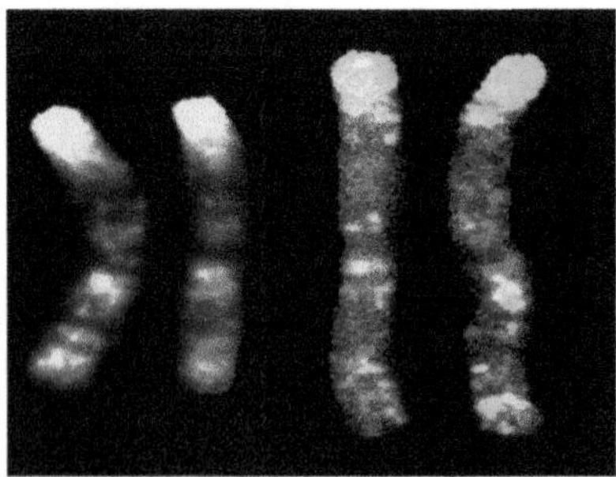

▲ **Figure 1** *The chromosomes on the left belong to 3-year-old identical twins – one chromosome from each twin. The chromosomes on the right belong to 50-year-old twins. Methylation differences are highlighted in red. Yellow means no epigenetic differences.*

Acetylation of histones

You saw in Topic 8.3 that, in eukaryotic cells, DNA is packaged and coiled around proteins called histones. Like DNA, histones can be chemically modified by the addition of **acetyl** groups. The addition of acetyl groups makes the DNA more accessible for transcription whereas decreased acetylation means less expression or even switching genes 'off'. Figure 2 summarises the effects of DNA methylation and histone acetylation on gene expression.

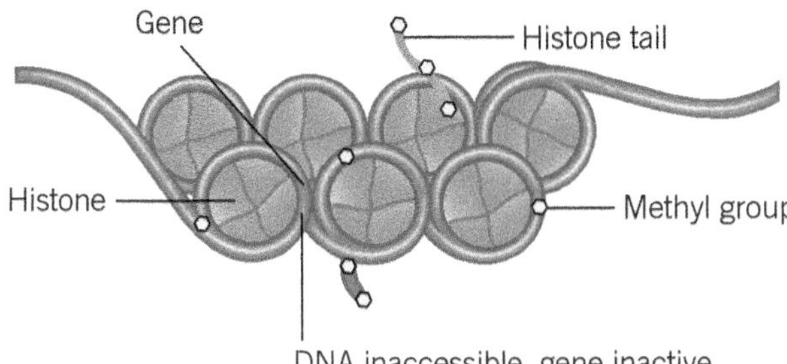

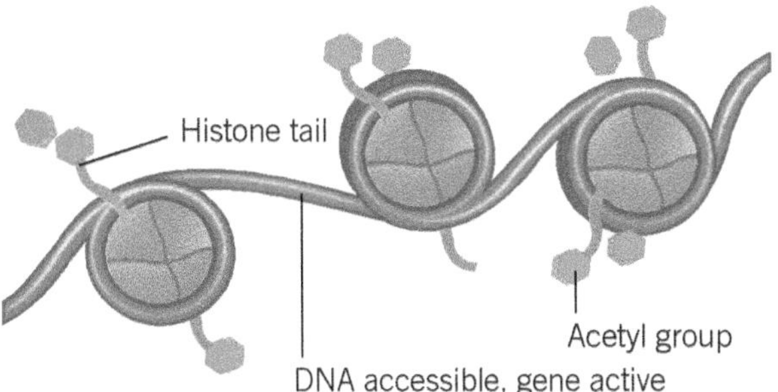

▲ **Figure 2** *Epigenetic changes to DNA and histones can alter gene expression*

Epigenetic modification and cancer

In healthy cells, the cell cycle is carefully regulated. In Topic 17.3, you learnt about the role of tumour suppressor genes and proto-oncogenes in the control of the cell cycle. Mutations in these genes can lead to the development of cancer. However we now know that epigenetic changes to DNA are also found in many types of cancer. Some of these changes are found in the **promotor** regions of tumour suppressor genes and proto-oncogenes. Promotors are regions close to the start of a gene where RNA polymerase will bind to initiate transcription of the gene. Too much methylation, or **hypermethylation**, of promotors for tumour suppressor genes means that these will not be expressed. Cell division will be unregulated leading to cancer. In Topic 17.3 you were told that mutations in proto-oncogenes can lead to them becoming oncogenes. Too little methylation, or **hypomethylation**, of promotors for oncogenes can lead to them being overexpressed, resulting in a more aggressive cancer.

Epigenetic therapies for cancer treatment

An understanding of how genes are switched on and switched off has opened the door to the potential for epigenetic therapies for cancers. The enzymes responsible for methylation of DNA (DNA methyltransferases) have been identified and some drugs that inhibit these enzymes (DNMTi) are now approved for use in cancer therapy. It is now possible to reactivate silenced tumour suppressor genes. The inhibitors bind to the enzyme permanently, making the active site unavailable. This means there is less active enzyme and, over successive rounds of cell division, less methylation occurs. Eventually the DNA in the cells shows the normal pattern of methylation and gene expression is restored. As shown in Figure 3, the cells, for example, white blood cells, go on to develop normally, with any abnormal cells being destroyed by programmed cell death(**apoptosis**).

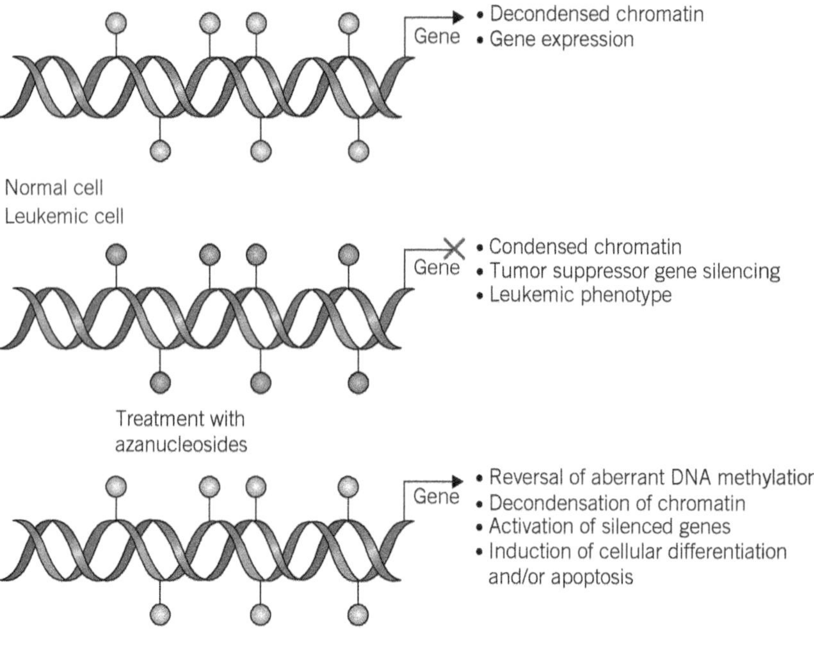

▲ **Figure 3** *In cells with a normal methylation pattern, the tumour suppressor gene is expressed. In leukaemia, abnormal methylation means that the DNA cannot be transcribed. DNMTi reverses the abnormal methylation, re-activating the genes.*

Summary questions

1 Explain what is meant by the term epigenetics.

2 In a CpG site, what is represented by 'p'?

3 Compare the effects of DNA methylation and histone acetylation on gene expression.

4 Explain the role of RNA polymerase in gene expression.

28.2 Hormones and RNA interference

In Topic 28.1, you saw that for genes to be expressed, RNA polymerase and transcription factors are necessary.

Hormones such as oestrogen can switch on a gene and thus start transcription by combining with a receptor on the transcription factor. This releases the inhibitor molecule. The process is illustrated in Figure 1 and is described below.

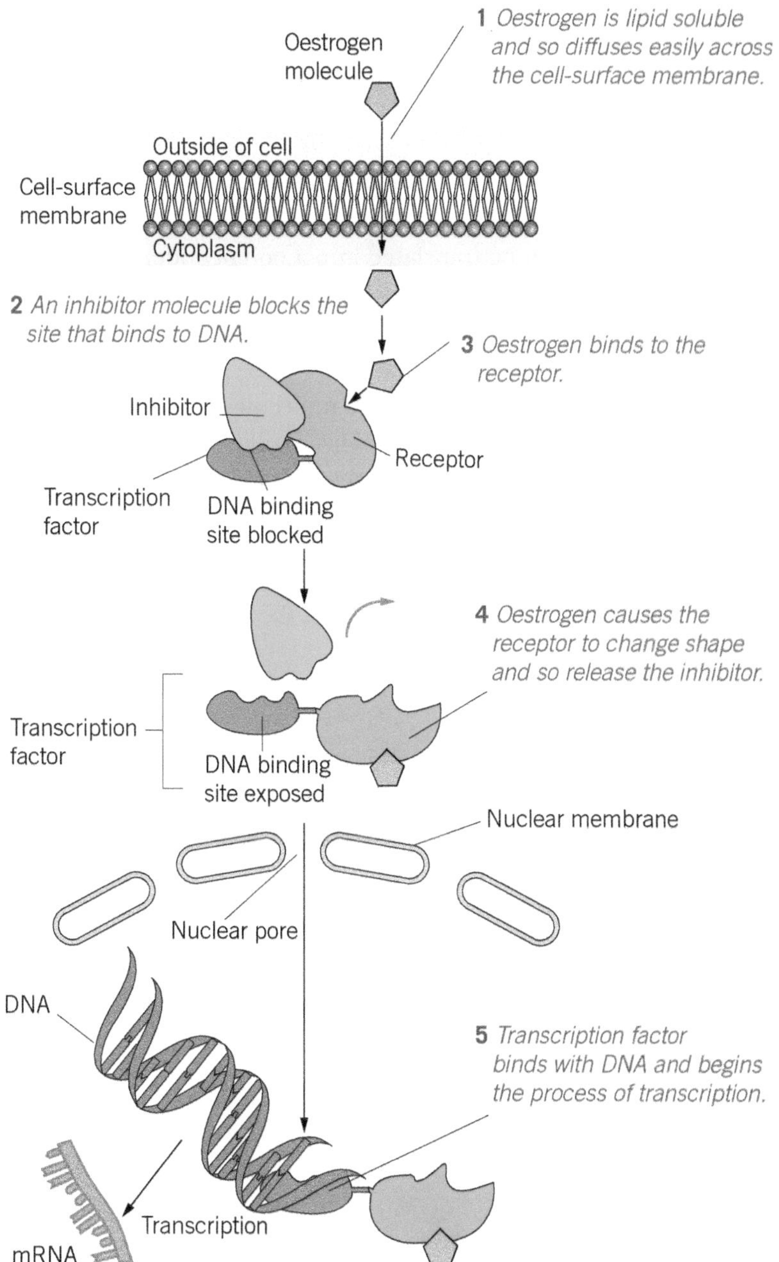

▲ **Figure 1** *The effect of oestrogen on gene transcription*

- Oestrogen is a lipid-soluble molecule and therefore diffuses easily through the **phospholipid** portion of cell-surface membranes.

Learning objectives:

- → Describe the role of transcription factors in regulation of gene expression.
- → Describe the effect of microRNA on gene expression.

Specification reference: 3.4.9.2

Synoptic link

The attachment of oestrogen to a receptor causes changes in the shape of the receptor in the same way as the attachment of a non-competitive inhibitor to an enzyme molecule changes its shape and also its active site. This involves the same basic mechanism, which is described in Topic 3.3. You are not required to learn about the action of oestrogen. The example is included as it illustrates how transcription factors themselves work.

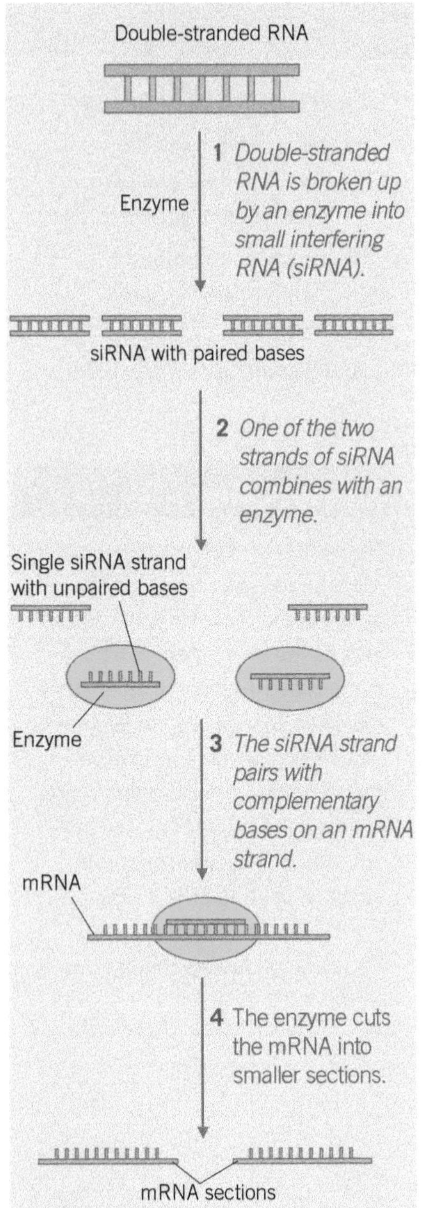

▲ **Figure 2** *The effect of siRNA on gene expression*

- Once inside the cytoplasm of a cell, oestrogen combines with a site on a receptor molecule of the transcription factor. The shape of this site and the shape of the oestrogen molecule complement one another (Figure 1, stage 3).
- By combining with the site, the oestrogen changes the shape of the receptor molecule. This change of shape releases the inhibitor molecule from the DNA binding site on the transcription factor (Figure 1, stage 4).
- The transcription factor can now enter the nucleus through a nuclear pore and combine with DNA (Figure 1, stage 5).
- The combination of the transcription factor with DNA stimulates transcription of the portion of DNA gene that makes up the gene (Figure 1, stage 5).

The effect of siRNA on gene expression

Gene expression can be prevented by breaking down messenger RNA before its genetic code can be translated into a polypeptide. Essential to this process are small double-stranded sections of RNA called **small interfering RNA (siRNA)**. The process (see Figure 2) operates as follows:

- An enzyme cuts large double-stranded molecules of RNA into smaller sections of siRNA (Figure 2, stage 1).
- One of the two siRNA strands combines with an enzyme (Figure 2, stage 2).
- The base sequence of a particular siRNA binds to a complementary base sequence on a messenger RNA (mRNA) molecule transcribed from a specific gene, so effectively guiding the enzyme into position (Figure 2, stage 3).
- Once in position the enzyme cuts the mRNA into smaller sections (Figure 2, stage 4).
- The mRNA is no longer capable of being translated into a polypeptide.
- This means that the gene has not been expressed, that is, it has been blocked.

The siRNA has a number of potential scientific and medical uses:

- It could be used to identify the role of genes in a biological pathway. Some siRNA that blocks a particular gene could be added to cells. By observing the effects (or lack of them) we could determine what the role of the blocked gene is.
- As some diseases are caused by genes, it may be possible to use siRNA to block the expression of the genes and so prevent the disease.

Cancer – the 'two-hit' hypothesis

You saw in Topic 17.3 that tumours can develop as a result of a mutation in proto-oncogenes that causes cells to divide more rapidly than normal. Tumours can also develop by a mutation in tumour suppressor genes that prevents them from inhibiting cell division. It only takes a single mutated allele to activate proto-oncogenes but it takes a mutation of both alleles to inactivate tumour suppressor genes (two hits). As natural mutation rates are slow, it takes a considerable time for both tumour suppressor alleles to mutate. This explains why the risk of many cancers increases as one gets older. It is thought that some people are born with one mutated allele. These people are at a greater risk of cancer as they need only one further mutation, rather than two, to develop the disease. This explains why certain cancers carry an inherited increased risk.

1 Explain why a doctor may enquire about a patient's family medical history before deciding on using X-ray analysis for a condition other than cancer.
2 Suggest a reason why a single mutant allele of a proto-oncogene can cause cancer, but it requires two mutant alleles of the tumour suppressor gene to do so.
3 One experimental treatment for cancer involves introducing tumour suppressor genes into rapidly dividing cells in order to arrest tumour growth. Explain how this treatment might work.
4 Another experimental treatment is the development of an anticancer drug that will destroy certain protein receptors on membranes of cancer cells. Explain how this treatment might be effective.

Gene expression in haemoglobin

You saw in Topic 5.7 that a haemoglobin molecule is made up of four polypeptide chains. Each is known as a globulin. In adult humans two of the polypeptides in a haemoglobin molecule are alpha-globulin and two are beta-globulin. In other words, 50 per cent of the total globulin in all haemoglobin is alpha and 50 per cent is beta. In a human fetus, however, the haemoglobin is different, with much of the beta-globulin being replaced by a third type, gamma-globulin.

Fetal haemoglobin has a greater affinity for oxygen than adult haemoglobin. The changes in the production of the three types of globulin during early human development are shown in Figure 3.

Summary questions

1 What is the role of a transcription factor?
2 Describe how oestrogen stimulates the expression of a gene.
3 One of the two strands of siRNA combines with an enzyme and guides it to an mRNA molecule, which it then cuts. Explain why the mRNA is unlikely to be cut if the other siRNA strand combines with the enzyme.

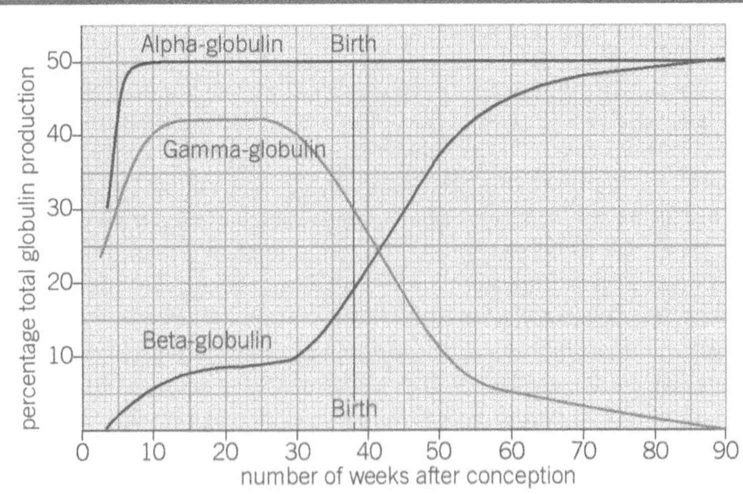

▲ **Figure 3** *Percentage total globulin production during early human development*

Humans have genes that code for the production of all three types of globulin. The production of the different haemoglobins depends upon which gene is expressed. The expression of these genes changes at different times during development.

1. Suggest an advantage of fetal haemoglobin having a greater affinity for oxygen than adult haemoglobin.
2. At birth, what percentage of the total globulin production is of each globulin type?
3. Describe the changes in gene expression that occur at 25 weeks.
4. Outline **two** possible explanations for the change in the expression of the gene for gamma-globulin after 25 weeks.
5. Sickle cell disease is the result of a mutant form of haemoglobin. In Saudia Arabia and India, some individuals have high levels of fetal haemoglobin in their blood, even as adults. When these individuals have sickle cell disease, their symptoms are much reduced. Suggest how controlling the expression of the genes for globulin might provide a therapy for sickle cell disease.

28.3 Totipotency and cell specialisation

All cells contain the same genes. Every cell is therefore capable of making everything that the body can produce. A cell in the lining of the small intestine has the gene coding for insulin just as a β cell of the pancreas has the gene coding for maltase. So why do the cells of the small intestine produce maltase rather than insulin and β cells of the pancreas produce insulin rather than maltase? The answer is that, although all cells contain all genes, only certain genes are expressed (switched on) in any one cell at any one time.

Some genes are permanently expressed (switched on) in all cells. For example, the genes that code for essential chemicals, such as the enzymes involved in respiration, are expressed in all cells. In some cells, other genes are permanently not expressed (switched off), for example, the gene for insulin in cells lining the small intestine. Further genes are switched on and off as and when they are needed. In this chapter you will learn how the expression of genes is controlled.

Differentiated cells differ from each other, often visibly so. This is mainly because they each produce different proteins. The proteins that a cell produces are coded for by the genes it possesses or, more accurately, by the genes that are expressed (switched on).

An organism develops from a single fertilised egg. A fertilised egg clearly has the ability to give rise to all types of cells. Cells such as fertilised eggs, which can mature into any and all body cell types including the extra-embryonic membranes, are known as **totipotent cells**. Totipotent cells can give rise to an entire new organism. The early cells that are derived from the fertilised egg are also totipotent. These later differentiate and become specialised for a particular function. For example, mesophyll cells become specialised for photosynthesis in plants and muscle cells in animals become specialised for contraction. This is because, during the process of cell specialisation, only some of the genes are expressed. This means that only part of the DNA of a cell is translated into proteins. The cell therefore only makes those proteins that it requires to carry out its specialised function. Although it is still capable of making all the other proteins, these are not needed and so it would be wasteful to produce them. Therefore, the genes for these other proteins are not expressed. The ways in which genes are prevented from expressing themselves include:

- preventing transcription and hence preventing the production of mRNA
- breaking down mRNA before its genetic code can be translated.

Details of these mechanisms were given in Topics 28.1 and 28.2.

If specialised cells still retain all the genes of the organism, can they still develop into any other cell? The answer is – it depends. There are no hard and fast rules. Xylem vessels, which transport water in plants, and red blood cells, which carry oxygen in animals, are so specialised that they lose their nuclei once they are mature. As the nucleus contains the genes, then clearly these cells cannot develop into other cells. In fact, specialisation is irreversible in most animal cells. Once cells have matured and specialised they can no longer develop into other cells or, in other words, they lose their totipotency. Only a few **multipotent** cells exist in mature animals. These are called **adult stem cells**. Multipotent cells can differentiate into some different cells but not others.

Learning objectives:

- → Explain what is meant by totipotency.
- → Describe the location of totipotent cells in different organisms.
- → Describe how cells lose their topipotency and become specialised.
- → Discuss the use of totipotent stem cells in the treatment of human disorders.

Specification reference: 3.4.9.3

Synoptic link

As a starting point for this topic, it would be useful to revise Topic 2.4.

Hint

Differentiation results from differential gene expression.

Adult stem cells are undifferentiated dividing cells that occur in adult animal tissues and need to be constantly replaced. They are found in the inner lining of the small intestine, in the skin, and also in the bone marrow, which produces red and white blood cells. Under certain conditions, adult stem cells can develop into any other types of cell. As a result they could theoretically be used to treat a variety of genetic disorders, such as the blood diseases thalassaemia and sickle cell anaemia. Unipotent cells are also found in animal tissues. As the name suggest these divide giving rise to just a single type of cell.

In addition to the adult stem cells found in mature organisms, stem cells also occur at the earliest stage of the development of an embryo, before the cells have differentiated. These are called **embryonic stem cells**. Embryonic stem cells are **pluripotent**. They can differentiate into most tissues present in the organism but not into the extra-embryonic membranes – so not into an entire new organism.

The situation in plants is different. Mature plants have many totipotent cells. Under the right conditions, many plant cells can develop into any other cell. For example, if we take a cell from the root of a carrot, place it in a suitable nutrient medium, and give it certain chemical stimuli at the right time, we can develop a complete new carrot plant. Growing cells outside of a living organism in this way is called *in vitro* development. Since this new carrot plant is genetically identical to the one from which the single root cell came, it is a **clone**. Cells from most plant species can be used to clone new plants in this way.

Summary questions

1. What are totipotent cells?
2. How does the distribution of totipotent cells in animals differ from that in plants?
3. All cells possess the same genes and yet a skin cell can produce the protein keratin but not the protein myosin, whilst a muscle cell can produce myosin but not keratin. Explain why.

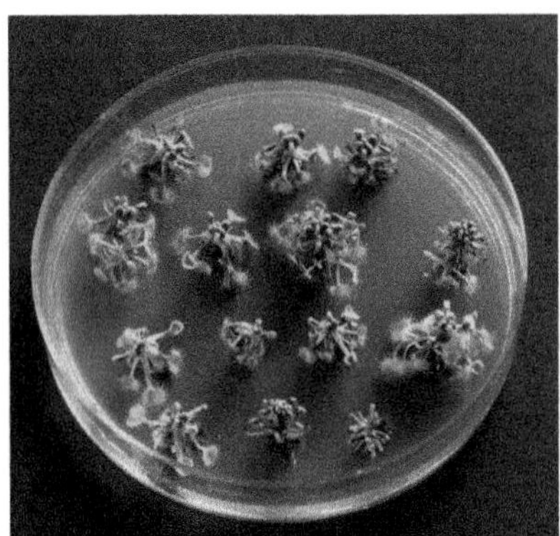

▲ **Figure 1** *Plants growing from tissue cultures in a petri dish*

Growth of plant tissue cultures

There are many factors that influence the growth of plant tissue cultures from totipotent cells. One group consists of plant growth factors, which are chemicals involved in the growth and development of plant tissues. Plant growth factors have a number of features:

- They have a wide range of effects on plant tissues.
- The effects on a particular tissue depend upon the concentration of the growth factor.
- The same concentration affects different tissues in different ways.
- The effect of one growth factor can be modified by the presence of another.

An experiment was carried out to investigate the effects on the development of a plant tissue culture of three growth factors: cytokinin, IAA, and 2,4-D. Samples of totipotent plant cells were grown on a basic growth medium in a series of test tubes. Each test tube contained a mixture of the three growth factors in different concentrations. After 2 weeks, the tubes were observed to see the effects of the growth factor mixtures on shoot and root growth. The results are shown in Table 1.

▼ **Table 1** *Effect of growth factors on shoot and root development*

Tube no.	Relative concentration of growth factors			Shoot development	Root development
	Cytokinin	IAA	2,4-D		
1	None	Low	None	Moderate	Little
2	Low	High	None	Extensive	Little
3	High	Low	None	Little	Moderate
4	None	High	High	Extensive	Extensive
5	None	None	None	Very little	Very little

1 Name the process by which the totipotent cells of the plant tissue culture change in appearance and develop into shoot or root cells.
2 From Table 1, state which **two** growth factors together produced the greatest development of both shoots and roots.
3 Describe one piece of evidence from Table 1 that supports the view that 'the effects of one growth factor can be modified by another'.

Human embryonic stem cells and the treatment of disease

Although there are a number of types of stem cell in the human body, it is the first few cells from the division of the fertilised egg that have the greatest potential to treat human diseases. As they come from the early stages of an embryo, they are called human embryonic stem cells. These cells can be grown *in vitro* and then induced to develop into a wide range of different human tissues. The process is illustrated in Figure 2.

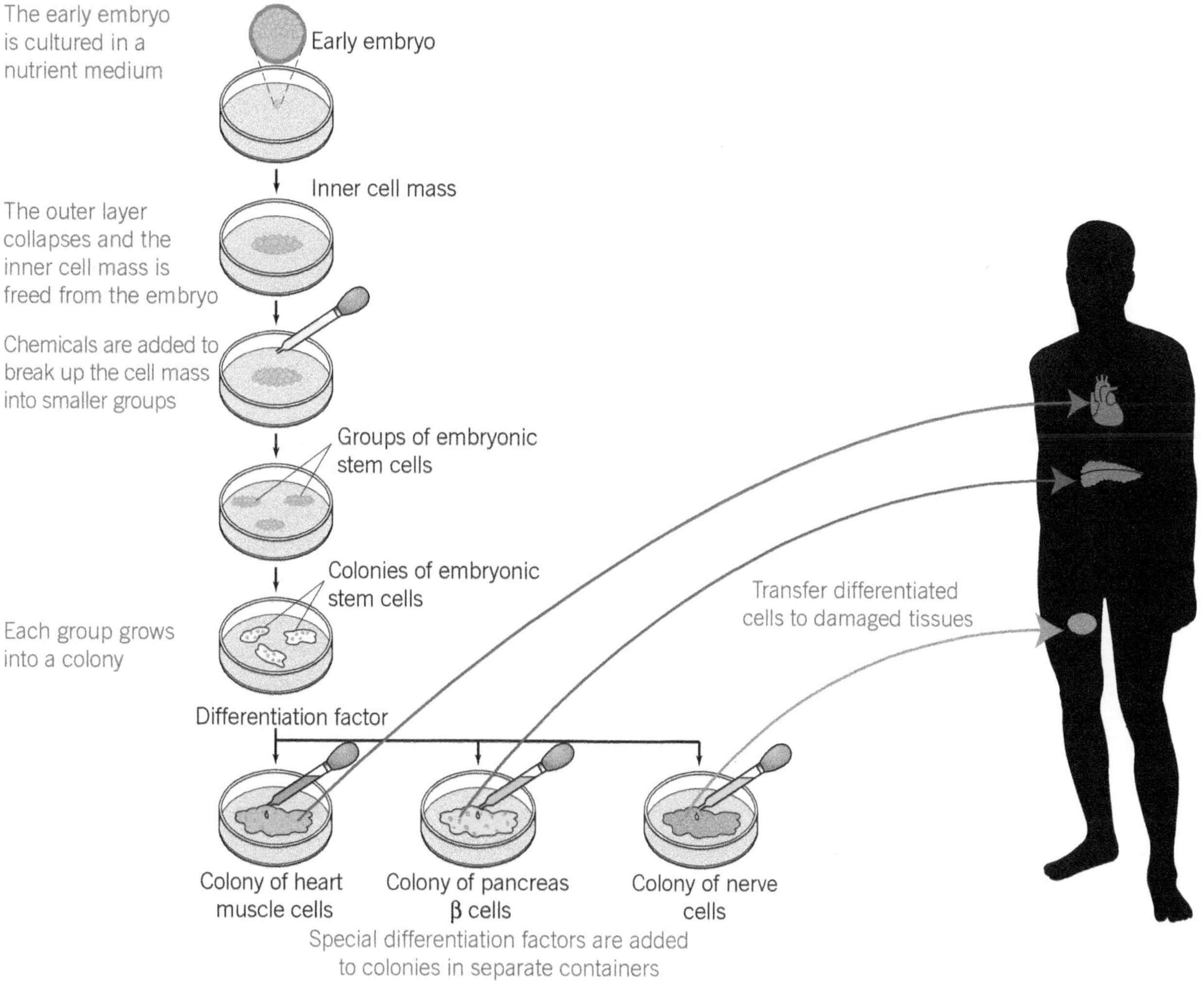

The early embryo is cultured in a nutrient medium
Early embryo
Inner cell mass
The outer layer collapses and the inner cell mass is freed from the embryo
Chemicals are added to break up the cell mass into smaller groups
Groups of embryonic stem cells
Colonies of embryonic stem cells
Each group grows into a colony
Transfer differentiated cells to damaged tissues
Differentiation factor
Colony of heart muscle cells
Colony of pancreas β cells
Colony of nerve cells
Special differentiation factors are added to colonies in separate containers

▲ **Figure 2** In vitro *culturing of human embryonic stem cells*

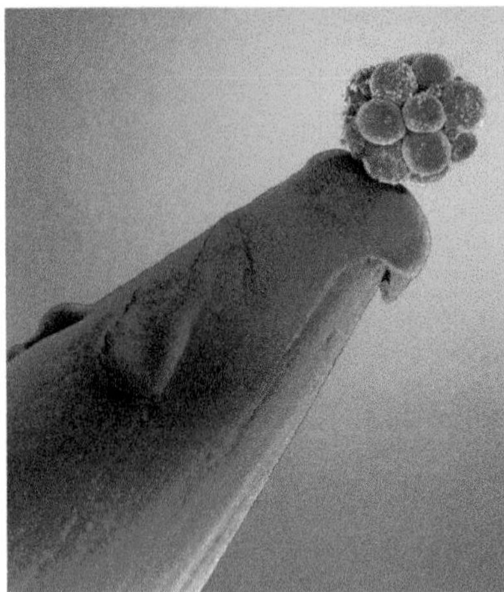

▲ **Figure 3** *Scanning electron micrograph (SEM) of a 3-day-old human embryo at the 16-cell stage on the tip of a pin*

▼ **Table 2** *Potential uses of human cells produced from stem cells*

Type of cell	Disease that could be treated
Heart muscle cells	Heart damage, e.g., as a result of a heart attack
Skeletal muscle cells	Muscular dystrophy
β cells of the pancreas	Type I diabetes
Nerve cells	Parkinson's disease, multiple sclerosis, strokes, Alzheimer's disease, paralysis due to spinal injury
Blood cells	Leukaemia, inherited blood diseases
Skin cells	Burns and wounds
Bone cells	Osteoporosis
Cartilage cells	Osteoarthritis
Retina cells of the eye	Macular degeneration

There are many possible uses of cells grown in this manner. The cells can be used to regrow tissues that have been damaged in some way, either by accident (e.g., skin grafts for serious burn damage) or as a result of disease (e.g., neurodegenerative diseases, such as Parkinson's disease). Table 2 lists some of the potential uses of human cells produced from stem cells. At present, embryonic stem cell research is only allowed in the UK under licence and specified conditions. These conditions include its use as a means of increasing knowledge about embryo development and serious diseases, including their treatment. Embryos used in this type of research are obtained from *in vitro* fertilisation. The process nevertheless presents a number of ethical issues.

One issue surrounds the argument as to whether a human embryo less than 14 days old should be afforded the same respect as a fetus or an adult person. Some people feel that using embryos in this way undermines our respect for human life and could progress to the use of fetuses, and even newborn babies, for research or the treatment of disease. They feel that it is a further move towards reproductive cloning and, even if this remains illegal in the UK, the information gained could be used to clone humans elsewhere. Others disagree, arguing that an embryo at such an early stage of development is just a ball of identical, undifferentiated cells, bearing no resemblance to a human being. They feel that the laws prohibiting cloning, in the UK and elsewhere, provide sufficient protection.

Supporters of human embryonic stem cell research contend that it is wrong to allow human suffering to continue when there is a possibility of alleviating it. They further argue that, since embryos are produced for other purposes, for example, fertility treatments, it makes no sense to destroy superfluous embryos that could be used in research. Opponents of embryonic stem cell research contend that it is wrong to use humans, including human embryos, as a means to an end, even if that end is the laudable one of alleviating human suffering.

However, human embryos are not the only source of stem cells. For example, they can be obtained from the bone marrow of adult humans. As long as a person gives consent, this source of stem cells raises no real ethical issues. At present, these cells have far more restricted medical applications but scientists hope, in time, to be able to make them behave more like embryonic stem cells.

1 Write **two** accounts, each of around 200 words, evaluating the case **a** for, and **b** against, the continued use of embryos for stem cell research.

Practice questions: Chapter 28

1 Plant physiologists attempted to produce papaya plants using tissue culture. They investigated the effects of different concentrations of two plant growth factors on small pieces of the stem tip from a papaya plant. Their results are shown in the table. Callus is a mass of undifferentiated plant cells. Plantlets are small plants

Concentration of auxin / $\mu mol\ dm^{-3}$	Concentration of cytokinin / $\mu mol\ dm^{-3}$		
	5	25	50
0	No effect	No effect	Leaves produced
1	No effect	Leaves produced	Leaves produced
5	No effect	Leaves produced	Leaves and some plantlets produced
10	Callus produced	Leaves and some plantlets produced	Plantlets produced
15	Callus produced	Callus and some leaves produced	Callus and some leaves produced

(a) Explain the evidence from the table that cells from the stem tip are totipotent. *(2 marks)*

(b) Calculate the ratio of cytokinin to auxin that you would recommend to grow papaya plants by this method. *(2 marks)*

(c) Papaya plants reproduce sexually by means of seeds. Papaya plants grown from seeds are very variable in their yield. Explain why. *(2 marks)*

AQA June 2011

2 Human immunodeficiency virus (HIV) particles have a specific protein on their surface. This protein binds to a receptor on the plasma membrane of a human cell and allows HIV to enter. This HIV protein is found on the surface of human cells after they have become infected with HIV. Scientists made siRNA to inhibit expression of a specific HIV gene inside a human cell. They attached this siRNA to a carrier molecule. The flow chart shows what happens when this carrier molecule reaches a human cell infected with HIV.

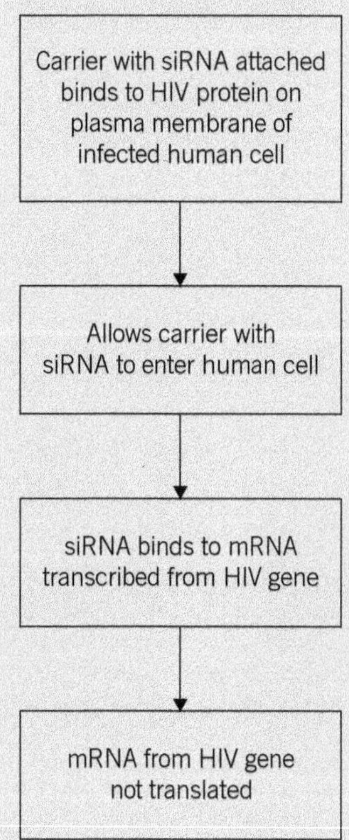

(a) When siRNA binds to mRNA, name the complementary base pairs holding the siRNA and mRNA together. One of the bases is named for you.

.... with ....

.... Adenine .... with .... *(1 mark)*

(b) This siRNA would only affect gene expression in cells infected with HIV. Suggest **two** reasons why. *(4 marks)*

(c) The carrier molecule on its own may be able to prevent the infection of cells by HIV. Explain how. *(2 marks)*

AQA June 2013

3 The figure shows one way in which methylation of DNA prevents the translation of a particular gene. A CpG island (step 1) consists of a number of cytosine-phosphate-guanine sequences.

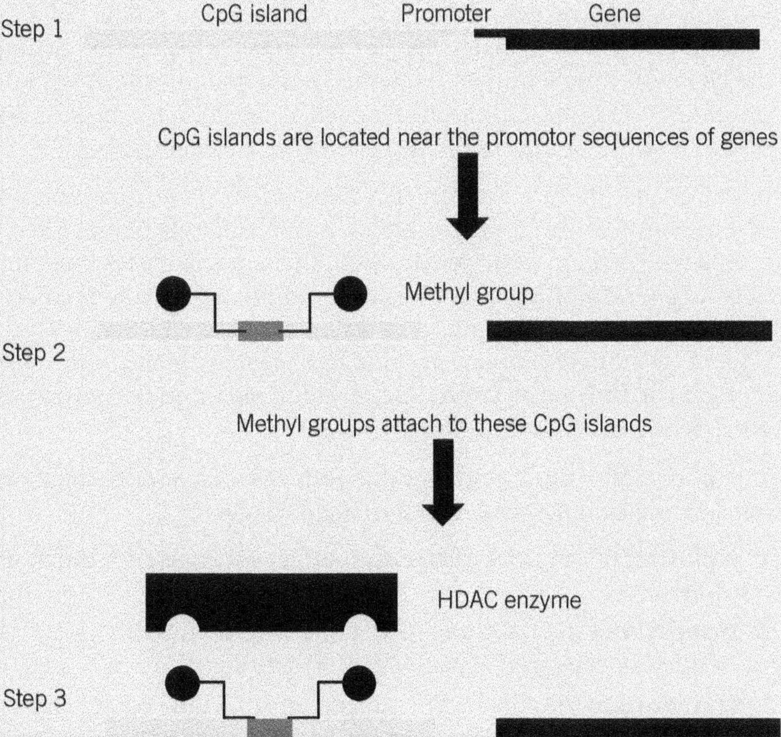

The enzyme HDAC binds, compresses the shape of the DNA, and prevents the gene from being transcribed.

(a) Describe the difference between a cytosine-phosphate-guanine sequence and a cytosine-guanine base pair. *(3 marks)*

(b) Explain why the enzyme HDAC only binds to the methylated CpG islands (step 3). *(3 marks)*

(c) Explain why tumour suppressor genes may prevent cancer. *(1 mark)*

(d) Use the figure to explain why HDAC inhibitors may be useful in treating cancer. *(3 marks)*

Oxford Int AQA specimen paper 4

Answers to the Practice Questions are available at
www.oxfordsecondary.com/oxfordaqaexams-alevel-biology

29 Recombinant DNA technology
29.1 Producing DNA fragments

Learning objectives:

- → Describe how complementary DNA is made using reverse transcriptase.
- → Describe how restriction endonucleases are used to cut DNA into fragments.

Specification reference: 3.4.10.1 and 3.4.10.2

Perhaps the most significant scientific advance in recent years has been the development of technology that allows genes to be manipulated, altered, and transferred from organism to organism – even to transform DNA itself. These techniques have enabled us to better understand how organisms work and to design new industrial processes and medical applications.

A number of human diseases result from individuals being unable to produce various metabolic chemicals for themselves. Many of these chemicals are proteins, such as insulin. They are therefore the product of a specific portion of DNA, that is, the product of a gene. Treatment of such deficiencies previously involved extracting the chemical from a human or animal donor and introducing it into the patient. This presents problems such as rejection by the immune system and risk of infection. The cost is also considerable.

It follows that there are advantages in producing large quantities of 'pure' proteins from other sources. As a result, techniques have been developed to isolate genes, clone them, and transfer them into microorganisms. These microorganisms are then grown to provide a 'factory' for the continuous production of a desired protein. The DNA of two different organisms that has been combined in this way is called **recombinant DNA**. The resulting organism is known as a **genetically modified organism** (**GMO**).

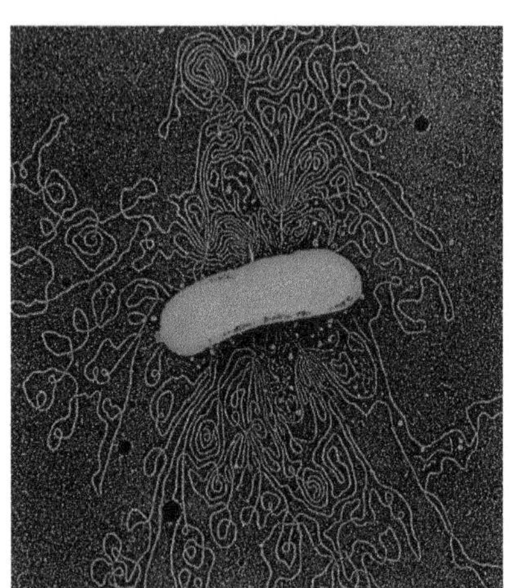

▲ **Figure 1** *An* Escherichia coli *bacterial cell that has been treated so that its DNA is ejected*

The process of making a protein using the DNA technology of gene transfer and cloning involves a number of stages:

1. **isolation** of the DNA fragments that have the gene for the desired protein
2. **insertion** of the DNA fragment along with promotor and terminator regions if appropriate into a vector
3. **transformation**, that is, the transfer of DNA into suitable host cells
4. **identification** of the host cells that have successfully taken up the gene by use of gene markers
5. **growth/cloning** of the population of host cells.

Let us consider each stage in detail.

Before a gene can be transplanted, it must be identified and isolated from the rest of the DNA. Given that the required gene may consist of a sequence of a few hundred bases amongst the many millions in human DNA, this is no small feat! Two of the methods employed use enzymes that have important roles in microorganisms: reverse transcriptase and restriction endonucleases.

Using reverse transcriptase

Retroviruses are a group of viruses of which the best known is human immunodeficiency virus (HIV). The genetic information of retroviruses is in the form of RNA. However, they are able to synthesise DNA from their RNA using an enzyme called reverse transcriptase. It is so named because it catalyses the production of DNA from RNA, which is the reverse of the more usual transcription of RNA from DNA. The process of using reverse transcriptase to isolate a gene is illustrated in Figure 2 and described below:

- A cell that readily produces the protein is selected (e.g., the β cells of the islets of Langerhans from the pancreas are used to produce insulin).
- These cells have large quantities of the relevant mRNA, which is therefore extracted.
- Reverse transcriptase is then used to make DNA from RNA. This DNA is known as **complementary DNA (cDNA)** because it is made up of the nucleotides that are complementary to the mRNA.
- To make the other strand of DNA, the enzyme DNA polymerase is used to build up the complementary nucleotides on the cDNA template. This double strand of DNA is the required gene.

Synoptic link

DNA polymerase acts in the same way when forming the second DNA strand during DNA replication, as described in Topic 8.4.

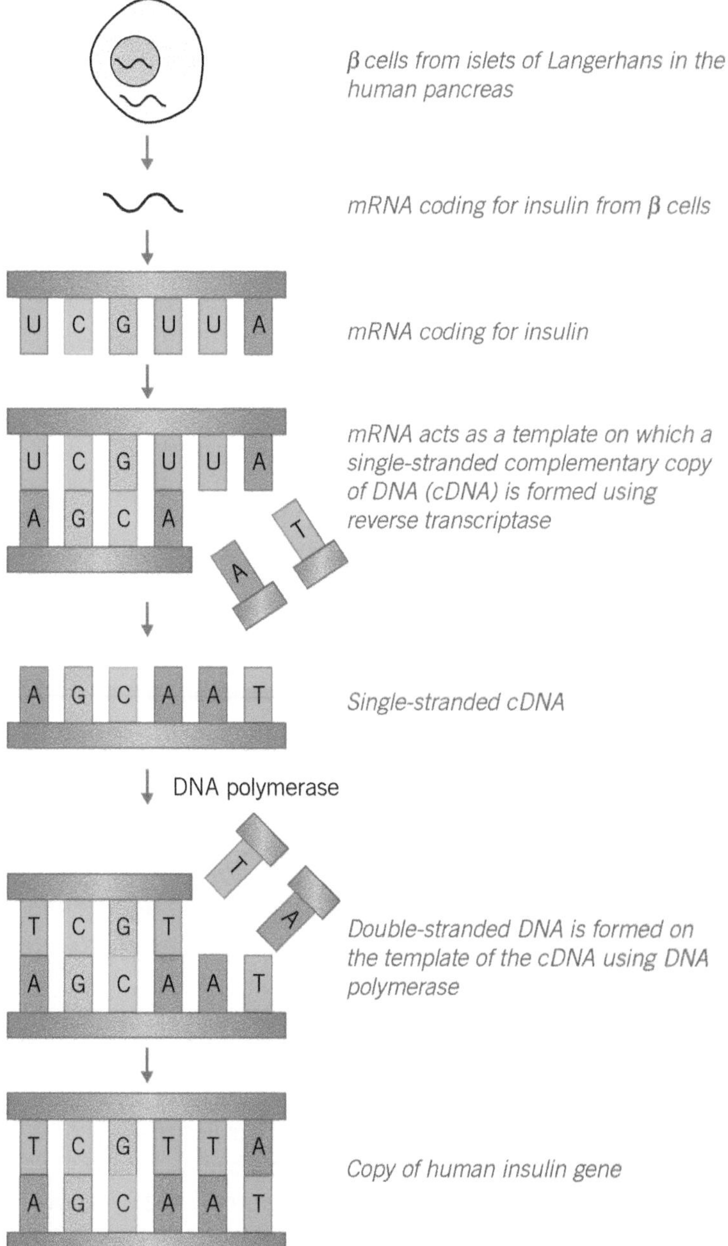

▲ **Figure 2** *The use of reverse transcriptase to isolate the gene that codes for insulin*

Hint

Each restriction endonuclease recognises and cuts DNA at a specific sequence of bases. These sequences occur in the DNA of all species of organisms – but not in the same places!

Using restriction endonucleases

All organisms use defensive measures against invaders. Bacteria are frequently invaded by viruses that inject their DNA into them in order to take over the cell. Some bacteria defend themselves by producing enzymes that cut up the viral DNA. These enzymes are called restriction endonucleases.

There are many types of restriction endonucleases. Each one cuts a DNA double strand at a specific sequence of bases called a recognition sequence. Sometimes, this cut occurs between two opposite base pairs. This leaves two straight edges known as blunt ends. For example, one restriction endonuclease cuts in the middle of the base recognition sequence GTTAAC (see Figure 3).

Other restriction endonucleases cut DNA in a staggered fashion. This leaves an uneven cut in which each strand of the DNA has exposed, unpaired bases. An example is a restriction endonuclease that recognises a six-base pair (or 6 bp) AAGCTT, as shown in Figure 3. In this figure, look at the sequence of unpaired bases that remain. If you read both the four unpaired bases at each end from left to right, the two sequences are opposites of one another, that is, they are a **palindrome**. The recognition sequence is therefore referred to as a 6 bp palindromic sequence. This feature is typical of the way restriction endonucleases cut DNA to leave 'sticky ends'. You shall look at the importance of these sticky ends in Topic 29.3.

a *HpaI restriction endonuclease has a recognition site GTTAAC, which produces a straight cut and therefore blunt ends*

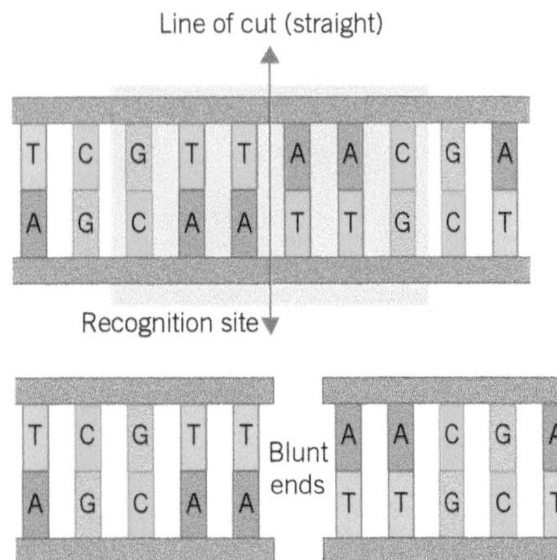

b *HindIII restriction endonuclease has the recognition site AAGCTT, which produces a staggered cut and therefore sticky ends*

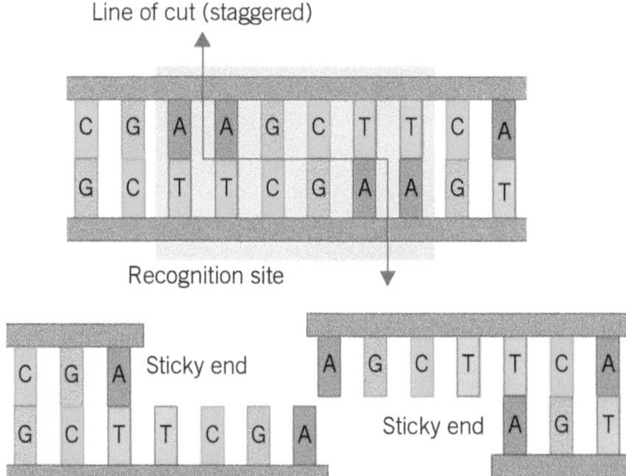

▲ **Figure 3** *Action of restriction endonucleases*

Summary question

1 In the following passage replace each of the letters, **a** to **h** with the most appropriate word or words.

Where the DNA of two different organisms is combined, the product is known as **a** DNA. One method of producing DNA fragments is to make DNA from RNA using an enzyme called **b**. This enzyme initially forms a single strand of DNA called **c** DNA. To form the other strand requires an enzyme called **d**. Another method of producing DNA fragments is to use enzymes called **e**, which cut up DNA. Some of these leave fragments with two straight edges, called **f** ends. Others leave ends with uneven edges, called **g** ends. If the sequence of bases on one of these uneven ends is GAATTC, then the sequence on the other end, if read in the same direction, will be **h**.

29.2 *In vitro* gene cloning – the polymerase chain reaction

You will be looking at *in vivo* gene cloning in Topic 29.3 but first we will consider *in vitro* gene cloning using the polymerase chain reaction.

Polymerase chain reaction

The polymerase chain reaction (PCR) is a method of copying fragments of DNA. The process is automated, making it both rapid and efficient. The process requires the following:

- **the DNA fragment** to be copied
- **DNA polymerase** – an enzyme capable of joining together tens of thousands of nucleotides in a matter of minutes. It is obtained from bacteria in hot springs and is therefore tolerant to heat (thermostable) and does not denature during the high temperatures of the process.
- **primers** – short sequences of nucleotides that have a set of bases complementary to those at one end of each of the two DNA fragments
- **nucleotides** – which contain each of the four bases found in DNA
- **thermocycler** – a computer-controlled machine that varies temperatures precisely over a period of time (Figure 1).

The PCR is illustrated in Figure 2 and is carried out in three stages:

1. **separation of the DNA strand**. The DNA fragments, primers, and DNA polymerase are placed in a vessel in the thermocycler. The temperature is increased to 95 °C, causing the two strands of the DNA fragments to separate.
2. **addition (annealing) of the primers**. The mixture is cooled to 55 °C, causing the primers to join (anneal) to their complementary bases at the end of the DNA fragment. The primers provide the starting sequences for DNA polymerase to begin DNA copying because DNA polymerase can only attach nucleotides to the end of an existing chain. Primers also prevent the two separate strands from simply rejoining.
3. **synthesis of DNA**. The temperature is increased to 72 °C. This is the optimum temperature for the DNA polymerase to add complementary nucleotides along each of the separated DNA strands. It begins at the primer on both strands and adds the nucleotides in sequence until it reaches the end of the chain.

Because both separated strands are copied simultaneously there are now two copies of the original fragment. Once the two DNA strands are completed, the process is repeated by subjecting them to the temperature cycle again, resulting in four strands. The whole temperature cycle takes around 2 minutes. More than a million copies of the DNA can be made in only 25 temperature cycles and

Learning objectives:

→ Describe how the polymerase chain reaction works.

Specification reference: 3.4.10.3

Study tip

Remember DNA polymerase causes nucleotides to join together as a strand, not complementary base pairing.

Hint

The polymerase chain reaction is not the same as semi-conservative replication of DNA in cells.

Study tip

Make sure you can describe the polymerase chain reaction, particularly the importance of the temperature changes.

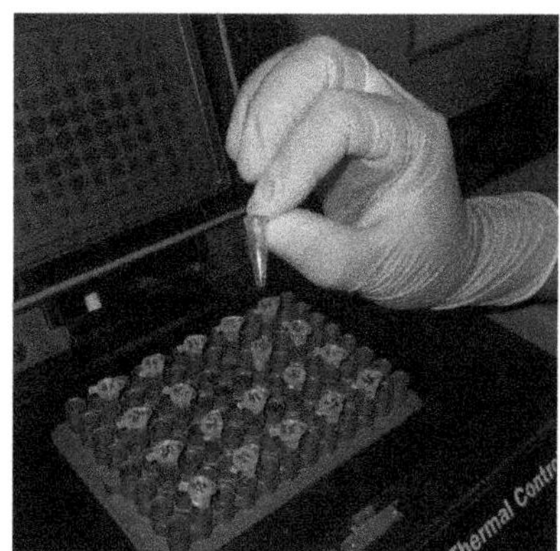

▲ **Figure 1** *This is a thermocycler, a machine that carries out the PCR*

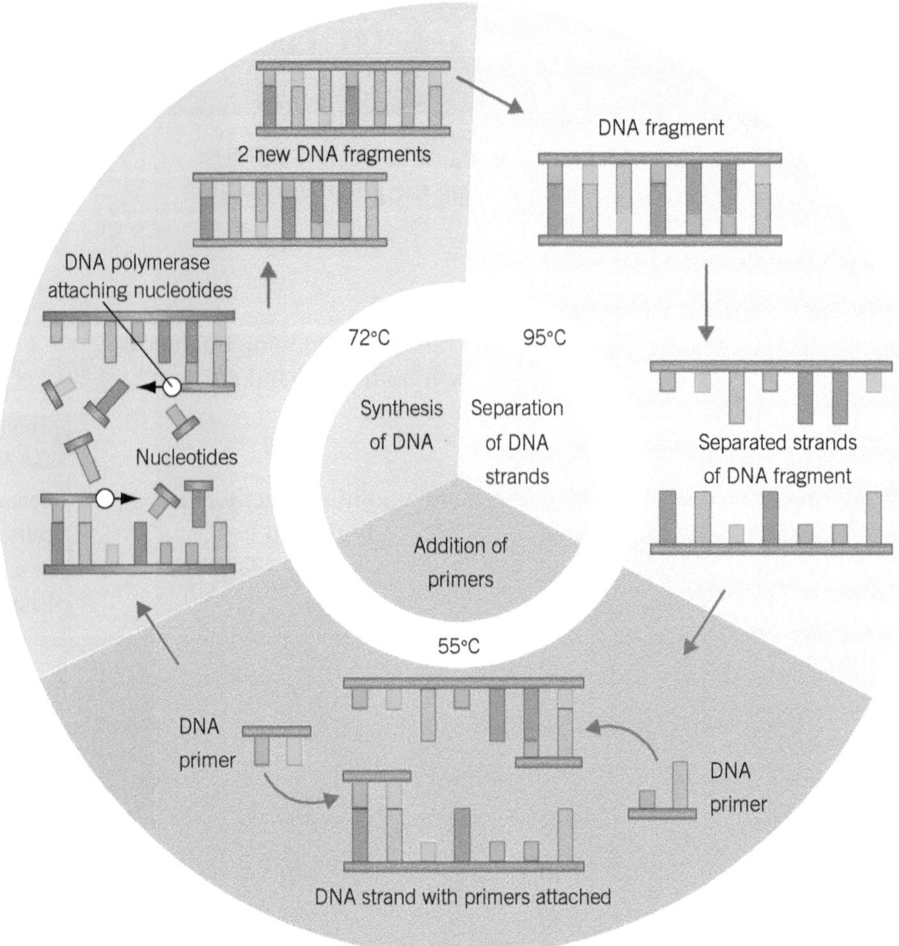

▲ **Figure 2** *The polymerase chain reaction showing a single cycle*

100 billion copies can be manufactured in just a few hours. The PCR has revolutionised many aspects of science and medicine. Even the tiniest sample of DNA from a single hair or a speck of blood can now be multiplied to allow forensic examination and accurate cross-matching.

Summary questions

1. In the polymerase chain reaction (PCR), what are primers?
2. What is the role of these primers?
3. Why are two different primers required?
4. When DNA strands are separated in PCR, what type of bond is broken?
5. It is important in PCR that the fragments of DNA used are not contaminated with any other biological material. Suggest a reason why.

29.3 *In vivo* gene cloning – the use of vectors

Once the fragments of DNA have been obtained, the next stage is to **clone** them so that there is a sufficient quantity for medical or commercial use. This can be achieved in two ways:

- *in vivo*, by transferring the fragments to a host cell using a vector
- *in vitro*, using the polymerase chain reaction (see Topic 29.2).

Before we consider how genes can be cloned within living organisms (*in vivo* cloning), let us look at the importance of the 'sticky ends' left when DNA is cut by **restriction endonucleases**.

Learning objectives:

- → Explain the significance of sticky ends in inserting DNA fragments into vectors.
- → Explain how the DNA of the vector is introduced into host cells.
- → Explain the role of gene markers in recombinant DNA technology.

Specification reference: 3.4.10.4

Importance of sticky ends

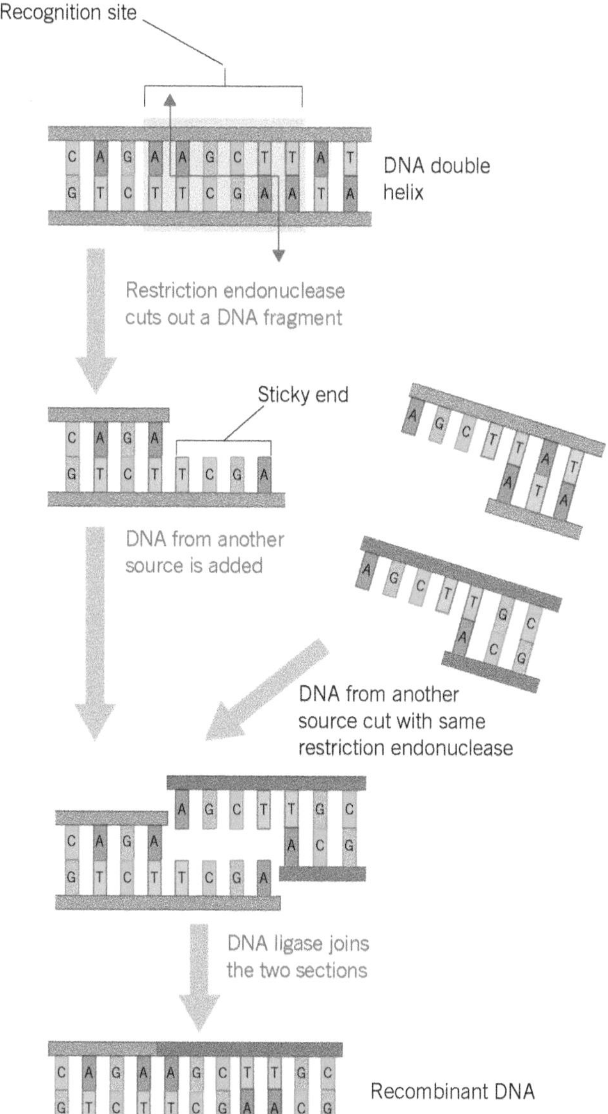

▲ **Figure 1** *The use of sticky ends to combine DNA from different sources*

The sequences of DNA that are cut by restriction endonucleases are called recognition sites. If the recognition site is cut in a staggered fashion, the cut ends of the DNA double strand are left with a single strand that is a few **nucleotide** bases long. The nucleotides on the single strand at one side of the cut are obviously complementary to those at the other side because they were previously paired together.

If the same restriction endonuclease is used to cut the DNA, then all the fragments produced will have ends that are complementary to one another. This means that the single-stranded end of any one fragment can be joined (stuck) to the single-stranded end of any other fragment. In other words, their ends are 'sticky'. Once the complementary bases of two sticky ends have paired up, an enzyme called **DNA ligase** is used to join the sugar-phosphate framework of the two sections of DNA and so unite them as one.

Sticky ends have considerable importance because, provided the same restriction endonuclease is used, we can combine the DNA of one organism with that of any other organism (see Figure 1).

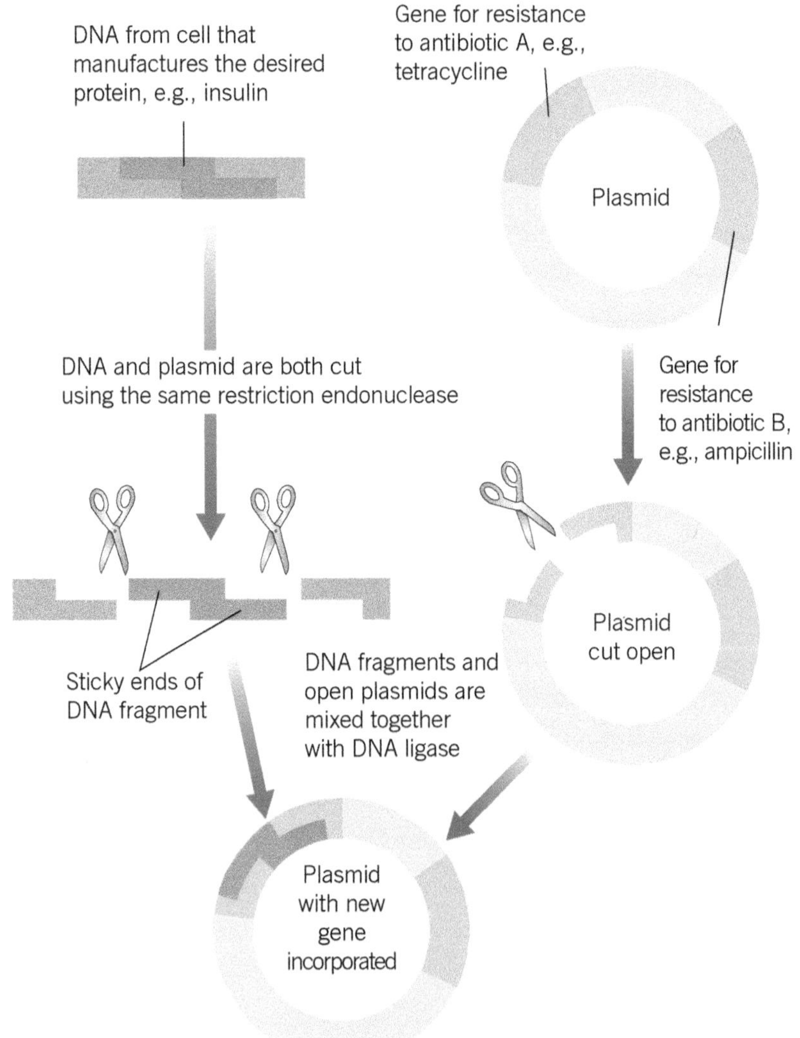

▲ **Figure 2** *Inserting a gene into a plasmid vector*

Insertion of DNA fragment into a vector

Once an appropriate fragment of DNA has been cut from the rest of the DNA, the next task is to join it into a carrying unit, known as a **vector**. This vector is used to transport the DNA into the host cell. There are different types of vector but the most commonly used is the **plasmid**. Plasmids are circular lengths of DNA, found in bacteria, which are separate from the main bacterial DNA. Plasmids almost always contain genes for antibiotic resistance, and restriction endonucleases are used at one of these antibiotic-resistance genes to break the plasmid loop.

The restriction endonuclease used is the same as the one that cut out the DNA fragment. This ensures that the sticky ends of the opened-up plasmid are complementary to the sticky ends of the DNA fragment. When the DNA fragments are mixed with the opened-up plasmids, they may become incorporated into them. Where they are incorporated, the join is made permanent using the enzyme DNA ligase. These plasmids now have recombinant DNA. These events are summarised in Figure 2.

To make sure the inserted gene is expressed, a promotor region is often introduced along with the gene (Topic 28.1). Alternatively, plasmids are constructed with a promotor already in place 'upstream' of where the gene will be inserted. In addition, some plasmids also contain a terminator sequence which signals the transciption process to stop.

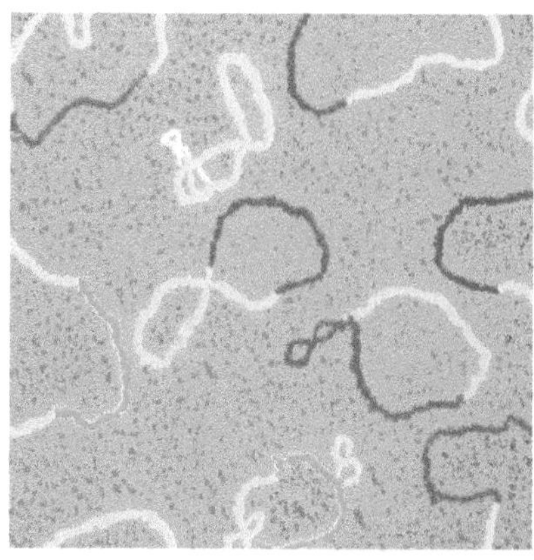

▲ **Figure 3** *False-colour transmission electron micrograph (TEM) of genetically engineered DNA plasmids from the bacterium* Escherichia coli. *The plasmids (yellow) have had different gene sequences (various colours) inserted into them.*

Introduction of DNA into host cells

Once the DNA has been incorporated into at least some of the plasmids, they must then be reintroduced into bacterial cells. This process is called **transformation** and involves the plasmids and bacterial cells being mixed together in a medium containing calcium ions. The calcium ions, and changes in temperature, make the bacteria permeable, allowing the plasmids to pass through the cell membrane into the cytoplasm. However, not all the bacterial cells will possess the DNA fragments. There are two main reasons for this:

- Only a few bacterial cells (as few as 1 per cent) take up the plasmids when the two are mixed together.
- Some plasmids will have closed up again without incorporating the DNA fragment.

The first task is to identify which bacterial cells have taken up the plasmid. To do so we use the fact that, over the years, bacteria have evolved mechanisms for resisting the effects of antibiotics, typically by producing an enzyme that breaks down the antibiotic before it can destroy the bacterium. The genes for the production of these enzymes are found in the plasmids.

Some plasmids carry genes for resistance to more than one antibiotic. One example is the R-plasmid, which carries genes for resistance to two antibiotics: ampicillin and tetracycline.

The task of finding out which bacterial cells have taken up the plasmids entails using the gene for antibiotic resistance, which is unaffected by the introduction of the new gene. In Figure 2, this is the gene for resistance to ampicillin. The process works as follows:

- All the bacterial cells are grown on a medium that contains the antibiotic ampicillin.

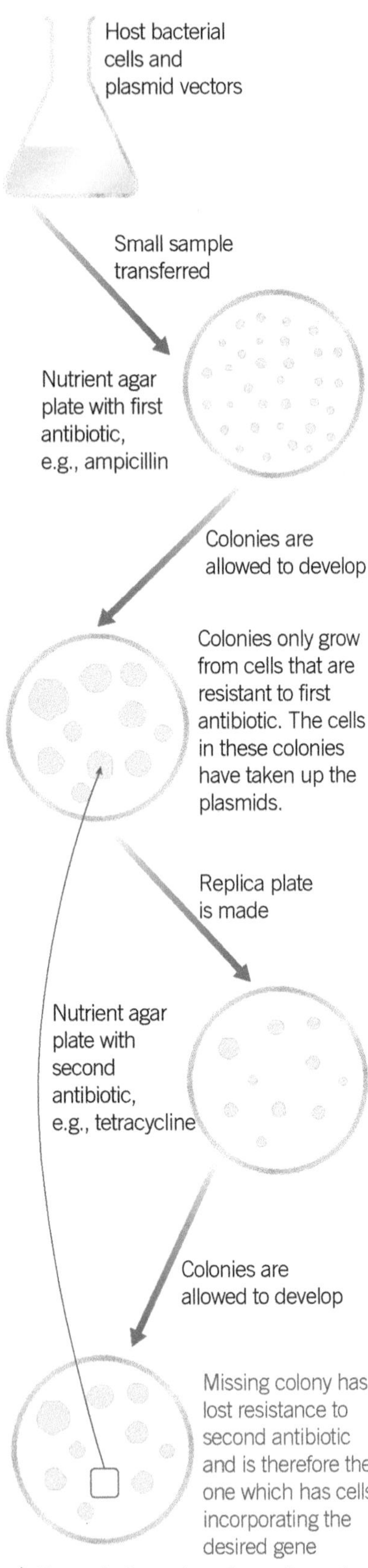

▲ **Figure 4** *Detection of transformed bacteria*

- Bacterial cells that have taken up the plasmids will have acquired the gene for ampicillin resistance.
- These bacterial cells are able to break down the ampicillin and therefore survive.
- The bacterial cells that have not taken up the plasmids will not be resistant to ampicillin and therefore die.

This is an effective method of showing which of the bacterial cells have taken up the plasmids. However, some bacteria will have taken up plasmids which closed up without incorporating the new gene, and these will also have survived. The next task is to identify these cells and eliminate them. This is achieved using gene markers.

Marker genes

There are a number of different ways of using gene markers to identify whether a gene has been taken up by bacterial cells. They all involve using a second, separate gene on the plasmid. This second gene is easily identifiable for one reason or another. For example:

- It may be resistant to an antibiotic.
- It may make a fluorescent protein that is easily seen.
- It may produce an enzyme whose action can be identified.

Antibiotic-resistance markers

The use of antibiotic-resistance genes as markers is a rather old technology and has been superseded by other methods. However, it is an interesting example of the way in which scientists use knowledge and understanding to solve new problems, use appropriate methodology, and carry out relevant experiments.

To identify those cells with plasmids that have taken up the new gene we use a technique called **replica plating**. This process uses the other antibiotic-resistance gene in the plasmid, the gene that was cut in order to incorporate the required gene. In Figure 2 this is the gene for resistance to tetracycline. As this gene has been cut, it will no longer produce the enzyme that breaks down tetracycline. In other words, the bacteria that have taken up plasmids with the required gene will no longer be resistant to tetracycline. We can therefore identify these bacteria by growing them on a culture that contains tetracycline.

The problem is that treatment with tetracycline will destroy the very cells that contain the required gene. This is where the technique of replica plating comes in. This works as follows:

- The bacterial cells that survived treatment with the first antibiotic (ampicillin) are known to have taken up the plasmid.
- These cells are cultured by spreading them very thinly on nutrient agar plates.
- Each separate cell on the plate will grow into a genetically identical colony.
- A tiny sample of each colony is transferred onto a second (replica) plate in exactly the same position as the colonies on the original plate.
- This replica plate contains the second antibiotic (tetracycline), against which the antibiotic-resistance gene will have been made useless if the new gene has been taken up.

- The colonies killed by the antibiotic must be the ones with the plasmids that have taken up the required gene.
- The colonies in exactly the same position on the original plate are the ones that possess the required gene. These colonies are therefore made up of bacteria that have been genetically modified, that is, they have been transformed.

The process of detecting transformed bacteria is summarised in Figure 4.

Fluorescent markers

A more recent and more rapid method is the transference of a gene from a jellyfish into the plasmid. The gene in question produces a green fluorescent protein (GFP). The gene to be cloned is transplanted into the centre of the GFP gene. Any bacterial cell that has taken up the plasmid with the gene that is to be cloned will not be able to produce GFP. Unlike the cells that have taken up the gene, these cells will not fluoresce. As the bacterial cells with the desired gene are not killed, there is no need for replica plating. Results can be obtained by simply viewing the cells under a microscope and retaining those that do not fluoresce. This makes the process more rapid.

Enzyme markers

Another gene marker is the gene that produces the enzyme lactase. Lactase will turn a particular colourless substrate blue. Again, the required gene is transplanted into the gene that makes lactase. If a plasmid with the required gene is present in a bacterial cell, the colonies grown from it will not produce lactase. Therefore, when these bacterial cells are grown on the colourless substrate they will be unable to change its colour. Where the gene has not transformed the bacteria, the colonies will turn the substrate blue.

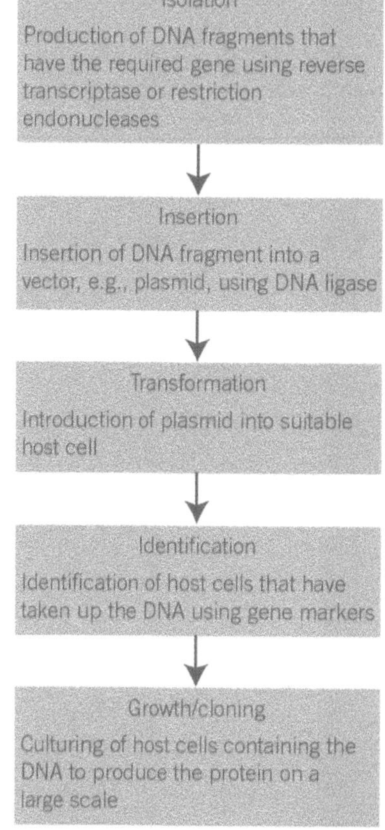

▲ **Figure 5** *Outline summary of gene transfer and* in vivo *cloning*

Advantages of *in vitro* and *in vivo* gene cloning

The advantages of *in vitro* gene cloning using the polymerase chain reaction (PCR) are:

- **It is extremely rapid**. Within a matter of hours a 100 billion copies of a gene can be made. This is particularly valuable where only a minute amount of DNA is available, for example, at the scene of a crime. This can quickly be increased using the polymerase chain reaction (PCR) and so there is no loss of valuable time before forensic analysis and matching can take place. *In vivo* cloning would take many days or weeks to produce the same quantity of DNA.
- **It does not require living cells**. All that is required is a base sequence of DNA that needs amplification. No complex culturing techniques, requiring time and effort, are needed.

The advantages of *in vivo* gene cloning are:

- **It is particularly useful where we wish to introduce a gene into another organism**. As it involves the use of **vectors**, once we have introduced the gene into a **plasmid**, this plasmid can be used to deliver the gene into another organism, such as a human being (i.e., it can transform other organisms).

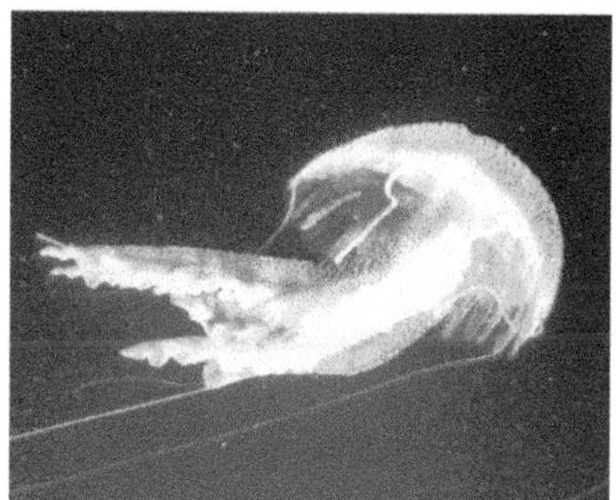

▲ **Figure 6** *The gene in this jellyfish that produces a green fluorescent protein can be transplanted into other organisms and used as a fluorescent marker*

Hint

Interestingly, the gene for the green fluorescent protein has itself been genetically modified by the same techniques it is used to support. As a result, varieties have been engineered that fluoresce more brightly and in a number of different colours.

- **It involves almost no risk of contamination**. This is because a gene that has been cut by the same **restriction endonuclease** can match the sticky ends of the opened-up plasmid. Contaminant DNA will therefore not be taken up by the plasmid. *In vitro* cloning requires a very pure sample because any contaminant DNA will also be multiplied and could lead to a false result.
- **It is very accurate**. The DNA copied has few, if any, errors. At one time, about 20 per cent of the DNA cloned *in vitro* by the PCR was copied inaccurately, but modern techniques have improved the accuracy of the process considerably. However, any errors in copying DNA or any contaminants in the sample will also be copied in subsequent cycles. This problem hardly ever arises with *in vivo* cloning because, although mutations can arise, these are very rare.
- **It cuts out specific genes**. It is therefore a very precise procedure as the culturing of transformed bacteria produces many copies of a specific gene and not just copies of the whole DNA sample.
- **It produces transformed bacteria that can be used to produce large quantities of gene products**. The transformed bacteria can produce proteins for commercial or medical use (e.g., hormones such as insulin).

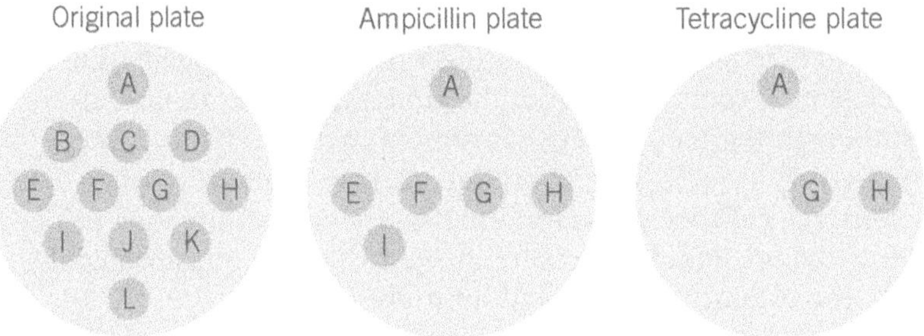

▲ **Figure 7**

Summary questions

1. What is the role of a vector during *in vivo* gene cloning?
2. Why are gene markers necessary during *in vivo* gene cloning?
3. Give **one** advantage of using fluorescent gene markers rather than antibiotic gene markers. Explain your answer.
4. Figure 7 shows the results of an experiment using antibiotic-resistance gene markers to find which bacterial cells have taken up a gene X. The circles within each plate represent a colony of growing bacteria. Which colonies on the original plate:

   a did not take up any plasmids

   b contained plasmids possessing gene X?

   Give reasons for your answers.

29.4 Use of recombinant DNA technology

Plants and animals have been used for many products since the earliest time of civilisation. Crop plants and domesticated animals have been genetically manipulated by **selective breeding** for thousands of years. This has produced crops with higher yields, meat and milk with a lower fat content, and more docile animals. It has only been with the rediscovery of Mendel's genetic work in 1900, and with modern DNA technology, that humans can now achieve, in weeks, genetic changes that once took hundreds of years.

Genetic modification

The genetic make-up of organisms can now be altered by transferring genes between individuals of the same species or between organisms of different species. These modifications can benefit humans in many ways including:

- increasing the yield from animals or plant crops
- improving the nutrient content of foods
- introducing resistance to disease and pests
- making crop plants tolerant to herbicides
- developing tolerance to environmental conditions, for example, extreme temperatures and drought
- making vaccines
- producing medicines for treating disease.

Let us look at some genetically modified microorganisms, plants, and animals that benefit humans.

Examples of genetically modified microorganisms

There are three main groups of substances produced using genetically modified bacteria.

- **Antibiotics** are produced naturally by bacteria. Although genetic engineering has not substantially improved the quality of antibiotics, it has produced bacteria that increase the quantity of the antibiotics produced and the rate at which they are made.
- **Hormones** – **insulin** is needed daily by more than 2 million diabetics, in order for them to lead normal lives. Previously, insulin extracted from cows or pigs was used, but this could produce side effects due to its rejection by the diabetic patient's immune system. With genetic engineering, bacterial cells have the human insulin gene incorporated into them and so the insulin produced is identical to human insulin and has no adverse effects on the patient. This method also avoids killing animals and the need to modify insulin before it is injected into humans. Other hormones produced in this way include human growth hormone, cortisone, and the sex hormones – testosterone and oestrogen.
- **Enzymes** – many enzymes used in the food industry are manufactured by genetically modified bacteria. These include amylases used to break down starch during beer production, lipases used to improve the flavour of cheeses, and proteases used to tenderise meat.

Learning objectives:

- → How has genetic modification of organisms benefited humans?
- → What roles have genetically modified microorganisms, plants, and animals played in the beneficial use of recombinant DNA technology?

Specification reference: 3.4.10.4

Synoptic link

Selective breeding is covered in Topic 10.1.

Hint

There is always a small risk that substances extracted from animals for human use may carry a disease. The use of the same substance produced by genetically modified microorganisms reduces this risk.

▲ **Figure 1** *Genetically modified maize resistant to a herbicide. This is a field trial in Lincolnshire where the first application of herbicide has killed the competing weeds and increased the crop's productivity.*

Examples of genetically modified plants

- **Genetically modified tomatoes** have been developed using the insertion of a gene. This gene has a base sequence that is complementary to that of the gene producing the enzyme that causes the tomatoes to soften. The mRNA transcribed from this inserted gene is therefore complementary to the mRNA of the original gene. The two therefore combine to form a double strand. This prevents the mRNA of the original gene from being translated. The softening enzyme is therefore not produced. This allows the tomatoes to develop flavour without the problems associated with harvesting, transporting, and storing soft fruit.
- **Herbicide-resistant crops** have a gene introduced that makes them resistant to a specific herbicide. When the herbicide is sprayed on the crops, the weeds that are competing with the crop plants for water, light, and minerals, are killed. The crop plants are resistant to the herbicide and so are unaffected.
- **Disease-resistant crops** have genes introduced that give resistance to specific diseases. Genetically modified rice, for example, can withstand infection by a particular virus.
- **Pest-resistant crops**, for example, maize, can have a gene added that allows the plant to make a toxin. This toxin kills insects that eat the maize, but is harmless to other animals including humans.
- **Plants that produce plastics** are a possibility currently being explored. It is hoped that we can genetically engineer plants that have the metabolic pathways necessary to make the raw material for plastic production.

Examples of genetically modified animals

An example of genetic modification in animals is the transfer of genes from an animal that has natural resistance to a disease into a totally different animal. This second animal is then made resistant to that disease. In this way, domesticated animals can be more economic to rear and hence help to reduce the price of food production.

Further examples are fast-growing food animals such as sheep and fish that have a growth hormone gene added so that, in the case of salmon, they can grow 30 times larger than normal and at 10 times the usual rate.

Another example is in the production of rare and expensive proteins for use in human medicine. Domesticated milk-producing animals such as goats can be used. The gene for the required protein is inserted alongside the gene that codes for proteins in goats' milk. In this way, the required protein is produced in the milk of the goat. The gene can be inserted into the fertilised egg of a goat, so that all the female offspring of that individual will be capable of producing the protein in their milk. One example of a protein made in this way is a protein that prevents blood from clotting (anticoagulant) called anti-thrombin.

Some individuals have an inherited disorder that affects one of the alleles that codes for the protein anti-thrombin. As a result, those affected are unable to produce sufficient quantities of anti-thrombin. These individuals are therefore at risk of blood clots. They are currently treated with drugs that thin the blood or are given anti-thrombin that has been extracted from donated blood.

Whilst small amounts of anti-thrombin can be extracted from human blood, far more can be produced in the milk of genetically transformed goats. The process, summarised in Figure 2, is as follows:

- Mature eggs are removed from female goats and fertilised by sperm.
- The normal gene for anti-thrombin production from a human is added to the fertilised eggs alongside the gene that codes for proteins in goats' milk.
- These genetically transformed eggs are implanted into female goats.
- Those resulting goats with the anti-thrombin gene are cross-bred, to give a herd in which goats produce milk rich in the protein anti-thrombin.
- The anti-thrombin is extracted from the milk, purified, and given to humans unable to manufacture their own anti-thrombin.

In its lifetime a single genetically modified goat can produce as much anti-thrombin in its milk as can be extracted from 90 000 blood donations. Anti-thrombin is sold under its commercial name Atryn®. It is the world's first drug from a genetically modified animal to be registered for general use.

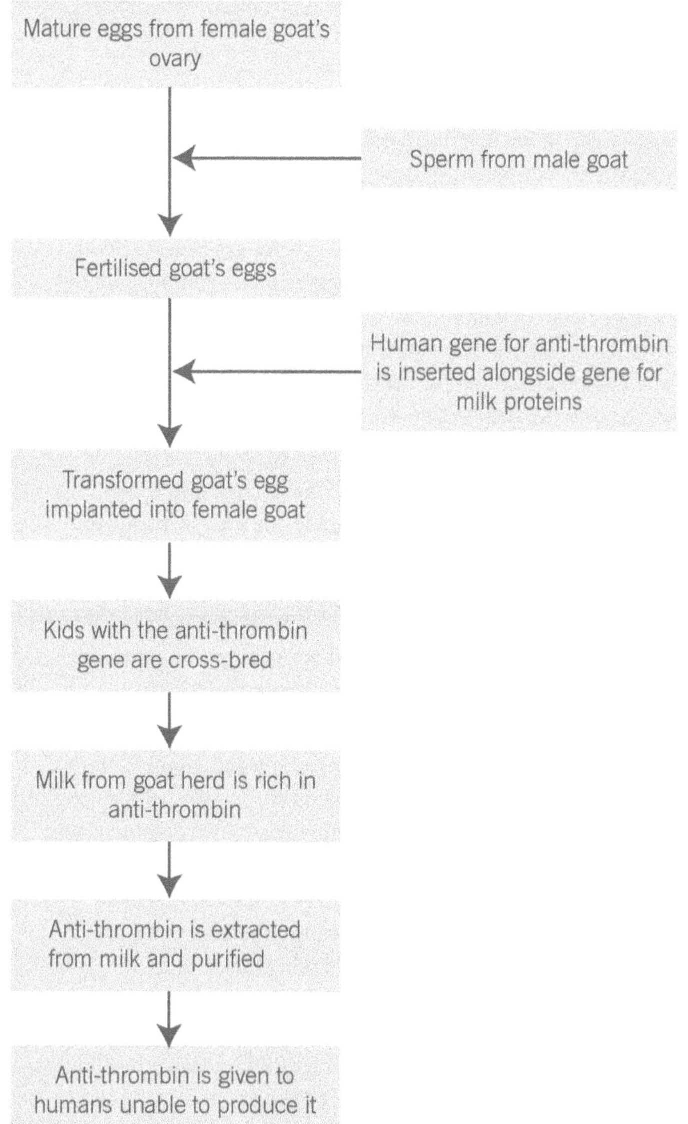

▲ **Figure 2** *Summary of genetic modification of goats in order to produce the protein anti-thrombin*

Summary questions

1 State **one** advantage to humans of genetically modified tomatoes.

2 Suggest **one** benefit and **one** possible disadvantage of using genetically modified herbicide-resistant crop plants together with the relevant herbicide. Explain your answer.

3 Why is insulin produced by recombinant DNA technology better than insulin extracted from animals?

Another means of producing drugs from genetically modified animals is being pioneered by the Edinburgh institute that produced Dolly the sheep. Domesticated chickens have had human genes for medicinal proteins added to their DNA. The eggs laid by these transgenic chickens contain the proteins in the white portion from which they can be easily extracted. The human genes are passed on from generation to generation. Large flocks can therefore be formed that offer a potentially unlimited source of cheap medicinal proteins. Drugs that have so far been produced in this way include a form of interferon used to treat multiple sclerosis and an **antibody** with the potential to treat skin cancer and arthritis. It could be 5 years before patient trials are carried out on these drugs and 10 years before the medicines are fully developed.

Evaluation of DNA technology

Genetic engineering undoubtedly brings many benefits to mankind, but it not without its risks. It is therefore important to evaluate the ethical, moral, and social issues associated with its use.

The benefits of recombinant DNA technology

- Microorganisms can be modified to produce a range of substances, for example, antibiotics, hormones, and enzymes, that are used to treat diseases and disorders.
- Microorganisms can be used to control pollution, for example, to break up and digest oil slicks or destroy harmful gases released from factories. Care needs to be taken to ensure that such bacteria do not destroy oil in places where it is required, for example, car engines. To do this, a 'suicide gene' can be incorporated that causes the bacteria to destroy themselves once the oil slick has been digested.
- Genetically modified plants can be transformed to produce a specific substance in a particular organ of the plant. These organs can then be harvested and the desired substance extracted. If a drug is involved, the process is called plant pharming. One promising application of this technique is in combating disease. This involves the production of plants that manufacture antibodies to pathogens and the toxins they produce. Alternatively the plants can be modified to manufacture **antigens** which, when injected into humans, induce natural antibody production.
- Genetically modified crops can be engineered to have economic and environmental advantages. These include making plants more tolerant to environmental extremes, for example, able to survive drought, cold, heat, salt, or polluted soils. This permits crops to be grown commercially in places where they are not at present. Globally, each year, an area of land equal to half the UK becomes unfit for normal crops because of increases in soil salt concentrations. Growing genetically modified plants, such as salt-tolerant tomatoes, could bring this land back into productivity. Other examples have been described earlier in this spread. In a world where millions lack a basic nutritious diet, and with a predicted 90 million more mouths to feed by 2025, can we ethically oppose the use of such plant crops?

Hint

MORALS are individual or group views about what is right or wrong. Such views refer to almost any subject, such as it is wrong to hunt foxes, work on a Sunday, to swear, or to tell lies. Morals vary from country to country and individual to individual, and change over time. Some of the accepted moral values of a hundred years ago in Britain, most people would now disagree with.

Hint

ETHICS is a narrower concept than morals. Ethics are a set of standards that are followed by a particular group of individuals and are designed to regulate their behaviour. They determine what is acceptable and legitimate in pursuing the aims of the group.

Hint

SOCIAL ISSUES relate to human society and its organisation. They concern the mutual relationships of human beings, their interdependence, and their cooperation for the benefit of all.

- Genetically modified crops can help prevent certain diseases. Rice can have a gene for vitamin A production added. Can we morally justify not developing more vitamin A-enriched crops when 250 million children worldwide are at risk from vitamin A deficiency, leading to 500 000 cases of irreversible blindness each year?
- Genetically modified animals are able to produce expensive drugs, antibiotics, hormones, and enzymes relatively cheaply. Details were given earlier in this topic.
- **Gene therapy** might be used to cure certain genetic disorders, such as cystic fibrosis. Details are given in Topic 16.5.
- Genetic fingerprinting can be used in forensic science. Details are given in Topic 16.8.

The risks of recombinant DNA technology

Against the benefits of genetic engineering, must be weighed the risks – both real and potential.

- It is impossible to predict with complete accuracy what the ecological consequences will be of releasing genetically engineered organisms into the environment. The delicate balance that exists in any habitat may be irreversibly damaged by the introduction of organisms with engineered genes. There is no going back once an organism is released.
- A recombinant gene may pass from the organism it was placed in to a completely different one. We know, for example, that viruses can transfer genes from one organism to another. What if a virus were to transfer genes for herbicide resistance and vigorous growth from a crop plant to a weed that competed with the crop plant? What if the same gene were transferred in pollen to other plants? How would we then be able to control this weed?
- Any manipulation of the DNA of a cell will have consequences for the metabolic pathways within that cell. We cannot be sure until after the event what unforeseen by-products of the change might be produced. Could these lead to metabolic malfunctions, cause cancer, or create a new form of disease?
- Genetically modified bacteria often have antibiotic resistance marker genes that have been added. These bacteria can spread antibiotic resistance to harmful bacteria.
- All genes mutate. What then, might be the consequences of our engineered gene mutating? Could it turn the organism into a **pathogen** which we have no means of controlling?
- What will be the long-term consequences of introducing new gene combinations? We cannot be certain of the effects on the future evolution of organisms. Will the **artificial selection** of 'desired' genes reduce the genetic variety that is so essential to evolution?
- What might be the economic consequences of developing plants and animals to grow in new regions? Developing bananas which grow in Britain could have disastrous consequences for the Caribbean economies that rely heavily on this crop for their income.

Hint

'Evaluate' always involves looking at the positives and negatives, i.e. the benefits and risks, of a particular issue.

- How far can we take gene therapy (Topic 16.5)? It may be acceptable to replace a defective gene to cure cystic fibrosis, but is it equally acceptable to introduce genes for intelligence, more muscular bodies, cosmetic improvements, or different facial features?
- Will knowledge of, and ability to change, human genes lead to eugenics, whereby selection of genes leads to a means of selecting one race rather than another?
- What will be the consequences of the ability to manipulate genes getting into the wrong hands? Will unscrupulous individuals, groups or governments use this power to achieve political goals, control opposition or gain ultimate power?
- Is the cost of genetic engineering justified, or would the money be better used fighting hunger and poverty, which are the cause of much human misery. Will sophisticated treatments, with their more high-profile images, be put before the everyday treatment of rheumatoid arthritis or haemorrhoids? Will such treatments only be within the financial reach of the better-off?
- Genetic fingerprinting (Topic 16.8), with its ability to identify an individual's DNA accurately, is a highly reliable forensic tool. How easy would it be for someone to exchange a DNA sample maliciously, leading to wrongful conviction?
- Is it immoral to tamper with genes at all? Should we let nature take its own course in its own time?
- How do we deal with the issues surrounding the **human genome project**? Is it right that an individual or company can patent, and therefore effectively own, a gene?

It is inevitable that we remain inquisitive about the world in which we live, and that we will seek to try to improve the conditions around us. Genetic research is bound to continue, but the challenge will be to develop the safeguards and ethical guidelines that will allow genetic engineering to be used in a safe and effective manner.

1. Take any **three** aspects of recombinant DNA technology that are beneficial to humans (as listed above) and present a reasoned argument in each case for the continued use of that technology.
2. Using the same three aspects, present a reasoned argument that an environmentalist or anti-globalisation activist might make against the continued use of that technology.

Practice questions: Chapter 29

1 The polymerase chain reaction (PCR) is a process which can be carried out in a laboratory to replicate DNA. Figure 1 shows the main stages involved in the PCR.

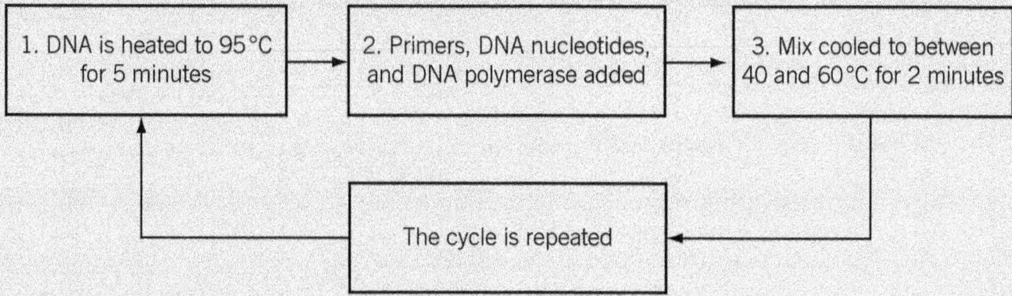

▲ **Figure 1**

(a) Explain why DNA is heated to 95 °C. *(1 mark)*

(b) What is the role of:
(i) a primer in this process
(ii) DNA polymerase? *(2 marks)*

(c) (i) How many DNA molecules will have been produced from one molecule of DNA after six complete cycles?
(ii) Suggest **one** use of the PCR. *(2 marks)*

(d) Give **two** ways in which the PCR differs from the process of transcription. *(2 marks)*

AQA 2005

2 Read the following passage:

The giant panda is one of the rarest animals in the world and is considered to be on the brink of extinction in the wild. Giant pandas have been kept and bred in zoos with the hope that they could be released in to the wild. One worry is that small populations, like those in zoos, reduce the genetic variation needed to allow species to adapt to changing situations.

Unfortunately, pandas find it difficult to reproduce in captivity. Fertilization of the females is guaranteed only by insemination with semen from several males. With so many potential fathers, the true paternity of the cubs is not clear. It is important to identify the fathers to maintain genetic variation.

Panda faeces can be collected in the wild. The faeces contain DNA from the panda, from the bamboo on which they feed, and bacteria. The DNA is subjected to the polymerase chain reaction (PCR). The primers used only attach to the panda DNA. The resulting DNA is subjected to genetic fingerprinting. This can help us count the number of individuals in the wild because it allows us to identify individual pandas.

Use information in the passage and your own biological knowledge to answer the questions.

(a) Describe how genetic fingerprinting may be carried out on a sample of panda DNA. *(6 marks)*

(b) (i) Explain how genetic fingerprinting allows scientists to identify the father of a particular panda club.
(ii) When pandas are bred in zoos, it is important to ensure only unrelated pandas breed. Suggest how genetic fingerprints might be used to do this. *(3 marks)*

(c) (i) Suggest why panda DNA is found in faeces.
(ii) Explain why the PCR is carried out on the DNA from the faeces.
(iii) Explain why the primers used in the PCR will bind to panda DNA, but not to DNA from bacteria or bamboo. *(4 marks)*

(d) DNA from wild pandas could also be obtained from blood samples. Suggest **two** advantages of using faeces, rather than blood samples, to obtain DNA from pandas. *(2 marks)*

AQA 2006

3 Haemophilia is a genetic condition in which blood fails to clot. Factor IX is a protein used to treat haemophilia. Sheep can be genetically engineered to produce Factor IX in the milk produced by their mammary glands. Figure 2 shows the stages involved in this process.

(a) Name the type of enzyme that is used to cut the gene for Factor IX from human DNA (stage 1). *(1 mark)*

(b) (i) The jellyfish gene attached to the human Factor IX gene (stage 2) codes for a protein that glows green under fluorescent light. Explain the purpose of attaching this gene *(2 marks)*

(ii) The promoter DNA from sheep (stage 3) causes transcription of genes coding for proteins found in sheep milk. Suggest the advantage of using this promoter DNA. *(2 marks)*

(c) (i) Many attempts to produce transgenic animals have failed. Very few live births result from the many embryos that are implanted. Suggest one reason why very few live births result from the many embryos that are implanted. *(2 marks)*

(ii) It is important that scientists still report the results from failed attempts to produce transgenic animals. Explain why. *(2 marks)*

AQA June 2012

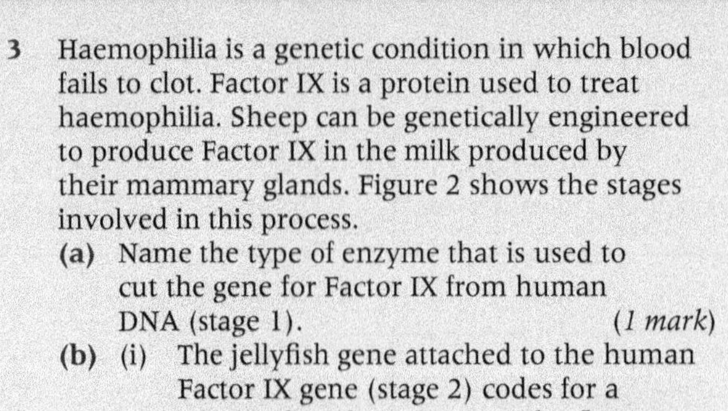

▲ **Figure 2**

4 Scientists wanted to measure how much mRNA was transcribed from allele A of a gene in a sample of cells. This gene exists in two forms, A and a. The scientists isolated mRNA from the cells. They added an enzyme to mRNA to produce cDNA.

(a) Name the type of enzyme used to produce the cDNA. *(1 mark)*

The scientists used the polymerase chain reaction (PCR) to produce copies of the cDNA. They added a DNA probe for allele A to the cDNA copies. This DNA probe had a dye attached to it. This dye glows with a green light only when the DNA probe is attached to its target cDNA.

(b) Explain why this DNA probe will only detect allele A. *(2 marks)*

(c) (i) The scientists used this method with cells from two people, H and G. One person was homozygous, AA, and the other was heterozygous, Aa. The scientists used the PCR and the DNA probe specific for allele A on the cDNA from both people. Explain the curve in Figure 3 for person H. *(3 marks)*

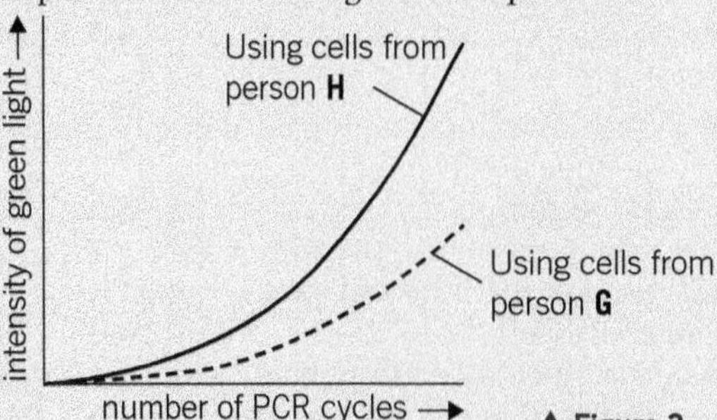

▲ **Figure 3**

(ii) Which person, H or G, was heterozygous, Aa? Explain your answer *(2 marks)*

AQA June 2014

Answers to the Practice Questions are available at www.oxfordsecondary.com/oxfordaqaexams-alevel-biology

Mathematical Skills in A level Biology

Biology students are often less comfortable with the application of mathematics than are students such as physicists, for whom complex maths is a more obvious everyday tool. Nevertheless, it is important to realise that biology does require competent maths skills in many areas. It is important to practise these skills so that you are familiar with them as part of your routine study of the subject.

Confidence with mental arithmetic is very helpful, but among the most important skills is that of taking care and checking calculations. You may not be required to understand the detailed theory of the maths you use, but you do need to be able to apply the skills accurately, whether simply calculating percentages or means, or substituting numbers into complex-looking algebraic equations, such as in statistical tests.

This chapter is designed to help with some of the regularly encountered mathematical problems in biology.

Working with the correct units

In biology it is very important to be secure in the use of correct units. These must always be written clearly in calculations.

Base units

The units we use are from the Système Internationale – the SI units. In biology we most commonly use the SI base units:

- metre (m) for length, height, distance
- kilogram (kg) for mass
- second (s) for time
- mole (mol) for the amount of a substance.

You should develop good habits right from the start, being careful to use the correct abbreviation for each unit used. For example, seconds should be abbreviated to s, not 'sec' or 'S'.

Derived units

Biologists also use SI derived units, such as:

- square metres (m^2) for area
- cubic metre (m^3) for volume
- cubic centimetre (cm^3), also written as millilitre (ml), for volume
- degree Celsius (°C) for temperature
- mole per litre (mol/L, mol dm^{-3}) is usually used for concentration of a substance in solution (although the official SI derived unit is moles per cubic metre)
- joule (J) for energy
- pascal (Pa) for pressure
- volt (V) for electrical potential.

Non-SI units

Although examination boards use SI units, you may also encounter non-SI units elsewhere, for example:

- litre (cubic decimetre) (l, L, dm^3) for volume
- minute (min) for time
- hour (h) for time
- svedberg (S) (for sedimentation rate), used for ribosome particle size.

Unit prefixes

To accommodate the huge range of dimensions in our measurements, units may be further modified using appropriate prefixes. For example, one thousandth of a second is a millisecond (ms). This is illustrated in Table 1.

▼ **Table 1**

Division	Factor	Prefix	Length		Mass		Volume		Time	
One thousand millionth	10^{-9}	nano	nanometre	nm	nanogram	ng	nanolitre	nl	nanosecond	ns
One millionth	10^{-6}	micro	micrometre	μm	microgram	μg	microlitre	μl	microsecond	μs
One thousandth	10^{-3}	milli	millimetre	mm	milligram	mg	millilitre	ml/cm^3	millisecond	ms
One hundredth	10^{-2}	centi	centimetre	cm						
Whole unit			metre	m	gram	g	litre	l/L/dm^3	second	s
One thousand times	10^3	kilo	kilometre	km	kilogram	kg				

Converting between units

You may need to convert between units in order to be able to scale and express numbers in sensible forms. For example, rather than refer to the width of a cell in metres you would use micrometres (μm). This allows your measurements to be understood within the relevant scale of the observation.

Divide by 1000 for each step to convert in this direction 

nano-	*micro-*	*milli-*	*whole unit*	*kilo-*
e.g., nm	*e.g., μm*	*e.g., mm*	*e.g., m*	*e.g., km*

Multiply by 1000 for each step to convert in this direction

▲ **Figure 1** *The effect on units of multiplying and dividing by 1000*

Examples:

Convert 1 m to mm: $1 \times 1000 = 1000$ mm

Convert 1 m to μm: $1 \times 1000 = 1000$ mm, then $1000 \times 1000 = 1\,000\,000$ μm

Convert 1 l to cm^3: $1 \times 1000 = 1000\ cm^3$

Convert 20 000 μm to mm: $20\,000 \div 1000 = 20$ mm

Converting between square or cube units requires a bit more care.

$1\,m^2 = 1000 \times 1000 = 1\,000\,000\,mm^2$, so your conversion factor becomes × or ÷ 1 000 000

$1\,m^3$ is $1000 \times 1000 \times 1000 = 1\,000\,000\,000\,mm^3$, so your conversion factor now becomes × or ÷ 1 000 000 000

Examples:

Convert $20\,m^2$ to km^2: $20 \div 1\,000\,000 = 0.000\,02\,km^2$

Convert $1\,m^2$ to mm^2: $1 \times 1\,000\,000 = 1\,000\,000\,mm^2$

Convert $5\,000\,000\,mm^3$ to m^3: $5\,000\,000 \div 1\,000\,000\,000 = 0.005\,m^3$

Convert $0.0000\,0007\,m^3$ to mm^3: $0.0000\,0007 \times 1\,000\,000\,000 = 70\,mm^3$

Decimals and standard form

When you are using numbers that are very small, such as dimensions of molecules and organelles, it is useful to use **standard form** to express them more easily. Standard form is also commonly called **scientific notation**.

Standard form is essentially expressing numbers in powers of 10. For example, 10 raised to the power 10 means 10 × 10, that is, 100. This may be written down as 10×10^1 or 1×10^2. To get to 1000 you use 10 × 10 × 10, which would be written as 1×10^3.

An easy way to look at this is to imagine the decimal point moving one place per power of 10. For example, to write down 58 900 000 000 as standard form, you would follow the steps below.

Step 1: write down the smallest number between 1 and 10 that can be derived from the number to be converted. In this case it would be 5.89.

Step 2: write the number of times the decimal place will have to shift to expand this to the original number as powers of ten. On paper this can be done by hopping the decimal over each number like this:

5.8900000000

▲ **Figure 2**

until the end of the number is reached. In this example, that requires 10 shifts, so the standard form should be written as 5.89×10^{10}.

Going the other way, for example, expressing 0.000 007 8 as standard form, write the number in terms of the number of places the decimal place would have to hop forward to make the smallest number between 1 and 10, so to get to 7.8 you would have to hop over six times, so this number is written as 7.8×10^{-6}.

Significant figures

There are some simple rules to use when working out significant figures.

Rule 1: All non-zero digits are significant.

For example, 78 has two significant figures, 9.543 has four significant figures and 340 has two significant figures.

Rule 2: Intermediate zeros are significant.

For example, 706 has three significant figures and 5.900 76 has six significant figures.

Rule 3: Any leading zeroes are not significant.

For example, 0.005 67 has three significant figures (5, 6, and 7; ignore the leading zeroes)

Rule 4: Zeroes at the ends of numbers containing decimal places are significant.

For example, 45.60 has four significant figures and 330.00 has five significant figures.

Significant figures and rounding

Table 2 shows the effect of rounding numbers to decimal places compared with significant figures. Remember that in rounding, when the next number is 5 or more round up, whereas if it is 4 or less don't round up. For example, 4.35 rounds to 4.4 and 4.34 rounds to 4.3.

Table 2 shows examples of rounding the number 23.336 00 to decimal places and to significant figures.

▼ **Table 2**

Measurements expressed by rounding to decimal places	Number of decimal places	Measurements expressed by rounding to significant figures	Number of significant figures	Measured to the nearest
23.336 00	5	23.336	5	100 thousandth
23.336 0	4	23.34	4	Ten thousandth
23.336	3	23.3	3	Thousandth
23.34	2	23	2	Hundredth
23.3	1	20	1	Tenth
23	0	—	—	Whole number

Significant figures and standard form

In standard form only the significant figures are written as digits, for example 5.600×10^3 has four significant figures. If this were written as a straight number it would be 5600. But according to the rules above, 5600 only has two significant figures – what does this mean?

In a given number, the significant figures are defined as the ones that contribute to its precision. Writing the number as 5600 implies precision only to the nearest whole hundred. The zeroes in the number could mean that it has simply been rounded, for example, from 5600.44 or even 5633. But if this number were actually more precise, for example, it had been measured with equipment genuinely sensitive to the nearest hundredth part (2 decimal places) then 5600.00 is actually very precise and the two zeros have significance because they tell us that the measurement is *exactly* 5600 with no tenths or hundredths at all. So using standard form allows this precision to remain clearly as part of the stated number, because all significant figures are written.

Averages

An average value is actually a measure of central tendency. The most familiar measurement is the arithmetic mean (mean for short), but median or modal values are sometimes more appropriate to the data.

The arithmetic mean

Usually referred to simply as the mean, this is a measure of central tendency that takes into account the number of times each measurement occurs together with the range of the measurements. When repeated measurements are averaged, the mean will more accurately approach the true value, which will lie somewhere in the middle of the observed range. This is why it is important to repeat experimental measurements, especially in biology where the natural unpredictability of living systems leads to inevitable fluctuations.

The mean is determined by adding together all the observed values and then dividing by the number of measurements made.

For a range of values of x, the mean $\bar{x} = \frac{\sum x}{n}$

$\bar{x}$ is the mean value.

$\sum x$ is the sum of all values of x.

n is the number of values of x.

For example, five mice were weighed, giving masses of 6.2 g, 7.7 g, 6.7 g, 7.1 g, and 6.3 g.

The mean mass is (6.2 + 7.7 + 6.7 + 7.1 + 6.3) ÷ 5 = 6.8 g

Be careful with your decimal places when calculating mean values. Your mean should normally have the same level of precision as the original measurements and therefore the same number of decimal places, otherwise you may be implying that the averaged measurements are more precise than they really are. For example, masses in whole grams would not average to a mass with one or more decimal places. Similarly averaging the numbers of whole objects should result in a whole number – if counts of bubbles in a pondweed experiment were averaged to a decimal place it implies you counted a fraction of one bubble, which is impossible!

The median

The median value in a set of data is calculated by placing the values in numerical order then finding the middle value in the range.

For example, the data set 12, 15, 10, 17, 9, 13, 13, 19, 10, 11 rearranges as 9, 10, 10, 11, 12, 13, 13, 15, 17, 19.

The middle of this range is 12.5.

The median value is very useful when data sets have a few values (outliers) at the extremes, which if included in an arithmetic mean could skew the data. It also allows comparison of data sets with similar means but a clear lack of overlap, skewed data, and when there are too few measurements to calculate a reliable mean value.

For example, in the data set 1, 3, 3, 11, 12, 12, 12, 13, 14, 15, the median value is 12, a sensible looking mid point, but the mean would be 9.6, skewed to the left by the numbers at the lower extreme.

The mode

The modal value is the most frequent value in a set of data. It is very useful when interpreting data that is qualitative or in situations where the distribution has more than one peak (bimodal).

For example, in the data set 9, 10, 11, 11, 12, 13, 13, 13, 14, 17, 18, 19, the modal value is 13.

In biology, caution should be used because the sets of data are usually small and can introduce confusion. For example, in the data set 9, 10, 11, 11, 12, 13, 13, 14, 17, 18 there are apparently two modal values, 11 and 13, whereas in the set 11, 12, 13, 14, 17 there is no most frequent number and the mode is effectively every number and therefore of no value at all.

The modal value is not used very often, but it can be usefully applied when data is collected in categories, for example, numbers of moths attracted to lights of different colours.

Percentages

A percentage is simply expressing a fraction as a decimal. It is important to be confident with calculating percentages, which although straightforward are commonly calculated incorrectly.

Percentages as proportions and fractions

For example, two shapes of primrose flowers exist depending on stigma length: 'pin eyed' and 'thrum eyed'. In a survey of two areas of grassland, one area had 323 pin and 467 thrum (total 790 plants), the other had 667 pin and 321 thrum (total 988 plants). The percentage of pin eyed plants in each area is calculated as follows:

Area 1: *fraction* = $\frac{323}{790}$ which gives *decimal* 0.41, which multiplied by 100 gives *percentage* 41%.

The percentage of pin eyed flowers in Area 2 is $\frac{667}{988} \times 100 = 68\%$.

Percentages as chance

In genetics the likelihood of different offspring phenotypes should always be expressed as a percentage. For example, in a simple genetic cross between two heterozygous parents carrying the cystic fibrosis allele, one out of every four possible children could potentially be affected by the disorder. The chance of a child with cystic fibrosis from these parents is therefore $\frac{1}{4} \times 100 = 25\%$.

Percentage change

This often comes up in osmosis experiments where samples (usually of potato tissue) gain and lose mass in different bathing solutions.

For example, a sample weighed 18.50 g at the start and at the end it weighed 11.72 g.

The actual loss in mass = 18.50 − 11.72 g = 6.78 g

The percentage change = $\frac{\text{mass change}}{\text{starting mass}} \times 100 = \frac{6.78}{18.50} \times 100 = -36.6\%$

Note the use of the minus sign to indicate that this is a loss.

Equations

Substituting into equations

There are several equations (mathematical formulae) that you will need to be able to use in A level biology. You do not need to learn the theoretical maths from which they are derived, but you do need to be able to put known numerical values in the right place (this is **substituting into the equation**) and then calculate the result of the equation by performing the different steps in the right order (this is **solving the equation**).

An example that you will encounter during ecology studies is called the Simpson's Index of Diversity, which has the formula:

$$D = \frac{N(N-1)}{\Sigma n(n-1)}$$

Each symbol (**term**) in the equation has a specific meaning. In this example:

N means the total number of all individual organisms in a survey.

n means the total number of each different species.

Σ means 'the total of' and requires you to add together all the indicated values.

Brackets indicate sub-calculations that must be done, for example $N - 1$ means the total of all species found minus 1.

The figures in brackets need to be multiplied by the figures outside them, for example, $N(N - 1)$ means $N \times (N - 1)$.

An example of the data to use could be counts of the plant species found in a certain area. To make life easy, use a table like Table 3.

▼ **Table 3**

Plant species	Number of plants of each species found (n)	(n – 1)	n(n – 1)
A	22	22 – 1 = 21	22 × 21 = 462
B	30	30 – 1 = 29	30 × 29 = 870
C	25	25 – 1 = 24	25 × 24 = 600
D	23	23 – 1 = 22	23 × 22 = 506
Totals of all plants = N	N = 100 N – 1 = 99		$\Sigma n(n-1) = 2438$

The brackets in equations always need to be solved first.

Begin by finding $n - 1$ for each plant (see column 3 in Table 3) and $N - 1$ (at the bottom of column 2).

Next work out $n(n - 1)$ (column 4 in Table 3).

Now find $\Sigma n(n - 1)$ by totalling the figures in column 4.

Substituting the known values into the equation works like this:

$D = \frac{N(N-1)}{\Sigma n(n-1)}$ becomes $D = \frac{100(99)}{2438}$ which calculates to $D = \frac{9900}{2438}$

which gives the result $D = 4.1$

Rearranging equations

The individual parts or terms in equations are all related, but sometimes you might know all the values of the terms except one. The equation can be re-written so that the unknown term can be calculated. This is called rearranging or **changing the subject of** an equation. A very useful example of this arises during the study of microscopy and magnification.

The different terms are magnification, size of image, and actual size of the object being observed. The equation that relates them together is:

$$\textit{magnification} = \frac{\textit{size of image}}{\textit{size of real object}}$$

You can use the equation to calculate magnification factors quite simply. For example, if you had a photograph of your pet dog, the magnification of the image would be the height of the image of that dog divided by its real height.

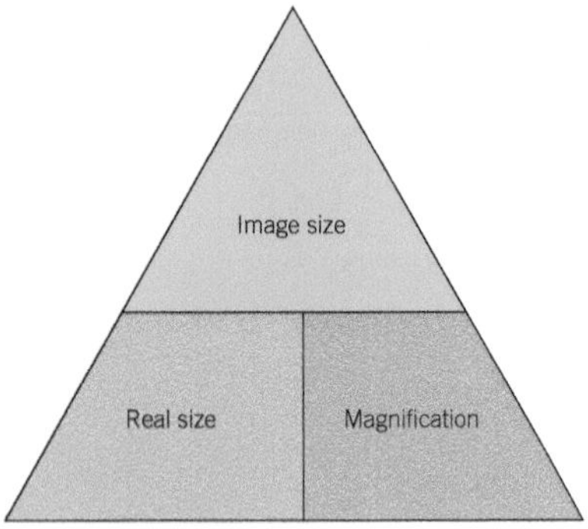

Be very careful to use the same units for each measurement! If the dog is 9 cm tall in the photograph and the real dog is 0.4 m tall you would have to convert the units before starting – 0.4 m is 40 cm, so the sum would be $9 \div 40 = 0.23$. You picture's magnification is ×0.23.

Suppose you only had the photo and the magnification. How would you find out how big the real object was? You may need to do this type of calculation on photomicrographs of cells or parts of cells.

For example, a photograph shows a mitochondrion which is 41 mm long in the picture and is taken at magnification ×34 000. How long is the original mitochondrion?

▲ **Figure 3** *The magnification triangle*

To find out, rearrange the equation. You might use an equation triangle to help.

On Figure 3 the horizontal line means divide and the vertical line means multiply.

So $\textit{magnification} = \dfrac{\textit{size of image}}{\textit{size of real object}}$

rearranges as $\textit{size of image} = \textit{magnification} \times \textit{real size}$

and $\textit{real size} = \dfrac{\textit{size of image}}{\textit{magnification}}$

You need to find the real size of the mitochondrion, so your sum will be:

$\textit{real size} = \dfrac{41}{34\,000} = 0.0012\,\text{mm}$

At this point you need to check that your units are sensible. A mitochondrion is so small that the appropriate unit of measurement is a micrometre (μm). The question may even ask you to use this unit. Earlier in this chapter you saw that 1 μm is 1/1000th of a mm, so to convert you need to multiply by 1000. The real mitochondrion is $0.0012 \times 1000 = 1.2\,\mu\text{m}$ long.

Gathering data and making measurements

Estimating results

When measuring and recording data it is useful to be able to make an estimate of the number you should be getting. This will allow you to judge whether the results you actually record seem believable. This is especially important when using a calculator, because it is easy to mistype an entry and get a wrong answer. An estimate is really a sensible guess. It is a good idea to practise this skill, for example, when collecting data from practical work in class.

Uncertainties in measurements

When making measurements, even using good quality instruments such as rulers and thermometers, there is a certain level of doubt in the precision of the measurement obtained. This is the **uncertainty of measurement**. The uncertainty can be stated, in which case a margin of error is identified. For example, measurements made using a good millimetre scale ruler may be reported as ±0.5 mm, which is the maximum error likely when using the ruler carefully.

Percentage error is a way of using the maximum error to calculate the possible total error in a given measurement. Some types of instrument have maximum errors written on them, for example a balance may state ±0.01 g. Other devices such as rulers and thermometers may rely on common sense, for example, ±0.5 mm or ±0.5 °C when recorded by eye. To find percentage error use the formula:

$$\textit{percentage error} = \frac{\textit{maximum error}}{\textit{measured value recorded}} \times 100$$

For example, with a ruler the maximum likely error is usually 0.5 mm. If an object is measured at 6 mm with the ruler:

percentage error $= \frac{0.5}{6} \times 100 = 8.3\%$.

A larger object will have a smaller percentage error because the ±0.5 mm is a lesser part of the total recorded. For example, an object measured at 87 mm:

percentage error $= \frac{0.5}{87} \times 100 = 0.6\%$.

Working with graphs and charts

Choosing the right type of graph or chart

During your course you will most commonly use line graphs, bar charts, and histograms. You need to be able to choose the right one to suit the data and be able to draw graphs accurately.

The first part of your decision depends on the type of independent variable that you have measured. When you have used an independent variable that has specific values on a continuous scale, such as temperature, you should use a line graph, for example, oxygen volume consumed by woodlice in a respirometer at a variety of temperatures. Alternatively your data might be in discrete categories, for example, the number of left-handed or right-handed people. For this data a bar chart should be used with a space between each bar. When your categoric data is in groups that can be arranged on a continuous scale, for example, height categories of plants such as 0 to 1 cm, >1 to 2 cm, >2 to 3 cm and so on, a histogram should be chosen, in which the bars are not separated by gaps.

Plotting the graph or chart

The rules when plotting the graph are:

- Ensure that the graph occupies the majority of the space available (this means more than half the space).

- Mark axes using a ruler and divide them clearly and equidistantly (i.e., 10, 20, 30, 40 not 10, 15, 20, 30, 45).
- Ensure that the dependent variable that you measured is on the *y* axis and the independent variable that you varied is on the *x* axis.
- Ensure that both axes have full titles and units clearly labelled, for example, pH of solution, not just 'pH', and mean height / m, not just 'height'.
- Plot the points accurately using a sharp pencil and '×' marks so the exact position of the point is obvious and is not obscured when you plot a trend line.
- Draw a neat best-fit line, either a smooth curve or a ruled line. It does not have to pass through all the points. Alternatively use a point to point ruled line, which is often used in biology where observed patterns do not necessarily follow mathematically predictable trends.
- Confine your line to the range of the points. Never extrapolate the line beyond the range within which you measured. Extrapolation is conjecture! A common mistake is to try and force the plotted line to go through the origin.
- Distinguish separate plotted trend lines using a key.
- Add a clear, concise title.
- Where data ranges fall a long way from zero, a broken axis will save space. For example, if the first value on the *y* axis is 36 it may be sensible to start the axis from 34 rather than zero. This will avoid leaving large areas of your graph blank.

You will be expected to follow these conventions. If you do, then questions that involve drawing a graph become easy.

Adding range bars and error bars to your plotted points

The position of the point on a graph is always subject to uncertainty. It may be a mean value, which will depend on the values averaged or whether you include or exclude any possible anomalies. A way of indicating the level of certainty in the positioning of your points is to use a range bar or error bar. These are ways of pictorially indicating the possible range of positions of the point and reflect the spread or variability in the original measurements that were averaged. The more spread the measurements, the less certain the position of the mean when plotted.

Table 4 shows some example data from an experiment on gas production by a photosynthesising plant at different temperatures, with which the different styles can be demonstrated.

▼ **Table 4**

Temperature / °C	Time taken to collect 10 cm^3 of gas / s						*s*	*mean +s*	*mean −s*
	1	2	3	4	5	mean			
15	**87**	95	102	**121**	117	104	14.4	118.4	89.6
20	67	78	**61**	**90**	86	76	12.3	88.3	63.7
25	57	59	**48**	**66**	51	56	7.0	63	49
30	**47**	45	39	42	**21**	39	10.4	49.4	28.6
35	**118**	123	**145**	136	132	131	10.7	141.7	120.3

Range bars

A range bar is the simplest way of showing the spread in the data. Look for the maximum and minimum values in each set of repeats, they are picked out in bold in the table. After plotting the points on your graph mark the positions of the maximum and minimum values above and below the point using a small bar. Join the two extremes with a neat ruler line running vertically through the plotted point (Figure 4).

Notice that the range bars are not always symmetrical above and below the plotted points. The tops and bottoms just show the largest or smallest values among the measurements made.

Error bars

To plot error bars you use the standard deviation (a calculated measure of the spread of the data), to indicate your ranges. This is better than using range bars because it reduces the effect of any extreme values in the dataset. In the table the values of standard deviation are shown in the column headed *s*.

Plot the bars by marking the top and lower limits as exactly plus and minus one standard deviation above and below the point. These values are also included in the table. The result is shown in Figure 5.

Notice that the error bars are symmetrical above and below the points, which are now indicated with a range of ± one standard deviation. The length of the bars now indicates not the maximum/minimum values but the mathematical spread in the data. The more the data spread out around the mean, the longer the error bar becomes, and the less certain you are that the mean is really accurate.

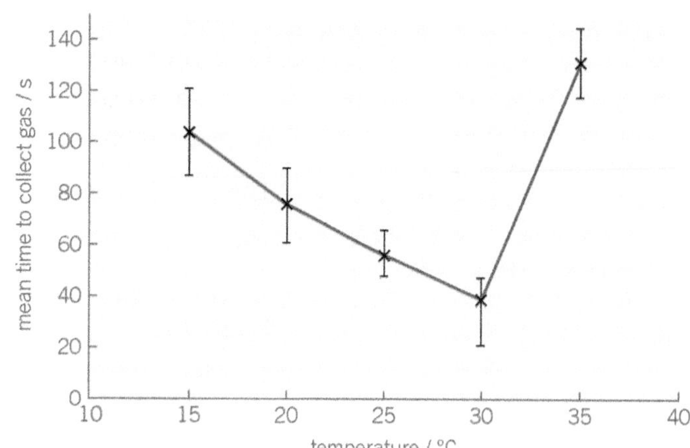

▲ **Figure 4** *Range bars*

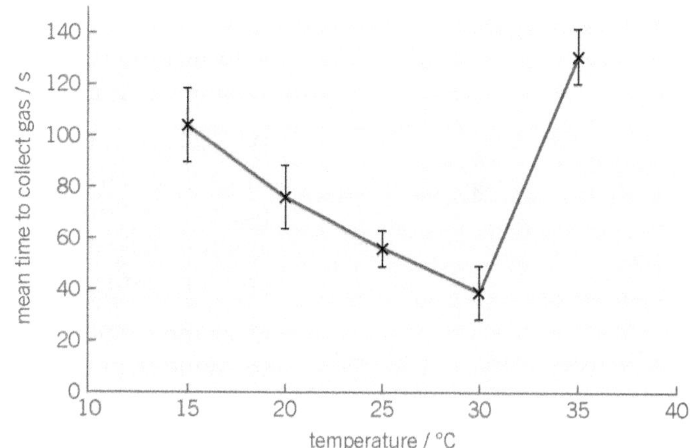

▲ **Figure 5** *Error bars*

Calculating rates from graphs

When data have been plotted on a line graph relating measured values on the y axis to time on the x axis, it is possible to calculate a rate of change for the y variable. There are two common graph forms that you will encounter, shown in Figure 6.

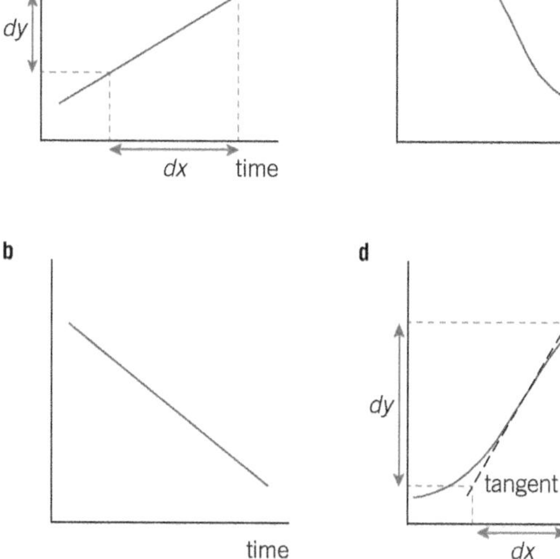

◀ **Figure 6** *Calculating rates from two common graph forms*

a might represent oxygen production by a photosynthesising plant.

b might represent oxygen consumption by a respiring organism.

c might represent pH change during a lipid digestion experiment.

d might represent growth of a bacterial population in a fermenter.

The rate of change is simply the gradient of the graph.

The formula that is used is $\frac{change\ in\ y}{change\ in\ x}$ or $\frac{dy}{dx}$

With a straight line graph follow these steps, which are marked on Figure 6a:

- select any two points on the plotted line
- use a ruler to mark construction lines from the two points to the *x* and *y* axes
- measure the difference between the two points on the *y* axis, this is *dy*
- measure the time difference between the points on the *x* axis, this is *dx*
- substitute the values into the equation and remember to quote suitable units for the result, for example, cm^3 O_2 per minute.

With a curved line the procedure is the same except that you need to start by marking a tangent against the curve, usually at its steepest point to find the maximum rate of change. This takes a bit of practise and is usually done by eye, although it is possible to calculate a position mathematically.

Once the tangent is drawn, select two points on it and proceed using the same steps that were applied to the straight line. Figure 6d has been marked with a tangent as an example.

Scatter diagrams

A scatter diagram is a method of plotting two variables in order to try and identify a correlation between them. The dependent variable is plotted on the *y* axis and the independent along the *x* axis. Once the points are plotted a trend line can be added to show a possible relationship. An example might be plotting incidence of lung cancer against number of cigarettes smoked per day. Once such a plot has been made the relationship can be tested using a statistical test, for example, the correlation coefficient, *r*.

Probability

When data appears to show a pattern, it is possible to determine whether the pattern is simply due to chance or whether it has an underlying cause. For example, 36 throws of die should give near enough six of each possible number. Throwing seven 4's and five 6's is likely to be a fluke, but if you threw twenty-three 6's then this is definitely against the rules of probability!

Probability is assessed using a statistical test, for example, chi-squared (χ^2), to test how closely the observed measurements fit with expectation, such as in genetic cross results, or the student's *t* test which compares the means of sets of data to assess whether they differ, for example, leaf widths of ivy plants grown in the sun or in the shade.

Such tests produce a calculated value that may be found in a table of probability. In these tables, it is the **probability that the data observed**

differ by chance alone that is being found. In biology we accept any probability greater than 5 per cent as likely to be just chance or fluke, but probabilities of 5 per cent or below show us that the data do differ significantly and there must be a cause influencing the outcome.

▼ **Table 5** *Formulae commonly used in biology*

Circumference of a circle	πd	d = diameter
Surface area of a cube or cuboid	$2(ab) + 2(ac) + 2(bc)$	a, b, and c are side lengths
Surface area of a sphere	$4\pi r^2$	r is the radius
Surface area of a cylinder	$2\pi r^2 + (\pi dh)$	h is the length or height of a cylinder
Volume of a cube or cuboid	$a \times b \times c$	a, b, and c are side lengths
Volume of a sphere	$\frac{4}{3}\pi r^3$	r is the radius
Volume of a cylinder	$\pi r^2 h$	r is the radius h is the height
Magnification	$magnification = \frac{image\ size}{real\ size}$	Rearrange to find the other quantities
pH	$pH = -log_{10}[H^+]$	$[H^+]$ is the concentration of the hydrogen ion in moles per litre
Pulmonary ventilation rate	$PVR = tidal\ volume \times breathing\ rate$	
Cardiac output	$CO = stroke\ volume \times heart\ rate$	
Species diversity index	$D = \frac{N(N-1)}{\Sigma n(n-1)}$	N is the grand total of all species sampled n is the number of each individual species sampled
Efficiency of energy transfer	$\frac{energy\ transferred}{energy\ intake} \times 100$	
Standard deviation, s Used to assess spread or dispersion in a set of data The higher the s, the more spread out your data is.	$s = \sqrt{\frac{\sum x^2 - \frac{(\sum x)^2}{n}}{n-1}}$	x refers to the values of the measurements taken n = the number of measurements $\sum$ means 'the sum of'
An alternative sum for standard deviation	$s = \sqrt{\frac{\Sigma(x-\bar{x})^2}{n-1}}$	$\bar{x}$ is the mean value of the values of x
Variance s^2, which is also a measure of dispersion in a set of data	$s^2 = \frac{\Sigma(x-\bar{x})^2}{n-1}$	
Standard error of the mean, SE, used to show how much your sample mean deviates from the actual population mean. The closer your mean is to the actual population mean, the smaller the SE.	$SE = \frac{s}{\sqrt{n}}$	s is the standard deviation n is your sample size
Chi² (χ^2) test, used to compare agreement between sample and expectation	$x^2 = \sum\frac{(O-E)^2}{E}$	O are the values you actually measure (observe) E are the values you expected to see
Spearman's rank coefficient, R, used to summarise the strength and direction (negative or positive) of a correlation between two variables.	$R = 1 - \frac{6\Sigma d^2}{(n^3 - n)}$	Σd^2 is the sum of the d^2 values n is the number of pairs of data collected and used

Glossary

A

acetylcholine: a chemical that functions in the brain and body of animals, including humans, as a chemical released by nerve cells to send signals to other cells.

action potential: the change in electrical potential associated with the passage of an impulse along the membrane of a muscle cell or nerve cell.

activation energy: energy required to bring about a reaction. The activation energy is lowered by the presence of *enzymes*.

active immunity: resistance to disease resulting from the activities of an individual's own immune system whereby an *antigen* induces plasma cells to produce *antibodies*.

active site: a group of amino acids that makes up the region of an *enzyme* into which the *substrate* fits in order to catalyse a reaction.

active transport: movement of a substance from a region where it is in a low concentration to a region where it is in a high concentration. The process requires the expenditure of *metabolic* energy.

aerobic: connected with the presence of free oxygen. Aerobic respiration requires free oxygen to release energy from glucose. See also *anaerobic*.

allele: one of a number of alternative forms of a *gene*. For example, the gene for the shape of pea seeds has two alleles: one for 'round' and one for 'wrinkled'.

anaerobic: Anaerobic exercise is a physical exercise intense enough to cause lactate to form.

antibiotic: a substance produced by living organisms that can destroy or inhibit the growth of microorganisms.

antibiotic resistance: the development in microorganisms of mechanisms that prevent *antibiotics* from killing them.

antibody: a protein produced by *lymphocytes* in response to the presence of the appropriate antigen.

antigen: a molecule that triggers an immune response by *lymphocytes*.

antioxidant: chemical which reduces or prevents *oxidation*. Often used as an additive to prolong the shelf-life of certain foods.

apoplastic pathway: route through the cell walls and intercellular spaces of plants by which water and dissolved substances are transported. See also *symplastic pathway*.

artificial selection: breeding of organisms by human selection of parents/gametes in order to perpetuate certain characteristics and/or to eliminate others.

atheroma: fatty deposits in the walls of arteries, often associated with high *cholesterol* levels in the blood.

ATP (adenosine triphosphate): *nucleotide* found in all living organisms, which is produced during respiration and is important in the transfer of energy.

axon: a projection of a nerve cell, or neuron, that conducts electrical impulses away from the neuron's cell body.

B

B cell (B lymphocyte): type of white blood cell that is produced and matures within the bone marrow. B lymphocytes produce *antibodies* as part of their role in *immunity*. See also *T cell*.

Benedict's test: a simple biochemical reaction to detect the presence of reducing sugars.

biodiversity: the range and variety of genes, species, and habitats within a particular region.

biomass: the total mass of living material, normally measured in a specific area over a given period of time.

Biuret test: a simple biochemical reaction to detect the presence of protein.

body mass index (BMI): a person's body mass in kilograms divided by the square of their height in metres.

C

cancer: a disease, resulting from mutations, that leads to uncontrolled cell division and the eventual formation of a group of abnormal cells called a *tumour*, from which cells may break away and form secondary tumours elsewhere in the body.

carcinogen: a chemical, a form of radiation, or other agent that causes *cancer*.

cardiac cycle: a continuous series of events which make up a single heart beat.

cardiac output: the total volume of blood that the heart can pump each minute. It is calculated as the volume of blood pumped at each beat (*stroke volume*) multiplied by the number of heart beats per minute (heart rate).

carrier molecule (carrier protein): a protein on the surface of a cell that helps to transport molecules and ions across plasma membranes.

Casparian strip: a distinctive band of suberin around the endodermal cells of a plant root that prevents water passing into xylem via the cell walls. The water is forced through the living part (*protoplast*) of the endodermal cells.

centrifugation: process of separating out particles of different sizes and densities by spinning them at high speed in a centrifuge.

cholesterol: lipid that is an important component of cell-surface membranes. Excess in the blood can lead to *atheroma*.

chromatid: one of the two copies of a *chromosome* that are joined together by a single centromere prior to cell division.

chromosome: a thread-like structure made of protein and DNA by which hereditary information is physically passed from one generation to the next.

clone: a group of genetically identical cells or organisms formed from a single parent as the result of asexual reproduction or by artificial means.

cohesion: attraction between molecules of the same type. It is important in the movement of water up a plant.

collagen: fibrous protein that is the main constituent of connective tissues such as tendons, cartilage, and bone.

community: all the living organisms present in an *ecosystem* at a given time.

complementary DNA: DNA that is made from messenger RNA in a process that is the reverse of normal transcription.

condensation: chemical process in which two molecules combine to form a more complex one with the elimination of a simple substance, usually water. Many biological *polymers*, such as polysaccharides and polypeptides, are formed by condensation. See also *hydrolysis*.

conjugation: the transfer of DNA from one cell to another by means of a thin tube between the two.

continuous variation: variation in which organisms do not fall into distinct categories but show gradations from one extreme to the other.

coronary arteries: arteries that supply blood to the cardiac muscle of the heart.

coronary heart disease (CHD): any condition, e.g., *atheroma* and *thrombosis*, affecting the coronary arteries that supply heart muscle.

correlation: when a change in one variable is reflected by a change in the second variable.

covalent bond: type of chemical bond in which two atoms share a pair of *electrons*, one from each atom.

crossing over: the process whereby a *chromatid* breaks during *meiosis* and rejoins to the chromatid of its *homologous chromosome* so that their *alleles* are exchanged.

D

denaturation: permanent changes due to the unravelling of the 3-D structure of a protein as a result of factors such as changes in temperature or pH.

diastole: the stage in the *cardiac cycle* when the heart muscle relaxes. See also *systole*.

dicotyledonous plants: any member of the class of flowering plants called Dicotyledonae. Their features include having two seed leaves (cotyledons) and broad leaves.

differentiation: the process by which cells become specialised for different functions.

diffusion: the movement of molecules or ions from a region where they are in high concentration to one where their concentration is lower.

diploid: a term applied to cells in which the nucleus contains two sets of *chromosomes*. See also *haploid*.

discontinuous variation: variation shown when the characters of organisms fall into distinct categories.

E

ecological niche: describes how an organism fits into its environment. It describes what a species is like, where it occurs, how it behaves, its interactions with other species, and how it responds to its environment.

ecosystem: all the living and non-living components of a particular area.

electron: negatively charged sub-atomic particle that orbits the positively charged nucleus of all atoms.

emphysema: a disease in which the walls of the alveoli break down, reducing the surface area for gaseous exchange, thereby causing breathlessness in the patient.

enzyme: a protein or RNA that acts as a catalyst and so alters the speed of a biochemical reaction.

epidemiology: the study of the spread of disease and the factors that affect this spread.

eukaryotic cell: a cell that has a membrane-bound nucleus and *chromosomes*. The cell also possesses a variety of other membranous organelles, such as mitochondria and endoplasmic reticulum. See also *prokaryotic cell*.

F

facilitated diffusion: diffusion involving the presence of protein *carrier molecules* to allow the passive movement of substances across plasma membranes.

G

gamete: reproductive (sex) cell that fuses with another gamete during fertilisation.

gene: section of DNA on a *chromosome* coding for one or more polypeptides.

gene pool: the total number of *alleles* in a particular population at a specific time.

H

habitat: the place where an organism normally lives and which is characterised by physical conditions and the types of other organisms present.

haemoglobin: globular protein in blood that readily combines with oxygen to transport it around the body. It comprises four polypeptide chains around an iron-containing haem group.

haploid: term referring to cells that contain only a single copy of each *chromosome*, e.g., the sex cells (*gametes*).

high-density lipoprotein (HDL): a compound of protein and lipid molecules found in blood plasma. It transports cholesterol from other cells to the liver. See also *low-density lipoprotein*.

histamine: substance released on tissue injury that causes dilation of blood vessels.

homologous chromosomes: a pair of *chromosomes*, one maternal and one paternal, that have the same gene *loci* and therefore determine the same features. They are not necessarily identical, however, as individual *alleles* of the same *gene* may vary, e.g., one chromosome may carry the allele for blue eyes, the other the allele for brown eyes. Homologous chromosomes are capable of pairing during *meiosis*.

human genome: the totality of the DNA sequences on the *chromosomes* of a single human cell.

hydrogen bond: chemical bond formed between the positive charge on a hydrogen atom and the negative charge on another atom of an adjacent molecule, e.g., between the hydrogen atom of one water molecule and the oxygen atom of an adjacent water molecule.

hydrolysis: the breaking down of large molecules into smaller ones by the addition of water molecules. See also *condensation*.

I

immunity: the means by which the body protects itself from infection.

interspecific variation: differences between organisms of different species.

intraspecific variation: differences between organisms of the same species.

ion: an atom or group of atoms that has lost or gained one or more *electrons*. Ions therefore have either a positive or negative charge.

ion channel: a passage across a cell-surface membrane made up of a protein that spans the membrane and opens and closes to allow *ions* to pass in and out of the cell.

isotonic: solutions that possess the same concentration of solutes and therefore have the same *water potential*.

isotope: variations of a chemical element that have the same number of *protons* and *electrons* but different numbers of neutrons. Whilst their chemical properties are similar they differ in mass. One example is carbon which has a relative atomic mass of 12 and an isotope with a relative atomic mass of 14.

K

kinetic energy: energy that an object possesses due to its motion.

L

locus: the position of a gene on a *chromosome*/DNA molecule.

low-density lipoprotein (LDL): a compound containing both protein and lipid molecules that occurs in blood plasma and *lymph*. It carries cholesterol from the liver to other cells in the body. See also *high-density lipoprotein*.

lumen: the hollow cavity inside a tubular structure such as the gut or a *xylem vessel*.

lymph: a slightly milky fluid found in lymph vessels and made up of *tissue fluid*, fats, and *lymphocytes*.

lymphocytes: types of white blood cell responsible for the immune response. They become activated in the presence of *antigens*. There are two types: *B lymphocytes* and *T lymphocytes*.

M

meiosis: the type of nuclear division in which the number of *chromosomes* is halved. See also *mitosis*.

mesophyll: tissue found between the two layers of epidermis in a plant leaf comprising an upper layer of *palisade cells* and a lower layer of spongy cells.

metabolism: all the chemical processes that take place in living organisms.

microvilli: tiny finger-like projections from the cell-surface membrane of some animal cells.

middle lamella: layer made up of pectins and other substances found between the walls of adjacent plant cells.

mitosis: the type of nuclear division in which the daughter cells have the same number of *chromosomes* as the parent cell. See also *meiosis*.

monomer: one of many small molecules that combine to form a larger one known as a *polymer*.

mono-unsaturated fatty acid: fatty acid that possesses a carbon chain with a single double bond. See also *polyunsaturated fatty acid*.

mutation: a sudden change in the amount or the arrangement of the genetic material in the cell.

myocardial infarction: otherwise known as a heart attack, results from the interruption of the blood supply to the heart muscle, causing damage to an area of the heart with consequent disruption to its function.

N

nucleotides: complex chemicals made up of an organic base, a sugar, and a phosphate. They are the basic units of which the nucleic acids DNA and RNA are made.

O

oral rehydration solution: means of treating dehydration involving giving, by mouth, a balanced solution of salts and glucose that stimulates the gut to reabsorb water.

osmosis: the passage of water from a region of high *water potential* to a region where its *water potential* is lower, through a partially permeable membrane.

osteoarthritis: degeneration of the cartilage of the joints, causing pain and stiffness of these joints.

oxidation: chemical reaction involving the loss of *electrons*.

P

palisade cells: long, narrow cells, packed with chloroplasts, that are found in the upper region of a leaf and which carry out photosynthesis.

passive immunity: resistance to disease that is acquired from the introduction of *antibodies* from another individual, rather than an individual's own immune system, e.g., across the placenta or in the mother's milk. It is usually short-lived.

pathogen: any microorganism that causes disease.

peptide bond: the chemical bond formed between two amino acids during *condensation*.

phagocytosis: mechanism by which cells engulf particles to form a vesicle or a vacuole.

phospholipid: triglycerides in which one of the three fatty acid molecules is replaced by a phosphate molecule. Phospholipids are important in the structure and functioning of the membranes.

photomicrograph: photograph of an image produced by a microscope.

plasmid: a small circular piece of DNA found in bacterial cells.

plasmodesmata: fine strands of cytoplasm that extend through pores in adjacent plant cell walls and connect the cytoplasm of one cell with another.

plasmolysis: the shrinkage of cytoplasm away from the cell wall that occurs as a plant cell loses water by *osmosis*.

polymer: large molecule made up of repeating smaller molecules (*monomers*).

polymerases: group of enzymes that catalyse the formation of long-chain molecules (*polymers*) from similar basic units (*monomers*).

polyunsaturated fatty acid: fatty acid that possesses carbon chains with many double bonds. See also *mono-unsaturated fatty acid*.

primary structure of a protein: the sequence of amino acids that makes up the polypeptides of a protein.

prokaryotic cell: a cell of an organism belonging to the kingdom Prokaryotae that is characterised by lacking a nucleus and membrane-bound organelles. Examples include bacteria. See also *eukaryotic cell*.

protoplast: the living portion of a plant cell, i.e., the nucleus and cytoplasm along with the organelles it contains.

Q

quaternary structure of a protein: a number of polypeptide chains linked together, and sometimes associated with non-protein groups, to form a protein.

R

reduction: chemical process involving the gain of *electrons*.

S

saturated fatty acid: a fatty acid in which there are no double bonds between the carbon atoms.

secondary structure of a protein: the way in which the chain of amino acids of the polypeptides of a protein is folded.

selective breeding: see *artificial selection*.

semi-conservative replication: the means by which DNA makes exact copies of itself by unwinding the double helix so that each chain acts as a template for the next. The new copies therefore possess one original and one new strand of DNA.

serum: clear liquid that is left after blood has clotted and the clot has been removed. It is therefore blood plasma without the clotting factors.

sinoatrial node (SAN): an area of heart muscle in the right atrium that controls and coordinates the contraction of the heart. Also known as the pacemaker.

species: a group of similar organisms that can breed together to produce fertile offspring.

stoma (plural stomata): pore, mostly in the lower epidermis of a leaf, through which gases diffuse in and out of the leaf.

stroke volume: the volume of blood pumped at each ventricular contraction of the heart.

substrate: a substance that is acted on or used by another substance or process. In microbiology, the nutrient medium used to grow microorganisms.

supernatant liquid: the liquid portion of a mixture left at the top of the tube when suspended particles have been separated out at the bottom during *centrifugation*.

symplastic pathway: route through the cytoplasm and *plasmodesmata* of plant cells by which water and dissolved substances are transported. See also *apoplastic pathway*.

systole: the stage in the *cardiac cycle* in which the heart muscle contracts. It occurs in two stages: atrial systole when the atria contract and ventricular systole when the ventricles contact. See also *diastole*.

T

T cell (T lymphocyte): type of white blood cell that is produced in the bone marrow but matures in the thymus gland. T lymphocytes coordinate the immune response and kill infected cells. See also *B cell*.

tertiary structure of a protein: the folding of a whole polypeptide chain in a precise way, as determined by the amino acids of which it is composed.

thrombosis: formation of a blood clot within a blood vessel that may lead to a blockage.

tidal volume: the volume of air breathed in and out during a single breath when at rest.

tissue: a group of similar cells organised into a structural unit that serves a particular function.

tissue fluid: fluid that surrounds the cells of the body. Its composition is similar to that of blood plasma except that it lacks proteins. It supplies nutrients to the cells and removes waste products.

transmission: the transfer of a *pathogen* from one individual to another.

transpiration: evaporation of water from a plant.

triglyceride: an individual lipid molecule made up of a glycerol molecule and three fatty acids.

tumour: a swelling in an organism that is made up of cells that continue to divide in an abnormal way.

turgid: a plant cell that is full of water. Additional entry of water is prevented by the cell wall stopping further expansion of the cell.

U

ultrafiltration: filtration assisted by blood pressure, e.g., in the formation of *tissue fluid*.

unsaturated fatty acid: a fatty acid in which there are one or more double bonds between the carbon atoms.

V

vaccination: the introduction of a vaccine containing appropriate disease *antigens* into the body, by injection or mouth, in order to induce artificial *immunity*.

W

water potential: the pressure created by water molecules. It is the measure of the extent to which a solution gives out water. The greater the number of water molecules present, the higher (less negative) the water potential. Pure water has a water potential of zero.

X

xerophyte: a plant adapted to living in dry conditions.

xylem vessels: dead, hollow, elongated tubes, with lignified side walls and no end walls, that transport water in most plants.

Answers to Summary Questions

Answers to the Practice Questions are available at
www.oxfordsecondary.com/oxfordaqaexams-alevel-biology

1.1

1 Carbon atoms readily link to one another to form a chain.

2 polymer

3 Sugar donates electrons that reduce blue copper(II) sulfate to orange copper(I) oxide.

Semi-quantitative nature of Benedict's test

1 B, E, A, D, C

2 Dry the precipitate in each sample and weigh it. The heavier the precipitate the more reducing sugar was present.

3 Once all the copper(II) sulfate has been reduced to copper(I) oxide, further amounts of reducing sugar cannot make a difference.

1.2

1 a glucose + galactose

b glucose + fructose

c glucose only

2 $C_{12}H_{22}O_{11}$ ($C_6H_{12}O_6 + C_6H_{12}O_6 - H_2O$)

3 Enzymes are denatured at higher temperatures and this prevents them functioning / enzymes lower the activation energy required.

1.3

1 starch

2 glycogen

3 α-glucose, β-glucose, starch, cellulose

4 starch, cellulose, glycogen

5 α-glucose

6 cellulose

7 starch, cellulose, glycogen

1.4

1 a triglycerides

b glycerol

c polyunsaturated

d two

e hydrophobic

2 triglyceride: three fatty acids / no phosphate group / non-polar; phospholipid: two fatty acids / one phosphate group / hydrophilic 'head' and hydrophobic 'tail'

3 Lipids provide more than twice as much energy as carbohydrate when they are oxidised. If fat is stored, the same amount of energy can be provided for less than half the mass. It is therefore a lighter storage product – a major advantage if the organism is motile.

1.5

1 peptide bond

2 condensation reaction

3 amino group (—NH_2), carboxyl group (—COOH), hydrogen atom (—H), R group

Protein shape and function

1 It has three polypeptide chains wound together to form a strong, rope-like structure that has strength in the direction of pull of a tendon.

2 They prevent the individual polypeptide chains from sliding past one another and so they gain strength because they act as a single unit.

3 The junctions between adjacent collagen are points of weakness. If they all occurred at the same point in a fibre, this would be a major weak point at which the fibre might break.

2.1

1 Magnification is how many times bigger the image is compared to the original object. Resolution is the minimum distance apart that two objects can be in order for them to appear as separate items.

2 200 times

3 10 mm

4 500 nm (0.5 μm)

5 Keep the plants in a cold, isotonic, buffered solution. Break up the cells using a mortar and pestle / homogeniser. Filter the homogenate to remove cell debris. Centrifuge the homogenate at 1000 times gravity and remove the supernatant liquid (leaving nuclei behind in the sediment). Then centrifuge the supernatant liquid at 2000–3000 times gravity. The sediment produced will be rich in chloroplasts.

6 a 1.6 μm

b 21 nm

7 a nuclei

b lysosomes

c mitochondria, lysosomes, and ribosomes

d ribosomes

2.2

1 The EM uses a beam of electrons that has a much smaller wavelength than light.

2 Electrons are absorbed by the molecules in air and, if present, this would prevent the electrons reaching the specimen.

3 a plant cell and bacteria

b all of them

c plant cell, bacterium, and virus

4 The preparation of the specimens may not be good enough.

5 The line X–Y is 25 mm (= 25 000 μm) long and represents 5 μm. Magnification is therefore 25 000 μm ÷ 5 μm = 5000 times.

2.3

1 protein synthesis

2 glucose, fructose, galactose

3 **a** mitochondrion **b** nucleus **c** Golgi apparatus **d** lysosome

4 **a** mitochondria, nucleus **b** Golgi apparatus, lysosomes **c** rough endoplasmic reticulum / ribosomes / mitochondria / smooth endoplasmic reticulum

2.4

1 a collection of similar cells aggregated together to perform a specific function

2 An artery is made up of more than one tissue (epithelial, muscle, connective tissues) whereas a blood capillary is made up of only one tissue (epithelial tissue).

3 a organ

b tissue

c organ

d tissue

2.5

1 A = absent; B = present; C = present; D = sometimes; E = sometimes; F = sometimes; G = present; H = present; I = sometimes; J = absent; K = present; L = present; M = absent; N = present

3.1

1 a substance that alters the rate of a chemical reaction without undergoing permanent change

2 They are not used up in the reaction and so can be used repeatedly.

3 The changed amino acid may no longer bind to the substrate, which will then not be positioned correctly, if at all, in the active site.

4 The changed amino acid may be one that forms hydrogen bonds with other amino acids. If the new amino acid does not form hydrogen bonds the tertiary structure of the enzyme will change, including the active site, so that the substrate may no longer fit.

3.2

1 To function enzymes must physically collide with their substrate. Lower temperatures decrease the kinetic energy of both enzyme and substrate molecules, which then move around less quickly. They hence collide less often and therefore react less frequently.

2 The heat causes hydrogen and other bonds in the enzyme molecule to break. The tertiary structure of the enzyme molecule changes, as does the active site. The substrate no longer fits the active site.

3 a High temperatures denature the enzyme and so they cannot spoil the food

b Vinegar is very acidic and the low pH will denature the enzymes and so preserve the food

3.3

1 Competitive inhibitors occupy the active site of an enzyme whereas non-competitive inhibitors attach to an enzyme at a site other than the active site.

2 Increase the substrate concentration. If the degree of inhibition is reduced it is a competitive inhibitor, if it stays the same, it is a non-competitive inhibitor.

Control of metabolic pathways

1 pH / substrate concentration (not temperature)

2 In a metabolic pathway, the product of one reaction acts as the substrate for the next reaction. By having the enzymes in appropriate sequence there is a greater chance of each enzyme coming into contact with its substrate than if the enzymes are floating freely in the organelle. This is a more efficient means of producing the end product.

3 a it would increase

b it would be unchanged

4 Advantage – the level of the end product does not fluctuate with changes in the level of substrate.
Explanation – Non-competitive inhibition occurs at a site on the enzyme other than the active site. Hence it is not affected by the substrate concentration. Therefore, in non-competitive inhibition, changes in the level of substrate do not affect the inhibition of the enzyme, nor the normal level of the end product.
Competitive inhibition involves competition for active sites. In this case the end product needs to compete with the substrate for the active sites of enzyme A. A change in the level of substrate would therefore affect how many end product molecules

combine with the active sites. As a result the degree of inhibition would fluctuate and so would the level of the end product.

4.1

1 to control the movement of substances in and out of the cell

2 hydrophobic tail

3 a phospholipid

b protein (intrinsic)

4 Any two from: lipid-soluble / small in size / have no electrical charge (or if it does, the charge should be opposite to that on the protein channels).

4.2

1 Any three from: concentration gradient / area over which diffusion takes place / thickness of exchange surface / temperature.

2 Facilitated diffusion only occurs at channels on the membrane where there are special protein carrier molecules.

3 There is no external energy used in the process. The only energy used is the in-built (kinetic) energy of the molecules themselves.

Diffusion in action

1 Only lipid-soluble substances diffuse across the phospholipid bilayer easily. Water-soluble substances like glucose diffuse only very slowly.

2 It could increase its surface area with microvilli and it could have more proteins with pores that span the phospholipid bilayer. (Note: the thickness of the cell-surface membranes does not vary to any degree.)

3 a increases two times / doubles

b no change

c decreases four times / it is one quarter

d increase two times / doubles (the CO_2 concentration is irrelevant)

4.3

1 a membrane that is permeable to water molecules (and a few other small molecules) but not to larger molecules.

2 zero

3 C, D, A, B

Osmosis and plant cells

1 Both cells have a lower water potential than pure water and so water enters them by osmosis. The animal cell is surrounded only by a thin cell-surface membrane and so it swells until it bursts. The plant cell is surrounded by a rigid cellulose cell wall. Assuming the cell is turgid, water cannot enter as the cellulose cell wall prevents the cell expanding and hence it bursting.

2 A = turgid, B = incipient plasmolysis, C = plasmolysed, D = turgid

3 solutions A, B, and D (all except C)

4.4

1 ATP releases its energy very rapidly. This energy is released in a single step and is transferred directly to the reaction requiring it.

2 ATP provides a phosphate that can attach to another molecule, making it more reactive and so lowering its activation energy. As enzymes work by lowering activation energy they have less 'work' to do and so function more readily.

3 Any three from: building up macromolecules (or named example of macromolecule) / active transport / secretions (formation of lysosomes) / activation of molecules.

4 Similarity – both use carrier proteins in the cell-surface membrane. Difference – active transport requires energy (ATP) / occurs against a concentration gradient.

5 Active transport requires energy in the form of ATP. Mitochondria supply ATP in cells and therefore they are numerous in cells carrying out active transport.

6 Diffusion, at best, can only reabsorb some of the glucose lost from the blood. The rest will be lost from the body. Active transport can absorb all the glucose, leaving none to be lost from the body.

5.1

1 respiratory gases, nutrients, excretory products, and heat

2 0.6

3 Any three from: surface area / thickness of cell-surface membrane / permeability of cell-surface membrane to the particular substance / concentration gradient of substance between inside and outside of cell / temperature.

Significance of surface area to volume ratio in organisms

1 They are very small and so have a very large surface area to volume ratio.

2 The blue whale has a very small surface area to volume ratio and so loses less heat to the water than it would if it were small.

Calculating a surface area to volume ratio

In making the calculation it is important to note that the cylinder sits on the rectangular box. This means that one end does not form part of the external surface. At the same time the equivalent area of the rectangular box also does not form part of the external surface. The area of these two discs is equal to $2 \times \pi r^2$ and must be subtracted from your calculation.

Since the surface area of the cylinder is calculated as $2\pi rh + 2\pi r^2$, you can therefore ignore the $2\pi r^2$ because this is the same as the area that must be subtracted from your calculation. The surface area of the cylinder is therefore taken to be $2\pi rh$ or $2 \times 3.14 \times 2 \times 8 = 100.48\,cm^2$. The surface area of the rectangular box is $2 \times (6 \times 5) + 2 \times (5 \times 12) + 2 \times (6 \times 12) = 324\,cm^2$. The total surface area = 100.48 + 324 = **424.48 cm²**.

The volume of the cylinder is calculated using the surface area of the base (πr^2) multiplied by its height (h). This equals $3.14 \times (2 \times 2) \times 8 = 100.48\,cm^3$. The volume of the rectangular box = $12 \times 6 \times 5 = 360\,cm^3$. The total volume is therefore 360 + 100.48 = **460.48 cm³**.

The surface area to volume ratio is 424.48 ÷ 460.48 = **0.92.**

5.2

1 diffusion over the body surface

2 by having valves that can close spiracles when the insect is inactive

3 Gas exchange requires a thin permeable surface with a large area. Conserving water requires thick, waterproof surfaces with a small area.

4 because it relies on diffusion to bring oxygen to the respiring tissues. If insects were large it would take too long for oxygen to reach the tissues rapidly enough to supply their needs.

Spiracle movements

1 It falls steadily and then remains at the same level

2 Cells use up oxygen during respiration and so it diffuses out of the tracheae and into these cells. With the spiracles closed, no oxygen can diffuse in from the outside to replace it. Ultimately, all the oxygen is used up and so the level ceases to fall.

3 the increasing level of carbon dioxide

4 It helps conserve water because the spiracles are not open continuously and therefore water does not diffuse out continuously

5 It contained more oxygen.

5.3

1 Any two from: no living cell is far from the external air / diffusion takes place in the gas phase / need to avoid excessive water loss / diffuse air through pores in their outer covering (can control the opening and closing of these pores).

2 Any two from: insects may create mass air flow – plants never do / insects have a smaller surface area to volume ratio than plants / insects have special structures (tracheae) along which gases can diffuse – plants do not / insects do not interchange gases between respiration and photosynthesis – plants do.

3 Helps to control water loss by evaporation/ transpiration.

Exchange of carbon dioxide

1 respiration

2 photosynthesis

3 At this light intensity the volume of carbon dioxide taken in during photosynthesis is exactly the same as the volume of carbon dioxide given out during respiration.

4 With stomata closed, there is little, if any, gas exchange with the environment. Whilst there will still be some interchange of gases produced by respiration and photosynthesis, neither process can continue indefinitely by relying exclusively on gases produced by the other. Some gases must be obtained from the environment. In the absence of this supply, both photosynthesis and respiration will ultimately cease and the plant will die.

5.4

1 Any two from: humans are large / have a large volume of cells; humans have a high metabolic rate / high body temperature.

2 alveoli, bronchioles, bronchus, trachea

3 The cells produce mucus that traps particles of dirt and bacteria in the air breathed in. The cilia on these cells move this debris up the trachea and into the stomach. The dirt / bacteria could damage / cause infection in the alveoli.

5.5

1 $0.48\,dm^3$

2 17.14 breaths min^{-1}. Measure the time interval between any two corresponding points on either graph that are at the same phase of the breathing cycle (e.g., two corresponding peaks on the volume graph or two corresponding troughs on the pressure graph). The interval is always 3.5 s. This is the time for one breath. The number of breaths in a minute (60 s) is therefore 60 s ÷ 3.5 s = 17.14.

3 It is essential to first convert all figures to the same units. For example $3000\,cm^3$ is equal to $3.0\,dm^3$. From the graph you can calculate that the exhaled volume is $0.48\,dm^3$ less than the maximum inhaled

volume. The exhaled volume is therefore 3.0 – 0.48 = 2.52 dm^3. If working in cm^3, the answer is 2520 cm^3.

4 The muscles of the diaphragm contract, causing it to move downwards. The external intercostals muscles contract, moving the rib cage upwards and outwards. Both actions increase the volume of the lungs. Consequently the pressure in the alveoli of the lungs is reduced.

5.6

1 a The rate of diffusion is more rapid the shorter the distance across which the gases diffuse.

b There is a very large surface area in 600 million alveoli (two lungs) and this makes diffusion more rapid.

c Diffusion is more rapid the greater the concentration gradient. Pumping of blood through capillaries removes oxygen as it diffuses from the alveoli into the blood. The supply of new carbon dioxide as it diffuses out of the blood helps to maintain a concentration gradient that would otherwise disappear as the concentrations equalised.

d Red blood cells are flattened against the walls of the capillaries to enable them to pass through. This slows them down, increasing the time for gas exchange and reducing the diffusion pathway, thereby increasing the rate of diffusion.

2 Four times greater.

Interpreting lung function measurements

1 Much air space within the lungs is occupied by fibrous tissue. This means that less air, and hence oxygen, is being taken into the lungs at each breath. In addition, the thickened epithelium of the alveoli means that the diffusion pathway is increased and so the diffusion of oxygen into the blood is extremely slow. The loss of elasticity makes ventilating the lungs very difficult. This makes it hard to maintain a diffusion gradient across the exchange surface and the patient becomes breathless in an attempt to compensate by breathing faster.

2 FEV will be lower/less because the expulsion of air when breathing out is due to the lungs springing back in the same way as a deflating balloon. To achieve this the lungs must be elastic. Fibrosis reduces elasticity and makes it difficult to breathe out.

5.7

1 Two pairs of polypeptides (α and β) link to form a spherical molecule. Each polypeptide has a haem group that contains a ferrous (Fe^{2+}) ion

2 Different base sequences in DNA – different amino acid sequences – different tertiary / quaternary structure and shape – different affinities for oxygen

3 If all oxygen molecules were released, there would be none in reserve to supply tissues when they were more active.

4 Carbon monoxide will gradually occupy all the sites on haemoglobin instead of oxygen. No oxygen will be carried to tissues such as the brain. These will cease to respire and to function, making the person lose consciousness.

Different lives – different haemoglobins: *Where you live is important*

1 At this partial pressure of oxygen, lugworm haemoglobin is 90% saturated, more than enough to supply sufficient oxygen to the tissues of a relatively inactive organism. Human haemoglobin, by contrast, is only 10% saturated – insufficient to supply enough oxygen to keep tissues alive.

2 The dissociation curve of the lugworm is shifted far to the left. This means it is fully loaded with oxygen even when there is very little available in its environment.

3 'The lugworm is not very active'. This means that it requires less oxygen and therefore what little there is in its burrow when the tide is out is sufficient to supply its needs so that it survives.

4 Respiration produces carbon dioxide. This builds up in the burrow when the tide is out. If lugworm haemoglobin exhibited the Bohr effect, it would not be able to absorb oxygen when it was present in only very low concentrations in the burrow.

5 The higher part of the beach is uncovered by the tide for a much longer period of time than the lower part. During this longer period all the oxygen in the burrow would be used up and the lugworm might die before the next tide.

6 It is shifted to the left.

Size matters

7 Unloading pressure of human haemoglobin is 5 kPa and of mouse haemoglobin is 9 kPa. Difference = 9 – 5 = 4 kPa.

8 a It unloads more readily.

b Oxygen is more readily released from haemoglobin to the tissues. This helps the tissues to respire more and so produce more heat, which helps to maintain the body temperature of the mouse.

c Even at an oxygen partial pressure of 21 kPa, mouse haemoglobin is still loaded to the maximum with oxygen.

9 Sigmoid-shaped curves, from left to right – elephant, human, shrew – because surface area to volume ratio increases in this order.

Activity counts

10 Shifted to the right because this means that oxygen is more readily released to the tissues and so the haemoglobin supplies more oxygen to enable the muscles to respire rapidly.

11 sigmoid-shaped curves, with plaice to the left of mackerel

12 The temperatures in Antarctic waters are very low. This means respiration rates of cold-blooded groups such as fish are also very low. As respiration needs oxygen, their oxygen requirements are also very low. Without haemoglobin, ice fish must rely on water alone to transport their oxygen.The amount of oxygen dissolved in water, whilst very little, is still adequate to supply their needs.

5.8

1 **a** 5 kPa **b** 90% **c** 70% (95% – 25%)

2 **a** the curve is shifted to the right

b haemoglobin has become less saturated

3 Exercising muscles release heat, shifting the curve to the right and causing the haemoglobin to release more oxygen to fuel the muscular activity.

6.1

1 mutation, meiosis, and fusion of gametes

2 mutation only

3 sampling bias, chance variation

4 by using random sampling – effectively using a computer to generate sampling sites

Investigating variation in plant pigments

1 This allows the atmosphere in the chamber to become saturated with solvent. Without this, the solvent would evaporate as it travelled up the strip.

2 Carotene

3 R_f values are characteristic for different pigments. The R_f values would be compared to those of know pigments such as anthocyanins.

6.2

1 **a** genetic **b** environmental **c** genetic

d genetic **e** environmental

f environmental

2 680

7.1

1 pentose(sugar), phosphate group, organic base

2 The bases are linked by hydrogen bonds. The molecular structures could be such that hydrogen bonds do not form between adenine and cytosine and between guanine and thymine.

3 ACCTCTGA

4 30.1%. If 19.9% is guanine then, as guanine always pairs with cytosine, it also makes up 19.9% of the bases in DNA, so together they make up 39.8%. This means the remaining 60.2% of DNA must be adenine and thymine and, as these also pair, each must make up half of this, i.e., 30.1%.

Unravelling the role of DNA

1 Alternative theories can be explored and investigated. As a result, new facts may emerge and so a new theory is put forward or the existing one is modified. In this way, scientific progress can be made.

2 A suggested explanation of something based on some logical scientific reasoning or idea.

3 The harmful bacteria in the sample could be tested to ensure they were dead, e.g., by seeing if they multiply when grown in ideal conditions. Dead bacteria cannot multiply.

4 The probability of the mutation happening once is very small. The probability of the same mutation occurring each time the experiment is repeated is so minute that it can be discounted.

5 Society will probably be affected by new discoveries and so is entitled to say how they can or cannot be used.

A prime location

1 It means 'in life'. In other words the synthesis of DNA by a living organism rather than in a laboratory.

2 Enzymes are very specific. They have active sites that are of a specific shape that fits their substrate. The shape of the 3′ end of the molecule with its hydroxyl group fits the active site of DNA polymerase whereas the shape of the 5′ end does not.

7.2

1 In prokaryotic cells the DNA is smaller, circular, and is not associated with proteins (i.e., does not have chromosomes). In a eukaryotic cell it is larger, linear, and associated with proteins / histones to form chromosomes.

2 It fixes the DNA into position.

3 It is looped and coiled a number of times.

7.3

1 TACGATGC

2 because half the original DNA is built into the new DNA strand

3 The linking together of the new nucleotides could not take place. Whilst the nucleotides would match up to their complementary nucleotides on the original DNA strand, they would not join together to form a new strand.

8.1

1 a base sequence of DNA that codes for the amino acid sequence of a polypeptide or functional RNA

2 18

3 A different base might code for a different amino acid. The sequence of amino acids in the polypeptide produced will be different. This change to the primary structure of the protein might result in a different shaped tertiary structure. The enzyme shape will be different and may not fit the substrate. The enzyme–substrate complex cannot be formed and so the enzyme is non-functional.

4 a 5

b the first and last (5th) / the two coded for by the bases TAC

c because some amino acids have up to six different codes, whereas others have just one triplet

Interpreting the genetic code

1 Trp – UGG and Met – AUG

2 a Leu

b Lys

c Asp

3 a Tyr-Ala-Ile-Pro-Ser

b Arg-Phe-Lys-Gly-Leu

8.2

1 The enzyme RNA polymerase moves along the template DNA strand, causing the bases on this strand to join with the individual complementary nucleotides from the pool that is present in the nucleus. The RNA polymerase adds the nucleotides one at a time, to build a strand of pre-RNA until it reaches a particular sequence of bases on the DNA that it recognises as a stop code.

2 DNA helicase – this acts on a specific region of the DNA molecule to break the hydrogen bonds between the bases, causing the two strands to separate and expose the nucleotide bases in that region.

3 Splicing is necessary because pre-mRNA has nucleotide sequences derived from introns in DNA. These introns are non-functional and, if left on the mRNA, would lead to the production of non-functional polypeptides or no polypeptides at all. Splicing removes these non-functional introns from pre-mRNA.

4 a UACGUUCAGGUC

b Four amino acids (one amino acid is coded for by three bases so 12 bases code for four amino acids)

5 Some of the base pairs in the genes are introns (non-functional DNA). These introns are spliced from pre-mRNA so the resulting mRNA has fewer nucleotides.

8.3

1 ribosome

2 a UAG on tRNA

b TAG on DNA

3 A tRNA molecule attaches an amino acid at one end and has a sequence of three bases, called an anticodon, at the other end. The tRNA molecule is transferred to a ribosome on an mRNA molecule. The anticodon on tRNA pairs with the complementary codon sequence on mRNA. Further tRNA molecules, with amino acids attached, line up along the mRNA in the sequence determined by the mRNA bases. The amino acids are joined by peptide bonds. Therefore the tRNA helps to ensure the correct sequence of amino acids in the polypeptide.

4 One of the codons is a stop codon that indicates the end of polypeptide synthesis. Stop codons do not code for any amino acid so there is one less amino acid than there are codons.

Protein synthesis

1 X = ribosome Y = mRNA

2 amino group

3 AUG

4 Val-Thr-Arg-Asp-Ser

5 CAATGGGCT

6 The mutation changes CAG to UAG. UAG is a stop codon that signifies the end of an amino acid sequence at which point the polypeptide is complete and is 'cast off'. The polypeptide chain is therefore shorter than it should be and may not function as normal.

7 a Glutamine has two codes GAG and GAA. The reversal of GAG produces the same codon and so still translates as glutamine and hence the polypeptide that is formed is unchanged. Reversal of GAA changes the codon to AAG which translates to a different amino acid – Lys. As a result, the polypeptide has a different primary structure which may affect bonding within the molecule and so change its tertiary structure also.

b Enzyme function depends on the substrate becoming loosely attached to an enzyme within its active site.

If the mutation has changed the amino acid from glutamine to Lys then the primary structure of the polypeptide will be different. Hydrogen and ionic bonds between the amino acids of the polypeptide may not be formed in the same way as before and so its tertiary shape may be changed. This change may alter the shape of the active site of the enzyme of which the polypeptide is a part and so the substrate no longer fits.

Glutamine may have been one of the amino acids in the active site to which the substrate normally attaches. Its replacement by another amino acid may mean that, although the shape of the active site is unchanged, the substrate cannot attach normally.

8.4

1 The active site of the molecule must have a specific 3-D shape which is complementary to its substrate or enzyme-substrate complexes cannot form.

2 They are no longer the right shape for enzymes to break down OR they are less soluble as the hydrophobic regions are in the wrong position.

3 Heat denatures proteins causing them to unfold. HSP 60 and 70 would make sure the cell proteins re-folded correctly so the proteins still function properly and the organism survives.

9.1

1 Haploid, because 27 is an odd number. Diploid cells have two sets of chromosomes and so their total must be an even number.

2 independent segregation of homologous chromosomes, recombination by crossing over

3 blue eyes and blood group A, brown eyes and blood group A

4 Gametes are produced by meiosis. In meiosis, homologous chromosomes pair up. With 63 chromosomes precise pairings are impossible. This prevents meiosis and hence gamete production, making them sterile.

9.2

1 They are similar to one another but different from members of other species. They are capable of breeding to produce offspring which themselves are fertile.

2 it is based on evolutionary relationships between organisms and their ancestors; it classifies species into groups using shared characteristics derived from their ancestors; it is arranged in a hierarchy in which groups are contained within larger composite groups with no overlap

3 1. phylum, 2. class, 3. order, 4. family, 5. *Rana*, 6. species, 7. *temporaria*

The difficulties of defining species

1 Fossil records are normally incomplete and not all features can be observed (there is no biochemical record) and so comparisons between individuals are hard to make. Fossil records can never reveal whether individuals could successfully mate.

2 Species change and evolve over time, sometimes developing into different species. There is considerable variety within a species. Fossil records are incomplete / non-existent. Current classifications only reflect current scientific knowledge and, as this changes, so does the naming and classifying of organisms.

3 During meiosis 1, chromosomes line up across the equator in their homologous pairs. With an odd number of chromosomes, exact pairings are not possible. This prevents meiosis occurring in the normal way.

4 No, it does not. Only fertile female mules are known, so interbreeding (a feature of any species) is impossible. The event is so rare that it can be considered abnormal and it would be wrong to draw conclusions from it. If a mule were a species, it would mean that the parents were the same species – however, donkeys and horses are sufficiently different to be recognised as separate species.

Phylogentic relationships

1 lizards

2 birds

3 Dinosaurs are extinct but all the other groups are still living and so they are shown extending further along the time line – as far as 'present'.

9.3

1 a to separate the two complementary strands of the DNA molecule

b because more complementary bases are joined together and therefore more energy is needed to break the hydrogen bonds linking them

c the more complementary bases that are joined the more similar are the DNA strands of the two species whose DNA forms the hybrid molecule. The more similar their DNA, the more closely they are related.

2 Humans and chimpanzees because the amount of precipitate formed is nearly the same showing that they have the greatest number of similar antigens.

Ethics of selective breeding in domesticated animals

1 Views of interested parties; technical feasibility; economic factors; safety issues; environmental issues; benefits versus risks, etc.

2 international and national laws / agreements / protocols

3 The process is largely a matter of luck. An organism may possess the desired features, but this does not mean that it will pass these onto the next generation. Equally an organism may possess the genes for the desired features, but it may not exhibit these characteristics and therefore may not be chosen to breed. This means the process is 'imprecise'. It is 'slow' because there will be many 'failures' along the way and it takes many generations to produce organisms that consistently exhibit the desired features. Also the number of offspring produced at each generation is small.

4 There is less genetic diversity. The male might carry a potentially harmful allele that is not apparent or not exhibited by the donor male. This harmful allele may be passed on to hundreds of offspring.

5 Any two from: cheaper milk / lower production costs; consistent quality and composition of milk; plentiful supply of milk.

6 Any two from: the welfare of cattle is at risk / cattle suffer distress / cattle are overworked; cattle are killed at a young age because they are no longer able to produce milk economically; it is unnatural / it interferes with the normal course of nature; it reduces genetic diversity / there is more uniformity / fewer alleles in the population – and potentially valuable alleles / characteristics are lost; cattle that require over-wintering in heated sheds contribute to global warming.

10.1

1 **a** increase **b** decrease **c** decrease **d** increase

2 Different DNA – different codes for amino acids – different amino acids – different protein shape – different protein function (e.g., non-functional enzyme) – change in a feature determined by that protein – altered appearance – greater genetic diversity.

3 Drop in population numbers due to chance event – few surviving individuals likely to have fewer, less diverse alleles – as population grows its alleles are equally less diverse – reduced genetic diversity.

10.2

1 the number of different species and the proportion of each species within a given area / community

2

Species	Numbers in salt marsh	$n(n-1)$
Salicornia maritima	24	24(23) = 552
Halimione portulacoides	20	20(19) = 380
Festuca rubra	7	7(6) = 42
Aster tripolium	3	3(2) = 6
Limonium humile	3	3(2) = 6
Suaeda maritima	1	1(0) = 0
	$\sum n(n-1)$	986

$$d = \frac{58(57)}{986} = \frac{3306}{986} = \mathbf{3.35}$$

3 It measures both the number of species and the number of individuals. It therefore takes account of species that are only present in small numbers.

Species diversity and ecosystems

1 Greenhouse gases lead to climate change. Communities with a high species diversity index are likely to include at least one species adapted to withstand the change and therefore survive. When the index is low, the community is less likely to include a species adapted to withstand the change and is therefore at greater risk of being damaged.

2 a The community fluctuates in line with environmental change – rising and falling in the same way but a little later in time.

b Communities with a high species diversity are more stable because they have a greater variety of species and therefore are more likely to have species that are adapted to the changed environment. Those with a low species diversity are less stable because they have fewer species and are less likely to include a species adapted to the change.

10.3

1 The few species possessing desirable qualities are selected for and bred. Other species are excluded, as far as possible, by culling or the use of pesticides. Many individuals of a few species = low species diversity.

2 Forests, with their many layers, have many habitats with many different species, i.e., a high species diversity. Grasslands have a single layer, fewer habitats, fewer species, and lower species diversity.

3 because tropical rainforests have a greater number and variety of species than any other ecosystem, i.e., the highest species diversity

Human activity and loss of species in the UK

1 500 000 km (350 000 × 100 ÷ 70)

2 Mixed woodlands comprise many species whereas the commercial conifer plantations that replace them are largely of a single predominant species

3 Any one from: benefit – cheaper grazing / fodder for animals and hence cheaper food / more efficient food production; risk – loss of species diversity / less stable ecosystem / more fertilisers and pesticides needed.

4 It provides evidence to inform and support decision-making. Data show where the most change has occurred and therefore the habitats most at risk. These can be prioritised and measures taken to conserve them, e.g., by giving them special protection. Funds can be directed towards reverting land to its former use, e.g., by grants to farmers to create hay meadows / convert to woodland / re-establish hedgerows. Helps decision makers form appropriate rules / legislation to prevent habitat destruction, e.g., ban on drainage of certain sites / rock removal. Informs decisions on planning applications for planting forests / reclaiming land.

5 Hedges provide more habitats / niches / food sources and therefore more species can survive. Species diversity is therefore increased.

11.1

1 a microorganism that causes disease

2 Diffusion takes place in these systems so they have a large surface area, are thin, moist, and well supplied with blood vessels. This makes it easy for pathogens to attach to and penetrate them.

3 by damaging host tissues, by producing toxins

4 A person with gastroenteritis has vomiting and diarrhoea. Both symptoms mean that the antibiotic is unlikely to remain in the body long enough to be absorbed.

11.2

1 correlation between the relative risk of lung cancer and the number of cigarettes smoked before stopping; correlation between the relative risk of lung cancer and the years since giving up smoking

2 There is no experimental evidence in the data provided to show that smoking causes cancer. Hence there is no causal link between the two variables.

3 the risk of getting lung cancer when compared to a non-smoker

Hill's Criteria of Causation

1 Some people develop CHD who do not smoke so temporality cannot always be established. There is a correlation between smoking and CHD but because there are other risk factors, the correlation is not a as strong as that between smoking and lung cancer. There is theoretical plausibility – nicotine increases heart rate and hence blood pressure. Higher blood pressure increases the risk of atherosclerosis. Nicotine also increases the stickiness of platelets increasing the risk of blood clots (thrombosis). However, CHD is a multifactorial disease with links to factors such as diet, exercise and genetic factors so criteria 7 - specificity – cannot be met.

Smoking and lung cancer

1 a 1940–50 b 1970–80

2 The lines for cigarettes smoked and deaths from lung cancer are a similar shape for both sexes (although separated in time).

3 Any three from: an increase in smoking / more air pollution / an overall increase in the population of the UK / any other reasonable answer.

4 Lung cancer develops slowly and so the patient dies many years after it has been first contracted.

11.3

1 giving up smoking / not starting smoking

2 by lowering blood pressure; by lowering blood cholesterol; by reducing the risk of obesity

3 drink less alcohol, eat more fruit and vegetables, reduce salt intake, eat less red and processed meat, reduce calorie intake if overweight (i.e., body mass index is above 25)

Smoking and disease

1 air pollution / inhaled substances (carcinogens), e.g., asbestos at work

2 a positive correlation between the incidence of lung cancer in men and the number of cigarettes smoked

3 It is unlikely that a coincidence would have occurred many times over.

4 No. Whilst the data clearly point to the likelihood that smoking causes lung cancer, they do not provide experimental evidence that specifically links smoking to lung cancer.

5 The experiment that showed that the derivative of benzopyrene caused changes to DNA at precisely the same three points as the mutations of the gene in a cancer cell.

6 This is a single case. The link between early death and lung cancer is about probabilities not certainties. Statistically it is unlikely, but not impossible, for smokers to live to be very old.

12.1

1 mouth – where teeth break up the food into small pieces, stomach – where the muscular wall churns the food

2 It releases enzymes via a duct into the duodenum and also hormones like insulin directly into the blood.

3 the expulsion of faeces from the rectum via the anus

12.2

1 the breakdown of molecules by the addition of water to the bonds that hold these molecules together

2 salivary glands, pancreas

3 Villi and microvilli increase surface area to speed up the absorption of soluble molecules. As the food in the stomach has not yet been hydrolysed into soluble molecules they cannot be absorbed and so villi and microvilli are unnecessary.

4 α-glucose

5 maltase, sucrase, lactase

Lactose intolerance

1 a respiration

b Carbon dioxide is formed in aerobic respiration, whereas conditions in the colon are anaerobic.

2 Modern storage and distribution methods mean that milk and milk products are readily available. Without these our ancestors rarely consumed milk as adults.

3 Low water potential in the colon causes water to move from epithelial cells into the lumen of the colon creating watery stools.

12.3

1 Endoplasmic reticulum to re-synthesise triglycerides from monoglycerides and fatty acids.

Golgi apparatus to form chylomicrons from triglycerides, cholesterol, and lipoproteins.

Mitochondria to provide ATP required for the co-transport of glucose and amino acid molecules.

2 sodium ions

3 an increase in the number/density of protein channels and carrier proteins

4 a active

b passive

c indirect active

Absorption of fatty acids

1 When glycerol or phosphate is added on its own, there is no increase in the absorption of fatty acids compared to when neither is present.

When glycerol and phosphate are both added together, the relative amount of absorption doubles compared to when neither is present.

When glycerol and phosphate are provided as a compound, the relative amount of absorption increases even further (to almost three times as much) compared to when neither is present.

2 There is no absorption when the inhibitor of phosphorylation (iodoacetate) is present.

13.1

1 DOMAIN – Eubacteriae

GENUS – *Vibrio*

2 Cholera toxin opens chloride channels in the cell-surface membrane of intestinal epithelial cells. Chloride ions flood into the lumen of the intestine. Water potential of epithelial cell rises whilst that of the lumen falls. Water moves into lumen by osmosis, causing watery stools (diarrhoea).

Transmission of cholera

1 Any three from: clean, uncontaminated water supplies / water treatment / chlorination/ proper sanitation and sewage treatment / personal hygiene (washing hands after using the toilet) / proper food hygiene.

2 Breast milk does not contain the bacterium that causes cholera (and contains antibodies that provide some resistance to it). Bottle-fed babies are given milk made up with water and this may be contaminated with bacteria that cause cholera.

3 It could prevent the bacterium penetrating the mucus barrier of the intestinal wall and so stop it reaching the epithelial cells.

4 The antibiotic may be digested by enzymes and therefore not function. / The antibiotic may be too large to diffuse across the intestinal epithelium. / Severe diarrhoea is a symptom of cholera. Therefore any antibiotic taken orally may pass through the intestines so rapidly that it is passed out of the body before it can come into contact with the cholera bacterium. Taken via the blood the antibiotic is not digested / does not have to diffuse into the body / can reach the bacterium and kill it.

13.2

1 a Glucose stimulates the uptake of sodium ions from the intestine and provides energy as it is a respiratory substrate.

b The sodium ions replace those lost from the body and encourage the use of the sodium-glucose transporter proteins to absorb more sodium ions.

c Boiling the water will kill any diarrhoeal pathogens that would otherwise make the patient's condition worse.

2 Potassium in the banana replaces the potassium ions that have been lost. It also stimulates the appetite and so aids recovery.

3 Banana improves the taste and so makes it easier for children to drink the mixture.

4 Too much glucose might lower the water potential within the intestine to a level below that within the epithelial cells. Water will then pass out of the cells by osmosis, increasing dehydration.

Developing and testing improved oral rehydration packs

1 Each species has different physical and chemical features and therefore may respond differently to the same drug. What is safe for some other animal may be harmful to a human.

2 It acts like a control experiment. Changes in the patients taking the real drug can be compared with patients taking the placebo to see whether they are due to the drug or to some other factor.

3 There is no risk of any deliberate or unwitting bias by the patients. Those knowing they are on the real drug might wrongly attribute changes in their symptoms to the drug.

13.3

1 It possesses RNA and the enzyme reverse transcriptase which can make DNA from RNA – a reaction that is the reverse of that carried out by transcriptase.

2 HIV is a virus – the human immunodeficiency virus – whilst AIDS (acquired immune deficiency syndrome) describes the condition caused by infection with HIV.

3 People with impaired immune systems, such as those with AIDS, are far less able to protect themselves from TB infections and so are more likely to contract and spread TB to others. Widespread use of condoms helps prevent HIV infection and so can reduce the number of people with impaired immune systems who are consequently more likely to contract TB.

4. reverse transcriptase inhibitors stop cDNA being formed, integrase inhibitors stop viral DNA being incorporated into the genomic DNA, and protease inhibitors stop viral particles being synthesised

14.1

1 A specific mechanism distinguishes between different pathogens but responds more slowly than a non-specific mechanism. A non-specific mechanism treats all pathogens in the same way but responds more rapidly than a non-specific mechanism.

2 The lymphocytes that will finally control the pathogen need to build up their numbers and this takes time.

3 The body responds immediately by 'recognising' the pathogen (and by phagocytosis); the delay is in building up numbers of lymphocytes and therefore controlling the pathogen.

14.2

1 **a** phagocytosis **b** phagosome **c** lysozymes **d** lysosomes

2 The protective covering of the eye, and especially the tear ducts, are potential entry points for pathogens. The eyes are vulnerable to infection because the coverings are thin to allow light through. Lysozyme will break down the cell walls of any bacterial pathogens and so destroy them before they can cause harm.

14.3

1 An organism or substance, usually a protein, that is recognised as foreign by the immune system and therefore stimulates an immune response.

2 Any two from: both are types of white blood cell / have a role in immunity / are produced from stem cells.

3 T cells mature in the thymus gland whereas B cells mature in the bone marrow; T cells are involved in cell-mediated immunity whereas B cells are involved in humoral immunity.

Bird flu

1 H5N1 infects the lungs, leading to a massive production of T cells. Accumulation of these cells may block the airways / fill the alveoli and cause suffocation.

2 Birds carry H5N1 virus. They can fly vast distances across the world in a very short space of time.

14.4

1 In the primary response, the antigens of the pathogen have to be ingested, processed, and presented by B cells. Helper T cells need to link with the B cells that then clone, with some of the cells developing into the plasma cells that produce antibodies. These processes occur consecutively and therefore take time. In the secondary response, memory cells are already present and the only processes are cloning and development into the plasma cells that produce antibodies. Fewer processes means a quicker response.

2 Examples of differences include:

Cell-mediated immunity	Humoral immunity
Involves T cells	Involves mostly B cells
No antibodies	Antibodies produced
First stage of immune response	Second stage of immune response after cell-mediated stage
Effective through cells	Effective through body fluids

3 rough endoplasmic reticulum – to make and transport the proteins of the antibodies; Golgi apparatus – to sort, process, and compile the proteins; mitochondria – to release the energy needed for such massive antibody production

14.5

1 Active immunity – individuals are stimulated to produce their own antibodies. Immunity is normally long-lasting.

Passive immunity – antibodies are introduced from outside rather than being produced by the individual. Immunity is normally only short-lived.

2 The influenza virus displays antigen variability. Its antigens change frequently and so antibodies no longer recognize the virus. New vaccines are required to stimulate the antibodies that complement the new antigens.

MMR vaccine

1 the MMR vaccine is given at 12–15 months – the same time as autism symptoms appear

2 It might: present the findings in an incomplete / biased fashion, ignore unfavourable findings, fund only further research that seems likely to produce the evidence that its seeks rather than investigating all possible outcomes, withdraw funding for research that seems likely to produce unfavourable findings.

15.1

1 **a** pulmonary artery **b** aorta
c hepatic vein **d** pulmonary vein
e aorta

2 low surface area to volume ratio; a high metabolic rate

3 It increases blood pressure and hence the rate of blood flow to the tissues.

15.2

1 **a** elastic tissue allows recoil and hence maintains blood pressure / smooth blood flow / constant blood flow

b muscle can contract, constricting the lumen of the arterioles and therefore controlling the flow of blood into capillaries

c valves prevent flow of blood back to the tissues and so keep it moving towards the heart / keep blood at low pressure flowing in one direction

d the wall is very thin, making the diffusion pathway short and exchange of material rapid

2 **a** C **b** B **c** E
d D **e** A

3 hydrostatic pressure (due to pumping of the heart)

4 via the capillaries and via the lymphatic system

Blood flow in various blood vessels

1 Rate of blood flow decreases gradually in the aorta and then very rapidly in the large and small arteries. It remains relatively constant in the arterioles and capillaries before increasing, at an increasing rate, in the venules and veins and vena cava.

2 Contraction of the left ventricle of the heart causes distension of the aorta. The elastic layer in the aorta walls creates a recoil action. There is therefore a series of pulses of increased pressure, each one the result of ventricle contraction.

3 Because the total cross-sectional area is increasing / there is increased frictional resistance from the increasing area of blood vessel wall.

4 Blood flow is slower, allowing more time for metabolic materials to be exchanged.

5 Capillaries have a large surface area and very thin walls (single cell thick) and hence a short diffusion pathway.

15.3

1 coronary artery

2 **a** deoxygenated **b** deoxygenated
c oxygenated

3 pulmonary vein left atrium left ventricle aorta vena cava right atrium right ventricle pulmonary artery

4 The mixing of oxygenated and deoxygenated blood would result in only partially oxygenated blood reaching the tissues and lungs. This would mean the supply of oxygen to the tissues would be inadequate and there would be a reduced diffusion gradient in the lungs, limiting the rate of oxygen uptake.

15.4

1 left ventricle

2 **a** true **b** true **c** false
d true **e** false

3 **a** atria **b** sinoatrial node
c bundle of His

4 Training builds up the muscles of the heart and so the stroke volume increases / more blood is pumped at each beat. This means that, if the cardiac output is the same, the heart rate / number of beats per minute decreases.

5 One complete cycle takes 0.8 s. Therefore the number of cycles in a minute = 60 ÷ 0.8 = 75. As there is one beat per cycle then there are 75 beats in a minute.

The effect of exercise on heart rate

1 Taking the pulse for longer could result in miscounting and lead to inaccuracies.

2 As muscles work harder there is increased demand for energy in the form of ATP. Most ATP comes from aerobic respiration, which requires oxygen. Heart rate increases to deliver more oxygen to respiring cells.

3 Subjects may have coronary heart disease. The increased heart rate during a step test might result in a myocardial infarction (a heart attack).

15.5

1 **a** myocardial infarction **b** atheroma
c thrombosis **d** aneurysm

2 the heart must work harder and is therefore prone to failure; an aneurysm may develop, causing a haemorrhage; the artery walls may thicken, restricting blood flow

A calculated risk

1 Reducing blood pressure – for any given cholesterol level, this will reduce the risk more than giving up smoking, e.g., at 8 mmol dm^{-3} the risk falls from 23% to 15% by giving up smoking but falls to 12% by lowering blood pressure.

2 At 5 mmol dm^{-3}, the risk is 5%. At 8 mmol dm^{-3}, the risk is 15%. The risk is therefore 15 ÷ 5 = 3 times greater.

3 The man who increases his blood cholesterol level is at greater risk. The risk of the man who starts to smoke increases from 2.5% to 3.5% = +1%. The risk of the man who increases his cholesterol level increases from 2.5% to 5.5% = +3%.

Electrocardiogram

1 a = normal – large peaks and small troughs repeated identically; b = heart attack – less pronounced peaks and smaller troughs repeated in a similar, but not identical, way; c = fibrillation – highly irregular pattern

16.1

1 Soil solution has a very high water potential, the root hair cell has a lower water potential (due to dissolved sugars, amino acids, and ions inside it). Water moves by osmosis from soil solution into the root hair cell.

2 apoplastic pathway takes place in cell walls and symplastic pathway in cytoplasm; apoplastic pathway occurs by cohesion and symplastic pathway by osmosis

3 because endodermal cells have a waterproof band / Casparian strip that prevents the passage of water

4 **a** passive **b** passive
c passive (the ions are actively transported but the **water** moves passively (osmosis))

5 The root hairs are delicate and may get broken. There may be air pockets rather than soil solution around the roots when transplanting.

16.2

1 **a** transpiration **b** stomata
c lower / reduced / more negative
d osmosis **e** cohesion
f increases **g** root pressure

Hug a tree

1 At 12.00 hours because this is when water flow is at its maximum. As transpiration creates most of the water flow they are both at a maximum at the same time.

2 Rate of flow increases from a minimum at 00.00 hours to a maximum at 12.00 hours and then decreases to a minimum again at 24.00 hours.

3 As evaporation / transpiration from leaves increases during the morning (due to higher temperature / higher light intensity) it pulls water molecules through the xylem because water molecules are cohesive / stick together. This transpiration pull creates a negative pressure / tension. The greater the rate of transpiration, the greater the water flow. The reverse occurs as transpiration rate decreases during the afternoon and evening.

4 As transpiration increases up to 12.00 hours, so there is a higher tension (negative pressure) in the xylem. This reduces the diameter of the trunk. As transpiration rate decreases, from 12.00 hours to 24.00 hours, the tension in the xylem reduces and the trunk diameter increases again.

5 Transpiration pull is a passive process / does not require energy. Xylem is non-living and so cannot provide energy. Although root cortex and leaf mesophyll cells are living, the movement of water across them uses passive processes, e.g., osmosis, and so continues at least for a while, even though the cells have been killed.

16.3

1 Plants photosynthesise and therefore have a large surface area to collect light and stomata through which carbon dioxide diffuses (necessary). Both features lead to a considerable loss of water by transpiration (evil).

2 a decreases b increases c increases
d increases e increases

3 Any two from: high humidity as water vapour cannot escape / still air / darkness – stomata close.

Measurement of water using a potometer

1 a As xylem is under tension, cutting the shoot in air would lead to air being drawn into the stem, which would stop transport of water up the shoot. Cutting under water means water, rather than air, is drawn in and a continuous column of water is maintained.

b Sealing prevents air being drawn into the xylem and stopping water flow up it / sealing prevents water leaking out which would produce an inaccurate result.

2 that all water taken up is transpired

3 Volume of water taken up in one minute: $3.142 \times (0.5 \times 0.5) \times 15.28 = 12.00\,mm^3$. Volume of water taken up in 1 hour: $12.00 \times 60 = 720\,mm^3$

4 their surface area / surface area of the leaves

5 An isolated shoot is much smaller than the whole plant / may not be representative of the whole plant / may be damaged when cut.

Conditions in the lab may be different from those in the wild, e.g., less air movement / greater humidity / more light (artificial lighting when dark).

16.4

a phloem b sources
c sinks d mass flow
e sieve tube f co-transport
g photosynthesising / chloroplast containing
h lower / more negative
i xylem j higher / less negative
k osmosis

16.5

1 a There would be a large swelling above the ring in summer but little, if any, swelling in winter.

b In summer the rate of photosynthesis, and therefore production of sugars, is greater due to higher temperatures, longer daylight, and higher light intensity. The translocation of these sugars leads to their accumulation, and therefore a swelling, above the ring. In winter lower temperatures, shorter daylight, and lower light intensity mean the rate of photosynthesis is less and any swelling is therefore smaller. In deciduous plants, the lack of leaves means there is no photosynthesis and therefore no swelling at all.

2 If the squirrel strips away the phloem around the whole circumference of the branch it may not have sufficient sugar for its respiration to release enough energy for survival as none can reach it from other parts of the plant.

3 If the branch has sufficient leaves to supply its own sugar needs from photosynthesis, rather than depending on supplies from elsewhere, it might survive for a while at least.

4 It is unlikely that squirrels would strip bark from around the whole circumference of a large tree trunk. Any intact phloem could still supply sufficient sugars to its roots to allow it to survive.

5 a It takes time for the sucrose from the leaves to be transported across the mesophyll of the leaf by diffusion and then to be actively transported into the phloem.

b The sucrose in the phloem is diluted with the water that enters it from the xylem. A little sucrose may be converted to glucose and used up by the leaves during respiration but this alone would not be sufficient to explain the reduction in concentration in the phloem.

Using radioactive tracers

1 The data suggest that the translocation of ^{42}K is almost entirely in the xylem with very little in the phloem.

2 The wax paper is 'impervious' which means that materials cannot pass across it. In the middle of the region where xylem and phloem are separated by wax paper, 99% of the ^{42}K is in the xylem. Even at the beginning and ends of the separated regions at least 85% of the ^{42}K is in the xylem. Where the two are in contact (control) the levels of ^{42}K are much more equal.

3 In sections 1 and 5, xylem and phloem are not separated by wax paper and so lateral movement of ^{42}K can take place. The ^{42}K therefore diffuses from the xylem into the phloem until the concentrations in both are similar.

4 The xylem and phloem could have been separated over the 225 mm portion of the control branch and then rejoined but without the wax paper. This is an improvement as it eliminates the physical disruption caused by separating xylem and phloem as the explanation for there being no lateral movement of ^{42}K.

16.6

1

Kingdom	Animalia
Phylum	Arthropoda
Class	Insecta
Order	Hemiptera
Family	Aphididae
Genus	Myzus
Species	persicae

2 They can survive periods of environmental stress as eggs. They can reproduce through parthenogenesis giving rise to large numbers of offspring in a short space of time. Winged females can move to a fresh plant and new colonies can start

3 Symptoms such as yellowing or distorted leaves reduce photosynthesis. In cereals this can mean a loss in size and numbers of grain. Fruits can be distorted or discoloured and these are less valuable.

17.1

1 a interphase b prophase
c spindle d nuclear envelope
e nucleolus f metaphase
g anaphase h growth / repair
i growth / repair

Recognising the stages of mitosis

1 A = telophase – chromosomes in two sets, one at each pole; B = prophase – chromosomes visible but randomly arranged; C = interphase – no chromosomes visible; D = metaphase – chromosomes lined up on equator; E = anaphase – chromosomes in two sets, each being drawn towards pole

2 24 minutes. Number of cells in metaphase ÷ total number of cells observed × time for one cycle (in minutes), i.e., 20 ÷ 1000 × 1200 = 24

3 11% chromosomes visible in prophase, metaphase, anaphase, and telophase (73 + 20 + 9 + 8) ÷ 1000 × 100

Calculating a mitotic index

1 It breaks down the material which holds together plant cell walls in a tissue so a single layer of cells can be produced.

2 Mitosis in plants is restricted to meristematic tissue. This is present in the growing point in the root. Behind the root tip, cells will elongate but will no longer be dividing

3 Divide the number of whole cells in a stage of mitosis by the total number of whole cells in the field of view. Remember some cells will be in prophase with chromosomes starting to condense. The mitotic index varies depending on the rate of root growth. mitotic index = 12 ÷ 84 = 0.14

17.2

1 It is circular as opposed to linear and the DNA is not attached to histone proteins and does not coil and condense prior to cell division.

2 horizontal gene transmission – transfer of DNA / genes from one species to another; vertical gene transmission – transfer of DNA / genes within a species from one generation to the next

3 This means that resistance can spread horizontally through a population of bacteria and may pass from one species of bacterium to another.

Discovering conjugation in bacteria

1 Both strains: 1 and 2

2 Neither would be expected to grow

3 Information on how to synthesise the two nutrients that it had previously been unable to synthesise for itself.

4 Any one from: mutation (leading to the ability to produce the nutrients they could not previously synthesise) / transformation (acquiring DNA with this information from the medium around it – see Topic 8.2) / transduction (via bacteriophages)

17.3

1 A deletion because the fifth nucleotide (A) has been lost. The sequence prior to and after this is the same. (Note: The last base in the mutant version was previously the 13th in the sequence and therefore not shown in the normal version.)

2 In a deletion, all codons after the deletion are affected (frame-shift). Therefore most amino acids coded for by these codons will be different and the polypeptide will be significantly affected. In a substitution, only a single codon, and therefore a single amino acid, will be affected. The effect on the polypeptide is likely to be less severe.

3 The mutation may result from the substitution of one base in the mRNA with another. Although the codon affected will be different, as the genetic code is degenerate the changed codon may still code for the same amino acid. The polypeptide will be unchanged and there will be no effect.

4 These errors may be inherited and may therefore have a permanent affect on the whole organism. Errors in transcription usually affect only specific cells, are temporary, and are not inherited. They are therefore less damaging.

5 Proto-oncogenes, which stimulate cell division, and tumour suppressor genes, which inhibit cell division.

Mutagenic agents

1 The codons in mRNA will be CAU AAA UAA (Note: In mRNA, guanine is coded for by cytosine in DNA, adenine by thymine, and uracil by adenine as usual, but after the change, cytosine becomes uracil in DNA and this codes for adenine in mRNA.)

2 substitution gene mutation

3 The active site of DNA polymerase can no longer fit the DNA molecule because the shapes of some DNA bases have been altered by X-rays.

4 The replication of DNA requires DNA polymerase and so the process cannot continue.

5 Public opinion, special interest groups such as the owners of shops selling or using sunbeds, manufacturers, consumers, professional bodies (e.g., members of the medical profession), the media, and other scientists.

Cancer and its treatment

1 0.2 million / 200,000

2 50%

3 8.33 times (0.5 ÷ 0.06)

4 More cancer cells are killed because they divide more rapidly than healthy cells and so are more susceptible to the drug.

5 Cancer cells take longer to recover. Cancer cells divide more slowly / rate of mitosis is reduced.

6 One dose of the drug does not kill all the cancer cells. Those that remain continue to divide and build up the number of cancer cells again.

18.1

a ecology

b biosphere

c biotic

d abiotic

e community

f population

g habitat

18.2

1 $\frac{100 \times 80}{5} = 1600$

2 a Population over-estimated (appears larger) as there will be proportionally fewer marked individuals in the second sample.

b Population over-estimated / appears larger as there will be proportionally fewer marked individuals in the second sample because all the 'new' individuals will be unmarked.

c No difference because the proportion of marked and unmarked individuals killed should be the same.

Ethics and fieldwork

1 They can be eaten by other organisms and so provide energy and nutrients to the ecosystem.

2 It allows the habitat to recover from any disturbance / removal of organisms. The results of a further study carried out too soon after may result in data that are not typical of the habitat under 'normal' conditions.

3 The organisms live beneath stones so they remain moist when not covered by the tide. If the stone is left upside down the organism may become desiccated and die.

4 A selection from each of the following:

For	Against
• Practical experience aids learning / is better than theoretical study	• Students are inexperienced and therefore more likely to damage habitats
• Experienced ecologists have to start somewhere	• Information could be provided by theoretical means / use of data obtained by experienced ecologists / videos
• Students may become ecologists and so aid conservation in the long term	• Large number of students puts pressure on / increases damage to popular sites
• Students can be taught conservation through ecology which makes people more aware of the environment and so more likely to support conservation	• Many students will not continue studies in biology / ecology and so may never use the information again

18.3

1 Certain factors limit growth, e.g., availability of food, accumulation of waste, disease.

2 Biotic factors involve the activities of living organisms.

Abiotic factors involve the non-living part of the environment.

3 a low light intensity

b lack of water

c low temperature

The influence of abiotic factors on plant populations

1 a 1

b 3

c 2

d 3

2 The pH is too high for species X and the temperature is too low for species Y.

18.4

1 Intraspecific competition occurs when individuals of the **same** species compete with one another for resources.

Interspecific competition occurs when individuals of **different** species compete for resources.

2 Any two from: food / water / breeding sites (or any other relevant factor, e.g., light, minerals).

The effects of interspecific competition on population size

1 After 1985 the rise in the grey squirrel population is mirrored by a fall in the red squirrel population.

2 Lack of food / adverse weather, e.g., cold winters / increase in number of squirrel predators / new disease.

3 Grey squirrels have more chance of finding fruits / nuts / seeds that have fallen to the ground as well as those that are still on the trees / bushes.

4 The sea presents a barrier to the grey squirrel reaching islands. The red squirrels already present on the islands have little or no competition from grey squirrels and so flourish.

Competing to the death

1 Population increases slowly at first and then at an accelerating / exponential / logarithmic rate to around 8 days. The growth rate then slows, reaching a maximum at around 12 days which is maintained at a constant level up to 20 days.

2 Population growth is faster initially. Maximum size is reached earlier. Maximum size is reduced to less than half. Size is not maintained at a constant level (it falls to zero).

3 *P. caudatum* is unsuccessful in competing with *P. aurelia* for yeast / food. Most available food is taken by *P. aurelia* and *P. caudatum* starves, leading to a population crash.

4 Some of the yeast / food is taken by *P. caudatum*, leaving less for the population growth of *P. aurelia*.

5 After 20 days all *P. caudatum* have died. *P. aurelia* has no competition for food and so it reaches its previous maximum. *P. aurelia* is in effect 'alone' again.

Investigating competition

1 The size of each pot, the type and quantity of compost or soil in each pot, the temperature and light exposure, and the watering regime.

2 Each species of seed may not be equally viable. If this is not the case, then fewer plants will germinate in one species. Sowing rates may need to be adjusted in order to obtain the right number of plants in each pot.

3 It is the wet weight that is being recorded. Any delay could result in water loss from the cut plants which could affect the results.

4 Interspecific competition occurs between marigold and zinnnia. Intraspecific competition occurs between marigold plants and between zinnia plants.

5 Intraspecific competition. Line X is much steeper than line Y. Plants of the same species are more likely to be competing in the same way for the same resource.

18.5

1 The range and variety of laboratory habitats is much smaller than in natural ones. This means that in nature there is a greater range of hiding places and so the prey has more space and places to escape the predator and survive.

2 With fewer predators, fewer prey are taken as food. The death rate of prey is reduced. Assuming the birth rate remains unchanged the population size increases.

3 Graph showing population fluctuations (peaks and troughs) of A. Species B mirrors these changes after a time lag. The population size of B is, for the most part, smaller than A.

B eats A → population of A falls → fewer A for B to eat → population of B falls → fewer B means fewer A are eaten → population of A rises → more A means more food for B → population of B rises.

The Canadian lynx and the snowshoe hare

1 The assumption is made that the relative numbers of each type of fur traded represents the relative size of each animal's population at the time.

2 The population size of the snowshoe hare fluctuated in a series of peaks and troughs. Each peak and trough was repeated about every 10 years. The population size of the Canadian lynx also fluctuated in a 10-year cycle of peaks and troughs. The relative pattern of peaks and troughs is similar for the lynx and the snowshoe hare. The rise in the population size of the lynx often (but not always) followed that of the snowshoe hare.

3 The snowshoe hare population increases due to the low numbers of Canadian lynx that feed on them → more hares mean more food for the lynx, whose population therefore increases as fewer starve / more are able to raise young → more lynx means there is more predation of hares, whose population therefore decreases → fewer hares means less food for the lynx, many of which starve and so their population decreases.

4 Four times

5 Addition of food – because the population increased more in every year that data were collected.

6 Both food supply and predation influence hare population size. Food supply has a greater influence than predation but a combination of both factors has an even greater influence than either of the other two separately.

18.6

1 pioneer species

2 primary colonisers (pioneer species) photosynthesise and fix nitrogen → these die and form a soil with nutrients → further colonisers can survive in this soil → environment is a little less hostile → more habitats and food sources available → other species are able to survive → increased biodiversity

3 climax community

Warming to succession

1 Biomass increases very slowly and so the line curves gently upwards at first (up to 60 years) because there is little nitrogen in the soil and therefore growth and hence net production of the pioneer species (*Dryas*) is small.

Biomass increases at a greater but constant rate and so the curve becomes a straight line with an upward gradient, from 60 to 120 years, as soil nitrogen levels rise. Increased levels of soil nitrogen remove this limit on growth (net productivity) therefore large species, such as alder, and later spruce, establish themselves and hence biomass increases more rapidly.

Biomass increase slows and finally stops and so the curve flattens out after 150 years because soil nitrogen levels fall as plants take it into their biomass – nitrogen again limits plant growth (net productivity).

2 a Nitrogen from the atmosphere is fixed into compounds, e.g., proteins and amino acids by the nitrogen-fixing species (lichens, *Dryas*, and alder). When these die or shed their leaves this nitrogen is released when decomposers break them down into ammonium compounds (ammonification), which are then broken down by nitrifying bacteria into nitrites and nitrates.

b More nitrogen is being absorbed by the increased biomass of the plants. The nitrogen-fixing lichens, *Dryas*, and alder have been replaced by spruce that does not fix nitrogen therefore less nitrogen is being added to the soil.

3 a (Pioneer) species are taking advantage of new habitats and lack of competition to rapidly colonise the empty land.

b Spruce is becoming dominant and outcompeting the other species, such as lichen, *Dryas*, and alder, for light, nutrients, etc. These other species are eliminated from the community.

4 Transects are better because there is a gradient of environmental factors that produce a series of changes over a long distance. Transects also ensure that every community is sampled, which may not be the case with random sampling.

18.7

1 The species within the habitat possess unique genes that at some point in the future may be useful. Conserving habitats maintains biodiversity. The greater the variety of habitats, the greater their potential to enrich our lives and provide enjoyment.

2 Cut back reeds to prevent them becoming dominant. Remove dead vegetation to prevent build-up and thus stop fens drying out. Pump water into fens to keep them waterlogged. Cut back grasses and shrubs to prevent succession.

Conflicting interests

1 96 (32% of 300)

2 It might increase the population of grouse as harriers would have alternative sources of food and therefore eat fewer grouse chicks. Alternatively it might lead to a large increase in harriers that then prey on grouse (especially once the supply of voles and meadow pipits has been exhausted). This would lead to a decrease in the grouse population.

3 The moorland would undergo secondary succession, finally reaching its climax community of deciduous (oak) woodland.

4 A selection from each of the following arguments:

For	Against
• The harrier is a very rare bird – there are only 750 pairs in the UK.	• The harrier is a major predator of grouse and so could threaten the already declining grouse population.
• Previous persecution led to its extinction on the UK mainland and this could happen again.	• If the grouse population is reduced / eliminated and / or the harrier population is not controlled, this could adversely affect the populations of alternative harrier prey, such as voles and meadow pipits.
• Harriers are part of our natural heritage and their population should not be controlled other than by natural means.	• Reduction / elimination of grouse population could make grouse shooting uneconomic and, unless money is found from elsewhere, the moorland habitat might be lost along with the species that live there, and so reduce biodiversity.

19.1

1 carbon dioxide and water

2 glucose and oxygen

3 a grana / thylakoids

b stroma

4 a reduced NADP, ATP, and oxygen

b sugars and other organic molecules

19.2

1 on the thylakoid membranes (of the grana in the chloroplast)

2 Water molecules are split to form electrons, protons, and oxygen, as a result of light exciting electrons / raising the energy levels of electrons in chlorophyll molecules.

3 a reduction

b reduction

c oxidation

Chloroplasts and the light-dependent reaction

1 A = (double) membrane of chloroplast / chloroplast envelope

C = granum

D = stroma

2 C

3 starch

4 Any two from: the light-dependent reaction does not produce sufficient ATP for the plants' needs / photosynthesis does not take place in the dark / cells without chlorophyll cannot produce ATP in this way and ATP cannot be transported around the plant.

5 Length X–Y on Figure 4 = 24mm (= 24000μm)

Actual length X–Y = 2μm

Magnification = $\frac{24\,000}{2}$ = 12000 times

19.3

1 It accepts / combines with a molecule of CO_2 (to produce two molecules of glycerate-3-phosphate).

2 It is used to reduce (donate hydrogen) glycerate-3-phosphate to triose phosphate.

3 ATP

4 Stroma of the chloroplasts.

5 The Calvin cycle requires ATP and reduced NADP in order to operate. Both are the products of the light-dependent reaction, which needs light. No light means no ATP or reduced NADP are produced and so the Calvin cycle cannot continue once any ATP or reduced NADP already produced have been used up.

Using a lollipop to work out the light-independent reaction

1 To allow the substances into which it becomes incorporated to be identified / to allow the sequence of substances produced to be identified.

2 The radioactive carbon is initially found in GP (5 seconds) and is next found in triose phosphate (10 seconds).

3 The high temperature and / or the methanol denature the enzymes that catalyse reactions.

4 The quantity of GP begins to decrease almost immediately. The rate of decrease becomes less until, after about 4.5 minutes, the quantity of GP becomes constant, but at around a quarter of its original level. The quantity of RuBP rises almost immediately. The rate of increase is steady at first, but then slows, peaking at 3.5 minutes. The quantity of RuBP then falls until it becomes constant at around 4.5 minutes, but at around double its original level. The quantities of GP and RuBP are the same after 2.5 minutes.

5 RuBP combines with CO_2 to form GP during the light-independent reaction / Calvin cycle of photosynthesis. GP is ultimately used to regenerate RuBP. When the CO_2 level is decreased, there is less to combine with RuBP and so less GP is formed but it is still being used up and so

its level falls. There is still some CO_2 and so some GP is made, but much less than originally. With less CO_2 to combine with, the RuBP accumulates because it cannot be converted to GP. Its quantity rises to a new higher level due to the lower level of CO_2.

19.4

1 Volume of oxygen produced / CO_2 absorbed.

2 Light intensity – because an increase in light intensity produces an increase in photosynthesis over this region of the graph.

3 Raising the CO_2 level to 0.1% – because this increases the rate of photosynthesis more than increasing the temperature to 35 °C.

4 Because light is limiting photosynthesis and so an increase in temperature will not increase the rate of photosynthesis.

5 More CO_2 is available to combine with RuBP to form more GP, then more triose phosphate and ultimately more glucose.

Measuring photosynthesis

1 Because any air escaping from or entering the apparatus will respectively decrease or increase the volume of gas measured, which will give an unreliable result.

2 So that any changes in the rate of photosynthesis can be said to be the result of changes in light intensity and not changes in temperature.

3 To ensure there is sufficient CO_2 and so it does not limit the rate of photosynthesis.

4 To prevent other light falling on the plant as this may fluctuate and will affect the light intensity and hence the rate of photosynthesis, leading to an unreliable result.

5 To prevent photosynthesis and to allow any oxygen produced before the experiment begins to disperse.

6 Because the volume of oxygen produced will be less than that produced by photosynthesis as some of the oxygen will be used up in cellular respiration / dissolved oxygen (and other gases) may be released from or absorbed by the water.

20.1

a cytoplasm
b glucose
c phosphate
d ATP
e phosphorylated glucose
f triose phosphate
g hydrogen
h NAD
i pyruvate
j ATP

20.2

1 a D
b A, C, D
c A, D
d A, B
e A, D
f B, C, D
g A

Investigating where certain respiratory pathways take place in cells

1 Homogenate is spun at slow speed. Heavier particles (e.g., nuclei) form a sediment. Supernatant is removed, transferred to another tube, and spun at a greater speed. Next heaviest particle is removed. Process is repeated.

2 Nuclei and ribosomes – because neither CO_2 nor lactate (products of respiration) are formed in any of the samples.

3 a mitochondria

b Krebs cycle produces CO_2 and results show that CO_2 is produced when mitochondria only are incubated with pyruvate.

4 remaining cytoplasm (Note: The complete homogenate is not a 'portion' of the homogenate.)

5 Cyanide prevents electrons passing down the transfer chain. Reduced NAD therefore accumulates and blocks Krebs cycle where CO_2 is produced. Glycolysis can still occur because the reduced NAD it produces is used to make lactate. Glucose can therefore be converted to lactate, but not into CO_2, in the presence of cyanide.

6 The conversion of glucose to CO_2 involves glycolysis (occurs in cytoplasm) and Krebs cycle (occurs in mitochondria). Only the complete homogenate contains both cytoplasm and mitochondria.

7 ethanol and CO_2

8 liver cell, epithelial cell, and muscle cell

20.3

1 Three

2 acetyl coenzyme A

3 matrix of mitochondria

4
1 = True 2 = True 3 = True
4 = True 5 = False 6 = False
7 = True 8 = True 9 = False
10 = False 11 = False 12 = True
13 = False 14 = True 15 = False
16 = False 17 = False 18 = False

Coenzymes in respiration

1 To show that the yeast suspension was responsible for any changes that occurred and the glucose did not change methylene blue nor did methylene blue change by itself.

2 a Yeast uses glucose as a respiratory substrate, producing hydrogen atoms that are taken up by methylene blue causing it to become reduced and changing from blue to colourless.

b As in (2a), except that the yeast uses stored carbohydrate as a respiratory substrate that has to be converted to glucose and so the production of hydrogen atoms is slower / reduced.

3 Contents of tube might have remained blue because the enzymes involved in respiration are denatured at 60°C and so respiration, and hence the reduction of methylene blue, ceases / the enzymes involved in hydrogen transport have been denatured and so the indicator is not reduced by hydrogen.

4 Air contains oxygen, which would re-oxidise methylene blue, turning it blue.

5 This is a single experiment. The same results would need to be obtained on many occasions.

20.4

1 The movement of electrons along the chain is due to oxidation. The energy from the electrons combines inorganic phosphate and ADP to form ATP = phosphorylation.

2 It provides a large surface area for the attachment of the coenzymes (NAD / FAD) and electron carriers that transfer the electrons along the chain.

3 Oxygen is the final acceptor of the electrons and hydrogen ions (protons) in the electron transfer chain. Without it the electrons would accumulate along the chain and respiration would cease.

4 water molecule

Sequencing the chain

1 Sequence – C, A, D, B

Explanation – Electron carriers become reduced by electrons from glycolysis and the Krebs cycle. Enzymes catalyse the transfer of these electrons to the next carrier. If an enzyme is inhibited all molecules prior to that enzyme will not be able to pass on their electrons and so will be reduced and those after it will be oxidised. The first molecule in the chain will be reduced with all inhibitors, the second with 2 out of 3 inhibitors, the third with 1 out of 3, and the last in the chain with none (i.e., it is always oxidised).

20.5

1 18/26 = 0.69

2 They do not have mitochondria so no aerobic respiration is possible. Anaerobic respiration uses glucose. Fatty acids and amino acids cannot be broken down anaerobically.

Calculating RQ values in germinating seeds

1 Temperature would affect the rate of respiration. The apparatus should be placed in a thermostatically controlled water bath at a temperature appropriate for the seeds such as 20 °C

2 The liquid will have moved towards tube B as oxygen is consumed in respiration but the carbon dioxide produced is absorbed by the potassium hydroxide. This reduces the pressure in that side of the apparatus.

3 This is less than 1 and indicates that some lipids must have been respired. However, it is higher than the 0.7 expected for lipids so a mixture of substrates is probably being respired including lipids.

4 The seeds could be respiring a mixture of lipids (an RQ below 1) and another substrate with an RQ value greater than 1. Equally, the seeds could be carrying out some anaerobic respiration of carbohydrates (RQ greater than 1) and aerobic respiration of lipids.

5 Measure the oxygen consumption over a fixed time period at a range of temperatures (e.g., every 5 degrees between 5 and 25°C. Plot the temperature (*x*-axis) against rate of oxygen consumption ($cm^3 min^{-1}$).

21.1

1 dragonfly nymphs

2 unicellular and filamentous algae

3 sticklebacks

4 the direction of energy flow

5 decomposers / saprobiotic (micro)organisms

21.2

1 Any 3 from: some of the organism is not eaten; some parts are not digested and so are lost as faeces; some energy is lost as excretory materials; some energy is lost as heat.

2 The proportion of energy transferred at each trophic level is small (less than 20%). After four trophic levels there is insufficient energy to support a large enough breeding population.

3 a $\frac{1250 \times 100}{6300} = 19.84\%$

b $\frac{50 \times 100}{42\,000} = 0.12\ (0.119)\%$

Adding up the totals

1 decomposers / saprobiotic (micro)organisms

2 insect-eating birds

3 $\frac{42\,500 \times 100}{1.7 \times 10^6} = 2.5\%$

4 Any three from: most (90%+) solar energy is reflected by clouds or dust or absorbed by the atmosphere / not all light wavelengths are used

in photosynthesis / much of the light does not fall on the chloroplast / chlorophyll molecule / factors may limit the rate of photosynthesis or photosynthesis is inefficient / respiration by producers means energy is lost (as heat).

5 $4120 - (1010 + 810) = 2300\,kJ\,m^{-2}\,year^{-1}$

21.3

1 In a pyramid of numbers: no account is taken of size / the number of individuals of one species may be so great that it is impossible to represent them on the same scale as other species in the food chain.

2 At certain times of year (e.g., spring) zooplankton consume phytoplankton so rapidly that their biomass temporarily exceeds that of phytoplankton.

3 grams per square metre ($g\,m^{-2}$) / grams per cubic metre ($g\,m^{-3}$)

A woodland food chain

1 moth caterpillars

2 $\frac{92 \times 100}{806} = 11.41\%$

3 $kJ\,m^{-2}\,year^{-1}$

4 Each beech tree has many caterpillar moths living on it, so the beech tree block is smaller in the number pyramid. Each beech tree is very large and so has a greater biomass and more energy than all the moth caterpillars together, so the beech tree block is larger in the biomass and energy pyramids.

5 Any two from: lost in respiration / lost as heat / lost to decomposers (in excretion, faeces, death, and decay) / part of robin is not eaten or not digested.

6 The parasitic flea block must be at the top of each pyramid. It is larger than the sparrowhawk block in the number pyramid but smaller in the biomass and energy pyramids.

21.4

1 gross productivity minus respiratory losses

2 $kJ\,m^{-2}\,year^{-1}$

3 In an agricultural ecosystem additional energy is put in to remove other species, add fertilisers, and pesticides. These reduce competition for light, water, CO_2, etc.; provide mineral ions; destroy pests; and reduce disease. All these increase photosynthesis and hence productivity.

4 Natural ecosystems use only solar energy, agricultural ecosystems use solar energy and additional energy from food (labour) and fossil fuels (machinery and transport).

Increasing productivity

1 On a cloudy day, light intensity will be the limiting factor and only an increase in this will increase the rate of photosynthesis. Increases in CO_2 concentration will have no effect and would be economically wasteful.

2 With the greenhouse open to the air any additional CO_2 would simply disperse outside. It therefore makes economic sense to set the level to around that found in the air, namely 400 ppm.

3 Long periods of CO_2 levels at 5000 ppm would cause stomata to close. Diffusion of gases, and therefore photosynthesis, would be considerably reduced. Keeping levels this high would be counterproductive. At 1000 ppm the higher level of CO_2 increases photosynthesis without causing stomata to close. For short durations there may be some benefit in bright sun of having a level of CO_2 as high as 5000 ppm as the plant takes some time to close its stomata. During this period 5000 ppm will increase photosynthesis as CO_2 is likely to be the limiting factor.

21.5

1 If the pesticide kills most of the pests then the population of organisms (predators) feeding on it will fall. With no predators controlling it, the pest population will increase again, possibly to a level higher than before. The crop will be even more affected by the pest, leading to lower productivity.

2 Advantages – any two from: highly specific, targeted only on pest / once introduced it reproduces itself and does not need to be re-applied / pests do not become resistant.

Disadvantages – any two from: effect is slow as there is a time lag between application and results / may itself become a pest or may disrupt the ecological balance.

3 Weeds compete for light, water, mineral ions, CO_2, space, etc. If these are in limited supply there will be less available to the crop plants. One or more may limit the rate of photosynthesis and hence productivity.

To weed or not to weed?

1 As the number of weeds increases, the productivity of wheat decreases. The reduction in productivity is initially large (between 0 and 40 weeds m^{-2}), but lessens as the number of weeds increases. From 40 to $50\,m^{-2}$, the loss in productivity levels out.

2 Soya bean because it has an increase in productivity of 50% (1000 to $1500\,kg\,ha^{-1}$) whilst wheat only increases by 33% (4500 to $6000\,kg\,ha^{-1}$).

3 No. Cost of herbicide per hectare = £100. Reducing weeds from 40 to $20\,m^{-1}$ increases wheat productivity from 4500 to $5000\,kg\,ha^{-1}$ – an increase of 500 kg or half a tonne. Wheat is sold at £150 per tonne, so increased income is £75 per hectare – £25 per hectare less than the cost of treating with herbicide.

A mighty problem

1 Description – In both experiments the spider mite populations rise slowly during the first 15 days and then very rapidly up to around 50 days.

In experiment 1 the spider mite population remains high up to 150 days but fluctuates (between 400 and 900). (Note: The scale is the square root of the numbers and so the figures on the *y*-axis need to be squared to give actual numbers.)

In experiment 2 the spider mite population falls over the period 50–150 days until it reaches the starting level.

Explanation – In experiment 1 the population of the spider mite increases until some factor (e.g., food supply) limits its size. It remains fairly constant as an equilibrium is reached with the limiting factor, fluctuating slightly as the factor fluctuates.

In experiment 2 the population of the spider mite increases up to 50 days, by which time the population of the predatory mite has increased considerably. The predatory mites feed on the spider mites, causing their population to drop to a very low level by 150 days.

2 Predatory mites are effective in controlling the population of spider mites as their presence reduces the spider mite population from around 400–900 when the predatory mite is absent to around 4 when the predatory mite is present.

3 The two populations will probably remain small as they remain in balance. They will fluctuate because, as the spider mite population falls, there will be less food for the predatory mite and so, a short time later, its population will also fall. The fall in the predatory mite's population means there will be less predation on the spider mite, whose population will then increase, followed in turn by an increase in the predatory mite's population.

21.6

1 Movement is restricted so less energy is expended in muscle contraction / heat loss is reduced so less energy is expended maintaining body temperature / the optimum amount and type of food for rapid growth can be provided / predators are excluded so no energy is lost to other organisms.

2 A longer dark period means more time is spent resting, less energy is expended, and more energy is converted into body mass.

Features of intensive rearing of livestock

1 The gene / allele / DNA for antibiotic resistance can pass from the animal disease-causing bacteria to the human disease-causing bacteria along a conjugation tube formed between the two (horizontal gene transmission).

2 A selection from each of the following:

For	Against
• Efficient energy conversion	• Lower quality eggs / less taste
• Economic – makes good use of resources	• Disease spreads more rapidly
• Cheaper eggs	• Antibiotic resistance – can spread to human disease-causing bacteria
• Less land required / more land for natural habitats	• Hens kept unnaturally – may be stressed or aggressive and therefore need to be de-beaked
• Safer / more easily regulated farms	• Restricted movement means more osteoporosis / joint pain
• Disease / predators more easily excluded	• More pollution / smells / harmful to environment
• Animals are warm and well fed	• Use of fossil fuels means more CO_2 / global warming
	• Hens require drugs as a consequence of the environment in which they are kept / to make them grow more rapidly

3 Any balanced discussion of issues such as crowded, confined conditions versus being warm, fed, kept healthy, free from predators / need for drugs versus becoming diseased / de-beaking versus harming one another / keeping animals unnaturally versus our demand for cheap food / treating animals badly (osteoporosis, boredom, frustration) versus animal welfare legislation to prevent cruelty.

22.1

1 Light intensity is greater / longer period of light in summer and the temperature is usually higher. Both factors increase the rate of photosynthesis (assuming there is no other limiting factor). More CO_2 is taken from the atmosphere during photosynthesis and so its concentration falls.

2 **A** = combustion (burning)

B = respiration

C = feeding (and digestion)

D = photosynthesis

3 d

22.2

1 Fluctuations are the result of annual changes in CO_2 concentration due to seasonal temperature changes. CO_2 levels fall in the summer as warmer temperatures and longer periods of more intense light lead to more photosynthesis and hence more CO_2 being fixed into organic molecules. In winter, colder temperatures, less light, and the loss of leaves in deciduous plants means less photosynthesis and less CO_2 absorption by plants. As respiration in organisms continues to produce CO_2, its level increases in winter.

2 Palm tree plantations – less productive than tropical rainforest therefore there is less photosynthesis → less CO_2 is absorbed → CO_2 level in the atmosphere increases → greenhouse gas → more global warming.

Burning of forest – heat directly contributes to global warming, the CO_2 produced does so indirectly (as above).

Palm oil extraction – manufacture of biofuel and its transport use fossil fuels that produce CO_2 and hence increase global warming.

Burning of biofuel – also produces CO_2 but as this has only recently been absorbed from the atmosphere by the palm trees it will have little or no effect on global warming.

3 Cattle produce methane, a greenhouse gas that contributes to global warming.

Digging into the past

1 There is a correlation because the graphs have similar shapes. Rises and falls in CO_2 concentration are reflected in similar rises and falls in temperature.

2 No. To prove this we would need to establish a causal link between changes in CO_2 concentration and temperature change. It is equally possible from the data to suggest that the temperature changes cause the changes in CO_2 concentration rather than the other way around.

Global warming and crop yields

1 The yield decreases slightly as the minimum temperature increases from 22.0 °C to 22.5 °C.

The yield decreases more significantly / to a greater extent as the minimum temperature increases from 22.5 °C to 24.0 °C.

The total decrease in yield is from 9.4 tonnes ha^{-1} at 22.0 °C down to 6.5 tonnes ha^{-1} at 24.0 °C.

2 Above-ground biomass at 22.0 °C = 1800 $g\,m^{-2}$

Above-ground biomass at 24.0 °C = 1500 $g\,m^{-2}$

Decrease in biomass = 1800 – 1500 = 300 $g\,m^{-2}$

% decrease = 300/1800 × 100 = 16.7%.

3 As the data are collected at night there is no photosynthesis but respiration is still taking place. An increase in temperature increases enzyme activity and hence respiration rate. More carbohydrate is used up in respiration leaving less available to form grain / biomass.

Global warming and insect pests

1 Description – The higher the mean winter temperature, the more rice stem-borer larvae there are per rice plant. For each 0.5 °C rise in mean winter temperature, the number of larvae per plant doubles. A temperature rise from 6.0 °C to 6.5 °C doubles the number of larvae from 10 to 20 and a rise from 6.5 °C to 7.0 °C doubles the number of larvae from 20 to 40.

Explanation – The larvae over-winter in the paddy fields and so the warmer the winter temperature the more are likely to survive and infect rice plants and the earlier they emerge in spring and so the longer they have to infect rice plants and the more larvae there are per plant.

2 From graph A – global warming from 1950 to 2000 has led to an increase in mean winter temperatures.

From graph B – this means more rice stem-borer larvae per rice plant.

From graph C – this means that the crop yield of rice is reduced.

3 10% – because in 1980 the mean winter temperature was 6.5 °C (graph A). The predicted number of larvae per plant when the mean winter temperature is 6.5 °C is 20 (graph B). When there are 20 larvae per plant, the loss in crop yield is 10% (graph C).

4 Larvae bore into stems of rice plants and so may block xylem / prevent transpiration so no water for photosynthesis reaches leaves, no photosynthesis, no sugars for respiration / products of photosynthesis / sugars cannot reach roots from leaves, no sugars for respiration and so roots, and hence plant, dies / damage to stem may cause plant to collapse / fall over.

22.3

a nitrogen fixation

b plants

c nitrate ions

d root hairs

e proteins / amino acids / nucleic acids

f decomposers / saprobiotic microorganisms

g ammonia / ammonium ions

h nitrifying

i nitrate

j denitrifying

22.4

1 Crops are grown repeatedly and intensively on the same area of land. Mineral ions are taken up by the crops, which are transported and consumed away from the land. The mineral ions they contain are not returned to the same area of land and so the levels in the soil are reduced, which can limit the rate of photosynthesis. Fertilisers need to be applied to replace them if photosynthesis / productivity is to be maintained.

2 $100\,kg\,ha^{-1}$ – although $150\,kg\,ha^{-1}$ gives a slightly better yield, this is marginal and the cost of using 50% more fertiliser makes it uneconomical.

3 Some other factor is limiting photosynthesis, e.g., light or CO_2, and only the addition of this factor will increase photosynthesis and hence productivity.

4 Natural fertilisers are organic and come from living organisms in the form of dead remains, urine, or faeces (manure).

Artificial fertilisers are inorganic and are mined from rocks and deposits.

Different forms of nitrogen fertilisers

1 manure, bone meal, and urea

2 To act as a control to show that any changes in productivity were the result of the nitrogen fertiliser being added.

3 Nitrogen is needed for proteins / amino acids / chlorophyll and DNA and therefore for plant growth. Nitrogen shortage may limit the production of proteins and DNA and hence growth. Its addition increases productivity.

4 Some forms of fertilisers contain more actual nitrogen than others and so different masses are added to ensure that the total nitrogen added was always the same ($140\,kg\,ha^{-1}$).

5 The data do not support the view. Whilst ammonium nitrate brings about the greatest increase in productivity, ammonium sulphate produces a smaller increase than both urea and bone meal. Therefore the investigation suggests that only some ammonium salts are better.

6 The farmer should spread the manure a few months before the main growing season for the crop.

22.5

1 Eutrophication is the process by which salts build up in bodies of water.

2 The concentration of algae near the surface becomes so dense that no light penetrates to deeper levels. No light means no photosynthesis and hence no carbohydrate for respiration and so plants at lower levels die.

3 Dead plants are used as food by saprobiotic microorganisms. With an increased supply of this food, the population of saprobiotic microorganisms increases exponentially. Being aerobic they use up the oxygen in the water leading to the death of the fish, which cannot respire without it.

Troubled waters

1 It has taken 10 days for the fertiliser that has dissolved in the rainwater to leach through the soil and into the lake.

2 In normal circumstances, a low level of nitrate (or other ions) is the limiting factor to algal growth. The fertiliser leaching into the lake contains nitrate (and other ions) and removes this limit on growth. The algal population grows rapidly, increasing in density.

3 Description – As the density of the algae increases so the clarity of the water decreases, i.e., there is a negative correlation. For the first 20 days the algal density (30 cells cm^{-3}) and water clarity (Secchi = 9 m) remain constant. From day 20 to day 100 the algal density increases from 30 to 120 cells cm^{-3} whilst the water clarity decreases from 9 to 1 m (Secchi depth). However, there is an anomaly between day 40 and day 50 when the water clarity suddenly falls from 7 to 4 m.

Explanation – As the density of algae increases, more light is absorbed / reflected by them and so less light penetrates / water clarity is reduced. Between day 40 and day 50 some factor (e.g., water turbulence stirring up sediment) other than algal density is reducing the water clarity.

4 Days 0–10: oxygen level is constant (at 10 ppm) because there is a balance between oxygen produced in photosynthesis of plants and algae, and oxygen used up in respiration of all organisms.

Days 10–25: oxygen level rises (up to around 13 ppm) due to increased photosynthesis by the larger population of algae.

Days 25–100: oxygen level decreases (more rapidly at first and then less so down to around 3 ppm) due to higher density of algae blocking out the light to lower depths and reducing the rate of photosynthesis of plants / algae at these depths. In time, light is blocked out altogether at lower depths → no photosynthesis → plants / algae die → saprobiotic microorganisms decompose them → their population increases → they use up much oxygen in respiration → oxygen levels fall.

23.1

a genotype

b mutation

c phenotype

d nucleotides/bases

e polypeptides
f locus
g homozygous
h heterozygous
i recessive
j co-dominant

23.2

1 Let allele for Huntington's disease = H

Let allele for normal condition = h

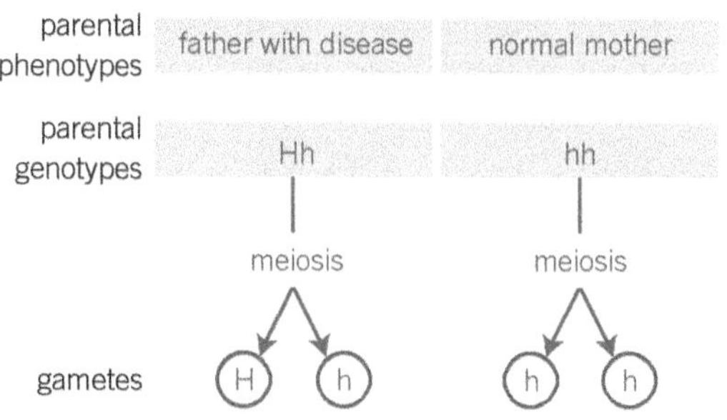

Offspring:

	Father's gametes	
Mother's gametes	H	h
h	Hh	hh
h	Hh	hh

Half (50%) of offspring will have Huntington's disease (Hh).

Half (50%) of offspring will be normal (hh).

2 a Let allele for black coat = B

Let allele for red coat = b

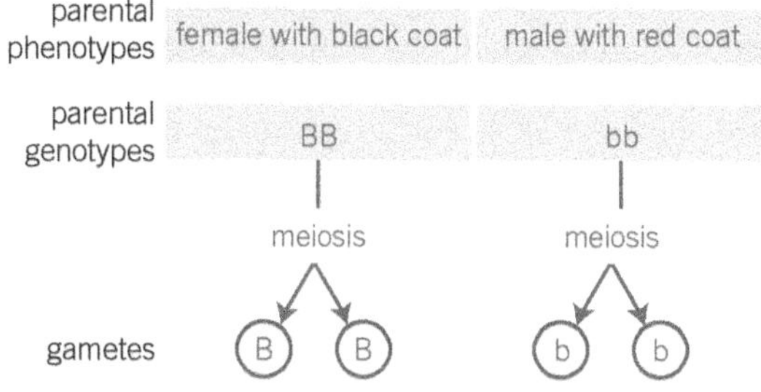

Offspring:

	Male gametes	
Female gametes	b	b
B	Bb	Bb
B	Bb	Bb

All (100%) offspring will have black coats (Bb).

b Let allele for black coat = B

Let allele for red coat = b

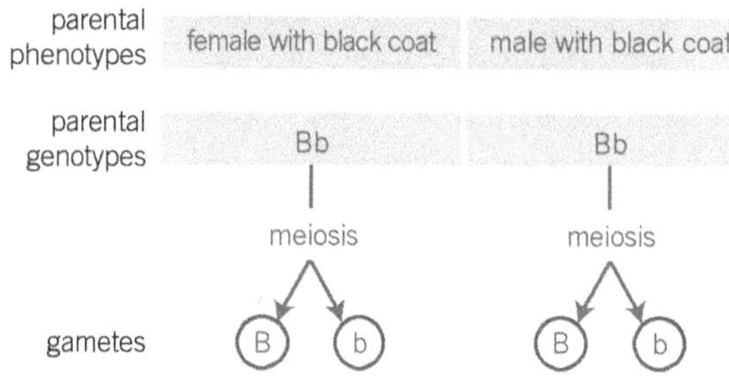

Offspring:

	Male gametes	
Female gametes	B	b
B	BB	Bb
b	Bb	bb

3 offspring (75%) with black coat (BB, Bb, and Bb).

1 offspring (25%) with red coat (bb).

Probability of offspring having red coat = 1 in 4 (25%/0.25)

Determining genotypes

1 Let allele for green pods = G

Let allele for yellow pods = g

a **Where a plant is homozygous dominant:**

parental phenotypes: green pods | yellow pods

parental genotypes: GG | gg

meiosis | meiosis

gametes: G G | g g

Offspring:

	♂ Gametes	
♀ Gametes	G	G
g	Gg	Gg
g	Gg	Gg

All (100%) green pods (Gg).

b **Where plant is heterozygous:**

parental phenotypes: green pods | yellow pods

parental genotypes: Gg | gg

meiosis | meiosis

gametes: G g | g g

Offspring:

Gametes	♂ Gametes G	g
g	Gg	gg
g	Gg	gg

Half (50%) green pods (Gg).

Half (50%) yellow pods (gg).

2 a homozygous dominant (GG)

b We cannot be absolutely certain because if the unknown genotype were heterozygous (Gg) the gametes produced would contain alleles of two types: either dominant (G) or recessive (g). It is a matter of chance which of these gametes fuses with those from our recessive parent – all these gametes have a recessive allele (g). It is just possible that, in every case, it is the gametes with the dominant allele that fuse and so all the offspring show the dominant character. Provided the sample of offspring is large enough, however, we can be reasonably sure that the unknown genotype is homozygous dominant.

3 a heterozygous (Gg)

b We can be certain because seven of the offspring display the recessive character (in our case yellow pods). These plants are homozygous recessive and must have obtained one recessive allele from each parent. Our unknown parental genotype must therefore have a recessive allele and be heterozygous (in our case Gg). It is theoretically possible that the plants with yellow pods were due to a mutation but this is most unlikely. The unexpectedly low number of plants with yellow pods is the result of random fusion of the gametes.

23.3

1 E = XX

F = XY

2 A = not colour blind/normal vision

B = not colour blind/normal vision

D = colour blind

3 G = X^RX^r

H = X^RY

I = X^RX^R

J = X^rY

4 0% – because sons inherit their X chromosome from their mother and she has only alleles for normal vision (X^R).

5 By mutation (of the R allele).

A right royal disease

1 Because the ancestors from whom they are descended (Edward VII and Victoria) did not have, or carry, alleles for haemophilia.

2 a The disease of haemophilia only occurs in males and not females.

b Parents without the disease are shown to have children with the disease. Alexandria and Tsar Nicholas II do not have the disease but their son Tsarevitch Alexis does. (Note: There are many other examples.)

3 a X^HX^H b X^hY c X^HX^h

4 Anastasia could have either genotype X^HX^h or X^HX^H, depending on whether she inherited an X^H or an X^h from her mother Alexandra. Waldemar's genotype must be X^hY. Therefore:

a Sons would inherit a Y from Waldemar and either an X^H or X^h from Anastasia (mother). Therefore the possible genotypes are X^HY or X^hY.

b Daughters must inherit X^h from Waldemar (father) and either X^H or X^h from Anastasia (mother). Therefore the possible genotypes are X^hX^h or X^HX^h.

23.4

1 C^AC^A C^AC^{Ch} C^AC^H C^AC^a

2 The man is not the father.

Reasons – child has blood group AB and therefore has alleles I^AI^B. The mother is blood group A and therefore either I^AI^O or I^AI^A. In either case she could have provided the I^A alleles to the child but not the I^B allele. The I^B allele must have come from the real father. The supposed father is blood group O and therefore has alleles I^OI^O. He cannot provide an I^B allele and so cannot be the father.

3

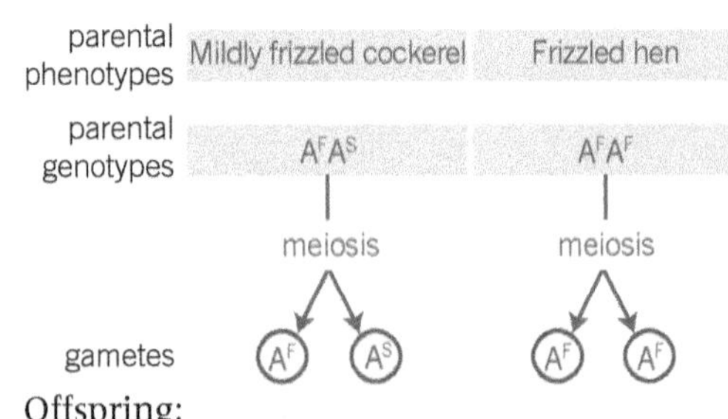

Offspring:

Hen gametes	Cockerel gametes A^F	A^S
A^F	A^FA^F	A^FA^S
A^F	A^FA^F	A^FA^S

Half (50%) frizzled fowl (A^FA^F)

Half (50%) mildly frizzled fowl (A^FA^S)

23.5

1 a

gametes	aB
Ab	AaBb – all offspring

b

gametes	GH	gH
GH	GGHH	GgHH
Gh	GGHh	GGHh
gH	GgHH	ggHH
gh	GgHh	ggHh

2 BBTT, BbTT, BBTt, BbTt

3

Genotype	Flower colour
aaBB	white
AaBB	orange
Aabb	yellow

Genetics in tomato plants

1

Phenotype	Genotype
Parent 1 – green hairless	pphh
Parent 2 – purple hairy	PPHH
F_1 – purple hairy	PpHh

2

Phenotype	Ratio
Purple hairy stem	9
Purple hairless stem	3
Green hairy stem	3
Green hairless stem	1

3 There is autosomal linkage. The dominant alleles P and H have been inherited together as have the recessive alleles p and h. There has been very little crossing over during meiosis in the production of gametes in F_1 generation.

23.6

1 a Gene pool is all the alleles of all the genes (or one particular gene) of all the individuals in a population at any one time.

b Allelic frequency is the number of times an allele occurs within the gene pool.

2 The proportion of dominant and recessive alleles of any gene in the population remains the same from one generation to the next.

3 No mutations arise.

The population is isolated / no flow of alleles into, or out of, the population.

No natural selection occurs / all alleles are equally advantageous.

The population is large.

Mating within the population is random.

4 $p + q = 1.0$ and $p = 0.942$

Therefore $q = 1.0 - 0.942 = 0.058$

Frequency of the heterozygous genotype = $2pq$

$= 2 \times 0.942 \times 0.058$

$= 0.109$

As a percentage = $0.109 \times 100 = 10.9\%$.

Not as black and white as it seems

1 It is not sex-linked because the number of males and females of each wing colour are approximately equal – the small difference is due to statistical error.

2 0.254 (25.4%)

Number of moths with dark-coloured wings (having two recessive alleles) = 562.
Total sample = 1653 + 562 = 2215.

Proportion with two recessive alleles = 562 ÷ 2215 = 0.254.

3 a 0.254 of the sample population has two recessive alleles

Therefore $q^2 = 0.254$ and $q = \sqrt{0.254} = 0.504$.

b $q = 0.504$ and $p + q = 1.0$.

Therefore $p = 1.0 - 0.504 = 0.496$.

c Frequency of heterozygotes = $2pq = 2 \times 0.496 \times 0.504 = 0.5$

Therefore % of heterozygotes = 50%

4 Capture a sample of moths and mark them in some way. Release them back into the population.

Some time later randomly recapture a given number of moths.

Record the number of marked and unmarked moths in this second sample.

Calculate the size of the population as follows:

$$\frac{\text{total number of moths in first sample} \times \text{total number of moths in second sample}}{\text{number of marked moths recaptured}}$$

23.7

1 Selection is the process by which organisms that are better adapted to their environment survive and breed, whilst those less well adapted fail to do so.

2

Directional selection	Stabilising selection
• Favours / selects phenotypes at one extreme of a population	• Favours / selects phenotypes around the mean of a population
• Changes the characteristics of a population	• Preserves the characteristics of a population
• Distribution curve remains the same shape but the mean shifts to the left or right	• Distribution curve becomes narrower and higher but the mean does not change

3 Directional selection – because birds to one side of the mean (heavier birds) were being selected for, whilst those to the other side of the mean (lighter birds) were being selected against. The population's characteristics are being changed, not preserved.

Early selection

1 a Few infants are born with a low birth mass (below 2 kg) and few infants are born with a high birth mass (above 4 kg). The majority of infants have a birth mass between 2 kg and 4 kg with the highest percentage having a birth mass of 3.25 kg.

b As infant birth mass increases up to 3.25 kg, the infant mortality rate decreases. At infant birth masses above 3.25 kg, the infant mortality rate increases again.

2 Stabilising selection – because the mortality rate is greater at the two extremes. The infants with the highest and lowest birth masses are more likely to die (are being selected against) whilst those around the mean are less likely to die (are being selected for / favoured). The population's characteristics are being preserved rather than changed.

They must be cuckoo!

1 Removing cuckoo eggs means there will be more food for the magpie's own chicks. These chicks have a greater probability of being successfully raised to adulthood.

2 Alleles for this type of behaviour are obviously present in the adult birds. There is a high probability that some of the chicks will inherit these alleles. Removing cuckoo eggs increases the probability of more of these chicks surviving to breed and therefore passing on the alleles for this behaviour to subsequent generations.

3 Displaying this behaviour has previously been of no advantage to magpies and so no selection for this behaviour has taken place. Although cuckoos have now arrived, it will take many generations for selection to operate and for allele frequencies to change.

4 Directional selection – because the population's characteristics are being changed, not preserved.

23.8

1 A species is a group of individuals that share similar genes and are capable of breeding with one another to produce fertile offspring. In other words they belong to the same gene pool.

2 Speciation is the evolution of new species from existing species.

3 Geographical isolation occurs when a physical barrier, such as mountains or oceans, prevents two populations from breeding with one another.

4 Allopatric speciation happens as a result of a geographical barrier. In sympatric speciation there is no such barrier but a breeding 'barrier' occurs within the population.

5 Geographically isolated populations may experience different environmental conditions. In each population, phenotypes that are best suited to the particular environmental conditions are selected. The composition of the alleles in each gene pool therefore changes as they pass to subsequent generations. The composition of the gene pool of each population becomes increasingly different over time. Being geographically isolated, individuals of each population cannot breed with one another and so the two gene pools remain separate and different.

24.1

1 (Negative chemo-) taxis – wastes are often removed from an organism because they are harmful. Moving away prevents the waste harming the organism and so increases its chance of survival.

2 (Positive chemo-) taxis – increases the chances of sperm cells fertilising the egg cells of other mosses and so helps to produce more moss plants / future generations. Cross-fertilisation increases genetic variability, making species better able to adapt to future environmental changes.

3 Kinesis – turning increases the chances of moving into an area with more nutrient

24.2

1 a special type of sodium channel that changes its permeability to sodium when it changes shape / is stretched

2 pressure on Pacinian corpuscle → corpuscle changes shape → stretches membrane of neurone → widens stretch-mediated sodium ion channels → allows sodium ions into neurone → changes potential of (depolarises) membrane → produces generator potential

3 Only rod cells are stimulated by low-intensity (dim) light. Rod cells cannot distinguish between different wavelengths / colours of light, therefore the object is perceived only in a mixture of black and white, i.e. grey.

4 Light reaching Earth from a star is of low intensity. Looking directly at a star, light is focused on to the fovea, where there are only cone cells. Cone cells respond only to high light intensity so they are not stimulated by the low light intensity from the star and it cannot be seen. Looking to one side of the star means that light from the star is focused towards the outer regions of the retina, where there are mostly rod cells. These are stimulated by low light intensity and therefore the star is seen.

24.3

a (nerve) impulses / action potentials
b nucleus
c rough endoplasmic reticulum
d dendrites
e Schwann cells
f insulation
g myelin
h motor
i sensory
j intermediate

24.4

1 Active transport of sodium ions out of the axon by sodium–potassium pumps is faster than active transport of potassium ions into the axon. Potassium ions diffuse out of the axon but few, if any, sodium ions diffuse into the axon because the sodium 'gates' are closed. Overall, there are more positive ions outside than inside and therefore the outside is positive relative to the inside.

2 A = closed B = open C = closed
D = closed E = closed F = open

Measuring action potentials

1 sodium and potassium ions

2 At resting potential (0.5 ms) there is a positive charge on the outside of the membrane and a negative charge inside, due to the high concentration of sodium ions outside the membrane. The energy of the stimulus causes the sodium voltage-gated channels in the axon membrane to open and therefore sodium ions diffuse in through the channels, along their electrochemical gradient. Being positively charged, they begin a reversal in the potential difference across the membrane. As sodium ions enter, so more sodium ion channels open, causing an even greater influx of sodium ions and an even greater reversal of potential difference, from −70 mv up to +40 mv at 2.0 ms.

3 Two action potentials take place in 10 ms.

Each action potential takes 10 ÷ 2 = 5 ms / action potentials are 5 ms apart.

There are 1000 ms in 1 second.

Therefore there are 1000 ÷ 5 = 200 action potentials in 1 second.

24.5

1 a node of Ranvier

b Because the remainder of the axon is covered by a myelin sheath that prevents ions being exchanged / prevents a potential difference being set up.

c It moves along in a series of jumps from one node of Ranvier to the next.

d saltatory (conduction)

e It is faster than in an unmyelinated axon.

2 It remains the same / does not change.

24.6

1 During the refractory period the sodium voltage-gated channels are closed so no sodium ions can move inwards and no action potential is possible. This means there must be an interval between one impulse and the next.

2 All-or-nothing principle – There is a particular level of stimulus that triggers an action potential. At any level above this threshold, a stimulus will trigger an action potential that is the same regardless of the size of the stimulus (the 'all' part). Below the threshold, no action potential is triggered (the 'nothing' part).

Different axons different speeds

1 The greater the diameter of an axon the faster the speed of conductance. Comparing the data for the two myelinated axons shows that the 20 μm diameter axon conducts at 120 m s^{-1} while the 10 μm diameter axon conducts at only 50 m s^{-1}. Likewise, the data for the two unmyelinated axons show that the 500 μm diameter axon conducts at 25 m s^{-1} while the 1 μm diameter axon conducts at 2 m s^{-1}.

2 In myelinated axons, the myelin acts as an electrical insulator. Action potentials can only form where there is no myelin (at nodes of Ranvier). The action potential therefore jumps from node to node (saltatory conduction) which makes its conductance faster.

3 Schwann cells

4 The presence of myelin has the greater effect because a myelinated human sensory axon conducts an action potential at twice the speed of the squid giant axon, despite being only 1/50th of its diameter. (Note: Similar comparisons can be made between other types of axon, e.g., squid and human motor axons.)

5 Temperature affects the speed of conductance of action potentials. The higher the temperature, the faster the conductance. The conductance of action potentials in the squid will therefore change as the environmental temperature changes. It will react more slowly at lower temperatures.

24.7

1 It possesses many mitochondria and large amounts of endoplasmic reticulum.

2 It has receptor molecules on its membrane.

3 Neurotransmitter is released from vesicles in the presynaptic neurone into the synaptic cleft when an action potential reaches the synaptic knob. The neurotransmitter diffuses across the synapse to receptor molecules on the postsynaptic neurone to which it binds, thereby setting up a new action potential.

4 Only one end can produce neurotransmitter and so this end alone can create a new action potential in the neurone on the opposite side of the synapse. At the other end there is no neurotransmitter that can be released to pass across the synapse and so no new action potential can be set up.

5 a The relatively quiet background noise of traffic produces a low-level frequency of action potentials in the sensory neurones from the ear. The amount of neurotransmitter released into the synapse is insufficient to exceed the threshold in the postsynaptic neurone and to trigger an action potential and so the noise is 'filtered out' / ignored. Louder noises create a higher frequency and the amount of neurotransmitter released is sufficient to trigger an action potential in the postsynaptic neurone and so there is a response. This is an example of temporal summation. (Note: An explanation in terms of spatial summation is also valid: many sound receptors with a range of thresholds → more receptors respond to the louder noise → more neurotransmitter → response.)

b Reacting to low-level stimuli (background traffic noise) that present little danger can overload the (central) nervous system and so organisms may fail to respond to more important stimuli. High-level stimuli (sound of horn) need a response because they are more likely to represent a danger.

6 As the inside of the membrane is more negative than at resting potential, more sodium ions must enter in order to reach the potential difference of an action potential, i.e., it is more difficult for depolarisation to occur. Stimulation is less likely to reach the threshold level needed for a new action potential.

24.8

1 a sodium ions

b acetylcholine

c ATP

d calcium ions

2 To recycle the choline and ethanoic acid; to prevent acetylcholine from continuously generating a new action potential in the postsynaptic neurone.

Effects of drugs on synapses

1 They will reduce pain.

2 They act like endorphins by binding to the receptors and therefore preventing action potentials being created in the neurones of the pain pathways.

3 Prozac might prevent the elimination of serotonin from the synaptic cleft (Note: any biologically accurate answer that results in more serotonin in the synaptic cleft is acceptable.)

4 Increasing the concentration of serotonin in the synaptic cleft increases, its activity, reducing depression, which is caused by reduced serotonin activity.

5 It will reduce muscle contractions (cause muscles to relax).

6 Valium increases the inhibitory effects of GABA so therefore there are fewer action potentials on the nerve pathways that cause muscles to contract.

7 The molecular structure of vigabatrin is similar to GABA so it may be a competitive inhibitor (compete) for the active site of the enzyme that breaks down GABA. As less GABA is broken down by the enzyme, more of it is available to inhibit neurone activity. Or vigabatrin might bind to GABA receptors on the neurone membrane and mimic its action, thereby inhibiting neuronal activity.

25.1

1 Muscles require much energy for contraction. Most of this energy is released during the Krebs cycle and electron transfer chain in respiration. Both these take place in mitochondria.

2 A = Z-line B = H-zone C = I-band (isotropic band) D = A-band (anisotropic band).

3 The actin and myosin filaments lie side by side in a myofibril and overlap at the edges where they meet. Where they overlap, both filaments can be seen. Where they do not overlap, we see one or other filament only.

4 Slow-twitch fibres contract more slowly and provide less powerful contractions over a longer period. Fast-twitch fibres contract more rapidly and produce powerful contractions but only for a short duration.

5 Slow-twitch fibres have myoglobin to store oxygen, much glycogen to provide a source of metabolic energy, a rich supply of blood vessels to deliver glucose and oxygen, and numerous mitochondria to produce ATP.

Fast-twitch fibres have thicker and more numerous myosin filaments, a high concentration of enzymes involved in anaerobic respiration, and a store of phosphocreatine to rapidly generate ATP from ADP in anaerobic conditions.

25.2

1 Myosin is made of two proteins. The fibrous protein is long and thin in shape, which enables it to combine with others to form a long thick filament along which the actin filament can move. The globular protein forms two bulbous structures (the head) at the end of a filament (the tail). This shape allows it to exactly fit recesses in the actin molecule,

to which it can become attached. Its shape also means it can be moved at an angle. This allows it to change its angle when attached to actin and so move it along, causing the muscle to contract.

2 Phosphocreatine stores the phosphate that is used to generate ATP from ADP in anaerobic conditions. A sprinter's muscles often work so strenuously that the oxygen supply cannot meet the demand. The supply of ATP from mitochondria during aerobic respiration therefore ceases. Sprinters with the most phosphocreatine have an advantage because ATP can be supplied to their muscles for longer, and so they perform better.

3 A single ATP molecule is enough to move an actin filament a distance of 40 nm.

Total distance moved by actin filament = 0.8 μm (= 800 nm).

Number of ATP molecules required = 800 ÷ 40 = 20.

4 One role of ATP in muscle contraction is to attach to the myosin heads, thereby causing them to detach from the actin filament and making the muscle relax. As no ATP is produced after death, there is none to attach to the myosin, which therefore remains attached to actin, leaving the muscle in a contracted state, i.e., *rigor mortis*.

Discovering the role of IAA in tropisms

1 experiment 1

2 As mica conducts electricity it will not prevent electrical messages passing from the shoot tip but it will prevent chemical messages passing. As there is no response, the message must be chemical and must pass down the shaded side.

3 Displacement of the tip means that the chemical initially only moves down the side of the shoot that is in contact with the tip. This side grows more rapidly, causing bending away from that side.

4 It prevents chemicals / IAA, but not light, passing from one side to the other.

5 Results support the hypothesis that IAA is transported from the lighter side to the darker side of the shoot.

Experiment 8 shows that the total IAA produced and collected is the same whether the shoot is in the light or the dark. This discounts the theory that light destroys IAA or inhibits its production.

Experiment 9 shows that the amount of IAA produced at either side of the tip is the same. The glass plate prevents any sideways transfer.

Experiment 10 shows that the IAA is transferred from the light to the dark side of the shoot soon after it is produced because more than twice as much IAA is found on the dark side of the shoot than on the light side.

26.1

1 Response ensures that roots grow downwards into the soil, thus anchoring the plant firmly and bringing them closer to water (needed for photosynthesis).

2 Animal hormones are made in particular organs and affect other organs some distance away.

Plant growth factors are made by cells located throughout the plant and have localised effects.

Investigating tropisms in seedlings

1 The seeds could have contaminants such as fungal spores or bacterial cells. The rate of growth of microorganisms is generally faster than that of plants so the germinating seedlings could be overcome by microbes and rot before they have chance to grow.

2 Conditions deep in the agar would be more anaerobic, reducing the amount of ATP production necessary for processes such as active transport or protein synthesis.

3 a temperature might vary in different locations

b seeds might vary in the thickness of the seed coat so water uptake and initial germination might be delayed in seeds with a thicker coat

4 The plate has been rotated at 90 degrees periodically / every few hours. The root has consistently shown positive geotropism and responds to each rotation by changing the direction of growth

26.2

1 The fruit will produce ethene which is also responsible for senescence – the flowers die sooner.

2 It is an ethene blocker – it prevents leaf and petal drop in response to ethene.

3 Both low water and high salt lower the water potential in the soil so it is more difficult to take up water. ABA reduces water loss through transpiration which is necessary to avoid the plant wilting.

27.1

1 Homeostasis is the maintenance of a constant internal environment in organisms.

2 Maintaining a constant temperature is important because enzymes function within a narrow range of temperatures. Fluctuations from the optimum temperature mean enzymes function less efficiently. If the variation is extreme, the enzyme may be denatured and cease to function altogether. A constant temperature means that reactions occur at a predictable and constant rate.

3 Maintaining a constant blood glucose concentration is important in ensuring a constant water potential. Changes to the water potential of the blood and tissue fluids may cause cells to shrink and expand (even to bursting point), due to water leaving or entering by osmosis. In both instances the cells

cannot operate normally. A constant blood glucose concentration also ensures a reliable source of glucose for respiration by cells.

27.2

1 Positive feedback occurs when the feedback causes the corrective measures to remain turned on. In doing so, it causes the system to deviate even more from the original (normal) level.

Negative feedback occurs when the feedback causes the corrective measures to be turned off. In doing so, it returns the system to its original (normal) level.

2 If the information is not fed back once an effector has corrected any deviation and returned the system to the set point, the receptor will continue to stimulate the effector and an over correction will lead to a deviation in the opposite direction from the original one.

3 It gives a greater degree of homeostatic control.

Control of blood water potential

1 As sweating involves a loss of water from the blood, its water potential will decrease (be lower or more negative).

2 a osmotic cells (in the hypothalamus)

b kidney

3 Being a hormone, it is transported in the blood plasma.

4 Absorption (taking in or consumption or drinking) of water because water has been lost during sweating. As the water potential of the blood returns to normal, the lost water must have been replaced. However, the kidney only excretes less water, it does not replace it. Therefore process X must be the way in which water is replaced.

5 negative feedback

27.3

1 Type I is caused by an inability to produce insulin. Type II is caused by receptors on body cells losing their responsiveness to insulin.

2 Type I is controlled by the injection of insulin. Type II is controlled by regulating the intake of carbohydrate in the diet and matching this to the amount of exercise taken.

3 Diabetes is a condition in which insulin is not produced by the pancreas. This leads to fluctuations in the blood glucose level. If the level is below normal, there may be insufficient glucose for the release of energy by cells during respiration. Muscle and brain cells in particular may therefore be less active, leading to tiredness.

4 Match your carbohydrate intake to the amount of exercise that you take. Avoid becoming overweight by not consuming excessive quantities of carbohydrate and by taking regular exercise.

Effects of diabetes on substance levels in the blood

1 adrenaline

2 The rise in insulin level is both greater and more rapid in group Y than in group X.

3 Glucose is removed from blood by cells using it during respiration.

4 Glucose level rises at first because the glucose that is drunk is absorbed into the blood (glucose line on the graph rises). This rise in blood glucose causes insulin to be secreted from cells (β cells) in the pancreas (insulin line rises steeply). Insulin causes increased uptake of glucose into liver and muscle cells, activates enzymes that convert glucose into glycogen and fat, and increases cellular respiration. The effect of all these actions is to reduce glucose levels (glucose line falls from 2.5 hours onwards). As the glucose level rises after 1 hour, so the glucagon level falls. The reduction in glucagon level decreases glucose production from other sources (glycogen, amino acids, and glycerol) and so also helps to reduce blood glucose levels. As the blood glucose level falls (after 2.5 hours) so the glucagon level increases to help maintain the blood glucose at its normal level.

5 Group X has diabetes and therefore the glucose intake does not stimulate insulin production (insulin level shown on the graph is low). The glucose level in the blood therefore continues to rise (glucose line rises steeply) as there is no insulin to reduce its level. Blood glucose level remains high, falling only slightly as it is respired by cells.

6 As it is respired by cells, the glucose level will decrease steadily until it falls below the normal level.

27.4

a respiration
b brain
c osmotic / water potential
d carbohydrate
e glycogen
f muscles
g amino acids
h gluconeogenesis
i glycogen
j respiration
k islets of Langerhans
l insulin
m glucagon
n adrenaline

27.5

1 Autonomic nervous system controls the involuntary activities of internal muscles and glands.

2 Sympathetic nervous system stimulates effectors and so speeds up an activity; prepares for stressful situations, e.g., the fight or flight response.

Parasympathetic nervous system inhibits effectors and slows down an activity; controls activities under resting conditions, conserving energy and replenishing the body's reserves.

3 Blood pressure remains high because the parasympathetic system is unable to transmit nerve impulses to the SAN, which decreases heart rate and so lowers blood pressure.

4 a Heart rate remains as it was before taking exercise – after exercise, blood pressure increases and CO_2 concentration of blood rises (causing blood pH to be lowered). The changes are detected by pressure and chemical receptors in the wall of the carotid arteries. As the nerve from here to the medulla oblongata is cut, no nerve impulse can be sent to the centres that control heart rate.

b Blood CO_2 concentration increases as a result of increased respiration during exercise.

28.1

1 Epigenetics means changes to gene expression which are not due to changes in the base sequence of the DNA.

2 'p' is the phosphate group. The C and G are adjacent to each other in the polynucleotide chain and would be joined through the 'sugar-phosphate' backbone.

3 DNA methylation prevents gene expression while histone acetylation is necessary for gene expression.

4 RNA polymerase binds to the gene promoter and synthesises the pre-mRNA molecule.

28.2

1 Transcription factors stimulate transcription of a gene.

2 Oestrogen diffuses through the phospholipid portion of a cell-surface membrane into the cytoplasm of a cell, where it combines with a site on a receptor portion of the transcription factor. Oestrogen changes the shape of the receptor molecule, releasing an inhibitor molecule from the DNA binding site on the transcription factor.

The transcription factor now enters the nucleus through a nuclear pore and combines with DNA, stimulating transcription of the gene that makes up that portion of DNA, i.e., it stimulates gene expression.

3 The other strand would have complementary bases (i.e., GCUA instead of CGAU respectively). It is unlikely that these opposite base pairings would complement a sequence on the mRNA. The siRNA, with enzyme attached, would therefore not bind to the mRNA and so would be unaffected.

Cancer – the 'two hit' hypothesis

1 A person with a family history of cancer may already have one mutated allele for the inactivation of the tumour suppressor gene. As X-rays increase mutation rates they might advance the likelihood of cancer in these patients. Patients with no family history of cancer are less at risk because they are less likely to have inherited a mutant allele.

2 The proto-oncogene mutant allele might be dominant whereas the tumour suppressor mutant allele might be recessive. If so, it requires just one dominant proto-oncogene allele to cause cancer where as it will take two recessive tumour suppressor alleles (homozygous state) to cause cancer.

3 Tumour repressor genes inhibit cell division. Mutated forms of these genes are inactive and so cell division increases and a tumour forms. The introduction of normal tumour repressor genes means that the inhibition of cell division will be resumed and the tumour growth will stop.

4 Oncogenes cause cancer by permanently activating protein receptors on cells and so they stimulate cell division. By destroying these receptors on cancer cells, division will be halted and tumour growth will stop.

Gene expression in haemoglobin

1 It allows the fetus to load its haemoglobin with oxygen from the mother's haemoglobin where the two blood supplies come close to each other (at the placenta).

2 alpha = 50% beta = 20% gamma = 30%

3 The gene for gamma-globulin is expressed less while the gene for beta-globulin is expressed more.

4 Expression of the gene for gamma-globulin is progressively reduced as a result of either preventing transcription, and hence preventing the production of mRNA, or by the breakdown of mRNA before its genetic code can be translated.

5 A possible therapy would be to express (switch on) the gene for gamma-globulin and prevent the expression of (switch off) the gene for beta-globulin. This would result in haemoglobin being of the fetal rather than the adult type.

28.3

1 Totipotent cells are cells with the ability to develop into any other cell of the organism.

2 In animals, only a few cells are totipotent. In humans these are known as stem cells and are found in the embryo, the inner lining of the intestine, skin, and bone marrow. In plants, many of the cells throughout a plant are totipotent.

3 In skin cells, the gene that codes for keratin is expressed, but not the gene for myosin. The genetic code for keratin is translated into the protein keratin, which the cell therefore produces, but the genetic code for myosin is not translated. In muscle cells, the gene for myosin is expressed but not the gene for keratin. In the same way, the genetic code for myosin rather than keratin is translated and so only myosin is produced.

Growth of plant tissue cultures

1 differentiation

2 IAA and 2,4-D

3 In test tube 1 the low concentration of IAA produces moderate shoot development but when a high concentration of cytokinin is added (test tube 3) the presence of cytokinin influences the effects of the IAA by reducing shoot development to a 'little'.

Human embryonic stem cells and the treatment of disease

1 Any properly structured and evaluated accounts that make scientifically accurate points in a reasoned fashion are acceptable, e.g.,:

For	Against
• Huge potential to cure many debilitating diseases	• It is wrong to use humans, including potential humans, as a means to an end
• Wrong to allow suffering when it can be relieved	• Embryos are human, they have human genes, and deserve the same respect and treatment as adult humans
• Embryos are created for other purposes (IVF) so why not stem cells	• It is a slippery slope to the use of older embryos and fetuses for research
• Embryos of less than 14 days are not recognisably human and so do not command the same respect as adults or fetuses	• It could lead to research and development of human cloning and, although banned in the UK, the information gained could be used elsewhere
• There is no risk of research escalating or including fetuses because current legislation prevents this	• It undermines respect for life
• Adult stem cells are not as suitable as embryonic stem cells and it may be many years before they are, in the meantime many people suffer unnecessarily	• Adult stem cells are an available alternative and energies should be directed towards developing these

29.1

a	recombinant	b	reverse transcriptase
c	complementary (cDNA)	d	DNA polymerase
e	restriction endonucleases	f	blunt
g	sticky	h	CTTAAG

29.2

1 Primers are short pieces of DNA that have a set of bases complementary to those at the end of the DNA fragment to be copied.

2 Primers attach to the end of a DNA strand that is to be copied and provide the starting sequences for DNA polymerase to begin DNA cloning. DNA polymerase can only attach nucleotides to the end of an existing chain. They also prevent the two separate strands from rejoining.

3 Because the sequences at the opposite ends of the two strands of DNA are different.

4 hydrogen bonds

5 Biological contaminants may contain DNA and this DNA would also be copied.

29.3

1 A vector transfers genes (DNA) from one organism into another.

2 To show which cells (bacteria) have taken up the plasmid (gene).

3 Results can be obtained more easily and more quickly because, with antibiotic-resistance markers, the bacterial cells with the required gene are killed, so replica plating is necessary to obtain the cells with the gene. With fluorescent gene markers, the bacterial cells are not killed and so there is no need to carry out replica plating.

4 a B, C, D, J, K, and L – because those that did not take up the plasmid will not have taken up the gene for ampicillin resistance and so will be the ones that are killed on the ampicillin plate, i.e., the colonies that have disappeared.

b E, F, and I – because those with the plasmid containing gene X will have lost the gene for tetracycline resistance and therefore the colonies will have been killed on the tetracycline plate, i.e., the colonies will have disappeared.

Index